PREFACE

I have written this introductory-level biostatistics text for upper-level undergraduate or graduate students interested in medicine or other health-related areas. This book requires no previous background in statistics, and its mathematical level assumes only a knowledge of algebra.

Fundamentals of Biostatistics evolved from a set of notes that I used in a course in biostatistics taught to Harvard University undergraduates and Harvard Medical School students over the past fifteen years. I wrote this book to help motivate students to master the statistical methods that are most often used in the medical literature. From the student's viewpoint, it is important that the example material used to develop these methods is representative of what actually exists in the literature. Therefore, most examples and exercises used in this book are either based on actual articles from the medical literature or on actual medical research problems I have encountered during my consulting experience at the Harvard Medical School.

Most other introductory statistics texts either use a completely nonmathematical, cookbook approach or develop the material in a rigorous, sophisticated mathematical framework. In this book I have attempted to follow an intermediate course, minimizing the amount of mathematical formulation and yet giving complete explanations of all the important concepts. *Every* new concept is developed systematically through completely worked out examples from current medical research problems. In addition, computer output is introduced where appropriate to illustrate these concepts.

The material in this book is suitable for either a one- or two-semester course in biostatistics. The material in Chapters 1 through 8 and Chapter 10 is suitable for a one-semester course. The instructor may select appropriate material from the other chapters as time permits.

The following changes have been made in the third edition:

- A set of computer exercises based on real data sets has been developed. The data sets are on a diskette and are available for instructors from the publisher.
- A new section on quantiles has been added.
- A separate section on permutations and combinations has been added.
- A new section concerning the mean and variance of linear combinations of normal random variables has been added.
- Separate sections on interval estimation for two-paired and independent samples, respectively, have been added to highlight the importance of interval estimation for two-sample problems.
- A brief introduction to survival methods, including a presentation of the log rank test and the Cox proportional hazards model, is given.

- New material concerning a comparison of incidence density measures has been added. This is an important area, since many epidemiological studies report results in terms of events per person-time, rather than events per person.
- A new section on power and sample size estimation for matched-pair binary data has been added.
- A new section on the analysis of covariance has been added, and PROC GLM of SAS is used to illustrate analysis of variance techniques for unbalanced designs.

Fundamentals of Biostatistics, third edition, is organized as follows:

Chapter 1 is an *introductory chapter* giving an outline of the development of an actual medical study I was involved with. It provides a unique sense of the role of biostatistics in the medical research process.

Chapter 2 concerns *descriptive statistics* and presents all the major numerical and graphical tools used for displaying medical data. This chapter is especially important for both consumers and producers of medical literature, since much of the actual communication of information is accomplished via descriptive material.

Chapters 3 through 5 discuss *probability*. The basic principles of probability are developed, and the most common probability distributions, such as the binomial and normal distributions, are introduced. These distributions are used extensively in the later chapters of the book.

Chapters 6 through 10 cover some of the basic methods of *statistical inference*.

Chapter 6 introduces the concept of drawing random samples from populations. The difficult notion of a sampling distribution is also developed, including an introduction to the most common sampling distributions, such as the t and chi-square distributions. The basic methods of *estimation* are also presented, including an extensive discussion of confidence intervals.

Chapters 7 and 8 contain the basic principles of *hypothesis testing*. The most elementary hypothesis test for normally distributed data, such as the t test, are also fully discussed for one- and two-sample problems.

Chapter 9 covers the basic principles of *nonparametric statistics*. The assumptions of normality are relaxed, and distribution-free analogues are developed for the tests in Chapters 7 and 8.

Chapter 10 contains the basic concepts of *hypothesis testing* as applied to categorical data, including some of the most widely used statistical procedures, such as the chi-square test and Fisher's exact test.

Chapter 11 develops the principles of *regression analysis*. The case of simple linear regression is thoroughly covered, and extensions are provided for the multiple regression case. An important section on the limitations of the use of regression analysis is also included.

Chapter 12 introduces the basic principles of the *analysis of variance* (ANOVA). The one-way and two-way analyses of variance are discussed as well as the analysis of covariance.

The elements of study design are not formally covered in this book but are informally introduced in much of the example material. The concepts of matching,

INDEX OF APPLICATIONS

FUNDAMENTALS
OF BIOSTATISTICS

THIRD EDITION

FUNDAMENTALS OF BIOSTATISTICS

BERNARD ROSNER

HARVARD UNIVERSITY

PWS-KENT PUBLISHING COMPANY

BOSTON, MASSACHUSETTS

PWS–KENT
Publishing Company

20 Park Plaza
Boston, Massachusetts 02116

**This book is dedicated to my wife Cynthia
and my children Sarah, David, and Laura.**

0534982140x

Copyright © 1990 by PWS-KENT Publishing Company. Copyright © 1986, © 1982 by PWS Publishers.

PWS-KENT Publishing Company is a division of Wadsworth, Inc.

Library of Congress Cataloging-in-Publication Data
Rosner, Bernard (Bernard A.)
   Fundamentals of biostatistics.
   Includes bibliographies and index.
1. Biometry.   2. Medical statistics.   I. Title
QH323.5.R674   1989   574'.072   89-13088
ISBN 0-534-91973-1

Printed in the United States of America
90 91 92 93 94–10 9 8 7 6 5 4 3 2 1

Sponsoring Editor: Michael R. Payne
Editorial Assistant: Susan Hankinson
Production Editor: S. London
Manufacturing Coordinator: Marcia Locke
Interior Designer: S. London
Cover Designer: Designworks
Typesetter: Composition House
Cover Printer: Arcata Graphics-Halliday Lithograph
Printer and Binder: Arcata Graphics-Halliday Lithograph

cohort studies, case-control studies, retrospective studies, prospective studies, and the sensitivity, specificity, and predictive value of screening tests are extensively discussed in the context of actual samples. In addition, specific sections on sample size estimation are provided for different statistical situations in Chapters 7, 8, and 10.

A flowchart of appropriate methods of statistical inference on pages 646–649 provides an easy reference to the methods developed in this book. This flowchart is referred to at the end of each of Chapters 7 through 12 to give the student some perspective on how the methods in a particular chapter fit in with the overall collection of statistical methods introduced in this book.

In addition, an index summarizing all examples and problems used in this book is provided, grouped by *medical specialty*.

I am grateful to the Literary Executor of the late Sir Ronald A Fisher, F.R.S., to Dr. Frank Yates, F.R.S., and to the Longman Group Ltd., London, for permission to reprint Table III from their book *Statistical Tables for Biological, Agricultural and Medical Research* (sixth edition, 1974).

I am indebted to Marie Sheehan and Harry Taplin, who have been invaluable in helping to type this manuscript. I am indebted to those who reviewed the manuscript among them: Charles S. Davis, University of Iowa; Donald J. Slymen, San Diego State University; Craig D. Turnbull, University of North Carolina at Raleigh; and William E. Strawderman, Rutgers University. I wish to thank Michael Payne, Susan London, and Sally Stickney who were instrumental in providing editorial advice in the preparation of the manuscript. I am indebted to my many colleagues at the Channing Laboratory, most notably Edward Kass, Frank Speizer, Charles Hennekens, Frank Polk, Ira Tager, Jerome Klein, James Taylor, Stephen Zinner, Scott Weiss, Frank Sacks, Walter Willett, and Alvaro Munoz and to my other colleagues at the Harvard Medical School, most notably Frederick Mosteller, Eliot Berson, Robert Ackerman, Mark Abelson, Arthur Garvey, Leo Chylack, Eugene Braunwald, and Arthur Dempster, who provided the inspiration for writing this book.

Bernard Rosner
Boston

CONTENTS

CHAPTER 5 **Continuous Probability Distributions** 107

CHAPTER 6 **Estimation** 141

CHAPTER 7 **Hypothesis Testing: One-Sample Inference** 186

CHAPTER 12   **Analysis of Variance 474**

APPENDIX 1   **Tables 527**

APPENDIX 2   **Index of Data Sets, 554**

CHAPTER ONE

GENERAL OVERVIEW

Statistics is the science whereby inferences are made about specific random phenomena on the basis of relatively limited sample material. The field of statistics can be subdivided into two main areas: mathematical statistics and applied statistics. **Mathematical statistics** concerns the development of new methods of statistical inference and requires detailed knowledge of abstract mathematics for its implementation. **Applied statistics** concerns the application of the methods of mathematical statistics to specific subject areas, such as economics, psychology, and public health. **Biostatistics** is the branch of applied statistics that concerns the application of statistical methods to medical and biological problems.

A good way to learn about biostatistics and its role in the research process is to follow the flow of a research study from its inception at the planning stage to its completion, which usually occurs when a manuscript reporting the results of the study is published. As an example, I will describe one such study in which I participated.

A friend called one morning and in the course of our conversation mentioned that he had recently used a new, automated blood-pressure device of the type seen in many banks, hotels, and department stores. The machine had read his average diastolic blood pressure on several occasions as 115 mm Hg; the highest reading was 130 mm Hg. I was horrified to hear of his experience, since if these readings were true, my friend might be in imminent danger of having a stroke or developing some other serious cardiovascular disease. I referred him to a clinical colleague of mine who, using a standard blood-pressure cuff, measured my friend's diastolic blood pressure as 90 mm Hg. The contrast in the readings aroused my interest, and I began to jot down the readings on the digital display every time I passed the machine at my local bank. I got the distinct impression that a large percentage of the reported readings were in the hypertensive range. Although one would expect that hypertensives would be more likely to use such a machine, I still believed that blood-pressure readings obtained with the machine might not be comparable with those obtained using standard methods of blood-pressure measurement. I spoke to Dr. B. Frank Polk about my suspicion and succeeded in interesting him in a small-scale evaluation of such machines. We decided to send a human observer who was well trained in blood-pressure measurement techniques to several of these machines. He would offer to pay subjects 50¢ for the cost of using the machine if they would agree to fill out a short questionnaire and have their blood pressure measured by both a human observer and the machine.

At this stage we had to make several important decisions, each of which would prove vital to the success of the study. The decisions were based on the following questions:

(1) How many machines should we test?

(2) How many people should we test at each machine?

(3) In what order should the measurements be taken—should the human observer or the machine be used first? Ideally, we would have preferred to avoid this problem by taking both the human and machine readings simultaneously, but this procedure was logistically impossible.

(4) What other data should we collect on the questionnaire that might influence the comparison between methods?

(5) How should the data be recorded to facilitate their computerization at a later date?

(6) How should the accuracy of the computerized data be checked?

We resolved these problems as follows:

(1) and (2) We decided to test more than one machine (four to be exact), since we were not sure if the machines were comparable in quality. However, we wanted to sample enough subjects from each machine so that we would have an accurate comparison of the standard and automated methods for each machine. We tried to predict how large a discrepancy there might be between the two methods. Using the methods of sample size determination discussed in this book, we calculated that we would need 100 subjects at each site to have an accurate comparison.

(3) We then had to decide in what order the measurements should be taken for each person. According to some reports, one problem that occurs with repeated blood-pressure measurements is that people tense up at the initial measurement, yielding higher blood pressures than at subsequent repeated measurements. Thus, we would not always want to use the automated or manual method first, since the effect of the method would get confused with the order-of-measurement effect. A conventional technique that we used here was to **randomize** the order in which the measurements were taken, so that for any person it was equally likely that the machine or the human observer would take the first measurement. This random pattern could be implemented by flipping a coin or, more likely, by using a table of **random numbers** as appears in Table 4 of Appendix 1.

(4) We felt that the major extraneous factor that might influence the results would be body size, since we might have more difficulty getting accurate readings from people with fatter arms than from those with leaner arms. We also wanted to get some idea of the type of people who use these machines; so we asked questions about age, sex, and previous hypertensive history.

(5) To record the data, we developed a coding form that could be filled out on site and from which data could be easily entered on a computer terminal for subsequent analysis. Each person in the study was assigned an identification (ID) number by which the computer could uniquely identify that person. The data on the coding forms were then keyed and verified. That is, the same form was entered twice, and a comparison was made between the two records to make sure they were the same. If the records were not the same, the form was reentered.

(6) After data entry we ran some editing programs to ensure that the data were accurate. Checking each item on each form was impossible because of the large amount of data. Alternatively, we checked that the values for individual variables were within specified ranges and printed out aberrant values for manual checking.

For example, we checked that all blood-pressure readings were at least 50 and no more than 300 and printed out all readings that fell outside this range.

After completing the data collection, data entry, and data editing phases, we were ready to look at the results of the study. The first step in this process is to get a general feel for the data by summarizing the information in the form of several descriptive statistics. This descriptive material can be numerical or graphical. If numerical, it can be in the form of a few summary statistics, which can be presented in tabular form or, alternatively, in the form of a **frequency distribution**, which lists each value in the data and how frequently it occurs. If graphical, the data are summarized pictorially and can be presented in one or more figures. The appropriate type of descriptive material will vary with the type of distribution considered. If the distribution is **continuous**, that is, if there are essentially an infinite number of possible values, as would be the case for blood pressure, then means and standard deviations might be the appropriate descriptive statistics. However, if the distribution is **discrete**, that is, if there are only a few possible values, as would be the case for sex, then percentages of people taking on each value would be the appropriate descriptive measure. In some cases both types of descriptive statistics are used for continuous distributions by condensing the range of possible values into a few groups and giving the percentage of people that fall into each group (e.g., the percentages of people that have blood pressures between 120 and 129 mm Hg and between 130 and 139 mm Hg).

In this study we decided first to look at mean blood pressure for each method at each of the four sites. Table 1.1 summarizes this information [1].

You might notice from this table that we did not obtain meaningful data from all of the 100 people interviewed at each site, since we could not obtain valid readings from the machine for many of the people. This type of missing data problem is very common in biostatistics and should be anticipated at the planning stage when deciding on sample sizes (which was not done in this study).

Our next step in the study was to determine whether the apparent differences in blood pressure between machine and human measurements at two of the locations (C, D) were "real" in some sense or were "due to chance." This type of question falls into the area of **inferential statistics**. We realized that although there was a 14-mm Hg difference in mean systolic blood pressure between the two

TABLE 1.1
Mean blood pressures and differences between machine and human readings at four locations

| | | Systolic blood pressure (mm Hg) | | | | | |
| | | Machine | | Human | | Difference | |
Location	Number of people	Mean	Standard deviation	Mean	Standard deviation	Mean	Standard deviation
A	98	142.5	21.0	142.0	18.1	0.5	11.2
B	84	134.1	22.5	133.6	23.2	0.5	12.1
C	98	147.9	20.3	133.9	18.3	14.0	11.7
D	62	135.4	16.7	128.5	19.0	6.9	13.6

(By permission of the American Heart Association, Inc.)

methods for the 98 people we interviewed at location C, this difference might not hold up if we interviewed 98 other people at a different time, and we wanted to have some idea as to the **error in the estimate** of 14 mm Hg. In technical jargon this group of 98 people represents a **sample** from the **population** of all people who use that machine. We were interested in the population and we wished to use the sample to help us learn something about the population. In particular, we wanted to know how different the **estimated** mean difference of 14 mm Hg in our sample was likely to be from the **true** mean difference in the population of all people who might use this machine. More specifically, we wanted to know if it was still possible that there was no underlying difference between the two methods and that our results were due to chance. The 14-mm Hg difference in our group of 98 people is referred to as an **estimator** of the true mean difference (d) in the population. The problem of inferring characteristics of a population from a sample is the central concern of statistical inference and is a major topic in this text. To accomplish this aim, we needed to develop a **probability model**, which would tell us how likely it is that we would obtain a 14-mm Hg difference between the two methods in a sample of 98 people if there were no real difference between the two methods over the entire population of users of the machine. If this probability were sufficiently small, then we would begin to believe that a real difference existed between the two methods. In this particular case, using a probability model based on the t distribution, we were able to conclude that this probability was less than 1 in 1000 for each of machines C and D. This probability was sufficiently small for us to conclude that there was a real difference between the automatic and manual methods of taking blood pressure for two of the four machines tested.

We used a statistical package to perform the preceding data analyses. A package is a collection of statistical programs that describe data and perform various statistical tests on the data. Currently the most widely used statistical packages include SAS, SPSS^X, BMDP, and Minitab.

The final step in this study, after completing the data analysis, was to compile the results in the form of a publishable manuscript. Inevitably, because of space considerations, much of the material developed during the data analysis phase was weeded out and only the essential items were presented for publication.

The review of this study should give you some idea of what medical research is about and what the role of biostatistics is in this process. The material in this text will parallel the description of the data analysis phase of the study described. Chapter 2 summarizes different types of descriptive statistics. In Chapters 3 through 5, some basic principles of probability and various probability models for use in later discussions of inferential statistics are presented. In Chapters 6 through 12, the major topics of inferential statistics as used in biomedical practice are discussed. Issues of study design or data collection are brought up only as they relate to other topics discussed in the text.

Reference

[1] Polk, B. F., Rosner, B., Feudo, R., & Vandenburgh, M. (1980). An evaluation of the Vita-Stat automatic blood pressure measuring device. *Hypertension*, *2*(2), 221–227.

CHAPTER TWO

DESCRIPTIVE STATISTICS

SECTION 2.1 **Introduction**

The first step in looking at data is to describe the data at hand in some concise way. In smaller studies this step can be accomplished by listing each data point. In general, however, this procedure is tedious or impossible and, even if it were possible, would not give an overall picture of what the data look like.

EXAMPLE 2.1 **Cancer, Nutrition** Some investigators have proposed that consumption of vitamin A prevents cancer. To test this theory, a dietary questionnaire to collect data on vitamin A consumption among 200 hospitalized cancer cases and 200 controls might be used. The controls would be matched on age and sex to the cancer cases and would be in the hospital at the same time for an unrelated disease. What should be done with these data after they are collected? ∎

Before any formal attempt to answer this question can be made, the vitamin A consumption among cases and controls must be described. Consider Figure 2.1. The **bar graphs** show visually that the controls have a higher vitamin A consumption than the cases do, particularly in doses higher than the recommended daily allowance (RDA).

EXAMPLE 2.2 **Pulmonary Disease** Medical researchers have often suspected that passive smokers—people who themselves do not smoke but who live or work in an environment where others smoke—might have impaired pulmonary function as a result. In 1980 a research group in San Diego published results indicating that passive nonsmokers who did not work in smoky environments [1]. As supporting evidence, the authors measured the carbon monoxide (CO) concentrations in the working environments of passive smokers and of nonsmokers (where no smoking was permitted in the workplace) to see if the relative CO concentration changed over the course of the day. These results are displayed in the form of a **scatter plot** in Figure 2.2. ∎

Figure 2.2 clearly shows that the CO concentrations in the two working environments are about the same early in the day but diverge widely in the middle of the day and then converge again after the working day is over at 7 P.M.

Graphical displays illustrate the important role of descriptive statistics, which is to quickly display data to give the researcher a clue as to the principal trends in the data and suggests hints as to where a more detailed look at the data, using the methods of inferential statistics, might be worthwhile. Descriptive statistics are also

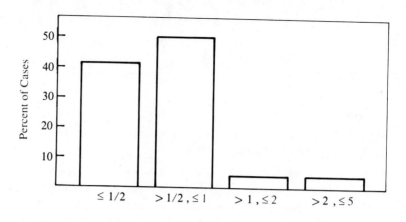

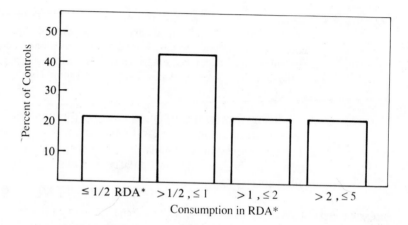

FIGURE 2.1
Daily vitamin A
consumption among
cancer cases and
controls

*RDA = Recommended Daily Allowance

crucially important in conveying the final results of studies in written publications. Unless it is one of their primary interests, most readers will not have time to critically evaluate the work of others but will be influenced mainly by the descriptive statistics presented.

What makes a good graphical or numerical display? The principal guideline is that the material should be as self-contained as possible and should be understandable without reading the text. These attributes require clear labeling. The captions, units, and axes on graphs should be clearly labeled, and the statistical terms used in tables and figures should be well defined. The quantity of material presented is equally important. If bar graphs are constructed, then care must be taken that neither too many nor too few groups be displayed. The same is true of tabular material.

Many methods are available for summarizing data in both numerical and graphical form. In this chapter the methods will be summarized and their strengths and weaknesses given.

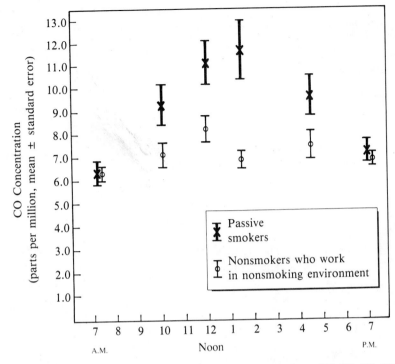

(Reproduced with permission of *The New England Journal of Medicine, 302,* 720–723, 1980.)

FIGURE 2.2
Mean carbon monoxide concentration (± standard error) by time of day as measured in the working environment of passive smokers and nonsmokers who work in nonsmoking environments

<u>SECTION 2.2</u> **Measures of Central Location**

The basic problem of statistics can be stated as follows: Consider a sample of data $x_1, \ldots, x_n$, where x_1 corresponds to the first sample point and x_n corresponds to the nth sample point. Presuming that the sample is drawn from some population P, what inferences or conclusions can be made about P from the sample?

Before this question can be answered, the data must be summarized as succinctly as possible, since the number of sample points is frequently large and it is easy to lose track of the overall picture by looking at all the data at once. One type of measure useful for summarizing data defines the center, or middle, of the sample. This type of measure is a **measure of central location**.

2.2.1 **The Arithmetic Mean**

How to define the middle of a sample may seem obvious, but the more you think about it, the less obvious it becomes. Suppose the sample consists of birthweights of all live-born infants born at a private hospital in San Diego, California, during a 1-week period. This sample is shown in Table 2.1.

One measure of central location for this sample is the arithmetic mean (colloquially referred to as the average). The arithmetic mean (or mean or sample mean) is usually denoted by $\bar{x}$.

TABLE 2.1
Sample of birthweights
of live-born infants born
at a private hospital in
San Diego, California,
during a 1-week
period (g)

i	x_i	i	x_i	i	x_i	i	x_i
1	3265	6	3323	11	2581	16	2759
2	3260	7	3649	12	2841	17	3248
3	3245	8	3200	13	3609	18	3314
4	3484	9	3031	14	2838	19	3101
5	4146	10	2069	15	3541	20	2834

DEFINITION 2.1

The **arithmetic mean** is the sum of all the observations divided by the number of observations. It is written in statistical terms as

$$\bar{x} = \frac{1}{n} \sum_{i=1}^{n} x_i$$

The sign Σ (sigma) in Definition 2.1 is referred to as a summation sign. The expression

$$\sum_{i=1}^{n} x_i$$

is simply a short way of writing the quantity $(x_1 + x_2 + \cdots + x_n)$.

If a and b are integers, then

$$\sum_{i=a}^{b} x_i$$

means that

(1) The value of a is substituted for i and x_a is used as the first term in the summation.
(2) The value of i is increased by 1 to $a + 1$ and x_{a+1} is added to x_a.
(3) Step 2 is repeated as many times as possible until $i = b$.

The term x_b is the last term in the summation. Thus,

$$\sum_{i=a}^{b} x_i = x_a + x_{a+1} + \cdots + x_b$$

If $a = b$, then $\sum_{i=a}^{b} x_i = x_a$. One fundamental property of summation signs is that if each term in the summation is a multiple of the same constant c, then c can be factored out from the summation, that is,

$$\sum_{i=1}^{n} cx_i = c\left(\sum_{i=1}^{n} x_i\right)$$

EXAMPLE 2.3 If

$$x_1 = 2 \qquad x_2 = 5 \qquad x_3 = -4$$

find

$$\sum_{i=1}^{3} x_i \quad \sum_{i=2}^{3} x_i \quad \sum_{i=1}^{3} x_i^2 \quad \sum_{i=1}^{3} 2x_i$$

SOLUTION

$$\sum_{i=1}^{3} x_i = 2 + 5 - 4 = 3 \qquad \sum_{i=2}^{3} x_i = 5 - 4 = 1$$

$$\sum_{i=1}^{3} x_i^2 = 4 + 25 + 16 = 45 \qquad \sum_{i=1}^{3} 2x_i = 2 \sum_{i=1}^{3} x_i = 6 \qquad \blacksquare$$

It is important to become familiar with summation signs, because they are used extensively in the remainder of this text.

EXAMPLE 2.4   What is the arithmetic mean for the sample of birthweights in Table 2.1?

$$\bar{x} = (3265 + 3260 + \cdots + 2834)/20 = 3166.9 \text{ g} \qquad \blacksquare$$

The arithmetic mean is, in general, a very natural measure of central location. One of its principal limitations, however, is that it is overly sensitive to extreme values. In this instance it may not be representative of the location of the great majority of the sample points. For example, if the first infant in Table 2.1 happened to be a premature infant weighing 500 g rather than 3265 g, then the arithmetic mean of the sample would be reduced to 3028.7 g. In this instance, 7 of the birthweights would be lower than the arithmetic mean, and 13 would be higher than the arithmetic mean. It is possible in extreme cases for all but one of the sample points to be on one side of the arithmetic mean. The arithmetic mean is a poor measure of central location in these types of samples, since it does not reflect the center of the sample. Nevertheless, the arithmetic mean is by far the most widely used measure of central location.

2.2.2   The Median

An alternative measure of central location, perhaps second in popularity to the arithmetic mean, is the **median** or, more precisely, the **sample median**.

Suppose there are n observations in a sample. If these observations are ordered from smallest to largest, then the median is defined as follows:

DEFINITION 2.2   ■■
The **sample median** is

(1) The $\left(\dfrac{n+1}{2}\right)$th largest observation if n is odd

(2) The average of the $\left(\dfrac{n}{2}\right)$th and $\left(\dfrac{n}{2}+1\right)$th largest observations if n is even   ■

The rationale for these definitions is to ensure an equal number of sample points on both sides of the sample median. The median is defined differently when n is even and odd because it is impossible to achieve this goal with one uniform definition. For samples with an odd sample size, there is a unique central point; for example, for samples of size 7, the fourth largest point is the central point in the sense that 3 points are both smaller and larger than it. For samples with an even sample size, there is no unique central point and the middle 2 values must be averaged. Thus, for samples of size 8, the fourth and fifth largest points would be averaged to obtain the median, since neither is the central point.

EXAMPLE 2.5   Compute the sample median for the sample in Table 2.1.

SOLUTION   First, arrange the sample in ascending order:

2069, 2581, 2759, 2834, 2838, 2841, 3031, 3101, 3200, 3245, 3248, 3260, 3265, 3314, 3323, 3484, 3541, 3609, 3649, 4146

Since n is even,

Sample median = average of the 10th and 11th largest observations

$$= (3245 + 3248)/2 = 3246.5 \text{ g}$$

∎

EXAMPLE 2.6   **Infectious Disease** Consider the data set in Table 2.2, which consists of white blood counts taken on admission of all patients entering a small hospital in Allentown, Pennsylvania, on a given day. Compute the median white blood count.

TABLE 2.2
Sample of admission white blood counts for all patients entering a hospital in Allentown, PA, on a given day (× 1000)

i	x_i	i	x_i
1	7	6	3
2	35	7	10
3	5	8	12
4	9	9	8
5	8		

SOLUTION   First, order the sample as follows: 3, 5, 7, 8, 8, 9, 10, 12, 35. Since n is odd, the sample median is given by the fifth largest point, which equals 8.

∎

The principal strength of the sample median is that it is insensitive to very large or very small values. In particular, if the second patient in Table 2.2 had a white count of 65,000 rather than 35,000, the sample median would remain unchanged, since the fifth largest value is still 8000. Conversely, the arithmetic mean would increase dramatically from 10,778 in the original sample to 14,111 in the new sample. The principal weakness of the sample median is that it is determined mainly by the middle points in a sample and is less sensitive to the actual numerical values of the remaining data points.

2.2.3   Comparison of the Arithmetic Mean and the Median

If a distribution is **symmetric**, then the relative position of the points on each side of the sample median will be the same. Examples of distributions expected to be roughly symmetric include the distribution of birthweights in Table 2.1 and the distribution of systolic blood-pressure measurements taken on all 30–39-year-old factory workers in a given workplace [Figure 2.3(a)].

If a distribution is **positively skewed** (or skewed to the right) then points above the median will tend to be farther from the median in absolute value than points below the median. An example of such a distribution would be the distribution of the number of years of oral contraceptive (OC) use by a group of women aged 20–29 years participating in a prepaid group health plan [Figure 2.3(b)]. Similarly, if a distribution is **negatively skewed** (or skewed to the left), then points below the

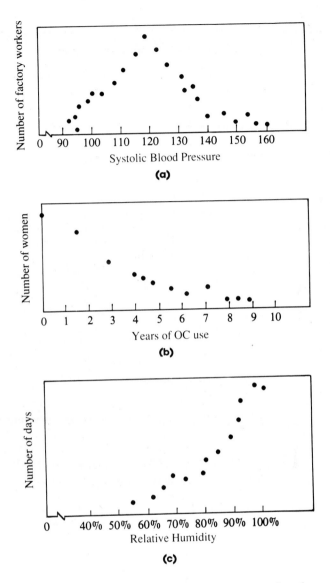

FIGURE 2.3

Graphical displays of (a) symmetric, (b) positively skewed, and (c) negatively skewed distributions

median will tend to be farther from the median in absolute value than points above the median. An example of such a distribution would be the distribution of relative humidities observed in a humid climate at the same time of day over a number of days. In this case, most of the humidities will be at or close to 100%, with a few very low humidities on dry days [Figure 2.3(c)].

In many samples, the relationship between the arithmetic mean and the sample median can be used to assess the symmetry of a distribution. In particular, for symmetric distributions, the arithmetic mean will be approximately the same as the median. For positively skewed distributions, the arithmetic mean will tend to be larger than the median; for negatively skewed distributions, the arithmetic mean will tend to be smaller than the median.

2.2.4 **The Mode**

Another widely used measure of central location is the mode.

DEFINITION 2.3 ■■■

The **mode** is the most frequently occurring value among all the observations in a sample.

■

EXAMPLE 2.7 **Family Planning** Consider the sample of time intervals between successive menstrual periods for a group of 500 college women aged 18–21, as shown in Table 2.3. The frequency column gives the number of women who reported each of the respective durations. Twenty-eight days is the mode, since it is the most frequently occurring value. ■

TABLE 2.3
Sample of time intervals between successive menstrual periods of college-aged women (days)

Value	Frequency	Value	Frequency	Value	Frequency
24	5	29	96	34	7
25	10	30	63	35	3
26	28	31	24	36	2
27	64	32	9	37	1
28	185	33	2	38	1

EXAMPLE 2.8 Compute the mode of the distribution in Table 2.2.

SOLUTION The mode is 8000 because it occurs more frequently than any other white count. ■

Some distributions have more than one mode. In fact, one useful method of classifying distributions is by the number of modes present. A distribution with one mode is referred to as **unimodal**; two modes, **bimodal**; three modes, **trimodal**; and so forth.

EXAMPLE 2.9 Compute the mode of the distribution in Table 2.1.

SOLUTION There is no mode, since all the values occur exactly once. ■

Example 2.9 illustrates a common problem with the mode: It is not a useful measure of location if there are a large number of possible values, each of which occurs infrequently. In such cases the mode will either be far from the center of the sample or, in extreme cases, will not exist, as in Example 2.9. The mode is not used much in this text because its mathematical properties are, in general, rather intractable, and in most common situations it is inferior to the arithmetic mean.

2.2.5 **The Geometric Mean**

Much laboratory data, specifically data in the form of concentrations of one substance in another, as assessed by serial dilution techniques, are either multiples of 2 or are a constant multiplied by a power of 2; that is, they can have outcomes only of the form $2^k c$, $k = 0, 1, \ldots$, for some constant c. For example, the data in Table 2.4 represent the minimal inhibitory concentration (MIC) of penicillin G in the urine for *N. gonorrhoeae* in 74 patients [2]. The arithmetic mean would not be appropriate as a measure of central location in this situation because the distribution is very skewed. However, the data do have a certain pattern since the

TABLE 2.4
Distribution of minimal
inhibitory concentration
(MIC) of penicillin G for
N. gonorrhoeae

(μg/mL) Concentration	Frequency	(μg/mL) Concentration	Frequency
$0.03125 = 2^0(0.03125)$	21	$0.250 = 2^3(0.03125)$	19
$0.0625\ = 2^1(0.03125)$	6	$0.50\ = 2^4(0.03125)$	17
$0.125\ = 2^2(0.03125)$	8	$1.0\ = 2^5(0.03125)$	3

(Reproduced with permission from *JAMA*, *220*, 205–208, 1972. Copyright 1972, American Medical Association.)

only possible values are of the form $2^k(0.03125)$ for $k = 0, 1, 2 \ldots$. One solution is to work with the distribution of the logs of the concentrations. The log concentrations have the property that successive possible concentrations differ by a constant; that is, $\log(2^{k+1}c) - \log(2^k c) = \log(2^{k+1}) + \log c - \log(2^k) - \log c = (k + 1)\log 2 - k \log 2 = \log 2$. Thus, the log concentrations are equally spaced from each other, and the resulting distribution is now not as skewed as the concentrations themselves. The arithmetic mean could then be computed in the log scale, that is,

$$\overline{\log x} = \frac{1}{n} \sum_{i=1}^{n} \log x_i$$

and used as a measure of location. However, it is usually preferable to work in the original scale by taking the antilogarithm of $\overline{\log x}$ to form the geometric mean, which leads to the following definition:

DEFINITION 2.4 ■■■
The **geometric mean** is the antilogarithm of $\overline{\log x}$, where

$$\overline{\log x} = \frac{1}{n} \sum_{i=1}^{n} \log x_i$$ ■

Any base can be used to compute logarithms for the geometric mean. The geometric mean will be the same regardless of which base is used. The only requirement is that the logs and antilogs in Definition 2.4 should be in the same base. Bases often used in practice are base 10 and base e; logs and antilogs using these bases can be computed using many pocket calculators.

EXAMPLE 2.10 **Infectious Disease** Compute the geometric mean for the sample in Table 2.4.

SOLUTION 1. For convenience, use base 10 to compute the logs and antilogs in this example.

2. Compute

$$\overline{\log x} = [21 \log(0.03125) + 6 \log(0.0625) + 8 \log(0.125)$$
$$+ 19 \log(0.250) + 17 \log(0.50) + 3 \log(1.0)]/74 = -0.846$$

3. The geometric mean = the antilogarithm of $-0.846 = 0.143$. ■

SECTION 2.3 **Some Properties of the Arithmetic Mean**

Consider a sample $x_1, \ldots, x_n$, which will be referred to as the original sample. To create a **translated sample** $x_1 + c, \ldots, x_n + c$, add a constant c to each data point.

Let $y_i = x_i + c$, $i = 1, \ldots, n$. Suppose we want to compute the arithmetic mean of the translated sample. We can show that the following relationship holds:

2.1 If
$$y_i = x_i + c, \quad i = 1, \ldots, n$$
then
$$\bar{y} = \bar{x} + c$$

Therefore, to find the arithmetic mean of the y's, compute the arithmetic mean of the x's and add the constant c.

This principle is useful because it is sometimes convenient to change the "origin" of the sample data, that is, compute the arithmetic mean after translation and transform back to the original origin.

EXAMPLE 2.11 In Table 2.3 it is more convenient to work with numbers that are near 0 than with numbers near 28 to compute the arithmetic mean of the time interval between menstrual periods. Thus, a translated sample might first be created by subtracting 28 days from each outcome in Table 2.3. The arithmetic mean of the translated sample could then be found and 28 added to get the actual arithmetic mean. The calculations are shown in Table 2.5.

TABLE 2.5
Translated sample for duration between successive menstrual periods in college-aged women

Value	Frequency	Value	Frequency	Value	Frequency
-4	5	1	96	6	7
-3	10	2	63	7	3
-2	28	3	24	8	2
-1	64	4	9	9	1
0	185	5	2	10	1

$\bar{y} = [(-4)(5) + (-3)(10) + \cdots + (10)(1)]/500 = 0.54$
$\bar{x} = \bar{y} + 28 = 0.54 + 28 = 28.54$ days ∎

Similarly, systolic blood-pressure scores are usually between 100 and 200. It is easy to subtract 100 from each blood-pressure score, find the mean of the translated sample, and add 100 to obtain the mean of the original sample.

What happens to the arithmetic mean if the units or scale being worked with are changed? A **rescaled sample** can be created:

$$y_i = cx_i, \quad i = 1, \ldots, n$$

The following result holds:

2.2 If
$$y_i = cx_i, \quad i = 1, \ldots, n$$
then
$$\bar{y} = c\bar{x}$$

Therefore, to find the arithmetic mean of the y's, compute the arithmetic mean of the x's and multiply it by the constant c.

EXAMPLE 2.12 Express the mean birthweight for the data in Table 2.1 (p. 8) in ounces rather than grams.

SOLUTION We know that 1 oz $= 28.35$ g and that $\bar{x} = 3166.9$ g. Thus, if the data were expressed in terms of ounces,

$$c = \frac{1}{28.35} \quad \text{and} \quad \bar{y} = \frac{1}{28.35}(3166.9) = 111.71 \text{ oz} \quad \blacksquare$$

Sometimes you want to change both the origin and the scale of the data at the same time. To do this, apply **(2.1)** and **(2.2)** as follows:

2.3 Let $x_1, \ldots, x_n$ be the original sample of data and let $y_i = c_1 x_i + c_2$, $i = 1, \ldots, n$, represent a transformed sample obtained by multiplying each original sample point by a factor c_1 and then shifting over by a constant c_2.

If $$y_i = c_1 x_i + c_2, \quad i = 1, \ldots, n$$

then $$\bar{y} = c_1 \bar{x} + c_2$$

EXAMPLE 2.13 If we have a sample of temperatures in °C with an arithmetic mean of 11.75°, then what is the arithmetic mean in °F?

SOLUTION Let y_i denote the °F temperature that corresponds to a °C temperature of x_i. Since the required transformation to convert the data to °F would be

$$y_i = \tfrac{9}{5}x_i + 32, \quad i = 1, \ldots, n$$

the arithmetic mean would be

$$\bar{y} = \tfrac{9}{5}(11.75) + 32 = 53.15°\text{F} \quad \blacksquare$$

SECTION 2.4 **Measures of Spread**

Consider the two samples shown in Figure 2.4. They represent two samples of cholesterol measurements, each on the same person, but using different measurement techniques. They appear to have about the same center, and whatever measure of central location is used will probably be about the same in the two

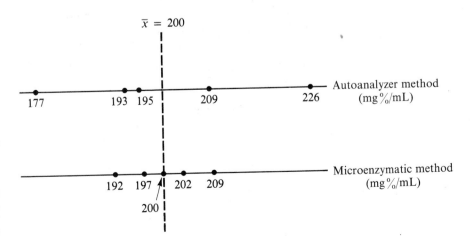

FIGURE 2.4
Two samples of
cholesterol
measurements on a
given person using an
Autoanalyzer and
a Microenzymatic
measurement
technique

samples. In fact, the arithmetic means are both 200 mg%/mL. However, the two samples visually appear to be radically different. This difference lies in the greater **variability**, or **spread**, of the Autoanalyzer method relative to the microenzymatic method. In this section, the notion of variability will be quantified. Many samples can be well described by the combination of a measure of central location and a measure of spread.

2.4.1 The Range

Several different measures can be used to describe the variability of a sample. Perhaps the simplest measure is the range.

DEFINITION 2.5 ■■■
The **range** is the difference between the largest and smallest observations in the sample. ■

EXAMPLE 2.14   The range in the sample of birthweights in Table 2.1 is

$$4146 - 2069 = 2077 \text{ g}$$ ■

EXAMPLE 2.15   Compute the ranges for the Autoanalyzer- and Microenzymatic-method data in Figure 2.4 and compare the variability of the two methods.

SOLUTION   The range for the Autoanalyzer method $= 226 - 177 = 49$ mg%/mL. The range for the Microenzymatic method $= 209 - 192 = 17$ mg%/mL. The Autoanalyzer method clearly seems more variable.
■

One advantage of the range is that it is very easy to compute once the sample points are ordered. One striking disadvantage is that it is very sensitive to extreme observations. Hence, if the lightest infant in Table 2.1 weighed 500 g rather than 2069 g, then the range would increase dramatically to $4146 - 500 = 3646$ g. Another disadvantage of the range is that it depends on the sample size (n). That is, the larger n is, the larger the range tends to be. This complication makes it difficult to compare ranges from different sized data sets.

2.4.2 Quantiles

Another approach that addresses some of the shortcomings of the range in quantifying the spread in a data set is the use of **quantiles** or **percentiles**. Intuitively, the pth percentile is the value V_p such that p percent of the sample points are less than or equal to V_p. The median, being the 50th percentile, is a special case of a quantile. As was the case for the median, a different definition is needed for the pth percentile, depending on whether $np/100$ is an integer or not.

DEFINITION 2.6 ■■■
The pth **percentile** is defined by
(1) The $(k + 1)$th largest sample point if $np/100$ is not an integer (where k is the largest integer less than $np/100$)
(2) The average of the $(np/100)$th and $(np/100 + 1)$th largest observations if $np/100$ is an integer.
■

The spread of a distribution can be characterized by specifying several percentiles. For example, the 10th and 90th percentiles are often used to characterize spread. Percentiles have the advantage over the range of being less sensitive to outliers and of not being much affected by the sample size (n).

EXAMPLE 2.16   Compute the 10th and 90th percentile for the birthweight data in Table 2.1 (p. 8).

SOLUTION   Since $20 \times .1 = 2$ and $20 \times 0.9 = 18$ are integers, the 10th and 90th percentiles are defined by

10th percentile: average of the 2nd and 3rd largest values = $(2581 + 2759)/2 = 2670$ g

90th percentile: average of the 18th and 19th largest values = $(3609 + 3649)/2 = 3629$ g

We would estimate that 80 percent of birthweights will fall between 2670 g and 3629 g, which gives us an overall feel for the spread of the distribution.   ■

EXAMPLE 2.17   Compute the 20th percentile for the white-count data in Table 2.2 (p. 10).

SOLUTION   Since $np/100 = 9 \times .2 = 1.8$ is not an integer, the 20th percentile is defined by the $(1 + 1)$th largest value = 2nd largest value = 5000.   ■

To compute percentiles, the sample points must be ordered. This can be difficult if n is even moderately large. An easy way to accomplish this is to use a stem-and-leaf plot (see Section 2.8.3).

There is no limit to the number of percentiles that can be computed. The most useful number is often determined by the sample size and by subject-matter considerations. Frequently used percentiles are quartiles (25th, 50th, and 75th percentiles), quintiles (20th, 40th, 60th, and 80th percentiles), and deciles (10th, 20th, ..., 90th percentiles). It is almost always instructive to look at some of the quantiles to get an overall impression of the spread and the general shape of a distribution.

2.4.3   **The Variance and Standard Deviation**

The principal difference between the Autoanalyzer- and Microenzymatic-method data in Figure 2.4 is that the Microenzymatic-method values are in some sense closer to the center of the sample than the Autoanalyzer-method values are. If the center of the sample is defined as the arithmetic mean, then a measure that can summarize the difference (or deviations) between the individual sample points and the arithmetic mean, that is,

$$x_1 - \bar{x}, x_2 - \bar{x}, \ldots, x_n - \bar{x}$$

is needed. One simple measure that would seem to accomplish this goal is

$$d = \frac{\sum_{i=1}^{n} (x_i - \bar{x})}{n}$$

Unfortunately, this measure will not work because of the following principle:

2.4   The sum of the deviations of the individual observations of a sample about the sample mean is always 0.

This can be seen as follows

$$\text{Sum of deviations} = \sum_{i=1}^{n} (x_i - \bar{x}) = \sum_{i=1}^{n} x_i - \sum_{i=1}^{n} \bar{x}$$

However, from the definition of the sample mean, $\sum_{i=1}^{n} x_i = n\bar{x}$. Furthermore, since $\bar{x}$ does not depend on i, $\sum_{i=1}^{n} \bar{x} = n\bar{x}$. Therefore,

$$\text{Sum of deviations} = n\bar{x} - n\bar{x} = 0$$

EXAMPLE 2.18   Compute the sum of the deviations about the mean for the Autoanalyzer- and Microenzymatic-method data in Figure 2.4.

SOLUTION   For the Autoanalyzer-method data,

$$d = (177 - 200) + (193 - 200) + (195 - 200) + (209 - 200) + (226 - 200)$$
$$= -23 - 7 - 5 + 9 + 26 = 0$$

For the Microenzymatic-method data,

$$d = (192 - 200) + (197 - 200) + (200 - 200) + (202 - 200) + (209 - 200)$$
$$= -8 - 3 + 0 + 2 + 9 = 0$$

Thus, d does not help distinguish the difference in spreads between the two methods.

A second idea is to use the squares of the deviations from the sample mean rather than the deviations themselves. The resulting measure of spread, denoted by s^2 is

$$s^2 = \frac{\sum_{i=1}^{n} (x_i - \bar{x})^2}{n}$$

The more usual form for this measure is with $(n - 1)$ in the denominator rather than with n. The resulting measure is called the sample variance (or variance).

DEFINITION 2.7 ■■
The **sample variance**, or **variance**, is defined as follows:

$$s^2 = \frac{\sum_{i=1}^{n} (x_i - \bar{x})^2}{(n - 1)}$$

A rationale for using $(n - 1)$ in the denominator rather than n is presented in the discussion of estimation in Chapter 6.

Another commonly used measure of spread is the sample standard deviation.

DEFINITION 2.8 ■■
The **sample standard deviation**, or **standard deviation**, is defined as follows:

$$s = \sqrt{\frac{\sum_{i=1}^{n} (x_i - \bar{x})^2}{(n - 1)}} = \sqrt{\text{sample variance}}$$

EXAMPLE 2.19   Compute the variance and standard deviation for the Autoanalyzer- and Microenzymatic-method data in Figure 2.4.

SOLUTION   **Autoanalyzer method**

$$s^2 = [(177 - 200)^2 + (193 - 200)^2 + (195 - 200)^2 + (209 - 200)^2 + (226 - 200)^2]/4$$

$$= (529 + 49 + 25 + 81 + 676)/4 = 1360/4 = 340$$

$$s = \sqrt{340} = 18.4$$

Microenzymatic method

$$s^2 = [(192 - 200)^2 + (197 - 200)^2 + (200 - 200)^2 + (202 - 200)^2 + (209 - 200)^2]/4$$

$$= (64 + 9 + 0 + 4 + 81)/4 = 158/4 = 39.5$$

$$s = \sqrt{39.5} = 6.3$$

Thus, the Autoanalyzer method has a standard deviation roughly three times as large as that of the Microenzymatic method.   ∎

One problem in using the variance is that it is difficult to compute in its original form, since the sample mean must first be computed, then the deviation of each sample point about the sample mean must be computed, and then the squares of these deviations about the sample mean must be summed. This procedure introduces two extra steps, which make the computation both more cumbersome and more error prone, especially because many pocket calculators can accumulate both the sum and sum of squares of a sample in one pass. This problem can be solved by using an alternative expression for the variance. To obtain this expression, recall from algebra that $(a + b)^2 = a^2 + 2ab + b^2$, and let $a = x_i$, $b = -\bar{x}$. Then $(x_i - \bar{x})^2$ can be written in the form $x_i^2 - 2x_i\bar{x} + \bar{x}^2$. Thus, s^2 can be rewritten as follows:

$$s^2 = \sum_{i=1}^{n} \frac{(x_i - \bar{x})^2}{(n - 1)} = \sum_{i=1}^{n} \frac{(x_i^2 - 2x_i\bar{x} + \bar{x}^2)}{(n - 1)}$$

$$= \frac{\sum_{i=1}^{n} x_i^2 + \sum_{i=1}^{n}(-2x_i\bar{x}) + \sum_{i=1}^{n} \bar{x}^2}{(n - 1)}$$

Since $-2\bar{x}$ and $\bar{x}^2$ are constants, $\sum_{i=1}^{n}(-2x_i\bar{x})$ can be written as $-2\bar{x}\sum_{i=1}^{n} x_i$ and $\sum_{i=1}^{n} \bar{x}^2$ as $n\bar{x}^2$, and the following expression can be obtained:

$$s^2 = \frac{\sum_{i=1}^{n} x_i^2 - 2\bar{x}\sum_{i=1}^{n} x_i + n\bar{x}^2}{(n - 1)}$$

$$= \frac{\sum_{i=1}^{n} x_i^2 - 2\bar{x}(n\bar{x}) + n\bar{x}^2}{(n - 1)}$$

$$= \frac{\left(\sum_{i=1}^{n} x_i^2 - n\bar{x}^2\right)}{(n - 1)}$$

It is sometimes more convenient to represent s^2 in terms of $\sum_{i=1}^{n} x_i$ rather than $\bar{x}$. To accomplish this, substitute $\left(\sum_{i=1}^{n} x_i/n\right)$ for $\bar{x}$ and obtain

$$s^2 = \frac{\sum_{i=1}^{n} x_i^2 - n\left(\sum_{i=1}^{n} x_i/n\right)^2}{(n - 1)}$$

$$= \frac{\sum_{i=1}^{n} x_i^2 - n\left(\sum_{i=1}^{n} x_i\right)^2/n^2}{(n - 1)} = \frac{\sum_{i=1}^{n} x_i^2 - \left(\sum_{i=1}^{n} x_i\right)^2/n}{(n - 1)}.$$

This gives the following two short forms for the variance:

2.5   The two short forms for the sample variance,

$$s^2 = \frac{\sum_{i=1}^n (x_i - \bar{x})^2}{(n-1)}$$

are given by $\dfrac{\sum_{i=1}^n x_i^2 - n\bar{x}^2}{(n-1)}$ and $\dfrac{\sum_{i=1}^n x_i^2 - \left(\sum_{i=1}^n x_i\right)^2/n}{(n-1)}$

The first form is useful when the sample mean has already been computed, whereas the second form is useful if the sum and sum of squares of the observations have been computed but the sample mean has not. Thus, the sample variance can be computed directly from the sum and the sum of squares of the individual observations. The second form is actually preferable from the standpoint of computational accuracy, since rounding error is often introduced in the computation of $\bar{x}$.

Similarly, the two short forms for the standard deviation can be written as follows:

2.6 $$s = \sqrt{s^2} = \sqrt{\frac{\sum_{i=1}^n x_i^2 - n\bar{x}^2}{(n-1)}} = \sqrt{\frac{\sum_{i=1}^n x_i^2 - \left(\sum_{i=1}^n x_i\right)^2/n}{(n-1)}}$$

EXAMPLE 2.20   Compute the variance and standard deviation for the Autoanalyzer and Microenzymatic data in Figure 2.4 using the short computational forms.

SOLUTION   **Autoanalyzer method**

$$\sum_{i=1}^5 x_i = 177 + 193 + 195 + 209 + 226 = 1000$$

$$\sum_{i=1}^5 x_i^2 = 177^2 + 193^2 + 195^2 + 209^2 + 226^2 = 201{,}360$$

$$s^2 = [201{,}360 - (1000)^2/5]/4 = (201{,}360 - 200{,}000)/4 = 1360/4 = 340$$

$$s = \sqrt{340} = 18.4$$

Microenzymatic method

$$\sum_{i=1}^5 x_i = 192 + 197 + 200 + 202 + 209 = 1000$$

$$\sum_{i=1}^5 x_i^2 = 192^2 + 197^2 + 200^2 + 202^2 + 209^2 = 200{,}158$$

$$s^2 = [200{,}158 - (1000)^2/5]/4 = (200{,}158 - 200{,}000)/4 = 158/4 = 39.5$$

$$s = \sqrt{39.5} = 6.3$$

Alternatively, s^2 could be represented in terms of $\bar{x}\,(=200)$ by writing

$$s^2 = \frac{200{,}158 - 5(200)^2}{4}$$

$$= \frac{(200{,}158 - 200{,}000)}{4} = \frac{158}{4} = 39.5$$

$$s = \sqrt{39.5} = 6.3 \qquad\blacksquare$$

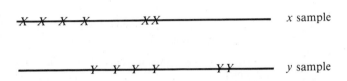

Some Properties of the Variance and Standard Deviation

The same questions can be asked of the variance and standard deviation as of the arithmetic mean: namely, How are the variance and standard deviation affected by a change in origin or a change in the units being worked with? Suppose there is a sample $x_1, \ldots, x_n$ and all data points in the sample are shifted by a constant c; that is, a new sample $y_1, \ldots, y_n$ is created such that $y_i = x_i + c$, $i = 1, \ldots, n$.

FIGURE 2.5
Comparison of the variances of two samples, where one sample has an origin shifted relative to the other

In Figure 2.5, we would clearly expect that the variance and standard deviation would remain the same, since the relationship of the points in the sample relative to one another remains the same. This property is stated as follows:

| 2.7 | Suppose there are two samples

$$x_1, \ldots, x_n \qquad \text{and} \qquad y_1, \ldots, y_n$$

where
$$y_i = x_i + c, \quad i = 1, \ldots, n$$

If the respective sample variances of the two samples are denoted by

$$s_x^2 \qquad \text{and} \qquad s_y^2$$

then
$$s_y^2 = s_x^2$$

EXAMPLE 2.21   Compare the variances and standard deviations for the menstural period data in Tables 2.3 (p. 12) and 2.5 (p. 14).

SOLUTION   The variance and standard deviation of the two samples are the same, since the second sample was obtained from the first by subtracting 28 days from each data value; that is,

$$y_i = x_i = 28 \qquad\blacksquare$$

Suppose the units are now changed so that a new sample $y_1, \ldots, y_n$ is created such that $y_i = cx_i$, $i = 1, \ldots, n$. The following relationship holds between the variances of the two samples.

| 2.8 | Suppose there are two samples |

$$x_1, \ldots, x_n \quad \text{and} \quad y_1, \ldots, y_n$$

where
$$y_i = cx_i, \quad i = 1, \ldots, n, \quad c > 0$$

Then
$$s_y^2 = c^2 s_x^2 \quad s_y = cs_x$$

This can be shown by noting that

$$s_y^2 = \frac{\sum_{i=1}^n (y_i - \bar{y})^2}{(n-1)} = \frac{\sum_{i=1}^n (cx_i - c\bar{x})^2}{(n-1)}$$

$$= \frac{\sum_{i=1}^n c[(x_i - \bar{x})]}{(n-1)} = \frac{\sum_{i=1}^n c^2(x_i - \bar{x})^2}{(n-1)}$$

$$= \frac{c^2 \sum_{i=1}^n (x_i - \bar{x})^2}{(n-1)} = c^2 s_x^2$$

$$s_y = \sqrt{c^2 s_x^2} = cs_x$$

EXAMPLE 2.22   Compute the variance and standard deviation of the birthweight data in Table 2.1 (p. 8) in both grams and ounces.

The original data are given in grams, so first compute the variance and standard deviation in these units.

$$\sum_{i=1}^{20} x_i = 63{,}338 \qquad \sum_{i=1}^{20} x_i^2 = 204{,}353{,}260$$

$$s^2 = [204{,}353{,}260 - (63{,}338)^2/20]/19 = 3{,}768{,}147.8/19 = 198{,}323.6 \ g^2$$

$$s = 445.3 \ g$$

To compute the variance and standard deviation in ounces, note that

$$1 \ oz = 28.35 \ g \quad \text{or} \quad y_i = \frac{1}{28.35} x_i$$

Thus
$$s^2 \ (oz) = \frac{1}{(28.35)^2} s^2(g) = 246.8 \ oz^2$$

$$s \ (oz) = \frac{1}{28.35} s \ (g) = 15.7 \ oz \qquad \blacksquare$$

Thus, if the sample points change in scale by a factor of c, the variance changes by a factor of c^2 and the standard deviation changes by a factor of c. This relationship is the main reason why the standard deviation is more often used than the variance as a measure of spread, since the standard deviation and the arithmetic mean are in the same units, whereas the variance and the arithmetic mean are not. Thus, as illustrated in Examples 2.12 and 2.22, both the mean and the standard deviation change by a factor of 28.35 in the birthweight data of Table 2.1 when the units are expressed in terms of ounces rather than grams.

The mean and standard deviation are the most widely used measures of location and spread in the literature. One of the principal reasons for this is that the normal (or bell-shaped) distribution is defined explicitly in terms of these two parameters, and this distribution has wide applicability in many biological and medical settings. The normal distribution is discussed extensively in Chapter 5.

SECTION 2.6   The Coefficient of Variation

It is useful to relate the arithmetic mean and the standard deviation together, since, for example, a standard deviation of 10 would mean something different conceptually if the arithmetic mean were 10 than if it were 1000. A special measure, called the coefficient of variation, is often used for this purpose.

DEFINITION 2.9 ■■

The **coefficient of variation** (CV) is defined by

$$100\% \times (s/\bar{x})$$ ■

This measure remains the same regardless of what units are used, because if the units are changed by a factor c, both the mean and standard deviation change by the factor c; the CV, which is the ratio between them, remains unchanged.

EXAMPLE 2.23   Compute the coefficient of variation for the data in Table 2.1 (p. 8) when the birthweights are expressed in either grams or ounces.

SOLUTION $$CV = 100\% \times (s/\bar{x}) = 100\% \times (445.3\ \text{g}/3166.9\ \text{g}) = 14.1\%$$

If the data were expressed in ounces, then

$$CV = 100\% \times (15.7\ \text{oz}/111.71\ \text{oz}) = 14.1\%$$ ■

The coefficient of variation is most useful in comparing the variability of several different samples, each with different arithmetic means. This is because a higher variability is usually expected when the mean increases, and the CV is a measure that accounts for this variability. Thus, if we are conducting a study where air pollution is measured at several sites and we wish to compare day-to-day variability at the different sites, we might expect a higher variability for the more highly polluted sites. A more accurate comparison could be made by comparing the CV's at different sites than by comparing the standard deviations.

SECTION 2.7   Grouped Data

Sometimes the sample size is prohibitively large to display all the raw data. Also, data are frequently collected in grouped form, since the required degree of accuracy to specify a measured quantity exactly is often lacking, because of either measurement error or imprecise patient recall. For example, systolic blood-pressure measurements taken with a standard cuff are usually specified to the nearest 2 mm Hg, since assessing them with any more precision is difficult using this instrument. Thus, a stated measurement of 120 mm Hg may actually imply that the reading is some number $\geqslant 119$ mm Hg and < 121 mm Hg. Similarly, because dietary recall is generally not very accurate, the most precise estimate of fish

TABLE 2.6									
58	118	92	108	132	32	140	138	96	161
120	86	115	118	95	83	112	128	127	124
123	134	94	67	124	155	105	100	112	141
104	132	98	146	132	93	85	94	116	113
121	68	107	122	126	88	89	108	115	85
111	121	124	104	125	102	122	137	110	101
91	122	138	99	115	104	98	89	119	109
104	115	138	105	144	87	88	103	108	109
128	106	125	108	98	133	104	122	124	110
133	115	127	135	89	121	112	135	115	64

TABLE 2.6
Sample of birthweights from 100 consecutive deliveries (oz)

consumption might take the following form: 2–3 servings per day, 1 serving per day, 5–6 servings per week, 2–4 servings per week, 1 serving per week, <1 serving per week and ⩾1 serving per month, never.

Consider the data set in Table 2.6, which represents the birthweights from 100 consecutive deliveries at a Boston hospital. Suppose we wish to display these data for publication purposes. How can we do this? If the data are on a computer, then the simplest way to display the data would be to generate a frequency distribution using one of the common statistical packages.

DEFINITION 2.10 ■■

A **frequency distribution** is an ordered display of each value in a data set together with its **frequency**, that is, the number of times that value occurs in the data set. In addition, the percentage of sample points that take on a particular value is also typically given. ■

A frequency distribution of the sample of 100 birthweights in Table 2.6 was generated using the Statistical Analysis System (SAS) package and is displayed in Table 2.7.

The SAS frequency-distribution program provides the frequency, cumulative frequency (CUM FREQ), relative frequency (PERCENT), and cumulative percent (CUM PERCENT) for each birthweight present in the sample. For any particular birthweight b, the cumulative frequency, or CUM FREQ, is the number of birthweights in the sample that are less than or equal to b. The PERCENT = $100 \times$ FREQUENCY/n, while the cumulative percent (CUM PERCENT) = $100 \times$ CUM FREQ/n = the percentage of birthweights less than or equal to b.

If the number of unique sample values is large, then a frequency distribution may still be too detailed a summary for publication purposes. Instead, the data could be grouped into broader categories. Some general instructions for categorizing the data are provided in the following guidelines:

1. Subdivide the data into k intervals, starting at some lower bound y_1 and ending at some upper bound y_{k+1}.

2. The first interval is from y_1 inclusive to y_2 exclusive; the second interval is from y_2 inclusive to y_3 exclusive; ...; the kth and last interval is from y_k inclusive to y_{k+1} exclusive. The rationale for this representation is to make certain that the group intervals include all possible values *and* do not overlap. These errors are common in the presentation of grouped data.

3. The group intervals are generally chosen to be equal, although the appropriateness of equal group sizes should be dictated more by subject-matter

TABLE 2.7

Frequency distribution of birthweight data in Table 2.6 using the Statistical Analysis System (SAS)

```
----------------------------------------------------------------
    SAMPLE OF BIRTHWEIGHTS FROM 100 CONSECUTIVE DELIVERIES (OZ.)
```

BIRTHWT	FREQUENCY	CUM FREQ	PERCENT	CUM PERCENT
32	1	1	1.000	1.000
58	1	2	1.000	2.000
64	1	3	1.000	3.000
67	1	4	1.000	4.000
68	1	5	1.000	5.000
83	1	6	1.000	6.000
85	2	8	2.000	8.000
86	1	9	1.000	9.000
87	1	10	1.000	10.000
88	2	12	2.000	12.000
89	3	15	3.000	15.000
91	1	16	1.000	16.000
92	1	17	1.000	17.000
93	1	18	1.000	18.000
94	2	20	2.000	20.000
95	1	21	1.000	21.000
96	1	22	1.000	22.000
98	3	25	3.000	25.000
99	1	26	1.000	26.000
100	1	27	1.000	27.000
101	1	28	1.000	28.000
102	1	29	1.000	29.000
103	1	30	1.000	30.000
104	5	35	5.000	35.000
105	2	37	2.000	37.000
106	1	38	1.000	38.000
107	1	39	1.000	39.000
108	4	43	4.000	43.000
109	2	45	2.000	45.000
110	2	47	2.000	47.000
111	1	48	1.000	48.000
112	3	51	3.000	51.000
113	1	52	1.000	52.000
115	6	58	6.000	58.000
116	1	59	1.000	59.000
118	2	61	2.000	61.000
119	1	62	1.000	62.000
120	1	63	1.000	63.000
121	3	66	3.000	66.000
122	4	70	4.000	70.000
123	1	71	1.000	71.000
124	4	75	4.000	75.000
125	2	77	2.000	77.000
126	1	78	1.000	78.000
127	2	80	2.000	80.000
128	2	82	2.000	82.000
132	3	85	3.000	85.000
133	2	87	2.000	87.000
134	1	88	1.000	88.000
135	2	90	2.000	90.000
137	1	91	1.000	91.000
138	3	94	3.000	94.000
140	1	95	1.000	95.000
141	1	96	1.000	96.000
144	1	97	1.000	97.000
146	1	98	1.000	98.000
155	1	99	1.000	99.000
161	1	100	1.000	100.000

```
----------------------------------------------------------------
```

considerations. Thus, equal intervals might be appropriate for the blood-pressure or birthweight data but not for the dietary-recall data, where the nature of the data dictates unequal group sizes corresponding to how most people remember what they eat.

4. A count is made of the number of units that fall in each interval, which is denoted by the frequency within that interval.

5. The midpoint of each group interval is computed for calculation of descriptive statistics. The midpoint of the first interval is denoted by

$$m_1 = \frac{y_1 + y_2}{2}$$

the midpoint of the second interval by

$$m_2 = \frac{y_2 + y_3}{2}, \ldots$$

and the midpoint of the last interval by

$$m_k = \frac{y_k + y_{k+1}}{2}.$$

The intervals and their midpoints are depicted in Figure 2.6.

FIGURE 2.6
Subdivision of the real
line for the purpose of
forming group intervals

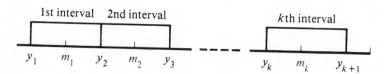

6. Finally, for the purpose of computing descriptive statistics, the group intervals and their midpoints, m_i, and frequencies, f_i, are then displayed concisely in a table such as Table 2.8.

For example, the raw data in Table 2.6 might be displayed according to the format in Table 2.9.

If we are confronted with grouped data either in the form of published data from a secondary source or from our own data, then we want to be able to compute

TABLE 2.8
General layout of
grouped data

Group interval	Midpoint of group interval	Frequency
$\geqslant y_1, < y_2$	m_1	f_1
$\geqslant y_2, < y_3$	m_2	f_2
$\vdots$	$\vdots$	$\vdots$
$\geqslant y_i, < y_{i+1}$	m_i	f_i
$\vdots$	$\vdots$	$\vdots$
$\geqslant y_k, < y_{k+1}$	m_k	f_k

TABLE 2.9

Grouped frequency distribution of birthweight (oz) from 100 consecutive deliveries

Group interval	Midpoint	Frequency
$\geqslant 29.5,\ < 69.5$	49.5	5
$\geqslant 69.5,\ < 89.5$	79.5	10
$\geqslant 89.5,\ < 99.5$	94.5	11
$\geqslant 99.5,\ < 109.5$	104.5	19
$\geqslant 109.5,\ < 119.5$	114.5	17
$\geqslant 119.5,\ < 129.5$	124.5	20
$\geqslant 129.5,\ < 139.5$	134.5	12
$\geqslant 139.5,\ < 169.5$	154.5	6
		100

grouped means and variances that are analogous to the arithmetic mean and variance. Suppose that f_i observations fall in the ith group interval, $i = 1, \ldots, k$, and that the midpoint of the ith interval is m_i, $i = 1, \ldots, k$, where $n = \sum_{i=1}^{k} f_i =$ total number of observations over all groups. The grouped mean is then defined as follows:

DEFINITION 2.11

The **grouped mean** is defined by

$$\bar{x}_g \equiv \frac{\sum_{i=1}^{k} f_i m_i}{\sum_{i=1}^{k} f_i}$$

EXAMPLE 2.24 Compute the grouped mean for the data in Table 2.9.

SOLUTION

$$\bar{x}_g = \frac{\sum_{i=1}^{k} f_i m_i}{\sum_{i=1}^{k} f_i}$$

$$= [5(49.5) + 10(79.5) + 11(94.5) + \cdots + 6(154.5)]/100$$

$$= (11{,}045)/100 = 110.45 \text{ oz}$$

DEFINITION 2.12

The **grouped variance** is defined by

$$s_g^2 \equiv \frac{\sum_{i=1}^{k} f_i (m_i - \bar{x}_g)^2}{\left(\sum_{i=1}^{k} f_i\right) - 1}$$

As for the ungrouped variance, the expression for the grouped variance can be simplified, yielding the following two short forms:

2.9 **Short Forms for the Grouped Variance**

$$\text{Grouped variance} = \frac{\sum_{i=1}^{k} f_i m_i^2 - n\bar{x}_g^2}{(n-1)} = \frac{\sum_{i=1}^{k} f_i m_i^2 - \left(\sum_{i=1}^{k} f_i m_i\right)^2/n}{(n-1)}$$

EXAMPLE 2.25 Compute the grouped variance for the data in Table 2.9.

SOLUTION

$$\sum_{i=1}^{k} f_i m_i^2 = 5(49.5)^2 + 10(79.5)^2 + \cdots + 6(154.5)^2 = 1{,}274{,}355$$

Thus,

$$s_g^2 = [1{,}274{,}355 - (11{,}045)^2/100]/99$$
$$= 54{,}434.75/99 = 549.85$$

and

$$s_g = \sqrt{549.85} = 23.45 \text{ oz} \qquad \blacksquare$$

SECTION 2.8 ## Graphical Methods for Grouped Data

In Section 2.7 we concentrated on methods for presenting grouped data in tabular form and on numerical measures for describing such data. In this section these techniques are supplemented by presenting certain commonly used graphical methods for displaying grouped data. The purpose of using graphical displays is to give a quick overall impression of the data, which is sometimes difficult to obtain with numerical measures.

2.8.1 ### Bar Graphs

One of the most widely used methods for displaying grouped data is the bar graph.

A **bar graph** can be constructed as follows:

(1) The data are divided into a number of groups using the guidelines provided in Section 2.7.

(2) For each group a rectangle is constructed with a base of a constant width and a height proportional to the frequency within that group.

(3) The rectangles are generally not contiguous and are equally spaced from each other.

A bar graph of daily vitamin A consumption among 200 cancer cases and 200 age- and sex-matched controls is presented in Figure 2.1 (p. 6).

2.8.2 ### Histograms

The bar graph tends to work well with grouped data when the groups are characterized by nonnumerical attributes, such as {current smoker/ex-smoker/ never smoker} or {patient gets worse/patient gets better/patient stays the same}. If the groups are characterized by a numerical attribute, such as systolic blood pressure or birthweight, then a histogram is preferable. For a histogram, the position of the rectangle will correspond to the location of the group interval along the x-axis, and the size of the rectangle will correspond to the frequency within the group.

A **histogram** is constructed as follows:

(1) The data are divided into groups as described in Section 2.7.

(2) A rectangle is constructed for each group. The location of the base of the rectangle corresponds to the position of the ends of the group interval along the x-axis, and the **area** of the rectangle is proportional to the frequency within the group.

(3) The scale used along either axis should allow all the rectangles to fit into the space allotted for the graph.

Note that the area, rather than the height, is proportional to the frequency. If the length of each group interval is the same, then the area and the height are in the same proportions and the height will be proportional to the frequency as well. However, if one group interval is 5 times as long as another and the two group intervals have the same frequency, then the first group interval should have a height $\frac{1}{5}$ as long as the second group interval so that the areas will be the same. A common mistake in the literature is to construct histograms with group intervals of different lengths but with the height proportional to the frequency. This representation gives a misleading impression of the data. A histogram for the birthweight data in Table 2.9 is given in Figure 2.7.

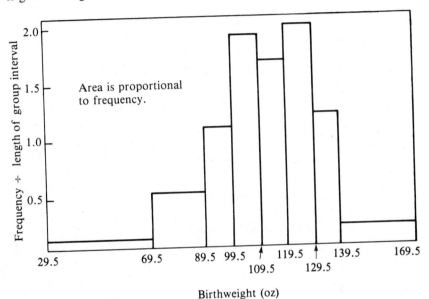

FIGURE 2.7
Histogram for the birthweight data in Table 2.9

2.8.3 Stem-and-Leaf Plots

Two problems with histograms are that (1) they are somewhat difficult to construct and (2) the sense of what the actual sample points are within the respective groups is lost. One type of graphical display that overcomes these problems is the stem-and-leaf plot.

A **stem-and-leaf** plot can be constructed as follows:

(1) Separate each data point into a stem component and a leaf component, respectively, where the stem component consists of the number formed by all but the rightmost digit of the number, and the leaf component consists of the rightmost digit. Thus, the stem of the number 483 is 48, and the leaf is 3.

(2) Write the smallest stem in the data set in the upper-left-hand corner of the plot.

(3) Write the second stem, which equals the first stem + 1, below the first stem.

(4) Continue with step 3 until you reach the highest stem in the data set.

(5) Draw a vertical bar to the right of the column of stems.

(6) For each number in the data set, find the appropriate stem and write the leaf to the right of the vertical bar.

The collection of leaves thus formed will take on the general shape of the distribution of the sample points. Furthermore, the actual sample values are preserved and yet there is a grouped display for the data, which is a distinct advantage over a histogram. Finally, a stem-and-leaf plot can usually be constructed more quickly than a histogram from raw data, since the number of data points in each group interval do not have to be counted. It is also easy to compute the median and the range from a stem-and-leaf plot. A stem-and-leaf plot is given in Figure 2.8 for the birthweight data in Table 2.6 (p. 24). Thus, the point 5|8 represents 58, 11|8 represents 118, and so forth. Notice how this plot gives an overall feel for the distribution without losing the individual values.

There are variations of stem-and-leaf plots where the leaf can consist of more than one digit. This variation might be appropriate for the birthweight data in Table 2.1 (p. 8), since the number of three-digit stems required would be very large relative to the number of data points. In this case the leaf would consist of the rightmost two digits and the stem the leftmost two digits, and the pairs of digits to the right of the vertical bar would be underlined to distinguish between two different leaves. The stem-and-leaf display for the data in Table 2.1 is presented in Figure 2.9.

Another common variation on the ordinary stem-and-leaf plot if the number of leaves is large is to allow more than one line for each stem. Similarly, one can position the largest stem at the top of the plot and the smallest stem at the bottom of the plot. In Figure 2.10 some graphical displays using the SAS UNIVARIATE procedure are given to illustrate this technique.

Notice that each stem is allowed two lines, with the leaves from 5 to 9 on the upper line and the leaves from 0 to 4 on the lower line. Furthermore, the leaves are ordered on each line, and a count of the number of leaves on each line is provided under the # column to allow easy computation of the median and other quantiles. Thus, the number 7 in the # column on the upper line for stem 12 indicates that

```
 3 │ 2
 4 │
 5 │ 8
 6 │ 7  8  4
 7 │
 8 │ 6  3  5  8  9  5  9  7  8  9
 9 │ 2  6  5  4  8  3  4  1  9  8  8
10 │ 8  5  0  4  7  8  4  2  1  4  9  4  5  3  8  9  6  8  4
11 │ 8  5  8  2  2  6  3  5  1  0  5  9  5  0  5  2  5
12 │ 0  8  7  4  3  4  1  2  6  1  4  5  2  2  8  5  2  4  7  1
13 │ 2  8  4  2  2  7  8  8  3  3  5  5
14 │ 0  1  6  4
15 │ 5
16 │ 1
```

FIGURE 2.8
Stem-and-leaf plot for the birthweight data (oz) in Table 2.6

```
20 | 69
21 |
22 |
23 |
24 |
25 | 81
26 |
27 | 59
28 | 41 38 34
29 |
30 | 31
31 | 01
32 | 65 60 45 00 48
33 | 23 14
34 | 84
35 | 41
36 | 49 09
37 |
38 |
39 |
40 |
41 | 46
```

FIGURE 2.9
Stem-and-leaf plot for the birthweight data (g) in Table 2.1

```
STEM LEAF                      #      BOXPLOT
  16 1                         1         |
  15 5                         1         |
  15                                     |
  14 6                         1         |
  14 014                       3         |
  13 557888                    6         |
  13 222334                    6         |
  12 5567788                   7         |
  12 0111222234444            13      +-----+
  11 5555556889               10      |     |
  11 0012223                   7      +--+--+
  10 5567888899               10      |     |
  10 012344444                 9      |     |
   9 568889                    6      +-----+
   9 12344                     5         |
   8 556788999                 9         |
   8 3                         1         |
   7                                     |
   7                                     |
   6 78                        2         |
   6 4                         1         |
   5 8                         1         0
   5
   4
   4
   3
   3 2                         1         0
     ----+----+----+----+
  MULTIPLY STEM.LEAF BY 10**+01
```

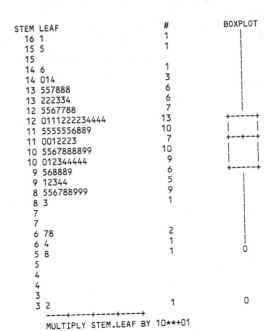

FIGURE 2.10
Stem-and-leaf and box plots for the birthweight data (oz) in Table 2.6 as generated by the SAS UNIVARIATE procedure

there are 7 birthweights from 125 to 129 oz in the sample, whereas the number 13 indicates that there are 13 birthweights from 120 to 124 oz. Finally, a multiplication factor is given in the bottom of the display to allow for the representation of decimal numbers in stem-and-leaf form. In particular, if no multiplication factor (m) is present, then it is assumed that all numbers have actual value stem.leaf; whereas if m is present, then the actual value of the number is assumed to be stem.leaf $\times 10^m$. Thus, for example, since the multiplication factor is 10^1, the value $6\,4$ on the stem-and-leaf plot represents the number $6.4 \times 10^1 = 64$ oz.

2.8.4 Box Plots

In Section 2.2.3 the comparison of the arithmetic mean and the median was discussed as a method for looking at the skewness of a distribution. This goal can also be accomplished by a graphical technique known as the **box plot**. To describe a box plot, the concept of the **hinges** of a sample must be introduced. To accomplish this, it is first necessary to understand the notion of the depth of the median.

DEFINITION 2.13

The **depth** (m) of the median for a sample of size n is

(1) $\dfrac{n}{2}$ if n is even

(2) $\dfrac{n+1}{2}$ if n is odd

The upper and lower hinges can be thought of conceptually as the approximate 75th and 25th percentiles of the sample, that is, the points $\frac{3}{4}$ and $\frac{1}{4}$ along the way in the ordered sample.

DEFINITION 2.14

The **upper hinge** of a sample is

(1) The $\dfrac{(m+1)}{2}$th largest point if m is odd

(2) The average of the $\dfrac{m}{2}$th and $\left(\dfrac{m}{2}+1\right)$th largest points if m is even

where $m =$ depth of the median. The **lower hinge** is defined similarly, starting from the smallest points in the sample.

EXAMPLE 2.26 Compute the upper and lower hinges for the birthweight data in Table 2.6 (p. 24).

SOLUTION Since $n = 100$, it follows that the depth of the median (m) = 50. Since m is even, the upper hinge is given by the average of the $\frac{50}{2}$th and $(\frac{50}{2}+1)$th largest sample values or the average of the 25th and 26th largest points in the sample. In the stem-and-leaf plot in Figure 2.10, counting down from the top, $1 + 1 + 1 + 3 + 6 + 6 + 7 = 25$ points are in the upper 12 row or above. Thus, the 25th largest point is the smallest number in the upper 12 row, which equals 125 oz. Also, the 26th largest point = largest number in the lower 12 row = 124 oz. Thus, the upper hinge = $(125 + 124)/2 = 124.5$ oz.

Similarly, the lower hinge = the average of the 25th and 26th smallest points in the sample. Counting up from the bottom, $1 + 1 + 1 + 2 + 1 + 9 + 5 = 20$ points are in the lower 9 row or below, and 26 points are in the upper 9 row or below. Thus, the 25th smallest point = the 2nd largest value in the upper 9 row = 98; the 26th smallest point = the largest value in the upper 9 row = 99. Therefore, the lower hinge = $(98 + 99)/2 = 98.5$ oz. ∎

How can hinges be used to judge the symmetry of a distribution?

(1) If the distribution is symmetric, then the upper and lower hinges should be approximately equally spaced from the median.

(2) If the upper hinge is farther from the median than the lower hinge, then the distribution is positively skewed.

(3) If the lower hinge is farther from the median than the upper hinge, then the distribution is negatively skewed.

These relationships are illustrated graphically in a box plot. In Figure 2.10 the top of the box corresponds to the upper hinge, whereas the bottom of the box corresponds to the lower hinge. A horizontal line is also drawn at the median value. Furthermore, in the SAS implementation of the box plot, the sample mean is indicated by a + sign.

EXAMPLE 2.27 What can be learned about the symmetry properties of the distribution of birthweights from the box plot in Figure 2.10?

SOLUTION In Figure 2.10, because the lower hinge is farther from the median than the upper hinge, the distribution is slightly negatively skewed. This pattern is true of many birthweight distributions. ∎

In addition to displaying the symmetry properties of a sample, a box plot can also be used to give a feel for the spread of a sample and can help identify possible outlying values, that is, values that seem inconsistent with the rest of the points in the sample. In the context of box plots, outlying values are defined as follows:

DEFINITION 2.15 ■■
An **outlying value** is a value x such that either

(1) $x >$ upper hinge + 1.5 × (upper hinge − lower hinge) or

(2) $x <$ lower hinge − 1.5 × (upper hinge − lower hinge) ■

DEFINITION 2.16 ■■
An **extreme outlying value** is a value x such that either

(1) $x >$ upper hinge + 3.0 × (upper hinge − lower hinge) or

(2) $x <$ lower hinge − 3.0 × (upper hinge − lower hinge)

The box plot is then completed by

(1) Drawing a vertical bar from the upper hinge to the largest nonoutlying value in the sample

(2) Drawing a vertical bar from the lower hinge to the smallest nonoutlying value in the sample

(3) Individually identifying the outlying and extreme outlying values in the sample by 0's and *'s, respectively ■

EXAMPLE 2.28   Using the box plot in Figure 2.10, comment on the spread of the sample in Table 2.6 (p. 24) and the presence of outlying values.

SOLUTION   Since the upper and lower hinges are 124.5 and 98.5 oz, respectively, an outlying value x must satisfy the following relations:

$$x > 124.5 + 1.5 \times (124.5 - 98.5) = 124.5 + 39.0 = 163.5$$

or $$x < 98.5 - 1.5 \times (124.5 - 98.5) = 98.5 - 39.0 = 59.5$$

Similarly, an extreme outlying value x must satisfy the following relations:

$$x > 124.5 + 3.0 \times (124.5 - 98.5) = 124.5 + 78.0 = 202.5$$

or $$x < 98.5 - 3.0 \times (124.5 - 98.5) = 98.5 - 78.0 = 20.5$$

Thus, the values 32 and 58 oz are outlying values but not extreme outlying values. These values are identified by 0's on the box plot. A vertical bar extends from 64 oz (the smallest nonoutlying value) to the lower hinge and from 161 oz (the largest nonoutlying value = the largest value in the sample) to the upper hinge. The accuracy of the two identified outlying values should probably be checked.   ∎

The methods used to identify outlying values are controversial, and the method given in Definitions 2.15 and 2.16 is not widely accepted by all statisticians. Nevertheless, it is given here to facilitate the understanding of box plots.

Many more details on stem-to-leaf plots, box plots, and other exploratory data methods are given in Tukey [3].

SECTION 2.9   ## Summary

In this chapter several **numerical and graphical methods for describing data** for the purpose of

(1) quickly summarizing a data set and for

(2) presenting results to others were presented.

In general, a data set can be described numerically in terms of a **measure of location** and a **measure of spread**. Several alternatives were introduced for each of these measures, including the **arithmetic mean, median, mode**, and **geometric mean** as possible choices for measures of location, and the **standard deviation, quantiles**, and **range** as possible choices for measures of spread. Criteria were discussed for choosing the appropriate measures in particular circumstances. Several graphical techniques for summarizing data, including traditional methods, such as the **bar graph** and **histogram**, and some more modern methods characteristic of exploratory data analysis (EDA), such as the **stem-and-leaf plot** and **box plot**, were introduced.

How do the descriptive methods in this chapter fit in with the methods of statistical inference discussed later in this book? Specifically, if, based on some prespecified hypotheses, some interesting trends using descriptive methods can be found, then we need some method to judge how "significant" these trends are. For this purpose several commonly used **probability models** are introduced in Chapters 3 through 5 and approaches for testing the validity of these models using the methods of **statistical inference** are explored in Chapters 6 through 12.

PROBLEMS

Pathology

The data in Table 2.10 are measurements from a group of 10 normal males and 11 males with left-heart disease taken at autopsy at a particular hospital. Measurements were made on several variables at that time, and the table presents the measurements on total heart weight (THW) and total body weight (BW). Assume that the diagnosis of left-heart disease is made independently of these variables.

2.1 Compute the mean and median for each variable in each disease group.

2.2 Compute the variance, standard deviation, range, and coefficient of variation for each variable in each disease group.

2.3 Group the data in some appropriate way and compute the grouped mean, grouped variance, and grouped standard deviation for each variable in each disease group.

2.4 Plot a histogram of each variable in each disease group with the groupings created in Problem 2.3.

2.5 Construct a stem-and-leaf plot for each variable in each disease group.

2.6 Can you qualitatively compare THW in the normal and abnormal groups from your answers to Problems 2.1 through 2.5? What about BW? (This topic will be covered formally in our later work on t tests.)

2.7 Is there any qualitative evidence of a relation between THW and BW *within* each of the disease groups? (*Hint*: A plot of THW versus BW may help here. This topic will be covered formally in our work on regression analysis.)

2.8 What are the principal differences (if any) between the groups as surmised from your answers to Problems 2.1 through 2.7?

Infectious Disease

The data in Table 2.11 are a sample from a larger data set collected on persons discharged from a selected Pennsylvania hospital as part of a retrospective chart review of antibiotic usage in hospitals [4].

2.9 Compute the mean and median for duration of hospitalization for the 25 patients.

2.10 Compute the standard deviation and range for the duration of hospitalization for the 25 patients.

2.11 It is of clinical interest to know if the duration of hospitalization is affected by whether or not a patient has received antibiotics. Can you answer this question using either numerical or graphical methods?

Suppose the origin for a data set is changed by adding a constant to each observation.

2.12 What is the effect on the median?

2.13 What is the effect on the mode?

TABLE 2.10 Autopsy data

Left-heart disease males			Normal males		
Observation number	THW(g)	BW(kg)	Observation number	THW(g)	BW(kg)
1	450	54.6	1	245	40.8
2	760	73.5	2	350	67.4
3	325	50.3	3	340	53.3
4	495	44.6	4	300	62.2
5	285	58.1	5	310	65.5
6	450	61.3	6	270	47.5
7	460	75.3	7	300	51.2
8	375	41.1	8	360	74.9
9	310	51.5	9	405	59.0
10	615	41.7	10	290	40.5
11	425	59.7			

TABLE 2.11 Hospital-stay data

ID no	Duration of hospital stay	Age	Sex (1 = M, 2 = F)	First temp. following admission	First WBC (× 10³) following admission	Received anti-biotic (1 = yes, 2 = no)	Received bacterial culture (1 = yes, 2 = no)	Service (1 = med., 2 = surg.)
1	5	30	2	99.0	8	2	2	1
2	10	73	2	98.0	5	2	1	1
3	6	40	2	99.0	12	2	2	2
4	11	47	2	98.2	4	2	2	2
5	5	25	2	98.5	11	2	2	2
6	14	82	1	96.8	6	1	2	2
7	30	60	1	99.5	8	1	1	1
8	11	56	2	98.6	7	2	2	1
9	17	43	2	98.0	7	2	2	1
10	3	50	1	98.0	12	2	1	2
11	9	59	2	97.6	7	2	1	1
12	3	4	1	97.8	3	2	2	2
13	8	22	2	99.5	11	1	2	2
14	8	33	2	98.4	14	1	1	2
15	5	20	2	98.4	11	2	1	2
16	5	32	1	99.0	9	2	2	2
17	7	36	1	99.2	6	1	2	2
18	4	69	1	98.0	6	2	2	2
19	3	47	1	97.0	5	1	2	1
20	7	22	1	98.2	6	2	2	2
21	9	11	1	98.2	10	2	2	2
22	11	19	1	98.6	14	1	2	2
23	11	67	2	97.6	4	2	2	1
24	9	43	2	98.6	5	2	2	2
25	4	41	2	98.0	5	2	2	1

2.14 What is the effect on the geometric mean?

2.15 What is the effect on the range?

Suppose the scale for a data set is changed by multiplying each observation by a positive constant.

2.16 What is the effect on the median?

2.17 What is the effect on the mode?

2.18 What is the effect on the geometric mean?

2.19 What is the effect on the range?

Renal Disease

For a study of kidney disease, the following measurements were made on a sample of women working in several factories in Switzerland. They represent concentrations of bacteria in a standard-size urine specimen. High concentrations of these bacteria may indicate possible kidney failure. The data are presented in Table 2.12.

TABLE 2.12 Concentration of bacteria in the urine in a sample of female factory workers in Switzerland

Concentration	Frequency
10^0	521
10^1	230
10^2	115
10^3	74
10^4	69
10^5	62
10^6	43
10^7	30
10^8	21
10^9	10
10^{10}	2

2.20 Compute the arithmetic mean for this sample.

2.21 Compute the geometric mean for this sample.

2.22 Which do you think is a more appropriate measure of location?

Ophthalmology

Table 2.13 comes from a paper giving the distribution of astigmatism in 1033 young men, aged 18–22, who were accepted for military service in Great Britain [5]. Assume that astigmatism is rounded to the nearest 10th of a diopter.

2.23 Compute the grouped mean.

2.24 Compute the grouped standard deviation.

2.25 Plot a histogram to properly illustrate these data.

Cardiovascular Disease

The mortality rates from heart disease (per 100,000 population) for each of the 50 states and the District of Columbia in 1973 are given in descending order in Table 2.14 [6].

TABLE 2.13 Distribution of astigmatism in 1033 young men aged 18–22

Degree of astigmatism (diopters)	Frequency
0.0 or less than 0.2	458
0.2–0.3	268
0.4–0.5	151
0.6–1.0	79
1.1–2.0	44
2.1–3.0	19
3.1–4.0	9
4.1–5.0	3
5.1–6.0	2
	1033

(Reprinted with permission of the Editor, the authors and the Journal from the *British Medical Journal*, May 7, 1394–1398, 1960.)

TABLE 2.14 Mortality rates from heart disease (per 100,000 population) for the 50 states and the District of Columbia in 1973

1	West Virginia	445.4	27	Louisiana	349.4
2	Pennsylvania	442.7	28	Connecticut	340.3
3	Maine	427.3	29	Oregon	338.7
4	Missouri	422.9	30	Washington	334.2
5	Illinois	420.8	31	Minnesota	332.7
6	Florida	417.4	32	Michigan	330.2
7	Rhode Island	414.4	33	Alabama	329.1
8	Kentucky	407.6	34	North Carolina	328.4
9	New York	406.7	35	DC	327.1
10	Iowa	396.9	36	South Carolina	322.4
11	Arkansas	396.8	37	Montana	319.1
12	New Jersey	395.2	38	Maryland	315.9
13	Massachusetts	394.0	39	Georgia	311.8
14	Kansas	391.7	40	Virginia	311.2
15	Oklahoma	391.0	41	California	310.6
16	Ohio	377.7	42	Wyoming	306.8
17	South Dakota	376.2	43	Texas	300.6
18	Wisconsin	369.8	44	Idaho	297.4
19	Vermont	369.2	45	Colorado	274.6
20	Nebraska	368.9	46	Arizona	265.4
21	Tennessee	361.4	47	Nevada	236.9
22	New Hampshire	358.2	48	Utah	214.2
23	Indiana	356.4	49	New Mexico	194.0
24	North Dakota	353.3	50	Hawaii	169.0
25	Delaware	351.6	51	Alaska	83.9
26	Mississippi	351.6			

Consider this data set as a sample of size 51 $(x_1, x_2, \ldots, x_{51})$. If

$$\sum_{i=1}^{51} x_i = 17,409 \qquad \left(\sum_{i=1}^{51} x_i\right)^2 = 303,073,281$$

$$\sum_{i=1}^{51} x_i^2 = 6,191,677$$

then:

2.26 Compute the arithmetic mean of this sample.

2.27 Compute the median of this sample.

2.28 Compute the standard deviation of this sample.

2.29 The national mortality rate for heart disease in 1973 was 360.8 per 100,000. Why does this figure *not* correspond to your answer for Problem 2.26?

2.30 Does the differential in raw rates between Florida (417.4) and Georgia (311.8) actually imply that the risk of dying from heart disease is greater in Florida than in Georgia? Why or why not?

Cardiovascular Disease

The data in Table 2.15 are a sample of cholesterol levels taken from 24 hospital employees who were on a standard American diet and who agreed to adopt a vegetarian diet for 1 month. Serum-cholesterol measurements were made before adopting the diet and 1 month after.

2.31 Compute the mean change in cholesterol.

2.32 Compute the standard deviation of the change in cholesterol levels.

2.33 Construct a stem-and-leaf plot of the cholesterol changes.

2.34 Compute the median change in cholesterol.

2.35 Construct a box plot of the cholesterol changes to the right of the stem-and-leaf plot.

2.36 Comment on the symmetry of the distribution of change scores based on your answers to Problems 2.31 through 2.35.

2.37 Some investigators feel that the effects of diet on cholesterol are more evident in people with high rather than low cholesterol levels. If you split the data in Table 2.15 according to whether baseline cholesterol is above or below the median, can you comment on this issue?

Some authors contend that cholesterol measurements are better expressed in the log scale, since they are skewed in the original scale.

TABLE 2.15 Serum cholesterol levels before and after adopting a vegetarian diet

Subject	Before	After	Before–after
1	195	146	49
2	145	155	−10
3	205	178	27
4	159	146	13
5	244	208	36
6	166	147	19
7	250	202	48
8	236	215	21
9	192	184	8
10	224	208	16
11	238	206	32
12	197	169	28
13	169	182	−13
14	158	127	31
15	151	149	2
16	197	178	19
17	180	161	19
18	222	187	35
19	168	176	−8
20	168	145	23
21	167	154	13
22	161	153	8
23	178	137	41
24	137	125	12

2.38 Compute the arithmetic mean and geometric mean of the "before" cholesterol measurements.

2.39 Draw stem-and-leaf and box plots of the "before" cholesterol measurements in raw and log scales.

2.40 Based on your answers to Problems 2.38 and 2.39, do you feel this distribution is more symmetric in the raw or log scale?

Hypertension

An experiment was performed to look at the effect of position on level of blood pressure [7]. In the experiment 32 subjects had their blood pressures measured while lying down with their arms at their sides and again standing with their arms supported at heart level. The data are given in Table 2.16.

2.41 Compute the arithmetic mean and median for the difference in systolic and diastolic blood pressure, respectively, between the positions (recumbent and standing).

2.42 Construct stem-and-leaf and box plots for each type of blood pressure in each position.

2.43 Based on your answers to Problems 2.41 and 2.42, comment on the effect of position on the levels of systolic and diastolic blood pressure.

TABLE 2.16 Effect of position on blood pressure

| | Blood pressure (mm Hg) | | | |
Subject	Recumbent, arm at side		Standing, arm at heart level	
B. R. A.	99*	71†	105*	79†
J. A. B.	126	74	124	76
F. L. B.	108	72	102	68
V. P. B.	122	68	114	72
M. F. B.	104	64	96	62
E. H. B.	108	60	96	56
G. C.	116	70	106	70
M. M. C.	106	74	106	76
T. J. F.	118	82	120	90
R. R. F.	92	58	88	60
C. R. F.	110	78	102	80
E. W. G.	138	80	124	76
T. F. H.	120	70	118	84
E. J. H.	142	88	136	90
H. B. H.	118	58	92	58
R. T. K.	134	76	126	68
W. E. L.	118	72	108	68
R. L. L.	126	78	114	76
H. S. M.	108	78	94	70
V. J. M.	136	86	144	88
R. H. P.	110	78	100	64
R. C. R.	120	74	106	70
J. A. R.	108	74	94	74
A. K. R.	132	92	128	88
T. H. S.	102	68	96	64
O. E. S.	118	70	102	68
R. E. S.	116	76	88	60
E. C. T.	118	80	100	84
J. H. T.	110	74	96	70
F. P. V.	122	72	118	78
P. F. W.	106	62	94	56
W. J. W.	146	90	138	94

* Systolic blood pressure
† Diastolic blood pressure
(Reprinted with permission of the *American Journal of Medicine*.)

Nutrition

Table 2.17 shows the distribution of dietary vitamin A intake as reported by 14 students who filled out a dietary questionnaire in class. The total intake is a combination of intake from individual food items and from vitamin pills. The units are in IU/100 (International Units/100).

TABLE 2.17 Distribution of dietary vitamin A intake as reported by 14 students

Student number	Intake (IU/100)	Student number	Intake (IU/100)
1	31.1	8	48.1
2	21.5	9	24.4
3	74.7	10	13.4
4	95.5	11	37.1
5	19.4	12	21.3
6	64.8	13	78.5
7	108.7	14	17.7

2.44 Compute the mean and median from these data.

2.45 Compute the standard deviation and coefficient of variation from these data.

2.46 Suppose the data are expressed in IU rather than IU/100. What are the mean, standard deviation, and coefficient of variation in the new units?

2.47 Construct a stem-and-leaf plot of the data on some convenient scale.

2.48 Do you think the mean or median is a more appropriate measure of location for this data set?

Pulmonary Disease

FEV (forced expiratory volume) is an index of pulmonary function that measures the volume of air expelled after one second of constant effort. The data set FEV.DAT contains determinations of FEV in 1980 on 654 children ages 6–22 who were seen in the Childhood Respiratory Disease Study (CRD Study) in East Boston, Massachusetts. These data are part of a longitudinal study to follow the change in pulmonary function over time in children [8].

The data in Table 2.18 are available for each child.

2.49 For each variable (other than ID), obtain appropriate descriptive statistics (both numerical and graphical).

2.50 Use both numerical and graphical measures to assess the relationship of FEV to age, height, and smoking status. (Do this separately for boys and girls.)

TABLE 2.18 Format for FEV.DAT

Column	Variable	Format or Code
1–5	ID number	
7–8	Age (years)	
10–15	FEV (liters)	X.XXX
17–20	Height (inches)	XX.X
22	Sex	0 = female/1 = male
24	Smoking status	0 = noncurrent smoker/1 = current smoker

2.51 Compare the pattern of growth of FEV by age for boys and girls. Are there any similarities? Any differences?

2.52 Answer Problem 2.51 for height rather than FEV.

Nutrition

The food frequency questionnaire (FFQ) is an instrument that is often used in dietary epidemiology to assess consumption of specific foods. A person is asked to write down the number of servings per day typically eaten in the past year of over 100 individual food items. A food composition table is then used to compute nutrient intakes (e.g., protein, fat, etc.), based on aggregating responses for individual foods. The FFQ is inexpensive to administer but is considered less accurate than the diet record (DR) (the gold standard of dietary epidemiology). For the diet record, a participant writes down the amount of each specific food eaten over the past week in a food diary and a nutritionist using a special computer program computes nutrient intakes from the food diaries. This is a much more expensive method of dietary recording. To validate the FFQ, 173 nurses participating in the Nurses Health Study completed 4 weeks of diet recording about equally spaced over a 12-month period and a FFQ at the end of diet recording [9]. Data are presented in the file VALID.DAT for saturated fat, total fat, total alcohol consumption, and total caloric intake for both the DR and FFQ. For the DR, average nutrient intakes were computed over the four weeks of diet recording. The format of this file is shown in Table 2.19.

2.53 Compute appropriate descriptive statistics for each nutrient for both DR and FFQ using both numerical and graphical measures.

2.54 Use descriptive statistics to relate nutrient intake for the DR and FFQ. Do you think that the FFQ is a reasonably accurate approximation to the DR? Why or why not?

2.55 A frequently used method for quantifying dietary intake is in the form of quintiles. Compute quintiles for each nutrient and each method of recording and relate the nutrient composition for DR and FFQ using the quintile scale. (That is, how do the quintile breakdowns based on DR relate to the quintile breakdowns based on FFQ for the same individual?) Do you get the same impression about the concordance between DR and FFQ using quintiles as in Problem 2.54, where raw (ungrouped) nutrient intake is considered?

TABLE 2.19 Format for VALID.DAT

Column	Variable	Format or code
1–6	ID number	
8–15	Saturated fat–DR	XXXXX.XX
17–24	Saturated fat–FFQ	XXXXX.XX
26–33	Total fat–DR	XXXXX.XX
35–42	Total fat–FFQ	XXXXX.XX
44–51	Alcohol consumption–DR	XXXXX.XX
53–60	Alcohol consumption–FFQ	XXXXX.XX
62–70	Total calories–DR	XXXXXX.XX
72–80	Total calories–FFQ	XXXXXX.XX

References

[1] White, J. R., & Froeb, H. E. (1980). Small-airways dysfunction in nonsmokers chronically exposed to tobacco smoke. *New England Journal of Medicine, 302*(33), 720–723.

[2] Pedersen, A., Wiesner, P., Holmes, K., Johnson, C., & Turck, M. (1972). Spectinomycin and Pencillin G in the treatment of gonorrhea. *JAMA, 220*(2), 205–208.

[3] Tukey, J. (1977). *Exploratory data analysis.* Reading, MA: Addison-Wesley.

[4] Townsend, T., Shapiro, M., Rosner, B., & Kass, E. H. (1979). Use of antimicrobial drugs in general hospitals I. Description of population and definition of methods. *Journal of Infectious Diseases, 139*(6), 688–697.

[5] Sorsby, A., Sheridan, M., Leary, G. A., & Benjamin, B. (1960). Vision, visual acuity and ocular refraction of young men in a sample of 1033 subjects. *British Medical Journal,* 1394–1398.

[6] National Center for Health Statistics. (1975, February 10). *Monthly vital statistics report, summary report, final mortality statistics (1973), 23*(11) (Suppl. 2).

[7] Kossmann, C. E. (1946). Relative importance of certain variables in the clinical determination of blood pressure. *American Journal of Medicine, 1,* 464–467.

[8] Tager, I. B., Weiss, S. T., Rosner, B., & Speizer, F. E. (1979). Effect of parental cigarette smoking on pulmonary function in children. *American Journal of Epidemiology, 110,* 15–26.

[9] Willett, W. C., Sampson, L., Stampfer, M. J., Rosner, B., Bain, C., Witschi, J., Hennekens, C. H., & Speizer, F. E. (1985). Reproducibility and validity of a semiquantitative food frequency questionnaire. *American Journal of Epidemiology, 122,* 51–65.

CHAPTER THREE

PROBABILITY

Introduction

In Chapter 2 various techniques for concisely describing data were outlined. But we usually want to do more with data than just describe it. In particular, we might want to test certain specific inferences about the behavior of the data.

EXAMPLE 3.1 **Cancer** One theory on the etiology of breast cancer states that women in a given age group who give birth to their first child relatively late in life (after 30) are at greater risk for eventually developing breast cancer over some time period t than are women who give birth to their first child early in life (before 20). Because women in the upper social classes tend to have children later, this theory has been used to explain why these women have a higher risk of developing breast cancer than women in the lower social classes. To test this hypothesis, we might identify 2000 women from a particular census tract who are currently aged 45–54 and have never had breast cancer, of whom 1000 had their first child before the age of 20 (call this group A) and 1000 after the age of 30 (call this group B). These 2000 women might be followed for 5 years and asked if they had a new case of breast cancer during this period. Suppose that there are 4 new cases of breast cancer out of 1000 in group A and 5 new cases out of 1000 in group B. ∎

Is this sufficient evidence to confirm a difference in risk between the two groups? Most people would feel uneasy about coming to this conclusion on the basis of such a limited amount of data.

Suppose we had a more ambitious plan and sampled 10,000 women from groups A and B, respectively, and found 40 new cases in group A and 50 new cases in group B and asked the same question. Although we might be more comfortable with the conclusion because of the larger sample size, we would still have to admit that there was some possibility that this apparent difference in the rates could be due to chance.

The problem is that we need a conceptual framework to make these decisions but have not explicitly stated what the framework is. This framework is provided by the underlying concept of **probability**. In this chapter probability is defined and some rules for working with probabilities are introduced. Understanding of probability is essential in the calculation and interpretation of p-values in the statistical tests of subsequent chapters. It also permits a discussion of sensitivity, specificity, and predictive values of screening tests which are discussed in Section 3.7.

Definition of Probability

EXAMPLE 3.2 **Obstetrics** Suppose we are interested in the probability of a male live childbirth (or livebirth) among all livebirths in the United States. Conventional wisdom tells us that this probability should be close to .5. We can explore this subject by looking at some vital-statistics data, as presented in Table 3.1 [1]. The probability of a male livebirth based on 1965 data is .51247, based on 1965–1969 data .51248, and based on 1965–1974 data .51268. These are **empirical** probabilities based on a finite amount of data. In principle, the sample size could be expanded indefinitely and an increasingly more precise estimate of this probability obtained.

TABLE 3.1

Probability of a male livebirth during the period 1965–1974

Time period	Number of male livebirths (a)	Total number of livebirths (b)	Empirical probability of a male livebirth (a/b)
1965	1,927,054	3,760,358	0.51247
1965–1969	9,219,202	17,989,361	0.51248
1965–1974	17,857,857	34,832,051	0.51268

This principle leads to the following definition of probability:

DEFINITION 3.1   ■■

The **sample space** is the set of all possible outcomes. In referring to probabilities of events, an **event** is any set of outcomes of interest. The **probability** of an event is the relative frequency (see p. 24) of this set of outcomes over an indefinitely large (or infinite) number of trials. ■

EXAMPLE 3.3   **Pulmonary Disease** The tuberculin skin test is a routine screening test used to detect tuberculosis. The results of this test can be categorized as either positive, negative, or uncertain. If the probability of a positive test is .1, it means that if a large number of such tests were performed, about 10% of them would be positive. The actual percentage of positive tests will be increasingly close to .1 the larger the number of tests performed.   ■

EXAMPLE 3.4   **Cancer** The probability of developing a new case of breast cancer in 1 year in 40-year-old women who have never had breast cancer is .001. This probability means that over a large sample of 40-year-old women who have never had breast cancer, approximately 1 in 1000 will develop the disease over 1 year, with this percentage becoming increasingly close to .001 as the number of women sampled increases.   ■

In real life, experiments cannot be performed an infinite number of times. Instead, probabilities of events are estimated from the empirical probabilities obtained from large samples (as was done in Examples 3.2–3.4). In other instances, theoretical probability models are constructed from which probabilities of many different kinds of events can be computed. One of the important issues in statistical inference is to compare empirical probabilities with theoretical probabilities, that is, to assess the goodness of fit of probability models. This topic is covered in Section 10.10.

EXAMPLE 3.5   **Cancer** The probability of developing a new case of stomach cancer over a 1-year period for 45–49-year-old women based on Connecticut Tumor Registry data from 1963–1965 is 14 per 100,000 [2]. Suppose we have studied cancer rates in a small group of Connecticut nurses over this period and wish to compare how close the rates from this limited sample are to the tumor registry figures. The figure 14 per 100,000 would be the best estimate of the probability prior to collecting any data, and we would then see how closely our new sample data conformed with this probability.   ■

From Definition 3.1 and from the preceding examples, we can deduce that probabilities have the following basic properties:

3.1

(1) The probability of any event E, denoted by $Pr(E)$, always satisfies $0 \leqslant Pr(E) \leqslant 1$.

(2) If outcomes A and B are two events that cannot both happen at the same time, then
$$Pr(A \text{ or } B \text{ occurs}) = Pr(A) + Pr(B).$$

EXAMPLE 3.6   **Hypertension** Let A be the event that a person has normotensive diastolic blood pressure (DBP) readings (i.e., DBP <90), and let B be the event that a person has borderline DBP readings (i.e., DBP $\geqslant 90$ and <95). Suppose that $Pr(A) = .7$, $Pr(B) = .1$. Let C be the event that a person has DBP <95. Then,

$$Pr(C) = Pr(A) + Pr(B) = .8$$

because the events A and B cannot occur at the same time.   ■

DEFINITION 3.2

Two events A and B are **mutually exclusive** if they cannot both happen at the same time.   ■

Thus, the events A and B in Example 3.6 are mutually exclusive.

EXAMPLE 3.7   **Hypertension** Let x be DBP, C be the event that $x \geqslant 90$, and D be the event that $75 \leqslant x \leqslant 100$. The events C and D are *not* mutually exclusive, since they both occur when $90 \leqslant x \leqslant 100$.   ■

SECTION 3.3   Some Useful Probabilistic Notation

DEFINITION 3.3

The symbol $\{\ \}$ is used as shorthand for the phrase "the event."   ■

DEFINITION 3.4

$A \cup B$ is the event that either A or B occurs or they both occur.   ■

Figure 3.1 diagrammatically depicts $A \cup B$ both for the case where A and B are and are not mutually exclusive.

EXAMPLE 3.8   **Hypertension** Let the events A and B be defined as in Example 3.6; that is, $A = \{x < 90\}$, $B = \{90 \leqslant x < 95\}$, where $x =$ DBP. Then, $A \cup B = \{x < 95\}$.   ■

EXAMPLE 3.9   **Hypertension** Let the events C and D be defined as in Example 3.7; that is,

$$C = \{x \geqslant 90\} \qquad D = \{75 \leqslant x \leqslant 100\}$$

Then
$$\{C \cup D\} = \{x \geqslant 75\}$$ ■

DEFINITION 3.5

$\{A \cap B\}$ is the event that both A and B occur simultaneously. $\{A \cap B\}$ is depicted diagrammatically in Figure 3.2.   ■

EXAMPLE 3.10   **Hypertension** Let the events C and D be defined as in Example 3.7; that is,

$$C = \{x \geqslant 90\} \qquad D = \{75 \leqslant x \leqslant 100\}$$

Then,
$$\{C \cap D\} = \{90 \leqslant x \leqslant 100\}$$ ■

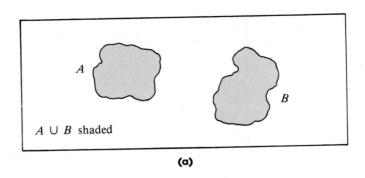

(a)

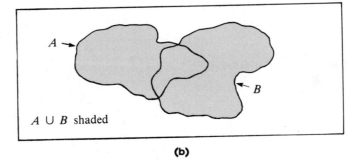

FIGURE 3.1
Diagrammatic
representation of $A \cup B$:
(a) A, B mutually
exclusive; (b) A, B not
mutually exclusive

(b)

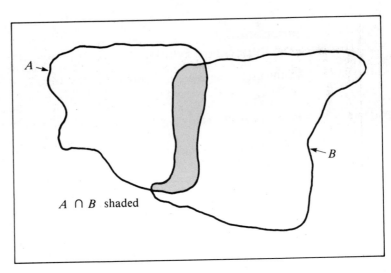

FIGURE 3.2
Diagrammatic
representation of $A \cap B$

Notice that $\{A \cap B\}$ is not well defined for the events A and B in Example 3.6, since both A and B cannot occur simultaneously. This situation is true for any mutually exclusive events.

DEFINITION 3.6 ■■■

$\bar{A}$ is the event that A does not occur. It is sometimes referred to as the **complement** of A. Notice that $Pr(\bar{A}) = 1 - Pr(A)$, since $\bar{A}$ occurs only when A does not occur. The event $\bar{A}$ is depicted diagrammatically in Figure 3.3.   ■

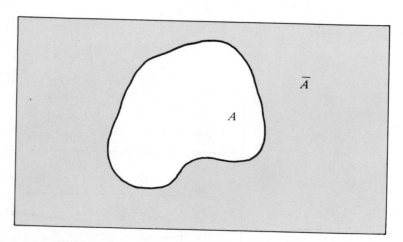

FIGURE 3.3
Diagrammatic
representation of $\bar{A}$

EXAMPLE 3.11   **Hypertension** Let the events A and C be defined as in Examples 3.6 and 3.7; that is,

$$A = \{x < 90\} \qquad C = \{x \geq 90\}$$

Then, $C = \bar{A}$, since C can only occur when A does not occur. Notice that

$$Pr(C) = Pr(\bar{A}) = 1 - .7 = .3$$

Thus, if 70% of people have DBP <90, then 30% of people must have DBP ≥ 90.   ∎

<u>SECTION 3.4</u>   **Independent
and Dependent Events**

In the preceding section events in general were described. In this section certain specific types of events are discussed.

EXAMPLE 3.12   **Hypertension, Genetics** Suppose we are conducting a hypertension screening program in the home. Consider all possible pairs of DBP measurements of the mother and father within a given family, assuming that the mother and father are not genetically related. This sample space consists of all pairs of numbers of the form (X, Y), where $X > 0$, $Y > 0$. Certain specific events might be of interest in this context. In particular, we might be interested in whether the mother or father is hypertensive, which is described, respectively, by the events $A = \{$mother's DBP $\geq 95\}$, $B = \{$father's DBP $\geq 95\}$. These events are depicted graphically in Figure 3.4.

Suppose we know that $Pr(A) = .1$, $Pr(B) = .2$. What can we say about $Pr(A \cap B) = Pr(\text{mother's DBP} \geq 95 \text{ and father's DBP} \geq 95) = Pr(\text{both mother and father are hypertensive})$? We can say nothing unless we are willing to make certain assumptions.   ∎

DEFINITION 3.7   ■■
Two events A and B are referred to as **independent events** if

$$Pr(A \cap B) = Pr(A) \times Pr(B)$$ ∎

EXAMPLE 3.13   **Hypertension, Genetics** Compute the probability that both the mother and father are hypertensive if the events in Example 3.12 are independent.

SOLUTION   If A and B are independent events, then

$$Pr(A \cap B) = Pr(A) \times Pr(B) = (.1)(.2) = .02$$ ∎

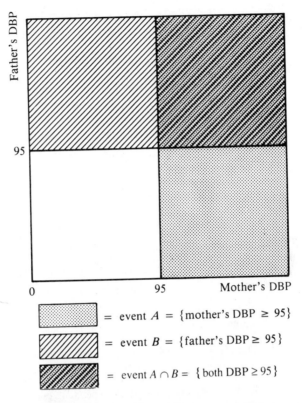

FIGURE 3.4
Possible diastolic blood-pressure measurements of the mother and father within a given family

= event A = {mother's DBP $\geq$ 95}

= event B = {father's DBP $\geq$ 95}

= event $A \cap B$ = {both DBP $\geq$ 95}

One way to interpret this example is to assume that the hypertensive status of the mother does not depend at all on the hypertensive status of the father. Thus, if these events are independent, then in 10% of all households where the father is hypertensive the mother is also hypertensive, and in 10% of all households where the father is *not* hypertensive the mother is hypertensive. We would expect these two events to be independent if the primary determinants of elevated blood pressure were genetic. However, if the primary determinants of elevated blood pressure were, to some extent, environmental, then we would expect that the mother would be more likely to have elevated blood pressure (A true) if the father had elevated blood pressure (B true) than if the father did not have elevated blood pressure (B not true). In this latter case the events would not be independent. The implications of this situation are discussed later in this chapter.

If two events are not independent, then they are said to be dependent.

DEFINITION 3.8 ■■■

Two events A, B are **dependent** if

$$Pr(A \cap B) \neq Pr(A) \times Pr(B)$$ ■

Example 3.14 is a classic example of dependent events.

EXAMPLE 3.14 **Hypertension, Genetics** Consider all possible diastolic blood-pressure measurements from a mother and her first-born child. Let

$$A = \{\text{mother's DBP} \geqslant 95\} \qquad B = \{\text{first-born child's DBP} \geqslant 80\}$$

Suppose $\qquad$ $Pr(A \cap B) = .05$ $\qquad$ $Pr(A) = .1$ $\qquad$ $Pr(B) = .2$

Then $\qquad\qquad\qquad$ $Pr(A \cap B) = .05 > Pr(A) \times Pr(B) = .02$

and the events A, B would be dependent. ∎

This outcome would be expected, since the mother and first-born child both share the same environment and are genetically related. In other words, the first-born child is more likely to have elevated blood pressure in households where the mother is hypertensive than in households where the mother is not hypertensive.

EXAMPLE 3.15 $\quad$ **Venereal Disease** Suppose two doctors, A and B, diagnose all patients coming into a VD clinic for syphilis. Let the events $A^+ = \{$doctor A makes a positive diagnosis$\}$, $B^+ = \{$doctor B makes a positive diagnosis$\}$. Suppose that doctor A diagnoses 10% of all patients as positive, doctor B diagnoses 17% of all patients as positive, and both doctors diagnose 8% of all patients as positive. Are the events A^+, B^+ independent?

SOLUTION $\quad$ We are given that

$$Pr(A^+) = .1 \qquad Pr(B^+) = .17 \qquad Pr(A^+ \cap B^+) = .08$$

Thus, $\qquad\qquad$ $Pr(A^+ \cap B^+) = .08 > Pr(A^+) \times Pr(B^+) = .1(.17) = .017$

and the events are dependent. This result would be expected, since there should be a similarity between how two doctors diagnose patients for syphilis. ∎

Definition 3.7 can be generalized to the case of $k(>2)$ independent events. This is often referred to as the multiplication law of probability.

3.2 $\quad$ If $A_1, \ldots, A_k$ are mutually independent events, then $Pr(A_1 \cap A_2 \cap \cdots \cap A_k) = Pr(A_1) \times Pr(A_2) \times \cdots \times Pr(A_k)$. This principle is referred to as the **multiplication law of probability**.

SECTION 3.5 $\quad$ **The Addition Law of Probability**

We have seen from the definition of probability that if A and B are mutually exclusive events, then $Pr(A \cup B) = Pr(A) + Pr(B)$. A more general formula for $Pr(A \cup B)$ can be developed when the events A and B are not necessarily mutually exclusive.

Specifically, the event $A \cup B$ can be subdivided into three mutually exclusive components, namely, $A \cap \bar{B}$, $\bar{A} \cap B$, $A \cap B$, that, in words, are the events A occurs and B does not occur, A does not occur and B occurs, and both A and B occur. If $A \cup B$ occurs, then exactly one of these events must occur. Therefore,

$$Pr(A \cup B) = Pr(A \cap \bar{B}) + Pr(\bar{A} \cap B) + Pr(A \cap B)$$

However, if A occurs, then it must occur either with B ($A \cap B$) or without B ($A \cap \bar{B}$) occurring. Therefore,

$$Pr(A) = Pr(A \cap B) + Pr(A \cap \bar{B})$$

If $Pr(A \cap B)$ is subtracted from both sides of the equation,

$$Pr(A \cap \bar{B}) = Pr(A) - Pr(A \cap B)$$

Similarly, if the roles of A and B are interchanged,

$$Pr(\bar{A} \cap B) = Pr(B) - Pr(A \cap B)$$

Finally, by substituting into the expression for $Pr(A \cup B)$,

$$Pr(A \cup B) = Pr(A) - Pr(A \cap B) + [Pr(B) - Pr(A \cap B)] + Pr(A \cap B)$$
$$= Pr(A) + Pr(B) - Pr(A \cap B)$$

This relationship is referred to as the addition law of probability.

| **3.3** | **Addition Law of Probability** |

If A and B are any events, then

$$Pr(A \cup B) = Pr(A) + Pr(B) - Pr(A \cap B)$$

This principle is depicted diagrammatically in Figure 3.5. Thus, to compute $Pr(A \cup B)$, add the probabilities of A and B separately and then subtract the overlap, which is $Pr(A \cap B)$.

EXAMPLE 3.16   **Venereal Disease** Consider the data given in Example 3.15. Suppose a patient is referred for further lab tests if either doctor A or B makes a positive diagnosis. What is the probability that a patient will be referred for further lab tests?

SOLUTION   The event that either doctor makes a positive diagnosis can be represented by $\{A^+ \cup B^+\}$. We know that

$$Pr(A^+) = .1 \qquad Pr(B^+) = .17 \qquad Pr(A^+ \cap B^+) = .08$$

Therefore, from the addition law of probability,

$$Pr(A^+ \cup B^+) = Pr(A^+) + Pr(B^+) - Pr(A^+ \cap B^+) = .1 + .17 - .08 = .19$$

Thus, 19% of all patients will be referred for further lab tests. ∎

There are special cases of the addition law that are of interest. First, if the events A and B are *mutually exclusive*, then $Pr(A \cap B) = 0$ and the addition law

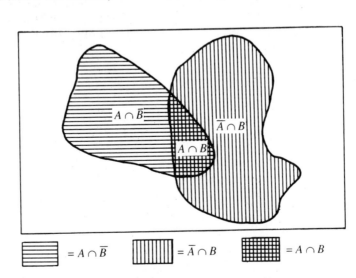

FIGURE 3.5
Diagrammatic
representation of the
addition law of
probability

reduces to $Pr(A \cup B) = Pr(A) + Pr(B)$. This property is given in **(3.1)** for probabilities over any two mutually exclusive events. Second, if the events A and B are *independent*, then by definition $Pr(A \cap B) = Pr(A) \times Pr(B)$ and $Pr(A \cup B)$ can be rewritten as $Pr(A) + Pr(B) - Pr(A) \times Pr(B)$. This leads to the following important special case of the addition law.

3.4 | **Addition Law of Probability for Independent Events**

If two events A and B are independent, then

$$Pr(A \cup B) = Pr(A) + Pr(B) \times [1 - Pr(A)]$$

This special case of the addition law can be interpreted as follows: The event $A \cup B$ can be separated into two mutually exclusive events: {A occurs} and {B occurs and A does not occur}. Furthermore, because of the independence of A and B, the probability of the latter event can be written as $Pr(B) \times [1 - Pr(A)]$. This probability is depicted diagrammatically in Figure 3.6.

EXAMPLE 3.17   **Hypertension**   Refer to Example 3.12, where

$$A = \{\text{mother's DBP} \geqslant 95\} \quad \text{and} \quad B = \{\text{father's DBP} \geqslant 95\}$$

$Pr(A) = .1$, $Pr(B) = .2$, and assume that A and B are independent events. Suppose a "hypertensive household" is defined as one in which either the mother or the father is hypertensive, and hypertension is defined for the mother and father, respectively, in terms of the events A and B. What is the probability of a hypertensive household?

SOLUTION Pr(hypertensive household) is

$$Pr(A \cup B) = Pr(A) + Pr(B) \times [1 - Pr(A)] = .1 + .2(.9) = .28$$

Thus, 28% of all households will be hypertensive.   ∎

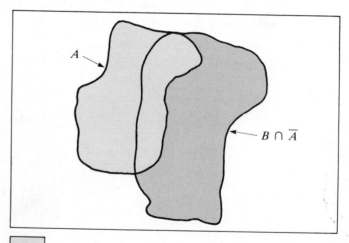

FIGURE 3.6
Diagrammatic representation of the addition law of probability for independent events

☐ = A

☐ = {B occurs and A does not occur} = $B \cap \bar{A}$

It is possible to extend the addition law to more than two events. In particular, if there are three events A, B, and C, then

$$Pr(A \cup B \cup C) = Pr(A) + Pr(B) + Pr(C) - Pr(A \cap B) - Pr(A \cap C) \\ - Pr(B \cap C) + Pr(A \cap B \cap C)$$

This result can be generalized to an arbitrary number of events, although this is beyond the scope of this text (see [3]).

SECTION 3.6 ## Conditional Probability

Suppose we want to compute the probability of several events occurring simultaneously. If the events are independent, then the multiplication law of probability can be used to accomplish this. If some of the events are dependent, then some quantitative measure of dependence is needed in order to extend the multiplication law to the case of dependent events. Consider the following example:

EXAMPLE 3.18 **Pulmonary Disease** In many places of employment, prospective employees are customarily given a screening test for tuberculosis (TB) before starting employment. The definitive test for the detection of TB is the chest X-ray. Unfortunately, the chest X-ray is somewhat expensive to administer and exposure to the radiation from the X-ray is an undesirable side effect of the test. A common procedure to avoid giving everyone a chest X-ray is to perform a less expensive test, the skin test, with the hope that only people who are positive on the skin test can possibly have TB. The ideal situation would be if the probability of having TB among all those with positive skin tests (SKT) were 1 and the probability of having TB among all those with negative skin tests were 0. The two events {SKT$^+$}, {TB} would then be completely dependent; that is, the result of the screening test would automatically determine the disease state. The opposite extreme is achieved when the events {SKT$^+$}, {TB} are completely independent. In this case the probability of TB is the same whether or not the skin test is positive, and the skin test would not be useful in screening for TB and should not be given. ∎

These concepts can be quantified in the following way. Let $A = \{\text{SKT}^+\}$, $B = \{\text{TB}\}$ and suppose that we are interested in the probability of TB (B) given that the skin test is positive (A). This probability can be written as $Pr(A \cap B)/Pr(A)$.

DEFINITION 3.9 The quantity $Pr(A \cap B)/Pr(A)$ is defined as the **conditional probability of B given A**, which is written as $Pr(B|A)$. ∎

However, from Section 3.4 we know that, by definition, if two events are independent, then $Pr(A \cap B) = Pr(A) \times Pr(B)$. If both sides are divided by $Pr(A)$, then $Pr(B) = Pr(A \cap B)/Pr(A) = Pr(B|A)$. Similarly, we can show that if A and B are independent events, then $Pr(B|\overline{A}) = Pr(B|A) = Pr(B)$. This relationship leads to the following alternative interpretation of independence in terms of conditional probabilities:

3.5 **(1)** If A and B are independent events, then $Pr(B|A) = Pr(B)$.
(2) If two events A, B are dependent, then $Pr(B|A) \neq Pr(B) \neq Pr(B|\overline{A})$ and $Pr(A \cap B) \neq Pr(A) \times Pr(B)$.

DEFINITION 3.10   ■■

The **relative risk (RR)** of B given A is

$$Pr(B|A)/Pr(B|\bar{A})$$ ■

Notice that if two events A, B are independent, then the relative risk will be 1. If two events A, B are dependent, then the relative risk will be different from 1. Heuristically, the dependence between events increases, the further the relative risk is from 1.

EXAMPLE 3.19   **Pulmonary Disease** Suppose that 1 person in 10,000 from those with negative skin tests has TB, or $Pr(B|\bar{A}) = .0001$, whereas 1 person in 100 from those with positive skin tests has TB, or $Pr(B|A) = .01$. The two events would be highly dependent here, since

$$RR = Pr(B|A)/Pr(B|\bar{A}) = .01/.0001 = 100$$

In words, people with positive skin tests are 100 times as likely to have TB as those with negative skin tests. This is the rationale for using the skin test as a screening test for TB. If the events A and B were independent, then the relative risk would be 1; that is, people with positive or negative skin tests would be equally likely to have TB and the test would not be useful as a screening test.   ■

EXAMPLE 3.20   **Venereal Disease** Using the data in Example 3.15 (p. 48), find the conditional probability that doctor B makes a positive diagnosis of syphilis given that doctor A makes a positive diagnosis. What is the conditional probability that doctor B makes a positive diagnosis of syphilis given that doctor A makes a negative diagnosis? What is the relative risk of $\{B^+\}$ given $\{A^+\}$?

SOLUTION
$$Pr(B^+|A^+) = Pr(B^+ \cap A^+)/Pr(A^+) = .08/.1 = .8$$

Thus, doctor B will confirm doctor A's positive diagnosis 80% of the time. Similarly,

$$Pr(B^+|A^-) = Pr(B^+ \cap A^-)/Pr(A^-) = Pr(B^+ \cap A^-)/.9$$

We must compute $Pr(B^+ \cap A^-)$. We know that if doctor B diagnoses a patient as positive, then doctor A either does or does not diagnose the patient as positive. Thus,

$$Pr(B^+) = Pr(B^+ \cap A^+) + Pr(B^+ \cap A^-)$$

since the events $\{B^+ \cap A^+\}$ and $\{B^+ \cap A^-\}$ are mutually exclusive. If we subtract $Pr(B^+ \cap A^+)$ from both sides of the equation, then

$$Pr(B^+ \cap A^-) = Pr(B^+) - Pr(B^+ \cap A^+) = .17 - .08 = .09$$

Therefore, $$Pr(B^+|A^-) = .09/.9 = .1$$

Thus, when doctor A diagnoses a patient as negative, doctor B will contradict the diagnosis 10% of the time. The relative risk of the event $\{B^+\}$ given $\{A^+\}$ is

$$Pr(B^+|A^+)/Pr(B^+|A^-) = .8/.1 = 8$$

This indicates that doctor B is 8 times as likely to diagnose a patient as positive when doctor A diagnoses the patient as positive than when doctor A diagnoses the patient as negative. These results quantify the dependence between the two doctors' diagnoses.   ■

The conditional ($Pr(B|A)$, $Pr(B|\bar{A})$) and unconditional ($Pr(B)$) probabilities mentioned previously can be related in the following way:

3.6 For any events A and B,

$$Pr(B) = Pr(B|A) \times Pr(A) + Pr(B|\overline{A}) \times Pr(\overline{A})$$

This formula tells us that the unconditional probability of B is the sum of the conditional probability of B given A *times* the unconditional probability of A *plus* the conditional probability of B given A *not* occurring *times* the unconditional probability of A *not* occurring.

EXAMPLE 3.21 **Pulmonary Disease** Let A and B be defined as in Example 3.19 and suppose that 1% of the general population will have a positive skin test. What is the probability of tuberculosis in the general population?

SOLUTION
$$Pr(B) = Pr(TB) = Pr(TB|SKT^+) \times Pr(SKT^+) + Pr(TB|SKT^-) \times Pr(SKT^-)$$
$$= (.01)(.01) + (10^{-4})(.99) = .000199 \simeq .0002 = 2 \times 10^{-4}$$

Thus, the unconditional probability of TB in the general population (2×10^{-4}) is a weighted average of the conditional probability of TB given a positive skin test ($.01 = 100 \times 10^{-4}$) and the conditional probability of TB given a negative skin test (10^{-4}). ∎

In **(3.6)** the probability of the event B is expressed in terms of the two events A and $\overline{A}$. In many instances the probability of an event B will need to be expressed in terms of more than two events, denoted by $A_1, A_2, \ldots, A_k$.

DEFINITION 3.11 ■■■
A set of events $A_1, \ldots, A_k$ are **exhaustive** if the least one of the events must occur. ■

Assume that the events $A_1, \ldots, A_k$ are mutually exclusive and exhaustive, that is, at least one of the events $A_1, \ldots, A_k$ must occur and no two events can occur simultaneously. Thus, exactly one of the events $A_1, \ldots, A_k$ must occur.

3.7 **Total Probability Rule**
Let $A_1, \ldots, A_k$ be mutually exclusive and exhaustive events. The unconditional probability of B ($Pr(B)$) can then be written as a weighted average of the conditional probabilities of B given A_i ($Pr(B|A_i)$) as follows:

$$Pr(B) = \sum_{i=1}^{k} Pr(B|A_i) \times Pr(A_i)$$

An application of the total probability rule is given in the following example:

EXAMPLE 3.22 **Ophthalmology** We are planning a 5-year study of cataract in a population of 5000 people 60 years of age and older. We know from census data that 45% of this population are ages 60–64, 28% are ages 65–69, 20% are ages 70–74, and 7% are age 75 or older. We also know from the Framingham Eye Study that 2.4%, 4.6%, 8.8%, and 15.3% of the people in those respective age groups will develop cataract over the next 5 years [4]. What percentage of our population will develop cataract over the next 5 years, and how many cataracts does this percentage represent?

SOLUTION Let $A_1 = \{$ages 60–64$\}$, $A_2 = \{$ages 65–69$\}$, $A_3 = \{$ages 70–74$\}$, $A_4 = \{$age 75+$\}$. These events are mutually exclusive and exhaustive, since exactly one event must occur for each person in our population. Furthermore, from the conditions of the problem, we know that $Pr(A_1) = .45$, $Pr(A_2) = .28$, $Pr(A_3) = .20$, $Pr(A_4) = .07$, $Pr(B|A_1) = .024$, $Pr(B|A_2) = .046$, $Pr(B|A_3) = .088$, and $Pr(B|A_4) = .153$. Finally, using the total probability rule,

$$Pr(B) = Pr(B|A_1) \times Pr(A_1) + Pr(B|A_2)$$
$$\times Pr(A_2) + Pr(B|A_3) \times Pr(A_3) + Pr(B|A_4) \times Pr(A_4)$$
$$= (.024)(.45) + (.046)(.28) + (.088)(.20) + (.153)(.07) = .052$$

Thus 5.2% of our population will develop cataract over the next 5 years, which represents a total of $5000 \times .052 = 260$ persons with cataract. ∎

The definition of conditional probability allows the multiplication law of probability to be extended to the case of dependent events.

3.8 | **Generalized Multiplication Law of Probability**

If $A_1, \ldots, A_k$ are an arbitrary set of events, then

$$Pr(A_1 \cap A_2 \cap \cdots \cap A_k)$$
$$= Pr(A_1) \times Pr(A_2|A_1) \times Pr(A_3|A_2 \cap A_1) \cdots \times Pr(A_k|A_{k-1} \cap \cdots \cap A_2 \cap A_1)$$

If the events are independent, then the conditional probabilities on the right-hand side of **(3.8)** reduce to unconditional probabilities and the generalized multiplication law reduces to the multiplication law for independent events given in **(3.2)**. Equation **(3.8)** also generalizes the relationship $Pr(A \cap B) = Pr(A) \times Pr(B|A)$ given in Definition 3.9 for two events to the case of more than two events.

SECTION 3.7   Bayes' Rule and Screening Tests

The tuberculosis skin test example given in Example 3.18 illustrates the general concept of the predictive value of a screening test, which can be defined as follows:

DEFINITION 3.12   ■■■

The **predictive value positive** (PV$^+$) of a screening test is the probability that a person has disease given that the test is positive

$$Pr(\text{disease}|\text{test}^+)$$

The **predictive value negative** (PV$^-$) of a screening test is the probability that a person does *not* have disease given that the test is negative

$$Pr(\text{no disease}|\text{test}^-)$$ ∎

EXAMPLE 3.23   **Pulmonary Disease** Find the predictive values positive and negative for the tuberculosis skin test given the data in Example 3.18.

SOLUTION   We see that
$$PV^+ = Pr(B|A) = .01$$

whereas
$$PV^- = Pr(\bar{B}|\bar{A}) = 1 - Pr(B|\bar{A}) = .9999$$

Thus, if the skin test is negative, the person is virtually certain not to have disease (PV$^- \approx 1$); whereas if the skin test is positive, the person still has only a small chance of having disease (PV$^+ = .01$). ∎

A symptom or a set of symptoms can also be regarded as a screening test for disease. The higher the predictive value of the screening test or symptoms, the more valuable the test. Ideally, we would like to find a set of symptoms such that both PV^+ and PV^- are 1. Then we would be able to accurately diagnose disease for each patient.

Clinicians often cannot directly measure the predictive value of a set of symptoms. However, they can measure how often specific symptoms occur in diseased and normal people. These measures are defined as follows:

DEFINITION 3.13 ■■

The **sensitivity** of a symptom (or set of symptoms or screening test) is the probability that the symptom is present given that the person has disease.   ■

DEFINITION 3.14 ■■

The **specificity** of a symptom (or set of symptoms or screening test) is the probability that the symptom is not present given that the person does not have disease.   ■

DEFINITION 3.15 ■■

A **false negative** is defined as a person who tests out as negative but who is actually positive. A **false positive** is defined as a person who tests out as positive but who is actually negative.   ■

It is important that both the sensitivity and specificity be high for a symptom to be effective in predicting disease.

EXAMPLE 3.24   **Cancer** Suppose that the disease is lung cancer and the symptom is cigarette smoking. If we assume that 90% of people with lung cancer and 50% of people without lung cancer (essentially the entire general population) are smokers, then the sensitivity and specificity are .9 and .5, respectively. Obviously cigarette smoking cannot be used by itself as a diagnostic tool for predicting lung cancer, because there will be too many false positives (normal people who are smokers).   ■

EXAMPLE 3.25   **Cancer** Suppose that the disease is breast cancer in women and the symptom is having a family history of breast cancer (i.e., either a mother or a sister with breast cancer). If we assume that 5% of people with breast cancer have a family history of breast cancer whereas only 2% of people without breast cancer have such a history, then the sensitivity is .05 and the specificity is .98 = (1 − .02). A family history of breast cancer cannot be used by itself to diagnose breast cancer because there will be too many false negatives (i.e., people without a family history who have the disease).   ■

How can the sensitivity and specificity of a symptom (or set of symptoms), which are quantities a physician can estimate, be used to compute predictive values, which are quantities a physician needs to make appropriate diagnoses?

Let A = symptom and B = disease. From Definitions 3.12, 3.13, and 3.14, we have

$$\text{Predictive value positive} = PV^+ = Pr(B|A)$$

$$\text{Predictive value negative} = PV^- = Pr(\bar{B}|\bar{A})$$

$$\text{Sensitivity} = Pr(A|B)$$

$$\text{Specificity} = Pr(\bar{A}|\bar{B})$$

Let $Pr(B)$ = probability of disease in the general population. We wish to compute $Pr(B|A)$ and $Pr(\bar{B}|\bar{A})$ in terms of the other quantities. From the definition of conditional probability,

$$PV^+ = Pr(B|A) = \frac{Pr(B \cap A)}{Pr(A)}$$

Also, from the definition of conditional probability,

$$Pr(B \cap A) = Pr(A|B) \times Pr(B)$$

Finally, from the total probability rule,

$$Pr(A) = Pr(A|B) \times Pr(B) + Pr(A|\bar{B}) \times Pr(\bar{B})$$

If the expressions for $Pr(B \cap A)$ and $Pr(A)$ are substituted into the equation for PV^+, we obtain

$$PV^+ = Pr(B|A) = \frac{Pr(A|B) \times Pr(B)}{Pr(A|B) \times Pr(B) + Pr(A|\bar{B}) \times Pr(\bar{B})}$$

This relationship is known as Bayes' rule.

3.9   **Bayes' Rule**

Let A = symptom and B = disease.

$$PV^+ = Pr(B|A) = \frac{Pr(A|B) \times Pr(B)}{Pr(A|B) \times Pr(B) + Pr(A|\bar{B}) \times Pr(\bar{B})}$$

In words, this can be written as

$$PV^+ = \frac{x \times \text{sensitivity}}{x \times \text{sensitivity} + (1 - x) \times (1 - \text{specificity})}$$

where $x = Pr(B)$ = prevalence of disease in the general population. Similarly

$$PV^- = \frac{(1 - x) \times \text{specificity}}{(1 - x) \times \text{specificity} + x \times (1 - \text{sensitivity})}$$

That is, predictive value can be expressed as a function of sensitivity, specificity, and probability of disease in the general population.

EXAMPLE 3.26   **Hypertension** Suppose that 84% of hypertensives and 23% of normotensives are classified as hypertensive by an automated blood-pressure machine. What is the predictive value positive and predictive value negative of the machine, assuming that 20% of the adult population is hypertensive?

SOLUTION   The sensitivity = .84 and specificity = $1 - .23 = .77$. Thus, from Bayes' rule it follows that

$$PV^+ = (.2)(.84)/[(.2)(.84) + (.8)(.23)]$$
$$= .168/.352 = .48$$

Similarly,

$$PV^- = (.8)(.77)/[(.8)(.77) + (.2)(.16)]$$
$$= .616/.648 = .95$$

Thus, a negative result from the machine is very predictive, since we are 95% sure that such a person is normotensive. However, a positive result is not very predictive, since we are only 48% sure that such a person is hypertensive. ∎

In Example 3.26 there were only two possible disease states: hypertensive and normotensive. In clinical medicine there are often more than two possible disease states. We would like to be able to predict the most likely disease state given a specific symptom (or set of symptoms). We will assume that the probability of having these symptoms for each disease state is known from clinical experience, as is the probability of each of the disease states in the general population. This objective leads us to the generalized Bayes' rule:

3.10 | **Generalized Bayes' Rule**

Let $B_1, B_2, \ldots, B_k$ be a set of mutually exclusive and exhaustive disease states, that is, at least one disease state must occur and no two disease states can occur at the same time. Let A represent the presence of a symptom or set of symptoms. Then

$$Pr(B_i|A) = Pr(A|B_i) \times Pr(B_i) \left/ \left[\sum_{j=1}^{k} Pr(A|B_j) \times Pr(B_j) \right] \right.$$

This result is obtained in a similar manner to that of Bayes' rule for two disease states in **(3.9)**. Specifically, from the definition of conditional probability, note that

$$Pr(B_i|A) = \frac{Pr(B_i \cap A)}{Pr(A)}$$

Also, from the definition of conditional probability,

$$Pr(B_i \cap A) = Pr(A|B_i) \times Pr(B_i)$$

From the total probability rule,

$$Pr(A) = Pr(A|B_1) \times Pr(B_1) + \cdots + Pr(A|B_k) \times Pr(B_k)$$

If the expressions for $Pr(B_i \cap A)$ and $Pr(A)$ are substituted we obtain

$$Pr(B_i|A) = \frac{Pr(A|B_i) \times Pr(B_i)}{\sum_{j=1}^{k} Pr(A|B_j) \times Pr(B_j)}$$

EXAMPLE 3.27 | **Pulmonary Disease** Suppose that a 60-year-old male who has never smoked cigarettes presents with symptoms consisting of a chronic cough and occasional breathlessness to a physician. The physician becomes concerned and orders the patient admitted to the hospital for a lung biopsy. Suppose that the results of the lung biopsy are consistent with either lung cancer or sarcoidosis, a fairly common, nonfatal lung disease. In this case

Symptoms A = {chronic cough, results of lung biopsy}

Disease state B_1 = normal

B_2 = lung cancer

B_3 = sarcoidosis

Suppose that $Pr(A|B_1) = .001$ $\quad$ $Pr(A|B_2) = .9$ $\quad$ $Pr(A|B_3) = .9$

and that in 60-year-old, never-smoking males

$Pr(B_1) = .99$ $\quad$ $Pr(B_2) = .001$ $\quad$ $Pr(B_3) = .009$

The first set of probabilities $Pr(A|B_i)$ could be obtained from clinical experience with the previous diseases, whereas the latter set of probabilities $Pr(B_i)$ would have to be obtained from age-sex-smoking specific prevalence rates for the diseases in question. The interesting question now is what are the probabilities $Pr(B_i|A)$ of the three disease states given the previous symptoms?

SOLUTION   Bayes' rule can be used to answer this question. Specifically,

$$Pr(B_1|A) = Pr(A|B_1) \times Pr(B_1) \Big/ \left[\sum_{j=1}^{3} Pr(A|B_j) \times Pr(B_j) \right]$$
$$= (.001)(.99)/[(.001)(.99) + .9(.001) + .9(.009)]$$
$$= .00099/.00999 = .099$$

$$Pr(B_2|A) = .9(.001)/[(.001)(.99) + .9(.001) + .9(.009)]$$
$$= .00090/.00999 = .090$$

$$Pr(B_3|A) = .9(.009)/[(.001)(.99) + .9(.001) + .9(.009)]$$
$$= .00810/.00999 = .811$$

Thus, although the unconditional probability of sarcoidosis is very low (.009), the conditional probability of the disease given these symptoms and this age-sex-smoking group is .811. Also, although the symptoms are consistent with both lung cancer and sarcoidosis, the latter is much more likely among patients in this age-sex-smoking group.   ∎

EXAMPLE 3.28   **Pulmonary Disease** Now, suppose that the patient in Example 3.27 was a smoker of two packs of cigarettes per day for 40 years. Then, assume that $Pr(B_1) = .98$, $Pr(B_2) = .015$, $Pr(B_3) = .005$ in this type of person. What are the probabilities of the three disease states given these symptoms for this type of patient?

SOLUTION
$$Pr(B_1|A) = (.001)(.98)/[(.001)(.98) + .9(.015) + .9(.005)]$$
$$= .00098/.01898 = .052$$
$$Pr(B_2|A) = .9(.015)/.01898 = .01350/.01898 = .711$$
$$Pr(B_3|A) = .9(.005)/.01898 = .237$$

Thus, in this type of patient, lung cancer is the most likely diagnosis.   ∎

SECTION 3.8   **Prevalence and Incidence**

In clinical medicine the terms *prevalence* and *incidence* are used to denote probabilities in a special context and are used frequently in this text.

DEFINITION 3.16 ■■■
The **prevalence** of a disease is the probability of currently having that disease regardless of the duration of time one has had the disease. It is obtained by dividing the number of people who currently have the disease by the number of people in the study population.   ∎

EXAMPLE 3.29   **Hypertension** The prevalence of hypertension in 1974 among all people 17 years of age and older was reported to be 15.7% as assessed by a government study [5]. It was computed by dividing the number of people who had elevated blood pressure and were 17 years of age and older (22,626) by the total number of people 17 years of age and older in the study population (144,380).   ∎

DEFINITION 3.17 ■■■
The **incidence** of a disease is the probability an individual with no prior disease will develop a new case of the disease over some specified time period.   ∎

EXAMPLE 3.30 **Cancer** The annual incidence rate of breast cancer in 40–44-year-old Connecticut women over the time period January 1, 1970, through December 31, 1970, was approximately 1 per 1000 [2]. This rate means that about 1 woman in 1000 of the 40–44-year-old women who had never had breast cancer on January 1, 1970, would have developed a new case of breast cancer by December 31, 1970. ∎

SECTION 3.9 **Summary**

In this chapter probabilities and how to work with them using the addition and multiplication laws were discussed. An important distinction was made between independent events, which are unrelated to each other, and dependent events, which tend to occur simultaneously. The general concepts of conditional probability and relative risk were introduced to quantify the dependence between two events. These ideas were then applied to the special area of screening populations for disease. In particular, the notions of sensitivity, specificity, and predictive value, which are used to define the accuracy of screening tests, were developed as applications of conditional probability. On some occasions only sensitivities and specificities are available and we wish to compute the predictive value of screening tests. This task can be accomplished using Bayes' rule. Indeed, Bayes' rule can be used generally to change the direction of conditional probabilities. Finally, prevalence and incidence, which are probabilistic parameters that are often used to describe the magnitude of disease in a population, were defined.

In the next two chapters, these general principles of probability are applied to derive some of the important probabilistic models often used in biomedical research, including the binomial, Poisson, and normal models. These models will be used eventually to test hypotheses about data.

PROBLEMS

Let $A = \{$serum cholesterol $= 250\text{–}299\}$, $B = \{$serum cholesterol $\geqslant 300\}$, $C = \{$serum cholesterol $\leqslant 280\}$.

3.1 Are the events A and B mutually exclusive?

3.2 Are the events A and C mutually exclusive?

3.3 Suppose $Pr(A) = .2$, $Pr(B) = .1$. What is $Pr($serum cholesterol $\geqslant 250)$?

3.4 What does $A \cup C$ mean?

3.5 What does $A \cap C$ mean?

3.6 What does $B \cup C$ mean

3.7 What does $B \cap C$ mean?

3.8 Are the events B and C mutually exclusive?

3.9 What does the event $\bar{B}$ mean? What is its probability?

Consider a family with a mother, father, and two children. Let $A_1 = \{$mother has influenza$\}$, $A_2 =$ $\{$father has influenza$\}$, $A_3 = \{$first child has influenza$\}$, $A_4 = \{$second child has influenza$\}$, $B = \{$at least one child has influenza$\}$, $C = \{$at least one parent has influenza$\}$, $D = \{$at least one person in the family has influenza$\}$.

3.10 What does $A_1 \cup A_2$ mean?

3.11 What does $A_1 \cap A_2$ mean?

3.12 Are A_3 and A_4 mutually exclusive?

3.13 What does $A_3 \cup B$ mean?

3.14 What does $A_3 \cap B$ mean?

3.15 Express C in terms of A_1, A_2, A_3, A_4.

3.16 Express D in terms of B and C.

3.17 What does $\bar{A}_1$ mean?

3.18 What does $\bar{A}_2$ mean?

3.19 Represent $\bar{C}$ in terms of A_1, A_2, A_3, A_4.

3.20 Represent $\bar{D}$ in terms of B and C.

Refer to Problem 3.10. Suppose that an influenza epidemic strikes a city. In 10% of families the mother has influenza; in 10% of families the father has influenza; and in 2% of families both the mother and father have influenza.

3.21 Are the events A_1, A_2 independent?

Suppose that the gender of successive offspring in the same family are independent events and that the probability of a male or female offspring is .5.

3.22 What is the probability of two successive female offspring?

3.23 What is the probability that exactly one of two successive children will be female?

3.24 Suppose that three successive offspring are male. What is the probability that a fourth child will be male?

Refer to Problem 3.21.

3.25 What is the probability that at least one parent will get influenza?

Suppose that there is a 20% chance that each child will get influenza, whereas in 10% of two-child families, both children get the disease.

3.26 What is the probability that at least one child will get influenza?

Hypertension

Multiple drugs are often used in treating hypertension. Suppose that 10% of patients taking antihypertensive agent A experience gastrointestinal (GI) side effects, whereas 20% of patients taking antihypertensive agent B experience such side effects.

3.27 If the side effects of the two agents are assumed to be independent events, then what is the probability that a patient taking the two agents simultaneously will experience GI side effects?

Refer to Problem 3.21.

3.28 What is the conditional probability that the father has influenza given that the mother has influenza?

3.29 What is the conditional probability that the father has influenza given that the mother does not have influenza?

Cardiovascular Disease

A survey was performed among people 65 years of age and older who underwent open-heart surgery. It was found that 30% of patients died within 90 days of the operation, whereas an additional 25% of those who survived 90 days died within the next 5 years.

3.30 What is the probability that a patient undergoing open-heart surgery will die within 5 years?

3.31 What is the mortality incidence (per patient month) in patients receiving this operation in the first 90 days after the operation?

3.32 Answer the same question as in Problem 3.31 for the period from 90 days to 5 years after the operation.

3.33 Can you tell if the operation prolongs life from the data presented? If not, then what additional data do you need?

Mental Health

Estimates of the prevalence of Alzheimer's disease have recently been provided by Pfeffer et al. [6]. The estimates are given in Table 3.2.

TABLE 3.2 Prevalence of Alzheimer's disease (cases per 100 population)

Age group	Males	Females
65–69	1.6	0.0
70–74	0.0	2.2
75–79	4.9	2.3
80–84	8.6	7.8
85+	35.0	27.9

Suppose an unrelated 77-year-old man, 76-year-old woman, and 82-year-old woman are selected from a community.

3.34 What is the probability that all three of these individuals have Alzheimer's disease?

3.35 What is the probability that at least one of the women has Alzheimer's disease?

3.36 What is the probability that at least one of the three individuals has Alzheimer's disease?

3.37 What is the probability that exactly one of the three individuals has Alzheimer's disease?

3.38 Suppose we know that one of the three individuals has Alzheimer's disease, but we don't know which one. What is the conditional probability that the affected individual is a woman?

3.39 Suppose we know that two of the three individuals have Alzheimer's disease. What is the conditional probability that they are both women?

3.40 Suppose we know that two of the three individuals have Alzheimer's disease. What is the conditional probability that they are both less than 80 years old?

Suppose the probability that both members of a married couple will have the disease, where each member is 75–79 years old, is .0015.

3.41 What is the conditional probability that the man will be affected given that the woman is affected? How does this value compare to the prevalence in Table 3.2? Why should it be the same (or different)?

3.42 What is the conditional probability that the woman will be affected given that the man is affected? How does this value compare to the prevalence in Table 3.2? Why should it be the same (or different)?

3.43 What is the probability that at least one member of the couple is affected?

Suppose a study of Alzheimer's disease is proposed in a retirement community with persons 65+ years of age, where the age-sex distribution is as shown in Table 3.3.

TABLE 3.3 Age-sex distribution of retirement community

	Male (%)*	Female (%)
65–69	5	10
70–74	9	17
75–79	11	18
80–84	8	12
85+	4	6

* Percent of total population

3.44 What is the expected overall prevalence of Alzheimer's disease in the community, if the prevalence estimates in Table 3.2 for specific age-sex groups holds?

3.45 If there are 1000 persons 65+ years of age in the community, then what is the expected number of cases of Alzheimer's disease in the community?

Cardiovascular Disease

A study relating smoking history to several measures of cardiopulmonary disability was recently reported [7]. The data in Table 3.4 were presented relating the number of people with different disabilities according to cigarette smoking status.

3.46 What is the prevalence of angina among light current smokers (<15 g/day)?

3.47 What is the relative risk of ex-smokers, light current smokers, and heavy current smokers, respectively, for shortness of breath as compared with nonsmokers?

3.48 Answer Problem 3.47 for angina.

3.49 Answer Problem 3.47 for possible infarction.

Occupational Health

A study is conducted in male workers 50–69 years old working in a chemical plant. We are interested in comparing the mortality experience of the workers in the plant with national mortality rates. Suppose that of the 500 workers in this age group in the plant, 35% are 50–54, 30% are 55–59, 20% are 60–64, and 15% are 65–69.

3.50 If the annual national mortality rates are 0.9% in 50–54-year-old men, 1.4% in 55–59-year-old men, 2.2% in 60–64-year-old men, and 3.3% in 65–69-year-old men, then what is the projected annual mortality rate in the plant as a whole?

The SMR (standardized mortality ratio) is often used in occupational studies as a measure of risk. It is defined as 100% *times* the observed number of events in the exposed group *divided by* the expected number of events in the exposed group (based on some reference population).

3.51 If 15 deaths are observed over 1 year among the 500 workers, then what is the SMR?

TABLE 3.4 Number of people with selected cardiopulmonary disabilities versus cigarette-smoking status

Disability	None ($n=656$)	Ex ($n=826$)	Current < 15 g/day ($n=955$)	Current ⩾ 15 g/day ($n=654$)
Shortness of breath	7	15	18	13
Angina	15	19	19	16
Possible infarction	3	7	8	6

Pulmonary Disease

Pulmonary embolism is a relatively common condition that necessitates hospitalization and also often occurs in patients hospitalized for other reasons. An oxygen tension (arterial P_{O_2}) < 90 mm Hg is one of the important criteria used in diagnosing this condition. Suppose that the sensitivity of this test is 95%, the specificity is 75%, and the estimated prevalence is 20% (i.e., a doctor estimates that a patient has a 20% chance of pulmonary embolism before performing the test).

3.52 What is the predictive value positive of this test? What does it mean in words?

3.53 What is the predictive value negative of this test? What does it mean in words?

3.54 Answer Problem 3.52 if the estimated prevalence is 80%.

3.55 Answer Problem 3.53 if the estimated prevalence is 80%.

Genetics

Suppose that a disease is inherited via a **dominant** mode of inheritance and that one of two parents is affected with the disease whereas one is not. The implications of this mode of inheritance are that the probability is $\frac{1}{2}$ that any particular offspring will get the disease.

3.56 What is the probability that in a family with two children, both siblings are affected?

3.57 What is the probability that exactly one sibling is affected?

3.58 What is the probability that neither sibling will be affected?

3.59 Suppose that the older child is affected. What is the probability that the younger child will be affected?

3.60 If A, B are two events such that $A = \{$older child is affected$\}$, $B = \{$younger child is affected$\}$, then are the events A, B independent?

Suppose that a disease is inherited via an **autosomal recessive** mode of inheritance. The implications of this mode of inheritance are that the children in a family each have a probability of $\frac{1}{4}$ of inheriting the disease.

3.61 What is the probability that in a family with two children, both siblings are affected?

3.62 What is the probability that exactly one sibling is affected?

3.63 What is the probability that neither sibling is affected?

Suppose that a disease is inherited via a **sex-linked** mode of inheritance. The implications of this mode of inheritance are that each male offspring has a 50% chance of inheriting the disease, whereas the female offspring have no chance of getting the disease.

3.64 In a family with one male and one female offspring, what is the probability that both siblings are affected?

3.65 What is the probability that exactly one sibling is affected?

3.66 What is the probability that neither sibling is affected?

3.67 Answer Problem 3.64 for families with two male siblings.

3.68 Answer Problem 3.65 for families with two male siblings.

3.69 Answer Problem 3.66 for families with two male siblings.

Suppose that in a family with two male siblings, both siblings are affected with a genetically inherited disease. Suppose also that, although the genetic history of the family is unknown, only a dominant, recessive, or sex-linked mode of inheritance is possible.

3.70 Assume that the dominant, recessive, and sex-linked modes of inheritance follow the probability laws given in Problems 3.56, 3.61, and 3.64 and that, without prior knowledge about the family in question, each is equally likely to occur. What is the probability of each mode of inheritance in this family?

3.71 Answer Problem 3.70 for a family with two male siblings where only one sibling is affected.

3.72 Answer Problem 3.70 for a family with one male and female sibling where both siblings are affected.

3.73 Answer Problem 3.72 where only the male sibling is affected.

Environmental Health, Pediatrics

3.74 Suppose that a company plans to build a lead smelter in a community and that the city council wishes to assess the health effects of the smelter. In particular, there is concern from previous literature that children living very close to the smelter will experience unusually high rates of lead poisoning in the first 3 years of life. Suppose that the projected rates of lead poisoning over this time period are 50 per 100,000 for those children living within 2 km of the smelter, 20 per 100,000 for children living >2 km but ≤5 km from the smelter, and 5 per 100,000 for children living > 5 km from the smelter. If 80% of the children live more than 5 km from the smelter, 15% live > 2 km but $\leqslant 5$ km from the smelter, and the remainder live $\leqslant 2$ km from the smelter, then what is the overall probability that a child from this community will get lead poisoning?

Obstetrics

The following data are derived from the 1973 Final Natality Statistics report issued by the National Center for Health Statistics [8]. These data are pertinent to live births only.

Suppose that infants are classified as low birthweight if they have a birthweight $\leqslant 2500$ g and as normal birthweight if they have a birthweight $\geqslant 2501$ g. Suppose that infants are also classified by period of gestation in the following four categories: < 20 weeks, 20–27 weeks, 28–36 weeks, > 36 weeks. Assume that the probabilities of the different periods of gestation are as given in Table 3.5.

TABLE 3.5 Relationship between birthweight and gestational age

Period of gestation	Probability low birthweight
< 20 weeks	.0004
20–27 weeks	.0059
28–36 weeks	.0855
> 36 weeks	.9082

Also assume that the probability of being low birthweight given that the period of gestation is < 20 weeks is .540, the probability of being low birthweight given that the period of gestation is 20–27 weeks is .813, the probability of being low birthweight given that the period of gestation is 28–36 weeks is .379, and the probability of being low birthweight given that the period of gestation is > 36 weeks is .035.

3.75 What is the probability of having a low birthweight infant?

3.76 Show that the events (period of gestation $\leqslant 27$ weeks) and (low birthweight) are not independent.

3.77 What is the probability of having a period of gestation $\leqslant 36$ weeks given that a child is low birthweight?

Cerebrovascular Disease

One problem with using the angiogram to diagnose stroke is the slight risk of mortality associated with this test ($< 1\%$). Some investigators have attempted to use the PET scanner (which measures blood flow in the brain) to detect stroke disease noninvasively as an alternative to the angiogram. A comparison was made on the same patients between these two methodologies for detecting stroke, with the results given in Table 3.6.

TABLE 3.6 Comparison of a noninvasive test for detecting stroke with an angiogram

Angiogram	Noninvasive test	Frequency
−	−	21
−	+	8
+	−	3
+	+	32

Regard the angiogram as the definitive test.

3.78 What is the sensitivity of the noninvasive test?

3.79 What is the specificity of the noninvasive test?

3.80 What is the predictive value positive of the noninvasive test if we assume that the patients in this series are typical of patients for whom the confirmation of a stroke diagnosis is necessary?

3.81 What is the predictive value negative of the noninvasive test under the same assumptions as in Problem 3.80?

Pulmonary Disease

A 1974 paper by Colley et al. looked at the relationship between parental smoking and the incidence of pneumonia and/or bronchitis in children in the first year of life [9]. One important finding of the paper was that 7.8% of children with nonsmoking parents had episodes of pneumonia and/or bronchitis in the first year of life, whereas, respectively, 11.4% of children with one smoking parent and 17.6% of children with two smoking parents had such an episode. Suppose that in the general population both parents are smokers in 40% of households, one parent smokes in 25% of households, and neither parent smokes in 35% of households.

3.82 What percentage of children in the general population will have pneumonia and/or bronchitis in the first year of life?

A group of families in which both parents smoke at the time of the first prenatal visit decide, after counseling by the nurse practitioner, to give up smoking. Suppose that in 10% of these families both parents resume smoking and in 30% of these families one parent resumes smoking. In the remainder of the families both parents have not resumed smoking at the time of birth of the child. Assume also that the smoking status of the parents at the time of the birth is maintained during the first year of life of the child.

3.83 What is the probability of pneumonia and/or bronchitis in children from families in this group?

3.84 What percentage of cases of pneumonia and/or bronchitis have been prevented by this type of counseling in families where both parents smoke?

Diabetes

The prevalence of diabetes in adults at least 20 years old has been studied in Tecumseh, Michigan [10]. The age-sex specific prevalence (per 1000) is given in Table 3.7.

TABLE 3.7 Age-sex specific prevalence of diabetes in Tecumseh, MI (per 1000)

Age group (years)	Sex	
	Male	Female
20–39	5	7
40–54	23	31
55+	57	89

(Reprinted with permission from the *American Journal of Epidemiology*, *116*(6), 971–980.)

3.85 Suppose we plan a new study in a town that consists of 48% males and 52% females. Of the males, 40% are ages 20–39, 32% are 40–54, and 28% are 55+. Of the females, 44% are ages 20–39, 37% are 40–54, and 19% are 55+. Assuming that the Tecumseh prevalence rates hold, what is the expected prevalence of diabetes in the new study?

3.86 What proportion of diabetics in the new study would be expected in each of the six age-sex groups?

Pulmonary Disease

The familial aggregation of respiratory disease is a well-established clinical phenomenon. However, whether this aggregation is due to genetic or environmental factors or both is somewhat controversial. An investigator wishes to study a particular environmental factor, namely, the relationship of cigarette smoking habits in the parents to the presence of absence of asthma in their oldest child living in the household in the 5–9-year-old age range (referred to below as their offspring). Suppose that the investigator finds that: (i) if both the mother and father are current smokers, then the probability of

their offspring having asthma is .15; (ii) if the mother is a current smoker and the father is not, then the probability of their offspring having asthma is .13; (iii) if the father is a current smoker and the mother is not, then the probability of their offspring having asthma is .05; (iv) if neither parent is a current smoker, then the probability of their offspring having asthma is .04.

3.87 Suppose that the smoking habits of the parents are independent and that the probability that the mother is a current smoker is .4, whereas the probability that the father is a current smoker is .5. What is the probability that both the father and the mother are current smokers?

3.88 What is the probability that the father is a current smoker if the mother is not a current smoker?

Suppose, alternatively, that if the father is a current smoker, then the probability that the mother is a current smoker is .6; whereas if the father is not a current smoker, then the probability that the mother is a current smoker is .2. Also assume that statements (i), (ii), (iii), and (iv) above hold.

3.89 If the probability that the father is a current smoker is .5, what is the probability that the father is a current smoker *and* that the mother is not a current smoker?

3.90 Are the current smoking habits of the father and the mother independent? Why or why not?

3.91 Find the unconditional probability that the offspring will have asthma under the assumptions in 3.89 and 3.90.

3.92 Suppose that a child has asthma. What is the probability that the father is a current smoker?

3.93 What is the probability that the mother is a current smoker if the child has asthma?

3.94 Answer Problem 3.92 if the child does not have asthma.

3.95 Answer Problem 3.93 if the child does not have asthma.

3.96 Are the child's asthma status and the father's smoking status independent? Why or why not?

3.97 Are the child's asthma status and the mother's smoking status independent? Why or why not?

Cancer

Table 3.8 shows the annual incidence rates for colon cancer, lung cancer, and stomach cancer in males ages 50 years and older from the Connecticut Tumor Registry, 1963–1965 [2].

TABLE 3.8 Average annual incidence per 100,000 males for colon, lung, and stomach cancer from the Connecticut Tumor Registry, 1963–1965 (2)

Type of cancer	Ages		
	50–54	55–59	60–64
Colon	35.7	60.3	98.9
Lung	76.1	137.5	231.7
Stomach	20.8	39.1	46.0

(Reprinted from *Cancer Incidence in Five Continents II*, 1970, with permission of Springer-Verlag, Berlin.)

3.98 What is the probability that a 57-year-old, disease-free male will develop lung cancer over the next year?

3.99 What is the probability that a 55-year-old, disease-free male will develop colon cancer over the next 5 years?

3.100 Suppose there is a cohort of 1000 50-year-old men who have never had cancer. How many colon cancers would be expected to develop in this cohort over a 15-year period?

3.101 Answer Problem 3.100 for lung cancer.

3.102 Answer Problem 3.100 for stomach cancer.

Pulmonary Disease

Smoking cessation is an important dimension in public health programs aimed at prevention of cancer and heart and lung diseases. For this purpose data were accumulated starting in 1962 on a group of current smoking men as part of the Normative Aging Study, a longitudinal study of the Veterans Administration in Boston. No interventions were attempted on this group of men, but the data in Table 3.9 were obtained as to annual quitting rates among initially healthy men who remained healthy during the entire period [11]:

TABLE 3.9 Annual quitting rates of men who smoked, from the Normative Aging Study, 1962–1975

Time period	Light smokers (≤ one pack per day) average annual quitting rate per 100 persons	Heavy smokers (> one pack per day) average annual quitting rate per 100 persons
1962–1966	3.1	2.0
1967–1970	7.1	5.0
1971–1975	4.7	4.1

Note that the quitting rates increased during the period of 1967 to 1970, which was around the time of the first Surgeon General's report on cigarette smoking.

3.103 Suppose a man was a light smoker on January 1, 1962. What is the probability that he quit smoking by the end of 1975 (a 14-year period)? (Assume that he remained a light smoker until just prior to quitting.)

3.104 Answer Problem 3.103 for a heavy smoker on January 1, 1962 (assume that he remained a heavy smoker until just prior to quitting).

Cardiovascular Disease

An experiment was set up by a group from the University of Utah to use Bayes' rule to help make clinical diagnoses [12]. In particular, a detailed medical history questionnaire and electrocardiogram were administered to each patient referred to a cardiovascular laboratory and suspected of having congenital heart disease. From the experience of this laboratory and from estimates based on other published data, two sets of probabilities were generated:

(1) The unconditional probability of each of several disease states (refer to prevalence column in Table 3.10)

(2) The conditional probability of specific symptoms given specific disease states (refer to the rest of Table 3.10)

Thus, the probability that a person has chest pain given that he or she is normal is .05. Similarly, the proportion of persons with isolated pulmonary hypertension is .020. A subset of the data is given in Table 3.10.

Assume that these diagnoses are the only ones possible and that a patient can have one and only one diagnosis.

3.105 What is the probability of having a symptom of chest pain given that you have isolated pulmonary hypertension?

3.106 What is the probability of being more than 20 years old in this clinic?

3.107 Suppose we assume that the probability of any set of symptoms are independent given a specific diagnosis (e.g., the probability of being >20 years old and having both chest pain and mild cyanosis given that one is normal is $.50 \times .05 \times .01 = .00025$). What is the probability of being diagnosed as normal given that you have the following symptoms: (i) age 1–20 years, (ii) repeated respiratory infections, and (iii) easy fatigue?

TABLE 3.10 Prevalence of symptoms and diagnoses for patients suspected of having congenital heart disease

Diagnosis	Prevalence	Symptoms						
		X_1	X_2	X_3	X_4	X_5	X_6	X_7
Y_1	.155	.49	.50	.01	.10	.05	.05	.01
Y_2	.126	.50	.50	.02	.50	.02	.40	.70
Y_3	.084	.55	.05	.25	.90	.05	.10	.95
Y_4	.020	.45	.45	.01	.95	.10	.10	.95
Y_5	.098	.10	.00	.20	.70	.01	.05	.40
Y_6	.391	.70	.15	.01	.30	.01	.15	.30
Y_7	.126	.60	.10	.30	.70	.10	.20	.70

Y_1 = normal
Y_2 = atrial septal defect without pulmonary stenosis or pulmonary hypertension*
Y_3 = ventricular septal defect with valvular pulmonary stenosis
Y_4 = isolated pulmonary hypertension*
Y_5 = transposed great vessels
Y_6 = ventricular septal defect without pulmonary hypertension*
Y_7 = ventricular septal defect with pulmonary hypertension*
X_1 = age 1–20 years old
X_2 = age > 20 years old
X_3 = mild cyanosis
X_4 = easy fatigue
X_5 = chest pain
X_6 = repeated respiratory infections
X_7 = EKG axis more than 110°

*Pulmonary hypertension is defined as pulmonary artery pressure ⩾ systematic arterial pressure. (Reprinted with permission of *The American Medical Association* from *The Journal of the American Medical Association, 177*(3), 177–183, 1961. Copyright 1961, American Medical Association.)

3.108 What is the *most likely* diagnosis given that you have all the following symptoms: (i) mild cyanosis, (ii) age > 20 years, (iii) EKG axis more than 110°? What is the second most likely diagnosis?

Suppose the symptom of an EKG axis of more than 110° is used as a screening criterion for diagnosing atrial septal defect without pulmonary stenosis or pulmonary hypertension.

3.109 What is the sensitivity of this test?

3.110 What is the specificity of this test?

3.111 Suppose we want to use another symptom in addition to EKG axis more than 110° to diagnose atrial septal defect without pulmonary stenosis or pulmonary hypertension. If we use the symptoms of age 1–20 years old and EKG axis more than 110°, then what is the predictive value positive?

3.112 Suppose we want to use two symptoms to diagnose atrial septal defect without pulmonary stenosis or pulmonary hypertension (not necessarily including EKG axis more than 110°). Which two symptoms can be used to maximize the predictive value positive? (Use a computer to answer this question.)

3.113 Answer Problem 3.112 using three symptoms rather than two. (Use a computer to answer this question.)

Pulmonary Disease
Research into cigarette smoking habits, smoking prevention, and cessation programs necessitates accurate measurement of smoking behavior. However, decreasing social acceptability of smoking appears to engender significant underreporting. Chemical markers for cigarette use can provide objective indicators of smoking

behavior. One widely used noninvasive marker is the level of saliva thiocyanate (SCN). In a Minneapolis school district, 1332 students in the eighth grade (ages 12–14) participated in a study [13] whereby they

(1) Viewed a film illustrating how recent cigarette use could be readily detected from small samples of saliva

(2) Provided a personal sample of saliva thiocyanate

(3) Provided a self-report on the number of cigarettes smoked per week

The results are given in Table 3.11.

TABLE 3.11 Relationship between saliva thiocyanate levels (SCN) and self-reported cigarettes smoked per week

Self-reported cigarettes smoked in last week	Number of students	Percent with SCN ≥ 100 µg/mL
None	1163	3.3
1–4	70	4.3
5–14	30	6.7
15–24	27	29.6
25–44	19	36.8
45+	23	65.2

(Reprinted with permission from the *American Journal of Public Health, 71*(12), 1320, 1981.)

Suppose the self-reports are completely accurate and are representative of the amount that eighth-grade students smoke in the general community. We are considering using an SCN level of $\geq 100\ \mu g/mL$ as a test criterion for identifying cigarette smokers. Regard a student as positive if he or she smokes 1 or more cigarettes per week.

3.114 What is the sensitivity of the test for light-smoking students (i.e., students who smoke ≤ 14 cigarettes per week)?

3.115 What is the sensitivity of the test for moderate-smoking students (i.e., students who smoke 15–44 cigarettes per week)?

3.116 What is the sensitivity of the test for heavy-smoking students (i.e., students who smoke ≥ 45 cigarettes per week)?

3.117 What is the specificity of the test?

3.118 What is the predictive value positive of the test?

3.119 What is the predictive value negative of the test?

Suppose we regard the self-reports of all students who report some cigarette consumption as valid but estimate that 10% of students who report no cigarette consumption actually smoke 1–4 cigarettes per week and an additional 2% smoke 5–14 cigarettes per week.

3.120 If we assume that the percentage of students with SCN $\geq 100\ \mu g/mL$ in these two subgroups is the same as in those who truly report 1–4 and 5–14 cigarettes per week, then what effect would this underreporting have on the predictive value positive of the test (i.e., would the true predictive value positive be the same, higher, or lower than that computed in 3.118)?

3.121 Compute the predictive value positive under these altered assumptions.

Cardiovascular Disease

The relationship between physical fitness and cardiovascular disease mortality was recently studied in a group of railroad working men, ages 22–79 [14]. Data were presented relating baseline exercise test heart rate and coronary heart disease mortality (Table 3.12).

TABLE 3.12 Relationship between baseline exercise test heart rate and coronary heart disease mortality

Exercise test heart rate (beats/min)	Coronary heart disease mortality (20 years) (per 100)
≤ 105	9.1
106–115	8.7
116–127	11.6
> 127	13.2

Suppose that 20%, 30%, 30%, and 20% of the population, respectively, have exercise test heart rates of ≤ 105, 106–115, 116–127, > 127. Suppose a test is positive if exercise test heart rate is > 127 beats/min and negative otherwise.

3.122 What is the probability of a positive test among men who have died over the 20-year period? Is there a name for this quantity?

3.123 What is the probability of a positive test among men who survived the 20-year period? Is there a name for this quantity?

3.124 What is the probability of death among men with a negative test? Is there a name for this quantity?

Hypertension

Laboratory measures of cardiovascular reactivity are receiving increasing attention. Much of the expanded interest is based on the belief that these measures obtained under challenge from physical and psychological stressors may yield a more biologically meaningful perspective of cardiovascular function than more traditional static measures. Typically, measurement of cardiovascular reactivity involves the use of an automated blood-pressure monitor to examine changes in blood pressure before and after a stimulating experience (such as playing a video game). For this purpose, BP measurements were made with the Vita-Stat machine both before and after playing a video game. Similar measurements were obtained using manual methods for obtaining blood pressure. A person was classified as a "reactor" if his or her diastolic blood pressure (DBP) increased by 10 mm Hg or more after playing the game and as a nonreactor otherwise. The results are given in Table 3.13.

TABLE 3.13 Classification of cardiovascular reactivity using an automated and manual sphygmomanometer

	Δ DBP, manual	
Δ DBP, automated	< 10	≥ 10
< 10	51	7
≥ 10	15	6

3.125 If the manual measurements are regarded as the "true" measure of reactivity, then what is the sensitivity of automated BP measurements?

3.126 What is the specificity of automated BP measurements?

3.127 If the population tested is representative of the general population, then what are the predictive values positive and negative using this test?

Otolaryngology

The data set in Table 3.14 is based on 214 children with acute otitis media (OME) who participated in a randomized clinical trial [15]. Each child had OME at the beginning of the study in either one (unilateral cases) or both (bilateral cases) ears. Each child was randomly assigned to receive a 14-day course of one of two antibiotics, either cefaclor (CEF) or amoxicillin (AMO). The focus here is on the 203 children whose middle-ear status was determined at a 14-day follow-up visit. The data in Table 3.14 are presented in the Data Set 6, EAR.DAT.

3.128 Does there seem to be any difference in the effect of the antibiotics on clearance of otitis media? Try to express your results in terms of relative risk. Consider separate analyses for unilateral and bilateral cases. Also consider an analysis combining the two types of cases.

3.129 The investigators recorded the age of the children because they felt this might be an important factor in determining outcome. Were they right? Try to express your results in terms of relative risk.

3.130 While controlling for age, propose an analysis comparing the effectiveness of the two antibiotics. Express your results in terms of relative risk.

3.131 Another issue in this trial is the possible dependence between ears for the bilateral cases. Can you comment on this issue based on the data collected?

The concept of a **randomized clinical trial** is discussed more completely in Chapter 6. The analysis of **contingency table data** is studied in Chapter 10, where many of the formal methods for analyzing this type of data are discussed.

Cardiovascular Disease

Exercise testing has sometimes been used to diagnose patients with coronary artery disease. One test criterion

TABLE 3.14 Format for EAR.DAT

Column	Variable	Format or code
1–3	ID	
5	Clearance by 14 days	1 = yes/0 = no
7	Antibiotic	1 = CEF/2 = AMO
9	Age	1 = <2 yrs/2 = 2–5 yrs/3 = 6+ yrs
11	Ear	1 = 1st ear/2 = 2nd ear

that has been used to identify those with disease is the abnormal ejection fraction criterion, that is, an absolute rise of less than .05 with exercise. The validity of this noninvasive test was assessed in 196 patients versus coronary angiography; the gold standard, a procedure that can unequivocally diagnose the disease but the administration of which carries some risk for the patient. A sensitivity of 79% and a specificity of 68% were found for the exercise test in this group

3.132 What does the sensitivity mean in words in this setting?

3.133 What does the specificity mean in words in this setting?

Suppose a new patient undergoes exercise testing and a physician feels before seeing the exercise test results that the patient has a 20% chance of having coronary artery disease.

3.134 If the exercise test is positive, then what is the probability that such a patient has disease?

3.135 Answer Problem 3.134 if the exercise test is negative.

Cardiovascular Disease
In Table 3.10 data on the relationship between various symptoms and disease states in patients suspected of having congenital heart disease were presented. In this table, for simplicity, only a subset of the symptoms (7) and disease states (7) were presented. In the original report [12], 50 symptoms and 33 disease states were considered. In Data Set 4, DISEASE.DAT, the prevalence of each of the disease states and the conditional probability of each of the symptoms given each of the disease states are presented. The documentation for this data set is given in Data Set 3, DISEASE.DOC.

3.136 Write a computer program to compute the probability of each of the disease states given the presence or absence of any combination of the 50 symptoms. Note that some of the symptoms are mutually exclusive and thus cannot occur simultaneously, for example, symptom 1 = age 1 month to 1 year and symptom 2 = age 1–20 years. Also, some of the symptoms have to be considered as a group. Read the original report for details concerning these points.

3.137 Test your program using some of the examples given in the article.

References

[1] National Center for Health Statistics (1976, February 13). *Monthly vital statistics report, advance report, final natality statistics (1974)*, *24*(11) (Suppl. 2).

[2] Doll, R., Muir, C., & Waterhouse, J. (Eds.) (1970). *Cancer incidence in five continents II.* Berlin: Springer-Verlag.

[3] Feller, W. (1960). *An introduction to probability theory and its applications.* New York: Wiley.

[4] Podgor, M. J., Leske, M. C., & Ederer, F. (1983). Incidence estimates for lens changes, macular changes, open-angle glaucoma, and diabetic retinopathy. *American Journal of Epidemiology, 118*(2), 206–212.

[5] National Center for Health Statistics (1976, November 8). *Advance data from vital and health statistics, 2.*

[6] Pfeffer, R. I., Afifi, A. A., & Chance, J. M. (1987). Prevalence of Alzheimer's Disease in a retirement community. *American Journal of Epidemiology, 125*(3), 420–436.

[7] Tenkanen, L., Teppo, L., & Hakulinen, T. (1987). Smoking and cardiac symptoms as predictors of lung cancer. *Journal of Chronic Disease, 40*(12), 1121–1128.

[8] National Center for Health Statistics (1975, January 30). *Monthly vital statistics report, final natality statistics (1973)*, *23*(11) (Suppl.).

[9] Colley, J. R. T., Holland, W. W., & Corkhill, R. T. (1974). Influence of passive smoking and parental phlegm on pneumonia and bronchitis in early childhood. *Lancet, II,* 1031.

[10] Butler, W. J., Ostrander, L. D., Jr., Carman, W. J., & Lamphiear, D. E. (1982). Diabetes mellitus in Tecumseh, Michigan: Prevalence, incidence and associated conditions. *American Journal of Epidemiology, 116*(6), 971–980.

[11] Garvey, A. J., Bossé, R., Glynn, R. J., & Rosner, B. (1983). Smoking cessation in a prospective study of healthy adult males: Effects of age, time period, and amount smoked. *American Journal of Public Health, 73*(4), 446–450.

[12] Warner, H., Toronto, A., Veasey, L. G., & Stephenson, R. (1961). A mathematical approach to medical diagnosis. *JAMA, 177*(3), 177–183.

[13] Luepker, R. V., Pechacek, T. F., Murray, D. M., Johnson, C. A., Hund, F., & Jacobs, D. R. (1981). Saliva thiocyanate: A chemical indicator of cigarette smoking in adolescents. *American Journal of Public Health, 71*(12), 1320.

[14] Slattery, M. L., & Jacobs, D. R., Jr. (1988). Physical fitness and cardiovascular disease mortality: The U.S. railroad study. *American Journal of Epidemiology, 127*(3), 571–580.

[15] Mandel, E., Bluestone, C. D., Rockette, H. E., Blatter, M. M., Reisinger, K. S., Wucher, F. P., & Harper, J. (1982). Duration of effusion after antibiotic treatment for acute otitis media: Comparison of cefaclor and amoxicillin. *Pediatric Infectious Diseases, 1,* 310–316.

CHAPTER FOUR

DISCRETE PROBABILITY DISTRIBUTIONS

Introduction

In Chapter 3 probability was defined and some of the basic tools used in working with probabilities were introduced. We now look at problems that can be put in a probabilistic framework. That is, by assessing the probabilities of certain events from actual past data, specific probability models that fit our problems can be used.

EXAMPLE 4.1   **Ophthalmology** Retinitis pigmentosa is a progressive ocular disease that in some cases eventually results in blindness. The three main genetic forms of the disease are the dominant mode, the recessive mode, and the sex-linked mode. Each mode has a different rate of progression, with the dominant mode being the slowest to progress and the sex-linked mode the fastest. Suppose a man does not have a clear idea of the prior history of disease in his family but he does know that 1 of his 2 male children is affected, whereas his 1 female child is not affected. Can this information help identify the genetic type? ∎

The **binomial distribution** can be applied to calculate the probability of this event occurring (1 out of 2 males affected, 0 out of 1 female affected) under each of the genetic modes mentioned, and these results can then be used to infer the most likely genetic mode. In fact, this distribution can be used to make an inference for any family where we know that k_1 out of n_1 male children are affected and k_2 out of n_2 female children are affected.

EXAMPLE 4.2   **Cancer** A second example of a commonly used probability model concerns a cancer scare in young children in Woburn, Massachusetts. A news story reported an "excessive" number of cancer deaths in young children in this town and speculated whether or not this high rate was due to the dumping of industrial wastes in the northeastern portion of town [1]. Suppose that 5 cases of leukemia were reported in a town where 1 would normally be expected. Is this difference sufficient evidence for concluding that there is an association between the industrial wastes and the cancer cases? ∎

The **Poisson distribution** can be used to calculate the probability of five or more cases if typical national rates for cancer were present in this town. If this probability were sufficiently small, then we would conclude that there was an association; otherwise, we would conclude that a longer surveillance of the town was necessary before arriving at a conclusion.

In this chapter the general concept of a discrete random variable is introduced and the binomial and Poisson distributions are described in-depth. This forms the basis for the discussion of hypothesis tests based on the binomial and Poisson distributions found in Chapters 7 and 10.

SECTION 4.2   ## Random Variables

In Chapter 3 we dealt with very specific events, such as the outcome of a tuberculin skin test or blood-pressure measurements taken on different members of a family. We now want to introduce ideas that will enable us to refer, in general terms, to different types of events having the *same probabilistic stucture*. For this purpose the concept of a random variable is introduced.

DEFINITION 4.1 ■■■

A **random variable** is a numerical quantity that takes different values with specified probabilities.
                                                                                ■

Two types of random variables are discussed in this text: discrete and continuous.

DEFINITION 4.2 ■■■

A random variable for which there exists a discrete set of values with specified probabilities is a **discrete random variable**.
                                                                                ■

EXAMPLE 4.3   **Otolaryngology** Otitis media is a disease of the middle ear and is one of the most frequent reasons for visiting a doctor in the first 2 years of life other than a routine well-baby visit. Let X be the random variable that represents the number of episodes of otitis media in the first 2 years of life. Then X is a discrete random variable, which takes on the values 0, 1, 2, ... ■

EXAMPLE 4.4   **Hypertension** Many new drugs have been introduced in the last decade to bring hypertension under control, that is, to reduce high blood pressure to normotensive levels. Suppose a physician agrees to use a new antihypertensive drug on a trial basis on the first 4 untreated hypertensives whom she encounters in her practice before deciding whether to adopt the drug for routine use. Let $X = $ the number of patients out of 4 who are brought under control. Then X is a discrete random variable, which takes on the values 0, 1, 2, 3, 4.   ■

DEFINITION 4.3 ■■■

A random variable whose values form a continuum (i.e., have no gaps), such that ranges of values occur with specified probabilities, is a **continuous random variable**.
                                                                                ■

EXAMPLE 4.5   **Environmental Health** The possible health effects on workers exposed to low levels of radiation over long periods of time are an issue of public health interest. One problem in assessing this situation is how to measure the cumulative exposure of a worker. A study was performed at the Portsmouth Naval Shipyard, whereby each exposed worker wore a badge, or dosimeter, which measured annual radiation exposure in rem [2]. The cumulative exposure over a worker's lifetime could then be obtained by summing the yearly exposures. The cumulative lifetime exposure is a good example of a continuous random variable, since it varied in this study from 0.000 rem to 91.414 rem, which would be regarded as taking on an essentially infinite number of values.
                                                                                ■

SECTION 4.3   **The Probability Mass Function for a Discrete Random Variable**

The values taken by a discrete random variable and its associated probabilities can be expressed by a rule, or relationship, which is called a probability mass function.

DEFINITION 4.4   ■■

A **probability mass function** is a mathematical relationship, or rule, that assigns to any possible value r of a discrete random variable X the probability $Pr(X = r)$. This assignment is made for all values r that have positive probability. The probability mass function is sometimes also referred to as a **probability distribution**.   ■

The probability mass function can be displayed in the form of a table giving the values and their associated probabilities and/or it can be expressed as a mathematical formula giving the probability of all possible values.

EXAMPLE 4.6   **Hypertension** Consider the situation in Example 4.4. Suppose that from previous experience with the drug, the drug company expects that for any clinical practice the probability that 0 patients out of 4 will be brought under control is .008, 1 patient out of 4 is .076, 2 patients out of 4 is .265, 3 patients out of 4 is .411, and all 4 patients is .240. This probability mass function, or probability distribution, is displayed in Table 4.1.   ■

TABLE 4.1
Probability mass function for the hypertension control example

$Pr(X = r)$	.008	.076	.265	.411	.240
r	0	1	2	3	4

Notice that for any probability mass function, the probability of any particular value must be between 0 and 1 and the sum of the probabilities of all values must exactly equal 1. Thus, $0 \leqslant Pr(X = r) \leqslant 1$, $\Sigma Pr(X = r) = 1$, where the summation is taken over all possible values that have positive probability.

EXAMPLE 4.7   **Hypertension** In Table 4.1, for any clinical practice, the probability that between 0 and 4 hypertensives are brought under control = 1; that is,

$$.008 + .076 + .265 + .411 + .240 = 1$$ ■

4.3.1   **Relationship of Probability Distributions to Sample Distributions**

In Chapters 1 and 2 the concept of a **frequency distribution** in the context of a sample was discussed. It was described as a list of each value in the data set and a corresponding count of how frequently the values occur. If each count is divided by the total number of points in the sample, then the frequency distribution can be considered as a sample analogue to a probability distribution. In particular, a probability distribution can be thought of as a model based on an infinitely large sample, giving the fraction of data points in a sample that *should* be allocated to each specific value. Since the frequency distribution gives the actual proportion of points in a sample that correspond to specific values, the appropriateness of the model can be validated by comparing the observed sample frequency distribution to the probability distribution. The formal statistical procedure for making this comparison is called a **goodness-of-fit test**, which is discussed in Chapter 10.

EXAMPLE 4.8

Hypertension How can the probability mass function in Table 4.1 be used to see if the drug behaves with the same efficacy in actual practice as predicted by the drug company? The drug company might distribute the drug to 100 physicians and ask each of them to treat their first 4 untreated hypertensives with it. Each physician would then report his or her results to the drug company, and the combined results could be compared with the expected results in Table 4.1. For example, suppose that out of 100 physicians who agree to participate, 19 are able to bring all of their first 4 untreated hypertensives under control, 48 are able to bring 3 of the 4 hypertensives under control, 24 are able to bring 2 out of 4 under control, 9 are able to bring only 1 of 4 under control, and none of the physicians brings 0 out of 4 hypertensives under control. The sample frequency distribution can be compared with the probability distribution given in Table 4.1. This comparison is shown in Table 4.2.

TABLE 4.2
Comparison of the sample frequency distribution and the theoretical probability distribution for the hypertension control example

Number of hypertensives under control = r	Probability distribution $Pr(X = r)$	Frequency distribution
0	.008	.000 = 0/100
1	.076	.090 = 9/100
2	.265	.240 = 24/100
3	.411	.480 = 48/100
4	.240	.190 = 19/100

The distributions look reasonably similar. The role of statistical inference is to compare the two distributions to judge if the differences between the two can be attributed to chance or whether real differences exist between the drug's performance in actual clinical practice and expectations from previous drug company experience. ∎

A question often asked is: Where does a probability mass function come from? In some instances previous data can be obtained on the same type of random variable being studied and the probability mass function can be computed from these data. In other instances previous data may not be available, but the probability mass function from some well-known distribution may be used to see how well it fits with some sample data. In fact, this approach was used in Table 4.2, where the probability mass function was derived from the binomial distribution and then compared with the frequency distribution from the sample of 100 physician practices.

SECTION 4.4

The Expected Value of a Discrete Random Variable

If a random variable has a large number of values with positive probability, then the probability mass function is not a useful summary measure. Indeed, we are faced with the same problem as in trying to summarize a sample by enumerating each data value.

Measures of location and spread can be developed for a random variable in much the same way as they were developed for samples. The analogue to the arithmetic mean $\bar{x}$ is referred to as the expected value of the random variable, or population mean, and is denoted by $E(X)$ or μ. The expected value represents the "average" value of the random variable. It is obtained by multiplying each possible value by its respective probability and summing over all the values that have positive (that is, non-zero) probability.

DEFINITION 4.5 ■■■

The **expected value of a discrete random variable** is defined as

$$E(X) \equiv \mu = \sum_{i=1}^{k} x_i Pr(X = x_i)$$

where the x_i's are the values the random variable assumes with positive probability.   ■

EXAMPLE 4.9   **Hypertension** Find the expected value for the random variable depicted in Table 4.1.

SOLUTION $E(X) = 0(.008) + 1(.076) + 2(.265) + 3(.411) + 4(.240) = 2.80$

Thus, on the average about 2.8 hypertensives would be expected to be brought under control for every 4 that are treated.   ■

EXAMPLE 4.10   **Otolaryngology** Consider the random variable mentioned in Example 4.3 representing the number of episodes of otitis media in the first 2 years of life. Suppose this random variable has a probability mass function as given in Table 4.3.

TABLE 4.3
Probability mass function for the number of episodes of otitis media in the first 2 years of life

r	0	1	2	3	4	5	6
Pr(X = r)	.129	.264	.271	.185	.095	.039	.017

What is the expected number of episodes of otitis media in the first 2 years of life?

SOLUTION $E(X) = 0(.129) + 1(.264) + 2(.271) + 3(.185) + 4(.095) + 5(.039) + 6(.017) = 2.04$

Thus, on the average a child would be expected to have 2 episodes of otitis media in the first 2 years of life.   ■

In Example 4.8 the probability mass function for the random variable representing the number of previously untreated hypertensives brought under control was compared with the actual number of hypertensives brought under control in 100 clinical practices. In much the same way, the expected value of a random variable can be compared with the actual sample mean in a data set ($\bar{x}$).

EXAMPLE 4.11   **Hypertension** Compare the number of hypertensives brought under control in the 100 clinical practices ($\bar{x}$) with the expected number of hypertensives brought under control (μ).

SOLUTION   From Table 4.2 we have

$$\bar{x} = [0(0) + 1(9) + 2(24) + 3(48) + 4(19)]/100 = 2.77$$

hypertensives controlled per clinical practice while $\mu = 2.80$. This agreement is rather good. The specific methods for comparing the observed average value and expected of a random variable ($\bar{x}$ and μ) will be covered in the material on statistical inference in Chapter 7. Notice that $\bar{x}$ could be written in the form

$$\bar{x} = 0(0/100) + 1(9/100) + 2(24/100) + 3(48/100) + 4(19/100)$$

that is, as a weighted average of observed probabilities. The expected value, in comparison, can be written as a weighted average of theoretical probabilities:

$$\mu = 0(.008) + 1(.076) + 2(.265) + 3(.411) + 4(.240)$$

Thus, the two quantities are actually obtained in the same way, one as a weighted average of "observed" probabilities and the other as a weighted average of "theoretical" probabilities.

■

The Variance of a Discrete Random Variable

The analogue to the sample variance (s^2) for a random variable is called the variance of the random variable, or population variance, and is denoted by $Var(X)$. The variance represents the spread of all values that have positive probability relative to the expected value. In particular, the variance is obtained by multiplying the squared distance of each possible value from the expected value by its respective probability and summing over all the values that have positive probability.

DEFINITION 4.6 ■■■

The **variance of a discrete random variable** denoted by X is defined by

$$Var(X) \equiv \sigma^2 = \sum_{i=1}^{k} (x_i - \mu)^2 Pr(X = x_i)$$

where the x_i's are the values for which the random variable takes on positive probability. The **standard deviation of a random variable** X, denoted by $sd(X)$ or σ, is defined by the square root of its variance.

■

There is also a short form for the population variance, which is similar to the equation presented for the sample variance.

This can be obtained by expanding $(x_i - \mu)^2$ in the form $x_i^2 - 2x_i\mu + \mu^2$ and rewriting $Var(X)$ as

$$Var(X) = \sum_{i=1}^{k} (x_i^2 - 2\mu x_i + \mu^2) Pr(X = x_i)$$

$$= \sum_{i=1}^{k} x_i^2 Pr(X = x_i) + \sum_{i=1}^{k} (-2\mu) x_i Pr(X = x_i) + \sum_{i=1}^{k} \mu^2 Pr(X = x_i)$$

Since -2μ and μ^2 are constants, this expression can be rewritten in the form

$$Var(X) = \sum_{i=1}^{k} x_i^2 Pr(X = x_i) - 2\mu \sum_{i=1}^{k} x_i Pr(X = x_i) + \mu^2 \sum_{i=1}^{k} Pr(X = x_i)$$

Since, by definition, $\sum_{i=1}^{k} x_i Pr(X = x_i) = E(X) = \mu$, and $\sum_{i=1}^{k} Pr(X = x_i) = 1$, it follows that

$$Var(X) = \sum_{i=1}^{k} x_i^2 Pr(X = x_i) - 2\mu^2 + \mu^2 = \sum_{i=1}^{k} x_i^2 Pr(X = x_i) - \mu^2$$

4.1   A short form for the **population variance** is given by

$$\sigma^2 = E(X - \mu)^2 = \sum_{i=1}^{k} x_i^2 \, Pr(x_i) - \mu^2$$

EXAMPLE 4.12   **Otolaryngology** Compute the variance and standard deviation for the random variable depicted in Table 4.3.

SOLUTION   We know from Example 4.10 that $\mu = 2.04$. Furthermore,

$$\sum_{i=1}^{k} x_i^2 \, Pr(x_i) = 0^2(.129) + 1^2(.264) + 2^2(.271) + 3^2(.185)$$
$$+ 4^2(.095) + 5^2(.039) + 6^2(.017)$$
$$= 0(.129) + 1(.264) + 4(.271) + 9(.185)$$
$$+ 16(.095) + 25(.039) + 36(.017)$$
$$= 6.12$$

Thus, $Var(X) = \sigma^2 = 6.12 - (2.04)^2 = 1.96$. The standard deviation of X is $\sigma = \sqrt{1.96} = 1.40$.   ∎

How can we get a feel for what the standard deviation of a random variable means? The following often-used principle is true for many, but not all, random variables:

4.2   **Approximately 95% of the probability mass falls within two standard deviations of the mean of a random variable.**

This statement holds exactly for normally distributed random variables, which are discussed in Chapter 5, and approximately for certain other random variables.

EXAMPLE 4.13   **Otolaryngology** Find a, b such that approximately 95% of infants will have between a and b episodes of otitis media in the first 2 years of life.

SOLUTION   The random variable depicted in Table 4.3 has mean $(\mu) = 2.04$ and standard deviation $(\sigma) = 1.40$. The interval $\mu \pm 2\sigma$ is given by

$$2.04 \pm 2(1.4) = 2.04 \pm 2.80$$

or from -0.76 to 4.84. Since only positive integer values are possible for this random variable, the valid range is from $a = 0$ to $b = 4$ episodes. In Table 4.3 the probability of having $\leqslant 4$ episodes is given as

$$.129 + .264 + .271 + .185 + .095 = .944$$ ∎

The rule allows us to quickly summarize the range of values that have most of the probability mass for a random variable without specifying each individual value. In Chapter 6 the type of random variable for which **(4.2)** applies is specified more precisely.

The Cumulative Distribution Function of a Discrete Random Variable

Many random variables are displayed in tables or figures in terms of a cumulative distribution function rather than a distribution of probabilities of individual values as in Table 4.1. The basic concept is to assign to each individual value the sum of probabilities of all values that are no larger than the value being considered. This function is defined as follows:

DEFINITION 4.7 ■■■

The **cumulative distribution function** of a discrete random variable X is denoted by $F(x)$ and is defined by $Pr(X \leqslant x)$.

EXAMPLE 4.14 **Otolaryngology** Compute the cumulative distribution function for the otitis media random variable in Table 4.3 (p. 75) and display it graphically.

SOLUTION The cumulative distribution function is given by

$$F(x) = 0 \qquad \text{if} \qquad x < 0$$
$$F(x) = .129 \qquad \text{if} \qquad 0 \leqslant x < 1$$
$$F(x) = .393 \qquad \text{if} \qquad 1 \leqslant x < 2$$
$$F(x) = .664 \qquad \text{if} \qquad 2 \leqslant x < 3$$
$$F(x) = .849 \qquad \text{if} \qquad 3 \leqslant x < 4$$
$$F(x) = .944 \qquad \text{if} \qquad 4 \leqslant x < 5$$
$$F(x) = .983 \qquad \text{if} \qquad 5 \leqslant x < 6$$
$$F(x) = 1.0 \qquad \text{if} \qquad x \geqslant 6$$

The function can be displayed as shown in Figure 4.1. ■

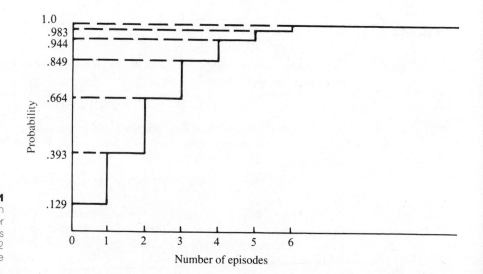

FIGURE 4.1
Cumulative distribution function for the number of episodes of otitis media in the first 2 years of life

The cumulative distribution for a discrete random variable looks like a series of steps. The steps become smaller as the number of values increases, and the function approaches that of a smooth curve.

Permutations and Combinations

In Sections 4.2 through 4.6 the concept of a discrete random variable was introduced in very general terms. In the remainder of this chapter, the focus is on some specific discrete random variables that occur frequently in medical and biological work. Consider the following example.

EXAMPLE 4.15   **Infectious Disease** One of the most common laboratory tests performed on any routine medical examination is a blood test. The two main aspects to a blood test are (1) counting the number of white blood cells (referred to as the "white count") and (2) differentiating the white blood cells that do exist into five categories, namely, neutrophils, lymphocytes, monocytes, eosinophils, and basophils (referred to as the "differential"). Both the white count and the differential are extensively used for clinical diagnoses. We will concentrate here on the differential, particularly on the distribution of the number of neutrophils k out of 100 white blood cells (which is the typical number counted). We will see that the number of neutrophils follows a binomial distribution. ∎

To study the binomial distribution, **permutations** and **combinations**, important topics in probability, must first be understood.

EXAMPLE 4.16   **Mental Health** Suppose we identify 5 male subjects aged 50–59 with schizophrenia in a community, and we wish to match these subjects with normal controls of the same sex and age living in the same community. Suppose we wish to employ a **matched pair design**, where each case is matched with a normal control of the same sex and age. Five psychologists are employed by the study, with each psychologist interviewing a single case and his matched control. If there are 10 eligible 50–59-year-old male controls in the community (labeled A, $B, \ldots, J$), then how many ways are there of choosing controls for the study if a control can never be used more than once?

SOLUTION   The first control can be any of $A, \ldots, J$ and thus can be chosen in 10 ways. Once the first control is chosen he can no longer be selected as the second control; therefore, the second control can be chosen in 9 ways. Thus, the first two controls can be chosen in any one of $10 \times 9 = 90$ ways. Similarly, the third control can be chosen in any one of 8 ways, the fourth control in 7 ways, and the fifth control in 6 ways. In total, there are $10 \times 9 \times 8 \times 7 \times 6 = 30{,}240$ ways of choosing the 5 controls. For example, one possible selection is $ACDFE$. This means that control A is matched to the first case, control C to the second case, and so on. The order of selection of the controls is important, since different psychologists may be assigned to interview each matched pair. Thus, the selection $ABCDE$ is different from $CBAED$, even though the same group of controls is selected. ∎

We can now ask the general question, How many ways can k objects be selected out of n where the order matters? Note that the first object has been selected in any one of $n = (n + 1) - 1$ ways. Given that the first object has been selected, the second object can be selected in any one of $n - 1 = (n + 1) - 2$ ways; $\ldots$; the kth object can be selected in any one of $n - k + 1 = (n + 1) - k$ ways.

DEFINITION 4.8 ▪▪

The number of **permutations** of n things taken k at a time is

$$_nP_k = n(n-1) \times \cdots \times (n-k+1)$$

It represents the number of ways of selecting k items out of n, where the order of selection is important.

▪

EXAMPLE 4.17 **Mental Health** Suppose there are 3 female schizophrenics aged 50–59 and 6 eligible controls living in the same community. How many ways are there of selecting three controls?

SOLUTION To answer this question, consider the number of permutations of 6 things taken 3 at a time.

$$_6P_3 = 6 \times 5 \times 4 = 120$$

Thus, there are 120 ways of choosing the controls. For example, one way would be to match control A to case 1, control B to case 2, and control C to case 3 (i.e., ABC). Another way would be to match control F to case 1, control C to case 2, and control D to case 3 (i.e., FCD). The order of selection is important, since, for example, the selection ABC is different than the selection BCA.

▪

In some instances we are interested in a special type of permutation: selecting n objects out of n, where the order of selection matters (i.e., ordering n objects). By the preceding principle,

$$_nP_n = n(n-1) \times \cdots \times [(n+1) - n] = n(n-1) \times \cdots \times 2 \times 1$$

The special symbol generally used for this quantity is $n!$, which is called n factorial and is defined as follows:

DEFINITION 4.9 ▪▪

$n! = n$ **factorial** is defined as $\quad n(n-1) \times \cdots \times 2 \times 1$ ▪

EXAMPLE 4.18 Evaluate 5 factorial.

SOLUTION
$$5! = 5 \times 4 \times 3 \times 2 \times 1 = 120$$
▪

The quantity $0!$ has no intuitive meaning, but for consistency it will be defined as 1.

EXAMPLE 4.19 **Mental Health** Consider a somewhat different study design for the situation described in Example 4.16. Suppose an **unmatched study design,** whereby *all* cases and controls will be interviewed by the same psychologist, is used. If there are 10 eligible controls, then how many ways are there of choosing 5 controls for the study?

SOLUTION In this case, since the same psychologist interviews all patients, what is important is which controls are selected, not the order of selection. Thus, the question is, How many ways can 5 out of 10 eligible controls be selected, where order is not important? Note that for each set of 5 controls (say A, B, C, D, E), there are $5 \times 4 \times 3 \times 2 \times 1 = 5!$ ways of ordering the controls among themselves (e.g., $ACBED$ and $DBCAE$ are two possible orders). Thus, the number of ways of selecting 5 out of 10 controls for the study without respect to order = (the number of ways of selecting 5 controls out of 10 where order is important)$/5! = {_{10}P_5}/5! = (10 \times 9 \times 8 \times 7 \times 6)/120 = 30{,}240/120 = 252$ ways. Thus, $ABCDE$ and $CDFIJ$ are two possible selections. Also, $ABCDE$ and $BCADE$ are *not* counted twice.

▪

The number of ways of selecting 5 objects out of 10 without respect to order is referred to as the number of **combinations** of 10 things taken at a time and is denoted by $_{10}C_5$ or $\binom{10}{5} = 252$.

This discussion can be generalized to evaluate the number of combinations of n things taken k at a time. Note that for every selection of k distinct items out of n, there are $k(k-1) \times \cdots \times (2) \times (1) = k!$ ways of ordering the items among themselves. Thus, we have the following definition:

DEFINITION 4.10 ■■■

The number of **combinations** of n things taken k at a time is

$$_nC_k = \binom{n}{k} = \frac{n(n-1) \times \cdots \times (n-k+1)}{k!}$$ ■

By multiplying the numerator and denominator of $_nC_k$ by

$$(n-k)! = (n-k)(n-k-1) \times \cdots \times 2 \times 1,$$

we obtain

$$_nC_k = \frac{n(n-1) \times \cdots \times (n-k+1)}{k!} \times \frac{(n-k)(n-k-1) \times \cdots \times 2 \times 1}{(n-k)(n-k-1) \times \cdots \times 2 \times 1}$$

Note that the numerator $= n!$ and the denominator $= k! \times (n-k)!$. Thus, we have the following alternative definition:

DEFINITION 4.11 ■■■

The number of **combinations** of n things taken k at a time is

$$_nC_k = \binom{n}{k} = \frac{n!}{k!\,(n-k)!}$$

It represents the number of ways of selecting k objects out of n where the order of selection does not matter. ■

EXAMPLE 4.20 Evaluate $_7C_3$.

SOLUTION
$$_7C_3 = \frac{7 \times 6 \times 5}{3 \times 2 \times 1} = 7 \times 5 = 35$$ ■

A special situation arises upon evaluating $\binom{n}{0}$. By definition, $\binom{n}{0} = n!/(0!\,n!)$, and 0! was defined as 1. Hence, $\binom{n}{0} = 1$ for any n.

Frequently, $\binom{n}{k}$ will need to be computed for $k = 0, 1, \ldots, n$. The combinatorials have the following symmetry property, which makes this calculation easier than it appears at first glance.

4.3 For any nonnegative integers n, k where $n \geqslant k$,

$$\binom{n}{k} = \binom{n}{n-k}$$

To see this, note from Definition 4.11 that

$$_nC_k = \frac{n!}{k!(n-k)!}$$

If k is substituted for $n - k$ in this expression, then we obtain

$$_nC_{n-k} = \frac{n!}{(n-k)![n-(n-k)]!} = \frac{n!}{(n-k)!k!} = {_nC_k}$$

Intuitively, this result makes sense, since $_nC_k$ represents the number of ways of selecting k objects out of n without regard to order. However, for every selection of k objects, we have also, in a sense, identified the other $n - k$ objects that were not selected. Thus, the number of ways of selecting k objects out of n without regard to order should be the same as the number of ways of selecting $n - k$ objects out of n without regard to order.

Hence we need only evaluate combinatorials $\binom{n}{k}$ for the integers $k \leqslant n/2$. If $k > n/2$, then the relationship $\binom{n}{n-k} = \binom{n}{k}$ can be used.

EXAMPLE 4.21   Evaluate

$$\binom{7}{0}, \binom{7}{1}, \dots, \binom{7}{7}$$

SOLUTION

$$\binom{7}{0} = 1 \qquad \binom{7}{1} = 7 \qquad \binom{7}{2} = \frac{7 \times 6}{2 \times 1} = 21 \qquad \binom{7}{3} = \frac{7 \times 6 \times 5}{3 \times 2 \times 1} = 35$$

$$\binom{7}{4} = \binom{7}{3} = 35 \qquad \binom{7}{5} = \binom{7}{2} = 21 \qquad \binom{7}{6} = \binom{7}{1} = 7 \qquad \binom{7}{7} = \binom{7}{0} = 1 \qquad ∎$$

SECTION 4.8   ## The Binomial Distribution

All examples involving the binomial distribution have a common stucture: a sample of n independent trials, each of which can have only two possible outcomes, which are denoted as "success" and "failure." Furthermore, the probability of a success at each trial is assumed to be some constant p, and hence the probability of a failure at each trial is $1 - p = q$. The term "success" is used in a general way, without any specific contextual meaning.

For Example 4.15, $n = 100$ and a "success" occurs when a cell is a neutrophil.

EXAMPLE 4.22   **Infectious Disease**   Reconsider Example 4.15 with 5 cells rather than 100 and ask the more limited question, What is the probability that the second and fifth cells considered will be neutrophils and the remaining cells nonneutrophils given that the probability that any one cell is a neutrophil is .6?

SOLUTION   If a neutrophil is denoted by an x and a nonneutrophil by an o, then the question being asked is, What is the probability of the outcome $oxoox = Pr(oxoox)$? Since the probabilities of success and failure are given respectively by .6 and .4, and the outcomes for different cells are presumed to be independent, then this probability is

$$q \times p \times q \times q \times p = p^2 q^3 = (.6)^2(.4)^3 \qquad ∎$$

EXAMPLE 4.23   **Infectious Disease** Now consider the more general question, What is the probability that any 2 cells out of 5 will be neutrophils?

SOLUTION   The arrangement *oxoox* is only one of many possible orderings that result in 2 neutrophils. The 10 possible orderings are given in Table 4.4.

TABLE 4.4
Possible orderings for 2
neutrophils out of 5 cells

xxooo	*oxxoo*	*ooxox*
xoxoo	*oxoxo*	*oooxx*
xooxo	*oxoox*	
xooox	*ooxxo*	

In terms of combinations, the number of orderings = the number of ways of selecting 2 cells to be neutrophils out of 5 cells = $_5C_2 = (5 \times 4)/(2 \times 1) = 10$.

The probability of any of the orderings in Table 4.4 is the same as that for the ordering *oxoox*, namely, $(.6)^2(.4)^3$. Thus, the probability of obtaining 2 neutrophils in 5 cells is $_5C_2(.6)^2(.4)^3 = 10(.6)^2(.4)^3 = .230$. ∎

Suppose the neutrophils problem is now considered more generally, with n trials rather than 5 trials, and the following question is asked: What is the probability of k successes (rather than 2 successes) in these n trials? The probability that the k successes will occur at k **specified** trials within the n trials and that the remaining trials will be failures is given by $p^k(1 - p)^{n-k}$. To compute the probability of k successes in any of the n trials, this probability must be multiplied by the number of ways in which k trials for the successes and $n - k$ trials for the failures $= \binom{n}{k}$ can be selected (as was done in Table 4.4). Thus, the probability of k successes in n trials, or k neutrophils in n cells, is

$$\binom{n}{k}p^k(1 - p)^{n-k} = \binom{n}{k}p^k q^{n-k}$$

4.4 | The distribution of the number of successes in n statistically independent trials, where the probability of success on each trial is p, is known as the **binomial distribution** and has a probability mass function given by

$$Pr(X = k) = \binom{n}{k}p^k q^{n-k}, \qquad k = 0, 1, \ldots, n$$

EXAMPLE 4.24   What is the probability of obtaining 2 boys out of 5 children if the probability of a boy is .51 at each birth and the sexes of successive children are considered independent random variables?

SOLUTION   Use a binomial distribution with $n = 5$, $p = .51$, $k = 2$. Compute

$$Pr(X = 2) = {_5C_2}(.51)^2(.49)^3 = \frac{5 \times 4}{2 \times 1}(.51)^2(.49)^3$$

$$= 10(.51)^2(.49)^3 = .306$$ ∎

4.8.1   **Using Binomial Tables**

Frequently, a number of binomial probabilities will need to be evaluated for the same n and p, which would be tedious if each probability had to be calculated from **(4.4)**. Instead, for small n ($n \leqslant 20$) and selected values of p, refer to Table 1 in the Appendix, where the individual binomial probabilities are calculated. In this table, the number of trials (n) is provided in the first column, the number of successes (k) out of the n trials is given in the second column, and the probability of success for an individual trial (p) is given in the first row. Binomial probabilities are provided for $n = 2, 3, \ldots, 20$, $p = .05, .10, \ldots, .50$.

EXAMPLE 4.25   **Infectious Disease** Evaluate the probability of 2 lymphocytes out of 10 white blood cells if the probability that any one cell is a lymphocyte is .2.

SOLUTION   Refer to Table 1 with $n = 10$, $k = 2$, $p = .20$. The appropriate probability, given in the $k = 2$ row and $p = .20$ column under $n = 10$, is .3020.   ∎

EXAMPLE 4.26   **Pulmonary Disease** An investigator notices that children develop chronic bronchitis in the first year of life in 3 of 20 households where both parents are chronic bronchitics, as compared with the national incidence rate of chronic bronchitis, which is 5 % in the first year of life. Is this difference "real" or can it be attributed to chance? Specifically, how likely are infants in at least 3 out of 20 households to develop chronic bronchitis if the probability of developing disease in any one household is .05?

SOLUTION   Suppose the underlying rate of disease in the offspring is .05. Under this assumption, the number of households where the infants develop chronic bronchitis will follow a binomial distribution with parameters $n = 20$, $p = .05$. Thus, the probability of observing k cases out of 20 with disease is given by

$$\binom{20}{k}(.05)^k(.95)^{20-k}, \qquad k = 0, 1, \ldots, 20$$

The question is, What is the probability of observing at least 3 cases? The answer is

$$Pr(X \geqslant 3) = \sum_{k=3}^{20} \binom{20}{k}(.05)^k(.95)^{20-k} = 1 - \sum_{k=0}^{2} \binom{20}{k}(.05)^k(.95)^{20-k}$$

These 3 probabilities in the sum can be evaluated using the binomial table (Table 1). Refer to $n = 20$, $p = .05$ and note that $Pr(X = 0) = .3585$, $Pr(X = 1) = .3774$, $Pr(X = 2) = .1887$. Thus,

$$Pr(X \geqslant 3) = 1 - (.3585 + .3774 + .1887) = .0754$$

Thus, $X \geqslant 3$ is an unusual event, but not very unusual. If 3 infants out of 20 were to develop the disease, it would be difficult to judge whether the familial aggregation was real until a larger sample was available.   ∎

One question that arises is how to use the binomial tables if the probability of success on an individual trial (p) is greater than .5. Recall that

$$\binom{n}{k} = \binom{n}{n-k}$$

and let X be a binomial random variable with parameters n and p, and Y be a binomial random variable with parameters n and $q = 1 - p$. Then **(4.4)** can be rewritten as

4.5
$$Pr(X = k) = \binom{n}{k} p^k q^{n-k} = \binom{n}{n-k} q^{n-k} p^k = Pr(Y = n - k)$$

In words, the probability of obtaining k successes for a binomial random variable X with parameters n and p is the same as the probability of obtaining $n - k$ successes for a binomial random variable Y with parameters n and q. Clearly, if $p > .5$, then $q = 1 - p < .5$, and Table 1 can be used with sample size n, referring to the $n - k$ row and the q column to obtain the appropriate probability.

EXAMPLE 4.27   **Infectious Disease** Evaluate the probabilities of obtaining k neutrophils out of 5 cells for $k = 0, 1, 2, 3, 4, 5$, where the probability that any one cell is a neutrophil is .6.

SOLUTION   Since $p > .5$, refer to the random variable Y with parameters $n = 5$, $p = 1 - .6 = .4$.

$$Pr(X = 0) = \binom{5}{0}(.6)^0(.4)^5 = \binom{5}{5}(.4)^5(.6)^0 = Pr(Y = 5) = .0102$$

upon referring to the $k = 5$ row and $p = .40$ column under $n = 5$. Similarly,

$Pr(X = 1) = Pr(Y = 4) = .0768$ upon referring to the 4 row and .40 column under $n = 5$

$Pr(X = 2) = Pr(Y = 3) = .2304$ upon referring to the 3 row and .40 column under $n = 5$

$Pr(X = 3) = Pr(Y = 2) = .3456$ upon referring to the 2 row and .40 column under $n = 5$

$Pr(X = 4) = Pr(Y = 1) = .2592$ upon referring to the 1 row and .40 column under $n = 5$

$Pr(X = 5) = Pr(Y = 0) = .0778$ upon referring to the 0 row and .40 column under $n = 5$ ∎

4.8.2   **Recursion Rule for Binomial Probabilities**

In many instances we will want to evaluate binomial probabilities for $n > 20$ and/or for values of p not given in Table 1 of Appendix 1. For sufficiently large n, the normal distribution can be used to approximate the binomial distribution, and tables of the normal distribution can be used to evaluate binomial probabilities. This procedure is usually less tedious than evaluating binomial probabilities directly using **(4.4)** and is studied in detail in Chapter 5. Alternatively, if the sample size is not large enough to use the normal approximation and if the value of p is not in Table 1, then a recursion rule can be used to evaluate binomial probabilities. This rule is particularly useful in evaluating many binomial probabilities for the same n and p. Using the recursion rule it is easy to evaluate $Pr(X = k + 1)$ once $Pr(X = k)$ is known. Thus, once the probability of 0 successes has been computed, the probability of 1 success, 2 successes, and so forth can easily be computed without computing any combinatorials. The recursion rule is given as follows:

4.6   **Recursion Rule for Binomial Probabilities**
$$Pr(X = k + 1) = [(n - k)/(k + 1)] \times (p/q) \times Pr(X = k), \qquad k = 0, 1, \ldots, n - 1$$

To see this, remember from **(4.4)** that

$$Pr(X = k + 1) = \binom{n}{k+1} p^{k+1} q^{n-(k+1)} = \frac{n!}{(k+1)!(n-k-1)!} p^{k+1} q^{n-k-1}$$

$$Pr(X = k) = \frac{n!}{k!(n-k)!} p^k q^{n-k}$$

Divide $Pr(X = k + 1)$ by $Pr(X = k)$ to obtain

$$\frac{Pr(X = k + 1)}{Pr(X = k)} = \frac{\{n!/[(k+1)!(n-k-1)!]\} p^{k+1} q^{n-k-1}}{\{n!/[k!(n-k)!]\} p^k q^{n-k}}$$

$$= \frac{k!}{(k+1)!} \times \frac{(n-k)!}{(n-k-1)!} \times \frac{p}{q}$$

However, since $k!/(k+1)! = 1/(k+1)$ and $(n-k)!/(n-k-1)! = n - k$, it follows that

$$\frac{Pr(X = k + 1)}{Pr(X = k)} = \frac{1}{k+1} \times (n-k) \times \frac{p}{q}$$

Upon multiplying both sides of the equation by $Pr(X = k)$, we have

$$Pr(X = k + 1) = \frac{(n-k)}{(k+1)} \times \frac{p}{q} \times Pr(X = k)$$

EXAMPLE 4.28   **Infectious Disease** Suppose that a group of 100 males aged 60–64 received a new flu vaccine in 1959 and that 5 of them died within the next year. Is this event unusual or can this kind of death rate be expected for people of this age-sex group? Specifically, how likely are at least 5 out of 100 60–64-year-old males who receive a flu vaccine to die in the next year?

SOLUTION   We first find the expected annual death rate in 60–64-year-old males. From a 1959 U.S. life table, we find that 60–64-year-old men have an approximate probability of death in the next year of .028 [3]. Thus, from the binomial distribution the probability that k out of 100 men will die during the next year is given by $\binom{100}{k}(.028)^k(.972)^{100-k}$. We want to know if 5 deaths in a sample of 100 men is an "unusual" event. One criterion for this evaluation might be to find the probability of getting at least 5 deaths in this group $= Pr(X \geq 5)$ given that the probability of death for an individual man is .028. This probability can be expressed as

$$\sum_{k=5}^{100} \binom{100}{k}(.028)^k(.972)^{100-k}$$

Because this sum of 96 probabilities is tedious to compute, we instead compute

$$Pr(X < 5) = \sum_{k=0}^{4} \binom{100}{k}(.028)^k(.972)^{100-k}$$

and then evaluate $Pr(X \geq 5) = 1 - Pr(X < 5)$. The binomial tables cannot be used because $n > 20$. Therefore, the sum of 5 binomial probabilities is evaluated using the recursion rule.

$$Pr(X = 0) = \binom{100}{0}(.028)^0(.972)^{100} = (.972)^{100} = .05843$$

$$Pr(X = 1) = \left(\frac{100 - 0}{0 + 1}\right)\left(\frac{.028}{.972}\right)(.05843) = .16832$$

$$Pr(X = 2) = \binom{99}{2}\left(\frac{.028}{.972}\right)(.16832) = .24001$$

$$Pr(X = 3) = \binom{98}{3}\left(\frac{.028}{.972}\right)(.24001) = .22585$$

$$Pr(X = 4) = \binom{97}{4}\left(\frac{.028}{.972}\right)(.22585) = .15777$$

Hence,

$$Pr(X < 5) = .05843 + .16832 + .24001 + .22585 + .15777 = .85038$$

and

$$Pr(X \geq 5) = 1 - Pr(X < 5) = .14962 \approx .15$$

Thus, 5 deaths in 100 is a slightly unusual, but not a very unusual, event. If there were 10 deaths rather than 5, then using the same approach,

$$Pr(X \geq 10) = 1 - Pr(X < 10) < .001$$

which is very unlikely and would probably be grounds for halting the use of the vaccine in the absence of any other evidence. ∎

SECTION 4.9   **Expected Value and Variance of the Binomial Distribution**

The expected value and variance of the binomial distribution are important both in terms of our general knowledge about the binomial distribution and for our later work on estimation and hypothesis testing. From Definition 4.5 we know that the general formula for the expected value of a discrete random variable is

$$E(X) = \sum_{i=1}^{k} x_i Pr(X = x_i)$$

In the special case of a binomial distribution, the only values that take on positive probability are $0, 1, 2, \ldots, n$, and these values occur with probabilities

$$\binom{n}{0}p^0q^n, \quad \binom{n}{1}p^1q^{n-1}, \ldots$$

Thus,

$$E(X) = \sum_{k=0}^{n} k\binom{n}{k}p^kq^{n-k}$$

It can be shown that this summation reduces to the simple expression np. Similarly, using Definition 4.6, we can show that

$$Var(X) = \sum_{k=0}^{n} (k - np)^2\binom{n}{k}p^kq^{n-k} = npq$$

which leads directly to the following result:

4.7   The **expected value and variance of a binomial distribution** are np and npq, respectively.

These results make good sense, since the expected number of successes in n trials is simply the probability of success on one trial multiplied by n, which equals np. Furthermore, for a given number of trials n, the binomial distribution has the highest variance when $p = \frac{1}{2}$, as shown in Figure 4.2. The variance of the distribution decreases as p moves away from $\frac{1}{2}$ in either direction, becoming 0 when $p = 0$ or 1. This result makes sense, since when $p = 0$ there must be 0 successes in n trials and when $p = 1$ there must be n successes in n trials, and there is no variability in either instance. Furthermore, when p is near 0 or near 1, the distribution of the number of successes is clustered near 0 and n, respectively, and there is comparatively little variability as compared with the situation when $p = \frac{1}{2}$. This point is depicted in Figure 4.3.

SECTION 4.10   **The Poisson Distribution**

The Poisson distribution is perhaps the second most frequently used discrete distribution after the binomial distribution. This distribution is usually associated with rare events.

EXAMPLE 4.29   **Infectious Disease** Consider the distribution of the number of deaths attributed to typhoid fever over a long period of time, for example, 1 year. Assuming that the probability of a new death from typhoid fever in any one day is very small and that the number of cases reported in any two distinct periods of time are independent random variables, then the number of deaths over a 1-year period will follow a Poisson distribution.   ∎

EXAMPLE 4.30   **Bacteriology** The preceding example concerns a rare event occurring over time. Rare events can also be considered not over time but on a surface area, such as the distribution of the number of bacterial colonies growing on an agar plate. Suppose we have a 100-cm^2 agar plate and that the probability of finding any bacterial colonies at any 1 point a (or more precisely in a small area around a) is very small and that the events of finding bacterial colonies at any 2 points a_1, a_2 are independent. The number of bacterial colonies over the entire agar plate will follow a Poisson distribution.   ∎

Consider Example 4.29. Ask the question, What is the distribution of the number of deaths due to typhoid fever from time 0 to time t (where t is some long period of time, such as 1 year or 20 years)?

Three assumptions must be made about the incidence of the disease. Consider any general *small* subinterval of the time period t, denoted by Δt.

ASSUMPTION 1   Assume that

(a) The probability of observing 1 death is directly proportional to the length of the time interval Δt. That is, $Pr(1 \text{ death}) \approx \lambda \Delta t$ for some constant λ.

(b) The probability of observing 0 deaths over Δt is approximately $1 - \lambda \Delta t$.

(c) The probability of observing more than 1 death over this time interval is essentially 0.   ∎

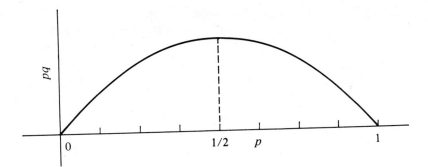

FIGURE 4.2
Plot of *pq* versus *p*

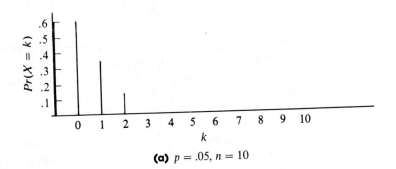

(a) $p = .05, n = 10$

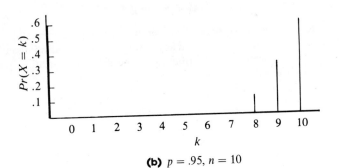

(b) $p = .95, n = 10$

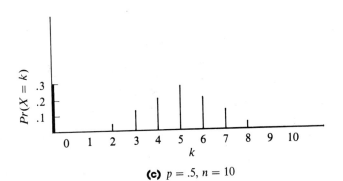

FIGURE 4.3
The binomial distribution
for various values of *p*
when $n = 10$

(c) $p = .5, n = 10$

ASSUMPTION 2   Assume that the number of deaths per unit time is the same thoughout the entire time interval t. Thus, an increase in the incidence of the disease as time goes on within the time period t would violate this assumption. Note that t should not be overly long, since this assumption is less likely to hold as t increases. ∎

ASSUMPTION 3   **Independence** If a death occurs within one time subinterval, it has no bearing on the probability of death in the next time subinterval. This assumption would be violated in an epidemic situation, because if a new case of disease occurs, then subsequent deaths are likely to build up over a short period of time until after the epidemic subsides. ∎

Based on these assumptions, the Poisson probability distribution can be derived:

4.8

The probability of k events occurring in a time period t for a Poisson random variable with parameter λ is

$$Pr(X = k) = e^{-\mu}\mu^k/k!, \qquad k = 0, 1, 2, \ldots$$

where $\mu = \lambda t$ and e is approximately 2.71828.

Thus, the Poisson distribution depends on one parameter $\mu = \lambda t$. Note that the parameter λ represents the *expected number of events per unit time*, whereas the parameter μ represents the *expected number of events over the time period t*. One important difference between the Poisson distribution and the binomial distribution concerns the numbers of trials and events. For a binomial distribution there are a finite number of trials n, and the number of events can be no larger than n. For a Poisson distribution the number of trials is essentially infinite and the number of events (or number of deaths) can be indefinitely large, although for very large k the probability of k events will get very small.

EXAMPLE 4.31   **Infectious Disease** Consider the typhoid-fever example. Suppose the number of deaths attributable to typhoid fever over a 1-year period is Poisson with parameter $\mu = 4.6$. What is the probability distribution of the number of deaths over a 6-month period? a 3-month period?

SOLUTION   Let $X =$ the number of deaths in 6 months. Since $\lambda = 4.6, t = 1$, it follows that $\mu = 4.6$. For a 6-month period we have that $\lambda = 4.6, t = .5$. Thus, $\mu = \lambda t = 2.3$. Therefore,

$$Pr(X = 0) = e^{-2.3} = .100$$

$$Pr(X = 1) = \frac{(2.3)}{1!} e^{-2.3} = .231$$

$$Pr(X = 2) = \frac{(2.3)^2}{2!} e^{-2.3} = .265$$

$$Pr(X = 3) = \frac{(2.3)^3}{3!} e^{-2.3} = .203$$

$$Pr(X = 4) = \frac{(2.3)^4}{4!} e^{-2.3} = .117$$

$$Pr(X = 5) = \frac{(2.3)^5}{5!} e^{-2.3} = .054$$

$$Pr(X \geqslant 6) = 1 - (.100 + .231 + .265 + .203 + .117 + .054) = .030$$

Let Y = the number of deaths in 3 months. For a 3-month period we have that $\lambda = 4.6$, $t = .25$, $\mu = \lambda t = 1.15$. Therefore,

$$Pr(Y = 0) = e^{-1.15} = .317$$

$$Pr(Y = 1) = \frac{1.15}{1!} e^{-1.15} = .364$$

$$Pr(Y = 2) = \frac{(1.15)^2}{2!} e^{-1.15} = .209$$

$$Pr(Y = 3) = \frac{(1.15)^3}{3!} e^{-1.15} = .080$$

$$Pr(Y \geqslant 4) = 1 - (.317 + .364 + .209 + .080) = .030$$

These distributions are plotted in Figure 4.4. Note that the distribution tends to become more symmetric as the time interval increases or, more specifically, as μ increases. ∎

The Poisson distribution can also be applied to Example 4.30, where the distribution of the number of bacterial colonies in an agar plate of area A is discussed. Assuming that the probability of finding 1 colony in an area of size ΔA at any point on the plate is $\lambda \Delta A$ for some λ and that the number of bacterial colonies found at 2 different points of the plate are independent random variables, then the probability of finding k bacterial colonies in an area of size A is given by $e^{-\mu}\mu^k/k!$, where $\mu = \lambda A$.

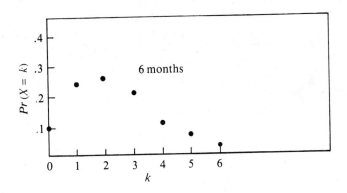

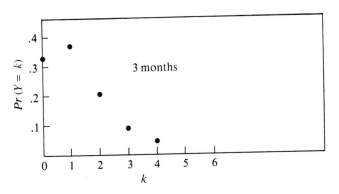

FIGURE 4.4
Distribution of the number of deaths attributable to typhoid fever over various time intervals

EXAMPLE 4.32   **Bacteriology** If $A = 100 \text{ cm}^2$, $\lambda = .02$, calculate the probability distribution of the number of bacterial colonies.

SOLUTION   We have that $\mu = \lambda A = 100(.02) = 2$. Let $X =$ the number of colonies.

$$Pr(X = 0) = e^{-2} = .135$$

$$Pr(X = 1) = e^{-2}(2)^1/1! = 2e^{-2} = .271$$

$$Pr(X = 2) = e^{-2}(2)^2/2! = 2e^{-2} = .271$$

$$Pr(X = 3) = e^{-2}(2)^3/3! = \frac{4}{3}e^{-2} = .180$$

$$Pr(X = 4) = e^{-2}(2)^4/4! = \frac{2}{3}e^{-2} = .090$$

$$Pr(X \geqslant 5) = 1 - (.135 + .271 + .271 + .180 + .090) = .053$$

Clearly, the larger λ is, the more bacterial colonies or deaths due to typhoid fever or whatever we would expect to find.   ∎

SECTION 4.11   **Computation of Poisson Probabilities**

4.11.1   **Using Poisson Tables**

A number of Poisson probabilities for the same parameter μ often need to be evaluated. This task would be tedious if **(4.8)** had to be applied repeatedly. Instead, for $\mu \leqslant 20$ refer to Table 2 in Appendix 1, in which individual Poisson probabilities are specifically calculated. In this table the Poisson parameter μ is given in the first row, the number of events (k) is given in the first column, and the corresponding Poisson probability is given in the k row and μ column.

EXAMPLE 4.33   Compute the probability of obtaining at least 5 events for a Poisson distribution with parameter $\mu = 3$.

SOLUTION   Refer to Table 2 under the 3.0 column. Let $X =$ the number of events.

$$Pr(X = 0) = .0498$$

$$Pr(X = 1) = .1494$$

$$Pr(X = 2) = .2240$$

$$Pr(X = 3) = .2240$$

$$Pr(X = 4) = .1680$$

Thus, $Pr(X \geqslant 5) = 1 - Pr(X \leqslant 4)$

$$= 1 - (.0498 + .1494 + .2240 + .2240 + .1680)$$

$$= 1 - .8152 = .1848$$ ∎

4.11.2   **Recursion Rule for Poisson Probabilities**

In many instances we will want to evaluate a collection of Poisson probabilities for the same μ, but μ will not be given in Table 2 of Appendix 1. For large μ ($\mu \geqslant 5$) a normal approximation, as given in Chapter 5, can be used. Otherwise, the following recursion rule, which is similar to that given for binomial probabilities, can be used:

4.9   **Recursion Rule for Poisson Probabilities**
If $Pr(X = k)$ is the Poisson probability of observing k events with underlying parameter μ, then

$$Pr(X = k + 1) = [\mu/(k + 1)]Pr(X = k)$$

EXAMPLE 4.34   **Infectious Disease** Apply the recursion rule to the distribution of deaths due to typhoid fever over a 3-month period given in Example 4.31.

SOLUTION   First, compute the probability of 0 deaths $= Pr(Y = 0) = e^{-1.15} = .3166$. Then,

$$Pr(Y = 1) = (1.15/1)Pr(Y = 0) = 1.15(.3166) = .3641$$
$$Pr(Y = 2) = (1.15/2)Pr(Y = 1) = (.575)(.3641) = .2094$$
$$Pr(Y = 3) = (1.15/3)Pr(Y = 2) = (1.15/3)(.2094) = .0803$$ ∎

SECTION 4.12   Expected Value and Variance of the Poisson Distribution

In many instances we cannot predict whether the assumptions for the Poisson distribution given in Section 4.10 are satisfied. Fortunately, the relationship between the expected value and variance of the Poisson distribution provides an important guideline that helps identify random variables that follow this distribution. This relationship can be stated as follows:

4.10   For a Poisson distribution with parameter μ, the mean and variance are both equal to μ.

This fact is useful to know, since if we have a data set from a discrete distribution where the *mean and variance are about the same*, then we can preliminarily identify it as a Poisson distribution and use various tests to confirm this hypothesis.

EXAMPLE 4.35   **Infectious Disease** The number of deaths attributable to polio during the years 1968–1976 are given in Table 4.5 [4, 5]. Comment on the applicability of the Poisson distribution to this data set.

SOLUTION   The mean and variance of the annual number of deaths due to polio during the period 1968–1976 are 11.3 and 51.5, respectively. The Poisson distribution clearly will not fit well here, since the variance is 4.6 times as large as the mean. The larger variance is probably due to the clustering of polio deaths at certain times and geographical locations, which leads to a violation of both the independence assumption and the assumption of constant incidence over time.   ∎

	Year	1968	1969	1970	1971	1972	1973	1974	1975	1976
TABLE 4.5 Number of deaths attributable to polio during the years 1968–1976	**Number of deaths**	24	13	7	18	2	10	3	9	16

Suppose we are studying a rare event phenomenon and wish to apply the Poisson distribution. A question that often arises is how to estimate the parameter μ of the Poisson distribution in this context. Since the expected value of the Poisson distribution is μ, μ can be estimated by the observed mean number of events, if such data are available. If the data are not available, other data sources can be used to estimate μ.

EXAMPLE 4.36   **Occupational Health** A public health issue arose concerning the possible carcinogenic potential of food ingredients containing ethylene dibromide (EDB). In some instances foods were removed from public consumption if they were shown to have excessive quantities of EDB. A study was previously performed looking at the mortality experience of 161 white male employees of two plants in Texas and Michigan who were exposed to EDB over the time period 1940–1975 [6]. Seven deaths due to cancer were observed among these employees. For this time period 5.8 cancer deaths were expected as calculated from overall mortality rates for U.S. white males. Assess if the observed number of cancer deaths was excessive in this group.

SOLUTION   Estimate the parameter μ from the expected number of cancer deaths from U.S. white male mortality rates; that is, $\mu = 5.8$. Then calculate $Pr(X \geq 7)$, where X is a Poisson random variable with parameter 5.8. Use the relationship

$$Pr(X \geq 7) = 1 - Pr(X \leq 6)$$

Since 5.8 is not in Table 2 of the Appendix, use the recursion rule.

$$Pr(X = 0) = \frac{e^{-5.8}(5.8)^0}{0!} = e^{-5.8} = .0030$$

$$Pr(X = 1) = \frac{5.8}{1} \times .0030 = .0174$$

$$Pr(X = 2) = \frac{5.8}{2} \times .0174 = .0505$$

$$Pr(X = 3) = \frac{5.8}{3} \times .0505 = .0976$$

$$Pr(X = 4) = \frac{5.8}{4} \times .0976 = .1415$$

$$Pr(X = 5) = \frac{5.8}{5} \times .1415 = .1641$$

$$Pr(X = 6) = \frac{5.8}{6} \times .1641 = .1586$$

Thus, $Pr(X \geq 7) = 1 - Pr(X \leq 6)$

$$= 1 - (.0030 + \cdots + .1586) = 1 - .6327 = .3673$$

Clearly, the observed number of cancer deaths is not excessive in this group.   ■

Poisson Approximation to the Binomial Distribution

As was seen in the preceding section, the Poisson distribution appears to fit well in some applications. Another important use for the Poisson distribution is as an approximation to the binomial distribution. Consider the binomial distribution for large n and small p. The mean of this distribution is given by np and the variance by npq. Note that $q \approx$ (is approximately equal to) 1 for small p, and thus $npq \approx np$. Therefore, the mean and variance of the binomial distribution are almost equal in this case, which suggests the following rule:

4.11	**Poisson Approximation to the Binomial Distribution**

The binomial distribution with large n and small p can be accurately approximated by a Poisson distribution with parameter $\mu = np$.

The rationale for using this approximation is that the Poisson distribution is easier to work with than the binomial distribution. The binomial distribution involves expressions such as $\binom{n}{k}$ and $(1 - p)^{n-k}$, which are cumbersome for large n.

EXAMPLE 4.37   **Cancer, Genetics**   Suppose we are interested in the genetic susceptibility to breast cancer. We find that 4 out of 1000 women aged 40–49 whose mothers have had breast cancer develop breast cancer over the next year of life. We would expect from large population studies that 1 in 1000 women of this age group will develop a new case of the disease over this period of time. How unusual is this event?

SOLUTION   The exact binomial probability could be computed by letting $n = 1000$, $p = 1/1000$. Hence,

$$Pr(X \geqslant 4) = 1 - Pr(X \leqslant 3)$$

$$= 1 - \left[\binom{1000}{0}(.001)^0(.999)^{1000} + \binom{1000}{1}(.001)^1(.999)^{999} \right.$$

$$\left. + \binom{1000}{2}(.001)^2(.999)^{998} + \binom{1000}{3}(.001)^3(.999)^{997} \right]$$

Instead, use the Poisson approximation with $\mu = 1000(.001) = 1$, which is obtained as follows:

$$Pr(X \geqslant 4) = 1 - [Pr(X = 0) + Pr(X = 1) + Pr(X = 2) + Pr(X = 3)]$$

Using Table 2 of the Appendix under the $\mu = 1.0$ column, we find that

$$Pr(X = 0) = .3679$$

$$Pr(X = 1) = .3679$$

$$Pr(X = 2) = .1839$$

$$Pr(X = 3) = .0613$$

Thus, $$Pr(X \geqslant 4) = 1 - (.3679 + .3679 + .1839 + .0613)$$

$$= 1 - .9810 = .0190$$

This event is indeed unusual and suggests a genetic susceptibility to breast cancer among female offspring of women who have had breast cancer.   ∎

How large should n be or how small should p be before the approximation is "adequate"? A conservative rule is to use the approximation when $n \geqslant 100$ and $p \leqslant .01$. As an example we give the exact binomial probability and the Poisson approximation for $n = 100$, $p = .01$, $k = 0, 1, 2, 3, 4, 5$ in Table 4.6. The two probability distributions agree to within .002 in all instances.

k	Exact binomial probability	Poisson approximation	k	Exact binomial probability	Poisson approximation
0	.366	.368	3	.061	.061
1	.370	.368	4	.015	.015
2	.185	.184	5	.003	.003

SECTION 4.14   **Summary**

In this chapter random variables were discussed and a distinction between discrete and continuous random variables was made. Specific attributes of random variables, including the notions of probability mass function (or probability distribution), cumulative distribution function, expected value, and variance were introduced. These notions were shown to be related to similar concepts for finite samples, which were discussed in Chapter 2. In particular, the sample frequency distribution is a sample realization of a probability distribution, whereas the sample mean ($\bar{x}$) and variance (s^2) are sample analogues of the expected value and variance, respectively, of a random variable. The relationship between attributes of probability models and finite samples is explored in more detail in Chapter 6.

Finally, some specific probability models were introduced, focusing on the binomial and Poisson distributions. The binomial distribution was shown to be applicable for binary outcomes, that is, if only two outcomes are possible, where outcomes on different trials are independent. These two outcomes are labeled as "success" and "failure," where the probability of success is the same for each trial. The Poisson distribution is a classic model used to describe the distribution of rare events.

The study of probability models continues in Chapter 5, where the focus is on continuous random variables.

PROBLEMS

Using the data in Problem 3.21 (p. 60), let X be the random variable representing the number of adults with influenza.

4.1 What is the probability mass function for this random variable?

4.2 What is its expected value?

4.3 What is its variance?

4.4 What is the cumulative distribution function?

Let X be the random variable representing the number of hypertensive adults in Example 3.13 (p. 46).

4.5 Derive the probability mass function for X.

4.6 What is its expected value?

4.7 What is its variance?

4.8 What is the cumulative distribution function?

Refer to Example 3.15 (p. 48). Let Y be a random variable representing the number of doctors who diagnose a patient as positive for syphilis.

4.9 What is the probability mass function for Y?

4.10 What is its expected value?

4.11 What is its variance?

4.12 What is the cumulative distribution function?

Suppose we wish to check the accuracy of self-reported diagnoses of angina by getting further medical records on a subset of the cases.

4.13 If we have 50 reported cases of angina and we wish to select 5 for further review, then how many ways can we select these cases if the order of selection matters?

4.14 Answer Problem 4.13 if the order of selection does not matter.

4.15 Evaluate the number of ways of selecting 4 objects out of 10 if the order of selection matters.

4.16 Evaluate the number of ways of selecting 4 objects out of 10 if the order of selection does *not* matter. What term is used to denote this quantity?

4.17 Evaluate $_{10}C_0, _{10}C_1, \ldots, _{10}C_{10}$.

4.18 Evaluate 9!.

Suppose that the probability that a person will develop hypertension over a lifetime is 20%.

4.19 What is the probability distribution of the number of hypertensives over a lifetime among 20 students graduating from the same high school class?

4.20 Suppose that 6 out of 15 students in a grade school class develop influenza, whereas nationwide 20% of grade school students develop influenza. Is there evidence of an excessive number of cases in the class? That is, what is the probability of obtaining at least 6 cases in this class if the nationwide rate holds true.

4.21 What is the expected number of students in the class who will develop influenza?

Refer to Example 4.28.

4.22 What is the probability that exactly 4 persons out of 50 aged 60–64 will die after receiving flu vaccine if the probability that 1 person will die is .028?

4.23 What is the probability that at least 4 persons will die after receiving the vaccine?

4.24 What is the expected number of deaths following the flu vaccine?

4.25 What is the standard deviation of the number of deaths following the flu vaccine?

4.26 What is the probability of obtaining exactly 6 events for a Poisson distribution with parameter $\mu = 4.0$?

4.27 What is the probability of obtaining at least 6 events for a Poisson distribution with parameter $\mu = 4.0$?

4.28 What is the expected value and variance for a Poisson distribution with parameter $\mu = 4.0$?

Mental Health
Refer to Table 3.2 (p. 60). Suppose 10 men are selected at random, 5 of whom are aged 80–84 and 5 are 85+.

4.29 What is the probability that exactly 3 80–84-year-olds have Alzheimer's disease?

4.30 What is the probability that at least 3 of the 5 80–84-year-olds have Alzheimer's disease?

4.31 What is the probability that exactly 3 85+-year-olds have Alzheimer's disease?

4.32 What is the probability that at least 3 85+-year-olds have Alzheimer's disease?

4.33 What is the probability that exactly 4 out of the 10 men have Alzheimer's disease?

4.34 What is the probability that at least 4 out of the 10 men have Alzheimer's disease?

Infectious Disease
Newborns were screened for human immunodeficiency virus (HIV or AIDS virus) in five Massachusetts hospitals. The data obtained [7] are shown in Table 4.7.

4.35 If 500 newborns are screened at the inner-city hospital, then what is the exact binomial probability of precisely 5 HIV-positive test results?

4.36 If 500 newborns are screened at the inner-city hospital, then what is the exact binomial probability of at least 5 HIV-positive test results?

TABLE 4.7 Seroprevalence of HIV antibody in newborn blood samples, according to hospital category

Hospital	Type	Number tested	Number positive	Number positive (per 1000)
A	Inner city	3,741	30	8.0
B	Urban/Suburban	11,864	31	2.6
C	Urban/Suburban	5,006	11	2.2
D	Suburban/Rural	3,596	1	0.3
E	Suburban/Rural	6,501	8	1.2

4.37 Answer Problems 4.35 and 4.36 using an approximation rather than an exact probability.

4.38 Answer Problem 4.35 for a mixed urban/suburban hospital (hospital C).

4.39 Answer Problem 4.36 for a mixed urban/suburban hospital (hospital C).

4.40 Answer Problem 4.37 for a mixed urban/suburban hospital (hospital C).

4.41 Answer Problem 4.35 for a mixed suburban/rural hospital (hospital E).

4.42 Answer Problem 4.36 for a mixed suburban/rural hospital (hospital E).

4.43 Answer Problem 4.37 for a mixed suburban/rural hospital (hospital E).

Health Services Administration

The in-hospital mortality rate for 16 clinical conditions at 981 hospitals was recently reported [8]. It was reported that in-hospital mortality was 10.5% for coronary bypass surgery and 5.0% for total hip replacement. Suppose an institution changes from an academic institution to a private for-profit institution. They find that after the change, of the first 20 patients receiving coronary bypass surgery, 5 die, while of 20 patients receiving total hip replacement, 4 die.

4.44 What is the probability that of 20 patients receiving coronary bypass surgery, exactly 5 will die in-hospital, if this hospital is representative of the total pool of 981 hospitals?

4.45 What is the probability of at least 5 deaths among the coronary bypass patients?

4.46 What is the probability of no more than 5 deaths among the coronary bypass patients?

4.47 What is the probability that exactly 4 will die among the hip replacement patients?

4.48 What is the probability that at least 4 will die among the hip replacement patients?

4.49 What is the probability of 4 or fewer deaths among the hip replacement patients?

4.50 Can you draw any conclusions based on the results in Problems 4.44–4.49 regarding any effects of the change in hospital administration on in-hospital mortality rates?

Occupational Health

Many investigators have suspected that workers in the tire industry have an unusual incidence of cancer.

4.51 Suppose the expected number of deaths due to bladder cancer for all workers in a tire plant on January 1, 1964, over the next 20 years (1/1/64–12/31/83) based on U.S. mortality rates is 1.8. If the Poisson distribution is assumed to hold and there are 6 reported deaths due to bladder cancer among the tire workers, then how unusual is this event?

4.52 Suppose a similar analysis is done for stomach cancer. In this plant 4 deaths due to stomach cancer are observed for the workers, whereas 2.5 are expected based on U.S. mortality rates. How unusual is this event?

Cardiovascular Disease

The rate of myocardial infarction (MI) in 50–59-year-old, disease-free women is approximately 2 per 1000 per year or 10 per 1000 over 5 years. Suppose that 3 MI's are reported over 5 years among 1000 women initially disease free who have been taking postmenopausal hormones.

4.53 Use the binomial distribution to see if this experience represents an unusually small number of events based on the overall rate.

4.54 Answer Problem 4.53 using the Poisson approximation to the binomial distribution.

4.55 Compare your answers in Problems 4.53 and 4.54.

Infectious Disease

One hypothesis is that gonorrhea tends to cluster in central cities.

4.56 Suppose that 10 gonorrhea cases are reported over a 3-month period among 10,000 people living in an urban county. The statewide incidence of gonorrhea is 50 per 100,000 over this period. Is the number of gonorrhea cases in this county unusual for this time period?

Cardiovascular Disease

4.57 A new hypothesis in the etiology of heart disease is that aspirin intake of 325 mg per day reduces subsequent cardiovascular mortality in men with a prior heart attack. Suppose that in a pilot study of 50 men who received 1 tablet per day (325 mg), only 2 die over a 3-year period from cardiovascular disease. How likely is it that not more than 2 men will die if the underlying 3-year mortality rate is 10% in such men?

Otolaryngology

Assume that the number of episodes per year of otitis media, a common disease of the middle ear in early childhood, follows a Poisson distribution with parameter $\lambda = 1.6$.

4.58 Find the probability of getting 3 or more episodes of otitis media in the first 2 years of life.

4.59 Find the probability of not getting any episodes of otitis media in the first year of life.

An interesting question in pediatrics is whether the tendency for children to have many episodes of otitis media is inherited in a family.

4.60 What is the probability that 2 siblings will both have 3 or more episodes of otitis media in the first 2 years of life?

4.61 What is the probability that exactly 1 of the siblings will have 3 or more episodes in the first 2 years of life?

4.62 What is the probability that neither sibling will have 3 or more episodes in the first 2 years of life?

4.63 What is the expected number of siblings in a 2-sibling family that will have 3 or more episodes in the first 2 years of life?

Pediatrics

A hospital administrator wants to construct a special-care nursery for low-birthweight infants ($\leqslant 2500$ g) and wants to have some idea as to the number of beds she should allocate to the nursery. She is willing to assume that the recovery period of each baby is exactly 4 days and thus is interested in the expected number of premature births over the period.

4.64 If the number of premature births in any 4-day period is binomially distributed with parameters $n = 25$ and $p = .1$, then find the probability of 0, 1, 2, ..., 7 premature births over this period.

4.65 The administrator wishes to allocate x beds where the probability of having more than x premature births over a 4-day period is less than 5%. What should x be?

4.66 Answer Problem 4.65 for 1%.

Hypertension

Hypertension has often been claimed to have a "familial aggregation." That is, if 1 person in a family is hypertensive, then his or her siblings are more likely to be hypertensive. Suppose that the prevalence of hypertension among 50–59-year-olds in the general population is 18%. Suppose we identify sibships of size 3 in a community where all members of the sibship are 50–59 years old.

4.67 What is the probability that 0, 1, 2, or 3 hypertensives will be identified in such sibships if the hypertensive status of 2 siblings in the same family are independent events?

4.68 Suppose that among 25 sibships of this type, 5 have at least 2 affected siblings. How does this situation agree with the independence assumption in Problem 4.67?

Cancer

The incidence rate of malignant melanoma is suspected to be increasing over time. To document this rate change, a questionnaire was mailed to 100,000 American nurses in 1976 and 1978, asking about any current or previous tumors. Thirty new cases of malignant melanoma were found to have developed over the 2-year period among women with no previous cancers in 1976.

4.69 If the annual incidence rate from cancer registry data is 10 per 100,000, then what is the expected number of new cases over 2 years?

4.70 Do the preceding results agree or disagree with the cancer registry data? Specifically, what is the probability of observing at least 30 new cases over a 2-year period if the cancer registry incidence rate is correct?

Environmental Health, Obstetrics

Suppose that the rate of major congenital malformations in the general population is 2.5 per 100 deliveries. A study is set up to investigate if the offspring of Vietnam veteran fathers are at special risk of having congenital malformations.

4.71 If 100 infants are identified in a birth registry as being offspring of a Vietnam veteran father and 4 have a major congenital malformation, then is there an excess risk of malformations in this group?

Using these same birth registry data, let us look at the effect of maternal use of marijuana on the rate of major congenital malformations.

4.72 Of 75 offspring of mothers who used marijuana, 8 are found to have a major congenital malformation. Is there an excess risk of malformations in this group?

Accident Epidemiology

Suppose the annual number of traffic fatalities at a given intersection follows a Poisson distribution with parameter $\mu = 10$.

4.73 What is the probability of observing exactly 10 traffic fatalities in 1989?

4.74 What is the probability of observing exactly 25 traffic fatalities over the 2-year period from January 1, 1987, to December 31, 1988?

4.75 Suppose that the traffic intersection is redesigned with better lighting, and 12 traffic fatalities are observed over the next 2 years. Is this rate a meaningful improvement over the previous rate of traffic fatalities?

Hypertension
A national study found that treating people appropriately for high blood pressure reduced their overall mortality by 20%. Treating people adequately for hypertension has been difficult, since it is estimated that 50% of hypertensives do not know they have high blood pressure; 50% of those that do know are inadequately treated by their physicians; and 50% that are appropriately treated fail to comply with this treatment by taking the appropriate number of pills.

4.76 What is the probability that among 10 true hypertensives at least 50% are being treated appropriately and are complying with this treatment?

4.77 What is the probability that at least 7 of the 10 hypertensives know they have high blood pressure?

4.78 If the preceding 50% rates were decreased to 40% by a massive education program, then what effect would this rate change have on the overall mortality rate among true hypertensives?

Pulmonary Disease, Environmental Health
Suppose the number of people seen for violent asthma attacks in the emergency ward of a hospital over a 1-day period is usually Poisson distributed with parameter $\lambda = 1.5$.

4.79 What is the probability of observing 5 or more cases over a 2-day period?

On a particular 2-day period, the air pollution levels increase dramatically and the distribution of attacks over a 1-day period is now estimated to be Poisson distributed with parameter $\lambda = 3$.

4.80 Answer Problem 4.79 under these assumptions.

4.81 If 10 days out of every year are high-pollution days what is the expected number of asthma cases seen in the emergency ward over a 1-year period?

Renal Disease
The presence of bacteria in a urine sample (bacteriuria) is sometimes associated with symptoms of kidney disease in women. Suppose that a determination of bacteriuria has been made over a large population of women at one point in time and that 5% of those sampled are positive for bacteriuria.

4.82 If a sample of size 5 is selected from this population, what would be the probability that 1 or more women would be positive for bacteriuria?

4.83 Suppose 100 women from this population are sampled. What is the probability that 3 or more women would be positive for bacteriuria?

One interesting phenomenon of bacteriuria is that there is a "turnover"; that is, if bacteriuria is measured on the same woman at 2 different points in time, the results are not necessarily the same. Assume that $\frac{1}{5}$ of all women who are bacteriuric at time 0 are again bacteriuric at time 1 (1 year later), whereas only 4.2% of women who were not bacteriuric at time 0 *are* bacteriuric at time 1. Let X be the random variable representing the number of bacteriuric events over the 2 time periods for 1 woman and still assume that the probability that a woman will be positive for bacteriuria at any one exam is 5%.

4.84 What is the probability distribution of X?

4.85 What is the mean of X?

4.86 What is the variance of X?

Demography
The data set in Table 4.8 is an example of current life-table data for males living in the United States in 1960 [3]. P_x represents the probability of living for the next year given that one is currently x years old. The ℓ_x column is obtained from the formula

$$\ell_0 = 100,000 \qquad \ell_x = \ell_0 \times P_0 \times P_1 \times \cdots \times P_{x-1},$$

$$x = 1, 2, \ldots, 100$$

Assume that the *current* death rates hold not only for the year 1960 but for all the subsequent years as well.

4.87 What is the probability of living to age 65 given that a man is 21 in 1960?

4.88 What is the probability of dying exactly between the ages of 56 and 57 given that a man is 21 in 1960?

4.89 Suppose 100 men of age 21 in 1960 live in a particular town and that 5 of them die before reaching the age of 30. Is this event unusual? Specifically, how likely are 5 or more men to die before reaching the age of 30?

4.90 What is the probability distribution and expected value of the lifetime of a man who is age 80 in 1960?

TABLE 4.8 Current life table for U.S. males in 1960

x	ℓ_x	x	ℓ_x	x	ℓ_x	x	ℓ_x
0	100,000	25	94,631	50	86,199	75	38,950
1	97,087	26	94,466	51	85,325	76	36,210
2	96,911	27	94,306	52	84,369	77	33,468
3	96,800	28	94,148	53	83,333	78	30,732
4	94,714	29	93,990	54	82,222	79	28,006
5	96,643	30	93,826	55	81,039	80	25,300
6	96,580	31	93,656	56	79,783	81	22,619
7	96,522	32	93,479	57	78,451	82	19,983
8	96,469	33	93,293	58	77,032	83	17,439
9	96,420	34	93,097	59	75,513	84	15,045
10	96,375	35	92,889	60	73,887	85	12,845
11	96,333	36	92,666	61	72,151	86	10,819
12	96,290	37	92,426	62	70,308	87	8980
13	96,242	38	92,166	63	68,361	88	7333
14	96,182	39	91,883	64	66,316	89	5876
15	96,107	40	91,572	65	64,177	90	4609
16	96,014	41	91,230	66	61,947	91	3534
17	95,905	42	90,854	67	59,631	92	2648
18	95,779	43	90,441	68	57,235	93	1939
19	95,641	44	89,988	69	54,770	94	1387
20	95,491	45	89,492	70	52,244	95	970
21	95,330	46	88,950	71	49,665	96	665
22	95,158	47	88,359	72	47,040	97	446
23	94,981	48	87,709	73	44,375	98	293
24	94,803	49	86,992	74	41,676	99	187
						100	0

4.91 Suppose we are not willing to assume that the P_x's remain constant in years subsequent to 1960. Can Problems 4.87–4.90 still be answered? If not, what additional information is needed?

Refer to the 11 males among the 25 people described in Table 2.11 (p. 36).

4.92 What is the expected number of deaths among the 11 males over the next year based on the life-table data? Refer to Table 4.8.

4.93 Answer Problem 4.92 for a 5-year period.

4.94 Answer Problem 4.92 for a 10-year period.

Use a computer, if necessary, to answer Problems 4.95–4.100.

4.95 What is the probability of exactly 2 deaths among the 11 males over the next year?

4.96 Answer Problem 4.95 for a 5-year period.

4.97 Answer Problem 4.95 for a 10-year period.

4.98 What is the probability of at least 4 deaths among the 11 males over the next year?

4.99 Answer Problem 4.98 for a 5-year period.

4.100 Answer Problem 4.98 for a 10-year period.

Pediatrics, Otolaryngology

Otitis media is a disease that occurs frequently in the first few years of life and is one of the most common reasons for physician visits after the routine check-up. A study was conducted to assess the frequency of otitis media in the general population in the first year of life. Table 4.9 gives the number of infants out of 2500 infants who were first seen at birth and who remained disease free by the end of the ith month of life, $i = 0, 1, \ldots, 12$. (Assume that no infants have been lost to follow-up.)

4.101 What is the probability that an infant will have 1 or more episodes of otitis media by the end of the 6th month of life? the first year of life?

TABLE 4.9 Number of infants (out of 2500) who remain disease free at the end of each month during the first year of life

i	Disease-free infants at the end of month i
0	2500
1	2425
2	2375
3	2300
4	2180
5	2000
6	1875
7	1700
8	1500
9	1300
10	1250
11	1225
12	1200

4.102 What is the probability that an infant will have 1 or more episodes of otitis media by the end of the 9th month of life given that no episodes have been observed by the end of the 3rd month of life?

4.103 Suppose an "otitis prone family" is defined as one where at least 3 siblings out of 5 develop otitis media in the first 6 months of life. What proportion of 5-sibling families are otitis prone if we assume that the disease occurs independently for different siblings in a family?

4.104 What is the expected number of otitis prone families out of 100 5-sibling families?

Pulmonary Disease

Each year approximately 4% of current smokers attempt to quit smoking, and 50% of those who try to quit are successful in the sense that they abstain from smoking for at least 1 year from the date they quit.

4.105 What is the probability that a current smoker will quit for at least 1 year?

4.106 What is the probability that among 100 current smokers, at least 5 will quit smoking for at least 1 year?

An educational program was conducted among smokers who attempt to quit to maximize the likelihood that such individuals would continue to abstain for the long term.

4.107 Suppose that of 20 people who enter the program when they first stop smoking, 15 will abstain from smoking 1 year later. Can the program be considered successful?

Cancer, Epidemiology

A frequent design for biomedical investigations is the case-control study. A group of **cases** is isolated on the basis of having a particular disease (e.g., lung cancer patients in a cancer registry), and a group of **controls** is chosen (e.g., patients on the same registry with cancer of the esophagus) such that every case is *matched* with 1 or more controls. That is, the case and control(s) are matched in every sense except that the controls do not have the disease trait. We can then look at whether or not some other factor (such as smoking) is associated with the disease trait. Obtaining "exact" matches is usually impossible, and several characteristics are selected to use for matching, such as age and sex. Suppose the group of cases and controls is as given in Table 4.10.

Suppose that the match is performed so that each control has the same age group and sex as its corresponding case. An example of a 1-to-1 match would be

Case number	1	2	3	4	5
Control number	24	51	18	14	22

An example of a 2-to-1 match would be

Case number	1	2	3	4	5
Control number	24, 26	51, 49	18, 21	14, 06	22, 27

Assume that the order within each group of matched controls in a many-to-one matching does not matter.

4.108 How many ways can 1-to-1 matches be assigned for age and sex?

4.109 Suppose the designers of the study get lazy and match only for age. How many ways can this matching be done?

4.110 If the groups were matched only on age, then what is the probability that the groups will be matched for sex as well if each match is equally likely?

4.111 Answer Problem 4.108 for 2-to-1 matches.

4.112 Answer Problem 4.109 for 2-to-1 matches.

4.113 Answer Problem 4.110 for 2-to-1 matches.

4.114 Answer Problem 4.108 for 3-to-1 matches.

TABLE 4.10 Selection of controls for a case-control study

Cases			Controls			
Case number	Age	Sex	Control number	Age group	Sex	Frequency
1	36	M	01–15	21–30	M	15
2	50	F	16–21	21–30	F	6
3	24	F	22–27	31–40	M	6
4	22	M	28–45	31–40	F	18
5	35	M	46–48	41–50	M	3
			49–54	41–50	F	6
			55–66	51–60	M	12
			67–69	51–60	F	3
			70–78	61–70	M	9
			79–84	61–70	F	6

4.115 Answer Problem 4.109 for 3-to-1 matches.

4.116 Answer Problem 4.110 for 3-to-1 matches.

An experiment is designed to test the potency of a drug on 20 rats. Previous animal studies have shown that a 10-mg dose of the drug is lethal 5% of the time within the first 4 hours; of the animals alive at 4 hours, 10% will die in the next 4 hours.

4.117 What is the probability that 3 or more rats will die in the first 4 hours?

4.118 Suppose 2 rats die in the first 4 hours. What is the probability that 2 or fewer rats will die in the next 4 hours?

4.119 What is the probability that 0 rats will die in the 8-hour period?

4.120 What is the probability that 1 rat will die in the 8-hour period?

4.121 What is the probability that 2 rats will die in the 8-hour period?

4.122 Can you write a general formula for the probability that x rats will die in the 8-hour period? Evaluate this formula for $x = 0, 1, \ldots, 10$.

Infectious Disease

An outbreak of acute gastroenteritis occurred at a nursing home in Baltimore, Maryland, in December 1980 [9]. A total of 46 out of 98 residents of the nursing home became ill. Persons living in the nursing home shared rooms: 13 rooms contained 2 occupants, 4 rooms contained 3 occupants, and 15 rooms contained 4 occupants. One question that arises is whether or not a geographical clustering of disease occurred for persons living in the same room.

4.123 If the binomial distribution holds, what is the probability distribution of the number of affected persons in rooms with 2 occupants? That is, what is the probability of finding 0 affected persons? 1 affected person? 2 affected persons?

4.124 Answer Problem 4.123 for the probability distribution of the number of affected persons in rooms with 3 occupants.

4.125 Answer Problem 4.123 for the probability distribution of the number of affected persons in rooms with 4 occupants.

4.126 One useful summary measure of geographical clustering is the number of rooms with 2 or more affected occupants. If the binomial distribution holds, what is the expected number of rooms with 2 or more affected occupants over the entire nursing home?

A summary of the number of affected persons and the total number of persons in a room is given in Table 4.11.

4.127 For rooms with 4 persons, compare the observed number of rooms with 2 or more affected occupants with the expected number of rooms. Does this comparison give any evidence that clustering of disease occurs within rooms?

TABLE 4.11 Number of affected persons and total number of persons in a room for an outbreak of acute gastroenteritis in a nursing home in Baltimore, Maryland

Persons in room	Total number of rooms	Number of rooms with				
		0 affected persons	1 affected person	2 affected persons	3 affected persons	4 affected persons
2	13	5	4	4	0	0
3	4	1	2	0	1	0
4	15	2	4	3	5	1

(Reprinted with permission of the *American Journal of Epidemiology*, *116*(6), 940-948, 1982.)

Environmental Health

One of the important issues in assessing nuclear energy is whether there are excess disease risks in the communities surrounding nuclear power plants. A study was undertaken in the community surrounding Hanford, Washington, looking at the prevalence of selected congenital malformations in the counties surrounding the nuclear test facility [10].

4.128 Suppose that 27 cases of Down's syndrome are found and only 19 are expected based on Birth Defects Monitoring Program prevalence estimates conducted in the states of Washington, Idaho, and Oregon. Is there a significant excess number of cases in the area surrounding the nuclear power plant?

Suppose that 12 cases of cleft palate are observed, while only 7 are expected based on Birth Defects Monitoring Program estimates.

4.129 What is the probability of observing exactly 12 cases of cleft palate if there is no excess risk of cleft palate in the study area?

4.130 Do you feel there is a meaningful excess number of cases of cleft palate in the area surrounding the nuclear power plant?

Cancer

The incidence rate of malignant melanoma in women ages 35–59 is approximately 11 new cases per 100,000 women per year. A study is planned to follow 10,000 women with excessive exposure to sunlight.

4.131 What is the expected number of cases among 10,000 women over 4 years? (Assume no excess risk due to sunlight exposure.)

4.132 Suppose that 9 new cases are observed in this period. How unusual a finding is this?

Health Promotion

A study was conducted among 234 people who had expressed a desire to stop smoking but who had not yet stopped. On the day they quit smoking, their carbon monoxide level (CO) was measured and the time was noted from the time they smoked their last cigarette to the time of the CO measurement. The CO level provides an "objective" indicator of the number of cigarettes smoked per day during the time immediately prior to the quit attempt. However, it is known to also be influenced by the time since the last cigarette was smoked. Thus, this time is provided as well as a "corrected CO level," which is adjusted for the time since last smoked. Information is also provided on age and sex of the subjects as well as the subject's self-report of number of cigarettes per day. The subjects were followed up for one year for the purpose of determining the number of days they remained abstinent. The number of days abstinent ranges from 0 days for those who quit for less than 1 day to 365 days for those who were abstinent for the full year. Assume that all persons were followed for the entire year.

The data are given in Data Set 24, SMOKE.DAT. The format of this file is given in Table 4.12.

4.133 Develop a life table similar to Table 4.9, giving the number of persons who remained abstinent at 1, 2, ..., 12 months of life (assume for simplicity that there are 30 days in each of the first 11 months after quitting and 35 days in the 12th month). Plot these data either by hand or on the computer. Compute the probability that a person will remain abstinent at 1, 3, 6, and 12 months after quitting.

4.134 Develop life tables for subsets of the data based on age, sex, number of cigarettes per day, and carbon monoxide level (one variable at a time). Based on these

TABLE 4.12 Format of SMOKE.DAT

Variable	Columns	Code
ID number	1–3	
Age	4–5	
Gender	6	1 = male, 2 = female
Cigarettes/day	7–8	
Carbon monoxide (CO) ($\times 10$)	9–11	
Minutes elapsed since the last cigarette smoked	12–15	
LogCOAdj* ($\times 1000$)	16–19	
Days abstinent†	20–22	

† Those abstinent less than 1 day were given a value of 0.

* This variable represents adjusted carbon monoxide (CO) values. CO values were adjusted for minutes elapsed since the last cigarette smoked using the formula, Log_{10}CO (adjusted) $= \text{Log}_{10}\text{CO} - (-0.000638)$ (min $- 80$), where min is the number of minutes elapsed since the last cigarette smoked.

data, do you feel that age, sex, number ̇garettes per day, or CO level are related to success in quitting? (Methods of analysis for life-table data are discu ̇ ̇ed in more detail in Chapters 10 and 11.)

Genetics

4.135 A topic of some interest in the genetic literature over at least the last 30 years has been the study of sex ratio data. In particular, one hypothesis that has been suggested is that there are a sufficient number of families with a preponderance of males (females) that the sexes of successive childbirths are not independent random variables but are related to each other. This hypothesis has been extended beyond just successive births so that some authors also consider relationships between offspring two birth orders apart (i.e., 1st and 3rd offspring, 2nd and 4th offspring, etc.). Sex ratio data in [11] from the first 5 births in 51,868 families are given

in Table 4.13. What are your conclusions concerning the above hypothesis based on your analysis of these data?

TABLE 4.13 Format of SEXRAT.DAT

Variable	Column
Number of children†	1
Sex of children*	3–7
Number of families	9–12

† For families with 5+ children, the sexes of the first 5 children are listed. The number of children is given as 5 for such families.

* The sex of successive births is given. Thus, MMMF means that the first 3 children were males and the 4th child was a female. There were 484 such families.

References

[1] *Boston Globe*, October 7, 1980.

[2] Rinsky, R. A., Zumwalde, R. O., Waxweiler, R. J., Murray, W. E., Bierbaum, P. J., Landrigan, P. J., Terpilak, M., & Cox, C. (1981, January 31). Cancer mortality at a naval nuclear shipyard. *The Lancet*, 231–235.

[3] U.S. Department of Health, Education, and Welfare. (1964). *United States life tables*: 1959–1961, *life tables*: 1956–1961, *1*(1).

[4] National Center for Health Statistics. (1974, June 27). *Monthly vital statistics report, annual summary for the United States (1973), 22*(13).

[5] National Center for Health Statistics. (1978, December 7). *Monthly vital statistics report, annual summary for the United States (1977), 26*(13).

[6] Ott, M. G., Scharnweber, H. C., & Langner, R. (1980). Mortality experience of 161 employees exposed to ethylene dibromide in two production units. *British Journal of Industrial Medicine, 37,* 163–168.

[7] Hoff, R., Berardi, V. P., Weiblen, B. J., Mahoney-Trout, L., Mitchell, M. L., & Grady, G. F. (1988). Seroprevalence of human immunodeficiency virus among childbearing women. *New England Journal of Medicine, 318*(9), 525–530.

[8] Shortell, S. M., & Hughes, E. F. X. (1988). The effects of regulation, competition, and ownership on mortality rates among hospital inpatients. *New England Journal of Medicine, 318*(17), 1100–1107.

[9] Kaplan, J. E., Schonberger, L. B., Varano, G., Jackman, N., Bied, J., & Gary, G. W. (1982). An outbreak of acute nonbacterial gastroenteritis in a nursing home: Demonstration of person-to-person transmission by temporal clustering of cases. *American Journal of Epidemiology, 116*(6), 940–948.

[10] Sever, L. E., Hessol, N. A., Gilbert, E. S., & McIntyre, J. M. (1988). The prevalence at birth of congenital malformations in communities near the Hanford site. *American Journal of Epidemiology, 127*(2), 243–254.

[11] Renkonen, K. O., Mäkelä, O., & Lehtovaara, R. (1961). Factors affecting the human sex ratio. *Annales Medicinae Experimentalis et Biologiae Fenniae, 39,* 173–184.

CHAPTER FIVE

CONTINUOUS PROBABILITY DISTRIBUTIONS

Introduction

In this chapter continuous probability distributions are discussed. In particular, the normal distribution, which is the most widely used distribution in statistical work, is explored in depth.

The normal, or Gaussian or "bell-shaped," distribution is the cornerstone of most of the methods of estimation and hypothesis testing that are developed in the rest of this text. Many random variables, such as the distribution of birthweights or blood pressures in the general population, tend to approximately follow a normal distribution. In addition, many random variables that are not themselves normal are closely approximated by a normal distribution when summed many times. In such cases, using the normal distribution is desirable, since tables for the normal distribution are more widely available than those for many other distributions.

EXAMPLE 5.1   **Infectious Disease** The number of neutrophils in a sample of 2 white blood cells is not normally distributed, but the number in a sample of 100 white blood cells is very close to being normally distributed.   ∎

General Concepts

We want to develop an analogue for a continuous random variable to the concept of a probability mass function, as was developed for a discrete random variable in Section 4.3. Thus, we would like to know which values are more probable than others and how probable they are.

EXAMPLE 5.2   **Hypertension** Consider the distribution of diastolic blood-pressure measurements in 35–44-year-old men. In actual practice this distribution is discrete because only a finite number of blood-pressure values are possible, since the measurement is only accurate to within 2 mm Hg or in some cases 5 mm Hg. However, assume that there is no measurement error and hence the random variable can take on a continuum of possible values. One consequence of this assumption is that the probabilities of specific blood-pressure measurement values such as 117.3 are 0 and, thus, the concept of a probability mass function cannot be used. The proof of this statement is beyond the scope of this text. Instead, we speak in terms of the probability that blood pressure falls within a range of values. Thus, the probabilities of blood pressures (denoted by X) falling in the ranges of $90 \leqslant X < 95$, $95 \leqslant X < 100$, and $X \geqslant 100$ might be 15%, 5%, and 2%, respectively. People whose blood pressures fall in these ranges might be denoted as borderline, mild hypertensive, and severe hypertensive, respectively.

∎

Although the probability of exactly obtaining any value in 0, we still have the intuitive notion that certain ranges of values occur more frequently than others. This notion can be quantified using the concept of a probability density function.

DEFINITION 5.1 ∎∎

The **probability density function** of the random variable X is a curve such that the area under the curve between any two points a and b is equal to the probability that the random variable X falls between a and b. Thus, the total area under the curve over the possible range of values for the random variable is 1.

∎

The probability density function takes on high values in regions of high probability and low values in regions of low probability.

EXAMPLE 5.3 **Hypertension** The probability density function for diastolic blood pressure in 35–44-year-old men is given in Figure 5.1. Areas A, B, and C correspond to the probabilities of being borderline, mild hypertensive, and severe hypertensive, respectively. Furthermore, the most likely range of values for diastolic blood pressure occurs around 80 mm Hg, with the values becoming increasingly unlikely as we move further away from 80.

∎

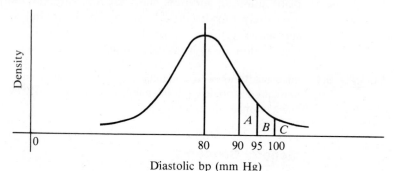

FIGURE 5.1
Probability density function for diastolic blood pressure in 35–44-year-old men

EXAMPLE 5.4 **Cardiovascular Disease** Serum triglycerides is an asymmetric, positively skewed continuous random variable whose probability density function appears in Figure 5.2.

∎

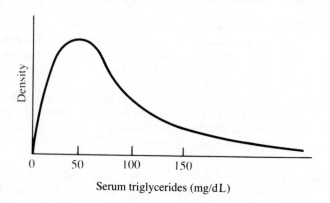

FIGURE 5.2
Probability density function for serum triglycerides

The cumulative distribution function is defined similarly to that for a discrete random variable (Section 4.6).

DEFINITION 5.2 ■■

The **cumulative distribution function** for the random variable X evaluated at the point a is defined as the probability that X will take on values $\leq a$. It is represented by the area under the probability density function to the left of a. ■

EXAMPLE 5.5 **Obstetrics** The probability density function for the random variable representing the distribution of birthweights (oz) in the general population is given in Figure 5.3. The cumulative distribution function evaluated at 88 oz $= Pr(X \leq 88)$ is represented by the area under this curve to the left of 88 oz. The region $X \leq 88$ oz has a special meaning in obstetrics, since 88 oz is the cutoff point usually used by obstetricians for identifying low-birthweight infants. Such infants are generally at higher risk for various unfavorable outcomes, such as mortality in the first year of life. ■

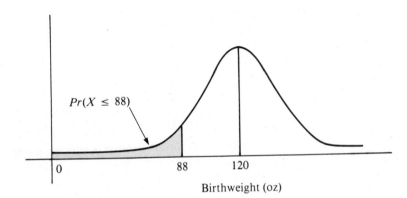

FIGURE 5.3
Cumulative distribution
function evaluated at
88 oz for the distribution
of birthweights in the
general population

Generally, a distinction will not be made between the probabilities $Pr(X \leq x)$ and $Pr(X < x)$ when X is a continuous random variable. The reason is that they represent the same quantity, because the probability of individual values is 0, that is, $Pr(X = x) = 0$.

The expected value and variance for continuous random variables have the same meaning as for discrete random variables (Sections 4.4 and 4.5). However, the mathematical definition of these terms is beyond the scope of this book.

DEFINITION 5.3 ■■

The **expected value** of a continuous random variable X, denoted by $E(X)$ or μ, is the average value taken on by the random variable. ■

DEFINITION 5.4 ■■

The **variance** of a continuous random variable X, denoted by $Var(X)$ or σ^2, is the average squared distance of each value of the random variable from its expected value. The standard deviation, or σ, is the square root of the variance, that is, $\sigma = \sqrt{Var(X)}$. ■

EXAMPLE 5.6 **Hypertension** The expected value and standard deviation of the distribution of diastolic blood pressures in 35–44-year-old men are 80 mm Hg and 12 mm Hg, respectively. ■

SECTION 5.3 ## The Normal Distribution

The normal distribution is the most widely used continuous distribution. It is also frequently referred to as the Gaussian distribution, after the well-known mathematician, Gauss.

EXAMPLE 5.7 **Hypertension** The distribution of body weights or of diastolic blood pressures for a group of 35–44-year-old males will follow a normal distribution. So will the distribution of tree diameters of a certain species of tree from some defined forest area. ∎

Many other distributions that are not themselves normal can be made normal by transforming the data onto a different scale.

EXAMPLE 5.8 **Cardiovascular Disease** The distribution of serum triglyceride concentrations from this same group of 35–44-year-old males is likely to be positively skewed. However, the log transformation of these measurements will usually follow a normal distribution. ∎

Generally speaking, any random variable that can be expressed as a sum of many other random variables can be well approximated by a normal distribution.

EXAMPLE 5.9 **Infectious Disease** The distribution of the number of lymphocytes in a differential of 100 white blood cells (refer to Example 4.15 for the definition of a differential) will tend to be normally distributed, since this random variable is a sum of 100 random variables, each representing whether or not an individual cell is a lymphocyte. ∎

Thus, because of its omnipresence, the normal distribution is vital to statistical work, and most of the estimation procedures and hypothesis tests that we will study are based on the assumption that the random variable being considered has an underlying normal distribution.

Another important area of application of the normal distribution is as an approximating distribution to other distributions. The normal distribution is generally more convenient to work with than any other distribution, particularly in hypothesis testing. Thus, if an accurate normal approximation to some other distribution can be found, then we will often use it.

DEFINITION 5.5 ∎∎

The **normal distribution** is defined by its **probability density function**, which is given as

$$f(x) = \frac{1}{(\sqrt{2\pi}\sigma)} \exp\left[-\frac{1}{2\sigma^2}(x - \mu)^2\right], \qquad -\infty < x < \infty$$

for some parameters μ, σ, where $\sigma > 0$. ∎

The exp function merely implies that the quantity to the right in brackets is the power to which "e" (≈ 2.71828) is raised. A plot of this probability density function is given in Figure 5.4.

The density function follows a bell-shaped curve, with the most frequently occurring value at μ. The curve is symmetric about μ, with points of inflection on each side of μ at $\mu - \sigma$ and $\mu + \sigma$, respectively. A point of inflection is a point where the slope of the curve changes direction. In Figure 5.4 the slope of the curve increases to the left of $\mu - \sigma$ and then starts to decrease to the right of $\mu - \sigma$ and

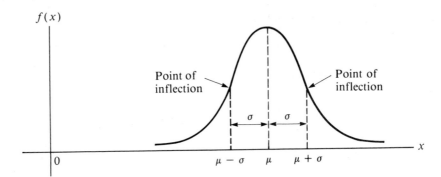

FIGURE 5.4
Probability density
function for a normal
distribution with mean μ
and variance σ^2

continues to decrease until reaching $\mu + \sigma$, after which it starts increasing again. Thus, the distances from μ to the points of inflection provide a good visual sense of the magnitude of the parameter σ.

You may wonder why the parameters μ and σ^2 have been used to define the normal distribution when the expected value and variance of an arbitrary distribution were previously defined as μ and σ^2. Indeed, from the definition of the normal distribution it can be shown, using calculus methods, that μ and σ^2 are, respectively, the expected value and variance of this distribution.

EXAMPLE 5.10 For diastolic blood pressure the parameters might be $\mu = 80$ mm Hg, $\sigma = 12$ mm Hg; for birthweight they might be $\mu = 120$ oz, $\sigma = 15$ oz; for tree diameters they might be $\mu = 8$ in., $\sigma = 2$ in. ∎

Interestingly, the entire shape of the normal distribution is determined by the two parameters μ and σ^2. If two normal distributions with the same variance σ^2 and different means μ_1, μ_2, where $\mu_2 > \mu_1$, are compared, then their density functions will appear as in Figure 5.5.

Similarly, two normal distributions with the same mean but different variances ($\sigma_2^2 > \sigma_1^2$) can be compared, as shown in Figure 5.6. Note that the area under any normal density function must be 1. Thus, the two normal distributions shown in Figure 5.6 must cross, since otherwise one curve would remain completely above the other and the areas under both curves could not simultaneously be 1.

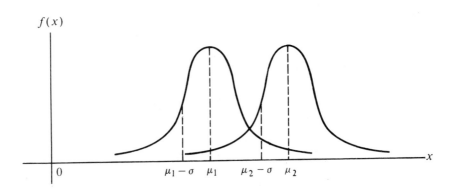

FIGURE 5.5
Comparison of two
normal distributions with
the same variance and
different means

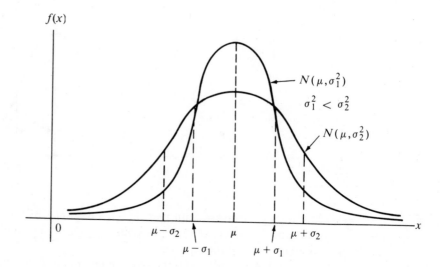

FIGURE 5.6
Comparison of two
normal distributions with
the same mean and
different variances

Another property of the normal distribution is that the height $= 1/(\sqrt{2\pi}\sigma)$. Thus, the height is inversely proportional to σ. This helps us to visualize σ, since the height of the $N(\mu, \sigma_1^2)$ distribution in Figure 5.6 is greater than the height of the $N(\mu, \sigma_2^2)$ distribution.

DEFINITION 5.6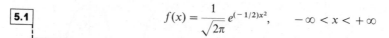

A **normal distribution with mean μ and variance σ^2** will generally be referred to as an $N(\mu, \sigma^2)$ distribution.

∎

Note that the second parameter is always the variance σ^2 and not the standard deviation σ.

DEFINITION 5.7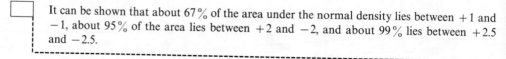

A normal distribution with mean 0 and variance 1 will be referred to as a **standard**, or **unit**, normal distribution. This distribution is denoted by $N(0, 1)$.

∎

We will see that any information concerning an $N(\mu, \sigma^2)$ distribution can be obtained from appropriate manipulations of an $N(0, 1)$ distribution.

SECTION 5.4   **Properties of the Standard Normal Distribution**

To become familiar with the $N(0, 1)$ distribution, some of its properties will be reviewed. First, the probability density function in this case reduces to

5.1
$$f(x) = \frac{1}{\sqrt{2\pi}} e^{(-1/2)x^2}, \qquad -\infty < x < +\infty$$

This distribution is symmetrical about 0, since $f(x) = f(-x)$, and is depicted in Figure 5.7.

It can be shown that about 67% of the area under the normal density lies between $+1$ and -1, about 95% of the area lies between $+2$ and -2, and about 99% lies between $+2.5$ and -2.5.

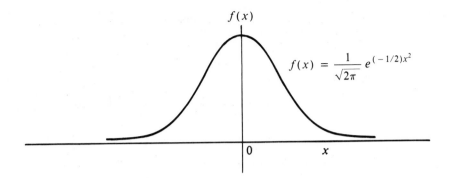

FIGURE 5.7
Probability density
function for a standard
normal distribution

$$f(x) = \frac{1}{\sqrt{2\pi}} e^{(-1/2)x^2}$$

These relationships can be expressed more precisely by saying that

$$Pr(-1 < X < +1) = .6827 \qquad Pr(-1.96 < X < +1.96) = .95$$

$$Pr(-2.576 < X < +2.576) = .99$$

Thus, the standard normal distribution slopes off very rapidly, and absolute values greater than 3 are unlikely. These relationships are depicted in Figure 5.8.

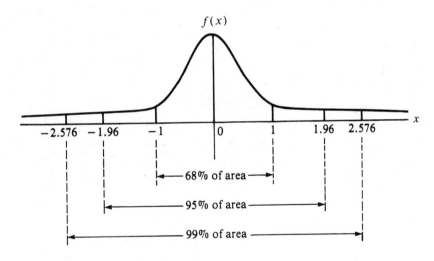

FIGURE 5.8
Empirical properties of
the standard normal
distribution

Tables of the area under the normal density function, or so-called normal tables, take advantage of the symmetry properties of the normal distribution and generally are concerned with areas for positive values of x.

DEFINITION 5.8 ■■■
The **cumulative distribution function for a standard normal distribution** is denoted by

$$\Phi(x) = Pr(X \leqslant x)$$

where X follows an $N(0, 1)$ distribution. This function is depicted in Figure 5.9. ■

DEFINITION 5.9 ■■■
The symbol $\sim$ is used as shorthand for the phrase "**is distributed as**". Thus, $X \sim N(0, 1)$ means that the random variable X is distributed as an $N(0, 1)$ distribution. ■

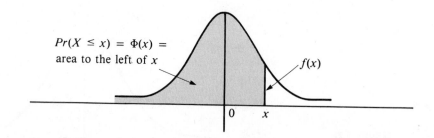

FIGURE 5.9
Cumulative distribution function ($\Phi(x)$) for a standard normal random variable

$$Pr(X \leq x) = \Phi(x) = \text{area to the left of } x$$

5.4.1 Using Normal Tables

Under column A in Table 3 of Appendix 1, $\Phi(x)$ for various positive values of x for a standard normal distribution are presented. This cumulative distribution function is depicted in Figure 5.10. Notice that the area to the left of 0 is 0.5. Furthermore, the area to the left of x approaches 0 as x becomes small and approaches 1 as x becomes large.

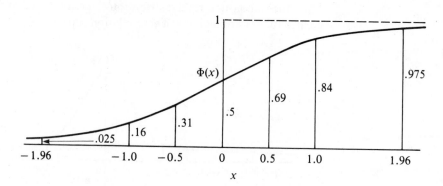

FIGURE 5.10
Cumulative distribution function for a standard normal distribution ($\Phi(x)$)

EXAMPLE 5.11 If

$$X \sim N(0, 1)$$

then find

$$Pr(X \leq 1.96) \quad \text{and} \quad Pr(X \leq 1)$$

SOLUTION   From Table 3, column A,

$$\Phi(1.96) = .975 \quad \text{and} \quad \Phi(1) = .8413 \qquad \blacksquare$$

5.2 **Symmetry Properties of the Standard Normal Distribution**
From the symmetry properties of the standard normal distribution,

$$\Phi(-x) = Pr(X \leq -x) = Pr(X \geq x) = 1 - Pr(X \leq x) = 1 - \Phi(x)$$

This symmetry property is depicted in Figure 5.11.

The right-hand tail of the standard normal distribution $= Pr(X \geq x)$ is provided in column B of Table 3.

EXAMPLE 5.12 Calculate

$$Pr(X \leq -1.96)$$

If

$$X \sim N(0, 1)$$

SOLUTION

$$Pr(X \leq -1.96) = Pr(X \geq 1.96) = .0250 \text{ from column B of Table 3} \qquad \blacksquare$$

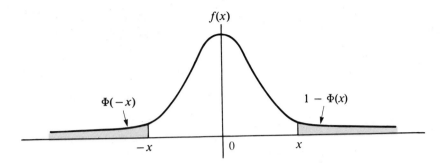

FIGURE 5.11
Illustration of the symmetry properties of the normal distribution

Furthermore, for any numbers a, b, we have $Pr(a \leqslant X \leqslant b) = Pr(X \leqslant b) - Pr(X \leqslant a)$ and thus can evaluate $Pr(a \leqslant X \leqslant b)$ for any a, b from Table 3.

EXAMPLE 5.13   Compute $\qquad\qquad Pr(-1 \leqslant X \leqslant 1.5)$

if $\qquad\qquad\qquad\qquad X \sim N(0, 1)$

SOLUTION $\qquad Pr(-1 \leqslant X \leqslant 1.5) = Pr(X \leqslant 1.5) - Pr(X \leqslant -1)$

$$= Pr(X \leqslant 1.5) - Pr(X \geqslant 1) = .9332 - .1587$$

$$= .7745 \qquad\qquad \blacksquare$$

EXAMPLE 5.14   **Pulmonary Disease** Forced Vital Capacity (FVC) is a standard measure of pulmonary function and represents the volume of air a person can expel in 6 seconds. A topic of current research interest is to look at potential risk factors, such as cigarette smoking, air pollution, or the type of stove used in the home, that may affect FVC in grade school children. One problem is that pulmonary function is affected by age, sex, and height, and these variables must be corrected for before looking at other risk factors. One way to make these adjustments for a particular child is to find the mean μ and standard deviation σ for the same age (in 1-year age groups), sex, and height (in 2-in. height groups) from large national surveys and compute a **standardized FVC**, which is defined as $(x - \mu)/\sigma$, where x is the original FVC. The standardized FVC would then approximately follow an $N(0, 1)$ distribution. Suppose that a child is considered in poor pulmonary health if his or her standardized FVC < -1.5. What percentage of children are in poor pulmonary health?

SOLUTION $\qquad\qquad\qquad Pr(X < -1.5) = Pr(X > 1.5) = .0668 \qquad\qquad \blacksquare$

Thus, about 7% of children are in poor pulmonary health.

In many instances we will be concerned with tail areas on either side of 0 for a standard normal distribution. For example, the *normal range* for a biological quantity is often defined by a range within x standard deviations of the mean for some specified value of x. The probability of a value falling in this range is given by $Pr(-x \leqslant X \leqslant x)$ for a standard normal distribution. This quantity is tabulated in column D of Table 3 for various values of x.

EXAMPLE 5.15   **Pulmonary Disease** Suppose a child is considered to have normal lung growth if his or her standardized FVC is within 1.5 standard deviations of the mean. What proportion of children are within the normal range?

SOLUTION   Compute $Pr(-1.5 \leqslant X \leqslant 1.5)$. Under 1.50 in Table 3, column D, this quantity is given as .8664. Thus, about 87% of children are considered to have normal lung growth using this definition. $\qquad\qquad \blacksquare$

Finally, in column C of Table 3, the area under the standard normal density from 0 to X is provided, since these areas will occasionally prove useful in work on statistical inference.

EXAMPLE 5.16   Find the area under the standard normal density from 0 to 1.45.

SOLUTION   Refer to column C of Table 3 under 1.45. The appropriate area is given by .4265. ∎

Of course, the areas given in columns A, B, C, and D are somewhat redundant in that *all* computations concerning the standard normal distribution could be performed using any one of these columns. In particular, we have seen that $B(x) = 1 - A(x)$. Also, from the symmetry of the normal distribution, we can easily show that $C(x) = A(x) - .5$, $D(x) = 2 \times C(x) = 2 \times A(x) - 1.0$. However, this redundancy is deliberate, since for some applications one or the other of these columns will be more convenient to use.

The percentiles of a standard normal distribution are often referred to in statistical inference. For this purpose the following definition is introduced:

DEFINITION 5.10 ▪▪

The **100% × uth percentile** of a standard normal distribution is denoted by z_u. It is defined by the relationship

$$Pr(X < z_u) = u, \quad \text{where } X \sim N(0, 1)$$ ∎

z_u is depicted graphically in Figure 5.12.

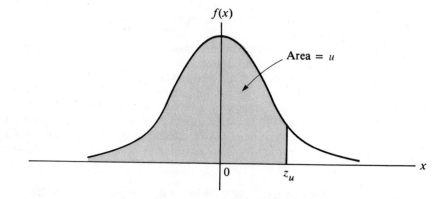

FIGURE 5.12
Graphical display of the *u*th percentile of a standard normal distribution (z_μ)

EXAMPLE 5.17   Compute $z_{.975}, z_{.95}, z_{.5}$, and $z_{.025}$

SOLUTION   From Table 3 we have that

$$\Phi(1.96) = .975$$
$$\Phi(1.645) = .95$$
$$\Phi(0) = .5$$
$$\Phi(-1.96) = 1 - \Phi(1.96) = 1 - .975 = .025$$

Thus,

$$z_{.975} = 1.96$$
$$z_{.95} = 1.645$$
$$z_{.5} = 0$$
$$z_{.025} = -1.96$$ ∎

SECTION 5.5 ## Conversion from an $N(\mu, \sigma^2)$ Distribution to an $N(0, 1)$ Distribution

EXAMPLE 5.18   **Hypertension** Suppose a borderline hypertensive is defined as a person whose diastolic blood pressure is between 90 and 100 mm Hg inclusive, and the subjects are 35–44-year-old males whose blood pressures are normally distributed with mean 80 and variance 144. What is the probability that a randomly selected person from this population will be a borderline hypertensive? This question can be restated more precisely:

If $$X \sim N(80, 144)$$

then what is $$Pr(90 < X < 100)$$

(The solution is given on page 118). ∎

More generally, the following question can be asked: If $X \sim N(\mu, \sigma^2)$, then what is $Pr(a < X < b)$ for any a, b? The basic idea is to convert a probability statement about an $N(\mu, \sigma^2)$ distribution to an equivalent probability statement about an $N(0, 1)$ distribution. Consider the random variable $Z = (X - \mu)/\sigma$. We can show that the following relationship holds:

5.3	If $\quad\quad X \sim N(\mu, \sigma^2) \quad$ and $\quad Z = (X - \mu)/\sigma$

then $Z \sim N(0, 1)$.

To see this, compute the expected value and variance of Z. To accomplish this, keep in mind that the expected value and variance have the same properties as the sample mean and variance upon addition of and/or multiplication by a constant. Specifically, for any constant c,

$$E(X + c) = E(X) + c$$

$$E(cX) = cE(X)$$

$$Var(X + c) = Var(X)$$

$$Var(cX) = c^2 Var(X)$$

Therefore, applying these principles,

$$E(Z) = E\left[\frac{(X - \mu)}{\sigma}\right] = \left(\frac{1}{\sigma}\right)E(X - \mu) = \left(\frac{1}{\sigma}\right)[E(X) - E(\mu)]$$

$$= \left(\frac{1}{\sigma}\right)[E(X) - \mu] = \left(\frac{1}{\sigma}\right)(\mu - \mu) = 0$$

$$Var(Z) = Var\left[\frac{(X - \mu)}{\sigma}\right] = \left(\frac{1}{\sigma^2}\right)Var(X - \mu)$$

$$= \left(\frac{1}{\sigma^2}\right)Var(X) = \left(\frac{1}{\sigma^2}\right)\sigma^2 = 1$$

Thus, the expected value of Z is 0 and its variance of Z is 1. It is also true that normality is preserved when converting from the random variable X to the random variable $Z = (X - \mu)/\sigma$, but to show this is beyond the scope of this book. Therefore, $Z \sim N(0, 1)$.

We now wish to convert $Pr(a < X < b)$ into a probability statement about Z, since Z is a standard normal random variable and tables are available only for the standard normal distribution.

We can easily show that $a < X < b$ if and only if $(a - \mu)/\sigma < Z < (b - \mu)/\sigma$. To see this, note that the inequality $a < X < b$ can be written as two inequalities, $a < X$ and $X < b$. If μ is subtracted from both sides of each inequality, we get

$$a - \mu < X - \mu \quad \text{and} \quad X - \mu < b - \mu$$

Also, if both sides of each inequality are divided by σ, we obtain

$$\frac{a - \mu}{\sigma} < \frac{X - \mu}{\sigma} \quad \text{and} \quad \frac{X - \mu}{\sigma} < \frac{b - \mu}{\sigma}$$

or upon rewriting,

$$\frac{a - \mu}{\sigma} < \frac{X - \mu}{\sigma} < \frac{b - \mu}{\sigma} \quad \text{or} \quad \frac{a - \mu}{\sigma} < Z < \frac{b - \mu}{\sigma}$$

Therefore, it follows that

$$Pr(a < X < b) = Pr\left[\frac{(a - \mu)}{\sigma} < Z < \frac{(b - \mu)}{\sigma}\right]$$

However, since $Z \sim N(0, 1)$,

$$Pr(a < X < b) = Pr\left[\frac{(a - \mu)}{\sigma} < Z < \frac{(b - \mu)}{\sigma}\right] = \Phi\left[\frac{(b - \mu)}{\sigma}\right] - \Phi\left[\frac{(a - \mu)}{\sigma}\right]$$

This procedure is known as **standardization of a normal variable**. This fact together with **(5.3)** leads to the following key relationship:

5.4   **Evaluation of Probabilities for Any Normal Distribution via Standardization**

If $\qquad\qquad X \sim N(\mu, \sigma^2) \qquad$ and $\qquad Z = (X - \mu)/\sigma$

then $\quad Pr(a < X < b) = Pr\left(\dfrac{a - \mu}{\sigma} < Z < \dfrac{b - \mu}{\sigma}\right) = \Phi[(b - \mu)/\sigma] - \Phi[(a - \mu)/\sigma]$

Since the Φ function, which is the cumulative distribution function for a standard normal distribution, is given in column A of Table 3 of Appendix 1, probabilities for *any* normal distribution can now be evaluated. This procedure is depicted in Figure 5.13.

SOLUTION TO   The probability of being a borderline hypertensive among the group of 35–44-year-old males
EXAMPLE 5.18   can now be calculated.

$$Pr(90 < X < 100) = Pr\left(\frac{90 - 80}{12} < Z < \frac{100 - 80}{12}\right)$$

$$= Pr(0.83 < Z < 1.67) = \Phi(1.67) - \Phi(0.83)$$

$$= .9525 - .7967 = .156$$

Thus, 15.6% of this population will be borderline hypertensive. ∎

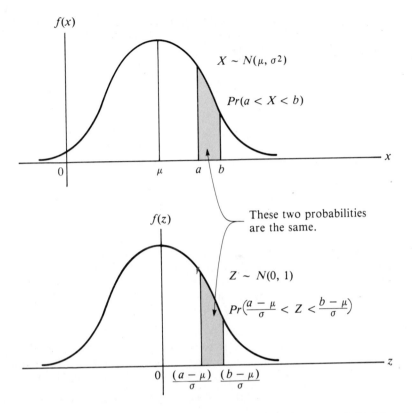

FIGURE 5.13
Evaluation of
probabilities for any
normal distribution via
standardization

EXAMPLE 5.19   **Botany** Suppose that tree diameters of a certain species of tree from some defined forest area are assumed to be normally distributed with mean 8 in. and standard deviation 2 in. Find the probability of a tree having an unusually large diameter, which is defined as >12 in.

SOLUTION   We have $X \sim N(8, 4)$ and require

$$Pr(X > 12) = 1 - Pr(X < 12) = 1 - Pr\left(Z < \frac{12 - 8}{2}\right)$$

$$= 1 - Pr(Z < 2.0) = 1 - .977 = .023$$

Thus, 2.3% of trees from this area have an unusually large diameter.   ∎

The general principle is that for any probability statement concerning normal random variables of the form $Pr(a < X < b)$, the population mean μ is subtracted from each boundary point and divided by the standard deviation σ to obtain an equivalent probability statement for the standard normal random variable Z,

$$Pr[(a - \mu)/\sigma < Z < (b - \mu)/\sigma]$$

The standard normal tables are then used to evaluate this latter probability.

EXAMPLE 5.20   **Cerebrovascular Disease** Diagnosing stroke strictly on the basis of clinical symptoms is difficult. A standard diagnostic test used in clinical medicine to detect stroke in patients is the angiogram. This test has some risks for the patient, and several noninvasive techniques have been developed that are hoped to be as effective as the angiogram. One such method utilizes the measurement of cerebral blood flow (CBF) in the brain, since stroke patients tend to have lower levels of CBF than normal. Assume that in the general population, CBF is normally distributed with mean 75 and standard deviation 17. A patient is classified as being at risk for stroke if his or her CBF is less than 40. What proportion of normal patients will be mistakenly classified as being at risk for stroke?

SOLUTION   Let X be the random variable representing CBF. Then $X \sim N(75, 17^2) = N(75, 289)$. We want to find $Pr(X < 40)$. We standardize the limit of 40 so as to use the standard normal distribution. The standardized limit is $(40 - 75)/17 = -2.06$. Thus, if Z represents the standardized normal random variable $= (X - \mu)/\sigma$, then

$$Pr(X < 40) = Pr(Z < -2.06)$$

$$= \Phi(-2.06) = 1 - \Phi(2.06) = 1 - .9803 \approx .020$$

Thus, about 2.0% of normal patients will be incorrectly classified as being at risk for stroke. ∎

EXAMPLE 5.21   **Ophthalmology** Glaucoma is a disease of the eye that is manifested by high intraocular pressure. The distribution of intraocular pressure in the general population is approximately normal with mean 16 mm Hg and standard deviation 3 mm Hg. If the normal range for intraocular pressure is considered to be between 12 mm Hg and 20 mm Hg, then what percentage of the general population would fall within this range?

SOLUTION   We wish to calculate $Pr(12 \leqslant X \leqslant 20)$, where $X \sim N(16, 9)$. The limits 12 and 20 must first be standardized so as to use the standard normal tables to evaluate this probability. The standardized limits are $(12 - 16)/3 = -1.33$ and $(20 - 16)/3 = +1.33$. Thus, we evaluate $Pr(-1.33 \leqslant Z \leqslant 1.33)$, where Z follows a standard normal distribution.

$$Pr(-1.33 \leqslant Z \leqslant 1.33) = Pr(Z \leqslant 1.33) - Pr(Z < -1.33)$$

$$= Pr(Z \leqslant 1.33) - [1 - Pr(Z \leqslant 1.33)]$$

$$= 2Pr(Z \leqslant 1.33) - 1 = 2(.9082) - 1 = .816$$

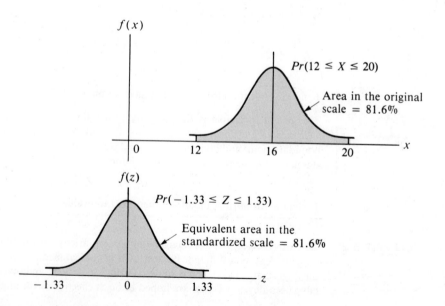

FIGURE 5.14
Calculation of the proportion of people with intraocular pressures in the normal range

Alternatively, $Pr(-1.33 \leqslant Z \leqslant 1.33)$ could be evaluated directly from the 1.33 row under column D of Table 3 of the Appendix, yielding a probability of .8165. Thus, 81.6% of the population have intraocular pressures in the normal range. These calculations are depicted in Figure 5.14.   ■

Normal Approximation to the Binomial Distribution

In Chapter 4 the binomial distribution was introduced to assess the probability of k successes in n independent trials, where the probability of success (p) is the same for each trial. If n is large, the binomial distribution is very cumbersome to work with and an approximation is easier to use rather than the exact binomial distribution. The normal distribution is often used to approximate the binomial since it is very easy to work with. The key question is, When will the normal distribution provide an accurate approximation to the binomial?

Suppose a binomial distribution has parameters n and p. If n is large and p is either near 0 or near 1, then the binomial distribution will be very positively or negatively skewed, respectively. See Figure 5.15(a) and (b). Similarly, when n is small, for any p, the distribution will tend to be skewed. See Figure 5.15(c). However, if n is moderately large and p is not too extreme, then the binomial distribution will tend to be symmetric and will be well approximated by a normal distribution. See Figure 5.15(d).

We know from Chapter 4 that the mean and variance of a binomial distribution are np and npq, respectively. A natural approximation to use is a normal distribution with the *same* mean and variance, that is, $N(np, npq)$. Suppose we want to compute $Pr(a \leqslant X \leqslant b)$ for some integers a, b, where X is binomially distributed with parameters n and p. This probability might be approximated by the area under the normal curve from a to b. However, we can show empirically that a better approximation to this probability is given by the area under the

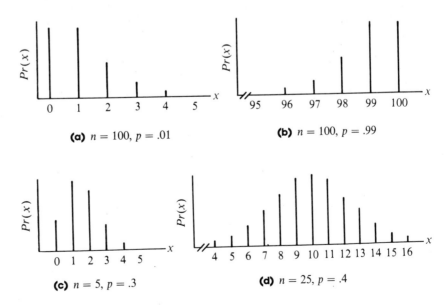

(a) $n = 100$, $p = .01$

(b) $n = 100$, $p = .99$

(c) $n = 5$, $p = .3$

(d) $n = 25$, $p = .4$

FIGURE 5.15
Symmetry properties of
the binomial distribution

normal curve from $a - \frac{1}{2}$ to $b + \frac{1}{2}$. This will generally be the case when any discrete distribution is approximated by the normal distribution. Thus the following rule applies:

5.5   **Normal Approximation to the Binomial Distribution**

If X is a binomial random variable with parameters n and p, then $Pr(a \leqslant X \leqslant b)$ is approximated by the area under an $N(np, npq)$ curve from $(a - \frac{1}{2})$ to $(b + \frac{1}{2})$. This rule implies that for the special case $a = b$, the binomial probability $Pr(X = a)$ is approximated by the area under the normal curve from $(a - \frac{1}{2})$ to $(a + \frac{1}{2})$. The only exception to this rule is that $Pr(X = 0)$ and $Pr(X = n)$ are approximated by the area under the normal curve to the left of $\frac{1}{2}$ and to the right of $n - \frac{1}{2}$, respectively.

EXAMPLE 5.22   Suppose a binomial distribution has parameters $n = 25, p = .4$. How can $Pr(7 \leqslant X \leqslant 12)$ be approximated?

SOLUTION   We have $np = 25(.4) = 10, npq = 25(.4)(.6) = 6.0$. Thus, this distribution is approximated by a normal random variable Y with mean 10 and variance 6. We specifically want to compute the area under this normal curve from 6.5 to 12.5. We have

$$Pr(6.5 \leqslant X \leqslant 12.5) = \Phi\left(\frac{12.5 - 10}{\sqrt{6}}\right) - \Phi\left(\frac{6.5 - 10}{\sqrt{6}}\right)$$

$$= \Phi(1.02) - \Phi(-1.43) = \Phi(1.02) - [1 - \Phi(1.43)]$$

$$= \Phi(1.02) + \Phi(1.43) - 1 = .8461 + .9236 - 1 = .770$$

This approximation is depicted in Figure 5.16.   ∎

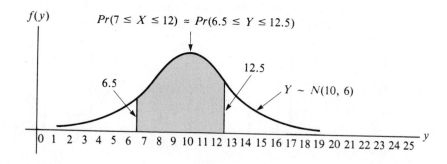

FIGURE 5.16
The approximation of the binomial random variable X with parameters $n = 25$, $p = .4$ by the normal random variable Y with mean 10 and variance 6

EXAMPLE 5.23   **Infectious Disease** Suppose we want to compute the probability that between 50 and 75 of 100 white blood cells will be neutrophils, where the probability that any one cell is a neutrophil is .6. These values are chosen as proposed limits to the range of neutrophils in normal people and we wish to predict what proportion of people will be in the normal range according to this definition.

SOLUTION   The exact probability is given by

$$\sum_{k=50}^{75} \binom{100}{k} (.6)^k (.4)^{100-k}$$

The normal approximation is used to approximate the exact probability. The mean of the binomial distribution in this case is $100(.6) = 60$, and the variance is $100(.6)(.4) = 24$. Thus, we find the area between 49.5 and 75.5 for an $N(60, 24)$ distribution. This area is

$$\Phi\left(\frac{75.5 - 60}{\sqrt{24}}\right) - \Phi\left(\frac{49.5 - 60}{\sqrt{24}}\right) = \Phi(3.16) - \Phi(-2.14)$$

$$= \Phi(3.16) + \Phi(2.14) - 1$$

$$= .9992 + .9838 - 1 = .983$$

Thus, 98.3% of the people will be normal.   ■

EXAMPLE 5.24   **Infectious Disease** Suppose a person is defined as abnormally high if the number of neutrophils is ≥ 76 and abnormally low if the number of neutrophils is ≤ 49. Calculate the proportion of people that are abnormally high and low.

SOLUTION   The probability of being abnormally high is given by $Pr(X \geq 76) \approx Pr(Y \geq 75.5)$, where X is a binomial random variable with parameters $n = 100$, $p = .6$ and $Y \sim N(60, 24)$. This latter probability is

$$1 - \Phi\left(\frac{75.5 - 60}{\sqrt{24}}\right) = 1 - \Phi(3.16) = .001$$

Similarly, the probability of being abnormally low is

$$Pr(X \leq 49) \approx Pr(Y \leq 49.5) = \Phi\left(\frac{49.5 - 60}{\sqrt{24}}\right)$$

$$= \Phi(-2.14) = 1 - \Phi(2.14)$$

$$= 1 - .9838 = .016$$

Thus, 0.1% of people will be abnormally high and 1.6% will be abnormally low. These probabilities are depicted in Figure 5.17.   ■

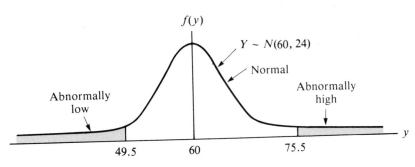

$f(y)$

$Y \sim N(60, 24)$

Normal

Abnormally low

Abnormally high

FIGURE 5.17
Normal approximation to the distribution of neutrophils

49.5   60   75.5 y

Under what conditions should this approximation be used?

> The normal distribution is used with mean np and variance npq to approximate a binomial distribution with parameters n and p when $npq \geq 5$.

This condition will be satisfied if n is moderately large and p is not too small. To illustrate this condition, the binomial probability distribution for $p = .1, n = 10,$ 20, 50, and 100 is plotted in Figure 5.18(a) through (d) and $p = .2, n = 10, 20, 50,$

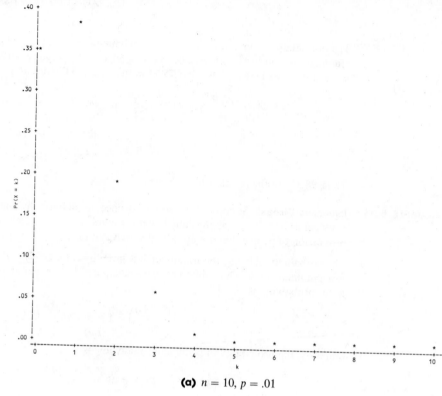

(a) $n = 10$, $p = .01$

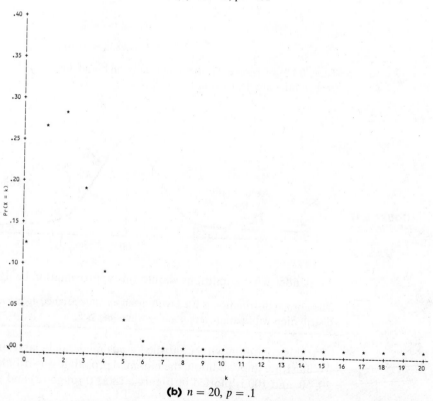

FIGURE 5.18(a)(b)
SAS plot of binomial
distribution

(b) $n = 20$, $p = .1$

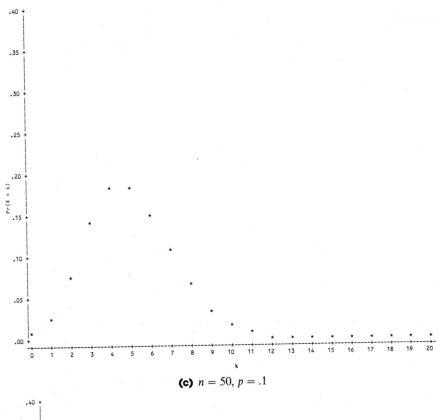

(c) $n = 50, p = .1$

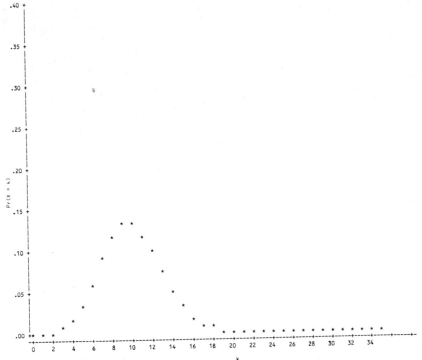

FIGURE 5.18(c)(d)
SAS plot of binomial
distribution

(d) $n = 100, p = .1$

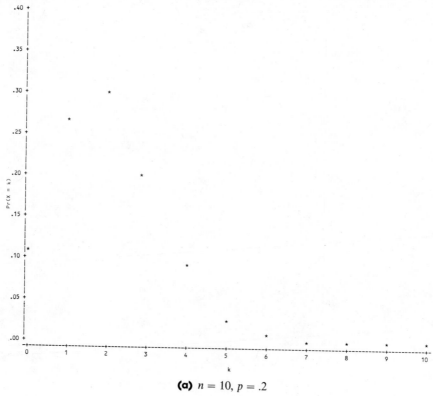

(a) $n = 10$, $p = .2$

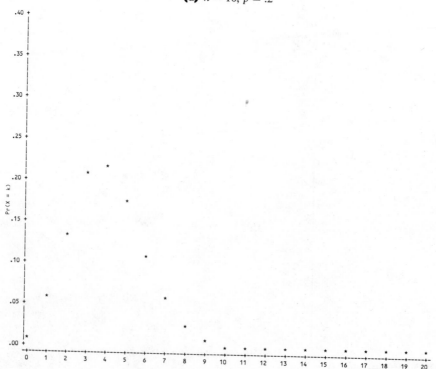

FIGURE 5.19(a)(b)
SAS plot of binomial
distribution

(b) $n = 20$, $p = .2$

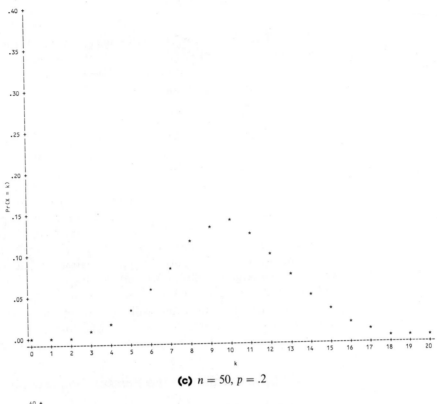

(c) $n = 50, p = .2$

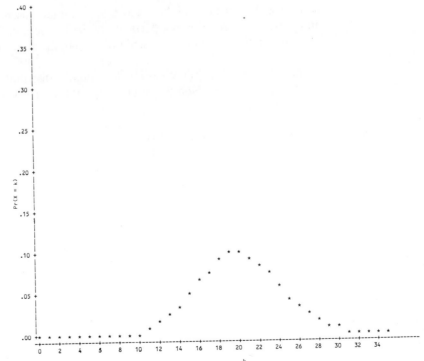

FIGURE 5.19(c)(d)
SAS plot of binomial
distribution

(d) $n = 100, p = .2$

and 100 is plotted in Figure 5.19(a) through (d) using a Statistical Analysis System (SAS) plotting routine (PROC PLOT). Notice that the normal approximation to the binomial distribution does not fit well in Figure 5.18(a), $n = 10$, $p = .1$ ($npq = 0.9$), or Figure 5.18(b), $n = 20$, $p = .1$ ($npq = 1.8$). The approximation is marginally adequate in Figure 5.18(c), $n = 50$, $p = .1$ ($npq = 4.5$), where the right-hand tail is only slightly longer than the left-hand tail. The approximation is quite good in Figure 5.18(d), $n = 100$, $p = .1$ ($npq = 9.0$), where the distribution appears to be quite symmetric. Similarly, for $p = .2$, although the normal approximation is not good for $n = 10$ [Figure 5.19(a), $npq = 1.6$], it becomes marginally adequate for $n = 20$ [Figure 5.19(b), $npq = 3.2$] and quite good for $n = 50$ [Figure 5.19(c), $npq = 8.0$] and $n = 100$ [Figure 5.19(d), $npq = 16.0$].

Note that the conditions under which the normal approximation to the binomial distribution works well (namely, $npq \geqslant 5$), which correspond to n moderate and p not too large or too small, are generally *not* the same as the conditions for which the Poisson approximation to the binomial distribution works well [n large ($\geqslant 100$) and p very small ($p \leqslant .01$)]. However, occasionally both of these criteria will be met. In such cases, for example, when $n = 1000$, $p = .01$, the two approximations will yield about the same results. The normal approximation is preferable because it is easier to apply.

SECTION 5.7   ## Normal Approximation to the Poisson Distribution

The normal distribution can be used to approximate discrete distributions other than the binomial distribution, particularly the Poisson distribution. The motivation for this is that the Poisson distribution is cumbersome to use for large values of μ.

The same technique is used as for the binomial distribution; that is, the means and variances of the Poisson distribution and the approximating normal distribution are equated.

5.6 | **Normal Approximation to the Poisson Distribution**

A Poisson distribution with parameter μ is approximated by a normal distribution with mean and variance both equal to μ. $Pr(X = x)$ is approximated by the area under an $N(\mu, \mu)$ density from $x - \frac{1}{2}$ to $x + \frac{1}{2}$ for $x > 0$ or by the area to the left of $\frac{1}{2}$ for $x = 0$. This approximation is used for $\mu \geqslant 5$.

The Poisson distribution for $\mu = 2$, 5, 10, and 20 is plotted using the SAS plotting program (PROC PLOT) in Figure 5.20(a) through (d), respectively. The normal approximation is clearly inadequate for $\mu = 2$ [Figure 5.20(a)], marginally adequate for $\mu = 5$ [Figure 5.20(b)], and adequate for $\mu = 10$ [Figure 5.20(c)] and $\mu = 20$ [Figure 5.20(d)].

EXAMPLE 5.25   **Bacteriology** Consider again the distribution of the number of bacteria in a petri plate of area A. Assume that the probability of observing x bacteria is given exactly by a Poisson distribution with parameter $\mu = \lambda A$, where $\lambda = 0.1$ and $A = 100 \text{ cm}^2$. Suppose 20 bacteria are observed in this area. How unusual is this event?

SOLUTION   Compute

$$Pr(X \geqslant 20) \approx Pr(Y \geqslant 19.5)$$

where

$$Y \sim N(\lambda A, \lambda A) = N(10, 10)$$

We have

$$Pr(Y \geqslant 19.5) = 1 - Pr(Y \leqslant 19.5) = 1 - \Phi\left(\frac{19.5 - 10}{\sqrt{10}}\right)$$

$$= 1 - \Phi\left(\frac{9.5}{\sqrt{10}}\right) = 1 - \Phi(3.00)$$

$$= 1 - .9987 = .0013$$

Thus, 20 or more colonies in 100 cm^2 would be expected only 1.3 times in 1000 plates, a rare event indeed. ∎

SECTION 5.8   Linear Combinations of Random Variables

In work on statistical inference, the use of sums or differences or more complicated linear functions of random variables (either continuous or discrete) will often arise. For this reason, the properties of linear combinations of random variables are important to discuss.

DEFINITION 5.11 ■■

A **linear combination** L of the random variables $X_1, \ldots, X_k$ is defined as any function of the form $L = c_1 X_1 + \cdots + c_k X_k$. ∎

EXAMPLE 5.26   **Renal Disease** Suppose X_1, X_2 represent serum creatinine levels for two different individuals with end-stage renal disease. Represent the sum, difference, and average of the random variables X_1, X_2 as linear combinations of the random variables X_1, X_2.

SOLUTION   The sum is $X_1 + X_2$, where $c_1 = 1$, $c_2 = 1$. The difference is $X_1 - X_2$, where $c_1 = 1$, $c_2 = -1$. The average is $(X_1 + X_2)/2$, where $c_1 = 0.5$, $c_2 = 0.5$. ∎

It will often be necessary to compute the expected value and variance of linear combinations of random variables. To find the expected value of L, the principle that the expected value of the sum of two random variables is the sum of the two respective expected values is used. Applying this principle,

$$E(L) = E(c_1 X_1 + \cdots + c_k X_k)$$

$$= E(c_1 X_1) + \cdots + E(c_k X_k) = c_1 E(X_1) + \cdots + c_k E(X_k)$$

5.7   **Expected Value of Linear Combinations of Random Variables**

The expected value of the linear combination $L = \sum c_i X_i$ is $E(L) = \sum c_i E(X_i)$.

- -

EXAMPLE 5.27   **Renal Disease** Suppose the expected values of serum creatinine for the two individuals in Example 5.26 are 1.5 and 1.3, respectively. What is the expected value of the average serum creatinine level of these two individuals?

SOLUTION   The expected value of the average serum creatinine level = $E(0.5X_1 + 0.5X_2) = 0.5E(X_1) + 0.5E(X_2) = 0.75 + 0.65 = 1.4$. ∎

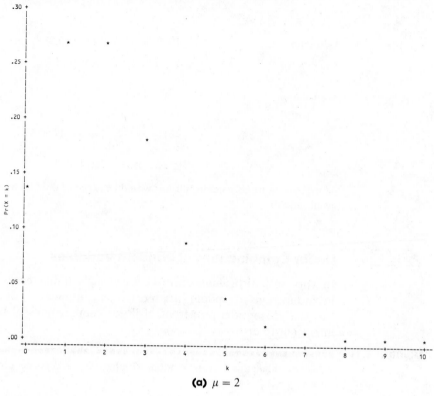

(a) $\mu = 2$

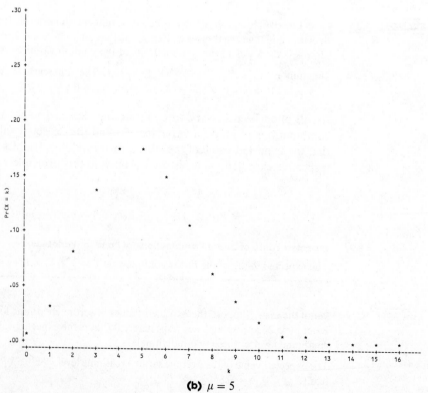

FIGURE 5.20(a)(b)
SAS plot of Poisson
distribution

(b) $\mu = 5$

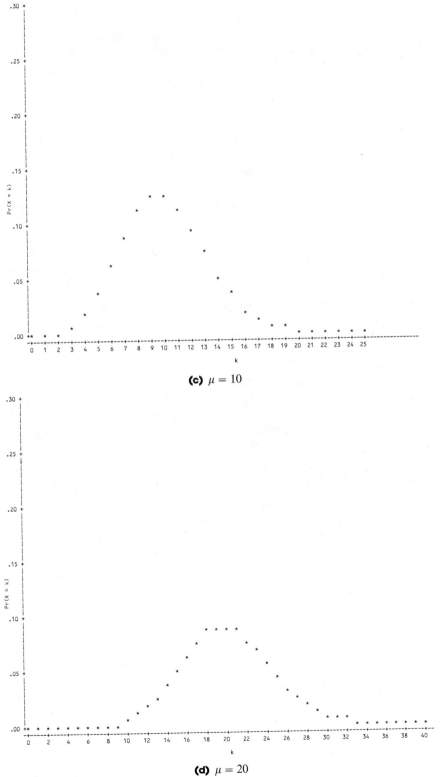

FIGURE 5.20(c)(d)
SAS plot of Poisson
distribution

(c) $\mu = 10$

(d) $\mu = 20$

To compute the variance of linear combinations of random variables, we assume that the random variables are independent. Under this assumption, it can be shown that the variance of the sum of two random variables is the sum of the respective variances. Applying this principle,

$$Var(L) = Var(c_1 X_1 + \cdots + c_k X_k)$$
$$= Var(c_1 X_1) + \cdots + Var(c_k X_k) = c_1^2\, Var(X_1) + \cdots + c_k^2\, Var(X_k)$$

5.8 | **Variance of Linear Combinations of Random Variables**

The variance of the linear combination $L = \sum_{i=1}^{k} c_i X_i$, where $X_1, \ldots, X_k$ are independent is $Var(L) = \sum_{i=1}^{k} c_i^2\, Var(X_i)$.

EXAMPLE 5.28   **Renal Disease** Suppose X_1, X_2 are defined as in Example 5.26. If we know that $Var(X_1) = Var(X_2) = 0.25$, then what is the variance of the average serum creatinine level over these two people?

SOLUTION   We wish to compute $Var(0.5X_1 + 0.5X_2)$. Applying **(5.8)**,

$$Var(0.5X_1 + 0.5X_2) = (0.5)^2 Var(X_1) + (0.5)^2 Var(X_2)$$
$$= 0.25(0.25) + 0.25(0.25) = 0.125 \qquad \blacksquare$$

The results for the expected value and variance of linear combinations in **(5.7)** and **(5.8)** *do not* depend on the assumption of normality. However, linear combinations of normal random variables are often of specific concern. It can be shown that any linear combination of normal random variables is itself normally distributed. This leads to the following important result:

5.9 | If $X_1, \ldots, X_k$ are independent normal random variables with expected values $\mu_1, \ldots, \mu_k$ and variances $\sigma_1^2, \ldots, \sigma_k^2$, and L is any linear combination $= \sum_{i=1}^{k} c_i X_i$, then L is normally distributed with

$$\text{expected value} = E(L) = \sum_{i=1}^{k} c_i \mu_i \quad \text{and} \quad \text{variance} = Var(L) = \sum_{i=1}^{k} c_i \sigma_i^2$$

EXAMPLE 5.29   **Renal Disease** If X_1, X_2 are defined as in Example 5.26 and are each normally distributed, then what is the distribution of the average $= 0.5X_1 + 0.5X_2$?

SOLUTION   Based on the solutions to Examples 5.27 and 5.28, we know that $E(L) = 1.4$, $Var(L) = 0.125$. Therefore, $(X_1 + X_2)/2 \sim N(1.4, 0.125)$. $\qquad \blacksquare$

SECTION 5.9   **Summary**

In this chapter continuous random variables were discussed. The concept of a probability density function which is the analogue to a probability mass function for discrete random variables, was introduced. In addition, generalizations of the concepts of expected value, variance, and cumulative distribution were presented for continuous random variables.

The normal distribution, the most important continuous distribution, was then studied in detail. The normal distribution is often used in statistical work, since many random phenomena follow this probability law, particularly those that can be expressed as a sum of many random variables. It was shown that the normal distribution is indexed by two parameters, the mean μ and the variance σ^2. Fortunately, all computations concerning any normal random variable can be accomplished using the standard, or unit, normal probability law, which has mean 0 and variance 1. Normal tables were introduced to use when working with the standard normal distribution. Also, since the normal distribution is easy to use, it is often employed to approximate other distributions. In particular, the focus was on the normal approximations to the binomial and Poisson distributions. Finally, the properties of linear combinations of random variables were discussed.

In the next three chapters, the normal distribution is used extensively as a foundation for work on statistical inference.

PROBLEMS

Cardiovascular Disease

Since serum cholesterol is related to age and sex, some investigators prefer to express it in terms of z-scores. If X = raw serum cholesterol, then $Z = \dfrac{X - \mu}{\sigma}$, where μ is the mean and σ is the standard deviation of serum cholesterol for a given age-sex group. Suppose Z is regarded as a standard normal distribution.

5.1 What is $Pr(z < 0.5)$?

5.2 What is $Pr(z > 0.5)$?

5.3 What is $Pr(-1.0 < z < 1.5)$?

Suppose a person is regarded as having high cholesterol if $z > 2.0$ and borderline cholesterol if $1.5 < z < 2.0$.

5.4 What proportion of people have high cholesterol?

5.5 What proportion of people have borderline cholesterol?

5.6 What are the deciles of the standard normal distribution, that is, the 10, 20, 30, ..., 90 percentiles?

5.7 What are the quartiles of the standard normal distribution?

Hypertension

Blood pressure in childhood tends to increase with age, but differently for boys and girls. Suppose that for both boys and girls, mean systolic blood pressure is 95 mm Hg at 3 years of age and increases 1.5 mm Hg per year up to the age of 13. Furthermore, starting at age 13, the mean increases by 2 mm Hg per year for boys and 1 mm Hg per year for girls up to the age of 18. Finally, assume that blood pressure is normally distributed and that the standard deviation is 12 mm Hg for all age-sex groups.

5.8 What is the probability that an 11-year-old boy will have a blood pressure greater than 130 mm Hg?

5.9 What is the probability that a 15-year-old girl will have a blood pressure between 100 and 120 mm Hg?

5.10 What proportion of 17-year-old boys have systolic blood pressure between 120 and 140 mm Hg?

5.11 What is the probability that of 200 15-year-old boys, at least 10 will have systolic blood pressure of 130 mm Hg or greater?

5.12 What level of systolic blood pressure is at the 80th percentile for 7-year-old boys?

5.13 What level of systolic blood pressure is at the 70th percentile for 12-year-old girls?

5.14 Suppose that a task force of pediatricians decides that children over the 95th percentile, but not over the 99th percentile, for their age-sex group should be encouraged to take preventive nonpharmacologic measures to reduce their blood pressure, whereas those children over the 99th percentile should receive antihypertensive drug therapy. Construct a table giving the appropriate levels of blood pressure to identify these groups for boys and girls for each year of age from 3 to 18.

Nutrition

Suppose that total carbohydrate intake in 12–14-year-old males is normally distributed with mean 124 g/1000 cal and standard deviation 20 g/1000 cal.

5.15 What percentage of boys in this age range have carbohydrate intake above 140 g/1000 cal?

5.16 What percentage of boys in this age range have carbohydrate intake below 90 g/1000 cal?

Suppose boys in this age range that live below the poverty level have a mean carbohydrate level of 121 g/1000 cal with a standard deviation of 19 g/1000 cal.

5.17 Answer Problem 5.15 for boys in this age range and economic environment.

5.18 Answer Problem 5.16 for boys in this age range and economic environment.

Cancer

The incidence of breast cancer in 40–59-year-old women is approximately 1 new case per 1000 women per year.

5.19 What is the incidence of breast cancer over 10 years in women initially 40 years old?

5.20 Suppose we are planning a study based on an enrollment of 10,000 women. What is the probability of obtaining at least 120 new breast cancer cases over a 10-year follow-up period?

Diabetes

A number of clinical characteristics were ascertained in a large group of subjects with insulin-dependent diabetes mellitus (IDDM). Suppose the distribution of percentage of ideal body weight in this group of patients is normal with mean 110 and standard deviation of 13.

5.21 What percentage of subjects with IDDM are above their ideal body weight, i.e., above 100% ideal body weight?

5.22 What percentage of subjects with IDDM are overweight (defined as 10% or more above ideal body weight)?

5.23 What percentage of subjects with IDDM are obese (defined as 20% or more above ideal body weight)?

5.24 What percentage of subjects with IDDM are underweight (defined as 10% or more below ideal body weight)?

5.25 What percentage of subjects with IDDM have normal body weight (within 10% of ideal body weight)?

Accident Epidemiology

Suppose the annual death rate from motor vehicle accidents in 1980 was 10 per 100,000 in urban areas of the United States.

5.26 In an urban state with a population of 2 million, what is the probability of observing not more than 150 traffic fatalities in a given year?

In rural areas of the United States the annual death rate from motor vehicle accidents is on the order of 100 per 100,000 population.

5.27 In a rurral state with a population of 100,000, what is the probability of observing not more than 80 traffic fatalities in a given year?

5.28 How large should x be so that the probability of observing not more than x traffic fatalities in a given year in a rural state with population 100,000 is 5%?

5.29 How large should x be so that the probability of observing x or more traffic fatalities in a given year, in an urban state with population 500,000 is 10%?

Pulmonary Disease

Many investigators have studied the relationship between asbestos exposure and death due to chronic obstructive pulmonary disease (COPD).

5.30 Suppose that among workers exposed to asbestos in a shipyard in 1960, 33 died over a 10-year period from COPD, whereas only 24 such deaths could be expected based on statewide mortality rates. Is the number of deaths due to COPD in this group excessive?

5.31 Twelve cases of leukemia are reported in people living in a particular census tract over a 5-year period. Is this number of cases abnormal if only 6.7 cases would be expected based on national cancer incidence rates?

Obstetrics

Assume that birthweights are normally distributed with a mean of 3400 g and a standard deviation of 700 g.

5.32 Find the probability of a low-birthweight child, where low birthweight is defined as ⩽ 2500 g.

5.33 Find the probability of a very low birthweight child, where very low birthweight is defined as ⩽2000 g.

5.34 Assuming that successive deliveries by the same woman have the same probability of being low birthweight, what is the probability that a woman with exactly 3 deliveries will have 2 or more low-birthweight deliveries?

Cardiovascular Disease, Pulmonary Disease

The duration of cigarette smoking has been linked to many diseases, including lung cancer and various forms of heart disease. Suppose we know that among men aged 30–34 who have ever smoked, the mean number of years they smoked is 12.8 with a standard deviation of 5.1 years. For women in this age group, the mean number of years they smoked is 9.3 with a standard deviation of 3.2.

5.35 Assuming that the duration of smoking is normally distributed, what proportion of men in this age group have smoked for more than 20 years?

5.36 Answer Problem 5.35 for women.

Renal Disease

The presence of bacteria in a urine sample (bacteriuria) is sometimes associated with symptoms of kidney disease. Assume that a determination of bacteriuria has been made over a large population at one point in time and that 5% of those sampled are positive for bacteriuria.

5.37 Suppose that 500 people from this population are sampled. What is the probability that 50 or more people would be positive for bacteriuria?

Cancer

Previous census data have indicated that approximately 0.2% of women aged 45–54 will have had cervical cancer at some point in their lives. However, the general feeling is that the rate of cervical cancer has increased.

5.38 If a new study by mail questionnaire is performed and it is found that 300 out of 100,000 women have had cervical cancer, then is this proportion consistent with the census rate?

Ophthalmology

A study was conducted of patients with retinitis pigmentosa, an ocular condition where pigment appears over the retina, resulting in substantial loss of vision in many cases [2]. The study was based on 94 patients who were seen annually at a baseline visit at three annual follow-up visits. In this study, 90 patients provided visual field measurements at each of the four examinations and are the subjects of the following data analyses. Visual field was transformed to the $\log_e$ scale to better approximate normality and yielded the data given in Table 5.1.

5.39 Assuming the change in visual field over 3 years is normally distributed when using the $\log_e$ scale, what is the proportion of patients who showed a decline in visual field over 3 years?

TABLE 5.1 Visual field measurements in retinitis pigmentosa patients

Year of examination	Mean*	Standard deviation	n
Year 0 (baseline)	8.15	1.23	90
Year 3	8.01	1.33	90
Year 0–year 3	0.14	0.66	90

* $\log_e$ (area of visual field) in degrees squared
(Reprinted with permission of the *American Journal of Ophthalmology*, 99, 240–251, 1985.)

5.40 What percentage of patients would be expected to show a decline of at least 20% in visual field over 3 years? (*Note:* In the $\log_e$ scale this is equivalent to a decline of at least $\log_e (1/0.8) = 0.223$).

5.41 Answer Problem 5.40 for a 50% decline over 3 years.

Cardiovascular Disease

Serum cholesterol is an important risk factor for coronary disease. We can show that $\log_e$ (serum cholesterol) is approximately normally distributed with mean 5.39 and standard deviation 0.23.

5.42 If the clinically normal range for cholesterol is 150–250 mg%/mL, then what proportion of people have abnormally low levels of cholesterol?

5.43 What proportion of people have abnormally high levels of cholesterol?

5.44 Some investigators feel that only cholesterol levels of over 300 mg%/mL indicate a high risk for heart disease. What proportion of the subpopulation with abnormally high levels does this group represent?

5.45 What proportion of the general population does the group with cholesterol levels over 300 mg%/mL represent?

Cancer

A study of the Massachusetts Department of Health found 46 deaths due to cancer among women in the city of Bedford over the period 1974–1978, where 30 deaths had been expected from statewide rates [2].

5.46 Write an expression for the probability of observing exactly k deaths over this period if the statewide rates are correct.

5.47 Can the occurrence of 46 deaths be attributed to chance? Specifically, what is the probability of observing at least 46 deaths if the statewide rates are correct?

Nutrition, Cancer

Beta carotene is a substance that is hypothesized to prevent cancer. A dietary survey was undertaken for the purpose of measuring the level of beta carotene intake in the typical American diet. Assume that the distribution of $\log_e$ carotene is normal with mean 8.34 and standard deviation 1.00. (Units are in $\log_e$ IU.)

5.48 What percentage of people have dietary carotene levels below 2000 IU? (*Note*: $\log_e 2000 = 7.60$.)

5.49 What percentage of people have dietary carotene levels below 1000 IU? (*Note*: $\log_e 1000 = 6.91$.)

5.50 Some studies suggest that carotene levels over 10,000 IU may protect against cancer. What percentage of people have a dietary intake of at least 10,000 IU?

Suppose that each person took a carotene supplement pill of dosage 5000 IU in addition to his or her normal diet. Assume that the resulting distribution of $\log_e$ carotene is normally distributed with mean 9.12 and standard deviation 1.00.

5.51 What percentage of people would have an intake from diet and supplements of at least 10,000 IU?

Hypertension

Suppose we want to recruit subjects for a hypertensive treatment study and we feel that 10% of the population to be sampled is hypertensive.

5.52 If 100 subjects are required for the study and perfect cooperation is assumed, then how many people need to be sampled to be 80% sure of ascertaining 100 hypertensives?

5.53 How many people need to be sampled to be 90% sure of ascertaining 100 hypertensives?

Hypertension

People are classified as hypertensive if their systolic blood pressure is higher than a specified level for their age group, according to the scheme in Table 5.2.

TABLE 5.2 Mean and standard deviation of systolic blood pressure (mm Hg) in specific age groups

Age group	Mean	Standard deviation	Specified hypertension level
1–14	105.0	5.0	115.0
15–44	125.0	10.0	140.0

Assume that systolic blood pressure is normally distributed with mean and standard deviation given in Table 5.2 for the age groups 1–14 and 15–44, re-

spectively. Define a *family* as a group of 2 people in the age group 1–14 and 2 people in the age group 15–44. A family is classified as hypertensive if *any one* family member is hypertensive.

5.54 What proportion of 1–14-year-olds are hypertensive?

5.55 What proportion of 15–44-year-olds are hypertensive?

5.56 What proportion of families are hypertensive? (Assume that the hypertensive status of different members of the family are independent random variables.)

5.57 Suppose an apartment building has 200 families living in it. What is the probability that between 10 and 25 families are hypertensive?

Nutrition

The distribution of serum levels of alpha tocopherol (serum vitamin E) is approximately normal with mean 860 $\mu g/dL$ and standard deviation 340 $\mu g/dL$.

5.58 What percentage of people have serum alpha tocopherol levels between 400 and 1000 $\mu g/dL$?

5.59 Suppose a person is identified as having toxic levels of alpha tocopherol if his or her serum level is > 2000 $\mu g/dL$. What percentage of people will be so identified?

5.60 A study is undertaken for evidence of toxicity of 2000 people who regularly take vitamin E supplements. The investigators found that 4 have serum alpha tocopherol levels > 2000 $\mu g/dL$. Is this an unusual number of people with toxic levels of serum alpha tocopherol?

Pulmonary Disease

Forced expiratory volume (FEV) is an index of pulmonary function that measures the volume of air expelled after 1 second of constant effort. FEV is known to be influenced by age, sex, and cigarette smoking. Assume that in 45–54-year-old nonsmoking males FEV is normally distributed with mean 4.0 liters and standard deviation 0.5 liter. In comparably aged currently smoking males FEV is normally distributed with mean 3.5 liters and standard deviation 0.6 liter.

5.61 If an FEV of less than 2.5 liters is regarded as showing some functional impairment (occasional breathlessness, inability to climb stairs, etc.), then what is the probability that a currently smoking male has functional impairment?

5.62 Answer Problem 5.61 for a nonsmoking male.

Many people are not functionally impaired now but their pulmonary function usually declines with age and they eventually will be functionally impaired. Assume that the *decline* in FEV over n years is normally distributed with mean $0.03n$ and standard deviation $0.02n$.

5.63 What is the probability that a 45-year-old man with FEV of 4.0 liters will be functionally impaired by the age of 75?

5.64 Answer Problem 5.63 for a 25-year-old man with FEV of 4.0 liters.

Epidemiology

A major problem in performing longitudinal studies in medicine is that people initially entered into a study are lost to follow-up for various reasons.

5.65 Suppose we wish to evaluate our data after 2 years and anticipate that the probability a patient will be available for study after 2 years is 90%. How many patients should be entered into the study to be 80% sure of having at least 100 patients left at the end of this period?

5.66 How many patients should be entered to be 90% sure of having at least 150 patients after 4 years if the probability of remaining in the study after 4 years is 80%?

Infectious Disease

The differential is a standard measurement made during a blood test. It consists of classifying white blood cells into the following 5 categories: (1) basophils, (2) eosinophils, (3) monocytes, (4) lymphocytes, and (5) neutrophils. The usual practice is to look at 100 randomly selected cells under a microscope and count the number of cells within each of the 5 categories. Assume that a normal adult will have the following proportions of the categories: basophils, 0.5%; eosinophils, 1.5%; monocytes, 4%; lymphocytes, 34%; and neutrophils, 60%.

5.67 An excess of eosinophils is sometimes consistent with a violent allergic reaction. What is the exact probability that a normal adult will have 5 or more eosinophils?

5.68 An excess of lymphocytes is consistent with various forms of viral infection, such as hepatitis. What is the probability that a normal adult will have 40 or more lymphocytes?

5.69 What is the probability that a normal adult will have 50 or more lymphocytes?

5.70 How many lymphocytes would have to appear in the differential before you would feel that the "normal" pattern was violated?

5.71 An excess of neutrophils is consistent with several types of bacterial infection. Suppose an adult has x neutrophils. How large would x have to be in order that the probability of a normal adult having x or more neutrophils was $\leqslant 5\%$?

5.72 How large would x have to be in order that the probability of a normal adult having x or more neutrophils was $\leqslant 1\%$?

Pulmonary Disease

5.73 The usual annual death rate from asthma in England over the period 1862–1962 for people aged 5–34 was approximately 1 per 100,000. Suppose that in 1963 twenty deaths were observed in a group of 1 million people in this age group living in Britain. Is this number of deaths inconsistent with the preceding 100-year rate? In particular, what is the probability of observing 20 or more deaths in 1 year in a group of 1 million people? (*Note*: This finding is both statistically and medically interesting, since it was found that the excess risk could be attributed to certain aerosols used by asthmatics in Britain during the period 1963–1967. The rate returned to normal during the period 1968–1972, when these types of aerosols were no longer used. For further information see Speizer *et al.* [3].)

Blood Chemistry

In pharmacologic research a variety of clinical chemistry measurements are routinely monitored closely for evidence of side effects of the medication under study. Suppose typical blood glucose levels are normally distributed with mean 90 mg/dL and standard deviation 38 mg/dL.

5.74 If the normal range is from 65–120 mg/dL, then what percentage of values will fall in the normal range?

5.75 In some studies only values that are at least 1.5 times as high as the upper limit of normal are identified as abnormal. What percentage of values would fall in this range?

5.76 Answer Problem 5.75 for 2.0 times the upper limit of normal.

5.77 Frequently, tests that yield abnormal results are repeated for confirmation. What is the probability that for a normal person a test will be at least 1.5 times as high as the upper limit of normal on two separate occasions?

5.78 Suppose that in a pharmacologic study involving 6000 patients, 75 patients have blood glucose levels at least 1.5 times the upper limit of normal on one occasion. What is the probability that this result could be due to chance?

Hypertension

Blood-pressure measurements are known to be variable, and repeated measurements are essential to accurately characterize a person's blood-pressure status. Suppose a person is measured on n visits with k measurements per visit and the average of all nk measurements ($\bar{x}$) is used to classify a person as to blood-pressure status. Specifically, if $\bar{x} \geq 95$ mm, then the person is classified hypertensive; if $\bar{x} < 90$ mm, then the person is classified normotensive; and if $\bar{x} \geq 90$ mm and < 95 mm, the person is classified borderline. It is also assumed that a person's "true" blood pressure is μ, representing an average over a large number of visits with a large number of measurements per visit, and that $\bar{x}$ is normally distributed with mean μ and variance $= 27.7/n + 7.9/(nk)$.

5.79 If a person's true diastalic blood pressure is 100 mm Hg, then what is the probability that the person will be classified accurately (as hypertensive) if a single measurement is taken at 1 visit?

5.80 Is the probability in Problem 5.79 a measure of sensitivity, specificity, or predictive value?

5.81 If a person's true blood pressure is 85 mm Hg, then what is the probability that the person will be accurately classified (as normotensive) if 3 measurements are taken at each of 2 visits?

5.82 Is the probability in Problem 5.81 a measure of sensitivity, specificity, or predictive value?

5.83 Suppose we decide to take 2 measurements per visit. How many visits are needed so that the reliability measures in Problems 5.79 and 5.81 would each be at least 95%?

Cancer

A treatment trial is proposed to test the efficacy of vitamin E as a preventive agent for cancer. One problem with such a study is how to assess compliance among study participants. A small pilot study is undertaken to establish criteria for compliance with the proposed study agents. In this regard, 10 patients are given 400 IU/day of vitamin E and 10 patients are given similar-sized tablets of placebo over a 3-month period. Their serum vitamin E levels are measured before and after the 3-month period and the change (3-month baseline) is shown in Table 5.3.

TABLE 5.3 Change in serum vitamin E(mg/dL) in pilot study

Group	Mean	sd	n
Vitamin E	0.80	0.48	10
Placebo	0.05	0.16	10

5.84 Suppose a change of 0.30 mg/dL in serum levels is proposed as a test criterion for compliance; that is, a patient who shows a change of ≥ 0.30 mg/dL is considered a compliant vitamin E taker. If normality is assumed, what percentage of the vitamin E group would be expected to show a change of at least 0.30 mg/dL?

5.85 Is the measure in Problem 5.84 a measure of sensitivity, specificity, or predictive value?

5.86 What percentage of the placebo group would be expected to show a change of not more than 0.30 mg/dL?

5.87 Is the measure in Problem 5.86 a measure of sensitivity, specificity, or predictive value?

5.88 Suppose a new threshold of change, Δ mg/dL, is proposed for establishing compliance. We wish to use a level of Δ such that the compliance measures in Problems 5.84 and 5.86 for the patients in the vitamin E and placebo groups are the same. What should Δ be? What would be the compliance in the vitamin E and placebo groups using this threshold level?

Hypertension

A study is planned to look at the effect of sodium restriction on lowering blood pressure. Nutritional counseling sessions are planned for the participants to encourage dietary sodium restriction. An important component in the study is validating the extent to which individuals comply with a sodium-restricted diet. This is usually accomplished by obtaining urine specimens and measuring sodium in the urine.

Assume that in free-living individuals (i.e., individuals with no sodium restriction) 24-hour urinary sodium excretion is normally distributed with mean 160.5 mEq/24 hrs and standard deviation $= 57.4$ mEq/24 hrs.

5.89 If 100 mEq/24 hr is the cutoff value chosen to measure compliance, then what percentage of noncompliant individuals, that is, individuals who do not restrict sodium, will have a 24-hour sodium level below this cutoff point?

Suppose that in an experimental study it is found that people who are on a sodium-restricted diet have 24-hour sodium excretion that is normally distributed with mean 57.5 mEq/24 hrs and standard deviation of 11.3 mEq/24 hrs.

5.90 What is the probability that a person who is on a sodium-restricted diet will have a 24-hour urinary sodium level above the cutoff point (i.e., 100 mEq/24 hrs)?

5.91 Suppose the investigators in the study wish to change the cutoff value (from 100 mEq/24 hrs) to another value such that the misclassification probabilities in Problems 5.89 and 5.90 are the same. What should the new cutoff value be, and what are the misclassification probabilities corresponding to your answers to Problems 5.89 and 5.90?

Mental Health
5.92 Refer to Tables 3.2 and 3.3 (pp. 60 and 61). Suppose a study of Alzheimer's disease is planned in more than one retirement community. How many retired people need to be studied to have a 90% chance of detecting at least 100 people with Alzheimer's disease, assuming that the age-sex-specific prevalence rates and age-sex distribution in Tables 3.2 and 3.3 hold?

5.93 Answer Problem 5.92 for 50 rather than 100 people.

Pulmonary Disease
Refer to the pulmonary function data in Data Set 10, FEV.DAT (see Problem 2.49, p. 39). We are interested in whether there is a relationship between smoking status and level of pulmonary function. However, FEV is affected by age, sex, and height; also, smoking children tend to be older than nonsmoking children. For these reasons, FEV should be standardized for age, sex, and height. To accomplish this, use the z-score approach outlined above in Problem 5.1, where the z-scores here are defined by age-sex-height groups.

5.94 Plot the distribution of z-scores for smokers and nonsmokers separately. Do these distributions look normal? Does there appear to be any relationship between smoking and pulmonary function in these data?

5.95 Repeat the analyses in Problem 5.94 for the subgroup of children 10+ years of age (since smoking is very rare prior to this age). Do you reach similar conclusions?

5.96 Repeat the analyses in Problem 5.95 separately for boys and girls. Are your conclusions the same in the two groups?
Note: Formal methods for comparing FEV's between smokers and nonsmokers are discussed in the material on statistical inference in Chapters 6 and 8.

Cardiovascular Disease
A clinical trial was conducted to test the efficacy of nifedipine, a new drug for stopping chest pain in patients with angina severe enough to require hospitalization. The duration of the study was 14 days in the hospital unless the patient was withdrawn prematurely from therapy, was discharged from the hospital, or died prior to this time. Patients were randomly assigned to either nifedipine or placebo and were given the same dosage of each drug in identical capsules at level 1 of therapy. If pain did not cease at this level of therapy, or if pain recurred after a period of pain cessation, then the patient progressed to level 2, whereby the dosage of each drug was increased according to a prespecified schedule. Similarly, if pain continued or recurred at level 2, then the patient progressed to level 3, whereby the dosage of the anginal drug was increased again.

TABLE 5.4 Format of NIFED.DAT

Column	Variable	Code
1–2	ID	
4	Treatment group	N = nifedipine/ P = placebo
6–8	Baseline heart rate*	beats/min
10–12	Level 1 heart rate†	beats/min
14–16	Level 2 heart rate	beats/min
18–20	Level 3 heart rate	beats/min
22–24	Baseline systolic bp	mm Hg
26–28	Level 1 systolic bp	mm Hg
30–32	Level 2 systolic bp	mm Hg
34–36	Level 3 systolic bp	mm Hg

* Immediately prior to randomization
† Highest heart rate and systolic bp at baseline and each level of therapy, respectively

Missing values indicate that either:
(a) the patient withdrew from the study prior to entering this level of therapy;
(b) the patient achieved pain relief prior to reaching this level of therapy; or,
(c) the patient encountered this level of therapy, but this particular piece of data was missing.

Patients randomized originally to either group were allowed to receive nitrates in any amount that was deemed clinically appropriate to help control pain.

The main objective of the study was to compare the degree of pain relief with nifedipine and placebo. A secondary objective was to better understand the effects of these agents on other physiologic parameters including heart rate and blood pressure. Data on these latter parameters are given NIFED.DAT; the format of this file is given in Table 5.4.

5.97 Describe the effect of each treatment regimen on changes in heart rate and blood pressure. Do the distribution of changes in these parameters look normal or not?

5.98 Compare graphically the effects of the treatment regimens on heart rate and blood pressure. Do you notice any difference between treatments?

(*Note*: Formal tests for comparing changes in heart rate and blood pressure in the two treatment groups are covered in Chapters 6 and 8.)

References

[1] Berson, L. L., Sandberg, M. A., Rosner, B., Birch, D. G., & Hanson, A. H. (1985). Natural cause of retinitis pigmentosa over a three-year interval. *American Journal of Ophthalmology, 94*, 240–251.

[2] Boston Globe, April 25, 1980.

[3] Speizer, F. E., Doll, R., Heaf, P., & Strang, L. B. (1968, February 8). Investigation into use of drugs preceding death from asthma. *British Medical Journal, 1*, 339–343.

CHAPTER SIX

ESTIMATION

Introduction

In Chapters 3 through 5 the properties of different probability models were explored. In doing this, it was always assumed that the specific probability distributions were known.

EXAMPLE 6.1 **Infectious Disease** We assumed that the number of neutrophils in a sample of 100 white blood cells was binomially distributed with parameter $p = .6$. ∎

EXAMPLE 6.2 **Bacteriology** We assumed that the number of bacterial colonies on a 100-cm² agar plate was Poisson distributed with parameter $\mu = 0.02$. ∎

EXAMPLE 6.3 **Hypertension** We assumed that the distribution of diastolic blood-pressure measurements in 35–44-year-old men was normal with mean $\mu = 80$ mm Hg and $\sigma = 12$ mm Hg. ∎

In general, we have been assuming that the properties of the underlying distributions from which our data are drawn are known and that the only question that remains is what can be predicted about the behavior of the data given a knowledge of these properties.

EXAMPLE 6.4 **Hypertension** Using the data in Example 6.3, we could predict that about 95% of all diastolic blood pressures from 35–44-year-old men should fall between 56 mm Hg and 104 mm Hg. ∎

The problem addressed in the remainder of this text, and the more basic statistical problem, is that we have a data set and we want to **infer** the properties of the underlying distribution from this data set. This inference usually involves **inductive** rather than **deductive** reasoning; that is, in principle a variety of different probability models must at least be tried to see which model best "fits" the data.

Statistical inference can be further subdivided into the two main areas of estimation and hypothesis testing. **Estimation** is concerned with predicting the values of specific population parameters; **hypothesis testing** is concerned with testing whether the value of a population parameter is equal to some specific value. Problems of estimation are covered in this chapter, while problems of hypothesis testing are discussed in Chapters 7 through 10.

Some typical problems that involve estimation follow.

EXAMPLE 6.5 **Hypertension** Suppose we measure the systolic blood pressures of a group of Polynesian villagers and we believe the underlying distribution is normal. How can the parameters of this distribution (μ, σ^2) be estimated if no previous data are available on these people? ∎

EXAMPLE 6.6   **Pulmonary Disease** Suppose we look at people living within a low-income census tract in some urban area and we wish to estimate the prevalence of tuberculosis (TB) in the community. We assume that the number of cases among n people sampled will be binomially distributed with some parameter p. How is the parameter p estimated?   ∎

In Examples 6.5 and 6.6, we were interested in obtaining specific numbers as estimates of our parameters. These numbers are often referred to as **point estimates**. Sometimes we want to specify a range within which the parameter values are likely to fall. If this range is narrow, then we may feel that our point estimate is a good one. This type of problem involves **interval estimation**.

EXAMPLE 6.7   **Ophthalmology** A study is proposed to screen a group of 1000 people ages 65 or older to identify those with "low vision," that is, a visual acuity of 20–50 or worse in both eyes, even with the aid of glasses. Suppose we assume that the number of such people ascertained in this manner is binomially distributed with parameters $n = 1000$ and unknown p. We would like to obtain a point estimate of p and to provide an interval about this point estimate to see how accurate our point estimate is. For example, we would feel better about a point estimate of 5% if this interval were .04–.06 than if it were .01–.10.   ∎

<u>SECTION 6.2</u>   **The Relationship Between Population and Sample**

EXAMPLE 6.8   **Obstetrics** Suppose we wish to characterize the distribution of birthweights of all liveborn infants that were born in the United States in 1988. Assume that the underlying distribution of these measurements has an expected value (or mean) μ and variance σ^2. We wish to estimate μ and σ^2 exactly. But this task is impossible with such a large group. Instead, we decide to select a random sample of n infants that are *representative of* this large group and use the birthweights $x_1, \ldots, x_n$ from this sample to help us estimate μ and σ^2. What is a random sample?   ∎

DEFINITION 6.1   ■■
A **random sample** is a selection of some of the members of a population such that each member is independently chosen and has a known non-zero probability of being selected.   ∎

DEFINITION 6.2   ■■
A **simple random sample** is a random sample in which each group member has the same probability of being selected.   ∎

DEFINITION 6.3   ■■
The **reference**, **target**, or **study**, population is the group that we wish to study. The random sample is selected from the study population.   ∎

For ease of discussion, the abbreviated term "random sample" will be used to denote a simple random sample.

EXAMPLE 6.9   **Epidemiology** The Nurses Health Study is a large epidemiologic study involving over 100,000 female nurses residing in 11 large states in the United States. The nurses were first contacted by mail in 1976 and have been followed every 2 years by mail since then. Suppose we want to select a test sample of 100 nurses to test a new procedure for obtaining serum samples by mail. One way of selecting the sample is to assign each nurse an ID number and then select the nurses with the lowest 100 ID numbers. This is definitely *not* a random sample since each nurse is not equally likely to be chosen. Indeed, since the first two digits of the ID number are assigned according to state, the 100 nurses with the lowest ID numbers would all come from the same state. An alternative method of selecting the sample is to have a computer generate a set of 100 **random numbers** (from among the numbers 1 to over 100,000), one to be assigned to each nurse. By doing this, each member is equally likely to be included in the sample. This would be a truly random sample. (More details on random numbers are given in Section 6.3.)   ∎

In practice, there is rarely an opportunity to enumerate each member of the reference population so as to select a random sample, and the assumption that the sample selected has all the properties of a random sample without formally being a random sample must be made.

In Example 6.8 the reference population is finite and well defined and can be enumerated. In many instances, the reference population is effectively infinite and is not well defined.

EXAMPLE 6.10   **Cancer** Suppose we wish to estimate the 5-year survival rate of women who are initially diagnosed as having breast cancer at the ages of 45–54 and who undergo radical mastectomy at this time. Our reference population is all women who have ever had a first diagnosis of breast cancer in the past when they were 45–54 years old or who ever will have such a diagnosis in the future when they are 45–54 years old and who receive radical mastectomies.   ∎

This population is effectively infinite. The population cannot be formally enumerated and thus a truly random sample cannot be selected from it. However, we will again assume that the sample we have selected behaves as if it were a random sample.

In this text we will assume that all reference populations discussed are **effectively infinite**, although, as in Examples 6.8 and 6.9, many of them are actually very large but finite. Sampling theory is the special branch of statistics that treats statistical inference for finite populations; it is beyond the scope of this text. See reference [1] for a good treatment of this subject.

SECTION 6.3   **Random Number Tables**

In this section practical methods for selecting random samples are discussed.

EXAMPLE 6.11   **Hypertension** Suppose we wish to study how effective a hypertension treatment program is in controlling the blood pressure of its participants. We are given a roster of all 1000 participants in the program but, due to limited resources, only 20 people can be surveyed. We would like the 20 people chosen to be a random sample from the population of all participants in the program. How should this random sample be selected?   ∎

A table of random numbers would probably be used to select this sample.

DEFINITION 6.4   ■■■
A **random number** (or **random digit**) is a random variable X that takes on the values 0, 1, 2,..., 9 with equal probability. Thus,

$$Pr(X = 0) = Pr(X = 1) = \cdots = Pr(X = 9) = \tfrac{1}{10}$$ ∎

DEFINITION 6.5 ■■

A **random number table** is a collection of digits that satisfies the following two properties:

(1) Each digit 0, 1, 2, ..., 9 is equally likely to occur.

(2) The value of any particular digit is independent of the value of any other digit in the table.
■

Table 4 in Appendix 1 lists 1000 random digits.

EXAMPLE 6.12 Suppose that a 5 appears as a digit in a random number table. Does this mean that 5's are more likely to occur in the next few digits in the table?

SOLUTION No. Each digit either after or before the 5 is still equally likely to be any of the digits 0, 1, 2, ..., 9.
■

Computer programs generate large quantities of random digits that approximately satisfy the conditions in Definition 6.5. Thus, the numbers in random number tables are sometimes referred to as **pseudorandom numbers**, since they are simulated to satisfy the properties in Definition 6.5.

EXAMPLE 6.13 **Hypertension** How can the random digits in Table 4 be used to select 20 random participants in the hypertension treatment program in Example 6.11?

SOLUTION A roster of the 1000 participants must be compiled and each participant must then be assigned a number from 000 to 999. Perhaps an alphabetical list of the participants already exists, which would make this task easy. Twenty groups of three digits would then be selected, starting at any position in the random number table. For example, if we start at the first row of Table 4, we have the numbers listed in Table 6.1.

TABLE 6.1
20 random participants chosen from 1000 participants in the hypertension treatment program

First 3 rows of random number table				Actual random numbers chosen				
32924	22324	18125	09077	329	242	232	418	125
54632	90374	94143	49295	090	775	463	290	374
88720	43035	97081	83373	941	434	929	588	720
				430	359	708	183	373

Therefore, our random sample would consist of the persons numbered 329, 242, ..., 373 in the alphabetical list. In this particular case there were no repeats in the 20 three-digit numbers selected. If there had been repeats, then more three-digit numbers would have been selected until 20 different numbers were selected. This process is referred to as **random selection**.
■

EXAMPLE 6.14 **Diabetes** Suppose we wish to conduct a clinical trial of a new treatment for diabetes and compare the new treatment with standard insulin therapy. A small study of this type will be conducted on 10 patients: 5 patients will be randomly assigned to the new therapy and 5 to insulin therapy. How can the table of random numbers be used to make the assignments?

SOLUTION The prospective patients are numbered from 0 to 9 and five unique random digits are selected from some arbitrary position in the random number table (e.g., from the 28th row).

TABLE 6.2 Sample of birthweights (oz) obtained from 1000 consecutive deliveries at Boston City Hospital

ID numbers	0	1	2	3	4	5	6	7	8	9	10	11	12	13	14	15	16	17	18	19
000–019	116	124	119	100	127	103	140	82	107	132	100	92	76	129	138	128	115	133	70	121
020–039	114	114	121	107	120	123	83	96	116	110	71	86	136	118	120	110	107	157	89	71
040–059	98	105	106	52	123	101	111	130	129	94	124	127	128	112	83	95	118	115	86	120
060–079	106	115	100	107	131	114	121	110	115	93	116	76	138	126	143	93	121	135	81	135
080–099	108	152	127	118	110	115	109	133	116	129	118	126	137	110	32	139	132	110	140	119
100–119	109	108	103	88	87	144	105	138	115	104	129	108	92	100	145	93	115	85	124	123
120–139	141	96	146	115	124	113	98	110	153	165	140	132	79	101	127	137	129	144	126	155
140–159	120	128	119	108	113	93	144	124	89	126	87	120	99	60	115	86	143	97	106	148
160–179	113	135	117	129	120	117	92	118	80	132	121	119	57	126	126	77	135	130	102	107
180–199	115	135	112	121	89	135	127	115	133	64	91	126	78	85	106	94	122	111	109	89
200–219	99	118	104	102	94	113	124	118	104	124	133	80	117	112	112	112	102	118	107	104
220–239	90	113	132	122	89	111	118	108	148	103	112	128	86	111	140	126	143	120	124	110
240–259	142	92	132	128	97	132	99	131	120	106	115	101	130	120	130	89	107	152	90	116
260–279	106	111	120	198	123	152	135	83	107	55	131	108	100	104	112	121	102	114	102	101
280–299	118	114	112	133	139	113	77	109	142	144	114	117	97	96	93	120	149	107	107	117
300–319	93	103	121	118	110	89	127	100	156	106	122	105	92	128	124	125	118	113	110	149
320–339	98	98	141	131	92	141	110	134	90	88	111	137	67	95	102	75	108	118	99	79
340–359	110	124	122	104	133	98	108	125	106	128	132	95	114	67	134	136	138	122	103	113
360–379	142	121	125	111	97	127	117	122	120	80	114	126	103	98	108	100	106	98	116	109
380–399	98	97	129	114	102	128	107	119	84	117	119	128	121	113	128	111	112	120	122	91
400–419	117	100	108	101	144	104	110	146	117	107	126	120	104	129	147	111	106	138	97	90
420–439	120	117	94	116	119	108	109	106	134	121	125	105	177	109	109	109	79	118	92	103
440–459	110	95	111	144	130	83	93	81	116	115	131	135	116	97	108	103	134	140	72	112
460–479	101	111	129	128	108	90	113	99	103	41	129	104	144	124	70	106	118	99	85	93
480–499	100	105	104	113	106	88	102	125	132	123	160	100	128	131	49	102	110	106	96	116
500–519	128	102	124	110	129	102	101	119	101	119	141	112	100	105	155	124	67	94	134	123
520–539	92	56	17	135	141	105	133	118	117	112	87	92	104	104	132	121	118	126	114	90
540–559	109	78	117	165	127	122	108	109	119	98	120	101	96	76	143	83	100	128	124	137
560–579	90	129	89	125	131	118	72	121	91	113	91	137	110	137	111	135	105	88	112	104
580–599	102	122	144	114	120	136	144	98	108	130	119	97	142	115	129	125	109	103	114	106
600–619	109	119	89	98	104	115	99	138	122	91	161	96	138	140	32	132	108	92	118	58
620–639	158	127	121	75	112	121	140	80	125	73	115	120	85	104	95	106	100	87	99	113
640–659	95	146	126	58	64	137	69	90	104	124	120	62	83	96	126	155	133	115	97	105
660–679	117	78	105	99	123	86	126	121	109	97	131	133	121	125	120	97	101	92	111	119
680–699	117	80	145	128	140	97	126	109	113	125	157	97	119	103	102	128	116	96	109	112
700–719	67	121	116	126	106	116	77	119	119	122	109	117	127	114	102	75	88	117	99	136
720–739	127	136	103	97	130	129	128	119	22	109	145	129	96	128	122	115	102	127	109	120
740–759	111	114	115	112	146	100	106	137	48	110	97	103	104	107	123	87	140	89	112	123
760–779	130	123	125	124	135	119	78	125	103	55	69	83	106	130	98	81	92	110	112	104
780–799	118	107	117	123	138	130	100	78	146	137	114	61	132	109	133	132	120	116	133	133
800–819	86	116	101	124	126	94	93	132	126	107	98	102	135	59	137	120	119	106	125	122
820–839	101	119	97	86	105	140	89	139	74	131	118	91	98	121	102	115	115	135	100	90
840–859	110	113	136	140	129	117	117	129	143	88	105	110	123	87	97	99	128	128	110	132
860–879	78	128	126	93	148	121	95	121	127	80	109	105	136	141	103	95	140	115	118	117
880–899	114	109	144	119	127	116	103	144	117	131	74	109	117	100	103	123	93	107	113	144
900–919	99	170	97	135	115	89	120	106	141	137	107	132	132	58	113	102	120	98	104	108
920–939	85	115	108	89	88	126	122	107	68	121	113	116	94	85	93	132	146	98	132	104
940–959	102	116	108	107	121	132	105	114	107	121	101	110	137	122	102	125	104	124	121	111
960–979	101	93	93	88	72	142	118	157	121	58	92	114	104	119	91	52	110	116	100	147
980–999	114	99	123	97	79	81	146	92	126	122	72	153	97	89	100	104	124	83	81	129

The first five unique digits are 6, 9, 4, 3, 7. Thus, the patients numbered 3, 4, 6, 7, 9 will be assigned to the new therapy and the remaining patients (numbered 0, 1, 2, 5, 8) to standard insulin therapy. In some studies the prospective patients are not known in advance and are recruited over time. In this case, if 00 is identified with the first patient recruited, 01 with the second patient recruited, ..., and 09 with the tenth patient recruited, then the new therapy would be assigned to the fourth (3 + 1), fifth (4 + 1), seventh (6 + 1), eighth (7 + 1), and tenth (9 + 1) patients recruited and the standard therapy to the first (0 + 1), second (1 + 1), third (2 + 1), sixth (5 + 1), and ninth (8 + 1). ∎

This process is referred to as **random assignment**. It is different from random selection (Example 6.13) in that, typically, the number, in this case, of patients to be assigned to each type of treatment (5), is fixed in advance. The random number table helps select the 5 patients who are to receive one of the two treatments (new therapy). By default, the patients not selected for new therapy are assigned to the alternative treatment (standard insulin therapy). No additional random numbers need to be chosen for the second group of 5 patients. If random selection were used instead, then one approach might be to draw a random digit for each patient. If the random digit is from 0 to 4, then the patient is assigned to new therapy; if from 5 to 9, then the patient is assigned to standard therapy. One problem with this approach is that in a finite sample equal numbers of patients will not necessarily be assigned to each therapy, which is usually the most efficient design. Indeed, referring to the first 10 digits in the 28th row of the random number table (69644   37198), we see that 4 patients would be assigned to new therapy (patients 4, 5, 6, and 8) and 6 patients would be assigned to standard therapy (patients 1, 2, 3, 7, 9, 10) if the method of random selection were used. Random assignment is preferable in this instance, since it ensures an equal number of patients assigned to each treatment group.

EXAMPLE 6.15   **Obstetrics** The birthweights from 1000 consecutive deliveries at Boston City Hospital (serving a low-income population) are enumerated in Table 6.2. For the purpose of this example, consider this population as effectively infinite. Suppose we wish to draw 5 random samples of size 10 from this population using the random numbers in Table 4. How can these samples be selected?

SOLUTION   Start anywhere in the table. Say we arbitrarily choose to start in the 17th row and read groups of three digits from left to right. The random numbers will thus range from 000 to 999. Suppose the three-digit random number selected is y. The appropriate row in Table 6.2

TABLE 6.3
17th and 18th rows of the random number table (Table 4 in the Appendix)

```
        41871     17566     61200     15994
        ⌞⌟⌞___⌟⌞___⌟⌞___⌟   ⌞⌟⌞
        25758     04625     43226     32986
        ⌞⌞⌞___⌟⌞___⌟⌞   ⌞⌟⌞___⌟⌞___⌟⌞
```

1st 3-digit no. = 418	Group = 400–419, birthweight = 97 oz	Column = 418 − 400 = 18
2nd 3-digit no. = 711	Group = 700–719, birthweight = 117 oz	Column = 711 − 700 = 11
3rd 3-digit no. = 756	Group = 740–759, birthweight = 140 oz	Column = 756 − 740 = 16

⋮

is then found by finding the group of ID numbers within which y falls and the appropriate column by subtracting the lower end of the group from y. For example, refer to the 17th and 18th rows of the random number table, which are reproduced in Table 6.3.

The random number selection process continues until 5 sets of 10 numbers are obtained, as shown in Table 6.4.   ∎

TABLE 6.4

5 random samples of size 10 from the population of infants whose birthweights (oz) appear in Table 6.2

Individual	Sample				
	1	2	3	4	5
1	97	177	97	101	137
2	117	198	125	114	118
3	140	107	62	79	78
4	78	99	120	120	129
5	99	104	132	115	87
6	148	121	135	117	110
7	108	148	118	106	106
8	135	133	137	86	116
9	126	126	126	110	140
10	121	115	118	119	98
$\bar{x}$	116.90	132.80	117.00	106.70	111.90
s	21.70	32.62	22.44	14.13	20.46

SECTION 6.4 **Estimation of the Mean of a Distribution**

Now that we understand the meaning of a random sample from a population and have explored some practical methods for selecting such samples using a random number table, we will move on to estimation. The question remains, How is a specific random sample $x_1, \ldots, x_n$ used to estimate μ and σ^2, the mean and variance of the underlying distribution. Estimating the mean is the focus of this section, and estimating the variance is covered in Section 6.5.

6.4.1 **Point Estimation**

A natural estimator to use for estimating the population mean μ is the sample mean

$$\bar{x} = \sum_{i=1}^{n} \frac{x_i}{n}$$

What are the properties of $\bar{x}$ that make it a desirable estimator of μ? Forget about our particular sample for the moment and consider the set of all possible samples of size n that could have been selected from the population. The values of $\bar{x}$ in each of these samples will, in general, be different. These values will be denoted by $\bar{x}_1, \bar{x}_2$, and so forth. The key conceptual point in this instance is to forget about our sample as a unique entity and to consider it instead as representative of all possible samples of size n that could have been drawn from the population. Stated another way, the specific value obtained for $\bar{x}$ from our sample is forgotten. Instead, $\bar{x}$ is regarded as a single realization of a random variable over all possible samples of size n that could have been selected from the population.

DEFINITION 6.6 ■■

The **sampling distribution** of $\bar{x}$ is the distribution of values of $\bar{x}$ over all possible samples of size n that could have been selected from the reference population.   ■

In Figure 6.1 an example of such a sampling distribution is provided. This example consists of a frequency distribution of the sample mean from 200 randomly selected samples of size 10 drawn from the distribution of 1000 birthweights given in Table 6.2, as generated by the Statistical Analysis System (SAS) procedure PROC CHART.

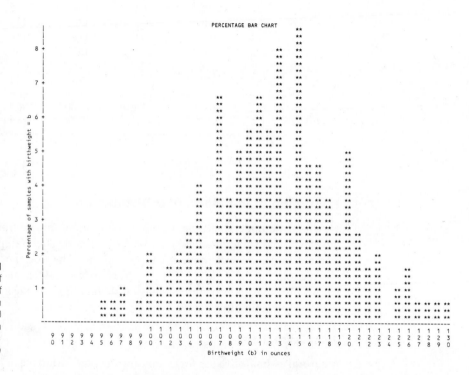

FIGURE 6.1
Sampling distribution of $\bar{x}$ over 200 samples of size 10 selected from the population of 1000 birthweights given in Table 6.2 (100 = 100.0–100.9, etc.)

We can show that the average of these sample means ($\bar{x}$'s) when taken over a large number of random samples of size n will approximate μ as the number of samples selected becomes large. In other words, the expected value of $\bar{x}$ over its sampling distribution is equal to μ. This result is summarized as follows:

6.1   Let $x_1, \ldots, x_n$ be a random sample drawn from some population with mean μ. Then for the sample mean $\bar{x}$, $E(\bar{x}) = \mu$.

Note that **(6.1)** holds for any population regardless of its underlying distribution. In other words, $\bar{x}$ is an unbiased estimator of μ.

DEFINITION 6.7 ■■

An **estimator** e of a parameter μ is unbiased if $E(e) = \mu$. This means that the average value of e over a large number of repeated samples of size n will be μ. ■

The unbiasedness of $\bar{x}$ is not a sufficient reason to use it as an estimator of μ. Many unbiased estimators of μ exist, including the sample median and the average value of the largest and smallest data points in a sample. Why is $\bar{x}$ chosen rather than any of the other unbiased estimators? The reason is that if the underlying distribution of the population is normal, then it can be shown that the unbiased estimator with the smallest variance is given by $\bar{x}$. Thus, $\bar{x}$ is referred to as the **minimum variance unbiased estimator** of μ.

This concept is illustrated in Figure 6.2(a) through (c), where for 200 random samples of size 10 drawn from the population of 1000 birthweights in Table 6.2, the sampling distribution of the sample mean ($\bar{x}$) is plotted in Figure 6.2(a), the sample median in Figure 6.2(b), and the average of the smallest and largest observations in the sample in Figure 6.2(c). Note that the variability of the distribution of sample means is slightly smaller than that of the sample median and considerably smaller than that of the average of the smallest and largest observations.

6.4.2 **Standard Error of the Mean**

From **(6.1)** we see that $\bar{x}$ will be an unbiased estimator of μ for any sample size n. Why then is it preferable to estimate parameters from large samples rather than from small ones? The intuitive reason is that the larger the sample size, the more accurate an estimator $\bar{x}$ will be.

EXAMPLE 6.16 **Obstetrics** Consider Table 6.4 (p. 147). Notice that the 50 individual birthweights range from 62 to 198 oz and have a sample standard deviation of 23.79 oz. The 5 sample means range from 106.7 to 132.8 oz and have a sample standard deviation of 9.77 oz. Thus, the sample means based on 10 observations are less variable from sample to sample than are the individual observations, which can be considered as sample means from samples of size 1. ■

Indeed, we would expect that the sample means from repeated samples of size 100 would be less variable than those from samples of size 10. We can show that this is true. By definition,

$$Var(\bar{x}) = \left(\frac{1}{n^2}\right) Var\left(\sum_{i=1}^{n} x_i\right)$$

Now, if a random sample $x_1, \ldots, x_n$ is drawn from the same population, then

$$Var\left(\sum_{i=1}^{n} x_i\right) = \sum_{i=1}^{n} Var(x_i)$$

Thus,
$$Var(\bar{x}) = \left(\frac{1}{n^2}\right) \sum_{i=1}^{n} Var(x_i)$$

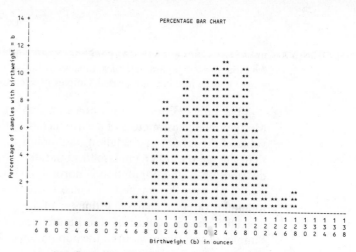

(a) Sampling distribution of the sample mean ($\bar{x}$)

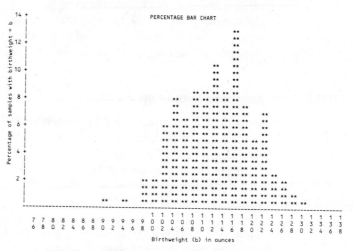

(b) Sampling distribution of the sample median

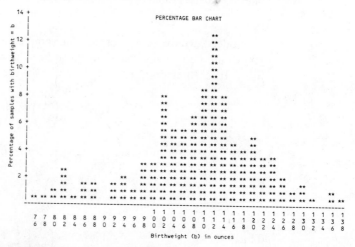

(c) Sampling distribution of the average of the smallest and largest observations

However, by definition $Var(x_i) = \sigma^2$. Therefore,

$$Var(\bar{x}) = (1/n^2)(\sigma^2 + \sigma^2 + \cdots + \sigma^2) = (1/n^2)(n\sigma^2) = \sigma^2/n$$

The standard deviation (sd) $= \sqrt{\text{variance}}$, $\text{sd}(\bar{x}) = \sigma/\sqrt{n}$. We have the following summary:

6.2 | Let $x_1, \ldots, x_n$ be a random sample from a population with underlying mean μ and variance σ^2. The set of sample means in repeated random samples of size n from this population has variance σ^2/n. The standard deviation of this set of sample means is thus $\sigma/\sqrt{n}$ and is referred to as the **standard error of the mean** (sem) or the **standard error**.

In practice, the population variance σ^2 is rarely known. We will see later in Section 6.5 that a reasonable estimator for the population variance σ^2 is the sample variance s^2, which leads to the following definition:

DEFINITION 6.8 ■■■

The **standard error of the mean (sem)**, or the **standard error**, is given by $\sigma/\sqrt{n}$ and is estimated by $s/\sqrt{n}$. It represents the estimated standard deviation obtained from a set of sample means from repeated samples of size n from a population with underlying variance σ^2. ■

Note that the standard error is *not* the standard deviation of an individual observation x_i but rather of the sample mean $\bar{x}$. The standard error of the mean is illustrated in Figure 6.3(a) through (c). In Figure 6.3(a) the frequency distribution of the sample mean is plotted for 200 samples of size 1 drawn from the collection of birthweights in Table 6.2. Similar frequency distributions are plotted for 200 sample means from samples of size 10 in Figure 6.3(b) and from samples of size 30 in Figure 6.3(c). Notice that the spread of the frequency distribution in Figure 6.3(a), corresponding to $n = 1$, is much larger than the spread of the frequency distribution in Figure 6.3(b), corresponding to $n = 10$. Furthermore, the spread of the frequency distribution in Figure 6.3(b), corresponding to $n = 10$, is much larger than the spread of the frequency distribution in Figure 6.3(c), corresponding to $n = 30$.

EXAMPLE 6.17   **Obstetrics** Compute the standard error of the mean for the third sample of birthweights in Table 6.4 (p. 147).

SOLUTION   The standard error of the mean is given by

$$s/\sqrt{n} = 22.44/\sqrt{10} = 7.10 \qquad ■$$

The standard error is a quantitative measure of the variability of sample means obtained from repeated random samples of size n drawn from the same population. Notice that the standard error is directly proportional to both $1/\sqrt{n}$ and to the population standard deviation of an individual observation σ. It justifies the concern with sample size in assessing the accuracy of our estimate $\bar{x}$ of the

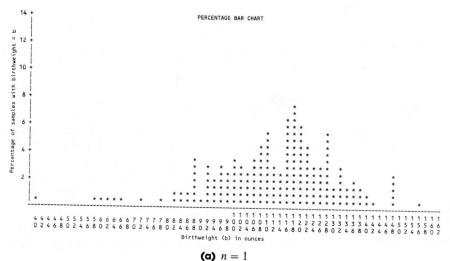

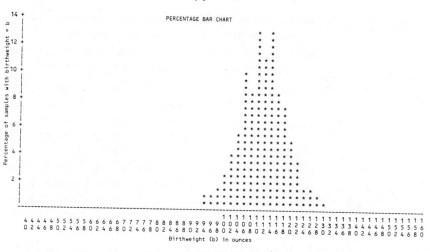

FIGURE 6.3(a)(b)
Illustration of the
standard error of the
mean (100 = 100.0–
101.9, etc.)

unknown population mean μ. The reason it is preferable to estimate μ from a sample size of 400 rather than from one of size 100 is that the standard error from the first sample will be $\frac{1}{2}$ as large as in the second sample. Thus, the larger sample should provide a more accurate estimate of μ. Notice that the accuracy of our estimate is also affected by the underlying variance σ^2, of individual observations from the population, a quantity which is unrelated to the sample size n.

EXAMPLE 6.18   **Gynecology** Suppose a woman wishes to estimate her exact day of ovulation for contraceptive purposes. A theory exists that at the time of ovulation the body temperature rises by an amount from 0.5°F to 1.0°F. Thus, changes in body temperature can be used to guess the day of ovulation. To use this method, a good estimate of basal body temperature during a period when ovulation is definitely not occurring is needed. Suppose that for this purpose a woman

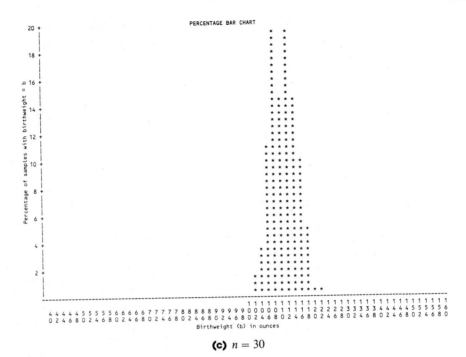

FIGURE 6.3(c)
Illustration of the
standard error of the
mean (100 = 100.0–
101.9, etc.)

(c) $n = 30$

measures her body temperature on awakening on the first 10 days after menstruation and obtains the following data: 97.2°, 96.8°, 97.4°, 97.4°, 97.3°, 97.0°, 97.1°, 97.3°, 97.2°, 97.3°. What is the best estimate of her underlying basal body temperature (μ)? How accurate is this estimate?

SOLUTION   Her best estimate of underlying body temperature during the nonovulation period (μ) is given by

$$\bar{x} = (97.2 + 96.8 + \cdots + 97.3)/10 = 97.20°$$

The standard error of this estimate is given by

$$s/\sqrt{10} = 0.189/\sqrt{10} = 0.06°$$

In our work on confidence intervals in Section 6.4.6 we will show that for many underlying distributions of temperature, we can be fairly certain that the true mean basal temperature μ is within two standard errors of $\bar{x}$, or within $97.20° \pm 2(0.06)° \approx (97.1°–97.3°)$. Thus, if the temperature is elevated by at least 0.5° above this range on a given day, then it might indicate that the woman was ovulating, and for contraceptive purposes, intercourse should not be attempted on that day. ∎

6.4.3 **Central Limit Theorem**

If the underlying distribution is normal, then it can be shown that the sample mean will itself be normally distributed with mean μ and variance σ^2/n (see Section 5.8). In other words, $\bar{x} \sim N(\mu, \sigma^2/n)$. If the underlying distribution is *not* normal, we would still like to make some statement about the sampling distribution of the sample mean. This statement is given by the following theorem:

6.3   **Central Limit Theorem**

Let $x_1, \ldots, x_n$ be a random sample from some population with mean μ and variance σ^2. Then for large n, $\bar{x} \sim N(\mu, \sigma^2/n)$ even if the underlying distribution of individual observations in the population is not normal. (The symbol $\sim$ is used to represent "approximately distributed.")

This theorem is very important because many of the distributions encountered in practice are not normal. In such cases the central limit theorem can often be applied; this will allow us to perform statistical inference based on the approximate normality of the sample mean, despite the nonnormality of the distribution of individual observations.

EXAMPLE 6.19   **Obstetrics**   The central limit theorem is illustrated by plotting, in Figure 6.4(a), the sampling distribution of mean birthweights obtained by drawing 200 random samples of size 1 from the collection of birthweights in Table 6.2. Similar sampling distributions of sample means are plotted from samples of size 5, in Figure 6.4(b), and samples of size 10, in Figure 6.4(c). Notice that the distribution of individual birthweights (i.e., sample means from samples of size 1) is slightly skewed to the left. However, the distribution of sample means becomes increasingly closer to bell shaped as the sample size increases to 5, in Figure 6.4(b), and 10, in Figure 6.4(c). ∎

EXAMPLE 6.20   **Cardiovascular Disease**   Serum cholesterol is an important risk factor for cardiovascular disease. Its distribution tends to be positively skewed or skewed to the right, with a few people with very high values, as is shown in Figure 6.5. However, hypothesis tests can be performed based on mean serum cholesterol over moderate samples of people, since from the central limit theorem the distribution of means will be approximately normal, even if the underlying distribution of individual measurements is not. ∎

EXAMPLE 6.21   **Obstetrics**   Compute the probability that the mean birthweight from a sample of 10 infants drawn from the Boston City Hospital population in Table 6.2 will fall between 98.0 and 126.0 oz (i.e., $98 \leqslant \bar{x} < 126$) if the mean birthweight for the 1000 birthweights from the Boston City Hospital population is 112.0 oz with a standard deviation of 20.6 oz.

SOLUTION   The central limit theorem is applied and it is assumed that $\bar{x}$ follows a normal distribution with mean $\mu = 112.0$ oz and standard deviation $\sigma/\sqrt{n} = 20.6/\sqrt{10} = 6.51$ oz. It follows that

$$Pr(98.0 \leqslant \bar{x} < 126.0) = \Phi\left(\frac{126.0 - 112.0}{6.51}\right) - \Phi\left(\frac{98.0 - 112.0}{6.51}\right)$$

$$= \Phi(2.15) - \Phi(-2.15)$$

$$= \Phi(2.15) - [1 - \Phi(2.15)] = 2\Phi(2.15) - 1$$

Refer to Table 3 in the Appendix and obtain

$$Pr(98.0 \leqslant \bar{x} < 126.0) = 2(.9842) - 1.0 = .968$$

Thus, 96.8% of the samples of size 10 would be expected to have mean birthweights between 98 and 126 oz if the central limit theorem holds. This value can be checked by

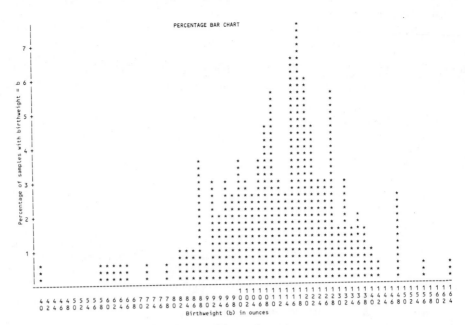

(a) $n = 1$

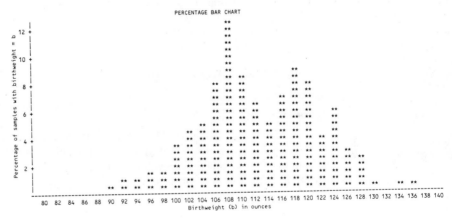

(b) $n = 5$

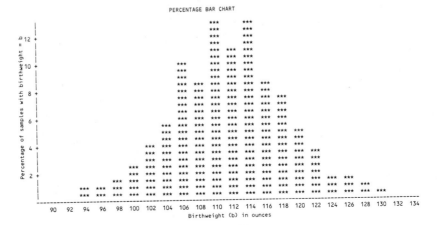

FIGURE 6.4(a)(b)(c)
Illustration of the central limit theorem (100 = 100.0–101.9, etc.)

(c) $n = 10$

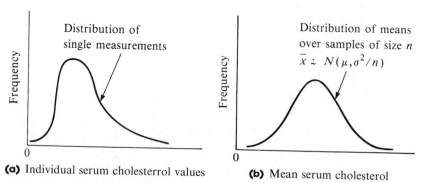

FIGURE 6.5
Distribution of single
serum cholesterol
measurements and of
means of such
measurements over
samples of size n

(a) Individual serum cholesterrol values

(b) Mean serum cholesterol

referring to Figure 6.2(a) (p. 150). Note that within a specific column 4 rows of *'s correspond to 2% of the distribution. Thus, for each column a row of *'s corresponds to 0.5% of the distribution. Furthermore, the 90 column corresponds to the birthweight interval 90.0–91.9, the 92 column to 92.0–93.9, and so forth. Note that one column of *'s is in the 90 column, one in the 94 column, two in the 96 column, three in the 128 column, and two in the 126 column. Thus, 4 rows of *'s (4 × 0.5% = 2% of the distribution) are less than 98.0 oz, and 5 rows of *'s (5 × 0.5% = 2.5% of the distribution) are greater than or equal to 126.0 oz. It follows that 100% − 4.5% = 95.5% of the distribution is actually between 98 and 126 oz. This value corresponds well to the 96.8% predicted by the central limit theorem, showing that the central limit theorem holds well for averages from samples of size 10 drawn from this population. ∎

6.4.4 Interval Estimation—Known Variance

We have been discussing the rationale for using $\bar{x}$ to estimate the mean of a distribution and have given a measure of variability of this estimate, namely, the standard error. These statements hold for any underlying distribution. However, we frequently wish to obtain an interval of plausible estimates of the mean as well as a best estimate of its precise value. Our interval estimates will hold exactly only if the underlying distribution is normal and only approximately if the underlying distribution is not normal, as stated in the central limit theorem.

EXAMPLE 6.22 **Obstetrics** Suppose the first sample of 10 birthweights given in Table 6.4 (p. 147) has been drawn. Our best estimate of the population mean μ would be the sample mean $\bar{x} = 116.9$ oz. Although 116.9 oz is our best estimate of μ, we still are not certain that μ is 116.9 oz. Indeed, if the second sample of 10 birthweights had been drawn, a point estimate of 132.8 oz would have been used. Our point estimate would certainly have a different meaning if we were quite certain in some sense that μ was within 1 oz of 116.9 rather than within 5 oz. ∎

We have assumed previously that the distribution of birthweights in Table 6.2 was normal with mean μ and variance σ^2. It follows from our previous discussion of the properties of the sample mean that $\bar{x} \sim N(\mu, \sigma^2/n)$. Thus, if μ and σ^2 were known, then the behavior of the set of sample means over a large number of samples of size n would be precisely known. In particular, 95% of all such sample means will fall within the interval $(\mu - 1.96\sigma/\sqrt{n}, \mu + 1.96\sigma/\sqrt{n})$. This statement can be written alternatively as follows:

6.4

$$Pr(\mu - 1.96\sigma/\sqrt{n} < \bar{x} < \mu + 1.96\sigma/\sqrt{n}) = .95$$

The inequality in **(6.4)** can actually be written as a set of two inequalities,

$$\mu - 1.96\sigma/\sqrt{n} < \bar{x} \quad \text{and} \quad \bar{x} < \mu + 1.96\sigma/\sqrt{n}$$

Suppose $1.96\sigma/\sqrt{n}$ is added to both sides of the first inequality and $1.96\sigma/\sqrt{n}$ is subtracted from both sides of the second inequality. The following inequalities are then obtained:

$$\mu < \bar{x} + 1.96\sigma/\sqrt{n} \quad \text{and} \quad \bar{x} - 1.96\sigma/\sqrt{n} < \mu$$

If these two inequalities are combined into one inequality, the result is

$$\bar{x} - 1.96\sigma/\sqrt{n} < \mu < \bar{x} + 1.96\sigma/\sqrt{n}$$

Thus, **(6.4)** can be rewritten in the following form:

6.5

$$Pr(\bar{x} - 1.96\sigma/\sqrt{n} < \mu < \bar{x} + 1.96\sigma/\sqrt{n}) = .95$$

DEFINITION 6.9

A **95 % confidence interval (CI)** for μ when σ^2 is known is defined by

$$(\bar{x} - 1.96\sigma/\sqrt{n}, \ \bar{x} + 1.96\sigma/\sqrt{n})$$

You may be puzzled at this point as to what the confidence interval means. The parameter μ is a fixed unknown constant. How can we state that the probability that it lies within some specific interval is 95%? The key point to understand is that the boundaries of the interval depend on the sample points chosen (or more precisely, on the sample mean) and will vary from sample to sample. Furthermore, 95% of such intervals that could be constructed from repeated random samples of size n will contain the parameter μ.

EXAMPLE 6.23

Obstetrics Consider the 5 samples of size 10 from the population of birthweights as shown in Table 6.4 (p. 147). Assume that σ is known to be 20. The interval

$$(\bar{x} - 1.96\sigma/\sqrt{n}, \ \bar{x} + 1.96\sigma/\sqrt{n}) = \left(\bar{x} - \frac{1.96(20)}{\sqrt{10}}, \ \bar{x} + \frac{1.96(20)}{\sqrt{10}} \right) = (\bar{x} - 12.4, \ \bar{x} + 12.4)$$

will be different for each sample and is given in Figure 6.6. A dashed line has been added to represent an imaginary value for μ. The idea is that over a large number of hypothetical samples of size 10, 95% of such intervals will contain the parameter μ. Any one interval from a particular sample *may* or *may not* contain the parameter μ. For example, in Figure 6.6 the first, third, fourth, and fifth intervals contain the parameter μ, whereas the second interval does not.

Therefore, we cannot say that there is a 95% chance that the parameter μ will fall within a particular 95% CI. However, we can say the following:

Over the collection of all 95% confidence intervals that could be constructed from repeated random samples of size n, 95% will contain the parameter μ.

The length of the confidence interval gives some idea of the precision of the point estimate $\bar{x}$. In this particular case the length of each confidence interval is about 25 oz, which makes the precision of the point estimate $\bar{x}$ doubtful and implies that a larger sample size is needed to get a more precise estimate of μ.

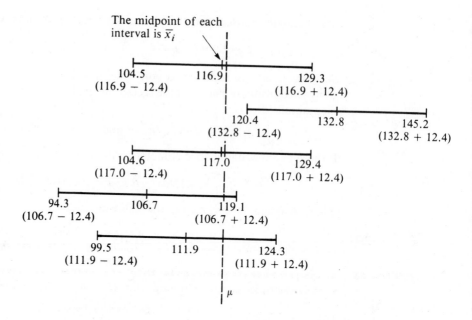

The midpoint of each
interval is $\bar{x}_i$

104.5   116.9| 129.3
(116.9 − 12.4)   (116.9 + 12.4)

120.4   132.8   145.2
(132.8 − 12.4)   (132.8 + 12.4)

104.6   117.0   129.4
(117.0 − 12.4)   (117.0 + 12.4)

94.3   106.7   119.1
(106.7 − 12.4)   (106.7 + 12.4)

99.5   111.9   124.3
(111.9 − 12.4)   (111.9 + 12.4)

μ

FIGURE 6.6
A collection of 95%
confidence intervals for
the mean μ as
computed from
repeated samples of
size 10 (see Table 6.4)
from the population of
birthweights given in
Table 6.2

EXAMPLE 6.24   **Gynecology** Compute a 95% CI for the underlying mean basal body temperature using the
data in Example 6.18 (p. 152), assuming that the standard deviation is 0.2°.

SOLUTION   The 95% CI is given by

$$\bar{x} \pm 1.96\sigma/\sqrt{n} = 97.2° \pm 1.96(0.2)/\sqrt{10} = 97.2° \pm 0.12°$$
$$= (97.08°, 97.32°)$$ ∎

We are frequently interested in obtaining confidence intervals with levels of
confidence other than 95%. In particular, we would like to develop confidence
intervals with confidence level $100\% \times (1 - \alpha)$ for any arbitrary α. This interval
can be developed in the same way as the 95% confidence interval in **(6.5)**. In
particular, if $\bar{x} \sim N(\mu, \sigma^2/n)$, then by definition, $100\% \times (1 - \alpha)$ of all sample
means will fall within the interval $(\mu - z_{1-\alpha/2}\sigma/\sqrt{n}, \mu + z_{1-\alpha/2}\sigma/\sqrt{n})$, or alternatively,

$$Pr(\mu - z_{1-\alpha/2}\sigma/\sqrt{n} < \bar{x} < \mu + z_{1-\alpha/2}\sigma/\sqrt{n}) = 1 - \alpha$$

This can be written as two inequalities

$$\mu - z_{1-\alpha/2}\sigma/\sqrt{n} < \bar{x} \quad \text{and} \quad \bar{x} < \mu + z_{1-\alpha/2}\sigma/\sqrt{n}$$

If $z_{1-\alpha/2}\sigma/\sqrt{n}$ is added to both sides of the first inequality and $z_{1-\alpha/2}\sigma/\sqrt{n}$ is
subtracted from both sides of the second inequality, we obtain

$$\mu < \bar{x} + z_{1-\alpha/2}\sigma/\sqrt{n} \quad \text{and} \quad \bar{x} - z_{1-\alpha/2}\sigma/\sqrt{n} < \mu$$

or $$Pr(\bar{x} - z_{1-\alpha/2}\sigma/\sqrt{n} < \mu < \bar{x} + z_{1-\alpha/2}\sigma/\sqrt{n}) = 1 - \alpha$$

DEFINITION 6.10 ■■

A **100% × (1 − α) confidence interval for** μ is defined by the interval

$$(\bar{x} - z_{1-\alpha/2}\sigma/\sqrt{n}, \; \bar{x} + z_{1-\alpha/2}\sigma/\sqrt{n})$$

where $z_{1-\alpha/2}$ equals the upper $\alpha/2$ percentile of an $N(0, 1)$ distribution. ∎

EXAMPLE 6.25 Suppose the first sample in Table 6.4 (p. 147) has been drawn. Compute a 99% CI for the underlying mean birthweight, assuming that $\sigma = 20$.

SOLUTION This value is given by

$$(116.9 - z_{.995}(20)/\sqrt{10}, \; 116.9 + z_{.995}(20)/\sqrt{10})$$

From Table 3 of the Appendix we see that $z_{.995} = 2.576$, and therefore the 99% CI is

$$(116.9 - 2.576(20)/\sqrt{10}, \; 116.9 + 2.576(20)/\sqrt{10}) = (100.6, 133.2)$$ ∎

Notice that the 99% confidence interval (100.6, 133.2) computed in Example 6.25 is wider than the corresponding 95% confidence interval (104.5, 129.3) computed for the first sample in Figure 6.6. The rationale for this difference is that the higher the level of confidence desired that μ lies within an interval, the wider the confidence interval must be. Indeed, for 95% confidence intervals the length was $2(1.96)\sigma/\sqrt{n}$; for 99% confidence intervals the length was $2(2.576)\sigma/\sqrt{n}$. In general, the length of the 100% × (1 − α) confidence interval is given by

$$2z_{1-\alpha/2}\sigma/\sqrt{n}$$

Therefore, we can see that the length of a confidence interval is governed by three variables: n, σ, and α.

The length of a 100% × (1 − α) confidence interval equals $2z_{1-\alpha/2}\sigma/\sqrt{n}$ and is determined by n, σ, and α.

n. As the sample size (n) increases, the length of the confidence interval decreases.

σ. As the standard deviation (σ), which reflects the variability of individual observations, increases, the length of the confidence interval increases.

α. As the confidence desired increases (α decreases), the length of the confidence interval increases.

EXAMPLE 6.26 **Gynecology** Compute a 95% CI for the underlying mean basal body temperature using the data in Example 6.18 (p. 152), assuming that the number of days sampled is 100 rather than 10 and the standard deviation = 0.2°.

SOLUTION The 95% CI is given by

$$97.2° \pm 1.96(0.2)/\sqrt{100} = 97.2° \pm 1.96(0.2)/10 = 97.2° \pm 0.04° = (97.16°, 97.24°)$$

Notice that this interval is much narrower than the corresponding interval (97.08°, 97.32°) based on a sample of 10 days given in Example 6.24. ∎

EXAMPLE 6.27 Compute a 95% CI for the underlying mean basal temperature using the data in Example 6.18, assuming that the standard deviation of basal body temperature is 0.4° rather than 0.2° with a sample size of 10.

SOLUTION The 95% CI is given by

$$97.2° \pm 1.96(0.4)/\sqrt{10} = 97.2° \pm 0.25° = (96.95°, 97.45°)$$

Notice that this interval is much wider than the corresponding interval (97.08°, 97.32°) based on a standard deviation of 0.2° with a sample size of 10. ∎

Usually only n and α can be controlled. σ is a function of the type of variable being studied, although σ itself can sometimes be decreased, if changes in technique can reduce the amount of measurement error, day-to-day variability, etc.

To this point confidence intervals have been used mainly as descriptive tools for characterizing the precision with which the parameters of a distribution can be estimated. Another use for confidence intervals is in making decisions on the basis of the data.

EXAMPLE 6.28 **Cardiovascular Disease, Pediatrics** Suppose we know from large studies that the mean cholesterol level in children ages 2–14 is 175 mg%/mL and the standard deviation is 30 mg%/mL. We wish to see if there is a familial aggregation of cholesterol levels. Specifically, we identify a group of fathers who have had a heart attack and who presumably have high cholesterol levels and measure the cholesterol levels of their offspring within the 2–14 age range.

Suppose we find that the mean cholesterol level in a group of 100 such children is 207.3 mg%/mL. Is this value sufficiently far from 175 mg%/mL for us to believe that the underlying mean cholesterol level in the population of all children selected in this way is greater than 175 mg%/mL?

SOLUTION One approach would be to construct a 95% confidence interval for μ on the basis of our sample data. We then could make the following decision: If the interval contains 175 mg%/mL, then we cannot say that the underlying mean for this group is any different than the mean for all children (175), because 175 is among the plausible values for μ provided by the 95% confidence interval. We would decide that there is no demonstrated familial aggregation of cholesterol levels. If the confidence interval does not contain 175, then we would conclude that the true underlying mean for this group is greater than 175 and therefore there is a demonstrated familial aggregation of cholesterol levels. The basis for this decision rule is discussed in the chapters on hypothesis testing.

The confidence interval in this case is given by

$$[207.3 - z_{.975}(30)/\sqrt{100}, 207.3 + z_{.975}(30)/\sqrt{100}]$$

$$= [207.3 - 1.96(30)/10, 207.3 + 1.96(30)/10] = (201.4, 213.2)$$

Clearly, 175 is far from the lower boundary of the interval, and we thus conclude that there is familial aggregation of cholesterol. ∎

6.4.5 *t* Distribution

In the previous section the problem of constructing confidence intervals for the mean of a normal distribution when the variance is known was discussed. This situation is somewhat artificial, since the population variance is seldom known when dealing with actual data. The first step in constructing confidence intervals in the previous section was to assume that if the individual observations came from an underlying normal distribution with mean μ and variance σ^2, then the quantity $(\bar{x} - \mu)/(\sigma/\sqrt{n}) \sim N(0, 1)$. Since σ is unknown, it is reasonable to estimate σ by the sample standard deviation s and to try to construct confidence intervals using the quantity $(\bar{x} - \mu)/(s/\sqrt{n})$. The problem is that this quantity is no longer normally distributed.

This problem was first solved in 1908 by a statistician named William Gossett. For his entire professional life, Gossett worked for the Guinness Brewery in Great Britain. He chose to identify himself by the pseudonym "Student," and thus the distribution of $(\bar{x} - \mu)/(s/\sqrt{n})$ is sometimes referred to as **Student's t distribution**. Gossett found that the shape of the distribution depended on the sample size n. Thus, the t distribution is not a unique distribution but is instead a family of distributions indexed by a parameter referred to as the **degrees of freedom** (*df*) of the distribution.

| 6.6 | If $x_1, \ldots, x_n \sim N(\mu, \sigma^2)$ and are independent, then $(\bar{x} - \mu)/(s/\sqrt{n})$ is distributed as a t distribution with $(n - 1)$ degrees of freedom (*df*). |

Once again, Student's t distribution is not a unique distribution but is a family of distributions indexed by the degrees of freedom d. The t distribution with d degrees of freedom is sometimes referred to as the t_d distribution.

DEFINITION 6.11 ■■
The **uth percentile of a t distribution with d degrees of freedom** is denoted by $t_{d,u}$, that is,

$$Pr(t_d < t_{d,u}) \equiv u$$ ■

EXAMPLE 6.29 What does $t_{20,.95}$ mean?

SOLUTION $t_{20,.95}$ is the 95th percentile or the upper 5th percentile of a t distribution with 20 degrees of freedom. ■

It is interesting to compare a t distribution with d degrees of freedom to an $N(0, 1)$ distribution. The density functions corresponding to these distributions are depicted in Figure 6.7.

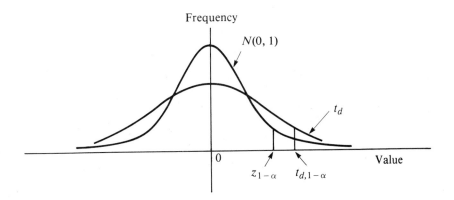

FIGURE 6.7
Comparison of
Student's *t* distribution
with *d* degrees of
freedom with an $N(0, 1)$
distribution

Notice that the t distribution is symmetric about 0 but is more spread out than the $N(0, 1)$ distribution. It can be shown that for any α, $t_{d, 1-\alpha}$ is always larger than the corresponding percentile for an $N(0, 1)$ distribution ($z_{1-\alpha}$). This relationship is depicted in Figure 6.7. However, as d becomes large, the t distribution converges to an $N(0, 1)$ distribution. An explanation for this convergence is that for finite samples the sample variance (s^2) is an approximation to the population variance (σ^2). This approximation gives the statistic $(\bar{x} - \mu)/(s/\sqrt{n})$ more variability than the corresponding statistic $(\bar{x} - \mu)/(\sigma/\sqrt{n})$. As n becomes large, this approximation gets better and s^2 will converge to σ^2 exactly. The two distributions thus get more and more alike as n becomes large. The upper 2.5th percentile of the t distribution for various degrees of freedom and the corresponding percentile for the normal distribution are given in Table 6.5 and are depicted in Figure 6.8.

TABLE 6.5
Comparison of the 97.5th percentile of the t distribution and the normal distribution

d	$t_{d,.975}$	$z_{.975}$	d	$t_{d,.975}$	$z_{.975}$
4	2.776	1.960	60	2.000	1.960
9	2.262	1.960	∞	1.960	1.960
29	2.045	1.960			

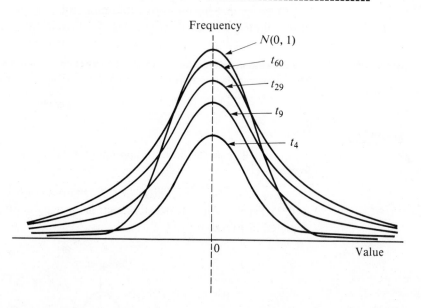

FIGURE 6.8
Comparison of various t distributions with different degrees of freedom with an $N(0, 1)$ distribution

The difference between the t distribution and the normal distribution is greatest for small values of n ($n < 30$). For most applications we can probably assume that a t distribution is approximately normal if the number of degrees of freedom is greater than 60, although an exact t table is always preferable if available. Also, many statistical packages such as Minitab, SPSS^X or SAS will compute exact probabilities associated with the t distribution. Table 5 in Appendix 1 gives the percentage points of the t distribution for various degrees of freedom. The degrees of freedom are given in the first column of the table, and the percentiles are given across the first row. The uth percentile of a t distribution with d degrees of freedom is found by reading across the row marked d and reading down the column marked u.

EXAMPLE 6.30 Find the upper 5th percentile of a t distribution with 23 df.

SOLUTION Find $t_{23,.95}$, which is given in row 23 and column 0.95 of Table 5 and is 1.714. ∎

6.4.6 **Interval Estimation—Unknown Variance**

6.7 Using similar logic to that in Section 6.4.4, we can show that a **100 % × (1 − α) confidence interval for the mean μ of a normal distribution with unknown variance** is given by

$$(\bar{x} - t_{n-1,\,1-\alpha/2}s/\sqrt{n}, \ \bar{x} + t_{n-1,\,1-\alpha/2}s/\sqrt{n})$$

To show this, we see that since $(\bar{x} - \mu)/(s/\sqrt{n})$ follows a t_{n-1} distribution, it follows that

$$Pr\left(t_{n-1,\alpha/2} < \frac{\bar{x} - \mu}{s/\sqrt{n}} < t_{n-1,\,1-\alpha/2}\right) = 1 - \alpha$$

that is, there is a probability of $1 - \alpha$ that a random variable that follows a t_{n-1} distribution will fall between the upper and lower $\alpha/2$ percentiles. This inequality can be written in the form of two inequalities:

$$t_{n-1,\alpha/2} < \frac{\bar{x} - \mu}{s/\sqrt{n}} \quad \text{and} \quad \frac{\bar{x} - \mu}{s/\sqrt{n}} < t_{n-1,\,1-\alpha/2}$$

Both sides of each inequality are now multiplied by $s/\sqrt{n}$ and μ is added to both sides to obtain

$$\mu + t_{n-1,\alpha/2}s/\sqrt{n} < \bar{x} \quad \text{and} \quad \bar{x} < t_{n-1,\,1-\alpha/2}s/\sqrt{n} + \mu$$

Finally, $t_{n-1,\alpha/2}s/\sqrt{n}$ is subtracted from both sides of the first inequality and $t_{n-1,\,1-\alpha/2}s/\sqrt{n}$ is subtracted from both sides of the second inequality, yielding

$$\mu < \bar{x} - t_{n-1,\alpha/2}s/\sqrt{n} \quad \text{and} \quad \bar{x} - t_{n-1,\,1-\alpha/2}s/\sqrt{n} < \mu$$

Expressed as one inequality, this is

$$\bar{x} - t_{n-1,\,1-\alpha/2}s/\sqrt{n} < \mu < \bar{x} - t_{n-1,\alpha/2}s/\sqrt{n}$$

From the symmetry of the t distribution, $t_{n-1,\alpha/2} = -t_{n-1,\,1-\alpha/2}$, and this inequality can be rewritten as

$$\bar{x} - t_{n-1,\,1-\alpha/2}s/\sqrt{n} < \mu < \bar{x} + t_{n-1,\,1-\alpha/2}s/\sqrt{n}$$

and we can say that

$$Pr(\bar{x} - t_{n-1,\,1-\alpha/2}s/\sqrt{n} < \mu < \bar{x} + t_{n-1,\,1-\alpha/2}s/\sqrt{n}) = 1 - \alpha$$

Thus, the interval $(\bar{x} - t_{n-1,\,1-\alpha/2}s/\sqrt{n}, \ \bar{x} + t_{n-1,\,1-\alpha/2}s/\sqrt{n})$ is a $100\% \times (1 - \alpha)$ confidence interval for μ.

EXAMPLE 6.31   **Obstetrics**   Now consider the birthweight data from the first sample in Table 6.4 (p. 147). Compute a 95% confidence interval for μ assuming that the variance is unknown.

SOLUTION   Assuming that the variance is unknown, a 95% confidence interval for μ is given as

$$[116.90 - t_{9,.975}(21.70)/\sqrt{10}, \ 116.90 + t_{9,.975}(21.70)/\sqrt{10}]$$
$$= [116.90 - 2.262(21.70)/\sqrt{10}, \ 116.90 + 2.262(21.70)/\sqrt{10}] = (101.38, 132.42) \quad \blacksquare$$

Generally, confidence intervals based on the t distribution (unknown variance) will be longer than confidence intervals based on the normal distribution (known variance). That is, the range of plausible values for μ will be wider, and it will be harder to rule out particular values, such as was attempted in Example 6.28. However, this principle does not always apply, since for a particular sample the sample variance s^2 may be considerably less than the population variance σ^2.

SECTION 6.5   **Estimation of the Variance of a Distribution**

6.5.1   **Point Estimation**

In Chapter 2 the sample variance was defined as

$$s^2 = \frac{1}{n-1} \sum_{i=1}^{n} (x_i - \bar{x})^2$$

This definition is somewhat counterintuitive, since the denominator would be expected to be n rather than $n-1$. A more formal justification for this definition is now given. If our sample $x_1, \ldots, x_n$ is considered as coming from some population with mean μ and variance σ^2, then how can the unknown population variance σ^2 be estimated from our sample? The following principle aids in deciding on a method of estimation:

6.8   Let $x_1, \ldots, x_n$ be a random sample from some population with mean μ and variance σ^2. The **sample variance s^2 is an unbiased estimator** of σ^2 over all possible random samples of size n that could have been drawn from this population; that is, $E(s^2) = \sigma^2$.

Therefore, if repeated random samples of size n are selected from the population, as was done in Table 6.4, and the sample variance s^2 is computed from each sample, then the average of these sample variances over a large number of such samples of size n will be the population variance σ^2. This statement holds for any underlying distribution.

EXAMPLE 6.32   **Gynecology**   Estimate the variance of the distribution of basal body temperatures using the data in Example 6.18 (p. 152).

SOLUTION   We have $$s^2 = \frac{1}{9} \sum_{i=1}^{n} (x_i - \bar{x})^2 = (0.189)^2 = 0.0357$$

which is an unbiased estimator of σ^2. $\blacksquare$

Note that the intuitive estimator for σ^2 with n in the denominator rather than $n - 1$, that is,

$$\frac{1}{n} \sum_{i=1}^{n} (x_i - \bar{x})^2$$

will tend to underestimate the underlying variance σ^2 by a factor of $(n - 1)/n$. This factor is considerable for small samples but tends to be negligible for large samples. A more complete discussion of the relative merits of different estimators for σ^2 is given in [2].

6.5.2 The Chi-Square Distribution

The problem of interval estimation of the mean of a normal distribution was discussed in Sections 6.4.4 and 6.4.6. We often want to obtain interval estimates of the variance as well. Once again, as was the case for the mean, the interval estimates will hold exactly only if the underlying distribution is normal. The interval estimates will perform much more poorly for the variance than for the mean if the underlying distribution is not normal, and they should be used with caution in this case.

EXAMPLE 6.33 **Hypertension** A new machine has been produced, called an arteriosonde machine, that "prints" blood-pressure recordings on a tape so that the measurements can be read rather than heard. A major argument for using such a machine is that the variability of measurements obtained by different observers on the same person will be lower than with a standard blood-pressure cuff.

Suppose we have the data presented in Table 6.6, consisting of systolic blood-pressure measurements obtained on 10 people and read by 2 observers. We will use the difference d_i between the first and second observer to assess interobserver variability. In particular, if we assume that the underlying distribution of these differences is normal with mean μ and variance σ^2, then it is of primary interest to estimate σ^2. The higher σ^2 is, the higher the interobserver variability.

TABLE 6.6

Systolic blood-pressure measurements (mm Hg) from an arteriosonde machine obtained from 10 people and read by 2 observers

	Observer		
Person (i)	1	2	Difference (d_i)
1	194	200	-6
2	126	123	$+3$
3	130	128	$+2$
4	98	101	-3
5	136	135	$+1$
6	145	145	0
7	110	111	-1
8	108	107	$+1$
9	102	99	$+3$
10	126	128	-2

We have seen previously that an unbiased estimate of the variance σ^2 is given by the sample variance s^2. In this case,

$$s^2 = \sum_{i=1}^{n} (d_i - \bar{d})^2/9 = \left[\sum_{i=1}^{n} d_i^2 - \left(\sum_{i=1}^{n} d_i \right)^2 \bigg/ 10 \right] \bigg/ 9$$

$$= \frac{[(-6)^2 + (3)^2 + \cdots + (-2)^2] - [(-6) + (3) + \cdots + (-2)]^2/10}{9} = 8.178$$

How can an interval estimate for σ^2 be obtained? ■

To obtain an interval estimate for σ^2, a new family of distributions, called chi-square (χ^2) distributions, must be introduced to enable us to find the sampling distribution of s^2 from sample to sample.

DEFINITION 6.12 ■■

If
$$G = \sum_{i=1}^{n} x_i^2$$

where
$$x_1, \ldots, x_n \sim N(0, 1),$$

and are independent, then G is said to follow a **chi-square distribution with n degrees of freedom (df)**. The distribution is often denoted by χ_n^2. ■

The chi-square distribution is actually a family of distributions indexed by the parameter n referred to, again, as the degrees of freedom, as was the case for the t distribution. Unlike the t distribution, which is always symmetric about 0 for any degrees of freedom, the chi-square distribution only takes on positive values and is generally skewed to the right, except for very large n ($n \geq 100$), where the distribution becomes more symmetric. The general shape of these distributions is indicated in Figure 6.9.

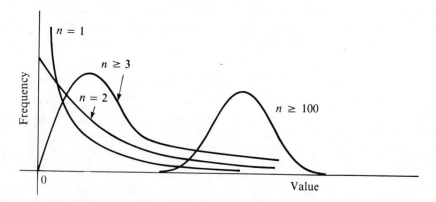

FIGURE 6.9
General shape of various χ^2 distributions with n df

For $n = 1, 2$, the distribution has a mode at 0 ([2]). For $n \geq 3$, the distribution has a mode greater than 0 and is skewed to the right. It can be shown that the expected value of a χ_n^2 distribution is n and the variance is $2n$. For large n ($n \geq 100$), the distribution tends to be roughly symmetric and can be approximated by a normal distribution with mean n and variance $2n$.

DEFINITION 6.13

The **uth percentile of a χ_n^2 distribution** is denoted by $\chi_{n,u}^2$, where $Pr(\chi_n^2 < \chi_{n,u}^2) \equiv u$. These percentiles are depicted in Figure 6.10 and appear in Table 6 in Appendix 1.

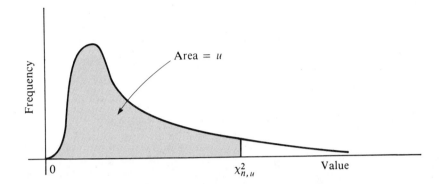

FIGURE 6.10
Graphical display of
the percentiles of a
χ_n^2 distribution

Table 6 is constructed similarly to the t table (Table 5), with the degrees of freedom (d) indexed in the first column and the percentile (u) indexed in the first row. The principal difference between the two tables is that both *lower* ($u \leqslant 0.5$) and *upper* ($u > 0.5$) percentiles are given for the chi-square distribution, whereas only upper percentiles are given for the t distribution. The t distribution is symmetric about 0, and therefore any lower percentile can be obtained as the negative of the corresponding upper percentile. Because the chi-square distribution is, in general, a skewed distribution, there is no simple relationship between the upper and lower percentiles.

EXAMPLE 6.34 Find the upper and lower 2.5th percentile of a chi-square distribution with 10 df.

SOLUTION According to Table 6, the upper and lower percentiles are given by

$$\chi_{10,.975}^2 = 20.48 \quad \text{and} \quad \chi_{10,.025}^2 = 3.25 \quad \text{respectively.}$$

6.5.3 **Interval Estimation**

To obtain an interval estimate of σ^2, we need to find the sampling distribution of s^2. Suppose we assume that $x_1, \ldots, x_n \sim N(\mu, \sigma^2)$. Then, it can be shown that

6.9

$$s^2 \sim \frac{\sigma^2 \chi_{n-1}^2}{(n-1)}$$

Thus, from **(6.9)** we see that s^2 follows a chi-square distribution with $n-1$ df multiplied by the constant $\sigma^2/(n-1)$. Manipulations similar to those given in Section 6.4.4 can now be used to obtain a $100\% \times (1-\alpha)$ confidence interval for σ^2.

In particular, from **(6.9)** it follows that

$$Pr\left(\frac{\sigma_{n-1,\alpha/2}^2}{n-1} < s^2 < \frac{\sigma^2 \chi_{n-1,1-\alpha/2}^2}{n-1}\right) = 1-\alpha$$

This inequality can be represented as two separate inequalities:

$$\frac{\sigma^2 \chi^2_{n-1,\alpha/2}}{n-1} < s^2 \quad \text{and} \quad s^2 < \frac{\sigma^2 \chi^2_{n-1,1-\alpha/2}}{n-1}$$

If both sides of the first inequality are multiplied by $(n-1)/\chi^2_{n-1,\alpha/2}$, and both sides of the second inequality are multiplied by $(n-1)/\chi^2_{n-1,1-\alpha/2}$, then

$$\sigma^2 < \frac{(n-1)s^2}{\chi^2_{n-1,\alpha/2}} \quad \text{and} \quad \frac{(n-1)s^2}{\chi^2_{n-1,1-\alpha/2}} < \sigma^2$$

or, upon combining these two inequalities,

$$\frac{(n-1)s^2}{\chi^2_{n-1,1-\alpha/2}} < \sigma^2 < \frac{(n-1)s^2}{\chi^2_{n-1,\alpha/2}}$$

It follows that

$$Pr\left[\frac{(n-1)s^2}{\chi^2_{n-1,1-\alpha/2}} < \sigma^2 < \frac{(n-1)s^2}{\chi^2_{n-1,\alpha/2}} \right] = 1 - \alpha$$

Thus, the interval $[(n-1)s^2/\chi^2_{n-1,1-\alpha/2}, (n-1)s^2/\chi^2_{n-1,\alpha/2}]$ is a $100\% \times (1-\alpha)$ confidence interval for σ^2.

6.10 | A $100\% \times (1-\alpha)$ **confidence interval for** σ^2 **is given by**

$$[(n-1)s^2/\chi^2_{n-1,1-\alpha/2}, (n-1)s^2/\chi^2_{n-1,\alpha/2}]$$

EXAMPLE 6.35   **Hypertension** We now return to the specific data set in Example 6.33. Suppose we wish to construct a 95% confidence interval for the interobserver variability as defined by σ^2.

SOLUTION   Since there are 10 people and $s^2 = 8.178$, the required interval is given by

$$(9s^2/\chi^2_{9,.975}, 9s^2/\chi^2_{9,.025}) = [9(8.178)/19.02, 9(8.178)/2.70] = (3.87, 27.26)$$

Similarly, a 95% confidence interval for σ is given by $(\sqrt{3.87}, \sqrt{27.26}) = (1.97, 5.22)$. Notice that the confidence interval for σ^2 is *not* symmetric about $s^2 = 8.178$, in contrast to the confidence intervals for μ, which *were* symmetric about $\bar{x}$. This characteristic is common in confidence intervals for the variance.

The utility of the confidence interval for σ^2 for decision-making purposes might be achieved if we had a good estimate of the interobserver variability of blood-pressure readings from a standard cuff. For example, suppose we know from previous work that if two people are listening to blood-pressure recordings from a standard cuff, then the interobserver variability as measured by the variance of the set of differences between the readings of the two observers is 35. This value is outside the range of the 95% confidence interval for σ^2 (3.87, 27.26), and we thus conclude that the interobserver variability is decreased by using an arteriosonde machine. Alternatively, if this prior variance were 15, then we could not say that the variances obtained from using the two methods are different.

Estimation for the Binomial Distribution

6.6.1 **Point Estimation**

Point estimation for the parameter p of a binomial distribution is discussed in this section.

EXAMPLE 6.36 **Cancer** Consider the problem of estimating the prevalence of malignant melanoma in 45–54-year-old women in the United States. Suppose that a random sample of 5000 women is selected from this age group and that 28 are found to have the disease. Let the random variable X for each of the 5000 women be 1 if a woman has the disease and 0 if she does not. Suppose that the prevalence rate of the disease in this age group is p. How can p be estimated? ∎

DEFINITION 6.14 ■■

Let X_i be a random variable that takes on the value 1 with probability p and the value 0 with probability $q = 1 - p$. This type of random variable is defined as a **Bernoulli trial**. This is a special case of a binomial random variable with $n = 1$. ∎

We know from the definition of an expected value that $E(X_i) = 1(p) + 0(q) = p$ and that $E(X_i^2) = 1^2(p) + 0^2(q) = p$. Therefore,

$$Var(X_i) = E(X_i^2) - [E(X_i)]^2 = p - p^2 = p(1 - p) = pq$$

Now consider the random variable

$$X = \sum_{i=1}^{n} X_i$$

In Example 6.36 this random variable simply represents the number of cases of malignant melanoma among n women.

$$E(X) = E\left(\sum_{i=1}^{n} X_i\right) = p + p + \cdots + p = np$$

and $$Var(X) = Var\left(\sum_{i=1}^{n} X_i\right) = \sum_{i=1}^{n} Var(X_i) = pq + pq + \cdots + pq = npq$$

Note that X can also be looked at as a binomial random variable with parameters n and p, since it represents the number of events in n independent trials.

Finally, consider the random variable $\hat{p}$ = sample proportion of events. In our example, $\hat{p}$ = proportion of women with malignant melanoma. Thus,

$$\hat{p} = \frac{1}{n} \sum_{i=1}^{n} X_i = \bar{x}$$

Since $\hat{p}$ is a sample mean, the results of **(6.1)** apply and we see that $E(\hat{p}) = E(X_i) \equiv \mu = p$. Furthermore, from **(6.2)** it follows that

$$Var(\hat{p}) = \sigma^2/n = pq/n \quad \text{and} \quad se(\hat{p}) = \sqrt{pq/n}$$

Thus, for any sample of size n, the sample proportion $\hat{p}$ is an unbiased estimator of the population proportion p. The standard error of this proportion is given exactly by $\sqrt{pq/n}$ and is estimated by $\sqrt{\hat{p}\hat{q}/n}$. These principles can be summarized as follows:

6.11	**Point Estimation of the Binomial Parameter p**

Let X be a binomial random variable with parameters n and p. An unbiased estimate of p is given by the sample proportion of events $\hat{p}$. Its standard error is given exactly by $\sqrt{pq/n}$ and is estimated by $\sqrt{\hat{p}\hat{q}/n}$.

EXAMPLE 6.37   Estimate the prevalence of malignant melanoma in Example 6.36 and give its standard error.

SOLUTION   Our best estimate of the prevalence rate of malignant melanoma among 45–54-year-old women is $28/5000 = .0056$. Its estimated standard error is

$$\sqrt{(.0056)(.9944)/5000} = .0011$$

■

6.6.2   Interval Estimation—Normal Theory Methods

The point estimation of the parameter p of a binomial distribution was covered in Section 6.6.1. How can an **interval estimate** of the parameter p be obtained?

EXAMPLE 6.38   **Cancer** Suppose we are interested in estimating the prevalence rate of breast cancer among 50–54-year-old women whose mothers have had breast cancer. Suppose that in a random sample of 10,000 such women, 400 are found to have had breast cancer at some point in their lives. We have shown that the best point estimate of the prevalence rate p is given by the sample proportion $\hat{p} = 400/10{,}000 = .040$. How can an interval estimate of the parameter p be obtained? (See the solution in Example 6.39 on page 172.)   ■

We will assume that the normal approximation to the binomial distribution is valid—whereby the number of events X observed out of n women will be approximately normally distributed with mean np and variance npq or, correspondingly, the proportion of women with events $= \hat{p} = X/n$ is normally distributed with mean p and variance pq/n.

The normal approximation can actually be justified on the basis of the central limit theorem. Indeed, in the previous section we showed that $\hat{p}$ could be represented as an average of n Bernoulli trials, each of which has mean p and variance pq. Thus, for large n, from the central limit theorem, we can see that $\hat{p} = \bar{x}$ is normally distributed with mean $\mu = p$ and variance $\sigma^2/n = pq/n$, or

6.12	$$\hat{p} \sim N(p, pq/n)$$

Alternatively, since the number of successes in n Bernoulli trials $= X = n\hat{p}$ (which is the same as a binomial random variable with parameters n and p), if **(6.12)** is multiplied by n,

6.13	$$X \sim N(np, npq)$$

This formulation is indeed the same as that for the normal approximation to the binomial distribution, which was given in Chapter 5. How large should n be before this approximation can be used? In Chapter 5 we said that the normal approximation to the binomial distribution is valid if $npq \geqslant 5$. However, in Chapter 5 we assumed that p was known, whereas here we assume that it is unknown. Thus, we shall estimate p by $\hat{p}$ and q by $\hat{q} = 1 - \hat{p}$ and will apply the normal approximation to the binomial if $n\hat{p}\hat{q} \geqslant 5$. Therefore, the results of this section should only be used if $n\hat{p}\hat{q} \geqslant 5$. An approximate $100\% \times (1 - \alpha)$ confidence interval for p can now be derived from **(6.12)** using methods similar to those given in Section 6.4.4. In particular, from **(6.12)**, we see that

$$Pr(p - z_{1-\alpha/2}\sqrt{pq/n} < \hat{p} < p + z_{1-\alpha/2}\sqrt{pq/n}) = 1 - \alpha$$

This inequality can be written in the form of two inequalities:

$$p - z_{1-\alpha/2}\sqrt{pq/n} < \hat{p} \quad \text{and} \quad \hat{p} < p + z_{1-\alpha/2}\sqrt{pq/n}$$

To explicitly derive a confidence interval based on these inequalities requires solving a quadratic equation for p in terms of $\hat{p}$. To avoid this complexity, it is customary to approximate $\sqrt{pq/n}$ by $\sqrt{\hat{p}\hat{q}/n}$ and rewrite the inequalities in the form

$$p - z_{1-\alpha/2}\sqrt{\hat{p}\hat{q}/n} < \hat{p} \quad \text{and} \quad \hat{p} < p + z_{1-\alpha/2}\sqrt{\hat{p}\hat{q}/n}$$

We now add $z_{1-\alpha/2}\sqrt{\hat{p}\hat{q}/n}$ to both sides of the first inequality and subtract this quantity from both sides of the second inequality, obtaining

$$p < \hat{p} + z_{1-\alpha/2}\sqrt{\hat{p}\hat{q}/n} \quad \text{and} \quad \hat{p} - z_{1-\alpha/2}\sqrt{\hat{p}\hat{q}/n} < p$$

Combining these two inequalities, we get

$$\hat{p} - z_{1-\alpha/2}\sqrt{\hat{p}\hat{q}/n} < p < \hat{p} + z_{1-\alpha/2}\sqrt{\hat{p}\hat{q}/n}$$

or

$$Pr(\hat{p} - z_{1-\alpha/2}\sqrt{\hat{p}\hat{q}/n} < p < \hat{p} + z_{1-\alpha/2}\sqrt{\hat{p}\hat{q}/n}) = 1 - \alpha$$

The approximate $100\% \times (1 - \alpha)$ confidence interval for p is thus given by

$$(\hat{p} - z_{1-\alpha/2}\sqrt{\hat{p}\hat{q}/n}, \; \hat{p} + z_{1-\alpha/2}\sqrt{\hat{p}\hat{q}/n})$$

6.14 **Normal Theory Method for Obtaining a Confidence Interval for the Binomial Parameter p**

An approximate $100\% \times (1 - \alpha)$ confidence interval for the binomial parameter p based on the normal approximation to the binomial distribution is given by

$$(\hat{p} - z_{1-\alpha/2}\sqrt{\hat{p}\hat{q}/n}, \; \hat{p} + z_{1-\alpha/2}\sqrt{\hat{p}\hat{q}/n})$$

This method of interval estimation should only be used if $n\hat{p}\hat{q} \geqslant 5$.

EXAMPLE 6.39   **Cancer** Using the data in Example 6.38, derive a 95% confidence interval for the prevalence rate of breast cancer among 50–54-year-old women whose mothers have had breast cancer.

SOLUTION
$$\hat{p} = .040 \qquad \alpha = .05 \qquad z_{1-\alpha/2} = 1.96 \qquad n = 10,000$$

Therefore, an approximate 95% confidence interval is given by

$$[.040 - 1.96\sqrt{(.04)(.96)/10,000}, \, .040 + 1.96\sqrt{(.04)(.96)/10,000}]$$
$$= (.040 - .004, .040 + .004) = (.036, .044)$$

Suppose we know that the prevalence rate of breast cancer among all 50–54-year-old American women is 2%. Since 2% does *not* fall in the preceding interval, we can be quite confident that the underlying rate for the group of women whose mothers have had breast cancer is higher than the rate in the general population. ∎

6.6.3 Interval Estimation—Exact Methods

The question remains, How is a confidence interval for the binomial parameter p obtained when either the normal approximation to the binomial distribution is not valid or a more exact confidence interval is desired?

EXAMPLE 6.40   **Cancer, Nutrition** Suppose we want to estimate the rate of bladder cancer in rats that have been fed a diet high in saccharin. We feed this diet to 20 rats and find that 2 develop bladder cancer. In this case our best point estimate of p is $\hat{p} = \frac{2}{20} = .1$. However, since

$$n\hat{p}\hat{q} = 20(2/20)(18/20) = 1.8 < 5$$

the normal approximation to the binomial distribution cannot be used and thus normal theory methods for obtaining confidence intervals are not valid. How can an interval estimate be obtained in this case? ∎

A small sample method for obtaining confidence limits will be presented.

6.15   **Exact Method for Obtaining a Confidence Interval for the Binomial Parameter p**

An exact $100\% \times (1 - \alpha)$ confidence interval for the binomial parameter p that is always valid is given by (p_1, p_2), where p_1, p_2 satisfy the equations

$$Pr(X \geqslant x \,|\, p = p_1) = \frac{\alpha}{2} = \sum_{k=x}^{n} \binom{n}{k} p_1^k (1 - p_1)^{n-k}$$

$$Pr(X \leqslant x \,|\, p = p_2) = \frac{\alpha}{2} = \sum_{k=0}^{x} \binom{n}{k} p_2^k (1 - p_2)^{n-k}$$

The main problem with using this method is the difficulty in computing expressions such as

$$\sum_{k=0}^{x} \binom{n}{k} p^k (1 - p)^{n-k}$$

Fortunately, special tables exist for the evaluation of such expressions, one of which is given in Table 7 in the Appendix. This table can be used as follows:

| 6.16 | **Exact Confidence Limits for Binomial Proportions** |

(1) The sample size (n) is given along each curve. Two curves should correspond to a given sample size. One curve is used to obtain the lower confidence limit and the other to obtain the upper confidence limit.

(2) If $0 \leqslant \hat{p} \leqslant .5$, then

(a) Refer to the lower horizontal axis and find the point corresponding to $\hat{p}$.

(b) Draw a line perpendicular to the horizontal axis and find the two points where this line intersects the two curves identified in **1**.

(c) Read across to the left vertical axis; the smaller value corresponds to the lower confidence limit and the larger value to the upper confidence limit.

(3) If $.5 < \hat{p} \leqslant 1.0$, then

(a) Refer to the upper horizontal axis and find the point corresponding to $\hat{p}$.

(b) Draw a line perpendicular to the horizontal axis and find the two points where this line intersects the two curves identified in **1**.

(c) Read across to the right vertical axis; the smaller value corresponds to the lower confidence limit and the larger value to the upper confidence limit.

EXAMPLE 6.41 **Cancer** Derive an exact 95% confidence interval from the rat bladder cancer data given in Example 6.40.

SOLUTION We refer to Table 7, $\alpha = 0.05$, and identify the two curves with $n = 20$. Since $\hat{p} = .1 \leqslant .5$, we refer to the lower horizontal axis and draw a vertical line at 0.10 until it intersects the two curves marked $n = 20$. We then read across to the left vertical axis and find the confidence limits of .01 and .32. Thus, the exact 95% confidence interval = (.01, .32). Notice that this confidence interval is *not* symmetric about $\hat{p} = .10$. ■

EXAMPLE 6.42 **Health Promotion** Suppose that as part of a program for counseling patients with many risk factors for heart disease, 100 smokers are identified. Of this group, 10 give up smoking for at least 1 month. After a 1-year follow-up, 6 of the 10 patients are found to have taken up smoking again. The proportion of ex-smokers who start smoking again is referred to as the *recidivism rate*. Derive a 99% confidence interval for the recidivism rate.

SOLUTION Exact binomial confidence limits must be used, since

$$n\hat{p}\hat{q} = 10(.6)(.4) = 2.4 < 5$$

We refer to the upper horizontal axis of the chart marked $\alpha = 0.01$ in Table 7 and note the point $\hat{p} = .60$. We then follow the vertical scale at .60 until it intersects the two curves marked $n = 10$. We then read across to the right vertical axis and find the confidence limits of .19 and .92. Thus, the exact 99% confidence interval = (.19, .92). ■

One-Sided Confidence Intervals

In the previous discussion of interval estimation, only what are known as *two-sided confidence intervals* have been described. Frequently, the following type of problem occurs.

EXAMPLE 6.43 **Cancer** A standard treatment exists for a certain type of cancer, and the patients receiving the treatment have a 5-year survival rate of 30%. A new treatment is proposed that has some unknown survival rate p. We would only be interested in using the new treatment if it were better than the standard treatment. Suppose that 40 out of 100 patients who receive the new treatment survive for 5 years. Can we say that the new treatment is better than the standard treatment? ∎

One way to assess these data is to construct a one-sided confidence interval, where we are interested in only *one* bound of the interval, in this case the lower bound. If 30% is below the lower bound, then it is an unlikely estimate of the 5-year-survival rate for patients getting the new treatment. We could reasonably conclude from this, that the new treatment is better than the standard treatment in this case.

6.17 **Upper One-Sided Confidence Interval**
for the Binomial Parameter p—Normal Theory Method

An **upper one-sided 100% × (1 − α) confidence interval** is of the form $p > p_1$ such that

$$Pr(p > p_1) = 1 - \alpha$$

If we assume that the normal approximation to the binomial holds true, then we can show that this confidence interval is given approximately by

$$p > \hat{p} - z_{1-\alpha}\sqrt{\hat{p}\hat{q}/n}$$

This interval estimator should only be used if $n\hat{p}\hat{q} \geqslant 5$.

To see this, note that if the normal approximation to the binomial distribution holds, then $\hat{p} \sim N(p, pq/n)$. Therefore, by definition

$$Pr(\hat{p} < p + z_{1-\alpha}\sqrt{pq/n}) = 1 - \alpha$$

We approximate $\sqrt{pq/n}$ by $\sqrt{\hat{p}\hat{q}/n}$ and subtract $z_{1-\alpha}\sqrt{\hat{p}\hat{q}/n}$ from both sides of the equation, yielding

$$\hat{p} - z_{1-\alpha}\sqrt{\hat{p}\hat{q}/n} < p$$

or $p > \hat{p} - z_{1-\alpha}\sqrt{\hat{p}\hat{q}/n}$ and $Pr(p > \hat{p} - z_{1-\alpha}\sqrt{\hat{p}\hat{q}/n}) = 1 - \alpha$

Therefore, if the normal approximation to the binomial distribution holds, then $p > \hat{p} - z_{1-\alpha}\sqrt{\hat{p}\hat{q}/n}$ is an approximate $100\% \times (1 - \alpha)$ one-sided confidence interval for p.

Notice that $z_{1-\alpha}$ is used in constructing one-sided intervals, whereas $z_{1-\alpha/2}$ was used in constructing two-sided intervals.

EXAMPLE 6.44   Suppose a 95% confidence interval for a binomial parameter p is desired. What percentile of the normal distribution should be used for a one-sided interval? a two-sided interval?

SOLUTION   For $\alpha = .05$, we use $z_{1-.05} = z_{.95} = 1.645$ for a one-sided interval and $z_{1-.05/2} = z_{.975} = 1.96$ for a two-sided interval. ∎

EXAMPLE 6.45   **Cancer** Construct an upper one-sided 95% confidence interval for the survival rate based on the cancer treatment data in Example 6.43.

SOLUTION First check that $n\hat{p}\hat{q} = 100(.4)(.6) = 24 \geqslant 5$. The confidence interval is then given by

$$Pr[p > .40 - z_{.95}\sqrt{(.4)(.6)/100}] = .95$$
$$Pr[p > .40 - 1.645(.049)] = .95$$
$$Pr(p > .319) = .95$$

Since .30 is not within the given interval, we would conclude that the new treatment is better than the standard treatment. ∎

If we were interested in 5-year death rates rather than survival rates, then a one-sided interval of the form $Pr(p < p_2) = 1 - \alpha$ would be appropriate, since we would only be interested in the new treatment if its death rate were lower than that of the standard treatment.

6.18 | **Lower One-Sided Confidence Interval**
for the Binomial Parameter p—Normal Theory Method

The interval $p < p_2$ such that

$$Pr(p < p_2) = 1 - \alpha$$

is referred to as a **lower one-sided 100% × (1 − α) confidence interval** and is given approximately by

$$p < \hat{p} + z_{1-\alpha}\sqrt{\hat{p}\hat{q}/n}$$

This expression can be derived in the same manner as in (**6.17**) by starting with the relationship

$$Pr(\hat{p} > p - z_{1-\alpha}\sqrt{pq/n}) = 1 - \alpha$$

If we approximate $\sqrt{pq/n}$ by $\sqrt{\hat{p}\hat{q}/n}$ and add $z_{1-\alpha}\sqrt{\hat{p}\hat{q}/n}$ to both sides of the equation, we get

$$Pr(p < \hat{p} + z_{1-\alpha}\sqrt{\hat{p}\hat{q}/n}) = 1 - \alpha$$

EXAMPLE 6.46 **Cancer** Compute a lower one-sided 95% confidence interval for the death rate using the cancer treatment data in Example 6.43.

SOLUTION We have that $\hat{p} = .6$. Thus, the 95% confidence interval is given by

$$Pr[p < .6 + 1.645\sqrt{(.6)(.4)/100}] = .95$$
$$Pr[p < .6 + 1.645(.049)] = .95$$
$$Pr(p < .681) = .95$$

Since 70% is not within this interval, we can conclude that the new treatment has a lower death rate than the old treatment does. ∎

Similar methods can be used to obtain one-sided confidence intervals for the mean and variance of a normal distribution and for the binomial parameter p using exact methods.

SECTION 6.8 ## Summary

In this chapter the concept of a sampling distribution was introduced. This concept is crucial to understanding the principles of statistical inference. The fundamental idea is to forget about our sample as a unique entity; instead, regard it as a random sample from all possible samples of size n that could have been drawn from the population under study. Using this concept, $\bar{x}$ was shown to be an unbiased estimator of the population mean μ; that is, the average of all sample means over all possible random samples of size n that could have been drawn will equal the population mean. Furthermore, if our population follows a normal distribution, then $\bar{x}$ has minimum variance among all possible unbiased estimators and is thus referred to as a minimum variance unbiased estimator of μ. Finally, if our population follows a normal distribution, then $\bar{x}$ will also follow a normal distribution. However, even if our population is not normal, the sample mean will still approximately follow a normal distribution for a sufficiently large sample size. This very important idea, which justifies many of the hypothesis tests we will study in the remainder of this book, is called the central limit theorem.

The idea of an interval estimate (or confidence interval) was then introduced. Specifically, a 95% confidence interval is defined as an interval that will contain the true parameter for 95% of all random samples that could have been obtained from the reference population. The preceding principles of point and interval estimation were applied to

(a) estimating the mean μ of a normal distribution when the variance is known

(b) estimating the mean μ of a normal distribution when the variance is unknown

(c) estimating the variance σ^2 of a normal distribution

(d) estimating the parameter p of a binomial distribution

The t and chi-square distributions were introduced to obtain interval estimates for (b) and (c), respectively.

In Chapters 7 through 10, the discussion of statistical inference continues, focusing primarily on testing hypotheses rather than on parameter estimation. In this regard some parallels between inference from the points of view of hypothesis testing and confidence intervals are discussed.

PROBLEMS

Suppose we wish to construct a list of treatment assignments for patients entering a study comparing different treatments for duodenal ulcer.

6.1 Anticipating that 20 patients will be entered in the study and 2 treatments will be used, construct a list of random treatment assignments starting in the 28th row of the random number table (Table 4 in the Appendix).

6.2 Count the number of people assigned to each treatment group. How does this number compare with the expected number in each group?

6.3 Suppose we change our minds and decide to enroll 40 patients and use 4 treatment groups. Start at the 12th row of Table 4 and construct the list of random treatment assignments referred to in Problem 6.1.

6.4 Answer Problem 6.2 for the list of treatment assignments derived in Problem 6.3.

Pulmonary Disease

The data in Table 6.7 concern the mean triceps skin fold thickness in a group of normal men and group of men with chronic airflow limitation [3].

TABLE 6.7 Triceps skin fold thickness in normal men and men with chronic airflow limitation

Group	Mean	sd	n
Normal	1.35	0.5	40
Chronic airflow limitation	0.92	0.4	32

(Reprinted with permission of *Chest*, 85(6), 585–595, 1984.)

6.5 What is the standard error of the mean for each group?

6.6 Assume that the central limit theorem is applicable. What does it mean in this context?

Cardiology

The data in Table 6.8 on left ventricular ejection fraction (LVEF) were collected from a group of 27 patients with acute dilated cardiomyopathy [4].

TABLE 6.8 Left ventricular ejection fraction (LVEF) for 27 patients with acute dilated cardiomyopathy

Patient number	LVEF	Patient number	LVEF
1	0.19	15	0.24
2	0.24	16	0.18
3	0.17	17	0.22
4	0.40	18	0.23
5	0.40	19	0.14
6	0.23	20	0.14
7	0.20	21	0.30
8	0.20	22	0.07
9	0.30	23	0.12
10	0.19	24	0.13
11	0.24	25	0.17
12	0.32	26	0.24
13	0.32	27	0.19
14	0.28		

Note: $\sum x_i = 6.05$, $\sum x_i^2 = 1.522$.
(Reprinted with permission of the *New England Journal of Medicine*, 312(14), 885–890, 1985.)

6.7 Calculate the standard deviation of LVEF for these patients.

6.8 Calculate the standard error of the mean for LVEF.

6.9 What is the difference in interpretation between standard deviation and standard error in this case?

6.10 What does the central limit theorem mean in this context?

6.11 Find the upper 1st percentile of a t distribution with 16 df.

6.12 Find the lower 10th percentile of a t distribution with 28 df.

6.13 Find the upper 2.5th percentile of a t distribution with 7 df.

6.14 Assuming that the standard deviation is known to be 0.1, compute a 95% confidence interval for the true mean LVEF for patients with acute dilated cardiomyopathy based on the data in Table 6.8.

6.15 Answer Problem 6.14 without assuming that the standard deviation is known.

6.16 Assuming that the standard deviation is known to be 6.0, compute a 95% confidence interval for the mean duration of hospitalization using the data in Table 2.11 (p. 36).

6.17 Compute a 95% confidence interval for the mean duration of hospitalization without assuming that the standard deviation is known.

6.18 Answer Problem 6.17 for a 90% confidence interval.

6.19 What is the relationship between your answers to Problems 6.17 and 6.18?

6.20 What is the upper 10th percentile of a chi-square distribution with 5 df?

6.21 What is the upper 1st percentile of a chi-square distribution with 3 df?

6.22 What are the upper and lower 2.5th percentiles for a chi-square distribution with 2 df? What notation is used to denote these percentiles?

6.23 What are the upper and lower 2.5th percentiles for a chi-square distribution with 140 df?

6.24 What is the best point estimate of the variance of LVEF in Table 6.8?

6.25 Construct a 95% confidence interval for the variance of LVEF in Table 6.8.

6.26 Using the data in Table 2.10 (p. 35), construct a 99% confidence interval for the variance of total heart weight for left heart disease males.

6.27 Using the data in Table 2.10 (p. 35), answer Problem 6.26 for normal males.

Gynecology

In a 1985 study 89 of 283 women with primary tubal infertility (cases) and 640 of 3833 control women reported ever having used an IUD [5].

6.28 What is the best point estimate of the rate of IUD use among case and control women, respectively?

6.29 Provide a 95% confidence interval for the estimates in Problem 6.28.

Refer to the data in Table 2.11 (p. 36). Regard this hospital as typical of Pennslyvania hospitals.

6.30 What is the best point estimate of the percentage of males among patients discharged from Pennslyvania hospitals?

6.31 What is the standard error of the estimate obtained in Problem 6.30?

6.32 Provide a 95% confidence interval for the percentage of males among patients discharged from Pennsylvania hospitals.

6.33 What is the best point estimate of the percentage of discharged patients 10 years of age and older who received antibiotics while in the hospital?

6.34 Using the normal approximation method, provide a 95% confidence interval for the estimate in Problem 6.33.

6.35 Answer Problem 6.34 using the exact method.

6.36 Compare your results in Problems 6.34 and 6.35.

6.37 What is the best point estimate of the percentage of discharged patients, exclusive of women of childbearing age (ages 18–45), who received a bacterial culture while in the hospital?

6.38 Provide a 95% confidence interval corresponding to the estimate in Problem 6.37.

6.39 Answer Problem 6.38 for a 99% confidence interval.

Cardiology

Suppose a drug to relieve anginal pain is effective within 8 hours in 30% of 100 patients studied.

6.40 Derive an upper one-sided 95% confidence interval for the percentage of patients who could get pain relief from the drug within 8 hours.

6.41 Suppose that if the patients are untreated, 15% will be free from pain within 8 hours. Assuming that the drug either benefits the patients or has no effect at all, what is your opinion on the effectiveness of the drug?

Suppose that in the same study 20% of the patients become pain-free within 4 hours of administration of the drug.

6.42 Derive an upper one-sided confidence interval for the percentage of patients who could get pain relief from the drug within 4 hours.

6.43 Suppose 10% of untreated patients will be pain-free within 4 hours. Assuming that the drug either benefits the patients or has no effect at all, what is your opinion on the effectiveness of the drug?

Microbiology

A nine-laboratory cooperative study was performed to evaluate quality control for susceptibility tests with 30 μg netilmicin disks [6]. Each laboratory tested 3 standard control strains on a different lot of Mueller–Hinton agar, with 150 tests performed per laboratory. For protocol control, each laboratory also performed 15 additional tests on each of the control strains using the *same* lot of Mueller–Hinton agar across laboratories. The mean zone diameters for each of the nine laboratories are given in Table 6.9.

6.44 Provide a point and interval estimate (95% confidence interval) for the mean zone diameter across laboratories for each type of control strain, if each laboratory uses different media to perform the susceptibility tests.

6.45 Answer Problem 6.44 if each laboratory uses a common medium to perform the susceptibility tests.

6.46 Provide a point and interval estimate (95% confidence interval) for the interlaboratory standard deviation of mean zone diameters for each type of control strain, if each laboratory uses different media to perform the susceptibility tests.

6.47 Answer Problem 6.46 if each laboratory uses a common medium to perform the susceptibility tests.

6.48 Are there any advantages to using a common medium versus using different media for performing the susceptibility tests with regards to standardization of results across laboratories?

Cancer

A case-control study of the effectiveness of the Pap test in preventing cervical cancer was performed [7]. It was found that 28.1% of 153 cervical cancer cases and 7.2% of 153 age-matched (± 5 years) controls had never had a PAP test prior to the time of the case's diagnosis.

6.49 Provide a 95% confidence interval for the percentage of cervical cancer cases who never had a PAP test.

TABLE 6.9 Mean zone diameters with 30 μg netilmicin disks tested in nine separate laboratories

| | Type of control strain | | | | | |
| | E. coli | | S. aureus | | P. aeruginosa | |
Laboratory	Different media	Common medium	Different media	Common medium	Different media	Common medium
A	27.5	23.8	25.4	23.9	20.1	16.7
B	24.6	21.1	24.8	24.2	18.4	17.0
C	25.3	25.4	24.6	25.0	16.8	17.1
D	28.7	25.4	29.8	26.7	21.7	18.2
E	23.0	24.8	27.5	25.3	20.1	16.7
F	26.8	25.7	28.1	25.2	20.3	19.2
G	24.7	26.8	31.2	27.1	22.8	18.8
H	24.3	26.2	24.3	26.5	19.9	18.1
I	24.9	26.3	25.4	25.1	19.3	19.2

6.50 Provide a 95% confidence interval for the percentage of controls who never had a PAP test.

6.51 Do you think the PAP test is helpful in preventing cervical cancer?

Renal Disease

A study of psychosocial and physiological changes in a cohort of dialysis patients with end-stage renal disease was conducted [8]. 102 patients were initially ascertained at baseline; 69 of the 102 patients were reascertained at an 18-month follow-up visit. The data in Table 6.10 were reported.

6.52 Provide a point and interval estimate (95% confidence interval) for the mean of each of the parameters at baseline and follow-up.

6.53 Do you have any opinion on the physiological and psychological changes in this group of patients?

Obstetrics

A new drug therapy is proposed for the prevention of low-birthweight deliveries. A pilot study undertaken, using the drug on 20 pregnant women, found that the mean birthweight in this group is 3500 g with a standard deviation of 500 g.

6.54 What is the standard error of the mean in this case?

6.55 What is the difference in interpretation between the standard deviation and standard error in this case (in words)?

Hypertension

In an effort to detect hypertension in young children, blood-pressure measurements were taken on 30 children aged 5–6 years living in a specific community. For these children the mean diastolic blood pressure was found to be 56.2 mm Hg with standard deviation 7.9

TABLE 6.10 Physiological and psychological parameters in patients with end-stage renal disease

| | Baseline ($n = 102$) | | 18-month follow-up ($n = 69$) | |
Variable	Mean	sd	Mean	sd
Serum creatinine (mmol/L)	0.97	0.22	1.00	0.19
Serum potassium (mmol/L)	4.43	0.64	4.49	0.71
Serum phosphate (mmol/L)	1.68	0.47	1.57	0.40
Psychosocial adjustment to illness scale (PAIS scale)	36.50	16.08	23.27	13.79

mm Hg. From a nationwide study we know that the mean diastolic blood pressure is 64.2 mm Hg for 5-6-year-old children.

6.56 Is there evidence that the mean diastolic blood pressure for the children in the community is different from the nationwide average of children of the same age group?

6.57 Provide a 95% confidence interval for the standard deviation of the diastolic blood pressure of 5-6-year-old children in this community based on the observed 30 children.

Infectious Disease, Pulmonary Disease, Hospital Epidemiology

A study was performed to relate reactivity to tuberculin to job activity within a hospital. In particular, suppose that according to statewide, age-specific rates, 31% of nurses in a hospital are expected to have positive tuberculin skin tests. In the hospital under study, 93 out of 221 nurses are tuberculin positive.

6.58 Give a 95% confidence interval for the proportion of positive tuberculin skin tests among nurses.

6.59 How does the nurses' observed rate compare with the expected rate of 31%?

Ophthalmology, Hypertension

A special study is conducted to test the hypothesis that people with glaucoma have higher blood pressure than average. In the study 200 people with glaucoma are recruited with a mean systolic blood pressure of 140 mm Hg and a standard deviation of 25 mm Hg.

6.60 Construct a 95% confidence interval for the mean systolic blood pressure among people with glaucoma.

6.61 If the average systolic blood pressure for people of comparable age is 130 mm Hg, then is there an association between gluacoma and blood pressure?

Hypertension

Hypertensive patients are screened at a neighborhood health clinic and are given methyl dopa, a strong antihypertensive medication for their condition. They are asked to come back 1 week later and have their blood pressures measured again. Suppose the initial and follow-up systolic blood pressures of the patients are given in Table 6.11.

To test the effectiveness of the drug, we want to measure the difference (D) between initial and follow-up blood pressures for each person.

6.62 What is the mean and sd of D?

6.63 What is the standard error of the mean?

TABLE 6.11 Initial and follow-up systolic bp (mm Hg) for hypertensive patients given methyl dopa

Patient no.	Initial systolic bp	Follow-up systolic bp
1	200.0	188.0
2	194.0	212.0
3	236.0	186.0
4	163.0	150.0
5	240.0	200.0
6	225.0	222.0
7	203.0	190.0
8	180.0	154.0
9	177.0	180.0
10	240.0	225.0

6.64 Assume that D is normally distributed with unknown mean μ and known standard deviation $= 20$; that is, $D \sim N(\mu, 400)$. Construct a 95% confidence interval for μ.

6.65 Do you have any opinion on the effect of methyl dopa from the results of these 10 patients?

Venereal Disease

Suppose a clinical trial is conducted to test the efficacy of a new drug, spectinomycin, in the treatment of gonorrhea for females. Forty-six patients are given a 4-g daily dose of the drug and are seen 1 week later, at which time 6 of the patients still have gonorrhea.

6.66 What is the best point estimate for p, the probability of a failure with the drug?

6.67 What is a 95% confidence interval for p?

6.68 Suppose we know that penicillin G at a daily dose of 4.8 mega units has a 10% failure rate. What can be said in comparing the two drugs?

Cancer

Data from American cancer tumor registries suggest that of all people with the type of lung cancer where surgery is the recommended therapy, 40% survive for 3 years from the time of diagnosis and 33% survive for 5 years.

6.69 Suppose that a group of patients who would have received standard surgery are assigned to a new type of surgery. Of 100 such patients, 55 survive for 3 years and 45 survive for 5 years. Can we say that the new form of surgery is better in any sense than the standard form of surgery?

Hepatic Disease

Suppose we are experimenting with a group of guinea pigs and inoculate them with a fixed dose of a particular toxin causing liver enlargement. We find that out of 40 guinea pigs, 15 actually have enlarged livers.

6.70 What is the best point estimate p of the probability of a guinea pig having an enlarged liver?

6.71 What is a two-sided 95% confidence interval for p assuming that the normal approximation is valid?

6.72 Answer Problem 6.71 if we do *not* assume that the normal approximation is valid.

Cardiovascular Disease

A recent hypothesis states that vigorous exercise is an effective preventive measure for subsequent cardiovascular death To test this hypothesis, a sample of 750 men aged 50–75 who report that they jog at least 10 miles per week is ascertained. After 6 years, 64 have died of cardiovascular disease.

6.73 Compute a 95% confidence interval for the incidence of cardiovascular death in this group.

6.74 If the expected death rate from cardiovascular disease over 6 years in 50–75-year-old men based on large samples is 10%, then can a conclusion be drawn concerning this hypothesis from these data?

Pharmacology

Suppose we wish to estimate the concentration (μg/mL) of a specific dose of ampicillin in the urine after various periods of time. We recruit 25 volunteers and find that they have a mean concentration of 7.0 μg/mL with a standard deviation of 2.0 μg/ mL. Assume that the underlying population distribution of concentrations is normally distributed.

6.75 Find a 95% confidence interval for the population mean concentration.

6.76 Find a 99% confidence interval for the population variance of the concentrations.

6.77 How large a sample would be needed to ensure that the length of the confidence interval in Problem 6.75 is 0.5 μg/mL if we assume that the sample standard deviation remains at 2.0 μg/mL?

Cardiovascular Disease

A group of 50 men under the age of 55 with a prior history of myocardial infarction are put on a strict vegetarian diet as part of an experimental program. After 5 years, 2 men from the group have died.

6.78 What is the best point estimate of the 5-year mortality rate in this group of men?

6.79 Derive a 95% confidence interval for the 5-year mortality rate.

6.80 Suppose that from a large sample of men under age 55 with a prior history of myocardial infarction, we know that the 5-year mortality rate is 18%. How does the observed mortality rate obtained in Problem 6.78 compare with the large sample rate of 18%?

Environmental Health

Much discussion has taken place concerning possible health hazards from exposure to anesthetic gases. In one study a group of 525 Michigan nurse anesthetists was ascertained by mail questionnaires and telephone interviews in 1972 to determine the incidence rate of cancer [9]. Of this group, 7 women reported having a new malignancy other than skin cancer during 1971.

6.81 What is the best estimate of the 1971 incidence rate from these data?

6.82 Provide a 95% confidence interval for the true incidence rate.

A comparison was made between the Michigan report and 1969 cancer incidence rates from the Connecticut tumor registry, where the expected incidence rate was determined to be 402.8 per 100,000.

6.83 Comment on the comparison between the observed incidence rate and the Connecticut tumor registry data.

Pulmonary Disease

A spirometric tracing is a standard device used to measure pulmonary function. These tracings represent plots of the volume of air expelled over a 6-second period and tend to look like Figure 6.11. One quantity of interest is the slope at various points along the curve. The slopes are referred to as **flow rates**. A problem that arises is that the flow rates cannot be accurately measured, and some observer error is always introduced. To quantify the observer error, an observer measures the flow at 50% of forced vital capacity (volume as measured at 6 seconds) twice on tracings from 10 different people. A machine called a digitizer can trace the curves automatically and can estimate the flow mechanically. Suppose the digitizer is also used to measure the flow twice on these 10 tracings. The data are given in Table 6.12.

6.84 Find a 95% confidence interval for the standard deviation of the difference between the first and second replicates using the manual method.

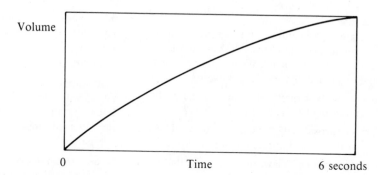

Volume

0   Time   6 seconds

FIGURE 6.11
A typical spirometric
tracing

TABLE 6.12 Estimation of flow rates (1) at 50% of forced vital capacity by a manual and a digitizer method

Person	Manual method replicate 1	Manual method replicate 2	Digitizer replicate 1	Digitizer replicate 2
1	1.80	1.84	1.82	1.83
2	2.01	2.09	2.05	2.04
3	1.63	1.52	1.62	1.60
4	1.54	1.49	1.49	1.45
5	2.21	2.36	2.32	2.36
6	4.16	4.08	4.21	4.27
7	3.02	3.07	3.08	3.09
8	2.75	2.80	2.78	2.79
9	3.03	3.04	3.06	3.05
10	2.68	2.71	2.70	2.70

6.85 Answer Problem 6.84 for the difference between the first and second replicates using the digitizer method.

Suppose we want to compare the variability of the two methods within the same person. Let x_{i1}, x_{i2} represent the 2 replicates on the ith person using the manual method, and let y_{i1}, y_{i2} represent the 2 replicates on the ith person using the digitizer method. Let

$$d_i = |x_{i1} - x_{i2}| - |y_{i1} - y_{i2}|$$

Then, d_i is a measure of the difference in variability using the two methods. Assume that d_i is normally distributed with mean μ_d and variance σ_d^2.

6.86 Find a 95% confidence interval for μ_d.

6.87 What is your opinion on the relative variability of the two methods?

Obstetrics, Serology

A new assay is developed to obtain the concentration of *M. Hominis* mycoplasma in the serum of pregnant women. The developers of this assay wish to make a

statement as to the variability of their laboratory technique. For this purpose, 10 subsamples of 1 ml each are drawn from a large serum sample from *one* woman, and the assay is performed on each subsample. The concentrations are given as follows: $2^4, 2^3, 2^5, 2^4, 2^5, 2^4, 2^3, 2^4, 2^4, 2^5$.

6.88 If the concentration is assumed to be normal in the log scale to the base 2, then give the best estimate of the variance of the method from these data.

6.89 Compute a 95% confidence interval for the variance of the method.

6.90 Assuming that the point estimate in Problem 6.88 is correct, what is the probability that a particular assay, when expressed in the log scale to the base 2, is no more than 1.5 log units off from its true value?

6.91 Answer Problem 6.90 for 2.5 log units.

Turn to the table of random digits (Table 4 in Appendix 1). Start at the top left. Reading across, record for each of the first 11 sets of 10 consecutive digits (1) the second digit, X_i, and (2) the mean of the 10 digits, Y_i.

6.92 Derive the theoretical mean and variance of the distribution of second digits.

6.93 How do the sample properties of the X_i compare with the results of Problem 6.92?

6.94 From your answer to Problem 6.92, what should the mean and variance of the Y_i be?

6.95 How do the actual Y_i's compare with the results of Problem 6.94?

6.96 Do the actual Y_i's relate to the central limit theorem in any way? Elaborate.

Hypertension

Suppose 100 hypertensive people are given an anti-hypertensive drug and the drug is *effective* in 20 of the people. By *effective*, we mean that their diastolic blood pressure is lowered by at least 10 mm Hg as judged from a repeat measurement 1 week after taking the drug.

6.97 What is the best point estimate of the probability p of the drug being effective?

6.98 Suppose we know that 10% of all hypertensive patients who are given a placebo will have their diastolic blood pressure lowered by 10 mm Hg. Can we carry out some procedure to be sure that we are not simply observing the placebo effect?

6.99 What assumptions have you made to carry out the procedure in Problem 6.98?

Suppose we decide that a better measure of the effectiveness of the drug is the absolute decrease in blood pressure rather than the measure of effectiveness used previously. Let $d_i = x_i - y_i$, $i = 1, \ldots, 100$, where $x_i =$ diastolic blood pressure on the ith person before taking the drug and $y_i =$ diastolic blood pressure on the ith person 1 week after taking the drug. Suppose that the sample mean of the d_i is $+5.3$ and the sample variance is 144.0.

6.100 What is the standard error of d?

6.101 What is a 95% confidence interval for the population mean of d?

6.102 Can we make a statement about the effectiveness of the drug?

6.103 What does a 95% confidence interval mean, in words, in this case?

Bacteriology

Suppose a group of mice are inoculated with a uniform dose of a specific type of bacteria and that all die within 24 days, with the distribution of survival times given in Table 6.13.

TABLE 6.13 Survival time of mice after inoculation with a specific type of bacteria

Survival time (days)	No. of mice
10	5
11	11
12	29
13	30
14	40
15	51
16	71
17	65
18	48
19	36
20	21
21	12
22	7
23	2
24	1

6.104 Assume that the underlying distribution of survival times is normal. Estimate the probability p that a mouse will survive for 20 or more days.

6.105 Suppose we are not willing to assume that the underlying distribution is normal. Estimate the probability p that a mouse will survive for 20 or more days.

6.106 Compute 95% confidence limits for the parameter estimated in Problem 6.105.

6.107 Compute 99% confidence limits for the parameter estimated in Problem 6.105.

Draw 6 random samples of size 5 from the data in Table 6.2 (p. 145).

6.108 Compute the mean birthweight for each of the 6 samples.

6.109 Compute the standard deviation based on the sample of 6 means.

6.110 Select the third point from each of the 6 samples and compute the sample standard deviation from this collection of 6 third points.

6.111 What theoretical relationship should there be between the standard deviation in Problem 6.109 and the standard deviation in Problem 6.110?

6.112 How do the actual sample results in Problems 6.109 and 6.110 compare?

Cardiovascular Disease

In Table 2.15 (p. 38) data on serum cholesterol levels of 24 hospital employees before and after they adopted a vegetarian diet were provided.

6.113 What is your best estimate of the effect of adopting a vegetarian diet on change in serum cholesterol levels?

6.114 What is the standard error of the estimate given in Problem 6.113?

6.115 Provide a 95% confidence interval for the estimate given in Problem 6.113.

6.116 What can you conclude from your results in Problem 6.115?

Some physicians consider only changes of at least 10 mg%/ml (the same units as in Table 2.15) to be clinically significant.

6.117 Among people with a clinically significant change in either direction, what is the best estimate of the proportion whose cholesterol levels have declined?

6.118 Provide a 95% confidence interval for the estimate in Problem 6.117.

6.119 What can you conclude from your results in Problem 6.118?

Obstetrics

In Figure 6.4(b) (p. 155) a plot of the sampling distribution of the sample mean from 200 samples of size 5 from the population of 1000 birthweights given in Table 6.2 was provided. The mean of the 1000 birthweights in Table 6.2 is 112.0 oz with standard deviation 20.6 oz.

6.120 If the central limit theorem holds, then what proportion of sample means should fall within 0.5 lb of the population mean (112.0 oz)?

6.121 Answer Problem 6.120 for 1 lb rather than 0.5 lb.

6.122 Compare your results in Problems 6.120 and 6.121 with the actual proportion of sample means that fall in these ranges.

6.123 Do you feel that the central limit theorem is applicable for samples of size 5 from this population?

Hypertension, Pediatrics

The etiology of high blood pressure remains a subject of active investigation. One widely accepted hypothesis is that excessive sodium intake adversely affects blood-pressure outcomes. To explore this hypothesis, an experiment was set up to measure the responsiveness to blood pressure. The protocol used involved testing 3-day-old infants in the newborn nursery by giving them a drop of various solutions and thus eliciting the sucking response and noting (a) the vigor with which they sucked—denoted by MSB = mean number of sucks per burst of sucking, and (b) the mean heart rate during sucking. The content of the solution was changed over 10 consecutive periods: (1) water, (2) water, (3) 0.1 molar salt + water, (4) 0.1 molar salt + water, (5) water, (6) water, (7) 0.3 molar salt + water, (8) 0.3 molar salt + water, (9) water, (10) water. In addition, as a control, the response of the baby to the taste of sugar was also measured after the salt-taste protocol was completed. In this experiment, the sucking response was measured over 5 different periods with the following stimuli: (1) nonnutritive sucking, that is, a pure sucking response was elucidated without using any external substance; (2) water; (3) 5% sucrose + water; (4) 15% sucrose + water; (5) nonnutritive sucking.

The data are given in Data Set 14, INFANTBP.DAT, Appendix 2. The format of the data are given in Data Set 13, INFANTBP.DOC, Appendix 2

6.124 Construct a descriptive table describing the relationship between blood-pressure level and responsiveness to salt. (*Hint*: Divide the infants into responders and nonresponders to the salt stimulus and use interval estimation methods to assess blood-pressure change within each subgroup.)

6.125 Answer Problem 6.124 regarding the sugar stimulus.

Genetics

In Data Set 22, SEXRAT.DAT, the sexes of children born in over 50,000 families with more than one child are listed.

6.126 Use interval estimation methods to determine if the sex of successive births is predictable from the sex of previous births.

Nutrition

In Data Set 26, VALID.DAT, estimated daily consumption of total fat, saturated fat, and alcohol as well as total caloric intake using two different methods of dietary assessment are provided for 173 subjects.

6.127 Use a computer to draw repeated samples of size 5 from this population. Does the central limit theorem seem to hold for these dietary attributes based on samples of size 5?

6.128 Answer Problem 6.127 for samples of size 10.

6.129 Answer Problem 6.127 for samples of size 20.

6.130 How do the sampling distributions compare based on samples of size 5, 10, and 20?

References

[1] Cochran, W. G. (1963). *Sampling techniques* (2nd ed.) New York: Wiley.

[2] Mood, A. & Graybill, F. (1973). *Introduction to the theory of statistics* (3rd ed.) New York: McGraw-Hill.

[3] Arora, N. S., & Rochester, D. F. (1984). Effect of chronic airflow limitation (CAL) on sternocleidomastoid muscle thickness. *Chest, 85*(6), 58S–59S.

[4] Dec, G. W., Jr., Palacios, I. F., Fallon, J. T., Aretz, H. T., Mills, J., Lee, D. C. S., & Johnson, R. A. (1985). Active myocarditis in the spectrum of acute dilated cardiomyopathies. *New England Journal of Medicine, 312*(14), 885–890.

[5] Cramer, D. W., Schiff, I., Schoenbaum, S. C., Gibson, M., Belisle, S., Albrecht, B., Stillman, R. J., Berger, M. J., Wilson, W., Stadel, B. V., & Seibel, M. (1985). Tubal infertility and the intrauterine device. *New England Journal of Medicine, 312*(15), 941–947.

[6] Barry, A. L., Gavan, T. L., & Jones, R. N. (1983). Quality control parameters for susceptibility data with 30 μg netilmicin disks. *Journal of Clinical Microbiology, 18*(5), 1051–1054.

[7] Celentano, D. D., Klassan, A. C., Weisman, C. S., & Rosenshein, N. S. (1988). Cervical cancer screening practices among older women: Results from the Maryland cervical cancer case-control study. *Journal of Clinical Epidemiology, 41*(6), 531–541.

[8] Oldenburg, B., Macdonald, G. J., & Perkins, R. J. (1988). Prediction of quality of life in a cohort of end-stage renal disease patients. *Journal of Clinical Epidemiology, 41*(6), 555–564.

[9] Corbett, T. H., Cornell, R. G., Leiding, K., & Endres, J. L. (1973). Incidence of cancer among Michigan nurse-anesthetists. *Anesthesiology, 38*(3), 260–263.

CHAPTER SEVEN

HYPOTHESIS TESTING: ONE-SAMPLE INFERENCE

Introduction

In Chapter 6 methods of point and interval estimation for parameters of various distributions were discussed. However, researchers often have preconceived ideas about what these parameters might be and wish to test whether the data conform with these ideas.

EXAMPLE 7.1   **Pediatrics, Genetics** A current area of research interest is the familial aggregation of cardiovascular risk factors in general and lipid levels in particular. Suppose it is known that the "average" cholesterol level in children is 175 mg%/mL. A group of men who have died from heart disease within the past year are identified and the cholesterol levels of their offspring are measured. Two hypotheses will be considered:

(1) The average cholesterol level of these children is 175 mg%/mL.

(2) The average cholesterol level of these children is greater than 175 mg%/mL. ∎

This type of question is formulated in a hypothesis-testing framework by specifying two hypotheses—a null and an alternative hypothesis. We wish to compare the relative probabilities of obtaining the sample data under each of these hypotheses. In Example 7.1 the null hypothesis is that the average cholesterol level of the children is 175 mg%/mL and the alternative hypothesis is that the average cholesterol level of the children is greater than 175 mg%/mL.

Why is hypothesis testing so important? Hypothesis testing provides an objective framework for making decisions using probabilistic methods, rather than relying on subjective impressions. People can form different opinions by looking at data, but a hypothesis test provides a uniform decision-making criterion that is consistent for all people.

In this chapter some of the basic concepts of hypothesis testing are developed and applied to one-sample problems of statistical inference. In a **one-sample problem** hypotheses are specified about a single distribution; in a **two-sample problem** two different distributions are compared.

SECTION 7.2   **General Concepts**

EXAMPLE 7.2   **Obstetrics**   Suppose we want to test the hypothesis that mothers with low socioeconomic status (SES) deliver babies whose birthweights are lower than "normal." To test this hypothesis, a list of birthweights from 100 consecutive, full-term, live-born deliveries from the maternity ward of a hospital in a low-SES area is obtained. The mean birthweight ($\bar{x}$) is found to be 115 oz with a sample standard deviation (s) of 24 oz. Suppose we know from nationwide surveys based on millions of deliveries that the mean birthweight in the United States is 120 oz with a standard deviation of 25 oz. Can we actually say that the underlying mean birthweight from this hospital is lower than the national average?   ∎

Assume that the 100 birthweights from this hospital come from an underlying normal distribution with unknown mean μ and known standard deviation $\sigma = 25$. The methods in Section 6.7 could be used to construct a 95% lower one-sided confidence interval for μ based on the sample data, that is, an interval of the form $\mu < c$. If this interval contains 120 oz (i.e., if $c \geqslant 120$), then the hypothesis that these birthweights are not different from the national average would be accepted. If it does not contain 120 oz ($c < 120$), then the hypothesis that these birthweights tend to be lower than the national average would be accepted.

Another way of looking at this problem is in terms of hypothesis testing. In particular, the hypotheses being considered can be formulated in terms of null and alternative hypotheses, which can be defined as follows:

DEFINITION 7.1   ■■■
The **null hypothesis**, denoted by H_0, is the hypothesis that is to be tested. The **alternative hypothesis**, denoted by H_1, is the hypothesis that in some sense contradicts the null hypothesis.   ∎

EXAMPLE 7.3   **Obstetrics**   In Example 7.2 the null hypothesis (H_0) is the hypothesis that the mean birthweight in the low-SES-area hospital (μ) is equal to the mean birthweight in the United States (μ_0). This is the hypothesis we want to test. The alternative hypothesis (H_1) is the hypothesis that the mean birthweight in this hospital (μ) is less than the mean birthweight in the United States (μ_0). We want to compare the relative probabilities of obtaining the sample data under each of these two hypotheses.   ∎

We also assume that the standard deviation is known to be 25 (σ_0) and that the underlying distribution is normal under either hypothesis. These hypotheses can be written more succinctly in the following form:

7.1

$$H_0: \mu = \mu_0, \sigma = \sigma_0 \quad \text{vs.} \quad H_1: \mu < \mu_0, \sigma = \sigma_0$$

Suppose that the only possible decisions are whether H_0 is true or H_1 is true. Actually, for ease of notation, all outcomes in a hypothesis-testing situation generally refer to the null hypothesis. Hence, if we decide that H_0 is true, then we say that we accept H_0. If we decide that H_1 is true, then we state that H_0 is not true or, equivalently, that we reject H_0. Thus four possible outcomes can occur:

(1) We accept H_0, and H_0 is in fact true.

(2) We accept H_0, and H_1 is in fact true.

(3) We reject H_0, and H_0 is in fact true.

(4) We reject H_0, and H_1 is in fact true.

These four possibilities are depicted in Table 7.1.

TABLE 7.1
Four possible outcomes
in hypothesis testing

Decision	H_0	H_1
Accept H_0	H_0 is true and H_0 is accepted	H_1 is true and H_0 is accepted
Reject H_0	H_0 is true and H_0 is rejected	H_1 is true and H_0 is rejected

In actual practice, it is impossible, using hypothesis-testing methods to *prove the null hypothesis*. Thus, in particular, if we *accept* H_0, then we have actually failed to reject H_0.

If H_0 is true and H_0 is accepted, or if H_1 is true and H_0 is rejected, then the correct decision has been made. If H_0 is true and H_0 is rejected or if H_1 is true and H_0 is accepted, then an *error* has been made. The two types of errors are generally treated differently.

DEFINITION 7.2 ■■■

The probability of a **type I error** is the probability of rejecting the null hypothesis given that H_0 is true.
■

DEFINITION 7.3 ■■■

The probability of a **type II error** is the probability of accepting the null hypothesis given that H_1 is true. This probability is a function of μ as well as other factors.
■

EXAMPLE 7.4   **Obstetrics** In the context of the birthweight data in Example 7.2, a type I error would be the probability of deciding that the mean birthweight in the hospital was less than 120 oz when in fact it was 120 oz. A type II error would be the probability of deciding that the mean birthweight was 120 oz when in fact it was less than 120 oz.
■

EXAMPLE 7.5   **Cardiovascular Disease, Pediatrics** What are the type I and type II errors for the cholesterol data in Example 7.1?

SOLUTION   The type I error is the probability of deciding that the offspring of men who have died from heart disease have an average cholesterol greater than 175 mg%/mL when in fact their average cholesterol level is 175. The type II error is the probability of deciding that the offspring have normal cholesterol levels when in fact their cholesterol levels are above average.
■

Type I and type II errors often result in monetary and nonmonetary costs.

EXAMPLE 7.6   **Obstetrics** The birthweight data in Example 7.2 might be used to decide if a special-care nursery for low-birthweight babies is needed in this hospital. If H_1 were true, that is, if the birthweights in this hospital did tend to be lower than the national average, then the hospital might be justified in having its own special-care nursery. If H_0 were true and the mean birthweight was no different from the U.S. average, then the hospital probably does not need such a nursery. If a type I error is made, then a special-care nursery will be recommended, with all the extra costs involved, when in fact it is not needed. If a type II error is made, a special-care nursery will not be funded, when in fact it is needed. The nonmonetary cost of this decision is that some low-birthweight babies may not survive without the unique equipment in a special-care nursery.
■

DEFINITION 7.4 ■■■
The probability of a type I error is usually denoted by α and is commonly referred to as the
significance level of a test.   ■

DEFINITION 7.5 ■■■
The probability of a type II error is usually denoted by β.   ■

DEFINITION 7.6 ■■■
The **power** of a test is defined as

$$1 - \beta = 1 - \text{probability of a type II error}$$ ■

The general aim in hypothesis testing is to use statistical tests that make α and
β as small as possible. This goal requires compromise, since making α small
involves rejecting the null hypothesis less often, whereas making β small involves
accepting the null hypothesis less often. These actions are contradictory; that is,
as α increases, β will decrease, while as α decreases, β will increase. Our general
strategy will be to fix α at some specific level, for example .10, .05, .01, ..., and to use
the test that minimizes β or, equivalently, maximizes the power.

SECTION 7.3   **One-Sample Test for the Mean of a Normal Distribution**
with Known Variance: One-Sided Alternatives

The appropriate hypothesis test for the birthweight data in Example 7.2 will now
be developed. The statistical model in this case is that the birthweights come from a
normal distribution with mean μ and variance σ^2. We assume that σ^2 is known to
be 625 and we wish to test the null hypothesis, H_0, that $\mu = 120$ oz versus the
alternative hypothesis, H_1, that $\mu < 120$ oz. Suppose a more specific alternative,
namely, $H_1: \mu = \mu_1 = 110$ oz, is selected. We will show that the nature of the best
test does not depend on the value chosen for μ_1 provided that μ_1 is less than 120 oz.
We will also fix the α level at .05 for concreteness.

EXAMPLE 7.7   A very simple test could be used by referring to the table of random digits in Table 4 in
Appendix 1. Suppose two digits are selected from this table and the null hypothesis is
rejected if these two digits are between 00 and 04 inclusive and is accepted if these two digits
are between 05 and 99. Clearly, from the properties of the random number table, the type I
error of this test = $\alpha = Pr(\text{rejecting the null hypothesis}|H_0 \text{ true}) = Pr(\text{drawing two random}$
digits between 00 and 04) $= \frac{5}{100} = .05$. Thus, the proposed test satisfies the α-level criterion
given previously. The problem with this test is that it has very low power. Indeed, the power
of the test = $Pr(\text{rejecting the null hypothesis}|H_1 \text{ true}) = Pr(\text{drawing two random digits}$
between 00 and 04) $= \frac{5}{100} = .05$.
   Note that the outcome of the test has nothing to do with the sample birthweights
drawn. H_0 will be rejected just as often when the sample mean birthweight ($\bar{x}$) is 110 oz as
when it is 120 oz. Thus, this test must be very poor, since we would expect to reject H_0 with
near certainty if $\bar{x}$ is sufficiently small and would expect never to reject H_0 if $\bar{x}$ is sufficiently
large.   ■

It can be shown that the best (most powerful) test in this situation is based on
the sample mean ($\bar{x}$) if the sample is a random sample. If $\bar{x}$ is sufficiently smaller

than μ_0, then H_0 is rejected; otherwise, H_0 is accepted. This test is reasonable, since if H_0 is true, then the most likely values of $\bar{x}$ will tend to cluster around μ_0, whereas if H_1 is true, the most likely values of $\bar{x}$ will tend to cluster around μ_1. The distributions of $\bar{x}$ under H_0 and H_1 are depicted in Figure 7.1.

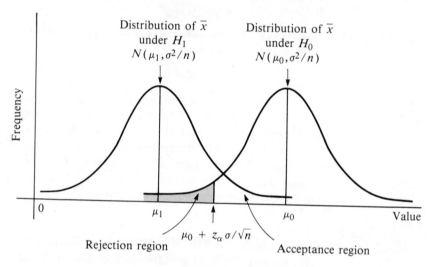

FIGURE 7.1
The distribution of x under the null (H_0) and alternative (H_1) hypotheses

Note that if H_0 is true, the distribution of $\bar{x}$ is depicted by the right-hand curve in Figure 7.1, whereby small values of $\bar{x}$ are unlikely. Similarly, if H_1 is true, then the distribution of $\bar{x}$ is depicted by the left-hand curve in Figure 7.1, whereby large values of $\bar{x}$ are unlikely. These distributions demonstrate why H_0 is rejected for small values of $\bar{x}$ and accepted for large values of $\bar{x}$.

DEFINITION 7.7   ■■
The **acceptance region** is the range of values of $\bar{x}$ for which H_0 is accepted.   ■

DEFINITION 7.8   ■■
The **rejection region** is the range of values of $\bar{x}$ for which H_0 is rejected.   ■

For the birthweight data in Example 7.2, the rejection region consists of small values of $\bar{x}$ because the underlying mean under the alternative hypothesis (μ_1) is less than the underlying mean under the null hypothesis. This type of test is called a one-tailed test.

DEFINITION 7.9   ■■
A **one-tailed test** is a test in which the values of the parameter being studied (in this case μ) under the alternative hypothesis are allowed to be either greater than or less than the values of the parameter under the null hypothesis (μ_0) *but not both.*   ■

EXAMPLE 7.8   **Cardiovascular Disease, Pediatrics** The hypotheses for the cholesterol data in Example 7.1 are $H_0: \mu = \mu_0$ versus $H_1: \mu > \mu_0$, where μ is the true mean cholesterol level for children of men who have died from heart disease. This test is also one-tailed, since the alternative mean is only allowed to be greater than the null mean.   ■

How small should $\bar{x}$ be for H_0 to be rejected? This issue can be settled by recalling that the significance level of the test is set at α. Suppose that H_0 is rejected for all values of $\bar{x} < c$ and accepted otherwise. The value c should be selected such that

7.2
$$\alpha = Pr(\text{type I error}|H_0) = Pr\left[\bar{x} < c | \bar{x} \sim N\left(\mu_0, \frac{\sigma^2}{n}\right)\right] = \Phi\left[\frac{c - \mu_0}{\sigma/\sqrt{n}}\right]$$

Thus, using the z notation for the percentiles of a normal distribution developed in Chapter 5,

7.3
$$z_\alpha = \frac{c - \mu_0}{\sigma/\sqrt{n}}$$

If both sides of **(7.3)** are multiplied by $\sigma/\sqrt{n}$ and μ_0 is added,

7.4
$$c = \mu_0 + \frac{z_\alpha \sigma}{\sqrt{n}}$$

Thus, the test takes on the following form:

7.5   **One-Sample Test for the Mean of a Normal Distribution with Known Variance (Alternative Mean < Null Mean)**

To test the hypothesis

$$H_0: \mu = \mu_0, \sigma = \sigma_0 \qquad \text{vs.} \qquad H_1: \mu < \mu_0, \sigma = \sigma_0$$

with a significance level of α, the best (most powerful) test is based on $\bar{x}$: if

$$\bar{x} < \mu_0 + z_\alpha \sigma/\sqrt{n}$$

then H_0 is rejected. If

$$\bar{x} \geqslant \mu_0 + z_\alpha \sigma/\sqrt{n}$$

then H_0 is accepted.

The acceptance and rejection regions for the birthweight data in Example 7.2 are depicted in Figure 7.1.

It can be shown that for a given α level, this test maximizes the power or, equivalently, minimizes the type II error.

EXAMPLE 7.9   **Obstetrics** If an α level of .05 is used for the birthweight data in Example 7.2, then $z_\alpha = -1.645$ and H_0 would be rejected if $\bar{x} < 120 - 1.645(25)/10$ or if $\bar{x} < 120 - 4.11 = 115.89$. H_0 would be accepted if $\bar{x} \geqslant 115.89$. Thus, the rejection region is $\bar{x} < 115.89$ and the acceptance region is $\bar{x} \geqslant 115.89$. H_0 would be rejected in this case, since $\bar{x}$ was in fact 115 oz. If, instead, an α level of .01 rather than .05 is used, H_0 would be rejected if

$$\bar{x} < 120 + z_{.01}(25)/10 = 120 - 2.326(25)/10 = 120 - 5.82 = 114.18$$

and H_0 would be accepted if $\bar{x} \geqslant 114.18$. In other words, the rejection region is $\bar{x} < 114.18$ and the acceptance region is $\bar{x} \geqslant 114.18$. Since $\bar{x} = 115$ oz, H_0 would be accepted in this case. ∎

How do we know what level of α to use? The actual α level used should depend on the relative importance of type I and type II errors, since, for a fixed sample size (n), the smaller α is made the larger β becomes. Most people feel uncomfortable with α levels much greater than .05. Traditionally, an α level of exactly .05 is used most frequently.

In general, a number of significance tests could be performed at different α levels, as was done in Example 7.9, and whether H_0 would be accepted or rejected in each instance could be noted. This can be somewhat tedious and is unnecessary since, instead, significance tests can be effectively performed *at all α levels* by obtaining the *p*-value for the test.

DEFINITION 7.10 ██

The *p*-value for any hypothesis test is the α level at which we would be indifferent to accepting or rejecting H_0 given the sample data at hand. That is, the *p*-value is the α level at which the given value of the statistic (such as $\bar{x}$) would be on the borderline between the acceptance and rejection regions. ■

According to the test criterion in **(7.5)**, if a significance level of p is used, then H_0 would be rejected if $\bar{x} < \mu_0 + z_p\sigma/\sqrt{n}$ and accepted if $\bar{x} \geq \mu_0 + z_p\sigma/\sqrt{n}$. Hence, we would be indifferent to accepting or rejecting H_0 if $\bar{x} = \mu_0 + z_p\sigma/\sqrt{n}$. By subtracting μ_0 from both sides of the equation and dividing by $\sigma/\sqrt{n}$, we obtain

7.6
$$z_p = (\bar{x} - \mu_0)/(\sigma/\sqrt{n})$$

or, alternatively,

7.7
$$p = \Phi[(\bar{x} - \mu_0)/(\sigma/\sqrt{n})]$$

The *p*-value can be depicted graphically as the area under an $N(\mu_0, \sigma^2/n)$ curve to the left of the sample mean $\bar{x}$, as is shown in Figure 7.2.

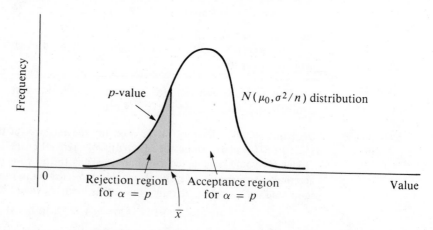

FIGURE 7.2
Graphical display of a
p-value

EXAMPLE 7.10   **Obstetrics** Compute the *p*-value for the birthweight data in Example 7.2.

SOLUTION   From **(7.7)** we have that the *p*-value is given by

$$\Phi\left[\frac{(115-120)}{(25/10)}\right] = \Phi(-2.0) = .023$$ ■

An alternative definition of a p-value that will be useful in other hypothesis-testing problems is given as follows:

DEFINITION 7.11 ■■

The **p-value** can also be thought of as the probability of obtaining a result as extreme or more extreme than the actual sample value obtained given that the null hypothesis is true.   ■

We know that under the null hypothesis, $\bar{x} \sim N(\mu_0, \sigma^2/n)$. Hence, the probability of obtaining a sample mean that is no larger than $\bar{x}$ under the null hypothesis is $\Phi[(\bar{x} - \mu_0)/(\sigma/\sqrt{n})] = p$-value, as shown in Figure 7.2.

EXAMPLE 7.11 **Cardiology** A topic of recent clinical interest is the possibility of using drugs to reduce infarct size in patients who have had a myocardial infarction within the past 24 hours. Suppose we know that in untreated patients the mean infarct size is 25 ($ck - g - EQ/m^2$) with a standard deviation of 10. Furthermore, in 8 patients treated with drug, the mean infarct size is 16. Is the drug effective in reducing infarct size?

SOLUTION The hypotheses are H_0: $\mu = 25$, $\sigma = 10$ versus H_1: $\mu < 25$, $\sigma = 10$. The p-value is computed using **(7.7)** as follows:

$$p = \Phi\left[\frac{(16 - 25)}{(10/\sqrt{8})}\right] = \Phi(-2.55) = 1 - \Phi(2.55) = 1 - .9946 \approx .005$$

Thus, H_0 is rejected and we conclude that the drug reduces infarct size (all other things being equal).

This can be interpreted as the probability that mean infarct size among a random sample of 8 patients will be no larger than 16, given that the null hypothesis is true. In this example, the null hypothesis is that the drug is ineffective, or in other words, that true mean infarct size for the population of all patients with myocardial infarction that are treated with drug = true mean infarct size for untreated patients = 25.   ■

The importance of the p-value is that it tells us *exactly* how significant the results are without performing repeated significance tests at different α levels. A question typically asked is, How small should the p-value be for results to be considered statistically significant? Although this question has no definite answer, some commonly used criteria follow:

7.8 **Guidelines for Judging the Significance of a p-value**

If $.01 \leqslant p < .05$, then the results are *significant*.
If $.001 \leqslant p < .01$, then the results are *highly significant*.
If $p < .001$, then the results are *very highly significant*.
If $p > .05$, then the results are considered *not statistically significant* (sometimes denoted by NS).
However,
If $.05 \leqslant p < .10$, then a trend toward statistical significance is sometimes noted.

Authors frequently do not specify the exact p-value beyond giving ranges of the type shown here, since whether the p-value is .024 or .016 is thought to be unimportant. Other authors give an exact p-value even for results that are not statistically significant so that the reader can appreciate how close to statistical significance the results have come. With the advent of statistics performed by personal computers, exact p-values are easy to obtain. These different approaches lead to the following general principle:

7.9	**Determination of Statistical Significance for Results from Hypothesis Tests**

Either of the following methods can be used to establish whether results from hypothesis tests are statistically significant:

(1) The test statistic $\bar{x}$ can be computed and compared with the critical value c at an α level of .05. Specifically, if $H_0: \mu = \mu_0$ versus $H_1: \mu < \mu_0$ are being tested and $\bar{x} < c = \mu_0 + z_{.05}\sigma/\sqrt{n}$, then H_0 is rejected and the results are declared *statistically significant* (i.e., $p < .05$). Otherwise, H_0 is accepted and the results are declared *not statistically significant* (i.e., $p \geq .05$).

(2) The exact p-value can be computed, and if $p < .05$, then H_0 is rejected and the results are declared *statistically significant*. Otherwise, if $p \geq .05$, then H_0 is accepted and the results are declared *not statistically significant*.

These two approaches are equivalent regarding the determination of statistical significance (i.e., whether $p < .05$ or $p \geq .05$). The second approach is somewhat more precise in that it yields an exact p-value. The two approaches in **(7.9)** can also be used to determine statistical significance in other hypothesis-testing problems.

EXAMPLE 7.12   **Obstetrics** Assess the statistical significance of the birthweight data in Example 7.2.

SOLUTION   Since the p-value is .023, the results would be considered statistically significant and we would conclude that the true birthweight is significantly lower in this hospital than in the general population.   ∎

EXAMPLE 7.13   **Cardiology** Assess the significance of the infarct-size data in Example 7.11.

SOLUTION   The p-value = .005, and thus the results are highly significant.   ∎

In writing up the results of a study, a distinction between scientific and statistical significance should be made, since the two terms do not necessarily coincide. The results of a study can be statistically significant but still not be scientifically important. This situation would occur if a small difference was found to be statistically significant because of a large sample size. Conversely, some statistically insignificant results can be scientifically important, encouraging researchers to perform larger studies to confirm the direction of the findings and possibly reject H_0 with a larger sample size.   ∎

EXAMPLE 7.14   **Obstetrics** Suppose the mean birthweight in Example 7.2 was 119 oz based on a sample of size 10,000. Assess the results of the study.

SOLUTION   The p-value would be given by

$$\Phi\left(\frac{119 - 120}{25/\sqrt{10,000}}\right) = \Phi(-4.00) < .001$$

The results are thus very highly significant but are clearly not very important because of the small difference in mean birthweight (1 oz) between this hospital and the national average.   ∎

EXAMPLE 7.15   **Obstetrics** Suppose that the mean birthweight in Example 7.2 was 110 oz based on a sample size of 10. Assess the results of the study.

SOLUTION   The p-value would be given by

$$\Phi\left(\frac{110 - 120}{25/\sqrt{10}}\right) = \Phi(-1.26) = 1 - \Phi(1.26) = 1 - .8962 \approx .104$$

These results are not statistically significant but could be important if the same trends were also apparent in a larger study.   ∎

The test criterion in **(7.5)** was based on an alternative hypothesis that $\mu < \mu_0$. In many situations we wish to use an alternative hypothesis that $\mu > \mu_0$. In this case H_0 would be rejected if $\bar{x}$ were large ($>c$) and accepted if $\bar{x}$ were small ($\leqslant c$). To ensure a type I error of α, find c such that

$$\alpha = Pr(\bar{x} > c | H_0) = Pr(\bar{x} > c | \mu = \mu_0)$$

$$= 1 - Pr(\bar{x} \leq c | \mu = \mu_0) = 1 - \Phi\left(\frac{c - \mu_0}{\sigma/\sqrt{n}}\right)$$

As in **(7.2)**, this equation can be rewritten as

$$\Phi\left(\frac{c - \mu_0}{\sigma/\sqrt{n}}\right) = 1 - \alpha \qquad \text{or} \qquad \frac{c - \mu_0}{\sigma/\sqrt{n}} = z_{1-\alpha}$$

Solving for c, we get $c = \mu_0 + z_{1-\alpha}\sigma/\sqrt{n}$. Thus, at level α, H_0 is rejected if $\bar{x} > \mu_0 + z_{1-\alpha}\sigma/\sqrt{n}$ and accepted otherwise. The p-value is the probability of observing a sample mean at least as large as $\bar{x}$ under the null hypothesis. Thus,

$$p = Pr(X > \bar{x} | X \sim N(\mu_0, \sigma^2/n)) = 1 - \Phi\left(\frac{\bar{x} - \mu_0}{\sigma/\sqrt{n}}\right)$$

7.10   **One-Sample Test for the Mean of a Normal Distribution with Known Variance (Alternative Mean > Null Mean)**

To test the hypothesis

$$H_0: \mu = \mu_0, \sigma = \sigma_0 \qquad \text{vs.} \qquad H_1: \mu > \mu_0, \sigma = \sigma_0$$

with a significance level of α, the best test is based on $\bar{x}$: If

$$\bar{x} > \mu_0 + z_{1-\alpha}\sigma/\sqrt{n}, \qquad \text{then } H_0 \text{ is rejected}$$

If

$$\bar{x} \leqslant \mu_0 + z_{1-\alpha}\sigma/\sqrt{n}, \qquad \text{then } H_0 \text{ is accepted}$$

The p-value for this test is given by

$$p = 1 - \Phi[(\bar{x} - \mu_0)/(\sigma/\sqrt{n})]$$

The acceptance and rejection regions for this test are depicted in Figure 7.3.

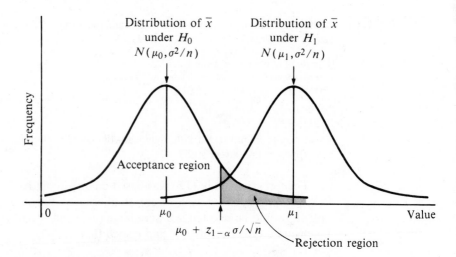

FIGURE 7.3
Acceptance and
rejection regions for the
one-sample normal test
when the alternative
mean (μ_1) > null mean
(μ_0)

EXAMPLE 7.16 **Cardiovascular Disease, Pediatrics** Suppose that the mean cholesterol level of 10 children in Example 7.1 (p. 186) is $200\,\mathrm{mg\%/mL}$ and the standard deviation is assumed to be 50 $\mathrm{mg\%/mL}$. Test the hypothesis that the mean cholesterol level is higher in this group than in the general population.

SOLUTION The hypothesis

$$H_0: \mu = 175, \sigma = 50 \quad \text{vs.} \quad H_1: \mu > 175, \sigma = 50$$

is tested using an α level of .05. H_0 is rejected if

$$\bar{x} > 175 + \frac{z_{.95}(50)}{\sqrt{10}} = \frac{175 + 1.645(50)}{\sqrt{10}} = 175 + 26.01 = 201.01$$

and accepted otherwise. Since $\bar{x} = 200 < 201.01$, H_0 is accepted. An exact p-value could also be computed, which is given by

$$p = 1 - \Phi[(200 - 175)/(50/\sqrt{10})] = 1 - \Phi(25/15.811)$$

$$= 1 - \Phi(1.58) = 1 - .9429$$

$$\approx .057$$

Since $p > .05$, we conclude that our results are not statistically significant, and the null hypothesis that the mean cholesterol level of these children is no different from that of an average child is accepted. ∎

SECTION 7.4 **One-Sample Normal Test: Two-Sided Alternatives**

In the previous section the alternative hypothesis was assumed to be in a *specific direction* relative to the null hypothesis.

EXAMPLE 7.17 **Obstetrics** In Example 7.2 it was assumed that birthweights of infants from a low-SES-area hospital were either the same or lower than average. In Example 7.1 it was assumed that cholesterol levels of children of men who died from heart disease were either the same or higher than average. ∎

In most instances this *prior knowledge* is unavailable. If the null hypothesis is not true, then we have no idea in which direction the alternative mean will fall.

EXAMPLE 7.18 **Cardiovascular Disease** Suppose we want to compare fasting serum cholesterol levels in people over 21 living in a group of islands in the South Pacific with typical levels found in the United States. Suppose we assume that levels in adults over 21 in the United States are approximately normally distributed with mean 190 mg/dL and standard deviation 40 mg/dL. We have no idea what the relative levels of serum cholesterol are on the islands as compared with the United States. We will assume that the levels on the islands are normally distributed with some unknown mean μ and standard deviation 40. Hence, we wish to test the null hypothesis $H_0: \mu = \mu_0 = 190$, $\sigma^2 = 1600$ versus the alternative hypothesis $H_1: \mu \neq \mu_0$, $\sigma^2 = 1600$. Blood tests are performed on 100 adults from the islands and the mean level ($\bar{x}$) is found to be 181.52 mg/dL. What can be concluded on the basis of this evidence? ∎

The type of alternative given in Example 7.18 is known as a *two-sided* alternative, since the alternative mean can be either less than or greater than the null mean.

DEFINITION 7.12 ■■■
A **two-tailed test** is a test in which the values of the parameter being studied (in this case μ) under the alternative hypothesis are allowed to be either *greater than or less than* the values of the parameter under the null hypothesis (μ_0). ∎

The best test here depends on the sample mean $\bar{x}$, as it did in the one-sided situation developed in Section 7.3. We showed in **(7.5)** that to test the hypotheses $H_0: \mu = \mu_0$ versus $H_1: \mu < \mu_0$, the best test was of the form: reject H_0 if $\bar{x} < c$ and accept H_0 if $\bar{x} \geqslant c$, where $c = \mu_0 + z_\alpha \sigma/\sqrt{n}$. This test is clearly only appropriate for alternatives on one side of the null mean, namely, $\mu < \mu_0$. We also showed in **(7.10)** that to test the hypothesis

$$H_0: \mu = \mu_0 \quad \text{vs.} \quad H_1: \mu > \mu_0$$

the best test was correspondingly of the form: reject H_0 if $\bar{x} > c = \mu_0 + z_{1-\alpha}\sigma/\sqrt{n}$ and accept H_0 if $\bar{x} \leqslant c$.

A reasonable decision rule to test for alternatives on *either* side of the null mean is to *reject H_0 if $\bar{x}$ is either too small or too large*. Another way of stating this rule is that H_0 will be rejected if $\bar{x}$ is either $<c_1$ or $>c_2$ for some constants c_1, c_2 and H_0 will be accepted if $c_1 \leqslant \bar{x} \leqslant c_2$.

The question remains, What are appropriate values for c_1 and c_2? These values are again determined by the type I error (α). The constants c_1, c_2 should be chosen such that

7.11
$$Pr(\text{reject } H_0 | H_0 \text{ true}) = Pr(\bar{x} < c_1 \text{ or } \bar{x} > c_2 | H_0 \text{ true})$$
$$= Pr(\bar{x} < c_1 | H_0 \text{ true}) + Pr(\bar{x} > c_2 | H_0 \text{ true}) = \alpha$$

Half of the type I error is arbitrarily assigned to each of the probabilities on the left-hand side of **(7.11)**. Thus, we wish to find c_1, c_2 such that

7.12
$$Pr(\bar{x} < c_1 | H_0 \text{ true}) = Pr(\bar{x} > c_2 | H_0 \text{ true}) = \alpha/2$$

$$Pr(\bar{x} < c_1 | H_0) = \Phi\left(\frac{c_1 - \mu_0}{\sigma/\sqrt{n}}\right) = \alpha/2$$

$$Pr(\bar{x} > c_2 | H_0) = 1 - \Phi\left(\frac{c_2 - \mu_0}{\sigma/\sqrt{n}}\right) = \alpha/2$$

Solving for c_1 and c_2 in the same manner as in the one-sided case,

$$c_1 = \mu_0 - z_{1-\alpha/2}\sigma/\sqrt{n}$$
$$c_2 = \mu_0 + z_{1-\alpha/2}\sigma/\sqrt{n}$$

This test procedure can be summarized as follows:

7.13   **One-Sample Test for the Mean of a Normal Distribution with Known Variance (Two-Sided Alternative)**

To test the hypothesis $H_0: \mu = \mu_0, \sigma = \sigma_0$ versus $H_1: \mu \neq \mu_0, \sigma = \sigma_0$ with a significance level of α, the best test is based on $\bar{x}$: If

$$\bar{x} < \mu_0 - z_{1-\alpha/2}\sigma/\sqrt{n} \quad \text{or} \quad \bar{x} > \mu_0 + z_{1-\alpha/2}\sigma/\sqrt{n}$$

then H_0 is rejected. If

$$\mu_0 - z_{1-\alpha/2}\sigma/\sqrt{n} \leqslant \bar{x} \leqslant \mu_0 + z_{1-\alpha/2}\sigma/\sqrt{n}$$

then H_0 is accepted.

The acceptance and rejection regions for this test are depicted in Figure 7.4.

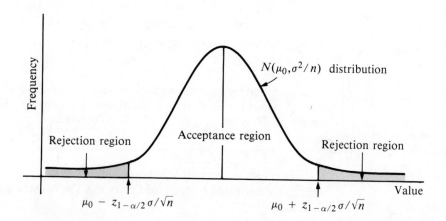

FIGURE 7.4
One-sample test for the mean of a normal distribution with known variance (two-sided alternative)

EXAMPLE 7.19   **Cardiovascular Disease** Test the hypothesis that cholesterol levels of adults living in the South Pacific are different from those in the United States using the data in Example 7.18.

SOLUTION   Compute c_1, c_2 as follows:

$$c_1 = 190 - z_{.975}(40)/\sqrt{100} = 190 - 1.96(4) = 182.16$$
$$c_2 = 190 + 1.96(4) = 197.84$$

Since $\bar{x} = 181.52 < c_1$, the null hypothesis would be rejected in favor of the alternative hypothesis that the underlying mean cholesterol level is different from 190. ■

Alternatively, we might want to compute a p-value as we did in the one-sided case. The p-value is computed in two different ways, depending on whether $\bar{x}$ is less than or greater than μ_0.

| 7.14 | p-value for the One-Sample Test for the Mean of a Normal Distribution with Known Variance (Two-Sided Alternative) |

$$p = \begin{cases} 2\Phi[(\bar{x} - \mu_0)/(\sigma/\sqrt{n})] & \text{if } \bar{x} \leqslant \mu_0 \\ 2\{1 - \Phi[(\bar{x} - \mu_0)/(\sigma/\sqrt{n})]\} & \text{if } \bar{x} > \mu_0 \end{cases}$$

Thus, in words, if $\bar{x} \leqslant \mu_0$, then $p = 2$ times the area under an $N(\mu_0, \sigma^2/n)$ curve to the left of $\bar{x}$; if $\bar{x} > \mu_0$, then $p = 2$ times the area under an $N(\mu_0, \sigma^2/n)$ curve to the right of $\bar{x}$. One way to interpret the p-value is as follows:

The **p-value** is the probability under the null hypothesis of obtaining a sample mean as extreme as or more extreme than the observed sample mean, where, because a two-sided alternative hypothesis is being used, extremeness is measured as the **absolute value** of the difference between $\bar{x}$ and μ_0.

Hence, if $\bar{x} \leqslant \mu_0$, the p-value is the area to the left of $\bar{x}$ plus the area to the right of $\mu_0 + (\mu_0 - \bar{x}) = 2\mu_0 - \bar{x}$ under an $N(\mu_0, \sigma^2/n)$ curve.

However, this area simply amounts to twice the left-hand tail area, since the normal curve is symmetric about μ_0. Similarly, if $\bar{x} > \mu_0$, then the p-value is the area to the right of $\bar{x}$ plus the area to the left of $\mu_0 - (\bar{x} - \mu_0) = 2\mu_0 - \bar{x}$ under an $N(\mu_0, \sigma^2/n)$ curve = twice the right-hand tail area.

These areas are illustrated in Figure 7.5.

EXAMPLE 7.20   **Cardiovascular Disease** Compute the p-value for the hypothesis test in Example 7.19.

SOLUTION   Since $\bar{x} = 181.52 < 190$, the p-value for the test would be twice the left-hand tail area, or

$$p = 2 \times \Phi[(181.52 - 190)/(40/10)] = 2 \times \Phi(-8.48/4) = 2 \times \Phi(-2.12)$$
$$= 2 \times [1 - \Phi(2.12)] = 2(1 - .983) = .034$$

Hence, the results are statistically significant with a p-value of .034. ■

When is a one-sided test more appropriate than a two-sided test? Generally, the sample mean falls in the expected direction from μ_0 and it is *easier* to reject H_0 using a one-sided test than using a two-sided test. Indeed, with the data in Example 7.18, the one-sided p-value to test $H_0 : \mu = \mu_0$ versus $H_1 : \mu < \mu_0$ would be

$$p = \Phi\left[(\bar{x} - \mu_0)\bigg/\left(\frac{\sigma}{\sqrt{n}}\right)\right] = \frac{1}{2}(.034) = .017 = \frac{1}{2} \text{ (two-sided } p\text{-value)}$$

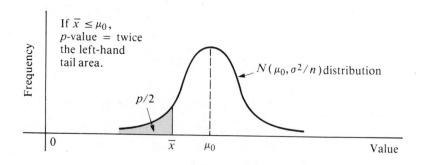

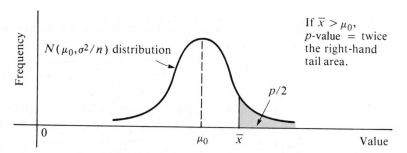

FIGURE 7.5
Illustration of the *p*-value for a one-sample test for the mean of a normal distribution with known variance (two-sided alternative)

Generally, a two-sided test is always appropriate, since then there can be no question about the conclusions. However, in certain situations only alternatives on one side of the null mean are of interest or are possible, and in this case a one-sided test is better because it has more power than its two-sided counterpart. In all instances, it is important to decide whether to use a one-sided or a two-sided test *before* data analysis (or preferably before data collection) begins so as not to bias conclusions based on results of hypothesis testing. In particular, you should not change from a two-sided to a one-sided test *after* looking at the data.

EXAMPLE 7.21   **Hypertension**   Suppose we are testing the efficacy of a drug to reduce blood pressure. We will assume that the change in blood pressure (baseline blood pressure minus follow-up blood pressure) is normally distributed with mean μ and variance σ^2. An appropriate hypothesis test might be $H_0: \mu = 0$ versus $H_1: \mu > 0$, since the drug interests us only if it reduces the level of blood pressure, not if it raises it.   ∎

<u>SECTION 7.5</u>   **One-Sample *t* Test**

In the previous sections of this chapter it has been assumed that the variance of the underlying distribution was known.

EXAMPLE 7.22   **Obstetrics, Cardiovascular Disease**   We assumed that the null distribution of birthweights in Example 7.2 and of cholesterol levels in Example 7.18 was the same, respectively, as distributions from much larger populations whose means and variances were known.   ∎

This type of information is usually unavailable. And even if it is available, its applicability to the study population is always questionable. Therefore, *it is usually assumed that the underlying variance of the population is unknown.* How then should the hypothesis tests be conducted?

It still makes good sense to base the significance tests on $\bar{x}$. In the two-sided normal test in (7.13), c_1 and c_2 were chosen such that $Pr(\bar{x} < c_1) = Pr(\bar{x} > c_2) = \alpha/2$ and H_0 was rejected if either $\bar{x} < c_1$ or $\bar{x} > c_2$ and accepted if $c_1 \leqslant \bar{x} \leqslant c_2$. We showed that if σ is known, then, because $(\bar{x} - \mu_0)/(\sigma/\sqrt{n})$ is distributed as an $N(0, 1)$ random variable under H_0, c_1 and c_2 can be derived as $\mu_0 - z_{1-\alpha/2}\sigma/\sqrt{n}$ and $\mu_0 + z_{1-\alpha/2}\sigma/\sqrt{n}$, respectively. However, if σ is unknown, c_1 and c_2 cannot be derived in this manner. We know from Section 6.4.5 that if $x_1, \dots, x_n$ are a random sample from an $N(\mu_0, \sigma^2/n)$ distribution and σ is estimated by s, then the random variable $(\bar{x} - \mu_0)/(s/\sqrt{n})$ follows a t distribution with $n - 1$ degrees of freedom. Hence, under the null hypothesis, that is, $\mu = \mu_0$

7.15
$$Pr\left[\frac{(\bar{x} - \mu_0)}{(s/\sqrt{n})} < t_{n-1, \alpha/2}\right] = Pr\left[\frac{(\bar{x} - \mu_0)}{(s/\sqrt{n})} > t_{n-1, 1-\alpha/2}\right] = \alpha/2$$

by the definition of the percentiles of a t distribution. Each inequality in (7.15) is multiplied by $s/\sqrt{n}$ and μ_0 is added:

$$Pr(\bar{x} < \mu_0 + t_{n-1, \alpha/2}s/\sqrt{n}) = Pr(\bar{x} > \mu_0 + t_{n-1, 1-\alpha/2}s/\sqrt{n}) = \alpha/2$$

Therefore,

7.16
$$c_1 = \mu_0 + t_{n-1, \alpha/2}s/\sqrt{n} = \mu_0 - t_{n-1, 1-\alpha/2}s/\sqrt{n}$$
$$c_2 = \mu_0 + t_{n-1, 1-\alpha/2}s/\sqrt{n}$$

This test can be summarized as follows:

7.17 **One-Sample *t* Test (Two-Sided Alternative)**

To test the hypothesis

$$H_0: \mu = \mu_0 \quad \text{vs.} \quad H_1: \mu \neq \mu_0$$

with significance level α assuming that σ^2 is the same under both hypotheses and is unknown, then the best test is based on $\bar{x}$: If

$$\bar{x} < \mu_0 - t_{n-1, 1-\alpha/2}s/\sqrt{n} \quad \text{or} \quad \bar{x} > \mu_0 + t_{n-1, 1-\alpha/2}s/\sqrt{n}$$

then H_0 is rejected. If

$$\mu_0 - t_{n-1, 1-\alpha/2}s/\sqrt{n} \leqslant \bar{x} \leqslant \mu_0 + t_{n-1, 1-\alpha/2}s/\sqrt{n}$$

then H_0 is accepted.

The acceptance and rejection regions for this test are illustrated in Figure 7.6.

We may also wish to compute a *p*-value for this test. The computation of the *p*-value will again depend on whether $\bar{x}$ is greater or less than μ_0. The *p*-value is given as in (7.18).

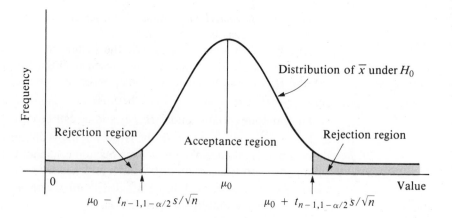

FIGURE 7.6
Illustration of the
acceptance and
rejection regions for the
one-sample *t* test (two-
sided alternative)

7.18 | **p-value for a One-Sample *t* Test for the Mean of a Normal Distribution (Two-Sided Alternative)**

If $\bar{x} < \mu_0$,

$$p = 2 \times [\text{area to the left of } (\bar{x} - \mu_0)/(s/\sqrt{n}) \text{ under a } t_{n-1} \text{ distribution}]$$

If $\bar{x} \geqslant \mu_0$,

$$p = 2 \times [\text{area to the right of } (\bar{x} - \mu_0)/(s/\sqrt{n}) \text{ under a } t_{n-1} \text{ distribution}]$$

This definition is similar to that of a *p*-value for a two-sided test when the variance is known, given in **(7.14)**, except that the t_{n-1} distribution takes the place of an $N(0, 1)$ distribution. It again corresponds to the probability of getting a sample mean as extreme or more extreme than the one obtained, relative to μ_0. The computation of the *p*-value is illustrated in Figure 7.7.

Note that if *n* is large ($n > 120$) and *s* is substituted for σ, then the one-sample normal test given in **(7.13)** and **(7.14)** will be virtually the same as the one-sample *t* test given in **(7.17)** and **(7.18)**. In this case the results will be similar using either test. If a computer is unavailable, then the one-sample normal test might be preferable, since the percentiles of the normal distribution are given in more detail in Table 3 in Appendix 1 than are the percentiles of the *t* distribution in Table 5, and thus more precise *p*-values can be obtained.

7.5.1 Evaluation of Statistical Significance Using the *t* Table

One problem in implementing the one-sample *t* test in **(7.17)** is that tables are in general not as complete for the *t* distribution as for the $N(0, 1)$ distribution. We have tables of the $N(0, 1)$ distribution in Table 3, which, for any argument (x) from 0.00 to 3.99 in intervals of 0.01, give various tail areas associated with the normal distribution. An equivalent table for the *t* distribution would require a comparable amount of information for each number of degrees of freedom, which would be very unwieldy. Instead, in Table 5 only selected percentiles of the t_d distribution for various values of *d* are given. Thus exact *p*-values cannot be computed; only a

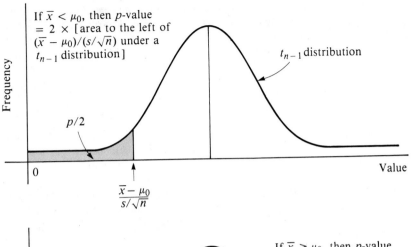

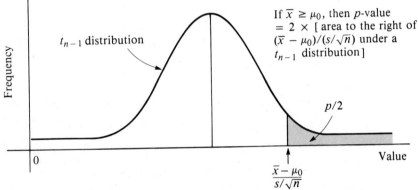

FIGURE 7.7
Computation of the *p*-value for a one-sample *t* test (two-sided alternative)

range within which the *p*-value lies can be specified. However, as previously mentioned, most statistical computer packages, and some hand-held calculators compute *p*-values exactly.

EXAMPLE 7.23   Suppose that the *t* value $= (\bar{x} - \mu_0)/(s/\sqrt{n})$ is -2.42 with a sample size of 8. Evaluate the *p*-value.

SOLUTION   Referring to Table 5 under the row for 7 degrees of freedom, we see that $t_{7,.975} = 2.365$, whereas $t_{7,.99} = 2.998$. Hence, since the *t* distribution is symmetric about 0, we know that the area to the left of -2.42 = the area to the right of $+2.42 = p/2$. Furthermore, since $2.365 < 2.42 < 2.998$, we know that $.01 < p/2 < .025$ or $.02 < p < .05$, and the results are statistically significant. ∎

EXAMPLE 7.24   Suppose that $(\bar{x} - \mu_0)/(s/\sqrt{n}) = 1.425$ based on a sample of size 14. Evaluate the *p*-value.

SOLUTION   Since

$$t_{13,.90} = 1.350 \quad \text{and} \quad t_{13,.95} = 1.771$$

and $1.350 < 1.425 < 1.771$, we have $.05 < p/2 < .10$ or $.1 < p < .2$, and the results are not statistically significant. ∎

EXAMPLE 7.25   **Occupational Medicine, Pulmonary Disease** Occupational medicine is a relatively new field in medicine, whereby specific health hazards are identified for particular occupations. One topic of recent interest is the effect of fire fighting on pulmonary function. Suppose a group of 25–34-year-old male fire fighters are identified and the change in their pulmonary function over a 5-year period is measured. Over 5 years it is found that 26 fire fighters have a mean decline in forced expiratory volume (FEV), which is the volume of air expelled in 1 second, of 0.27 liters with a sample standard deviation of 0.32 liters. Can any conclusions be drawn about the occupational exposure if the expected change over 5 years is 0.10 liters in normal males in this age group?

SOLUTION   A two-sided test will be used, since pulmonary function of fire fighters may decline either more than expected because of their occupational exposure or less than expected because of the likelihood of their being healthier than the general population (they must initially pass a rigorous physical examination). Assume that the decline in FEV is normally distributed with mean μ and variance σ^2. A one-sample t test is used since σ^2 is unknown.

To test $H_0: \mu = 0.10$ versus $H_1: \mu \neq 0.10$, we compute the test statistic

$$t = \frac{(\bar{x} - \mu_0)}{(s/\sqrt{n})} = \frac{(0.27 - 0.10)}{(0.32/\sqrt{26})} = \frac{0.17}{0.063} = 2.70$$

Under H_0, t follows a t distribution with 25 degrees of freedom. Referring to Table 5, we see that $t_{25,.99} = 2.485$, $t_{25,.995} = 2.787$ and therefore the p-value is between $2(1 - .995) = .01$ and $2(1 - .99) = .02$. The results are statistically significant with $.01 < p < .02$, and we conclude that the pulmonary function of fire fighters declines significantly faster than the typical 25–34-year-old male.

To obtain a more precise p-value, a computer program that can evaluate areas under the t distribution for any specified degrees of freedom must be used. The HP-41C hand-held calculator t distribution program has been used to evaluate the exact p-value. The program computes the left-hand tail area, which is given by .994 in this case. This value is subtracted from 1 to obtain the right-hand tail area and then multiplied by 2 to obtain a p-value of .012. The details are shown in Table 7.2. ∎

TABLE 7.2
Computation of the exact p-value for the pulmonary function data in Example 7.25 using a one-sample t test with the HP-41C t distribution program

```
    ----------------------------------------------------------------
                            XEQ "T"            (a)  Degrees of freedom
                25.00000000  XEQ A
    (a)  V=25.00000000                          (b)  t

    (b)      2.700000000  XEQ C
    (c)  P=0.993870881                          (c)  Left-hand tail area

                                                (d)  Exact two-sided p-value
                            CHS
             1.000000000     +
             2.000000000     *
    (d)      0.012258239    ***

    ----------------------------------------------------------------
```

Another problem in implementing the one-sample t test using a t table is that not all degrees of freedom are given in the table. How are **(7.17)** and **(7.18)** used if the degrees of freedom are between two listed degrees of freedom? If a computer is unavailable, then a simple approximate rule often used in this case is to interpolate the percentiles of a t distribution using **harmonic interpolation**, which is given as follows:

| 7.19 | **Interpolation for the t Table** |

To compute $t_{d,p}$ when only $t_{d_1,p}$ and $t_{d_2,p}$ are available in the t table, where $d_1 < d < d_2$, a good approximation is given by

$$t_{d,p} \approx \frac{w_1 t_{d_1,p} + w_2 t_{d_2,p}}{w_1 + w_2} \qquad \text{where} \qquad w_1 = \frac{1}{d} - \frac{1}{d_2}, \quad w_2 = \frac{1}{d_1} - \frac{1}{d}$$

EXAMPLE 7.26   Evaluate the 97.5th percentile of a t distribution with 99 degrees of freedom.

SOLUTION   Compute $t_{99,.975}$. Referring to Table 5, we see that 99 degrees of freedom is not listed in the table, but 60 degrees of freedom and 120 degrees of freedom are listed. Therefore, from **(7.19)** we have $d = 99$, $d_1 = 60$, $d_2 = 120$ and therefore

$$t_{99,.975} \approx \frac{(\frac{1}{99} - \frac{1}{120})t_{60,.975} + (\frac{1}{60} - \frac{1}{99})t_{120,.975}}{(\frac{1}{60} - \frac{1}{120})}$$

$$= \frac{(0.0101 - 0.0083)(2.000) + (0.0167 - 0.0101)(1.980)}{(0.0167 - 0.0083)}$$

$$= \frac{(0.0018)(2.000) + (0.0066)(1.980)}{0.0084} = \frac{18(2.000) + 66(1.980)}{84} = 1.984 \qquad \blacksquare$$

EXAMPLE 7.27   **Cardiovascular Disease** Refer to the cholesterol data in Example 7.18 (p. 197). Suppose the sample standard deviation is 48.23 and we *do not* assume that the variance is known. Evaluate the statistical significance of the results.

SOLUTION   Test the hypothesis $H_0: \mu = 190$ versus $H_1: \mu \neq 190$, where σ^2 is assumed unknown. We know that the mean cholesterol level in the sample of 100 adults is 181.52 and that the sample standard deviation (s) is 48.23. Then, from **(7.18)**, we compute the test statistic

$$t = (\bar{x} - \mu_0)/(s/\sqrt{n}) = \frac{(181.52 - 190)}{(48.23/\sqrt{100})} = -8.48/4.823 = -1.758$$

Since from Example 7.26 we know that $t_{99,.975} = 1.984$, it follows that $p > 2(.025) = .05$, and the results are not statistically significant (NS). Actually, the interpolation was probably unnecessary in this case, since from Table 5 we see that $t_{120,.975} = 1.980$ and $t_{99,.975} > t_{120,.975} = 1.980 > 1.758$. Thus, $p > 2(.025) = .05$ and the results are again not statistically significant.

Another approach to this problem is to use the critical value method given in **(7.17)**. H_0 will be rejected if

$$\bar{x} < \mu_0 - t_{n-1,1-\alpha/2} s/\sqrt{n} = 190 - t_{99,.975}(48.23)/\sqrt{100}$$
$$= 190 - 1.984(48.23)/10 = 190 - 9.57 = 180.43$$

or $$\bar{x} > \mu_0 + t_{n-1,1-\alpha/2} s/\sqrt{n} = 190 + 9.57 = 199.57$$

Since $\bar{x} = 181.52 > 180.43$ and $\bar{x} = 181.52 < 199.57$, the results are not statistically significant and $p \geq .05$. $\qquad \blacksquare$

The two-sided version of the one-sample t test has been discussed in this section. The test procedure for the one-sided case is similar to the one-sample normal test when the variance is known.

To test the hypothesis $H_0: \mu = \mu_0$ versus $H_1: \mu < \mu_0$, we wish to reject H_0 if $\bar{x} < c_1$ and accept H_0 otherwise, where c_1 is chosen such that $Pr(\bar{x} < c_1 | H_0) = \alpha$ or, equivalently, that

$$Pr\left(\frac{\bar{x} - \mu_0}{s/\sqrt{n}} < \frac{c_1 - \mu_0}{s/\sqrt{n}}\right) = \alpha$$

Since $t = (\bar{x} - \mu_0)/(s/\sqrt{n})$ follows a t distribution with $n - 1$ degrees of freedom,

$$\frac{c_1 - \mu_0}{s/\sqrt{n}} = t_{n-1,\alpha}$$

Solving for c_1, we get

$$c_1 = \mu_0 + t_{n-1,\alpha}s/\sqrt{n} = \mu_0 - t_{n-1,1-\alpha}s/\sqrt{n}$$

The p-value for this test is the probability of obtaining a sample mean no larger than $\bar{x}$ under H_0 = the area to the left of $(\bar{x} - \mu_0)/(s/\sqrt{n})$ under a t_{n-1} distribution.

To test the hypothesis $H_0: \mu = \mu_0$ versus $H_1: \mu > \mu_0$, H_0 is rejected if $\bar{x} > c_2$ and accepted otherwise, where c_2 is chosen such that $Pr(\bar{x} > c_2 | H_0) = \alpha$ or, equivalently, that

$$Pr\left(\frac{\bar{x} - \mu_0}{s/\sqrt{n}} > \frac{c_2 - \mu_0}{s/\sqrt{n}}\right) = \alpha$$

Since $(\bar{x} - \mu_0)/(s/\sqrt{n})$ follows a t distribution with $n - 1$ degrees of freedom,

$$\frac{c_2 - \mu_0}{s/\sqrt{n}} = t_{n-1,1-\alpha}$$

Solving for c_2, we get

$$c_2 = \mu_0 + t_{n-1,1-\alpha}s/\sqrt{n}$$

The p-value for this test is the probability of obtaining a sample mean at least as large as $\bar{x}$ under H_0 = the area to the right of $(\bar{x} - \mu_0)/(s/\sqrt{n})$ under a t_{n-1} distribution.

These procedures are summarized as follows:

7.20 | **One-Sample t Test (One-Sided Alternative)**

To test the hypothesis $H_0: \mu = \mu_0$ versus $H_1: \mu < \mu_0$ with significance level α assuming that σ^2 is the same under both hypotheses and is unknown, proceed as follows: If

$$\bar{x} < \mu_0 - t_{n-1,1-\alpha}s/\sqrt{n}$$

then H_0 is rejected. If

$$\bar{x} \geqslant \mu_0 - t_{n-1,1-\alpha}s/\sqrt{n}$$

then H_0 is accepted. The p-value for this test is given by the area to the left of $(\bar{x} - \mu_0)/(s/\sqrt{n})$ under a t_{n-1} distribution.

To test the hypothesis $H_0: \mu = \mu_0$ versus $H_1: \mu > \mu_0$ with significance level α assuming that σ^2 is the same under both hypotheses and is unknown, then the best test is based on $\bar{x}$: If

$$\bar{x} > \mu_0 + t_{n-1, 1-\alpha} s/\sqrt{n}$$

then H_0 is rejected. If

$$\bar{x} \leqslant \mu_0 + t_{n-1, 1-\alpha} s/\sqrt{n}$$

then H_0 is accepted. The p-value for this test is given by the area to the right of $(\bar{x} - \mu_0)/(s/\sqrt{n})$ under a t_{n-1} distribution.

SECTION 7.6   The Power of a Test

7.6.1   One-Sided Alternatives

In Section 7.3 the appropriate hypothesis test was derived to test

$$H_0: \mu = \mu_0 \quad \text{vs.} \quad H_1: \mu = \mu_1 < \mu_0$$

where the underlying distribution is assumed to be normal and the population variance is assumed to be known. The best test was based on the sample mean $\bar{x}$. In particular, from **(7.5)** for a type I error of α, H_0 is rejected if $\bar{x} < \mu_0 + z_\alpha \sigma/\sqrt{n}$ and H_0 is accepted if $\bar{x} \geqslant \mu_0 + z_\alpha \sigma/\sqrt{n}$. The form of the best test *does not depend on the alternative mean chosen* (μ_1) as long as this mean is less than the null mean μ_0.

Hence, if we were interested in an alternative mean of $\mu_1 = 115$ oz rather than $\mu_1 = 110$ oz, then the same test procedure would still be used. However, what differs for the two alternative means is the power of the test $= 1 - Pr$(type II error). Recall from Definition 7.6 that

$$\text{Power} = Pr(\text{rejecting } H_0 | H_0 \text{ false}) = Pr(\bar{x} < \mu_0 + z_\alpha \sigma/\sqrt{n} | \mu = \mu_1)$$

We know that under H_1, $\bar{x} \sim N(\mu_1, \sigma^2/n)$. Hence, upon standardization of limits,

$$\text{Power} = \Phi[(\mu_0 + z_\alpha \sigma/\sqrt{n} - \mu_1)/(\sigma/\sqrt{n})] = \Phi\left[z_\alpha + \frac{(\mu_0 - \mu_1)}{\sigma} \sqrt{n} \right]$$

This power is depicted graphically in Figure 7.8.

Note that the area to the left of $\mu_0 + z_\alpha \sigma/\sqrt{n}$ under the H_0 distribution is the significance level α, whereas the area to the left of $\mu_0 + z_\alpha \sigma/\sqrt{n}$ under the H_1 distribution is the power $= 1 - \beta$.

Why should power concern us? The power of a test tells how likely it is that a significant difference will be found given that the alternative hypothesis is true, that is, given that the true mean μ is different from the mean under the null hypothesis (μ_0). If the power is too low, then there is little chance of finding a significant difference and nonsignificant results are likely even if real differences exist between the true mean μ of the group being studied and the null mean μ_0. An inadequate sample size is almost always the cause of low power to detect a scientifically meaningful difference.

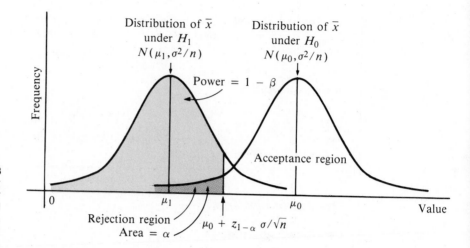

FIGURE 7.8
Illustration of power for
the one-sample test for
the mean of a normal
distribution with known
variance $(\mu_1 < \mu_0)$

EXAMPLE 7.28   **Obstetrics** Compute the power of the test for the birthweight data in Example 7.2 (p. 187) with an alternative mean of 115 oz and $\alpha = .05$.

SOLUTION   We have $\mu_0 = 120$ oz, $\mu_1 = 115$ oz, $\alpha = .05$, $\sigma = 25$, $n = 100$. Thus,

$$\text{Power} = \Phi[z_{.05} + (120 - 115)\sqrt{100}/25] = \Phi[-1.645 + 5(10)/25] = \Phi(.355) = .639$$

Therefore, there is about a 64% chance of detecting a significant difference using a 5% significance level with this sample size. ∎

We have focused on the situation where $\mu_1 < \mu_0$. We are also interested in power when testing the hypothesis

$$H_0: \mu = \mu_0 \quad \text{vs.} \quad H_1: \mu = \mu_1 > \mu_0$$

as was the case with the cholesterol data in Example 7.1. The best test for this situation was presented in **(7.10)**, where H_0 is rejected if $\bar{x} > \mu_0 + z_{1-\alpha}\sigma/\sqrt{n}$ and accepted if $\bar{x} \leqslant \mu_0 + z_{1-\alpha}\sigma/\sqrt{n}$. The power of the test is given by

$$\text{Power} = Pr(\bar{x} > \mu_0 + z_{1-\alpha}\sigma/\sqrt{n}|\mu = \mu_1) = 1 - Pr(\bar{x} < \mu_0 + z_{1-\alpha}\sigma/\sqrt{n}|\mu = \mu_1)$$

$$= 1 - \Phi\left(\frac{\mu_0 + z_{1-\alpha}\sigma/\sqrt{n} - \mu_1}{\sigma/\sqrt{n}}\right) = 1 - \Phi\left[z_{1-\alpha} + \frac{(\mu_0 - \mu_1)\sqrt{n}}{\sigma}\right]$$

Using the relationships $\Phi(-x) = 1 - \Phi(x)$ and $z_\alpha = -z_{1-\alpha}$, this expression can be rewritten as

$$\Phi\left[-z_{1-\alpha} + \frac{(\mu_1 - \mu_0)\sqrt{n}}{\sigma}\right] = \Phi\left[z_\alpha + \frac{(\mu_1 - \mu_0)\sqrt{n}}{\sigma}\right] \quad \text{if} \quad \mu_1 > \mu_0$$

This power is depicted graphically in Figure 7.9.

EXAMPLE 7.29   **Cardiovascular Disease, Pediatrics** Using a 5% level of significance and a sample of size 10, compute the power of the test for the cholesterol data in Example 7.16 (p. 196) with an alternative mean of 190 mg%/mL.

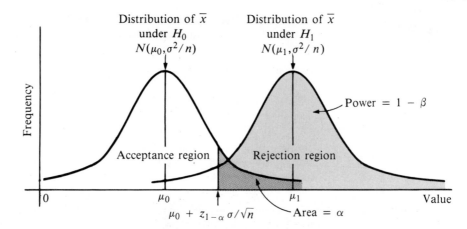

FIGURE 7.9
Illustration of power for the one-sample test for the mean of a normal distribution with known variance $(\mu_1 > \mu_0)$

SOLUTION   We have $\mu_0 = 175$, $\mu_1 = 190$, $\alpha = .05$, $\sigma = 50$, $n = 10$. Thus,

$$\text{Power} = \Phi[-1.645 + (190 - 175)\sqrt{10}/50]$$
$$= \Phi[-1.645 + 15\sqrt{10}/50] = \Phi(-0.70)$$
$$= 1 - \Phi(0.70) = 1 - .758 = .242$$

Therefore, the chance of finding a significant difference in this case is only 24%. Thus, it is not surprising that a significant difference was not found in Example 7.16, since the sample size was too small. ∎

The power formulas presented in this section can be summarized as follows:

7.21 | **Power for the One-Sample Test for the Mean of a Normal Distribution with Known Variance (One-Sided Alternative)**

The power of the test for the hypothesis

$$H_0: \mu = \mu_0 \quad \text{vs.} \quad H_1: \mu = \mu_1$$

where the underlying distribution is normal and the population variance (σ^2) is known is given by

$$\Phi[z_\alpha + (\mu_0 - \mu_1)\sqrt{n}/\sigma] \quad \text{if} \quad \mu_1 < \mu_0$$

$$\Phi[z_\alpha + (\mu_1 - \mu_0)\sqrt{n}/\sigma] \quad \text{if} \quad \mu_1 > \mu_0$$

Notice from **(7.21)** that the power depends on four factors: α, $|\mu_0 - \mu_1|$, n, and σ.

7.22 | **Factors Affecting the Power**

(1) If the significance level is made smaller (α decreases), z_α decreases and hence the power decreases.

(2) If the alternative mean is shifted further away from the null mean ($|\mu_0 - \mu_1|$ increases), then the power increases.

(3) If the standard deviation of an individual observation increases (σ increases), then the power decreases.

(4) If the sample size increases (n increases), then the power increases.

EXAMPLE 7.30   **Cardiovascular Disease, Pediatrics** Compute the power of the test for the cholesterol data in Example 7.16 (p. 196) with a significance level of .01 versus an alternative mean of 190 mg%/mL.

SOLUTION   If $\alpha = .01$, then the power is given by

$$\Phi[z_{.01} + (190 - 175)\sqrt{10}/50] = \Phi[-2.326 + 15\sqrt{10}/50]$$
$$= \Phi(-1.38) = 1 - \Phi(1.38) = 1 - .9162 \approx 8\%$$

which is lower than the power of 24% for $\alpha = .05$, as computed in Example 7.29. What does this mean? It means that if the α level is lowered from .05 to .01, the β error will be higher or, equivalently, the power, which decreases from .24 to .08, will be lower. ∎

EXAMPLE 7.31   **Obstetrics** Compute the power of the test for the birthweight data in Example 7.2 (p. 187) with $\mu_1 = 110$ oz rather than 115 oz.

SOLUTION   If $\mu_1 = 110$ oz, then the power is given by

$$\Phi[-1.645 + (120 - 110)10/25] = \Phi(2.355) = .991 \approx 99\%$$

which is higher than the power of 64%, as computed in Example 7.28 for $\mu_1 = 115$ oz. What does this mean? It means that if the alternative mean changes from 115 oz to 110 oz, then the chance of finding a significant difference increases from 64% to 99%. ∎

EXAMPLE 7.32   **Cardiology** Compute the power of the test for the infarct size data in Example 7.11 (p. 193) with $\sigma = 10$ and $\sigma = 15$ using an alternative mean of 20 ($ck - g - EQ/m^2$).

SOLUTION   If $\sigma = 10$, then

$$\text{Power} = \Phi[-1.645 + (25 - 20)\sqrt{8}/10] = \Phi(-0.23)$$
$$= 1 - \Phi(0.23) = 1 - 0.591 = .409 \approx 41\%$$

whereas if $\sigma = 15$, then

$$\text{Power} = \Phi[-1.645 + (25 - 20)\sqrt{8}/15] = \Phi(-0.70)$$
$$= 1 - 0.758 = .242 \approx 24\%$$

What does this mean? It means that the chance of finding a significant difference declines from 41% to 24% if σ increases from 10 to 15. ∎

EXAMPLE 7.33   **Obstetrics** Assuming a sample size of 10 rather than 100, compute the power for the birthweight data in Example 7.2 with an alternative mean of 115 oz.

SOLUTION   We have $\mu_0 = 120$ oz, $\mu_1 = 115$ oz, $\alpha = .05$, $\sigma = 25$, and $n = 10$. Thus,

$$\text{Power} = \Phi[z_{.05} + (120 - 115)\sqrt{10}/25] = \Phi(-1.645 + 5\sqrt{10}/25)$$
$$= \Phi(-1.01) = 1 - .8438 = .156$$

What does this mean? It means there is only a 16% chance of finding a significant difference with a sample size of 10, whereas there was a 64% chance with a sample size of 100. These results imply that if 10 infants were sampled, we would have virtually no chance of finding a significant difference and would almost surely report a false negative result. ∎

For given levels of α, σ, and n, a **power curve** can be drawn for the power of a test for various alternatives μ_1. Such a power curve is shown in Figure 7.10 for the birthweight data in Example 7.2.

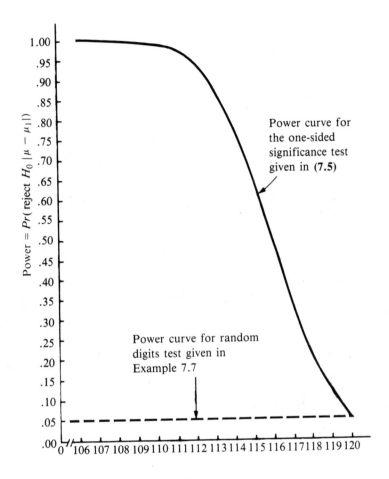

Power curve for the one-sided significance test given in **(7.5)**

Power curve for random digits test given in Example 7.7

FIGURE 7.10
Power curve for the birthweight data in Example 7.2

The power curve for the random digits test described in Example 7.7 (p. 189), which is just a straight line parallel to the x-axis at .05, has also been drawn. The point is that the latter power curve is always below the power curve for the test based on the sample mean, since the test we have derived has the most power for a given significance level for any particular alternative μ_1 if the observations are normally distributed.

7.6.2   **Two-Sided Alternatives**

The power formula given in **(7.21)** is appropriate for a one-sided significance test at level α for the mean of a normal distribution with known variance. Using a two-sided test with hypotheses $H_0: \mu = \mu_0$ versus $H_1: \mu \neq \mu_0$, from **(7.13)** H_0 is rejected if $\bar{x} < c_1 = \mu_0 - z_{1-\alpha/2}\sigma/\sqrt{n}$ or if $\bar{x} > c_2 = \mu_0 + z_{1-\alpha/2}\sigma/\sqrt{n}$. The power of the test versus the specific alternative hypothesis $\mu = \mu_1$ is given by

$$\text{Power} = Pr(\bar{x} < c_1 | \mu = \mu_1) + Pr(\bar{x} > c_2 | \mu = \mu_1)$$

$$= \Phi\left(\frac{c_1 - \mu_1}{\sigma/\sqrt{n}}\right) + 1 - \Phi\left(\frac{c_2 - \mu_1}{\sigma/\sqrt{n}}\right)$$

$$= \Phi\left(\frac{\mu_0 - z_{1-\alpha/2}\sigma/\sqrt{n} - \mu_1}{\sigma/\sqrt{n}}\right) + 1 - \Phi\left(\frac{\mu_0 + z_{1-\alpha/2}\sigma/\sqrt{n} - \mu_1}{\sigma/\sqrt{n}}\right)$$

$$= \Phi\left[-z_{1-\alpha/2} + \frac{(\mu_0 - \mu_1)\sqrt{n}}{\sigma}\right] + 1 - \Phi\left[z_{1-\alpha/2} + \frac{(\mu_0 - \mu_1)\sqrt{n}}{\sigma}\right]$$

Using the relationship $1 - \Phi(x) = \Phi(-x)$, the last two terms can be combined as follows:

7.23

$$\text{Power} = \Phi\left[-z_{1-\alpha/2} + \frac{(\mu_0 - \mu_1)\sqrt{n}}{\sigma}\right] + \Phi\left[-z_{1-\alpha/2} + \frac{(\mu_1 - \mu_0)\sqrt{n}}{\sigma}\right]$$

Equation 7.23 is somewhat more tedious to use than is usually necessary. Specifically, if $\mu_1 < \mu_0$, then the second term is usually negligible relative to the first term. On the other hand, if $\mu_1 > \mu_0$, then the first term is usually negligible relative to the second term. Therefore, the following approximate power formula is used for a two-sided test:

If $\mu_1 < \mu_0$, then

$$\text{Power} = \Phi\left[-z_{1-\alpha/2} + \frac{(\mu_0 - \mu_1)\sqrt{n}}{\sigma}\right]$$

If $\mu_1 > \mu_0$, then

$$\text{Power} = \Phi\left[-z_{1-\alpha/2} + \frac{(\mu_1 - \mu_0)\sqrt{n}}{\sigma}\right]$$

If $-z_{1-\alpha/2}$ is replaced by $z_{\alpha/2}$, then this can be summarized as follows:

7.24   **Power for the One-Sample Test for the Mean of a Normal Distribution with Known Variance (Two-Sided Alternative)**

The power of the two-sided test $H_0: \mu = \mu_0$ versus $H_1: \mu \neq \mu_0$ for the specific alternative $\mu = \mu_1$, where the underlying distribution is normal and the population variance (σ^2) is known, is given by

$$\Phi\left[z_{\alpha/2} + \frac{(\mu_0 - \mu_1)\sqrt{n}}{\sigma}\right] \qquad \text{if } \mu_1 < \mu_0$$

$$\Phi\left[z_{\alpha/2} + \frac{(\mu_1 - \mu_0)\sqrt{n}}{\sigma}\right] \qquad \text{if } \mu_1 > \mu_0$$

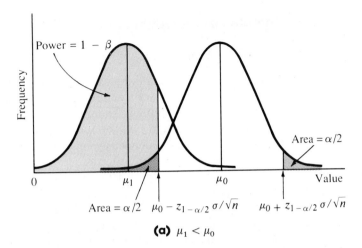

(a) $\mu_1 < \mu_0$

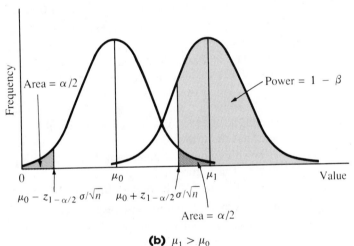

FIGURE 7.11
Illustration of power for the two-sample test for the mean of a normal distribution with known variance

(b) $\mu_1 > \mu_0$

The power is depicted graphically in Figure 7.11. Note that the power formula for the two-sided test in **(7.24)** is the same as the formula for the one-sided test in **(7.21)**, with α replaced by $\alpha/2$.

EXAMPLE 7.34   **Cardiology** A new drug in the class of calcium channel blockers is to be tested for the treatment of patients with unstable angina, a severe type of angina. The effect this drug will have on heart rate is unknown. Suppose that 20 patients are to be studied and the change in heart rate after 48 hours is known to have a standard deviation of 10 beats per minute. What power would such a study have of detecting a significant difference in heart rate over 48 hours if it is hypothesized that the true change in heart rate from baseline to 48 hours could be 5 beats per minute in either direction?

SOLUTION   Use **(7.24)** with $\sigma = 10$, $|\mu_0 - \mu_1| = 5$, $\alpha = .05$, $n = 20$. We have

$$\text{Power} = \Phi(z_{.05/2} + 5\sqrt{20}/10) = \Phi(-1.96 + 2.236) = \Phi(0.276) = .609 \approx .61$$

Thus, the study would have a 61% chance of detecting a significant difference. ∎

<u>SECTION 7.7</u>   **Sample Size Determination**

7.7.1   **One-Sided Alternatives**

Frequently, for planning purposes, we need to have some idea of an appropriate sample size for investigation before a study actually begins. One possible result of making these calculations is finding out that the appropriate sample size is far beyond the financial means of the investigator(s) and thus abandoning the proposed investigation. Obviously, reaching this conclusion before a study starts is far better than after it is in progress.

What does "an appropriate sample size for investigation" actually mean? Consider the birthweight data in Example 7.2. We are testing the null hypothesis $H_0: \mu = \mu_0$ versus the alternative hypothesis $H_1: \mu = \mu_1$, assuming that the distribution of birthweights is normal in both cases and that the standard deviation σ is known. We are presumably going to conduct a test with significance level α and have some idea of what the magnitude of the alternative mean μ_1 is likely to be. If the test procedure in **(7.5)** is used, then H_0 would be rejected if $\bar{x} < \mu_0 + z_\alpha \sigma/\sqrt{n}$ and accepted if $\bar{x} \geqslant \mu_0 + z_\alpha \sigma/\sqrt{n}$. Suppose the alternative hypothesis is actually true. The investigator should have some idea as to what he or she would like the probability of rejecting H_0 to be in this instance. This probability is, of course, nothing other than the power, or $1 - \beta$. Typical values for the desired power are 80%, 90%, ..., and so forth. The problem of determining **sample size** can be summarized as follows: Given that a significance test will be conducted at level α and that the true alternative mean is expected to be μ_1, what sample size is needed to be able to detect a significant difference with probability $1 - \beta$? The situation is depicted in Figure 7.12.

In Figure 7.12 the underlying distribution of $\bar{x}$ has been drawn under the null and alternative hypotheses, respectively, and the point $\mu_0 + z_\alpha \sigma/\sqrt{n}$ has been identified. H_0 will be rejected if $\bar{x} < \mu_0 + z_\alpha \sigma/\sqrt{n}$. Hence, the area to the left of

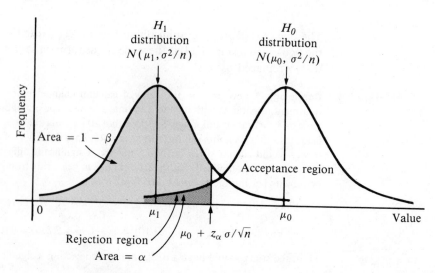

FIGURE 7.12
Requirements for
appropriate sample
size

$\mu_0 + z_\alpha \sigma/\sqrt{n}$ under the rightmost curve is α. However, we also want the area to the left of $\mu_0 + z_\alpha \sigma/\sqrt{n}$ under the leftmost curve, which represents the power, to be $1 - \beta$. These requirements will be met if n is made sufficiently large, since the variance of each curve (σ^2/n) will decrease as n increases and thus the curves will separate. From the power formula in **(7.21)**, we see that

$$\text{Power} = \Phi[z_\alpha + (\mu_0 - \mu_1)\sqrt{n}/\sigma] = 1 - \beta$$

We wish to solve for n in terms of α, β, $\mu_0 - \mu_1$ and σ. To accomplish this, recall that $\Phi(z_{1-\beta}) = 1 - \beta$, and, therefore,

$$z_\alpha + (\mu_0 - \mu_1)\sqrt{n}/\sigma = z_{1-\beta}$$

Subtract z_α from both sides of the equation and multiply by $\sigma/(\mu_0 - \mu_1)$ to obtain

$$\sqrt{n} = \frac{(-z_\alpha + z_{1-\beta})\sigma}{\mu_0 - \mu_1}$$

Replace $-z_\alpha$ by $z_{1-\alpha}$ and square both sides of the equation to obtain

$$n = \frac{(z_{1-\alpha} + z_{1-\beta})^2 \sigma^2}{(\mu_0 - \mu_1)^2}$$

Similarly, if we were to test the hypothesis

$$H_0: \mu = \mu_0 \quad \text{vs.} \quad H_1: \mu = \mu_1 > \mu_0$$

as was the case with the cholesterol data in Example 7.1, using a significance level of α and a power of $1 - \beta$, then, from **(7.21)**, the same sample size formula would hold. This procedure can be summarized as follows:

7.25 | **Sample Size Estimation When Testing for the Mean of a Normal Distribution (One-Sided Alternative)**

Suppose we wish to test

$$H_0: \mu = \mu_0 \quad \text{vs.} \quad H_1: \mu = \mu_1$$

where the data are normally distributed with mean μ and known variance σ^2. The **sample size** needed to conduct a one-sided test with significance level α and probability of detecting a significant difference $= 1 - \beta$ is

$$n = \frac{\sigma^2(z_{1-\beta} + z_{1-\alpha})^2}{(\mu_0 - \mu_1)^2}, \quad \text{one-sided}$$

EXAMPLE 7.35   **Obstetrics** Consider the birthweight data in Example 7.2. Suppose that $\mu_0 = 120$ oz, $\mu_1 = 115$ oz, $\sigma^2 = 625$, $\alpha = .05$, $1 - \beta = .80$. Compute the appropriate sample size needed to conduct the test.

SOLUTION
$$n = \frac{625(z_{.8} + z_{.95})^2}{25} = 25(0.84 + 1.645)^2 = 25(6.175) = 154.4$$

Thus, a sample size of 155 is needed to have an 80% chance of detecting a significant difference at the 5% level if the alternative mean is 115 oz.   ∎

Notice that the sample size is very sensitive to the alternative mean chosen. We see from **(7.25)** that the sample size is inversely proportional to $(\mu_0 - \mu_1)^2$. Thus, if the distance between the null and alternative means is halved, then the sample size needed is 4 times as large. Similarly, if the distance between the null and alternative means is doubled, then the sample size needed is $\frac{1}{4}$ as large.

EXAMPLE 7.36   **Obstetrics** Compute the sample size for the birthweight data in Example 7.2 if $\mu_1 = 110$ oz rather than 115 oz.

SOLUTION   The required sample size would be $\frac{1}{4}$ as large, since $(\mu_0 - \mu_1)^2 = 100$ rather than 25. Thus, $n = 38.6$, or 39 people would be needed. ∎

EXAMPLE 7.37   **Cardiovascular Disease, Pediatrics** Consider the cholesterol data in Example 7.1. Suppose that the null mean is $175\,mg\%/mL$, the alternative mean is $190\,mg\%/mL$, the standard deviation is 50, and we wish to conduct a significance test at the 5% level with a power of 90%. How large should the sample size be?

SOLUTION
$$n = \frac{\sigma^2(z_{1-\beta} + z_{1-\alpha})^2}{(\mu_0 - \mu_1)^2} = \frac{(50)^2(z_{.9} + z_{.95})^2}{(190 - 175)^2}$$

$$= \frac{2500(1.28 + 1.645)^2}{15^2} = \frac{2500(8.556)}{225} = 95.1$$

Thus, 96 people are needed to achieve a power of 90% using a 5% significance level. We should not be surprised that we did not find a significant difference with a sample size of 10 in Example 7.16 (p. 196). ∎

Clearly, the required sample size is related to the following four quantities:

7.26 | **Factors Affecting the Sample Size**

(1) The sample size increases as σ^2 increases.
(2) The sample size increases as the significance level is made smaller (α decreases).
(3) The sample size increases as the required power increases ($1 - \beta$ increases).
(4) The sample size decreases as the distance between the null and alternative means increases ($\mu_0 - \mu_1$) increases.

EXAMPLE 7.38   **Obstetrics** What would happen to the sample size estimate in Example 7.35 if σ were increased to 30? If α were reduced to .01? If the required power were increased to 90%? If the alternative mean were changed to 110 oz (keeping all other parameters the same in each instance)?

SOLUTION   From Example 7.35 we see that 155 infants need to be sampled to achieve 80% power using a 5% significance level with a null mean of 120 oz, an alternative mean of 115 oz, and a standard deviation of 25 oz.

If σ increases to 30, then we need

$$n = (30)^2(z_{.8} + z_{.95})^2/(120 - 115)^2 = 900(0.84 + 1.645)^2/25 = 222.3,\ \text{or } 223 \text{ infants}$$

If α were reduced to .01, then we need

$$n = (25)^2(z_{.8} + z_{.99})^2/(120 - 115)^2 = 625(0.84 + 2.326)^2/25 = 250.6,\ \text{or } 251 \text{ infants}$$

If $1 - \beta$ were increased to 0.9, then we need

$$n = (25)^2(z_{.9} + z_{.95})^2/(120 - 115)^2 = 625(1.28 + 1.645)^2/25 = 213.9, \text{ or } 214 \text{ infants}$$

If μ_1 is decreased to 110 or, equivalently, if $(\mu_0 - \mu_1)$ is increased from 5 to 10, then we need

$$n = (25)^2(z_{.8} + z_{.95})^2/(120 - 110)^2 = 625(0.84 + 1.645)^2/100 = 38.6, \text{ or } 39 \text{ infants}$$

Thus the required sample size increases if σ increases, α decreases, or $1 - \beta$ increases, respectively. The required sample size decreases if the distance between the null and alternative means increases. ∎

One question that arises is how to estimate the parameters necessary to compute sample size. It usually is easy to specify the magnitude of the null mean (μ_0). Similarly, by convention the type I error (α) is usually set at .05. What the level of the power should be is somewhat less clear, although most investigators seem to feel uncomfortable with powers of less than .80. The appropriate values for μ_1 and σ^2 are usually unknown. The parameters μ_1, σ^2 may be obtained from previous work, similar experiments, or prior knowledge of the underlying distribution. In the absence of such information, the parameter μ_1 is sometimes estimated by assessing what a *scientifically important difference* $|\mu_0 - \mu_1|$ would be in the context of the problem being studied. Conducting a small *pilot study* is sometimes valuable. Such a study is generally inexpensive, and its sole aim is to obtain estimates or μ_1 and σ^2 for the purpose of estimating the sample size needed to conduct the major investigation.

Keep in mind that most sample size estimates are very rough because of the inaccuracy in estimating μ_1 and σ^2. These estimates are often used merely to check that the proposed sample size of a study is in the vicinity of what is actually needed rather than to identify a precise sample size.

7.7.2   Sample Size Determination (Two-Sided Alternatives)

The sample size formula given in (**7.25**) was appropriate for a one-sided significance test at level α for the mean of a normal distribution with known variance. If it is not known whether the alternative mean (μ_1) is greater or less than the null mean (μ_0), then a two-sided test is appropriate, and the corresponding sample size needed to conduct a study with power $1 - \beta$ is given by

$$n = \frac{\sigma^2(z_{1-\beta} + z_{1-\alpha/2})^2}{(\mu_0 - \mu_1)^2}$$

To see this, suppose that $\mu_1 < \mu_0$. Use the power formula in (**7.24**) and solve for n in terms of the other parameters, whereby

$$\Phi\left[z_{\alpha/2} + \frac{(\mu_0 - \mu_1)\sqrt{n}}{\sigma}\right] = 1 - \beta$$

or

$$z_{\alpha/2} + \frac{(\mu_0 - \mu_1)\sqrt{n}}{\sigma} = z_{1-\beta}$$

If $z_{\alpha/2}$ is subtracted from both sides of the equation and multiplied by $\sigma/(\mu_0 - \mu_1)$, we get

$$\sqrt{n} = \frac{(z_{1-\beta} - z_{\alpha/2})\sigma}{\mu_0 - \mu_1}$$

If $-z_{\alpha/2}$ is replaced by $z_{1-\alpha/2}$ and both sides of the equation are squared, we get

$$n = \frac{(z_{1-\beta} + z_{1-\alpha/2})^2\sigma^2}{(\mu_0 - \mu_1)^2}$$

If $\mu_1 > \mu_0$, we get the same formula since $(\mu_0 - \mu_1)^2 = (\mu_1 - \mu_0)^2$. This procedure can be summarized as follows:

7.27 | **Sample Size Estimation When Testing for the Mean of a Normal Distribution (Two-Sided Alternative)**

Suppose we wish to test $H_0: \mu = \mu_0$ versus $H_1: \mu = \mu_1$, where the data are normally distributed with mean μ and known variance σ^2. The **sample size** needed to conduct a two-sided test with significance level α and power $1 - \beta$ is

$$n = \frac{\sigma^2(z_{1-\beta} + z_{1-\alpha/2})^2}{(\mu_0 - \mu_1)^2}, \qquad \text{two-sided}$$

Note that this sample size is always larger than the corresponding sample size for a one-sided test, given in **(7.25)**, since $z_{1-\alpha/2}$ is larger than $z_{1-\alpha}$.

EXAMPLE 7.39   **Cardiology** Consider a study of the effect of a calcium channel blocking agent on heart rate for patients with unstable angina, as described in Example 7.34 (p. 213). Suppose we want to have at least 80% power for detecting a significant difference if the effect of the drug is to change heart rate by 5 beats per minute over 48 hours in either direction. How many patients should be enrolled in such a study?

SOLUTION   We assume that $\alpha = .05$ and $\sigma = 10$ beats per minute, as in Example 7.34. We intend to use a two-sided test, since we are not sure in what direction heart rate will change after using the drug. Therefore, the sample size is estimated using the two-sided formulation in **(7.27)**:

$$n = \frac{\sigma^2(z_{1-\beta} + z_{1-\alpha/2})^2}{(\mu_0 - \mu_1)^2}$$

$$= \frac{10^2(z_{.8} + z_{.975})^2}{5^2} = \frac{100(0.84 + 1.96)^2}{25}$$

$$= 4(7.84) = 31.36, \text{ or } 32 \text{ patients}$$

Thus, 32 patients need to be studied to have at least an 80% chance of finding a significant difference using a two-sided test with $\alpha = .05$ if the true mean change in heart rate from using the drug is 5 beats per minute. Note that in Example 7.34 the investigators proposed a study with 20 patients, which would have provided only 61% power for testing the preceding hypothesis.

If the direction of effect of the drug on heart rate were well known, then a one-sided test might be justified. In this case the appropriate sample size could be obtained from the one-sided formulation in **(7.25)**:

$$n = \frac{\sigma^2(z_{1-\beta} + z_{1-\alpha})^2}{(\mu_0 - \mu_1)^2} = \frac{(10)^2(z_{.8} + z_{.95})^2}{5^2}$$

$$= \frac{100(0.84 + 1.645)^2}{25} = 4(6.175) = 24.7, \text{ or } 25 \text{ patients}$$

Thus, only 25 patients would need to be studied for a one-sided test instead of the 32 patients needed for a two-sided test. ∎

SECTION 7.8

The Relationship Between Hypothesis Testing and Confidence Intervals

A test procedure was presented in **(7.13)** for testing the hypothesis $H_0: \mu = \mu_0$, $\sigma^2 = \sigma_0^2$ versus $H_1: \mu \neq \mu_0$, $\sigma^2 = \sigma_0^2$. Similarly, in Section 6.4.4 a method for obtaining a two-sided confidence interval for the parameter μ of a normal distribution when the variance is assumed known was discussed. The relationship between these two procedures can be stated as follows:

7.28 | **The Relationship Between Hypothesis Testing and Confidence Intervals (Two-Sided Case)**

Suppose we are testing $H_0: \mu = \mu_0$, $\sigma^2 = \sigma_0^2$ versus $H_1: \mu \neq \mu_0$, $\sigma^2 = \sigma_0^2$. H_0 is rejected with a two-sided level α test if and only if the two-sided $100\% \times (1 - \alpha)$ confidence interval for μ *does not* contain μ_0. H_0 is accepted if and only if the two-sided $100\% \times (1 - \alpha)$ confidence interval for μ *does* contain μ_0. This relationship is illustrated in Figure 7.13.

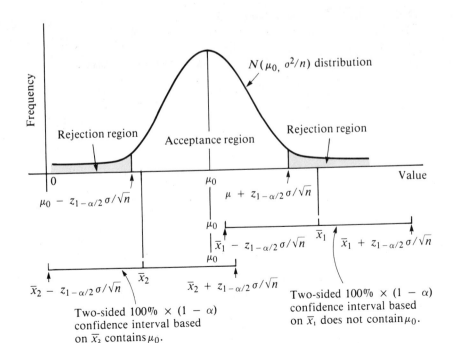

FIGURE 7.13
The relationship between a two-sided hypothesis test and a two-sided confidence interval

Suppose that the sample mean $\bar{x}_1$ is in the rejection region. The two-sided $100\% \times (1 - \alpha)$ confidence interval for μ, which ranges from $\bar{x}_1 - z_{1-\alpha/2}\sigma/\sqrt{n}$ to $\bar{x}_1 + z_{1-\alpha/2}\sigma/\sqrt{n}$, *does not* include μ_0 as depicted in the figure. Similarly, the sample mean $\bar{x}_2$ is in the acceptance region, and correspondingly the two-sided $100\% \times (1 - \alpha)$ confidence interval from $\bar{x}_2 - z_{1-\alpha/2}\sigma/\sqrt{n}$ to $\bar{x}_2 + z_{1-\alpha/2}\sigma/\sqrt{n}$ *does* contain μ_0.

Hence, this relationship is the rationale for using confidence intervals in Chapter 6 to decide on the reasonableness of specific values for the parameter μ. If any specific proposed value μ_0 did not fall in the two-sided $100 \times (1 - \alpha)$ confidence interval for μ, then we said that it was an unlikely value for the parameter μ. Equivalently, we could have tested the hypothesis $H_0: \mu = \mu_0$ versus $H_1: \mu \neq \mu_0$ and rejected H_0 at significance level α.

EXAMPLE 7.40   **Cardiovascular Disease** Consider the cholesterol data in Example 7.18 (p. 197). The two-sided 95% confidence interval for μ is given by

$$(\bar{x} - z_{1-\alpha/2}\sigma/\sqrt{n}, \bar{x} + z_{1-\alpha/2}\sigma/\sqrt{n})$$

$$= \left[181.52 - \frac{1.96(40)}{10}, 181.52 + \frac{1.96(40)}{10} \right]$$

$$= (181.52 - 7.84, 181.52 + 7.84) = (173.68, 189.36)$$

This 95% confidence interval does *not* contain 190, which corresponds to the statement that the two-sided p-value (.034) as computed in Example 7.20 (p. 199) is less than .05.   ∎

EXAMPLE 7.41   **Cardiovascular Disease** Suppose the sample mean for cholesterol was 185 mg/dL for the cholesterol data in Example 7.18. The 95% confidence interval would be

$$(185 - 7.84, 185 + 7.84) = (177.16, 192.84)$$

which contains the null mean (190). The p-value for the hypothesis test would be

$$p = 2 \times \Phi[(185 - 190)/4] = 2 \times \Phi(-1.25) = 2[1 - \Phi(1.25)]$$

$$= 2(1 - .8944) = 2(.1056) = .2112 > .05$$

Thus, the conclusions based on the confidence interval and hypothesis-testing approaches are also the same here.   ∎

A similar relationship exists between the one-sided hypothesis test developed in Section 7.3 and a one-sided confidence interval for the parameter μ, which can be stated as follows:

7.29 | **The Relationship Between Hypothesis Testing and Confidence Intervals (One-Sided Alternative)**

To test the hypothesis

$$H_0: \mu = \mu_0, \sigma^2 = \sigma_0^2 \quad \text{vs.} \quad H_1: \mu < \mu_0, \sigma^2 = \sigma_0^2$$

with significance level α, H_0 is rejected if and only if the lower one-sided $100\% \times (1 - \alpha)$ confidence interval $\mu \leqslant \bar{x} + z_{1-\alpha}\sigma/\sqrt{n}$ *does not contain* μ_0; H_0 is accepted if this interval *does contain* μ_0.

To test the hypothesis

$$H_0: \mu = \mu_0, \sigma^2 = \sigma_0^2 \quad \text{vs.} \quad H_1: \mu > \mu_0, \sigma^2 = \sigma_0^2$$

with significance level α, H_0 is rejected if and only if the upper one-sided $100\% \times (1 - \alpha)$ confidence interval $\mu \geqslant \bar{x} - z_{1-\alpha}\sigma/\sqrt{n}$ *does not contain* μ_0; H_0 is accepted if this interval *does* contain μ_0.

Equivalent confidence interval statements can also be made about any of the other hypothesis tests covered in this text.

Since the hypothesis-testing and confidence-interval approaches yield the same conclusions, is there any advantage to using one method over the other? The *p*-value from a hypothesis test tells precisely how statistically significant the results are. However, often results that are statistically significant are in fact not very important in the context of the subject matter, since the actual difference between $\bar{x}$ and μ_0 may not be very large, although the results are statistically significant because of a large sample size. A 95% confidence interval for μ would give additional information in this regard, since it would give a range of values within which μ is likely to fall. Conversely, the 95% confidence interval does not contain all the information contained in a *p*-value: It does not tell precisely how significant the results are but merely tells whether or not they are significant at the 5% level. Hence, it is good practice to compute both a *p*-value and a 95% confidence interval for μ.

Unfortunately, some researchers have become polarized on this issue, with some statisticians favoring only the hypothesis-testing approach, and some epidemiologists favoring only the confidence-interval approach. These issues have correspondingly influenced editorial policy, with some journals *requiring* that results be presented in one format or the other. The crux of the issue is that, traditionally, results need to be statistically significant (at the 5% level) in order to demonstrate the validity of a particular finding. One advantage of this approach is that a uniform statistical standard is provided (the 5% level) for all researchers in order to demonstrate evidence of an association. This protects the research community against scientific claims not based on any statistical or empirical criteria whatsoever (such as on the sole basis of clinical case reports). Advocates of the confidence-interval approach contend that the width of the confidence interval provides information on the strength of association, regardless of the level of significance. The author's opinion is that both significance levels and confidence limits provide complementary information and should both be reported, where possible.

EXAMPLE 7.42   **Cardiovascular Disease**   Consider the cholesterol data in Examples 7.20 and 7.40. The *p*-value of .034 computed in Example 7.20 tells precisely how significant the results are. The 95% confidence interval for $\mu = (173.68, 189.36)$ computed in Example 7.40 gives a range of possible values that μ might assume. The two types of information are complementary. ∎

One-Sample χ^2 Test for the Variance of a Normal Distribution

EXAMPLE 7.43   **Hypertension** Consider the data in Example 6.33 (p. 165) concerning the variability of blood-pressure measurements taken on an arteriosonde machine. We were concerned with the difference between measurements taken by two observers on the same person $= d_i = x_{1i} - x_{2i}$, where $x_{1i} =$ the measurement on the ith person by the first observer and $x_{2i} =$ the measurement on the ith person by the second observer. We will assume that this difference is a good measure of interobserver variability, and we wish to compare this variability with comparable data using a standard blood-pressure cuff. We have reason to believe that the variability of the arteriosonde machine might be different from that of a standard cuff. Intuitively, the variability of the new method should be lower. However, since the new method is not as widely used, the observers are probably less experienced at using the method, and the variability of the new method could possibly be higher than that of the old method. Thus, a two-sided test will be used to study this question. Suppose we know from previously published work that $\sigma^2 = 35$ for d_i obtained from the standard cuff. We wish to test the hypothesis $H_0: \sigma^2 = \sigma_0^2 = 35$ versus $H_1: \sigma^2 \neq \sigma_0^2$. How should this test be performed?   ∎

If $x_1, \ldots, x_n$ are a random sample, then we can reasonably base the test on s^2, since it is an unbiased estimator of σ^2. Specifically, if we wish to conduct a two-sided test with significance level α, then we wish to find c_1, c_2 such that

$$Pr(s^2 \leqslant c_1 | H_0) = Pr(s^2 \geqslant c_2 | H_0) = \alpha/2$$

We know from **(6.9)** that if $x_1, \ldots, x_n$ are a random sample from an $N(\mu, \sigma^2)$ distribution, then under H_0,

$$s^2 \sim \frac{\sigma_0^2 \chi_{n-1}^2}{n-1}$$

Therefore,

$$Pr\left[s^2 < \frac{\sigma_0^2 \chi_{n-1,\alpha/2}^2}{n-1}\right] = Pr\left[s^2 > \frac{\sigma_0^2 \chi_{n-1,1-\alpha/2}^2}{n-1}\right] = \frac{\alpha}{2}$$

It follows that

$$c_1 = \frac{\sigma_0^2 \chi_{n-1,\alpha/2}^2}{n-1}$$

$$c_2 = \frac{\sigma_0^2 \chi_{n-1,1-\alpha/2}^2}{n-1}$$

Hence, the test procedure is given as follows:

7.30 **One-Sample χ^2 Test for the Variance of a Normal Distribution (Two-Sided Alternative)**

If
$$s^2 < \frac{\sigma_0^2 \chi_{n-1,\alpha/2}^2}{n-1} \quad \text{or} \quad s^2 > \frac{\sigma_0^2 \chi_{n-1,1-\alpha/2}^2}{n-1}$$

then H_0 is rejected. If

$$\frac{\sigma_0^2 \chi_{n-1,\alpha/2}^2}{n-1} \leqslant s^2 \leqslant \frac{\sigma_0^2 \chi_{n-1,1-\alpha/2}^2}{n-1}$$

then H_0 is accepted.
The acceptance and rejection regions for this test are depicted in Figure 7.14.

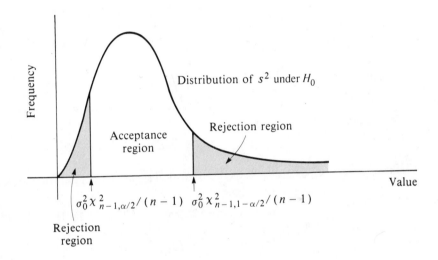

FIGURE 7.14
Acceptance and
rejection regions for the
one-sample χ^2 test for
the variance of a
normal distribution (two-
sided alternative)

Alternatively, we may wish to compute a p-value for our experiment. The computation of the p-value will again depend on whether $s^2 \leqslant \sigma_0^2$ or $s^2 > \sigma_0^2$. The rule is given as follows:

7.31 **p-value for a One-Sample χ^2 Test for the Variance of a Normal Distribution (Two-Sided Alternative)**

If $s^2 \leqslant \sigma_0^2$,

p-value = 2 × [area to the left of $(n-1)s^2/\sigma_0^2$ under a χ_{n-1}^2 distribution]

If $s^2 > \sigma_0^2$,

p-value = 2 × [area to the right of $(n-1)s^2/\sigma_0^2$ under a χ_{n-1}^2 distribution]

These p-values are illustrated in Figure 7.15.

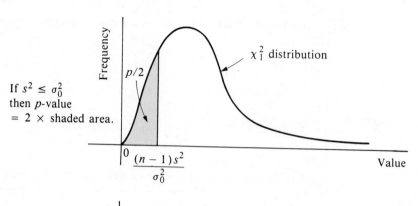

If $s^2 \le \sigma_0^2$
then p-value
$= 2 \times$ shaded area.

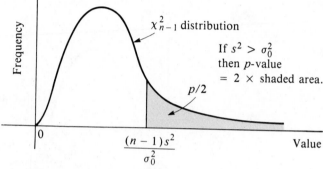

If $s^2 > \sigma_0^2$
then p-value
$= 2 \times$ shaded area.

FIGURE 7.15
Illustration of the p-value for a one-sample χ^2 test for the variance of a normal distribution (two-sided alternative)

EXAMPLE 7.44   **Hypertension** Assess the statistical significance of the arteriosonde machine data in Example 7.43.

SOLUTION   We know from Example 6.33 that $s^2 = 8.178$, $n = 10$. Thus, from **(7.30)** the critical values are given by

$$c_1 = \frac{\sigma_0^2 \chi_{n-1,\alpha/2}^2}{n-1} = \frac{35\chi_{9,.025}^2}{9} = \frac{35(2.70)}{9} = 10.50$$

$$c_2 = \frac{\sigma_0^2 \chi_{n-1,1-\alpha/2}^2}{n-1} = \frac{35\chi_{9,.975}^2}{9} = \frac{35(19.02)}{9} = 73.97$$

Since $s^2 = 8.178 < c_1 = 10.50$, H_0 is rejected using a two-sided test with $\alpha = .05$. To obtain the p-value, refer to **(7.31)** and compute the test statistic

$$X^2 = \frac{(n-1)s^2}{\sigma_0^2} = \frac{9(8.178)}{35} = 2.103$$

Under H_0, X^2 follows a χ^2 distribution with nine degrees of freedom. Thus, since $s^2 = 8.178 < 35 = \sigma_0^2$, the p-value is computed as follows:

$$p = 2 \times Pr(\chi_9^2 < 2.103)$$

From Table 6 of Appendix 1 we see that

$$\chi_{9,.025}^2 = 2.70, \quad \chi_{9,.01}^2 = 2.09$$

Thus, since $2.09 < 2.103 < 2.70$, we have $(.01) < p/2 < (.025)$ or $.02 < p < .05$.

To obtain the exact p-value, use the HP-41C chi-square distribution calculator program to evaluate areas under the χ^2 distribution. The program computes left-hand tail areas. Thus, multiply by 2 to obtain the exact two-sided p-value = .021. The details are given in Table 7.3.

TABLE 7.3
Computation of the exact p-value for the arteriosonde machine data in Example 7.44 using a one-sample χ^2 test with the HP-41C chi-square distribution program

```
ΣCHISQD

(a)               9.00   XEQ A          (a)  Degrees of freedom
(b)               2.103  XEQ E
                                        (b)  X

(c)       0.010267996    ***            (c)  Left-hand tail area
           2.000000000     *
(d)       0.020535992    ***            (d)  Exact two-sided p-value
```

Therefore, the results are statistically significant, and we conclude that the interobserver variance using the arteriosonde machine is significantly different from the interobserver variance using the standard cuff. To quantify how different the two variances are, a two-sided 95% confidence interval for σ^2 could be obtained, as was done in Example 6.35 (p. 168). This interval was (3.87, 27.26). Of course, it does not contain 35 because the p-value is less than .05. ∎

For a one-sided test, follow a similar procedure:

7.32

One-Sample χ^2 Test for the Variance of a Normal Distribution (One-Sided Alternative)

To test the hypothesis $H_0: \sigma^2 = \sigma_0^2$ versus $H_1: \sigma^2 < \sigma_0^2$ with significance level α, proceed as follows: If

$$s^2 < \frac{\sigma_0^2 \chi_{n-1,\alpha}^2}{n-1}$$

then H_0 is rejected. If

$$s^2 \geqslant \frac{\sigma_0^2 \chi_{n-1,\alpha}^2}{n-1}$$

then H_0 is accepted. The p-value for this test is given by the area to the left of $(n-1)s^2/\sigma_0^2$ under a χ_{n-1}^2 distribution.

To test the hypothesis $H_0: \sigma^2 = \sigma_0^2$ versus $H_1: \sigma^2 > \sigma_0^2$ with significance level α, proceed as follows: If

$$s^2 > \frac{\sigma_0^2 \chi_{n-1,1-\alpha}^2}{n-1}$$

then H_0 is rejected. If

$$s^2 \leqslant \frac{\sigma_0^2 \chi_{n-1,1-\alpha}^2}{n-1}$$

then H_0 is accepted.

The p-value for this test is given by the area to the right of $(n-1)s^2/\sigma_0^2$ under a χ_{n-1}^2 distribution.

In general, the assumption of normality is particularly important for hypothesis testing and confidence-interval estimation for variances. If this assumption is not satisfied, then the critical regions and p-values in **(7.30)** through **(7.32)** and the confidence limits in **(6.10)** will not be valid.

<u>SECTION 7.10</u>   **One-Sample Test for a Binomial Proportion**

7.10.1   **Normal Theory Methods**

EXAMPLE 7.45   **Cancer** Consider the breast cancer data in Example 6.38 (p. 170). In that example we were interested in the effect of having a family history of breast cancer on the incidence of breast cancer. Suppose that out of 10,000 50–54-year-old women sampled whose mothers had breast cancer, 400 had breast cancer at some time in their lives. Based on large studies, assume that the prevalence rate of breast cancer for American women in this age group is about 2%. The question is, How compatible is the sample rate of 4% with a population rate of 2%?

Another way of asking this question is to restate it in terms of hypothesis testing: If p = prevalence rate of breast cancer in 50–54-year-old women whose mothers have had breast cancer, then we wish to test the hypothesis $H_0: p = .02 = p_0$ versus $H_1: p \neq .02$. How can this be done? ∎

The significance test will be based on the sample proportion of cases $\hat{p}$. We wish to find c_1, c_2 such that H_0 is rejected if $\hat{p} < c_1$ or $\hat{p} > c_2$ and accepted if $c_1 \leqslant \hat{p} \leqslant c_2$. For a given significance level α, we must find c_1, c_2 such that

7.33
$$Pr(\hat{p} < c_1 | H_0) = Pr(\hat{p} > c_2 | H_0) = \alpha/2$$

We will assume that the normal approximation to the binomial distribution is valid. This assumption is reasonable when $np_0 q_0 \geqslant 5$. Therefore, from **(6.12)** we know that under H_0

$$\hat{p} \sim N\left(p_0, \frac{p_0 q_0}{n}\right)$$

It follows that
$$\frac{\hat{p} - p_0}{\sqrt{p_0 q_0/n}} \sim N(0, 1)$$

Thus,
$$Pr\left(\frac{\hat{p} - p_0}{\sqrt{p_0 q_0/n}} < z_{\alpha/2}\right) = Pr\left(\frac{\hat{p} - p_0}{\sqrt{p_0 q_0/n}} > z_{1-\alpha/2}\right) = \frac{\alpha}{2}$$

If both sides of each inequality are multiplied by $\sqrt{p_0 q_0/n}$ and p_0 is added, we obtain

$$Pr(\hat{p} < p_0 + z_{\alpha/2}\sqrt{p_0 q_0/n}) = Pr(\hat{p} > p_0 + z_{1-\alpha/2}\sqrt{p_0 q_0/n}) = \frac{\alpha}{2}$$

If $z_{\alpha/2}$ is replaced by $-z_{1-\alpha/2}$, it follows that

$$c_1 = p_0 - z_{1-\alpha/2}\sqrt{\frac{p_0 q_0}{n}}$$

$$c_2 = p_0 + z_{1-\alpha/2}\sqrt{\frac{p_0 q_0}{n}}$$

Thus, the test takes the following form:

7.34 | **One-Sample Test for a Binomial Proportion—Normal Theory Method (Two-Sided Alternative)**

If
$$\hat{p} < p_0 - z_{1-\alpha/2}\sqrt{\frac{p_0 q_0}{n}} \quad \text{or} \quad \hat{p} > p_0 + z_{1-\alpha/2}\sqrt{\frac{p_0 q_0}{n}}$$

then H_0 is rejected. If

$$p_0 - z_{1-\alpha/2}\sqrt{\frac{p_0 q_0}{n}} \leqslant \hat{p} \leqslant p_0 + z_{1-\alpha/2}\sqrt{\frac{p_0 q_0}{n}}$$

then H_0 is accepted. The acceptance and rejection regions are depicted in Figure 7.16.

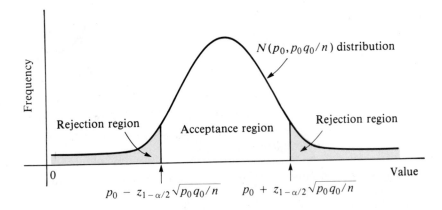

FIGURE 7.16
Acceptance and rejection regions for the one-sample binomial test—normal theory test (two-sided alternative)

Alternatively, a p-value could be computed. The computation of the p-value will again depend on whether $\hat{p} \leqslant p_0$ or $\hat{p} > p_0$. If $\hat{p} \leqslant p_0$, then

$$p\text{-value} = 2 \times \text{area to the left of } \hat{p} \text{ under an } N(p_0, p_0 q_0/n) \text{ curve}$$

If $\hat{p} > p_0$, then

$$p\text{-value} = 2 \times \text{area to the right of } \hat{p} \text{ under an } N(p_0, p_0 q_0/n) \text{ curve}$$

This is summarized as follows:

7.35 | **Computation of the *p*-value for the One-Sample Binomial Test—Normal Theory Method (Two-Sided Alternative)**

$$p\text{-value} = 2 \times \Phi\left[(\hat{p} - p_0)\Big/\sqrt{\frac{p_0 q_0}{n}}\right] = \text{twice the area to the left of } \hat{p} \text{ under an } N(p_0, p_0 q_0/n) \text{ curve if } \hat{p} \leqslant p_0$$

$$p\text{-value} = 2 \times \left\{1 - \Phi\left[(\hat{p} - p_0)\Big/\sqrt{\frac{p_0 q_0}{n}}\right]\right\} = \text{twice the area to the right of } \hat{p} \text{ under an } N(p_0, p_0 q_0/n) \text{ curve if } \hat{p} > p_0$$

The calculation of the p-value is illustrated in Figure 7.17.

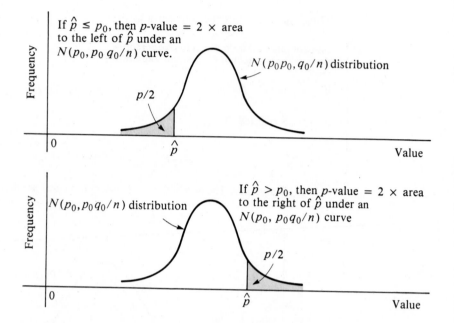

FIGURE 7.17
Illustration of the p-value for a one-sample binomial test—normal theory method (two-sided alternative)

These definitions of a p-value are again compatible with the idea of a p-value as the probability of obtaining results as extreme as or more extreme than the results in our particular sample.

EXAMPLE 7.46 **Cancer** Assess the statistical significance of the data in Example 7.45.

SOLUTION Using the critical value method,

$$c_1 = .02 - 1.96 \sqrt{\frac{(.02)(.98)}{10,000}} = .02 - .0027 = .0173$$

$$c_2 = .02 + 1.96 \sqrt{\frac{(.02)(.98)}{10,000}} = .02 + .0027 = .0227$$

Since $\hat{p} = .04 > c_2$, H_0 can be rejected using a two-sided test with $\alpha = .05$. To use the p-value method, note that $\hat{p} = .04 > p_0 = .02$. Hence, the p-value is given by

$$\text{p-value} = 2 \times \left\{ 1 - \Phi \left[\frac{.04 - .02}{\sqrt{(.02)(.98)/10,000}} \right] \right\}$$

$$= 2 \times \left\{ 1 - \Phi \left[\frac{.02}{.0014} \right] \right\} = 2 \times [1 - \Phi(14.3)] < .001$$

Thus, the results are very highly significant. ∎

7.10.2 **Exact Methods**

The test procedure presented in **(7.34)** to test the hypothesis $H_0: p = p_0$ versus $H_1: p \neq p_0$ depends on the assumption that the normal approximation to the binomial distribution is valid. This assumption will only be true if $np_0 q_0 \geq 5$. How can the preceding hypothesis be tested if this criterion is not satisfied?

Our approach will be to base our test on *exact* binomial probabilities. In particular, let X be a binomial random variable with parameters n and p_0 and let $\hat{p} = x/n$, where x is the observed number of events. The computation of the p-value depends on whether $\hat{p} \leqslant p_0$ or $\hat{p} > p_0$. If $\hat{p} \leqslant p_0$, then

$$p/2 = Pr(\leq x \text{ successes in } n \text{ trials} | H_0)$$

$$= \sum_{k=0}^{x} \binom{n}{k} p_0^k (1 - p_0)^{n-k}$$

If $\hat{p} > p_0$, then

$$p/2 = Pr(\geqslant x \text{ successes in } n \text{ trials} | H_0)$$

$$= \sum_{k=x}^{n} \binom{n}{k} p_0^k (1 - p_0)^{n-k}$$

This is summarized as follows:

7.36 | **Computation of the *p*-value for the One-Sample Binomial Test— Exact Method (Two-Sided Alternative)**

If $\hat{p} \leqslant p_0$,

$$p = 2 \times Pr(X \leqslant x) = 2 \sum_{k=0}^{x} \binom{n}{k} p_0^k (1 - p_0)^{n-k}$$

If $\hat{p} > p_0$,

$$p = 2 \times Pr(X \geqslant x) = 2 \sum_{k=x}^{n} \binom{n}{k} p_0^k (1 - p_0)^{n-k}$$

The computation of the p-value is depicted in Figure 7.18.

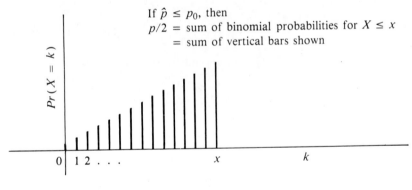

If $\hat{p} \leq p_0$, then
$p/2 = $ sum of binomial probabilities for $X \leq x$
$= $ sum of vertical bars shown

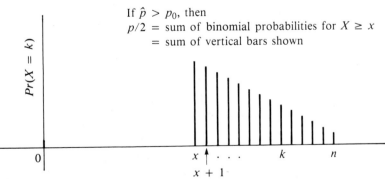

If $\hat{p} > p_0$, then
$p/2 = $ sum of binomial probabilities for $X \geq x$
$= $ sum of vertical bars shown

FIGURE 7.18

Illustration of the p-value for a one-sample binomial test—exact method (two-sided alternative)

In either case the *p*-value corresponds to the sum of the probabilities of all events that are as extreme as or more extreme than the sample result obtained.

EXAMPLE 7.47   **Occupational Medicine, Cancer**   The safety of those who work at or live close to nuclear power plants has been the subject of widely publicized debate in recent years. One possible health hazard due to radiation exposure is an excess of cancer deaths among those exposed. One problem with studying this question is that the number of deaths attributable to either cancer in general or specific types of cancer is small, and reaching statistically significant conclusions is difficult, except after long periods of follow-up. An alternative approach is to perform a *proportional mortality study*, whereby the proportion of deaths attributed to a specific cause in an exposed group is compared with the corresponding proportion in a large population. Suppose, for example, that 13 deaths have occurred among 55–64-year-old male workers in a nuclear power plant and that the cause of death was cancer in 5 of them. Assume, based on vital statistics reports, that approximately 20% of all deaths can be attributed to some form of cancer. Is this result significant?

SOLUTION   We wish to test the hypothesis $H_0: p = .20$ versus $H_1: p \neq .20$, where $p =$ probability that the cause of death was cancer in nuclear power workers. The normal approximation to the binomial cannot be used, since

$$np_0 q_0 = 13(.2)(.8) = 2.1 < 5$$

However, the exact procedure in **(7.36)** can be used:

$$\hat{p} = \frac{5}{13} = .38 > .20$$

Therefore, $$p = 2 \sum_{k=5}^{13} \binom{13}{k} (.2)^k (.8)^{13-k} = 2 \times \left[1 - \sum_{k=0}^{4} \binom{13}{k} (.2)^k (.8)^{13-k} \right]$$

From Table 1 in Appendix 1 we have

$$Pr(0) = .0550$$
$$Pr(1) = .1787$$
$$Pr(2) = .2680$$
$$Pr(3) = .2457$$
$$Pr(4) = .1535$$

Therefore, $$p = 2 \times [1 - (.0550 + .1787 + .2680 + .2457 + .1535)]$$
$$= 2 \times (1 - .9009) = .198$$

In summary, the results are *not* statistically significant and an excess cancer proportional mortality cannot be attributed to the nuclear power plant workers. ∎

7.10.3   **Power and Sample Size Estimation**

The power of the one-sample binomial test can also be considered using the large sample test procedure given in Section 7.10.1. Suppose we are conducting a two-tailed test at level α, where $p = p_0$ under the null hypothesis. Under the alternative hypothesis of $p = p_1$, the power is given by the following formula:

7.37 | **Power for the One-Sample Binomial Test (Two-Sided Alternative)**

The **power** of the one-sample binomial test for the hypothesis

$$H_0: p = p_0 \quad \text{vs.} \quad H_1: p \ne p_0$$

for the specific alternative $p = p_1$ is given by

$$\Phi\left\{ \sqrt{\frac{p_0 q_0}{p_1 q_1}} \left[z_{\alpha/2} + \frac{(p_0 - p_1)\sqrt{n}}{\sqrt{p_0 q_0}} \right] \right\} \quad \text{if } p_1 < p_0$$

and $\quad \Phi\left\{ \sqrt{\frac{p_0 q_0}{p_1 q_1}} \left[z_{\alpha/2} + \frac{(p_1 - p_0)\sqrt{n}}{\sqrt{p_0 q_0}} \right] \right\} \quad \text{if } p_1 > p_0$

We assume that $np_0 q_0 \ge 5$ so that the normal theory methods in Section 7.10.1 are valid.

EXAMPLE 7.48   **Cancer** Suppose we wish to test the hypothesis that women with a sister history of breast cancer are at higher risk of developing breast cancer themselves. Suppose we assume, as in Example 7.45, that the prevalence rate of breast cancer is 2% among 50–54-year-old American women, whereas it is 5% among women with a sister history. We propose to interview 500 50–54-year-old women with a sister history of the disease. What is the power of such a study assuming that we conduct a two-sided test with $\alpha = .05$?

SOLUTION   We have $\alpha = .05$, $p_0 = .02$, $p_1 = .05$, $n = 500$. The power, as given by (**7.37**), is:

$$\text{Power} = \Phi\left[\sqrt{\frac{(.02)(.98)}{(.05)(.95)}} \left(z_{.025} + \frac{.03\sqrt{500}}{\sqrt{.02(.98)}} \right) \right]$$

$$= \Phi[(.642)(-1.96 + 4.792)] = \Phi(1.818) = .965$$

Thus, there should be a 96.5% chance of finding a significant difference if the true rate of breast cancer among women with a sister history is 2.5 times as high as that of typical 50–54-year-old women. ∎

Similarly, we can consider the issue of appropriate sample size if the one-sample binomial test for a given α, p_0, p_1, and power is being used. The sample size is given by the following formula:

7.38 | **Sample Size Estimation for the One-Sample Binomial Test (Two-Sided Alternative)**

Suppose we wish to test $H_0: p = p_0$ versus $H_1: p \ne p_0$. The sample size needed to conduct a two-sided test with significance level α and power $1 - \beta$ versus the specific alternative hypothesis $p = p_1$ is

$$n = \frac{p_0 q_0 \left(z_{1-\alpha/2} + z_{1-\beta} \sqrt{\dfrac{p_1 q_1}{p_0 q_0}} \right)^2}{(p_1 - p_0)^2}$$

EXAMPLE 7.49 **Cancer** How many women should be interviewed in the study proposed in Example 7.48 to achieve 90% power if a two-sided significance test with $\alpha = .05$ is being used?

SOLUTION We have $\alpha = .05$, $1 - \beta = .90$, $p_0 = .02$, $p_1 = .05$. The sample size is given by **(7.38)**:

$$n = \frac{(.02)(.98)\left[z_{.975} + z_{.90}\sqrt{\dfrac{(.05)(.95)}{(.02)(.98)}} \right]^2}{(.03)^2}$$

$$= \frac{(.0196)[1.96 + 1.28(1.557)]^2}{.0009} = \frac{(.0196)(15.626)}{.0009} = 340.3, \text{ or } 341 \text{ women}$$

Thus, 341 women need to be interviewed to have a 90% chance of detecting a significant difference using a two-sided test with $\alpha = .05$ if the true rate of breast cancer among women with a sister history is 2.5 times as high as that of a typical 50–54-year-old woman. ∎

Note that if we wish to perform a one-sided test rather than a two-sided test at level α, then α is substituted for $\alpha/2$ in the power formula in **(7.37)** and the sample size formula in **(7.38)**. These formulas are summarized as follows:

7.39 | **Power for the One-Sample Binomial Test (One-Sided Alternative)**

To test the hypothesis $H_0: p = p_0$ versus $H_1: p < p_0$ for the specific alternative $p_1 < p_0$, the power is given by

$$\Phi\left\{ \sqrt{\frac{p_0 q_0}{p_1 q_1}}\left[z_\alpha + \frac{(p_0 - p_1)\sqrt{n}}{\sqrt{p_0 q_0}} \right] \right\}$$

To test the hypothesis $H_0: p = p_0$ versus $H_1: p > p_0$ for the specific alternative $p_1 > p_0$, the power is given by

$$\Phi\left\{ \sqrt{\frac{p_0 q_0}{p_1 q_1}}\left[z_\alpha + \frac{(p_1 - p_0)\sqrt{n}}{\sqrt{p_0 q_0}} \right] \right\}$$

7.40 | **Sample Size Estimation for the One-Sample Binomial Test (One-Sided Alternative)**

Suppose we wish to test the hypothesis $H_0: p = p_0$ versus $H_1: p < p_0$. The sample size needed to conduct a one-sided test with significance level α and power $1 - \beta$ versus the specific alternative $p = p_1 < p_0$ is

$$n = \frac{p_0 q_0 \left(z_{1-\alpha} + z_{1-\beta}\sqrt{\dfrac{p_1 q_1}{p_0 q_0}} \right)^2}{(p_1 - p_0)^2}$$

The same formula is used to test the hypothesis $H_0: p = p_0$ versus $H_1: p > p_0$ for the specific alternative $p = p_1 > p_0$.

7.10.4   **The Standardized Mortality Ratio**

EXAMPLE 7.50   **Occupational Health** Many studies have looked at possible heath hazards of rubber workers. In one such study, a group of 8418 white male workers ages 40–84 (either active or retired) on January 1, 1964, were followed for 10 years for various mortality outcomes [1]. Their mortality rates were then compared with U.S. white male mortality rates in 1968. In one of the reported findings, 21 deaths due to bladder cancer were observed compared with 18.1 deaths expected from U.S. mortality rates. Is this difference significant?   ∎

One problem with this type of study is that workers of different ages in 1964 have very different mortality risks over time. Thus the test procedures in **(7.34)** and **(7.36)**, which assume a constant p for all persons in the sample, are not applicable. However, these procedures can be generalized to take account of the different mortality risks of different individuals. Let

O = total observed number of deaths for members of the study population

p_i = probability of death for the ith individual

Under the null hypothesis that the death rates for the study population are the same as those for the general U.S. population, the expected number of events E is given by

$$E = \sum_{i=1}^{n} p_i$$

DEFINITION 7.13   ∎∎
The **standardized mortality ratio (SMR)** is defined by $100\% \times O/E = 100\% \times$ the observed number of deaths in the study population divided by the expected number of deaths in the study population under the assumption that the mortality rates for the study population are the same as those for the general population. For nonfatal conditions the standardized mortality ratio is sometimes known as the **standardized morbidity ratio**.   ∎

Thus,

If SMR $> 100\%$, there is an excess risk in the study population relative to the general population.

If SMR $< 100\%$, there is a reduced risk in the study population relative to the general population.

If SMR $= 100\%$, there is neither an excess nor a deficit of risk in the study population relative to the general population.

EXAMPLE 7.51   **Occupational Health** What is the standardized mortality ratio for bladder cancer using the data in Example 7.50?

SOLUTION   SMR $= 100\% \times 21/18.1 = 116\%$.   ∎

The significance test will be based on the SMR. If the SMR is far from 100%, then H_0 will be rejected; otherwise, H_0 will be accepted. We need to compute

Var(SMR) under H_0. We can show that under H_0 Var(SMR/100) approximately equals $1/E$ for large E. This result leads to the following test procedure:

7.41 **Generalized One-Sample Binomial Test (Large-Sample Test)**

To test the hypothesis H_0: SMR $= 100$ versus H_1: SMR $\neq 100$,

(1) Compute

$$O = \text{observed number of deaths in the study population}$$

$$E = \text{expected number of deaths in the study population} = \Sigma p_i$$

where p_i is the probability of death for the ith member of the study population under H_0, that is, under the assumption that the general population death rates apply to the study population.

$$\text{SMR} = 100\% \times O/E$$

(2) Compute the test statistic

$$X^2 = \frac{(O-E)^2}{E} = E \times \left(\frac{\text{SMR}}{100} - 1\right)^2 \sim \chi^2_1 \text{ under } H_0$$

(3) For a two-sided test at level α, H_0 is rejected if

$$X^2 > \chi^2_{1,1-\alpha}$$

and H_0 is accepted if

$$X^2 \leqslant \chi^2_{1,1-\alpha}$$

(4) The exact p-value is given by

$$Pr(\chi^2_1 > X^2)$$

(5) This test should only be used if $E \geqslant 5$.

The acceptance and rejection regions for this test are depicted in Figure 7.19. The computation of the exact p-value is given in Figure 7.20.

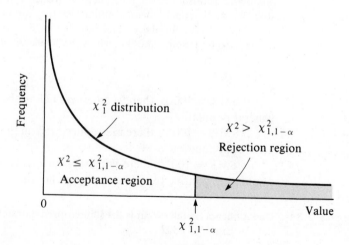

FIGURE 7.19
Acceptance and rejection regions for the generalized one-sample binomial test (large-sample test)

EXAMPLE 7.52   **Occupational Health** Assess the statistical significance of the bladder cancer data in Example 7.50.

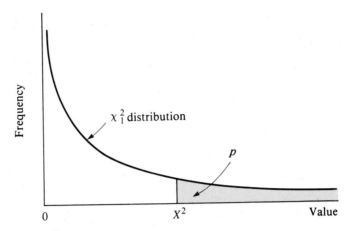

FIGURE 7.20
Computation of the p-value for the generalized one-sample binomial test (large-sample test)

SOLUTION $O = 21$, $E = 18.1$, SMR $= 116$. Therefore, we have the test statistic

$$X^2 = \frac{(21 - 18.1)^2}{18.1} \quad \text{or} \quad 18.1 \times (1.16 - 1)^2$$

$$= \frac{8.41}{18.1} = 0.46 \sim \chi_1^2 \text{ under } H_0$$

Since $\chi_{1,.95}^2 = 3.84 > X^2$, $p > .05$ and H_0 is accepted. Furthermore, from Table 6 in Appendix 1 we note that $\chi_{1,.50}^2 = 0.45$, $\chi_{1,.75}^2 = 1.32$, and $0.45 < X^2 < 1.32$. Thus, $1 - .75 < p < 1 - .50$, or $.25 < p < .50$. Therefore, the rubber workers in this plant do not have a significantly increased or decreased risk of bladder cancer relative to the general population. ∎

In many applications involving the SMR, the expected number of events will be less than 5. In these cases the large-sample test procedure in **(7.41)** is not applicable, and the following small-sample test procedure should be used:

7.42 | **Generalized One-Sample Binomial Test (Small-Sample Test)**

To test the hypothesis H_0: SMR $= 100$ versus H_1: SMR $\neq 100$,

(1) Compute

O = observed number of events in the study population

E = expected number of events in the study population $= \Sigma p_i$

where p_i is the probability of death for the ith member of the study population under H_0, that is, under the assumption that the general population death rates apply to the study population.

(2) The observed number of deaths (O) will follow a Poisson distribution with parameter E. Thus, the exact p-value is given by

$$2 \times \sum_{k=0}^{O} \frac{e^{-E}E^k}{k!} \qquad \text{if } O < E$$

$$2 \times \left(1 - \sum_{k=0}^{O-1} \frac{e^{-E}E^k}{k!} \right) \qquad \text{if } O \geqslant E$$

These computations are depicted in Figure 7.21.

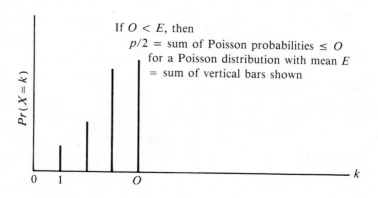

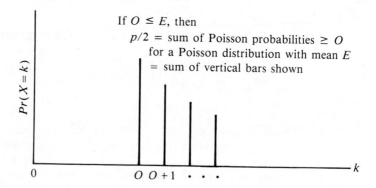

FIGURE 7.21
Computation of the
exact *p*-value for the
generalized one-
sample binomial test
(small-sample case)

EXAMPLE 7.53   **Occupational Health** The observed number of deaths due to Hodgkin's disease in the sample in Example 7.50 was 4 with SMR = 123. Test for the significance of these findings.

SOLUTION

$$E = 100 \times \frac{O}{\text{SMR}} = 100 \times \frac{4}{123} = 3.3 < 5.0$$

Thus, the small-sample procedure in **(7.42)** must be used. Since $O \geq E$,

$$p = 2 \times \left[1 - \sum_{k=0}^{3} \frac{e^{-3.3}(3.3)^k}{k!} \right]$$

From the recursion rule for the Poisson distribution,

$$Pr(0) = e^{-3.3} = .0369$$

$$Pr(1) = \frac{3.3}{1} \times .0369 = .1218$$

$$Pr(2) = \frac{3.3}{2} \times .1218 = .2010$$

$$Pr(3) = \frac{3.3}{3} \times .2010 = .2211$$

Thus
$$p = 2 \times [1 - (.0369 + .1218 + .2010 + .2211)]$$
$$= 2 \times (1 - .5808) = .838$$

Thus there is no significant excess or deficit of Hodgkin's disease in this population. ∎

Summary

In this chapter some of the fundamental ideas of hypothesis testing were introduced: (1) specification of the null (H_0) and alternative (H_1) hypotheses; (2) type I error (α), type II error (β), and power of a hypothesis test; (3) the p-value of a hypothesis test; and (4) the distinction between one-sided and two-sided tests. Methods for estimating the appropriate sample size for a proposed study as determined by the prespecified null and alternative hypotheses and the type I and type II errors were also discussed.

These general concepts were applied to several one-sample hypothesis-testing situations:

(1) The mean of a normal distribution with known variance (one-sample normal test)

(2) The mean of a normal distribution with unknown variance (one-sample t test)

(3) The variance of a normal distribution (one-sample chi-square test)

(4) The parameter p of a binomial distribution (one-sample binomial test)

Each of the hypothesis tests can be conducted in one of two ways:

(1) Specify critical values to determine the acceptance and rejection regions.

(2) Compute p-values

These methods were shown to be equivalent in the sense that they yield the same inferences regarding the acceptance and rejection of the null hypothesis.

Finally, the relationship between the hypothesis-testing methods in this chapter and the confidence-interval methods in Chapter 6 was explored. We showed that the inferences that can be drawn from using these methods are the same.

Many hypothesis tests are covered in this book. A flowchart is provided at the back of the book to help clarify the decision process in selecting the appropriate test. The flowchart can be used to choose the proper test by answering a series of yes/no questions. The specific hypothesis tests covered in this chapter have been shaded in an excerpt from the flowchart shown in Figure 7.22. For example, if we are interested in performing hypothesis tests concerning the mean of a normal distribution with known variance, then, beginning at the start box of the flowchart, we would answer *yes* to each of the following questions: (1) only one variable of interest? (2) one-sample problem? (3) underlying distribution normal? (4) inference concerning μ? (5) σ known? The flowchart leads us to the box on the lower left of the figure, indicating that the one-sample normal test should be used.

The study of hypothesis testing is extended in Chapter 8 to situations in which two different samples are compared. This topic corresponds to the answer *yes* to the questions (1) only one variable of interest? and *no* to (2) one-sample problem?

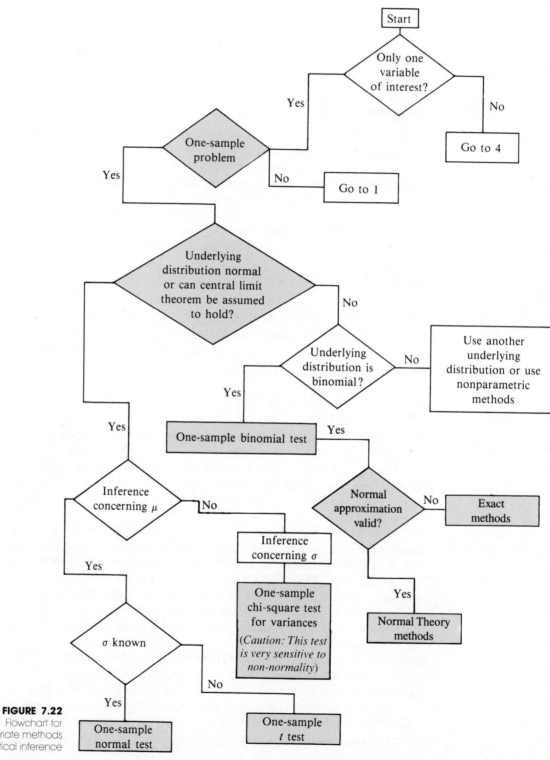

FIGURE 7.22
Flowchart for
appropriate methods
of statistical inference

PROBLEMS

Nutrition

As part of a dietary instruction program, ten 25–34-year-old males adopted a vegetarian diet for 1 month. During the diet, the average daily intake of linoleic acid was 13 g.

7.1 If the average daily intake among 25–34-year-old males in the general population is 15 g with standard deviation 4 g, then, using a significance level of .05, test the hypothesis that the intake of linoleic acid in this group is lower than that in the general population.

7.2 Compute a p-value for the hypothesis test in Problem 7.1.

As part of the same program, eight 25–34-year-old females report an average daily intake of saturated fat of 11 g.

7.3 If the average daily intake of saturated fat among 25–34-year-old females in the general population is 24 g with standard deviation 11 g, then, using a significance level of .01, test the hypothesis that the intake of saturated fat in this group is lower than that in the general population.

7.4 Compute a p-value for the hypothesis test in Problem 7.3.

7.5 What is the relationship between your answers to Problems 7.3 and 7.4?

Suppose we are uncertain what effect a vegetarian diet will have on the level of linoleic acid intake in Problem 7.1.

7.6 What are the null and alternative hypotheses in this case?

7.7 Compare the mean level of linoleic acid in this vegetarian population with that of the general population under the hypotheses in Problem 7.6. Report a p-value.

Infectious Disease

The mean serum creatinine level measured in 12 patients 24 hours after they received a newly proposed antibiotic was 1.2 mg/dL.

7.8 If the mean and standard deviation of serum creatinine in the general population are 1.0 and 0.4 mg/dL, respectively, then, using a significance level of .05, test if the mean serum creatinine level in this group is different from that of the general population.

7.9 What is the p-value for the test?

7.10 Suppose $\dfrac{\bar{x} - \mu_0}{s/\sqrt{n}} = 2.73$ and a one-sample t test is performed based on 20 subjects. What is the two-tailed p-value?

7.11 Suppose $\dfrac{\bar{x} - \mu_0}{s/\sqrt{n}} = -1.52$ and a one-sample t test is performed based on 7 subjects. What is the two-tailed p-value?

7.12 What is the approximate 95th percentile of a t distribution with 35 df (i.e., $t_{35,.95}$)?

7.13 What is the approximate 99th percentile of a t distribution with 75 df?

7.14 Suppose the sample standard deviation of linoleic acid is 6 g in Problem 7.7. Assume that the standard deviation for the vegetarian population is not known, and perform the hypothesis test for the hypotheses in Problem 7.6. Report a p-value.

7.15 Suppose the sample standard deviation of serum creatinine in Problem 7.8 is 0.6 mg/dL. Assume that the standard deviation of serum creatinine is not known, and perform the hypothesis test in Problem 7.8. Report a p-value.

7.16 Suppose a larger study is planned with 50 patients to be enrolled. If the alternative mean is 1.2 mg/dL and the standard deviation is 0.4 mg/dL, then what is the power of such a study using a one-sided test with $\alpha = .05$?

7.17 Answer Problem 7.16 if a two-sided test is used.

7.18 How large a study would be needed in Problem 7.16 to achieve 80% power?

7.19 How large a study would be needed in Problem 7.17 to achieve 80% power?

7.20 Answer Problem 7.19 for 90% power.

Diabetes

Plasma glucose levels are used to determine the presence of diabetes. Suppose the mean $\log_e$ plasma glucose concentration (mg/dL) in 35–44 year olds is 4.86 with standard deviation 0.54. A study of 100 sedentary persons in this age group is planned to test if they have a higher or lower level of plasma glucose than the general population.

7.21 If the expected difference is 0.10 $\log_e$ units, then what is the power of such a study if a two-sided test is to be used with $\alpha = .05$?

7.22 Answer Problem 7.21 if the expected difference is $0.20 \log_e$ units.

7.23 How many people would need to be studied to have 80% power under the assumptions in Problem 7.21?

7.24 Compute a lower one-sided 95% confidence interval for the true mean intake of linoleic acid in the vegetarian population depicted in Problem 7.1. Assume that the standard deviations of the vegetarian population and the general population are the same.

7.25 How does your answer to Problem 7.24 relate to your answer to Problem 7.1?

7.26 Compute a two-sided 95% confidence interval for the true mean serum creatinine level in Problem 7.8. (Assume that the standard deviation is known to be 0.4 mg/dL.)

7.27 How does your answer to Problem 7.26 relate to your answer to Problem 7.8?

Nutrition

A food frequency questionnaire was mailed to 20 subjects to assess the intake of various food groups. The sample standard deviation of vitamin C intake over the 20 subjects was 15 (exclusive of vitamin C supplements). Suppose we know from using an in-person diet interview method in a large previous study that the standard deviation is 20.

7.28 What hypotheses can be used to test if there are any differences between the standard deviations of the two methods?

7.29 Perform the test described in Problem 7.28 and report a p-value.

The sample standard deviation for $\log_e$ (vitamin A intake) exclusive of supplements based on the 20 subjects using the food frequency questionnaire was 0.016. Suppose the standard deviation from the diet interview method is known to be 0.020 based on a large previous study.

7.30 Test the hypothesis that the variances using the two methods are the same. Use the critical value method with $\alpha = .05$.

7.31 Report a p-value corresponding to the test in Problem 7.30.

Cardiovascular Disease

Suppose the incidence rate of MI per year was 5 per 1000 among 45–54-year-old males in 1970. To look at changes in incidence over time, 5000 45–54-year-old

men were followed for 1 year starting in 1980. Fifteen new cases of MI were found.

7.32 Using the critical value method with $\alpha = .05$, test the hypothesis that incidence rates of MI changed from 1970 to 1980.

7.33 Report a p-value to correspond to your answer to Problem 7.32.

Suppose that 25% of MI cases in 1970 died within 24 hours. This proportion is called the 24-hour case fatality rate.

7.34 Of the 15 new MI cases in the preceding study, 5 died within 24 hours. Test if the 24-hour case fatality rate changed from 1970 to 1980.

7.35 Suppose we eventually plan to accumulate 50 MI cases during the period 1980–1985. Assume that the 24-hour case fatality rate is truly 20% during this period. How much power would such a study have in distinguishing between case fatality rates in 1970 and 1980–1985 if a two-sided test with significance level .05 is planned?

7.36 How large a sample is needed in Problem 7.35 to achieve 90% power?

Occupational Health

Suppose that 28 cancer deaths are noted among workers exposed to asbestos in a building materials plant from 1981–1985. Only 20.5 cancer deaths are expected from statewide cancer mortality rates.

7.37 What is the estimated SMR for total cancer mortality?

7.38 Is there a significant excess or deficit of total cancer deaths among these workers?

In the same group of workers, 7 deaths due to leukemia are noted. Only 4.5 are expected from statewide rates.

7.39 What is the estimated SMR for leukemia?

7.40 Is there a significant excess or deficit of leukemia deaths among these workers?

Pulmonary Disease

Suppose the annual incidence of asthma in the general population among children 0–4 years of age is 1.4% for boys and 1% for girls.

7.41 If 10 cases are observed among 500 boys 0–4 years of age with smoking mothers, then test if there is a significant difference in asthma incidence between this group and the general population using the critical value method with a two-sided test.

7.42 Report a p-value corresponding to your answer to Problem 7.41.

7.43 Suppose that 4 cases are observed among 300 girls 0–4 years of age with smoking mothers. Answer Problem 7.41 based on these data.

7.44 Report a p-value corresponding to your answer to Problem 7.43.

Infectious Disease

Suppose the annual incidence of diarrhea (defined as 1+ episodes per year) in a Third-World country is 5% in children under the age of 2.

7.45 If 10 children out of 108 under the age of 2 in a poor rural community in the country have 1+ episodes of diarrhea in a year, then test if this represents a significant departure from the overall rate for the country using the critical value method.

7.46 Report a p-value corresponding to your answer to Problem 7.45.

Molecular Genetics

Ribosomal 5S RNA can be represented as a sequence of 120 nucleotides. Each nucleotide can be represented by one of four characters: A(adenine), G(guanine), C(cytosine), or U(uracil). The characters occur with different probabilities for each position. We wish to test if a new sequence is the same as ribosomal 5S RNA. For this purpose, we replicate the new sequence 100 times and find that there are 60 A's in the 20th position.

7.47 If the probability of an A in the 20th position is .79 in ribosomal 5S RNA, then test the hypothesis that the new sequence is the same as ribosomal 5S RNA using the critical value method.

7.48 Report a p-value corresponding to your results in Problem 7.47.

Ophthalmology

Suppose the distribution of systolic blood pressure in the general population is normal with a mean of 130 mm Hg and a standard deviation of 20 mm Hg. In a special subgroup of 85 persons with glaucoma, we find that the mean systolic blood pressure is 135 mm Hg with a standard deviation of 22 mm Hg.

7.49 Assuming that the standard deviation of the glaucoma patients is the same as that of the general population, test for an association between glaucoma and high blood pressure.

7.50 Answer Problem 7.49 without making the assumption concerning the standard deviation.

Cancer

7.51 Suppose we identify fifty 50–54-year-old women who have both a mother *and* a sister with a history of breast cancer. Five of these women themselves have developed breast cancer at some time in their lives. If we assume that the expected prevalence rate of breast cancer in women whose mothers have had breast cancer is 4%, then does having a sister with the disease add to the risk?

Occupational Health

7.52 Suppose it is known that the average life expectancy of a 50-year-old man in 1945 was 18.5 years. Twenty men aged 50 who have been working for at least 20 years in a potentially hazardous industry were ascertained in 1945. Upon follow-up in 1985 all the men have died, with an average lifetime of 16.2 years and a standard deviation of 7.3 years since 1945. Assuming that the life expectancy of 50-year-old men is approximately normally distributed, test if the underlying life expectancy for workers in this industry is shorter than for comparably aged men in the general population.

Obstetrics

7.53 The probability of having twins in the United States is approximately 1 in 90. This proportion is thought to be affected by a number of factors, including age, race, and parity. To test the effect of age, hospital records are abstracted. Of 538 deliveries for women under 20, 2 were found to have resulted in twins. What can be said about the effect of age on having twins?

Cancer

7.54 An area of current interest in cancer epidemiology is the possible role of oral contraceptives (OC's) in the development of breast cancer. Suppose that in a group of 1000 premenopausal women ages 40–49 who are current users of OC's, 15 subsequently develop breast cancer over the next 5 years. If the expected incidence of breast cancer in this group is 1.2% based on national incidence rates, then test the hypothesis that there is an association between current OC use and the subsequent development of breast cancer.

Obstetrics

Erythromycin is a drug that has been proposed to possibly lower the risk of premature delivery. A related area of interest is its association with the incidence of side effects during pregnancy. Assume that 30% of all pregnant women complain of nausea between the 24th and 28th week of pregnancy. Furthermore, suppose that of 200 women who are taking erythromycin regularly during this period, 110 complain of nausea.

7.55 Test the hypothesis that the incidence rate of nausea for the erythromycin group is the same as that for a typical pregnant woman.

Cancer

A group of investigators wishes to explore the relationship between the use of hair dyes and the development of breast cancer in females. A group of 1000 beauticians 40–49 years of age is identified and followed for 5 years. After 5 years, 20 new cases of breast cancer have occurred. Assume that breast cancer incidence over this time period for an average American woman in this age group is $\frac{7}{1000}$. We wish to test the hypothesis that using hair dyes increases the risk of breast cancer.

7.56 Is a one-sided or two-sided test appropriate here?

7.57 Test the hypothesis.

Epidemiology

One hundred volunteers agree to participate in a clinical trial involving a dietary intervention. The investigators want to check how representative this sample is of the general population. One interesting finding is that 10 of the volunteers are current cigarette smokers.

7.58 Assuming that 30% of the general population of adults are current smokers, state the hypotheses needed to test whether the volunteer group is representative of the general population regarding cigarette smoking.

7.59 Carry out the test described in Problem 7.58 and report a *p*-value.

Obstetrics

Assume that the distribution of birthweights in the general population is normal with mean 120 oz and standard deviation 20 oz. We wish to test a drug that, when administered to mothers in the prenatal period, will reduce the number of low-birthweight infants. We anticipate that the mean birthweight of the infants whose mothers are on the drug will be $\frac{1}{2}$ lb heavier than that for the general newborn population.

7.60 How large a sample is needed to have an 80% chance of finding a significant difference if a one-sided test with significance level .05 is used?

7.61 Answer Problem 7.60 for a 90% chance.

Cardiovascular Disease, Nutrition

Much discussion has appeared in the medical literature in recent years on the role of diet in the development of heart disease. The serum cholesterol levels of a group of

people who eat a primarily macrobiotic diet are measured. Among 24 of them, aged 20–39, the mean cholesterol level was found to be 175 mg/dL with a standard deviation of 35 mg/dL.

7.62 If the mean cholesterol level in the general population in this age group is 230 mg/dL and the distribution is assumed to be normal, then test the hypothesis that the group of people on a macrobiotic diet have cholesterol levels different from those of the general population.

7.63 Compute a 95% confidence interval for the true mean cholesterol level in this group.

7.64 What type of complementary information is provided by the hypothesis test and confidence interval in this case?

Renal Disease

The level of serum creatinine in the blood is considered a good indicator of the presence or absence of kidney disease. Normal people generally have low concentrations of serum creatinine, whereas diseased people have high concentrations. Suppose we want to look at the relation between analgesic abuse and kidney disorder. In particular, suppose we look at 15 people working in a factory who are known to be "analgesic abusers" (i.e., they take more than 10 pills per day) and we measure their creatinine levels. The creatinine levels are

$$0.9, 1.1, 1.6, 2.0, 0.8,$$
$$0.7, 1.4, 1.2, 1.5, 0.8,$$
$$1.0, 1.1, 1.4, 2.2, 1.4$$

7.65 If we assume that creatinine levels for normal people are normally distributed with mean 1.0 and standard deviation 0.40, then can we make any comment about the levels for analgesic abusers via some statistical test?

7.66 Suppose we are skeptical about assuming that the standard deviation is known. Can we test the validity of this assumption?

7.67 If we do not want to assume that the standard deviation is known, then answer the question in Problem 7.65 in some other way.

Cardiovascular Disease

High cholesterol levels have for some time been suspected as predictors of future heart attacks. The problem is that other factors such as age, smoking, body weight, family history, and so forth enter into the picture, and the different factors are difficult to separate out. Suppose we isolate a group of 100 men who have "high" cholesterol levels and can predict, on the basis of other factors, that 10% of these men will have a heart attack in the next 5 years.

7.68 Suppose that the incidence rate of heart attacks in the next 5 years in this group is 13%. Is this finding indicative of anything about cholesterol?

7.69 Suppose we had 1000 men instead of 100 men and also found an incidence rate of 13%. Is this finding indicative of anything about cholesterol?

7.70 How large a sample would be needed to have an 80% chance of finding a significant difference if the true rate of heart disease in this group is 13%?

Otolaryngology

Otitis media is an extremely common disease of the middle ear in children under 2 years of age. It can cause prolonged hearing loss during this period and may result in subsequent defects in speech and language. Suppose we wish to design a study to test the latter hypothesis, and we set up a study group consisting of children with 3 or more episodes of otitis media in the first 2 years of life. We have no idea what the size of the effect will be. Thus, we set up a pilot study with 20 cases, whereby we find that 5 of the cases have speech and language defects at age 3.

7.71 If we regard this experience as representative of what would occur in a large study and if we *know* that 15% of all normal children have speech and language defects by age 3, then how large a study group is needed to have an 80% chance of detecting a significant difference using a one-sided test at the 5% level?

7.72 Suppose only 50 cases can be recruited for the study group. How likely are we to find a significant difference if the true proportion of affected children with speech and language defects at age 3 is the same as that in the pilot study?

Hypertension

A pilot study of a new antihypertensive agent is performed for the purpose of planning a larger study. Five patients who have a mean diastolic blood pressure of at least 95 mm Hg are recruited for the study and are kept on the agent for 1 month. After 1 month the observed mean decline in diastolic blood pressure in these 5 patients is 4.8 mm Hg with a standard deviation of 9 mm Hg.

7.73 If μ_d = true mean difference in diastolic blood pressure between baseline and 1 month, then how many patients would be needed to have a 90% chance of

detecting a significant difference using a one-tailed test with a significance level of 5%? Assume that the true mean and standard deviation of the blood-pressure difference was the same as that observed in the pilot study.

7.74 Suppose we conduct a study of the preceding hypothesis based on 20 subjects. What is the probability that we will be able to reject H_0 using a one-sided test at the 5% level if the true mean and standard deviation of the blood-pressure difference is the same as that in the pilot study?

Hypertension

Suppose we are interested in investigating the effect of race on level of blood pressure. A study was conducted in Evans County, Georgia, comparing the mean level of blood pressure among whites and blacks for different age-sex groups [2]. The mean and standard deviation of systolic blood pressure among 25–34-year-old white males were reported as 128.6 mm Hg and 11.1 mm Hg, respectively, based on a large sample.

7.75 Suppose the mean bp of 38 25–34-year-old black males is reported as 135.7 mm Hg. If the standard deviation of white males is assumed to hold for black males as well, then test if the underlying mean systolic blood pressures are the same in the two groups.

7.76 The actual reported standard deviation among 25–34-year-old black males is 12.5 mm Hg. Test the hypothesis that the underlying variance of blood pressure for white and black males is the same.

7.77 Suppose the actual underlying mean for black males is 135 mm Hg. What is the power of the test in Problem 7.75, in this case if we assume that the standard deviations for white males and black males are the same?

Suppose we do *not* assume that the standard deviation of 11.1 mm Hg is correct for 25–34-year-old black males.

7.78 Test the hypothesis that the underlying mean systolic blood pressures are the same in the two groups.

7.79 Derive a 95% confidence interval for the mean systolic blood pressure for 25–34-year-old black males under the assumptions in Problem 7.78.

7.80 Relate your answers to Problems 7.78 and 7.79.

Occupational Health

The proportion of deaths due to lung cancer in males aged 15–64 in England and Wales during the period 1970–1972 was 12%. Suppose that of 20 deaths that occur among male workers in this age group who have worked for at least 1 year in a chemical plant, 5 are due to lung cancer. We wish to determine if there is a difference between the proportion of lung cancer deaths in this plant and the proportion in the general population.

7.81 State the hypotheses to be used in answering this question.

7.82 Is a one-sided or two-sided test appropriate here?

7.83 Perform the hypothesis test and report a *p*-value. After reviewing the results from 1 plant, the company decides to expand its study to include results from 3 additional plants. They find that of 90 deaths occurring among 15–64-year-old male workers who have worked for a least 1 year in these 4 plants, 19 are due to lung cancer.

7.84 Answer Problem 7.83 using the data from 4 plants and report a *p*-value.

One criticism of studies of this type is that they are biased because of the "healthy worker" effect. That is, workers in general are healthier than the general population, particularly regarding cardiovascular endpoints, which makes the proportion of deaths due to noncardiovascular causes seem abnormally high.

7.85 If the proportion of deaths due to ischemic heart disease (IHD) is 40% for all 15–64-year-old men in England and Wales, whereas 18 of the preceding 90 deaths are attributed to IHD, then answer Problem 7.84 if deaths due to IHD are *excluded* from the total.

Epidemiology

Height and weight are often used in epidemiological studies as possible predictors of disease outcomes. If the people in the study are assessed in a clinic, then heights and weights are usually measured directly. However, if the people are interviewed at home or by mail, then a person's self-reported height and weight are often used instead. Suppose we conduct a study on 10 people to test the comparability of these two methods. The data for weight are given in Table 7.4.

7.86 Should a one-sided or two-sided test be used here?

7.87 Which test procedure should be used to test the preceding hypothesis?

7.88 Conduct the test in Problem 7.87 using the critical value method with $\alpha = .05$.

7.89 Compute the *p*-value for the test in Problem 7.87.

TABLE 7.4 A comparison of self-reported and measured weight (lb) for 10 subjects

Person number	Self-reported weight	Measured weight	Difference
1	120	125	−5
2	120	118	+2
3	135	139	−4
4	118	120	−2
5	120	125	−5
6	190	198	−8
7	124	128	−4
8	175	176	−1
9	133	131	+2
10	125	125	0

7.90 Is there evidence of digit preference among the self-reported weights? Specifically, compare the observed proportion of reported weights whose last digit is 0 or 5 with the expected proportion based on chance and report a *p*-value.

Nutrition

Iron deficiency anemia is an important nutritional health problem in the United States. A dietary assessment was performed in 51 9–11-year-old male children whose families were below the poverty level. The mean daily iron intake among these children was found to be 12.50 mg with standard deviation 4.75 mg. Suppose that the mean daily iron intake among a large population of 9–11-year-old boys from all income strata is 14.44 mg. We wish to test if the mean iron intake among the low-income group is different from that of the general population.

7.91 State the hypotheses that can be used to consider this question.

7.92 Carry out the hypothesis test in Problem 7.91 using the critical value method with an α level of .05 and summarize your findings.

7.93 What is the *p*-value for the test conducted in Problem 7.92?

The standard deviation of daily iron intake in the larger population of 9–11-year-old boys was 5.56 mg. We wish to test if the standard deviation from the low-income group is comparable to that of the general population.

7.94 State the hypotheses that can be used to answer this question.

7.95 Carry out the test in Problem 7.94 using the critical value method with an α level of .05 and summarize your findings.

7.96 What is the p-value for the test conducted in Problem 7.95?

7.97 Compute a 95% confidence interval for the underlying variance of daily iron intake in the low-income group. What can you infer from the confidence interval?

7.98 Compare the inferences you made from the procedures in Problems 7.95, 7.96, and 7.97.

Hypertension

Suppose we wish to design an experiment to assess the effectiveness of a new drug for treating hypertensive patients. We decide to declare the drug "effective" if the mean diastolic blood pressure of the 50 patients participating in the study drops by at least 2 mm Hg after using the drug daily for 1 month.

Let us set up this study as a hypothesis-testing problem and assume that the population standard deviation of the before-after blood pressure differences is *known* to be 10 mm Hg.

7.99 If we identify declaring the drug effective with rejecting the null hypothesis, then what is the significance level of the test?

7.100 Suppose that the underlying population before-after diastolic blood pressure decrease from using this drug is 3 mm Hg. What is the power of the test in Problem 7.99?

Gynecology

A survey of contraceptive methods conducted by the National Center for Health Statistics in 1965 indicated that among 30–39-year-old, married, nonpregnant women who practiced contraception, 20% used some form of permanent contraception (i.e., either tubal ligation for the women or vasectomy for their spouses). The frequency of use of permanent contraception was suspected to have changed in the next decade. To test this hypothesis, records were selected of 50 women subscribing to a prepaid health plan in 1975 who satisfied the preceding demographic criteria. From the records, 35 of the women were found to have practiced some form of contraception, of whom 10 used the method of permanent contraception. Suppose we wish to test the hypothesis that a change has occurred in the percentage of contraceptors who use permanent con-

traception, without specifying the direction of the change.

7.101 Specify the hypotheses needed to perform this test.

7.102 Use the critical value method to conduct the hypothesis test with an α level of .01.

7.103 What is the exact p-value for this test?

The investigators are encouraged by the results of the small study and wish to enlarge the study.

7.104 How large a sample would be needed to test the preceding hypotheses if the following assumptions hold:

(a) The true rate of permanent contraception among contraceptors is 30%.

(b) Seventy percent of the women use some form of contraception.

(c) We wish to have a 90% chance of finding a significant difference using a two-sided test with an α level of .05.

Hypertension

Several studies have been performed relating urinary potassium excretion to blood-pressure level. These studies have tended to show an inverse relationship between these two variables, with the higher the level of potassium excretion, the lower the blood-pressure level. Therefore, a treatment trial is planned to look at the effect of potassium intake in the form of supplement capsules on changes in blood-pressure level. Suppose that in a pilot study, 20 people are given potassium supplements for 1 month. The data are as follows:

mean change	−3.2
sd change	8.5
n	20

7.105 What test should be used to assess if the potassium supplements have any effect on blood-pressure level?

7.106 Perform the test in Problem 7.105 using a two-sided test and report the p-value.

7.107 Derive a 95% confidence interval for the true mean change based on the above data. What is the relationship of your results here and in Problem 7.106?

7.108 How many subjects need to be studied to have an 80% chance of detecting a significant treatment effect using a two-sided test with an α level of .05 if the results (i.e., mean, sd) of the pilot study are assumed to be the population mean and sd?

Demography

A study is performed using census data to look at various health parameters for a group of 10,000 Americans of Chinese descent living in the Chinatown area of New York and San Francisco in 1970. The comparison group for this study is the total U.S. population in 1970.

Suppose it is found that 100 of the Chinese have died over a 1-year period and that this represents a 15% decline from the expected mortality rate based on 1970 U.S. age-sex specific mortality rates. What is the standardized mortality ratio for this group?

7.109 Test if the total mortality experience in this group is significantly different from that found in the total U.S. population. Report an exact p-value.

Suppose that 8 deaths due to tuberculosis are found in this group, which is twice the rate expected based on the total U.S. population in 1970.

7.110 Test if the mortality experience due to tuberculosis in this group is significantly different from that found in the total U.S. population. Report an exact p-value.

Occupational Health

To assess quantitatively the association between benzene exposure and cancer incidence, a study was conducted among 1165 men who worked in a rubber hydrochloride plant with occupational exposure to benzene. A comparison was made between observed cancer mortality during the period 1950–1981 and expected cancer mortality as derived from U.S. white male mortality rates over the same time period. The data in Table 7.5 were obtained [3].

TABLE 7.5 Mortality data for workers exposed to benzene

| Cause of death | Number of deaths | |
	Observed	Expected
Total cancer	69	66.8
Multiple myeloma	4	1.0

7.111 What are the standardized mortality ratios for total cancer and multiple myeloma in this group?

7.112 Perform a test to compare observed and expected total cancer mortality in this cohort. Report a p-value.

7.113 Perform a test comparing observed and expected multiple myeloma rates in this cohort. Report a p-value.

Occupational Health

The mortality experience of 8146 male employees of a research, engineering, and metal fabrication plant in Tonawanda, New York, was studied from 1946 to 1981. [4]. Potential workplace exposures included welding fumes, cutting oils, asbestos, organic solvents, and environmental ionizing radiation, as a result of disposal of wastes during the Manhattan Project of World War II. Comparisons were made for specific causes of death between mortality rates in workers and U.S. white male mortality rates from 1950–1978.

Suppose that among workers who were hired prior to 1946 and who had worked in the plant for 10 or more years that 17 deaths due to cirrhosis of the liver were observed, while 6.3 were expected based on U.S. white male mortality rates.

7.114 What is the standardized mortality ratio for this group?

7.115 Perform a significance test to assess whether there is an association between long duration of employment and mortality due to cirrhosis of the liver in the group hired prior to 1946. Report a p-value.

7.116 A similar analysis was performed among workers who were hired after 1945 and who were employed for 10 or more years. It was found that there were 4 deaths due to cirrhosis of the liver, while only 3.4 were expected.

7.117 Perform a significance test to assess whether there is an association between mortality due to cirrhosis of the liver and duration of employment in the group hired after 1945. Report a p-value.

Hospital Epidemiology

A study was conducted to identify characteristics that would predict 1-year survival for patients admitted to the medical service at New York Hospital [5]. One factor that was considered was the physician's estimate of the patients' severity of illness at the time of admission.

Suppose that it is expected, based on previous studies in this hospital, that $\frac{2}{3}$ of admitted patients will survive for at least 1 year.

7.118 If 47% of 136 patients deemed severely ill survive at least 1 year, then what test can be used to test the hypothesis that the severity-of-illness rating is predictive of 1-year mortality?

7.119 Perform the test in Problem 7.118 using the critical value method.

7.120 Provide a 95% confidence interval for the 1-year survival rate among severely ill patients.

Suppose that it is expected, based on previous studies, that among *all patients who survive hospitalization,* 75% will survive for 1 year. Furthermore, of the 136 severely ill patients, 33 die during hospitalization, and an additional 39 will die during the 1st year, but after hospitalization.

7.121 Test the hypothesis that severity of illness is predictive of 1-year mortality among patients who are discharged from the hospital. Report a *p*-value.

Nutrition
Refer to Data Set 26, VALID.DAT, Appendix 2.

7.122 Assess if reported nutrient consumption (i.e., saturated fat, total fat, alcohol consumption, total ca-

loric intake) is comparable for the diet record and the food frequency questionnaire. Use either hypothesis testing and/or confidence interval methodology.

7.123 Answer Problem 7.122 for the percentage of calories due to fat (separately for total fat and saturated fat) as reported on the diet record and the food frequency questionnaire. Assume there are 9 calories due to fat for every gram of fat consumed.

Demography
Refer to Data Set 22, SEXRAT.DAT, Appendix 2.

7.124 Apply hypothesis-testing methods to answer the questions posed in Problem 4.135 (p. 105).

Cardiovascular Disease
Refer to Data Set 18, NIFED.DAT. Appendix 2.

7.125 Use hypothesis-testing methods to assess if either treatment has an effect on blood pressure and heart rate in patients with severe angina.

References

[1] Andjelkovic, D., Taulbee, J., & Symons, M. (1976). Mortality experience of a cohort of rubber workers, 1964–1973. *Journal of Occupational Medicine, 18*(6), 387–394.

[2] McDonough, J. R., Garrison, G. E., & Hames, C. G. (1967). Blood pressure and hypertensive disease among negroes and whites in Evans County, Georgia. In J. Stamler, R. Stamler, & T. N. Pullman (eds.), *The epidemiology of hypertension.* New York: Grune and Stratton.

[3] Rinsky, R., Smith, A. B., Hornung, R., Filloon, T. G., Young, R. J., Okun, A. H., & Landrigan, P. J. (1987). Benzene and Leukemia: An epidemiologic risk assessment. *New England Journal of Medicine, 316*(17), 1044–1050.

[4] Teta, M. J., & Ott, M. G. (1988). A mortality study of a research, engineering and metal fabrication facility in western New York state. *American Journal of Epidemiology, 127*(3), 540–551.

[5] Pompei, P., Charlson, M. E., & Douglas, R. G., Jr. (1988). Clinical assessments as predictors of one-year survival after hospitalization: Implications for prognostic stratification. *Journal of Clinical Epidemiology, 41*(3), 275–284.

CHAPTER EIGHT

HYPOTHESIS TESTING: TWO-SAMPLE INFERENCE

SECTION 8.1 **Introduction**

All the tests introduced in Chapter 7 were one-sample tests. The underlying parameters of the population from which the sample is drawn have been compared with comparable values from other generally large populations *whose parameters are assumed to be known.*

EXAMPLE 8.1 **Obstetrics** In the birthweight data in Example 7.2, the underlying mean birthweight in one hospital was compared with the underlying mean birthweight in the United States, *whose value was assumed known.* ∎

A more frequently encountered situation is the two-sample hypothesis-testing problem.

DEFINITION 8.1 ■■■
In a **two-sample** hypothesis-testing problem, the underlying parameters of two different populations, *neither of whose values is assumed known*, are compared. ∎

EXAMPLE 8.2 **Cardiovascular Disease, Hypertension** We might be interested in the relationship between the use of oral contraceptives (OC) and the level of blood pressure (bp) in women. ∎

Two different experimental designs can be used to assess this relationship. One method would involve the following design:

8.1 **Longitudinal Study**

(1) Identify a group of nonpregnant, premenopausal women of childbearing age (16–49) from a prepaid health plan who are not currently OC users and measure their bp, which will be referred to as baseline bp.

(2) Rescreen these women 1 year later to ascertain a subgroup who have remained nonpregnant throughout the year and have become OC users. This subgroup will be the study population.

(3) Measure the bp of the study population at the follow-up visit.

(4) Compare the baseline and follow-up bp of the women in the study population to determine the difference between the bp of women when they *were* using the pill at follow-up and when they *were not* using the pill at baseline.

Another method would involve the following design:

8.2 | **Cross-Sectional Study**

(1) Identify both a group of OC users and a group of non-OC users among nonpregnant, premenopausal women of childbearing age (16–49) from a prepaid health plan and measure their bp.

(2) Compare the bp of the OC users and nonusers.

DEFINITION 8.2   ■■
The first type of study is called a **longitudinal** or **follow-up study**, since the same group of women are followed *over time*.

DEFINITION 8.3   ■■
The second type of study is called a **cross-sectional study**, since the women are seen at only one point in time.

There is another important difference between these two designs. The first study represents a *paired-sample* design, since each woman is used as her own control. The second study represents an *independent-sample* design, since two completely different groups of women are being compared.

DEFINITION 8.4   ■■
Two samples are said to be **paired** when each data point of the first sample is matched and is related to a unique data point of the second sample.

EXAMPLE 8.3   The paired samples may represent two sets of measurements on the same people. In this case each person is serving as his or her own control, as is the case in (8.1). The paired samples may also represent measurements on different people who are chosen on an individual basis using matching criteria, such as age and sex, to be very similar to each other.

DEFINITION 8.5   ■■
Two samples are said to be **independent** when the data points in one sample are unrelated to the data points in the second sample.

EXAMPLE 8.4   The samples in (8.2) are completely independent, since the data are obtained from unrelated groups of women.

Which type of study is better in this case? The first type of study is probably more definitive, since most confounding factors that influence the women's blood pressure at the first screening will also be present at the second screening and will not influence the comparison of blood-pressure levels at the first and second screenings. The second type of study by itself can only be considered suggestive, since other confounding factors may influence blood pressure in the two samples and cause an apparent difference to be found where none is actually present.

For example, OC users are known to weigh less than non-OC users. Since low weight tends to be associated with low bp, OC users' blood-pressure levels as a group would appear lower than non-OC users'.

On the other hand, a follow-up study is more expensive than a cross-sectional study. Therefore, a cross-sectional study may be the only financially feasible way of doing the study.

In this chapter the appropriate methods of hypothesis testing for both the paired-sample and independent-sample situations are studied.

SECTION 8.2   **The Paired _t_ Test**

Suppose the paired-samples study design in **(8.1)** is adopted and the sample data in Table 8.1 are obtained. The systolic bp level of the _i_th woman is denoted at baseline by x_{i1} and at follow-up by x_{i2}.

TABLE 8.1
Systolic blood-pressure levels (mm Hg) in 10 women while not using (baseline) and while using (follow-up) oral contraceptives

_i_	Systolic blood-pressure level while not using OC's (x_{i1})	Systolic blood-pressure level while using OC's (x_{i2})	$d_i (x_{i2} - x_{i1})$
1	115	128	13
2	112	115	3
3	107	106	-1
4	119	128	9
5	115	122	7
6	138	145	7
7	126	132	6
8	105	109	4
9	104	102	-2
10	115	117	2

8.3   Assume that the systolic bp of the _i_th woman is normally distributed at baseline with mean μ_i and variance σ^2 and at follow-up with mean $\mu_i + \Delta$ and variance σ^2.

We are thus assuming that the mean difference in bp between follow-up and baseline is Δ, which is constant for all women. If $\Delta = 0$, then there is no difference between baseline and follow-up bp. If $\Delta > 0$, then the use of OC pills is associated with an increase in bp. If $\Delta < 0$, then the use of OC pills is associated with a decline in bp.

We wish to test the hypothesis $H_0: \Delta = 0$ versus $H_1: \Delta \neq 0$. How should this be done? The problem is that μ_i is unknown, and we are assuming, in general, that it is different for each woman. However, consider the difference $d_i = x_{i2} - x_{i1}$. From **(8.3)** we know that d_i is normally distributed with mean Δ and a variance that shall be denoted by σ_d^2. Thus, although bp levels μ_i are different for each woman, the differences in bp between baseline and follow-up have the same underlying mean (Δ) and variance (σ_d^2) over the entire population of women. The hypothesis-testing problem can thus be considered a _one-sample t test based on the differences_ (d_i). From our work on the one-sample _t_ test in Section 7.5, we know that the best test of the hypothesis $H_0: \Delta = 0$ versus $H_1: \Delta \neq 0$, when the variance is unknown, is based on the mean difference

$$\bar{d} = (d_1 + d_2 + \cdots + d_n)/n$$

Specifically, from **(7.17)** (p. 201) for a two-sided level α test, we have the following test procedure, which is referred to as the paired _t_ test:

| **8.4** | **Paired *t* Test** |

Denote the test statistic $\bar{d}/(s_d/\sqrt{n})$ by t, where s_d is the sample standard deviation of the observed differences:

$$s_d = \sqrt{\left[\sum_{i=1}^{n} d_i^2 - \left(\sum_{i=1}^{n} d_i\right)^2 \bigg/ n\right] \bigg/ (n-1)}$$

$$n = \text{number of matched pairs}$$

If

$$t > t_{n-1,1-\alpha/2} \qquad \text{or} \qquad t < -t_{n-1,1-\alpha/2}$$

then H_0 is rejected. If

$$-t_{n-1,1-\alpha/2} \leqslant t \leqslant t_{n-1,1-\alpha/2}$$

then H_0 is accepted. The acceptance and rejection regions for this test are depicted in Figure 8.1.

Distribution of t in **(8.4)** under $H_0 = t_{n-1}$ distribution

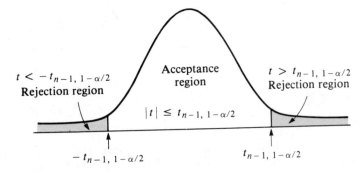

FIGURE 8.1
Acceptance and rejection regions for the paired *t* test

Similarly, from **(7.18)** (p. 202) a p-value for the test can be computed as follows:

| **8.5** | **Computation of the *p*-value for the Paired *t* Test** |

If $t < 0$,

$$p = 2 \times [\text{the area to the left of } t = \bar{d}/(s_d/\sqrt{n}) \text{ under a } t_{n-1} \text{ distribution}]$$

If $t \geqslant 0$,

$$p = 2 \times [\text{the area to the right of } t \text{ under a } t_{n-1} \text{ distribution}]$$

The computation of the p-value is illustrated in Figure 8.2.

EXAMPLE 8.5   **Cardiovascular Disease, Hypertension**   Assess the statistical significance of the OC-bp data in Table 8.1.

SOLUTION

$$\bar{d} = (13 + 3 + \cdots + 2)/10 = 4.80$$

$$s_d^2 = \{[(13)^2 + (3)^2 + \cdots + (2)^2] - 10(4.80)^2\}/9 = 20.844$$

$$s_d = \sqrt{20.844} = 4.566$$

$$t = (4.80)/(4.566/\sqrt{10}) = 4.80/1.444 = 3.32$$

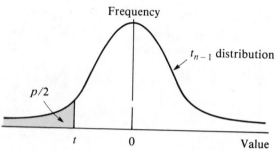

If $t = \bar{d}/(s_d/\sqrt{n}) < 0$, then $p = 2 \times$ (area to the left of t under a t_{n-1} distribution).

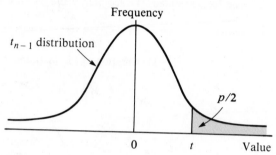

FIGURE 8.2
Computation of the
p-value for the paired
t test

If $t = \bar{d}/(s_d/\sqrt{n}) \geqslant 0$, then $p = 2 \times$ (area to the right of t under a t_{n-1} distribution).

The critical value method is first used to perform the significance test. There are $10 - 1 = 9$ degrees of freedom, and from Table 5 in Appendix 1 we see that $t_{9,.975} = 2.262$. Since $t = 3.32 > 2.262$, it follows from **(8.4)** that H_0 can be rejected using a two-sided significance test with $\alpha = .05$. To compute an approximate p-value, refer to Table 5 and note that $t_{9,.9995} = 4.781$, $t_{9,.995} = 3.250$. Thus, since $3.25 < 3.32 < 4.781$, it follows that $.0005 < p/2 < .005$ or $.001 < p < .01$. To compute a more exact p-value, a computer program must be used. The results in Table 8.2 were obtained using the SPSS^x/PC paired T-TEST program.

The output provides the mean, standard deviation, and standard error for systolic blood pressure at each of the baseline and follow-up visits. In addition, the mean, standard deviation, and standard error of the difference scores are provided along with the paired t statistic $= t$ (labeled t value), the degrees of freedom, and the two-tailed p-value (labeled 2-Tail Prob.). Finally, the **correlation coefficient** between baseline and follow-up blood pressures is given (labeled Corr.) along with a two-tailed p-value associated with this correlation. The meaning of a correlation coefficient is discussed in detail in Chapter 11.

Note from Table 8.2 that the exact two-sided p-value $= .009$. Therefore, we can conclude that starting oral contraceptive use is associated with a significant increase in blood pressure. ∎

Example 8.5 is a classic example of a paired study, since each woman is used as her own control. In many other paired studies, different people are used for the two groups, but they are matched individually on the basis of specific matching characteristics.

TABLE 8.2
Use of the SPSS^X/PC
paired T-TEST program
to analyze the blood-
pressure data in
Table 8.1

```
------------------------------------------------------------------
                          SPSSX/PC  Release 1.0


Paired samples t-test:  FUP       SYS BP WHILE USING OCS
                        BASE      SYS BP WHILE NOT USING OCS


Variable    Number               Standard   Standard
           of Cases    Mean      Deviation    Error

  FUP         10     120.4000     13.226      4.183
  BASE        10     115.6000     10.309      3.260


(Difference) Standard   Standard  |   2-Tail   |   t     Degrees of  2-Tail
   Mean      Deviation    Error   | Corr. Prob.| Value   Freedom     Prob.

  4.8000      4.566       1.444   | 0.955 0.000|  3.32      9        0.009
------------------------------------------------------------------
```

EXAMPLE 8.6   **Family Planning, Gynecology** A topic of recent clinical interest is the effect of different contraceptive methods on fertility. In particular, suppose we wish to compare how long it takes users of oral contraceptives and diaphragms, respectively, to become pregnant after stopping contraception. A study group of 20 oral contraceptive users is formed and diaphragm users who match each OC user in age (within 5 years), race, parity (number of previous pregnancies), and socioeconomic status (SES) are found. The differences in time to fertility between previous OC and diaphragm users are computed and it is found that the mean difference $\bar{d}$ (OC minus diaphragm) in time to fertility is 4 months with a standard deviation (s_d) of 8 months. What can we conclude from these data?

SOLUTION   Perform the paired t test. We have

$$t = \bar{d}/(s_d/\sqrt{n}) = 4/(8/\sqrt{20}) = 4/1.789 = 2.24 \sim t_{19}$$

under H_0. Referring to Table 5 in Appendix 1, we find that

$$t_{19,.975} = 2.093 \qquad \text{and} \qquad t_{19,.99} = 2.539$$

Then, since $2.093 < 2.24 < 2.539$, it follows that $.01 < p/2 < .025$ or $.02 < p < .05$. Therefore, previous OC users take a significantly longer time to become pregnant than do previous diaphragm users. ∎

SECTION 8.3   **Interval Estimation for the Comparison of Means from Two Paired Samples**

In the previous section, methods of hypothesis testing for comparing means from two paired samples were discussed. It is also useful to construct confidence limits for the true mean difference (μ_d). The observed difference scores $= d_i$ are normally distributed with mean Δ and variance σ_d^2. Thus, the sample mean difference $= \bar{d}$ is normally distributed with mean Δ and variance σ_d^2/n, where σ_d^2 is unknown. The methods of confidence interval estimation in **(6.7)** (p. 163) can be used to derive a $100\% \times (1 - \alpha)$ confidence interval for Δ, which is given by

$$\left(\frac{\bar{d} - t_{n-1,1-\alpha/2}s_d}{\sqrt{n}}, \bar{d} + t_{n-1,1-\alpha/2}s_d/\sqrt{n} \right)$$

| 8.6 | **Confidence Interval for the True Difference (Δ) Between the Underlying Means of Two Paired Samples (Two-Sided)** |

A two-sided $100\% \times (1 - \alpha)$ confidence interval for the true mean difference (Δ) for two paired samples is given by

$$\left(\frac{\bar{d} - t_{n-1,1-\alpha/2} s_d}{\sqrt{n}}, \bar{d} + t_{n-1,1-\alpha/2} s_d / \sqrt{n} \right)$$

EXAMPLE 8.7   **Cardiovascular Disease, Hypertension** Using the data in Table 8.1, compute a 95% confidence interval for the true increase in mean systolic blood pressure after starting oral contraceptives.

SOLUTION   From Example 8.5 we have $\bar{d} = 4.80$, $s_d = 4.566$, $n = 10$. Thus, from **(8.6)**, a 95% confidence interval for the true mean blood-pressure change is given by

$$\bar{d} \pm t_{n-1,.975} s_d/\sqrt{n} = 4.80 \pm t_{9,.975}(1.444)$$
$$= 4.80 \pm 2.262(1.444) = 4.80 \pm 3.27 = (1.53, 8.07)$$

Thus, the true change in mean bp is between 1.5 and 8 mm Hg. ∎

EXAMPLE 8.8   **Family Planning, Gynecology** Using the data in Example 8.6, compute a 95% confidence interval for the true mean difference in time to fertility between OC users and diaphragm users.

SOLUTION   From Example 8.6 we have that $\bar{d} = 4$ months, $s_d = 8$ months, $n = 20$. Thus, the 95% confidence interval for μ_d is given by

$$\bar{d} \pm \frac{t_{n-1,.975} s_d}{\sqrt{n}} = 4 \pm \frac{t_{19,.975} 8}{\sqrt{20}}$$
$$= 4 \pm \frac{2.093(8)}{\sqrt{20}} = 4 \pm 3.74 = (0.26, 7.74) \text{ months}$$

Thus, the true lag in time to fertility can be anywhere from about 0.25 months to nearly 8 months. A much larger study is needed to narrow the width of this confidence interval. ∎

SECTION 8.4   **Two-Sample *t* Test for Independent Samples with Equal Variances**

The question posed in Example 8.2 will now be discussed, assuming that the cross-sectional study defined in **(8.2)** rather than the longitudinal study defined in **(8.1)** is being performed.

EXAMPLE 8.9   **Cardiovascular Disease, Hypertension** Suppose a random sample of 8 women is selected from the group of 35–39-year-old nonpregnant premenopausal OC users in the prepaid health plan who have mean systolic blood pressure of 132.86 mm Hg and sample standard deviation of 15.34 mm Hg. A random sample of 21 women is also selected from the group of 35–39-year-old nonpregnant premenopausal non-OC users in the prepaid health plan who have mean systolic blood pressure of 127.44 mm Hg and sample standard deviation of 18.23 mm Hg. What can be said about the underlying mean difference in blood pressure between the two populations? ∎

Assume that the blood pressure is normally distributed in the first group with mean μ_1 and variance σ_1^2 and in the second group with mean μ_2 and variance σ_2^2. We want to test the hypothesis $H_0: \mu_1 = \mu_2$ versus $H_1: \mu_1 \neq \mu_2$. Assume in this section that the underlying variances in the two groups are the same (i.e., $\sigma_1^2 = \sigma_2^2 = \sigma^2$). The means and variances in the two samples are denoted by $\bar{x}_1$, $\bar{x}_2$, s_1^2, s_2^2, respectively.

It seems reasonable to base the significance test on the difference between the two sample means, $\bar{x}_1 - \bar{x}_2$. If this difference is far from 0, then H_0 will be rejected; otherwise, it will be accepted. Thus, we wish to study the behavior of $\bar{x}_1 - \bar{x}_2$ under H_0. We know that $\bar{x}_1$ is normally distributed with mean μ_1 and variance σ^2/n_1 and that $\bar{x}_2$ is normally distributed with mean μ_2 and variance σ^2/n_2. Hence, from **(5.9)**, since the two samples are independent, $\bar{x}_1 - \bar{x}_2$ is normally distributed with mean $\mu_1 - \mu_2$ and variance $\sigma^2(1/n_1 + 1/n_2)$. In symbols,

8.7
$$\bar{x}_1 - \bar{x}_2 \sim N\left[\mu_1 - \mu_2, \sigma^2\left(\frac{1}{n_1} + \frac{1}{n_2}\right)\right]$$

Under H_0 we know that $\mu_1 = \mu_2$. Thus, **(8.7)** reduces to

8.8
$$\bar{x}_1 - \bar{x}_2 \sim N\left[0, \sigma^2\left(\frac{1}{n_1} + \frac{1}{n_2}\right)\right]$$

If σ^2 were known, then $\bar{x}_1 - \bar{x}_2$ could be divided by $\sigma\sqrt{1/n_1 + 1/n_2}$. From **(8.8)**,

8.9
$$\frac{\bar{x}_1 - \bar{x}_2}{\sigma\sqrt{\dfrac{1}{n_1} + \dfrac{1}{n_2}}} \sim N(0, 1)$$

and the test statistic in **(8.9)** could be used as a basis for the hypothesis test. Unfortunately, σ^2 in general is unknown, and it must be estimated from the data. How can σ^2 be best estimated in this situation?

From the first and second sample, the sample variances are s_1^2, s_2^2, respectively, each of which could be used to estimate σ^2. The average of s_1^2 and s_2^2 could simply be used as the estimate of σ^2. However, this average will weight the sample variances equally even if the sample sizes are very different from each other. The sample variances should not be weighted equally, since the sample variance from the larger sample is probably more precise and should be weighted more heavily. The best estimate of the population variance σ^2, which is denoted by s^2, is given by a weighted average of the two sample variances, where the weights are the number of degrees of freedom in each sample.

8.10 The **pooled estimate of the variance** from two independent samples is given by

$$s^2 = \frac{[(n_1 - 1)s_1^2 + (n_2 - 1)s_2^2]}{n_1 + n_2 - 2}$$

In particular, s^2 will then have $n_1 - 1$ df from the first sample and $n_2 - 1$ df from the second sample, or

$$(n_1 - 1) + (n_2 - 1) = n_1 + n_2 - 2 \ df$$

overall. s can then be substituted for σ in **(8.9)**, and the resulting test statistic can then be shown to follow a t distribution with $n_1 + n_2 - 2 \ df$ rather than an $N(0, 1)$ distribution, since σ^2 is unknown. Thus the following test procedure is used:

8.11 | **Two-Sample *t* Test for Independent Samples with Equal Variances**

Suppose we wish to test the hypothesis $H_0: \mu_1 = \mu_2$ versus $H_1: \mu_1 \neq \mu_2$ with a significance level of α for two normally distributed populations, where σ^2 is assumed to be the same for each population.

Compute the test statistic

$$t = \frac{\bar{x}_1 - \bar{x}_2}{\left(s\sqrt{\dfrac{1}{n_1} + \dfrac{1}{n_2}}\right)}$$

where

$$s = \sqrt{[(n_1 - 1)s_1^2 + (n_2 - 1)s_2^2]/(n_1 + n_2 - 2)}$$

If

$$t > t_{n_1 + n_2 - 2, 1 - \alpha/2} \quad \text{or} \quad t < -t_{n_1 + n_2 - 2, 1 - \alpha/2}$$

then H_0 is rejected. If

$$-t_{n_1 + n_2 - 2, 1 - \alpha/2} \leqslant t \leqslant t_{n_1 + n_2 - 2, 1 - \alpha/2}$$

then H_0 is accepted.

The acceptance and rejection regions for this test are depicted in Figure 8.3.

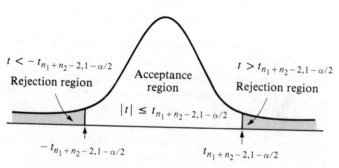

$t < -t_{n_1 + n_2 - 2, 1 - \alpha/2}$

Rejection region

Acceptance region

$t > t_{n_1 + n_2 - 2, 1 - \alpha/2}$

Rejection region

$|t| \leq t_{n_1 + n_2 - 2, 1 - \alpha/2}$

$-t_{n_1 + n_2 - 2, 1 - \alpha/2}$

$t_{n_1 + n_2 - 2, 1 - \alpha/2}$

FIGURE 8.3
Acceptance and rejection regions for the two-sample *t* test for independent samples with equal variances

Distribution of t in **(8.11)** under $H_0 = t_{n_1 + n_2 - 2}$ distribution

Similarly, a p-value can be computed for the test. The computation of the p-value will depend on whether $\bar{x}_1 \leqslant \bar{x}_2 (t \leqslant 0)$ or $\bar{x}_1 > \bar{x}_2 (t > 0)$. In each case, the p-value corresponds to the probability of obtaining a test statistic at least as extreme as the observed value t. This is given as follows:

8.12 | **Computation of the *p*-value for the Two-Sample *t* Test for Independent Samples with Equal Variances**

If $t \leq 0$,

$$p = 2 \times \text{area to the left of } t = \frac{\bar{x}_1 - \bar{x}_2}{s\sqrt{1/n_1 + 1/n_2}} \text{ under a } t_{n_1 + n_2 - 2} \text{ distribution}$$

where

$$s = \sqrt{[(n_1 - 1)s_1^2 + (n_2 - 1)s_2^2]/(n_1 + n_2 - 2)}$$

If $t > 0$,

$$p = 2 \times [\text{area to the right of } t \text{ under a } t_{n_1 + n_2 - 2} \text{ distribution}]$$

The computation of the *p*-value is illustrated in Figure 8.4.

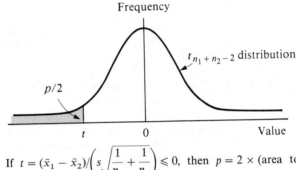

If $t = (\bar{x}_1 - \bar{x}_2) / \left(s\sqrt{\dfrac{1}{n_1} + \dfrac{1}{n_2}} \right) \leq 0$, then $p = 2 \times$ (area to

the left of t under a $t_{n_1 + n_2 - 2}$ distribution)

If $t = (\bar{x}_1 - \bar{x}_2) / \left(s\sqrt{\dfrac{1}{n_1} + \dfrac{1}{n_2}} \right) > 0$, then $p = 2 \times$ (area to

the right of t under a $t_{n_1 + n_2 - 2}$ distribution)

FIGURE 8.4
Computation of the
p-value for the two-
sample *t* test for
independent samples
with equal variances

EXAMPLE 8.10   **Cardiovascular Disease, Hypertension** Assess the statistical significance of the data in Example 8.9.

SOLUTION   The common variance is first estimated:

$$s^2 = \frac{7(15.34)^2 + 20(18.23)^2}{27} = \frac{8293.87}{27} = 307.18$$

or $s = 17.527$. The following test statistic is then formed:

$$t = \frac{132.86 - 127.44}{17.527\sqrt{1/8 + 1/21}} = \frac{5.42}{17.527 \times 0.415} = \frac{5.42}{7.274} = 0.75$$

If the critical value method is used, then note that under H_0, t comes from a t_{27} distribution. Referring to Table 5 in Appendix 1, we see that $t_{27,.975} = 2.052$. Since $-2.052 \leqslant 0.75 \leqslant 2.052$, it follows that H_0 is accepted using a two-sided test at the 5% level, and we conclude that the mean blood pressures of the two groups of OC users and non-OC users are not significantly different from each other. In a sense this result shows the superiority of the longitudinal design in Example 8.5. Despite the similarity in the magnitudes of the blood-pressure differences between users and nonusers in the two studies, significant differences could be detected in Example 8.5 in contrast to the nonsignificant results that were obtained using the preceding cross-sectional design. The longitudinal design is more efficient because it uses people as their own controls.

To compute an approximate p-value, note from Table 5 that $t_{27,.75} = 0.684$, $t_{27,.80} = 0.855$. Since $0.684 < 0.75 < 0.855$, it follows that $.2 < p/2 < .25$ or $.4 < p < .5$. The exact p-value is given by the HP-41C t distribution program as $2 \times Pr(t_{27} > .75) = 2 \times .230 = .46$. ∎

EXAMPLE 8.11   **Environmental Health, Mental Health** A topic of ongoing interest is the short- and long-term health effects of exposure to various agents in the environment. In 1975 a study was published that looked at the effect of chronic exposure to low levels of lead in children who lived near a lead smelter in El Paso, Texas [1]. A lead absorption group and a control group of children were identified by the amount of lead in blood samples, and various neurological and psychological tests were performed to compare the two groups. In particular, the WISC performance IQ test was performed on 34 children, 5 years of age and older, in the lead absorption group and 63 comparably aged children in the control group. Assess the statistical significance of the IQ test results, which are given in Table 8.3.

TABLE 8.3
Results of WISC
performance IQ test for
exposed and control
children

	Mean	sd	n
Lead absorption	96.44	13.74	34
Control	103.29	17.87	63

(Reprinted with permission of *The Lancet*, March 29, 708–715, 1975.)

SOLUTION   The two-sample t test for independent samples with equal variances is performed. A pooled estimate of the variance is first computed:

$$s^2 = [33(13.74)^2 + 62(17.87)^2]/95 = 273.99 \quad \text{or} \quad s = \sqrt{273.99} = 16.55$$

Thus the test statistic

$$t = (96.44 - 103.29)/[16.55\sqrt{(1/34) + (1/63)}] = -6.85/3.522 = -1.945$$

is computed, which follows a t distribution with $34 + 63 - 2 = 95$ df under H_0.

The critical value is given by $t_{95,.975}$, which is not given in Table 5. However, since $t_{60,.975} = 2.000$, $t_{120,.975} = 1.980$, we see that $1.980 < $ critical value < 2.000. Since $1.945 = |t| < 1.980 <$ critical value, H_0 can be accepted using a two-sided test with $\alpha = .05$. To compute an approximate p-value, note from Table 5 that $t_{60,.95} = 1.671$. Thus, since

$1.671 < 1.945 < 2.000$, it follows that if there were 60 df, then $.025 < p/2 < .05$ or $.05 < p <$ $.10$. Similarly, since $t_{120,.95} = 1.658$ and $1.658 < 1.945 < 1.980$, it follows that if there were 120 df, then $.025 < p/2 < .05$ or $.05 < p < .10$. Since the results agree for 60 and 120 df and the df (95) is between 60 and 120, we must reach the same conclusion; that is, $.05 < p < .10$. Thus, there is a trend toward statistical significance, but the results are not quite significant using a two-sided test ($.05 < p < .10$).

The exact p-value is given by the HP-41C t distribution program as $2 \times Pr(t_{95} > 1.945) = .055$. Actually, an argument can be made for using a one-sided test here, since the presence of high lead levels would be expected to worsen IQ. Thus, we would be testing the hypothesis $H_0: \mu_1 = \mu_2$ versus $H_1: \mu_1 < \mu_2$, in which case the one-sided p-value would be half of the two-sided p-value, or $.027$. This is what the authors of the study actually did. ∎

SECTION 8.5

Interval Estimation for the Comparison of Means from Two Independent Samples (Equal Variance Case)

In the previous section, methods of hypothesis testing for the comparison of means from two independent samples were discussed. It is also useful to compute $100\% \times (1 - \alpha)$ confidence limits for the true mean difference between the two groups $= \mu_1 - \mu_2$. From **(8.7)**, if σ is known, then $\bar{x}_1 - \bar{x}_2 \sim N[\mu_1 - \mu_2, \sigma^2(1/N_1 + 1/N_2)]$, or, equivalently,

$$\frac{\bar{x}_1 - \bar{x}_2 - (\mu_1 - \mu_2)}{\sigma\sqrt{\dfrac{1}{n_1} + \dfrac{1}{n_2}}} \sim N(0, 1)$$

If σ is unknown, then σ is estimated by s from **(8.10)** and

$$\frac{\bar{x}_1 - \bar{x}_2 - (\mu_1 - \mu_2)}{s\sqrt{\dfrac{1}{n_1} + \dfrac{1}{n_2}}} \sim t_{n_1 + n_2 - 2}$$

To construct a two-sided $100\% \times (1 - \alpha)$ confidence interval, note that

$$Pr\left[-t_{n_1 + n_2 - 2, 1 - \alpha/2} \leqslant \frac{\bar{x}_1 - \bar{x}_2 - (\mu_1 - \mu_2)}{s\sqrt{\dfrac{1}{n_1} + \dfrac{1}{n_2}}} \leqslant t_{n_1 + n_2 - 2, 1 - \alpha/2} \right] = 1 - \alpha$$

This can be written in the form of two inequalities:

$$-t_{n_1 + n_2 - 2, 1 - \alpha/2} \leqslant \frac{\bar{x}_1 - \bar{x}_2 - (\mu_1 - \mu_2)}{s\sqrt{\dfrac{1}{n_1} + \dfrac{1}{n_2}}}$$

and

$$\frac{\bar{x}_1 - \bar{x}_2 - (\mu_1 - \mu_2)}{s\sqrt{\dfrac{1}{n_1} + \dfrac{1}{n_2}}} \leqslant t_{n_1 + n_2 - 2, 1 - \alpha/2}$$

Each inequality is multiplied by $s\sqrt{\dfrac{1}{n_1} + \dfrac{1}{n_2}}$ and $\mu_1 - \mu_2$ is added to obtain

$$\mu_1 - \mu_2 - t_{n_1 + n_2 - 2, 1 - \alpha/2} \, s \sqrt{\frac{1}{n_1} + \frac{1}{n_2}} \leqslant \bar{x}_1 - \bar{x}_2$$

and

$$\bar{x}_1 - \bar{x}_2 \leqslant \mu_1 - \mu_2 + t_{n_1 + n_2 - 2, 1 - \alpha/2} \, s \sqrt{\frac{1}{n_1} + \frac{1}{n_2}}$$

Finally, $t_{n_1 + n_2 - 2, 1 - \alpha/2} \, s \sqrt{\dfrac{1}{n_1} + \dfrac{1}{n_2}}$ is added to both sides of the first inequality and subtracted from both sides of the second inequality to obtain

$$\mu_1 - \mu_2 \leqslant \bar{x}_1 - \bar{x}_2 + t_{n_1 + n_2 - 2, 1 - \alpha/2} \, s \sqrt{\frac{1}{n_1} + \frac{1}{n_2}}$$

$$\bar{x}_1 - \bar{x}_2 - t_{n_1 + n_2 - 2, 1 - \alpha/2} \, s \sqrt{\frac{1}{n_1} + \frac{1}{n_2}} \leqslant \mu_1 - \mu_2$$

If these two inequalities are combined, the required confidence interval is obtained.

$$\left(\bar{x}_1 - \bar{x}_2 - t_{n_1 + n_2 - 2, 1 - \alpha/2} \, s \sqrt{\frac{1}{n_1} + \frac{1}{n_2}}, \quad \bar{x}_1 - \bar{x}_2 + t_{n_1 + n_2 - 2, 1 - \alpha/2} \, s \sqrt{\frac{1}{n_1} + \frac{1}{n_2}} \right)$$

This is summarized as follows:

8.13 | **Confidence Interval for the Underlying Mean Difference ($\mu_1 - \mu_2$) Between Two Groups (Two-Sided) ($\sigma_1^2 = \sigma_2^2$)**

A two-sided $100\% \times (1 - \alpha)$ confidence interval for the true mean difference $\mu_1 - \mu_2$ based on two independent samples is given by

$$\left(\bar{x}_1 - \bar{x}_2 - t_{n_1 + n_2 - 2, 1 - \alpha/2} \, s \sqrt{\frac{1}{n_1} + \frac{1}{n_2}}, \quad \bar{x}_1 - \bar{x}_2 + t_{n_1 + n_2 - 2, 1 - \alpha/2} \, s \sqrt{\frac{1}{n_1} + \frac{1}{n_2}} \right)$$

EXAMPLE 8.12   **Cardiovascular Disease, Hypertension** Using the data in Example 8.9, compute a 95% confidence interval for the true mean difference in blood pressure between the two OC-use groups.

SOLUTION   A 95% confidence interval for the underlying mean difference in systolic blood pressure between the population of 35–39-year-old OC users and non-OC users is given by

$$[5.42 - t_{27, .975}(7.274), \, 5.42 + t_{27, .975}(7.274)]$$

$$= [5.42 - 2.052(7.274), \, 5.42 + 2.052(7.274)] = (-9.51, 20.35)$$

This interval is rather wide and indicates that a much larger sample is needed to accurately assess the true mean difference. ∎

EXAMPLE 8.13   **Environmental Health, Mental Health** Compute a 95% confidence interval for the difference in performance IQ between the lead absorption group and the control group in Example 8.11.

SOLUTION   A 95% confidence interval for $\mu_1 - \mu_2$ is given by

$$[-6.85 - t_{95,.975}(3.522), \; -6.85 + t_{95,.975}(3.522)]$$

We need to interpolate to obtain $t_{95,.975}$:

$$t_{95,.975} = \frac{(1/60 - 1/95)t_{120,.975} + (1/95 - 1/120)t_{60,.975}}{1/60 - 1/120}$$

$$= \frac{0.0061(1.980) + 0.0022(2.000)}{0.0083} = 1.985$$

Thus, the 95% confidence interval is given by

$$[-6.85 - 1.985(3.522), \; -6.85 + 1.985(3.522)] = (-13.84, 0.14)$$

This is also a rather wide interval.   ∎

SECTION 8.6   Testing for the Equality of Two Variances

In Section 8.4, when a two-sample t test for independent samples was conducted, it was assumed that the underlying variances of the two samples were the same. The common variance was then estimated using a weighted average of the individual sample variances. In this section a significance test to validate this assumption is developed. In particular, we want to test the hypothesis H_0: $\sigma_1^2 = \sigma_2^2$ versus H_1: $\sigma_1^2 \neq \sigma_2^2$, where the two samples are assumed to be independent random samples from an $N(\mu_1, \sigma_1^2)$ and $N(\mu_2, \sigma_2^2)$ distribution, respectively.

EXAMPLE 8.14   **Cardiovascular Disease, Pediatrics** Consider a problem that was discussed earlier, namely, the familial aggregation of cholesterol levels. In particular, suppose cholesterol levels are assessed in 100 2–14-year-old children of men who have died from heart disease and it is found that the mean cholesterol level in this group ($\bar{x}_1$) is 207.3 mg%/mL. Suppose that the sample standard deviation in this group (s_1) is 35.6. Previously, the cholesterol levels in this group of children were compared with the baseline level of 175 mg%/mL, which was assumed to be the underlying mean level in children in this age group based on previous large studies.

A better experimental design would be to select a group of control children whose fathers are alive and do not have heart disease and who are from the same census tract as the case children and then compare their cholesterol levels with those of the case children. If the case fathers are identified by a search of death records from the census tract, then control children who live in the same census tract as the case families but whose fathers have no history of heart disease can be selected. The case and control children come from the same census tract but are *not* individually matched. Thus, they are considered as two independent samples rather than as two paired samples. The cholesterol levels in these children can then be measured. Suppose this procedure is done and it is found that among 74 control children, the mean cholesterol level ($\bar{x}_2$) is 193.4 mg%/mL with a sample standard deviation (s_2) of 17.3 mg%/mL. We would like to compare the means of these two groups using the two-sample t test for independent samples given in **(8.11)**, but we are hesitant to assume equal variances because the sample variance of the case group is about 4 times as large as that of the control group:

$$(35.6)^2/(17.3)^2 = 4.23$$

What should be done?   ∎

What is needed is a significance test to determine if the underlying variances are in fact equal; that is, we wish to test the hypothesis $H_0: \sigma_1^2 = \sigma_2^2$ versus $H_1: \sigma_1^2 \neq \sigma_2^2$. It seems reasonable to base the significance test on the relative magnitudes of the sample variances (s_1^2, s_2^2). The best test in this case is based on the ratio of the sample variances (s_1^2/s_2^2) rather than on the difference between the sample variances $(s_1^2 - s_2^2)$. Thus, H_0 would be rejected if the variance ratio is either too large or too small and accepted otherwise. To accomplish this end, the sampling distribution of s_1^2/s_2^2 under the null hypothesis must be found.

8.6.1   The *F* Distribution

The distribution of the variance ratio (s_1^2/s_2^2) was studied by the statisticians R. A. Fisher and G. Snedecor. It can be shown that the variance ratio follows an **F distribution** under the null hypothesis that $\sigma_1^2 = \sigma_2^2$. There is no unique *F* distribution but instead a family of *F* distributions. This family is indexed by two parameters termed the *numerator* and *denominator degrees of freedom (df)*, respectively. Specifically, if the sample sizes of the first and second samples are n_1 and n_2, respectively, then the variance ratio follows an *F* distribution with $n_1 - 1$ (numerator *df*) and $n_2 - 1$ (denominator *df*), which is denoted by F_{n_1-1, n_2-1}.

The *F* distribution is generally positively skewed, with the skewness dependent on the relative magnitudes of the two degrees of freedom. If the numerator degree of freedom is 1, then the distribution has a mode at 0; otherwise, it has a mode at some point greater than 0. This distribution is illustrated in Figure 8.5. Table 8 in the Appendix gives the percentiles of the *F* distribution.

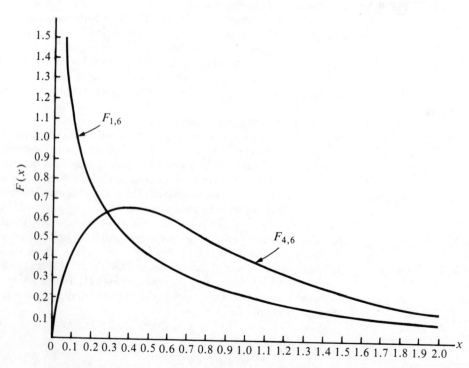

FIGURE 8.5
Probability density for
the *F* distribution

DEFINITION 8.6 ■■

The *p*th **percentile of an** *F* **distribution** with d_1 and d_2 degrees of freedom is denoted by $F_{d_1, d_2, p}$. Thus

$$Pr(F_{d_1, d_2} \leqslant F_{d_1, d_2, p}) = p \qquad \blacksquare$$

The *F* table is organized such that the different numerator *df* (d_1) are listed in the first row, the different denominator *df* (d_2) are listed in the first column, and the various percentiles (p) are listed in the second column.

EXAMPLE 8.15   Find the upper 1st percentile of an *F* distribution with 5 and 9 degrees of freedom.

SOLUTION $F_{5.9, .99}$ must be found. Look in the 5 column, the 9 row, and the sub row marked .99 to obtain

$$F_{5.9, .99} = 6.06 \qquad \blacksquare$$

Generally, *F* distribution tables give only upper percentage points because the symmetry properties of the *F* distribution make it possible to derive the lower percentage points of any *F* distribution from the corresponding upper percentage points of an *F* distribution with the appropriate degrees of freedom. Specifically, note that under H_0, s_2^2 / s_1^2 follows an F_{d_2, d_1} distribution. Therefore,

$$Pr(s_2^2 / s_1^2 \geqslant F_{d_2, d_1, 1-p}) = p$$

By taking the inverse of each side and reversing the direction of the inequality, we get

$$Pr\left(\frac{s_1^2}{s_2^2} \leqslant \frac{1}{F_{d_2, d_1, 1-p}}\right) = p$$

However, under H_0, s_1^2 / s_2^2 follows an F_{d_1, d_2} distribution. Therefore,

$$Pr\left(\frac{s_1^2}{s_2^2} \leqslant F_{d_1, d_2, p}\right) = p$$

It follows from the last two inequalities that

$$F_{d_1, d_2, p} = \frac{1}{F_{d_2, d_1, 1-p}}$$

This principle is summarized as follows:

8.14 | **Computation of the Lower Percentiles of an** *F* **Distribution**

The **lower** *p*th **percentile** of an *F* distribution with d_1 and d_2 *df* is the reciprocal of the **upper** *p*th **percentile** of an *F* distribution with d_2 and d_1 *df*. In symbols,

$$F_{d_1, d_2, p} = 1/F_{d_2, d_1, 1-p}$$

Thus, from **(8.14)** we see that the lower *p*th percentile of any *F* distribution is the same as the inverse of the upper *p*th percentile of an *F* distribution with the degrees of freedom reversed.

EXAMPLE 8.16   Estimate $F_{6,8,.05}$.

SOLUTION   From **(8.14)**, $F_{6,8,.05} = 1/F_{8,6,.95} = 1/4.15 = 0.241$ ∎

8.6.2   **Evaluation of Statistical Significance Using the *F* Table**

Frequently, either the numerator *df* or the denominator *df* does not appear in the *F* tables. The percentiles then need to be estimated by performing some type of interpolation based on the percentiles given in the table. **Harmonic interpolation** is a useful method of interpolation whereby we interpolate linearly using the inverse of both the numerator and denominator degrees of freedom. This method of interpolation is similar to that presented for the *t* distribution in Section 7.5.1 and is shown as follows:

8.15   **Interpolation for the *F* Table**

Suppose we wish to estimate the *p*th percentile of an *F* distribution with d_1 and d_2 *df* ($F_{d_1,d_2,p}$), where neither d_1 nor d_2 appears in the table.

(1) Find a_1 and b_1 such that $a_1 \leqslant d_1 \leqslant b_1$, and a_1 and b_1 are given in the table for the numerator *df*.

(2) Find a_2 and b_2 such that $a_2 \leqslant d_2 \leqslant b_2$, and a_2 and b_2 are given in the table for the denominator *df*.

(3) Estimate $F_{d_1,a_2,p}$ by

$$F_{d_1,a_2,p} \approx \frac{[(1/d_1) - (1/b_1)]F_{a_1,a_2,p} + [(1/a_1) - (1/d_1)]F_{b_1,a_2,p}}{[(1/a_1) - (1/b_1)]}$$

(4) Estimate $F_{d_1,b_2,p}$ by

$$F_{d_1,b_2,p} \approx \frac{[(1/d_1) - (1/b_1)]F_{a_1,b_2,p} + [(1/a_1) - (1/d_1)]F_{b_1,b_2,p}}{[(1/a_1) - (1/b_1)]}$$

(5) Then interpolate linearly in the inverse of the denominator *df* as follows:

$$F_{d_1,d_2,p} \approx \frac{[(1/d_2) - (1/b_2)]F_{d_1,a_2,p} + [(1/a_2) - (1/d_2)]F_{d_1,b_2,p}}{[(1/a_2) - (1/b_2)]}$$

EXAMPLE 8.17   Estimate the 99.9th percentile of an *F* distribution with 99 and 73 *df*.

SOLUTION   Note that 99 is not listed for the numerator *df* nor is 73 listed for the denominator *df*. However, 24 and ∞ are listed for the numerator *df*, where $24 \leqslant 99 \leqslant \infty$, and 60 and 120 are listed for the denominator *df*, where $60 \leqslant 73 \leqslant 120$. Thus, set $a_1 = 24$, $b_1 = \infty$, $a_2 = 60$, $b_2 = 120$, $d_1 = 99$, $d_2 = 73$ and apply **(8.15)**. First, estimate $F_{99,60,.999}$ as follows, noting that $1/\infty = 0$:

$$F_{99,60,.999} = \frac{[(1/99) - (1/\infty)]F_{24,60,.999} + [(1/24) - (1/99)]F_{\infty,60,.999}}{[(1/24) - (1/\infty)]}$$

$$= \frac{(1/99)(2.69) + [(1/24) - (1/99)](1.89)}{1/24} = \frac{0.0868}{0.0417} = 2.082$$

Second, estimate $F_{99,120,.999}$ as follows:

$$F_{99,120,.999} = \frac{[(1/99) - (1/\infty)]F_{24,120,.999} + [(1/24) - (1/99)]F_{\infty,120,.999}}{[(1/24) - (1/\infty)]}$$

$$= \frac{(1/99)(2.40) + [(1/24) - (1/99)](1.54)}{1/24} = \frac{0.0729}{0.0417} = 1.748$$

Third, estimate $F_{99,73,.999}$ as follows:

$$F_{99,73,.999} = \frac{[(1/73) - (1/120)]F_{99,60,.999} + [(1/60) - (1/73)]F_{99,120,.999}}{[(1/60) - (1/120)]}$$

$$= \frac{[(1/73) - (1/120)](2.082) + [(1/60) - (1/73)](1.748)}{[(1/60) - (1/120)]}$$

$$= 0.01636/0.00833 = 1.96 \qquad \blacksquare$$

8.6.3 The *F* Test

We now return to the significance test for the equality of two variances. We wish to test the hypothesis $H_0: \sigma_1^2 = \sigma_2^2$ versus $H_1: \sigma_1^2 \neq \sigma_2^2$. We stated that the test would be based on the variance ratio s_1^2/s_2^2, which under H_0 follows an F distribution with $n_1 - 1$ and $n_2 - 1$ df. Since this is a two-sided test, we wish to reject H_0 for both small and large values of s_1^2/s_2^2. This procedure can be made more specific, as follows:

8.16 | **F Test for the Equality of Two Variances**

In conducting a test of the hypothesis $H_0: \sigma_1^2 = \sigma_2^2$ versus $H_1: \sigma_1^2 \neq \sigma_2^2$ with significance level α, if

$$F = s_1^2/s_2^2 > F_{n_1 - 1, n_2 - 1, 1 - \alpha/2} \qquad \text{or} \qquad F < F_{n_1 - 1, n_2 - 1, \alpha/2}$$

then H_0 is rejected. If

$$F_{n_1 - 1, n_2 - 1, \alpha/2} \leqslant F \leqslant F_{n_1 - 1, n_2 - 1, 1 - \alpha/2}$$

then H_0 is accepted. The acceptance and rejection regions for this test are depicted in Figure 8.6.

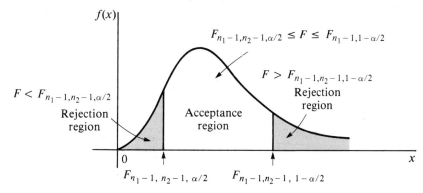

FIGURE 8.6
Acceptance and
rejection regions for the
F test for the equality of
two variances

$F_{n_1 - 1, n_2 - 1}$ distribution = distribution of $F = s_1^2/s_2^2$ under H_0

Alternatively, the exact p-value is given by:

8.17 | **Computation of the *p*-value for the *F* Test for the Equality of Two Variances**

$$p = 2 \times Pr(F_{n_1-1,n_2-1} > F) \quad \text{if} \quad F = s_1^2/s_2^2 \geq 1$$

$$p = 2 \times Pr(F_{n_1-1,n_2-1} < F) \quad \text{if} \quad F < 1$$

This computation is illustrated in Figure 8.7.

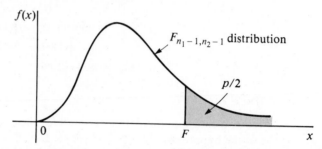

If $F = s_1^2/s_2^2 \geq 1$, then $p = 2 \times$ (area to the right of F under an F_{n_1-1,n_2-1} distribution)

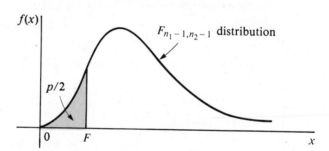

FIGURE 8.7
Computation of the *p*-value for the *F* test for the equality of two variances

If $F = s_1^2/s_2^2 < 1$, then $p = 2 \times$ (area to the left of F under an F_{n_1-1,n_2-1} distribution)

EXAMPLE 8.18 **Cardiovascular Disease, Pediatrics** Test for the equality of the two variances given in Example 8.14.

SOLUTION
$$F = s_1^2/s_2^2 = (35.6)^2/(17.3)^2 = 4.23$$

Since the two samples have 100 and 74 people, respectively, we know from **(8.16)** that under H_0, $F \sim F_{99,73}$. Thus, H_0 is rejected if

$$F > F_{99,73,.975} \quad \text{or} \quad F < F_{99,73,.025}$$

Note from Example 8.17 that $F_{99,73,.999} = 1.96$. Since $F = 4.23 > 1.96$, it follows that $p/2 < 1 - .999 = .001$ or $p < .002$. Thus the two sample variances are significantly different. The two-sample t test with equal variances, as given in Section 8.4, cannot be used, since this test depends on the assumption that the variances are equal. ∎

A satisfactory approximate p-value for the F test can often be computed using the df that appear in Table 8 in Appendix 1, without the need for interpolating.

The general rule is to find a numerator (a_1) df and denominator (a_2) df in the table that are either both less than or both greater than the df we are interested in (say, d_1, d_2):

8.18 If $p > .50$

$$a_1 \leqslant d_1 \quad \text{and} \quad a_2 \leqslant d_2$$

Then,
$$F_{d_1, d_2, p} \leqslant F_{a_1, a_2, p}$$

Similarly, if
$$a_1 \geqslant d_1 \quad \text{and} \quad a_2 \geqslant d_2$$

then
$$F_{d_1, d_2, p} \geqslant F_{a_1, a_2, p}$$

From (**8.18**), if $a_1 \leqslant d_1$, $a_2 \leqslant d_2$, and $F = s_1^2/s_2^2 > F_{a_1, a_2, 1-\alpha/2}$, then $F > F_{d_1, d_2, 1-\alpha/2}$, and H_0 can be rejected using a two-sided test with significance level α. Similarly, if $a_1 \geqslant d_1$, $a_2 \geqslant d_2$, and $F = s_1^2/s_2^2 < F_{a_1, a_2, 1-\alpha/2}$, then $F < F_{d_1, d_2, 1-\alpha/2}$, and H_0 can be accepted using a two-sided test with significance level α. The only counterexample to this principle is if $a_2 = d_2$ and $d_2 \leqslant 2$, which is a very unusual situation.

EXAMPLE 8.19 Evaluate the significance of the test statistic in Example 8.18 without using interpolation.

SOLUTION Percentiles for an $F_{99, 73}$ distribution are needed, but 99 does not appear under the numerator df and 73 does not appear under the denominator df. However, 24 does appear under numerator df, and 60 appears under denominator df. Thus from (**8.18**)

$$F_{99, 73, .999} \leqslant F_{24, 60, .999} = 2.69 < 4.23 = s_1^2/s_2^2 = F$$

Thus, without interpolation we can conclude that $p/2 < .001$ or $p < .002$ and that the two variances are significantly different. ∎

A question often asked about the F test is whether or not it makes a difference which sample is selected as the numerator sample and which as the denominator sample. The answer is that, for a two-sided test, it *does not* make a difference, because of the rules for calculating lower percentiles given in (**8.14**). A variance ratio > 1 is usually more convenient, so there is no need to use (**8.14**). Thus, the larger variance is usually put in the numerator and the smaller variance in the denominator.

EXAMPLE 8.20 **Cardiovascular Disease, Hypertension** Using the data in Example 8.9 (p. 254), test whether or not the variance of blood pressure is significantly different among OC users and non-OC users.

SOLUTION The sample standard deviation of blood pressure for the 8 OC users was 15.34 and for the 21 non-OC users was 18.23. Hence the variance ratio is

$$F = (18.23/15.34)^2 = 1.41$$

Under H_0, F follows an F distribution with 20 and 7 df. From (**8.18**),

$$F_{20, 7, .975} \geqslant F_{24, 7, .975} = 4.42 > 1.41$$

It follows that $p > 2(.025) = .05$, and the underlying variances of the two samples are not significantly different from each other. Thus it was correct to use the two-sample t test for independent samples with *equal variances* for these data, where the variances were assumed to be the same.

Suppose the numerator and denominator samples were changed so that the variance for OC users was in the numerator and the variance for non-OC users was in the denominator. The variance ratio = F would then be $(15.34/18.23)^2 = 1/1.41 = 0.71$. Under H_0, F follows an $F_{7,20}$ distribution. Thus, from (8.16), since $F < 1$, we wish to compare 0.71 with $F_{7,20,.025}$ to test for statistical significance at the 5% level. However, from (8.14),

$$F_{7,20,.025} = 1/F_{20,7,.975} \leqslant 1/F_{24,7,.975} = 1/4.42 = 0.23 < 0.71 = F$$

and it follows that F falls in the acceptance region at the 5% level or $p/2 > .025$ or $p > .05$. Notice that the arithmetic is equivalent whether 1.41 is compared with 4.42 when the non-OC users were in the numerator or 1/1.41 is compared with 1/4.42 when the OC users were in the numerator. For a two-sided test, which sample is placed in the numerator is simply a matter of convenience. ∎

To compute an exact p-value and/or avoid using the preceding interpolation methods, a computer program must be used to evaluate the area under the F distribution. The exact p-value for Example 8.20 has been evaluated using the HP-41C F distribution program, with the results given in Table 8.4. The program evaluates the right-hand tail area = $Pr(F_{20,7} > 1.41) = .335$. Thus, the two-tailed p-value = $2 \times .335 = .670$.

TABLE 8.4
Computation of the exact p-value for the blood-pressure data in Example 8.20 using the F test for the equality of two variances with the HP-41C F distribution program

```
----------------------------------------------------------------
                    XEQ "FDIST"
            20.00000   XEQ A
(a)   V1=20.00000                        (a)  Numerator degrees of freedom
             7.00000   XEQ B
(b)   V2=7.00000                         (b)  Denominator degrees of freedom
(c)          1.41000   XEQ C             (c)  F
                       RUN
(d)   P=0.33516                          (d)  Right-hand tail area
             2.00000      *
(e)          0.67031    ***              (e)  Exact two-tailed p-value
----------------------------------------------------------------
```

SECTION 8.7

Two-Sample *t* Test for Independent Samples with Unequal Variances

The F test for the equality of two variances from two independent, normally distributed samples was presented in (8.16). If the two variances *are not* significantly different, then the two-sample t test for independent samples with *equal variances* outlined in Section 8.4 can be used. If the two variances *are* significantly different, then a two-sample t test for independent samples with *unequal variances*, which is presented in this section, must be used.

Specifically, assume that there are two normally distributed samples, where the first sample is a random sample of size n_1 from an $N(\mu_1, \sigma_1^2)$ distribution, the second sample is a random sample from an $N(\mu_2, \sigma_2^2)$ distribution, and $\sigma_1^2 \neq \sigma_2^2$. We again wish to test the hypothesis $H_0: \mu_1 = \mu_2$ versus $H_1: \mu_1 \neq \mu_2$. Statisticians refer to this problem as the **Behrens-Fisher problem**.

It still makes sense to base the significance test on the difference between the sample means $\bar{x}_1 - \bar{x}_2$. Under either hypothesis, $\bar{x}_1$ is normally distributed with mean μ_1 and variance σ_1^2/n_1, and $\bar{x}_2$ is normally distributed with mean μ_2 and variance σ_2^2/n_2. Hence it follows that

8.19
$$\bar{x}_1 - \bar{x}_2 \sim N\left(\mu_1 - \mu_2, \frac{\sigma_1^2}{n_1} + \frac{\sigma_2^2}{n_2}\right)$$

Under H_0, $\mu_1 - \mu_2 = 0$. Thus, from **(8.19)**,

8.20
$$\bar{x}_1 - \bar{x}_2 \sim N\left(0, \frac{\sigma_1^2}{n_1} + \frac{\sigma_2^2}{n_2}\right)$$

If σ_1^2 and σ_2^2 were known, then the test statistic

8.21
$$(\bar{x}_1 - \bar{x}_2)\bigg/ \sqrt{\frac{\sigma_1^2}{n_1} + \frac{\sigma_2^2}{n_2}}$$

could be used for the significance test, which under H_0 would be distributed as an $N(0, 1)$ distribution. However, σ_1^2 and σ_2^2 are usually unknown and are estimated by s_1^2 and s_2^2, respectively (i.e., the sample variances in the two samples). Notice that a pooled estimate of the variance was not computed as in **(8.10)**, because the variances (σ_1^2, σ_2^2) are assumed to be different. If s_1^2 is substituted for σ_1^2 and s_2^2 for σ_2^2 in **(8.21)**, then the following test statistic is obtained:

8.22
$$t = (\bar{x}_1 - \bar{x}_2)/\sqrt{s_1^2/n_1 + s_2^2/n_2}$$

The exact distribution of t under H_0 is difficult to derive. However, several approximate solutions have been proposed that have appropriate type I error. The Satterthwaite approximation is presented here. Its advantage is its easy implementation using the ordinary t tables and the interpolation formula for the t distribution given in **(7.19)** [2].

8.23 **Two-Sample t Test for Independent Samples with Unequal Variances (Satterthwaite's Method)**

(1) Compute the test statistic

$$t = \frac{\bar{x}_1 - \bar{x}_2}{\sqrt{\dfrac{s_1^2}{n_1} + \dfrac{s_2^2}{n_2}}}$$

(2) Compute the approximate degrees of freedom d', where

$$d' = \frac{(s_1^2/n_1 + s_2^2/n_2)^2}{(s_1^2/n_1)^2/(n_1 - 1) + (s_2^2/n_2)^2/(n_2 - 1)}$$

(3) Round d' down to the nearest integer d''. If

$$t > t_{d'', 1-\alpha/2} \quad \text{or} \quad t < -t_{d'', 1-\alpha/2}$$

then reject H_0. If

$$-t_{d'', 1-\alpha/2} \leq t \leq t_{d'', 1-\alpha/2}$$

then accept H_0.

The acceptance and rejection regions for this test are illustrated in Figure 8.8.

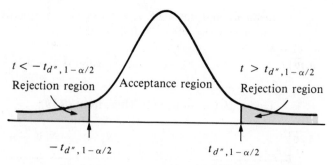

FIGURE 8.8
Acceptance and
rejection regions for the
two-sample t test for
independent samples
with unequal variances

$t_{d''}$ distribution = distribution of t in **(8.23)** under H_0

Similarly, the exact p-value for the hypothesis test can be computed as follows:

8.24 **Computation of the p-value for the Two-Sample Test for Independent Samples with Unequal Variances (Satterthwaite Approximation)**

If

$$t = \frac{\bar{x}_1 - \bar{x}_2}{\sqrt{\frac{s_1^2}{n_1} + \frac{s_2^2}{n_2}}} \leqslant 0$$

then

$$p = 2 \times (\text{area to the left of } t \text{ under a } t_{d''} \text{ distribution})$$

If $t > 0$, then

$$p = 2 \times (\text{area to the right of } t \text{ under a } t_{d''} \text{ distribution})$$

The computation of the p-value is illustrated in Figure 8.9.

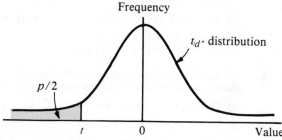

If $\quad t = (\bar{x}_1 - \bar{x}_2)/(\sqrt{s_1^2/n_1 + s_2^2/n_2}) \leqslant 0, \quad$ then $\quad p = 2 \times$ (area to the left of t under a $t_{d''}$ distribution)

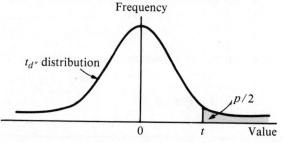

FIGURE 8.9
Computation of the
p-value for the two-
sample t test for
independent samples
with unequal variances

If $\quad t = (\bar{x}_1 - \bar{x}_2)/(\sqrt{s_1^2/n_1 + s_2^2/n_2}) > 0, \quad$ then $\quad p = 2 \times$ (area to the right of t under a $t_{d''}$ distribution)

EXAMPLE 8.21 **Cardiovascular Disease, Pediatrics** Consider the cholesterol data in Example 8.14 (p. 261). Test for the equality of the mean cholesterol levels of the children whose fathers have died from heart disease and whose fathers do not have a history of heart disease.

SOLUTION We have already tested for the equality of the two variances in Example 8.18 and found them to be significantly different. Thus, the t test for unequal variances in **(8.23)** must be used. The test statistic is

$$t = \frac{207.3 - 193.4}{\sqrt{35.6^2/100 + 17.3^2/74}} = \frac{13.9}{4.089} = 3.40$$

The approximate degrees of freedom are now computed:

$$d' = \frac{(s_1^2/n_1 + s_2^2/n_2)^2}{(s_1^2/n_1)^2/(n_1 - 1) + (s_2^2/n_2)^2/(n_2 - 1)}$$

$$= \frac{(35.6^2/100 + 17.3^2/74)^2}{(35.6^2/100)^2/99 + (17.3^2/74)^2/73} = \frac{(16.718)^2}{1.8465} = 151.4$$

Therefore, the approximate degrees of freedom $= d'' = 151$. If the critical value method is used, note that $t = 3.40 > t_{120,.975} = 1.980 > t_{151,.975}$. Therefore, H_0 can be rejected using a two-sided test with $\alpha = .05$. Furthermore, $t = 3.40 > t_{120,.9995} = 3.373 > t_{151,.9995}$, which implies that the p-value $< 2 \times (1.0 - .9995) = .001$. We conclude that the cholesterol levels in children of fathers who have died from heart disease is greater than the cholesterol levels in children of fathers without heart disease. It would be of great interest to identify the source of this difference, that is, whether it is due to genetic factors, environmental factors such as diet, or both. ∎

In this chapter, two procedures for comparing two means from independent, normally distributed samples have been presented. The first step in this process is to test for the equality of the two variances using the F test in **(8.16)**. If this test is not significant, then the t test with equal variances is used; otherwise, the t test with unequal variances is used. This overall strategy is illustrated in Figure 8.10.

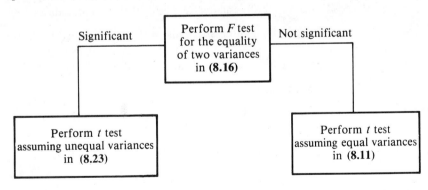

FIGURE 8.10
Strategy for testing for the equality of means in two independent, normally distributed samples

EXAMPLE 8.22 **Infectious Disease** Using the data in Table 2.11 (p. 36), compare the duration of hospitalization among antibiotic users and nonantibiotic users.

SOLUTION Refer to Table 8.5, where the SPSS/PC independent samples T-TEST program was used to analyze these data. Among the 7 antibiotic users (ANTIB = 1), the mean duration of hospitalization was 11.57 days with standard deviation 8.81 days; among the 18 nonantibiotic users (ANTIB = 2), the mean duration of hospitalization was 7.44 days with standard deviation 3.70 days. Both the F test and the t test with equal and unequal variances are displayed in this program. Using Figure 8.10, note that the first step in comparing the two

TABLE 8.5
Use of the SPSS^X/PC
independent samples
T-TEST program to
analyze the association
between antibiotic use
and duration of
hospitalization (raw
data presented in
Table 2.11)

```
-------------------------------------------------------------------------

                          SPSSX/PC   Release 1.0

Independent samples of  ANTIB      received antibiotic

Group 1:  ANTIB  EQ    1.00            Group 2:  ANTIB  EQ    2.00

t-test for:  DUR        duration of hospitalization

                    Number               Standard    Standard
                   of Cases     Mean     Deviation     Error

      Group 1         7       11.5714     8.810        3.330
      Group 2        18        7.4444     3.698        0.872


           | Pooled Variance Estimate | Separate Variance Estimate
   F   2-Tail |    t   Degrees of 2-Tail |   t   Degrees of 2-Tail
 Value  Prob. | Value  Freedom   Prob.   | Value  Freedom   Prob.

  5.68  0.004 |  1.68    23      0.106    |  1.20    6.84    0.270
-------------------------------------------------------------------------
```

means is to perform the F test for the equality of two variances in order to decide whether to use the t test with equal or with unequal variances. The F statistic is given in Table 8.5 by F Value = 5.68, with p-value (labeled 2-Tail Prob.) = .004. Thus, the variances are significantly different, and a two-sample t test with unequal variances should be used. Therefore, refer to the output labeled Separate Variance Estimate, where the t statistic [as given in **(8.23)**] is found to be 1.20 with degrees of freedom $d' = 6.84$. The corresponding two-tailed p-value (labeled 2-Tail Prob.) = .270. Thus, there is no significant difference between the mean duration of hospitalization in these two groups.

If the results of the F test had revealed a nonsignificant difference between the variances of the two samples, then the t test with equal variances would have been used, which is provided in the output labeled Pooled Variance Estimate. In this case, considerable differences are present in both the test statistics (1.68 vs. 1.20), and the two-tailed p-values (.106 vs. .270) resulting from using these two procedures. ∎

Using similar methods to those developed in Section 8.5, we can show that a two-sided $100\% \times (1 - \alpha)$ confidence interval for the underlying mean difference $\mu_1 - \mu_2$ in the case of unequal variances is given as follows:

8.25 | **Two-Sided 100% × (1 − α) Confidence Interval for $\mu_1 - \mu_2 (\sigma_1^2 \neq \sigma_2^2)$**

$$\left(\bar{x}_1 - \bar{x}_2 - t_{d'', 1 - \alpha/2}\sqrt{s_1^2/n_1 + s_2^2/n_2}, \quad \bar{x}_1 - \bar{x}_2 + t_{d'', 1 - \alpha/2}\sqrt{s_1^2/n_1 + s_2^2/n_2}\right)$$

EXAMPLE 8.23 **Cardiovascular Disease, Pediatrics** Using the data in Table 8.5, compute a 95 % confidence interval for the mean difference in duration of hospital stay between patients who do and patients who do not receive antibiotics.

SOLUTION Using Table 8.5, the 95 % confidence interval is given by

$$[11.571 - 7.444 - t_{6, .975}\sqrt{(8.810)^2/7 + (3.698)^2/18},$$

$$11.571 - 7.444 + t_{6, .975}\sqrt{(8.810)^2/7 + (3.698)^2/18}]$$

$$= [4.127 - 2.447(3.442), 4.127 + 2.447(3.442)]$$

$$= (4.127 - 8.423, 4.127 + 8.423) = (-4.30, 12.55)$$
∎

<u>SECTION 8.8</u> **Estimation of Sample Size and Power for Comparing Two Means**

8.8.1 **Estimation of Sample Size**

Methods of sample size estimation for the one-sample normal test were presented in Section 7.7. Estimates of sample size that are useful in planning studies in which *two* samples are to be compared are covered in this section.

EXAMPLE 8.24 **Cardiovascular Disease, Hypertension** Consider the blood-pressure data for OC users and non-OC users in Example 8.9 (p. 254) as a pilot study conducted to obtain parameter estimates to plan for a larger study. Suppose we assume that the true blood-pressure distribution of 35–39-year-old OC users is normal with mean μ_1 and variance σ_1^2. Similarly, for non-OC users we assume that the distribution is normal with mean μ_2 and variance σ_2^2. We wish to test the hypothesis $H_0: \mu_1 = \mu_2$ versus $H_1: \mu_1 \neq \mu_2$. How can the sample size needed for the larger study be estimated? ∎

Suppose we assume that σ_1^2 and σ_2^2 are known and we anticipate equal sample sizes in the two groups. To conduct a two-sided test with significance level α and a power of $1 - \beta$, the appropriate sample size for *each* group is as follows:

| 8.26 | **Sample Size Needed for Comparing the Means of Two Normally Distributed Samples of Equal Size Using a Two-Sided Test with Significance Level α and Power $1 - \beta$** |

$$n = \frac{(\sigma_1^2 + \sigma_2^2)(z_{1-\alpha/2} + z_{1-\beta})^2}{\Delta^2} = \text{sample size for each sample}$$

where $\Delta = \mu_2 - \mu_1$. The means and variances of the two respective samples are (μ_1, σ_1^2) and (μ_2, σ_2^2).

In words, n is the appropriate sample size in each group to have a probability of $1 - \beta$ of finding a significant difference if the true difference in means between the two groups is $\mu_2 - \mu_1$ based on a two-sided level α significance test.

EXAMPLE 8.25 **Cardiovascular Disease, Hypertension** Determine the appropriate sample size for the large study proposed in Example 8.24 using a two-sided test with a significance level of .05 and a power of .80.

SOLUTION In the small study,

$$\bar{x}_1 = 132.86$$
$$s_1 = 15.34$$
$$\bar{x}_2 = 127.44$$
$$s_2 = 18.23$$

If the sample data $(\bar{x}_1, s_1^2, \bar{x}_2, s_2^2)$ are used as estimates of the population parameters $(\mu_1, \sigma_1^2, \mu_2, \sigma_2^2)$, then ensuring an 80% chance of finding a significant difference using a two-sided significance test with $\alpha = .05$ would require a sample size of

$$n = (15.34^2 + 18.23^2)(1.96 + 0.84)^2/(132.86 - 127.44)^2 = 151.5$$

or 152 people in each group. It is not surprising that a significant difference was not found with sample sizes of 8 and 21 in the two groups, respectively. ∎

In many instances an imbalance between the groups can be anticipated and it can be predicted in advance that the number of people in one group will be k times the number in the other group for some number $k \neq 1$. In this case, where $n_2 = kn_1$, the appropriate sample size in the two groups for achieving a power of $1 - \beta$ using a two-sided level α significance test is given by the following formulas:

8.27   **Sample Size Needed for Comparing the Means of Two Normally Distributed Samples of Unequal Size Using a Two-Sided Test with Significance Level α and Power $1 - \beta$**

$$n_1 = \frac{(\sigma_1^2 + \sigma_2^2/k)(z_{1-\alpha/2} + z_{1-\beta})^2}{\Delta^2} = \text{sample size of first sample}$$

$$n_2 = \frac{(k\sigma_1^2 + \sigma_2^2)(z_{1-\alpha/2} + z_{1-\beta})^2}{\Delta^2} = \text{sample size of second sample}$$

where $\Delta = \mu_2 - \mu_1$; (μ_1, σ_1^2), (μ_2, σ_2^2) are the means and variances of the two respective samples; and $k = n_2/n_1 =$ the projected ratio of the two sample sizes.

Note that if $k = 1$, the sample size estimates given in **(8.27)** are the same as those in **(8.26)**.

EXAMPLE 8.26   **Cardiovascular Disease** Suppose we anticipate two times as many non-OC users as OC users entering the study proposed in Example 8.24. Project the required sample size if a two-sided test is used with a 5% significance level and an 80% power is desired.

SOLUTION   If **(8.27)** is used with

$$\mu_1 = 132.86$$
$$\sigma_1 = 15.34$$
$$\mu_2 = 127.44$$
$$\sigma_2 = 18.23$$
$$k = 2$$
$$\alpha = .05$$
$$1 - \beta = .8$$

then to achieve an 80% power in the study using a two-sided significance test with $\alpha = .05$, we need to enroll

$$n_1 = \frac{(15.34^2 + 18.23^2/2)(1.96 + 0.84)^2}{(132.86 - 127.44)^2} = 107.1, \text{ or } 108 \text{ OC users}$$

and

$$n_2 = 2(108) = 216 \text{ non-OC users} \qquad \blacksquare$$

If the variances in the two groups are the same, then for a given α, β, the smallest total sample size needed is achieved by the *equal sample size allocation rule* in **(8.26)**. Thus in the case of equal variances, the sample sizes in the two groups should be as nearly equal as possible.

Finally, to perform a one-sided rather than a two-sided test, the following sample size estimates are used:

| 8.28 | **Sample Size Needed for Comparing the Means of Two Normally Distributed Samples Using a One-Sided Test with Significance Level α and Power $1 - \beta$**

To test the hypothesis $H_0: \mu_1 = \mu_2$ versus $H_1: \mu_1 < \mu_2$ for the specific alternative $\mu_1 = \mu_2 - \Delta$, where $\Delta > 0$, with significance level α and power $1 - \beta$, a sample size of

$$n_1 = \frac{(\sigma_1^2 + \sigma_2^2/k)(z_{1-\alpha} + z_{1-\beta})^2}{\Delta^2} = \text{sample size of first sample}$$

$$n_2 = \frac{(k\sigma_1^2 + \sigma_2^2)(z_{1-\alpha} + z_{1-\beta})^2}{\Delta^2} = \text{sample size of second sample}$$

is needed, where (μ_1, σ_1^2), (μ_2, σ_2^2) are the means and variances of the two respective samples and $k = n_2/n_1$ = the projected ratio of the two sample sizes.

To test the hypothesis $H_0: \mu_1 = \mu_2$ versus $H_1: \mu_1 > \mu_2$ for the specific alternative $\mu_1 = \mu_2 + \Delta$, where $\Delta > 0$, with significance level α and power $1 - \beta$, the same sample size estimates are used.

8.8.2 Estimation of Power

In many situations, a predetermined sample size is available for study and how much power the study will have for detecting specific alternatives needs to be determined.

EXAMPLE 8.27   **Cardiovascular Disease** Suppose 100 OC users and 100 non-OC users are available for study and a true difference in systolic bp of 5 mm Hg is anticipated, with OC users having the higher blood pressure. How much power would such a study have assuming that the variance estimates in the pilot study in Example 8.9 (p. 254) are correct?   ■

Assuming that σ_1^2 and σ_2^2 are known, to conduct a two-sided test with significance level α, the power is given by:

| 8.29 | **Power for Comparing the Means of Two Normally Distributed Samples Using a Two-Sided Test with Significance Level α**

To test the hypothesis $H_0: \mu_1 = \mu_2$ versus $H_1: \mu_1 \neq \mu_2$ for the specific alternative $\mu_1 = \mu_2 - \Delta$ (or $\mu_1 = \mu_2 + \Delta$), where $\Delta > 0$, with significance level α,

$$\text{Power} = \Phi\left(-z_{1-\alpha/2} + \frac{\sqrt{n_1}\Delta}{\sqrt{\sigma_1^2 + \sigma_2^2/k}}\right)$$

where (μ_1, σ_1^2), (μ_2, σ_2^2) are the means and variances of the two respective samples and $k = n_2/n_1$ = the projected ratio of the two sample sizes.

EXAMPLE 8.28   **Cardiovascular Disease** Estimate the power available for the study proposed in Example 8.27 using a two-sided test with significance level = .05.

SOLUTION   From Example 8.27, $n_1 = n_2 = 100$, $k = n_2/n_1 = 1$, $\Delta = 5$, $\sigma_1 = 15.34$, $\sigma_2 = 18.23$, and $\alpha = .05$. Therefore, from **(8.29)**,

$$\text{Power} = \Phi\left(-z_{.975} + \frac{\sqrt{100}(5)}{\sqrt{15.34^2 + 18.23^2/1}}\right) = \Phi\left[-1.96 + \frac{10(5)}{23.83}\right]$$

$$= \Phi(-1.96 + 2.098) = \Phi(0.138) = .555$$

Thus, there is a 55.5% chance of detecting a significant difference using a two-sided test with significance level = .05. ∎

To perform a one-sided rather than a two-sided test, the following power formula is used:

8.30 | **Power for Comparing the Means of Two Normally Distributed Samples Using a One-Sided Test with Significance Level α**

To test the hypothesis $H_0: \mu_1 = \mu_2$ versus $H_1: \mu_1 < \mu_2$ for the specific alternative $\mu_1 = \mu_2 - \Delta$, where $\Delta > 0$, with significance level α,

$$\text{Power} = \Phi\left(-z_{1-\alpha} + \frac{\sqrt{n_1}\Delta}{\sqrt{\sigma_1^2 + \sigma_2^2/k}}\right)$$

where (μ_1, σ_1^2), (μ_2, σ_2^2) are the means and variances of the two respective samples and $k = n_2/n_1 =$ the projected ratio of the two sample sizes.

To test the hypothesis $H_0: \mu_1 = \mu_2$ versus $H_1: \mu_1 > \mu_2$ for the specific alternative $\mu_1 = \mu_2 + \Delta$, where $\Delta > 0$, with significance level α, the same power estimate is used.

SECTION 8.9   **Summary**

In this chapter methods of hypothesis testing for comparing the means and variances of two samples that are assumed to be normally distributed were studied. The basic strategy is outlined in the shaded boxes of the flowchart in Figure 8.11, which is an extract from the larger flowchart in the back of this book. Referring to ① in the upper left, first note that we are dealing with the case of a two-sample problem in which either the underlying distributions are normal or the central limit theorem can be assumed to hold. If we are interested in comparing the means of the two samples, then we refer to box ③. If our two samples are paired, that is, if each person is used as his or her own control or if the samples consist of different people who are matched on a one-to-one basis, then the paired *t* test is appropriate. If the samples are independent, then the *F* test for the equality of two variances is used to decide whether or not the variances are significantly different. If the variances are not significantly different, then the two-sample *t* test with equal variances is used; if the variances are significantly different, then the two-sample *t* test with unequal variances is used. If we are only comparing the variances of the two samples, then only the *F* test for comparing variances is used, as indicated in the lower left of Figure 8.11.

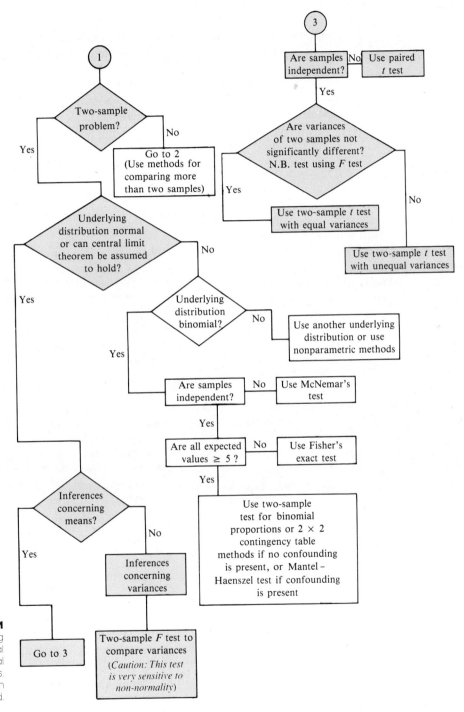

FIGURE 8.11

Flowchart summarizing two-sample statistical inference—normal theory methods. Material covered in Chapter 8 is shaded.

The chapter concluded by presenting appropriate sample size and power formulas for planning investigations in which the goal is to compare the means from two independent samples. In Chapter 9 alternative test procedures to those presented in this chapter, that are useful when the assumptions of normality are no longer valid, are looked at. These procedures are called nonparametric tests as opposed to the parametric tests presented in Chapters 7 and 8, where specific assumptions are made about the underlying distributions.

PROBLEMS

Ophthalmology

In a study of the natural history of retinitis pigmentosa (RP), 94 RP patients were followed for 3 years [3]. Among 90 patients with complete follow-up, the mean ± 1 se of $\log_e$ (visual field loss) over 1, 2, and 3 years was 0.02 ± 0.04, 0.08 ± 0.05, and 0.14 ± 0.07, respectively.

8.1 What test procedure can be used to test for changes in $\log_e$ (visual field) over any given time period?

8.2 Implement the procedure in Problem 8.1 to test for significant changes in visual field over 1 year. Report a p-value.

8.3 Answer Problem 8.2 for changes over 2 years.

8.4 Answer Problem 8.2 for changes over 3 years.

Cardiovascular Disease

Twenty volunteers adopt a low-cholesterol diet for 3 months. The mean ± 1 sd of changes in serum cholesterol over the 3-month period was 20.0 ± 35.0 (mg/dL).

8.5 Test for significant changes in cholesterol over 3 months.

One important component of cholesterol, which is widely believed to have a beneficial effect on heart disease, is HDL cholesterol. The mean ± 1 sd of changes in HDL cholesterol over the 3-month period was 3.0 ± 12.0 (mg/dL).

8.6 Test for significant changes in HDL cholesterol over 3 months.

The mean ± 1 sd weight loss over the 3-month period was 5.2 ± 8.0 (lb).

8.7 Test for significant changes in weight over 3 months.

8.8 Find the upper 5th percentile of an F distribution with 24 and 30 df.

8.9 Find the lower 2.5th percentile of an F distribution with 14 and 7 df.

8.10 Find the upper 5th percentile of an F distribution with 22 and 14 df.

8.11 Find the upper 1st percentile of an F distribution with 22 and 35 df.

8.12 Find the lower 2.5th percentile of an F distribution with 50 and 10 df.

8.13 Suppose we have two normally distributed samples of sizes 9 and 15 with sample standard deviations of 13.7 and 7.2, respectively. Test if the variances are significantly different in the two samples.

8.14 Suppose we have two normally distributed samples of sizes 16 and 12 with sample standard deviations of 12.6 and 6.2, respectively. Test if the variances are significantly different in the two samples.

Nutrition

The mean ± 1 sd of $\log_e$ [calcium intake (mg)] among 25 12–14-year-old females below the poverty level is 6.56 ± 0.64. Similarly, the mean ± 1 sd of $\log_e$ [calcium intake (mg)] among 40 12–14-year-old females above the poverty level is 6.80 ± 0.76.

8.15 Test for a significant difference between the variances of the two groups.

8.16 What is the appropriate procedure to test for a significant difference in means between the two groups?

8.17 Implement the procedure in Problem 8.16 using the critical value method.

8.18 What is the p-value corresponding to your answer to Problem 8.17?

8.19 Compute a 95% confidence interval for the difference in means between the two groups.

Refer to the data in Table 2.11 on page 36.

8.20 Test for a significant difference in the variances of the initial white blood count (WBC) between patients who did and patients who did not receive a bacterial culture.

8.21 What is the appropriate test procedure to test for significant differences in mean WBC between people who do and people who do not receive a bacterial culture?

8.22 Perform the procedure in Problem 8.21 using the critical value method.

8.23 What is the *p*-value corresponding to your answer to Problem 8.22?

8.24 Compute a 95% confidence interval for the true difference in mean WBC between the two groups.

Health Services Administration

A comparison is made of demographic characteristics of patients using fee-for-service practices and prepaid group health plans. Suppose the data presented in Table 8.6 are found.

TABLE 8.6 Characteristics of patients using fee-for-service practices and prepaid group health plans

Characteristic	Fee-for-service			Prepaid group health plans		
	Mean	sd	n	Mean	sd	n
Age (years)	58.1	6.2	57	52.6	4.3	48
Education (years)	11.8	0.7	57	12.7	0.8	48

8.25 Test for a significant difference in the variance of age between the two groups.

8.26 What is the appropriate test to compare the mean ages of the two groups?

8.27 Perform the test in Problem 8.26 and report a *p*-value.

8.28 Compute a 95% confidence interval for the mean age difference between the two groups.

8.29 Answer Problem 8.25 for number of years of education.

8.30 Answer Problem 8.26 for number of years of education.

8.31 Answer Problem 8.27 for number of years of education.

8.32 Answer Problem 8.28 for number of years of education.

Refer to Problem 8.15.

8.33 Suppose an equal number of 12–14-year-old girls below and above the poverty level are recruited to study differences in mean $\log_e$ (calcium intake). How many girls should be recruited to have an 80% chance

of detecting a significant difference using a two-sided test with $\alpha = .05$?

8.34 Answer Problem 8.33 if a one-sided rather than a two-sided test is used.

8.35 Using a two-sided test with $\alpha = .05$, answer Problem 8.33 anticipating that 2 girls above the poverty level will be recruited for every girl below the poverty level.

8.36 Suppose 50 girls above the poverty level and 50 girls below the poverty level are recruited for the study. How much power will the study have of finding a significant difference using a two-sided test with $\alpha = .05$ assuming that the population parameters are the same as the sample estimates in Problem 8.15?

8.37 Answer Problem 8.36 if a one-sided rather than a two-sided test is used.

8.38 Suppose that 50 girls above the poverty level and 25 girls below the poverty level are recruited for the study. How much power will the study have if a two-sided test is used with $\alpha = .05$?

8.39 Answer Problem 8.38, if a one-sided test is used with $\alpha = .05$.

For Problems 8.40 and 8.41 refer to the data on hospital stays in Table 2.11 (page 36).

8.40 Test whether or not the mean duration of hospitalization is the same among those patients under 50 years old and those patients 50 years old or older.

8.41 Test whether or not the mean duration of hospitalization is the same among those patients using the medical service and those patients using the surgical service.

Gynecology

A study was conducted to compare the age at menarche (the age at the first menstrual period) of girls entering the first-year class of a small American private college in the year 1975 with that of girls entering the first-year class of the same college in 1985. This study is in response to reports of differences over time in other countries. Suppose that 30 girls in the 1975 class have a mean age at menarche of 12.78 years with a standard deviation of 0.43 year and 40 girls in the class of 1985 have a mean age at menarche of 12.42 years with a standard deviation of 0.67 year.

8.42 What are the appropriate null and alternative hypotheses to test whether or not the ages at menarche are comparable in the two groups? Justify your choice of the appropriate hypotheses.

8.43 Perform the significance test indicated in Problem 8.42 and state your conclusions.

8.44 What is the advantage of comparing two different classes of the same school as opposed to comparing the 1975 entering class of one school with the 1985 entering class of another school?

Pathology

Refer to the heart weight data in Table 2.10 (page 35).

8.45 Test for a significant difference in total heart weight between the diseased and normal groups.

8.46 Test for a significant difference in body weight between the diseased and normal groups.

Ophthalmology

Diflunisal is a drug used to treat mild to moderate pain, osteoarthritis, or rheumatoid arthritis. Diflunisal's ocular effects had not been studied until a study on its effect on intraocular pressure in glaucoma patients who were already receiving maximal therapy for glaucoma was conducted [4].

8.47 Suppose the change (mean $\pm$ sd) in ocular pressure after administration of diflunisal among 10 patients whose standard therapy was methazolamide and topical glaucoma medications was -1.6 ± 1.5 mm Hg. Assess the statistical significance of the results.

8.48 The change in ocular pressure after administration of diflunisal among 30 patients whose standard therapy was topical drugs only was -0.7 ± 2.1 mm Hg. Assess the statistical significance of these results.

8.49 Compute 95% confidence limits for the change in pressure in each of the two groups identified in Problems 8.47 and 8.48.

8.50 Compare the change in ocular pressure in the two groups identified in Problems 8.47 and 8.48 using hypothesis-testing methods.

Psychiatry, Renal Disease

Severe anxiety often occurs in patients who must undergo chronic hemodialysis. A set of progressive relaxation exercises was shown on videotape to a group of 38 experimental subjects, while a set of neutral videotapes was shown to a control group of 23 patients who were also on chronic hemodialysis [5]. The results of a psychiatric questionnaire (the State-Trait Anxiety Inventory) are presented in Table 8.7.

8.51 Perform a statistical test to compare the experimental and control groups' pretest scores.

8.52 Perform a statistical test to compare the experimental and control groups' posttest scores.

TABLE 8.7 Pretest and posttest State-Trait Anxiety means and standard deviations for the experimental and control groups of hemodialysis patients

	Pretest			Posttest		
	Mean	sd	n	Mean	sd	n
Experimental	37.51	10.66	38	33.42	10.18	38
Control	36.42	8.59	23	39.71	9.16	23

A lower score on the test corresponds to less anxiety.
(Reprinted with permission of the *Journal of Chronic Diseases*, *35*(10), 797–802.)

Cardiovascular Disease, Pediatrics

A study in Pittsburgh looked at various cardiovascular risk factors in children, as measured at birth and during their first 5 years of life [6]. In particular, heart rate was assessed at birth, 5 months, 15 months, 24 months, and annually thereafter until 5 years of age. Heart rate was related to age, sex, race, and socioeconomic status. The data in Table 8.8 were presented relating heart rate to race among newborns.

TABLE 8.8 Relationship of heart rate to race among newborns

Race	Mean (beats per minute)	sd	n
White	125	11	218
Black	133	12	156

(Reprinted with permission of the *American Journal of Epidemiology*, *119*(4), 554–563.)

8.53 Test for a significant difference in heart rates between white and black newborns.

8.54 Report a *p*-value for the test performed in Problem 8.53.

Hypertension

An investigator wishes to determine if sitting upright in a chair versus lying down on a bed will affect a person's blood pressure. The investigator decides to use each of 10 patients as his or her own control and collects systolic blood-pressure data in both the sitting and lying positions, as given in Table 8.9.

8.55 What is the distinction between a one-sided and a two-sided hypothesis test in this problem?

8.56 Which hypothesis test is appropriate here? Why?

TABLE 8.9 Effect of position on level of bp (mm Hg)

Patient	Sitting upright	Lying down
1	142	154
2	100	106
3	112	110
4	92	100
5	104	112
6	100	100
7	108	120
8	94	90
9	104	104
10	98	114

8.57 Using an α level of .05, test the hypothesis that position affects the level of blood pressure.

8.58 What is the p-value of the preceding test? Compute either the specific p-value or a range within which the p-value lies.

Pharmacology

One method for assessing the effectiveness of a drug is to note its concentration in blood and/or urine samples at certain periods of time after giving the drug. Suppose we wish to compare the concentrations of two types of aspirin (types A and B) in urine specimens taken from the same person, 1 hour after he or she has taken the drug. Hence a specific dosage of either type A or type B aspirin is given at one time and the 1-hour urine concentration·is measured. One week later, after the first aspirin has presumably been cleared from the system, the same dosage of the other aspirin is given to the same person and the 1-hour urine concentration is noted. Since the order of giving the drugs may affect the results, a table of random numbers is used to decide which of the two types of aspirin to give first. This experiment is performed on 10 people; the results are given in Table 8.10.

Suppose we wish to test the hypothesis that the concentrations of the two drugs are the same in urine specimens.

8.59 What are the appropriate hypotheses?

8.60 What are the assumptions behind the test used?

8.61 Conduct the test mentioned in Problem 8.59.

8.62 What is the best point estimate of the difference in concentrations between the two drugs?

8.63 What is a 95% confidence interval for the mean difference?

TABLE 8.10 Concentration of aspirin in urine samples

Person	Aspirin A 1-hour concentration (mg%)	Aspirin B 1-hour concentration (mg%)
1	15	13
2	26	20
3	13	10
4	28	21
5	17	17
6	20	22
7	7	5
8	36	30
9	12	7
10	18	11
Mean	19.20	15.60
sd	8.63	7.78

8.64 Suppose an α level of .05 is used for the test in Problem 8.61. What is the relationship between the decision reached with the test procedure in Problem 8.61 and the nature of the confidence interval in Problem 8.63?

Hypertension

Blood-pressure measurements taken on the left and right arms of a person are assumed to be comparable. To test this assumption, 10 volunteers are obtained and systolic blood-pressure readings are taken simultaneously on both arms by two different observers, Mr. Jones for the left arm and Mr. Smith for the right arm. The data are given in Table 8.11.

TABLE 8.11 Effect of arm on level of blood pressure (mm Hg)

Patient	Left arm	Right arm
1	130	126
2	120	124
3	135	127
4	100	95
5	98	102
6	110	109
7	123	124
8	136	132
9	140	137
10	155	156

8.65 Assuming that the two observers are comparable, test whether or not the two arms give comparable readings.

8.66 Suppose we do *not* assume that the two observers are comparable. Can the experiment as it is defined detect differences between the two arms? If not, can you design the experiment differently so as to achieve this aim?

Nutrition

A hypothesis of ongoing clinical interest is that vitamin C prevents the common cold. A study is organized to test this hypothesis using 20 prisoners as participants. In the study 10 are randomly allocated to receive vitamin C capsules and 10 are randomly allocated to receive placebo capsules. The number of colds over a 12-month period for each participant is given in Table 8.12. We wish to test the hypothesis that vitamin C prevents the common cold.

8.67 Is a one-sample or two-sample test needed here?

8.68 Is a one-sided or two-sided test needed here?

8.69 Which of the following test procedures should be used to test this hypothesis? (More than one may be necessary.)

(a) Paired t test

(b) Two-sample t test with equal variances

(c) Two-sample t test with unequal variances

(d) Two-sample F test

(e) One-sample t test

8.70 Carry out the test procedure(s) in Problem 8.69 and report a p-value.

8.71 Derive a lower one-sided 95% confidence interval for the mean difference (vitamin C − placebo) in number of colds per year between the two groups.

8.72 What is the relationship between your answers to Problems 8.70 and 8.71?

Ophthalmology

A topic of current interest in ophthalmology is whether or not spherical refraction is different between the left and right eyes. For this purpose refraction is measured in both eyes of 17 people. The data are given in Table 8.13.

8.73 Is a one-sample or two-sample test needed here?

8.74 Is a one-sided or two-sided test needed here?

8.75 Which of the following test procedures is appropriate to use on these data? (More than one may be necessary.)

(a) Paired t test

(b) Two-sample t test for independent samples with equal variances

TABLE 8.12 Number of colds over a 12-month period for people taking vitamin C and placebo capsules

Vitamin C		Placebo		
i	x_{i1}	i	x_{i2}	$d_i(x_{i1} - x_{i2})$
1	4	1	7	−3
2	0	2	8	−8
3	3	3	4	−1
4	4	4	6	−2
5	4	5	6	−2
6	3	6	4	−1
7	4	7	6	−2
8	3	8	4	−1
9	2	9	6	−4
10	6	10	6	0
$\sum_{i=1}^{10} x_{i1} = 33$		$\sum_{i=1}^{10} x_{i2} = 57$		$\sum_{i=1}^{10} d_i = -24$
$\sum_{i=1}^{10} x_{i1}^2 = 131$		$\sum_{i=1}^{10} x_{i2}^2 = 341$		$\sum_{i=1}^{10} d_i^2 = 104$

TABLE 8.13 Spherical refraction in the right and left eye

(i) Person	(x_i) Spherical refraction OD (right eye) (diopters)	(y_i) Spherical refraction OS (left eye) (diopters)	$(d_i = x_i - y_i)$ Difference (OD − OS)	(i) Person	(x_i) Spherical refraction OD	(y_i) Spherical refraction OS	$(d_i = x_i - y_i)$ Difference (OD − OS)
1	+1.75	+2.00	−0.25	9	0	0.50	−0.50
2	−4.00	−4.00	0	10	−1.00	−1.25	+0.25
3	−1.25	−1.00	−0.25	11	+0.50	−1.75	+2.25
4	+1.00	+1.00	0	12	−8.50	−5.00	−3.50
5	−1.00	−1.00	0	13	+0.50	+0.50	0
6	−0.75	+0.25	−1.00	14	−5.25	−4.75	−0.50
7	−2.25	−2.25	0	15	−2.25	−2.50	+0.25
8	+0.25	+0.25	0	16	−6.50	−6.25	−0.25
				17	+1.75	+1.75	0

$$\sum_{i=1}^{17} x_i = -27.00 \qquad \sum_{i=1}^{17} x_i^2 = 180.00 \qquad \sum_{i=1}^{17} y_i = -23.50 \qquad \sum_{i=1}^{17} y_i^2 = 129.25 \qquad \sum_{i=1}^{17} d_i = -3.50 \qquad \sum_{i=1}^{17} d_i^2 = 19.13$$

(c) Two-sample t test for independent samples with unequal variances

(d) One-sample t test

8.76 Carry out the hypothesis test in Problem 8.75 and report a p-value.

8.77 Estimate a 90% confidence interval for the mean difference in spherical refraction between the two eyes.

Pulmonary Disease

A 1980 study attempted to compare the working environment in offices where smoking was permitted with that in offices where smoking was not permitted [7]. Measurements were made of carbon monoxide (CO) at 1:20 PM in 40 work areas where smoking was permitted and 40 work areas where smoking was not permitted. Where smoking was permitted, the mean CO = 11.6 parts per million (ppm) and the standard deviation CO = 7.3 ppm. Where smoking was not permitted, the mean CO = 6.9 ppm and the standard deviation CO = 2.7 ppm.

8.78 Test for whether or not the standard deviation of CO is significantly different in the two types of working environments.

8.79 Test for whether or not the mean CO is significantly different in the two types of working environments.

8.80 Does your answer to Problem 8.78 affect your answer to Problem 8.79? If so, in what way?

Cardiovascular Disease

A study was performed in 1976 to relate the use of oral contraceptives with the levels of various lipid fractions in a group of 163 nonpregnant, premenopausal women ages 21–39. The mean serum cholesterol among 66 current users of oral contraceptives was 201 ± 37 (mg%/mL) (mean ± sd), whereas for 97 nonusers it was 193 ± 37 (mg%/mL).

8.81 Test for significant differences in cholesterol levels between the two groups.

8.82 Report a p-value based on your hypothesis test in Problem 8.81.

8.83 Derive a 95% confidence interval for the true mean difference in cholesterol levels between the groups.

8.84 Suppose we compute the two-tailed p-value in Problem 8.82 and obtain .03 and compute the two-sided 95% confidence interval in Problem 8.83 and obtain (−0.6, 7.3). Do these two results contradict each other? Why or why not? (*Note*: These values are not necessarily the actual results in Problems 8.82 and 8.83.)

Ophthalmology

A camera has been developed to detect the presence of cataract more accurately. Using this camera, the gray level of each point (or pixel) in the lens of a human eye can be characterized into 256 gradations, where a gray level of 1 represents black and a gray level of 256 represents white. To test the camera, photographs were taken of 6 randomly selected normal eyes and 6 randomly selected cataractous eyes (the two groups consist of different people). The median gray level of each eye was computed over the 10,000 + pixels in the lens. The data are given in Table 8.14.

TABLE 8.14 Median gray level for cataractous and normal eyes

Patient number	Cataractous median gray level	Normal median gray level
1	161	158
2	140	182
3	136	185
4	171	145
5	106	167
6	149	177
$\bar{x}$	143.8	169.0
s	22.7	15.4

8.85 What statistical procedure can be used to test if there is a significant difference in the median gray levels between the cataractous and normal eyes?

8.86 Carry out the test procedure mentioned in Problem 8.85 and report a *p*-value.

8.87 Provide a 99% confidence interval for the mean difference in gray levels between cataractous and normal eyes.

Hypertension

A 1982 study by the Lipid Research Clinics looked at the relationship between alcohol consumption and level of systolic blood pressure in women not using oral contraceptives [8]. Alcohol consumption was categorized as follows: no alcohol use; ≤10 oz per week alcohol consumption; >10 oz per week alcohol consumption. The results for women 30–39 years of age are given in Table 8.15.

Suppose we wish to compare the levels of systolic blood pressure of groups A and B and we have no prior information regarding which group has higher blood pressure.

8.88 Should a one-sample or two-sample test be used here?

8.89 Should a one-sided or two-sided test be used here?

8.90 Which test procedure(s) should be used to test the preceding hypotheses?

TABLE 8.15 Relationship of systolic blood pressure and alcohol consumption in 30–39-year-old women not using oral contraceptives

	Systolic blood pressure (mm Hg)		
	Mean	sd	n
A. No alcohol use	110.5	13.3	357
B. ≤10 oz per week alcohol consumption	109.1	13.4	440
C. >10 oz per week alcohol consumption	114.5	14.9	23

(Reprinted with permission of the *Journal of Chronic Diseases, 35*(4), 251–257.)

8.91 Carry out the test in Problem 8.90 and report a *p*-value.

8.92 Compute a 95% confidence interval for the mean difference in blood pressure between the two groups.

8.93 Answer Problem 8.90 for a comparison of groups A and C.

8.94 Answer Problem 8.91 for a comparison of groups A and C.

8.95 Answer Problem 8.92 for a comparison of groups A and C.

Obstetrics

A clinical trial is conducted at the gynecology unit of a major hospital to determine the effectiveness of drug A in preventing premature birth. In the trial, 30 pregnant women are to be studied, 15 assigned to a treatment group that will receive drug A and 15 assigned to a control group that will receive a placebo. The patients are to take a fixed dose of each drug on a one-time-only basis between the 24th and 28th weeks of pregnancy. The patients are assigned to groups using a random number table, whereby for every 2 patients eligible for the study, one is assigned randomly to the treatment group and the other to the control group.

8.96 Suppose *you* are conducting the study. What would be a reasonable way of allocating women to the treatment and control groups?

Suppose the weights of the babies are those given in Table 8.16.

8.97 How would you assess the effects of drug A in light of your answer to Problem 8.96? Specifically, would you use a paired or an unpaired analysis, and of what type?

TABLE 8.16 Birthweights in a clinical trial to test a drug for preventing low birthweights

Number of pregnant women	Treatment group baby weight (lb)	Control group baby weight (lb)
1	6.9	6.4
2	7.6	6.7
3	7.3	5.4
4	7.6	8.2
5	6.8	5.3
6	7.2	6.6
7	8.0	5.8
8	5.5	5.7
9	5.8	6.2
10	7.3	7.1
11	8.2	7.0
12	6.9	6.9
13	6.8	5.6
14	5.7	4.2
15	8.6	6.8

8.98 Perform both a paired and an unpaired analysis of the data. Does the type of analysis affect the assessment of the results?

8.99 Suppose patient 3 in the control group subsequently moves to another city before giving birth and her child's weight is unknown. Does this event affect the analyses in Problem 8.98?

Pulmonary Disease

Forced expiratory volume (FEV) is a standard measure of pulmonary function. We would expect that any reasonable measure of pulmonary function would reflect that a person's pulmonary function declines with age after 20. Suppose we test this hypothesis by looking at 10 nonsmoking males ages 35–39, heights 68–72 inches, and measuring their FEV (L) initially and then once again 2 years later. The data in Table 8.17 are obtained.

8.100 What are the appropriate null and alternative hypotheses in this case?

8.101 In words, what is the meaning of a type I and a type II error here?

8.102 Carry out the test in Problem 8.100. What are your conclusions?

Another aspect of the preceeding study involves looking at the effect of smoking on baseline pulmonary function and on change in pulmonary function over time. We must be careful, since FEV depends on many factors, particularly age and height. Suppose we have a comparable group of 15 men in the same age and height group who are smokers, and we measure their FEV at year 0 and year 2. The data are given in Table 8.18.

8.103 What are the appropriate null and alternative hypotheses to compare the smokers and nonsmokers at baseline?

8.104 Carry out the procedure(s) necessary to conduct the test in Problem 8.103.

8.105 Suggest a procedure for testing whether or not the *change* in pulmonary function over 2 years is the same in the two groups.

Pharmacology

In a pediatric clinic a study is carried out to see how effective aspirin is in reducing temperature. Twelve 5-year-old girls suffering from influenza had their temperatures taken immediately before and 1 hour after

TABLE 8.17 Pulmonary function in nonsmokers at two points in time

Person	FEV year 0 (l)	FEV year 2 (l)	Person	FEV year 0 (l)	FEV year 2 (l)
1	3.22	2.95	6	3.25	3.20
2	4.06	3.75	7	4.20	3.90
3	3.85	4.00	8	3.05	2.76
4	3.50	3.42	9	2.86	2.75
5	2.80	2.77	10	3.50	3.32
			Mean	3.43	3.28
			sd	0.485	0.480

TABLE 8.18 Pulmonary function in smokers at two points in time

Person	FEV year 0 (I)	FEV year 2 (I)	Person	FEV year 0 (I)	FEV year 2 (I)
1	2.85	2.88	9	2.76	3.02
2	3.32	3.40	10	3.00	3.08
3	3.01	3.02	11	3.26	3.00
4	2.95	2.84	12	2.84	3.40
5	2.78	2.75	13	2.50	2.59
6	2.86	3.20	14	3.59	3.29
7	2.78	2.96	15	3.30	3.32
8	2.90	2.74			
			Mean	2.98	3.03
			sd	0.279	0.250

administration of aspirin. The results are given in Table 8.19. Suppose we assume normality and want to test the hypothesis that aspirin is reducing the temperature.

TABLE 8.19 Body temperature (°F) before and after taking aspirin

Patient	Before	After
1	102.4	99.6
2	103.2	100.1
3	101.9	100.2
4	103.0	101.1
5	101.2	99.8
6	100.7	100.2
7	102.5	101.0
8	103.1	100.1
9	102.8	100.7
10	102.3	101.1
11	101.9	101.3
12	101.4	100.2

8.106 What are the null and alternative hypotheses in this situation?

8.107 In words, what is meant by a type I error in this situation?

8.108 Suppose the alternative that aspirin reduces the temperature by 1 degree is considered. What is meant by the power of the test against this specific alternative?

8.109 How would the power in Problem 8.108 change if the alternative were a temperature reduction of 2 degrees?

8.110 Compute the power for the alternatives mentioned in Problems 8.108 and 8.109.

8.111 Perform a significance test for the hypotheses in Problem 8.106.

Hypertension

A study of the relationship between salt intake and blood pressure of infants is in the planning stages. A pilot study is done, comparing five one-year-old infants on a high-salt diet with five one-year-old infants on a low-salt diet. The results are given in Table 8.20.

TABLE 8.20 Relationship between salt intake and level of blood pressure

	High-salt diet	Low-salt diet
Mean systolic bp	90.8	87.2
sd systolic bp	10.3	9.2
n	5	5

8.112 If the means and standard deviations are considered to be true population parameters, then, using a one-sided test with significance level = .05, how many infants are needed in each group to have an 80% chance of detecting a significant difference?

8.113 Suppose it is easier to recruit high-salt-diet infants and the investigators decide to enroll twice as many high-salt-diet infants as low-salt-diet infants. How many infants are needed in each group to have a 90% chance of detecting a significant difference using a one-sided test with an α level of .05?

8.114 Suppose the budget will only allow for recruiting 50 high-salt-diet and 50 low-salt-diet infants into the study. How much power would such a study have of detecting a significant difference using a one-sided test with significance level = .05 if the true difference between the groups is 5 mm Hg?

8.115 Answer Problem 8.114 for a true difference of 2 mm Hg.

8.116 Answer Problem 8.114 if a two-sided test is used and the true difference is 5 mm Hg.

8.117 Answer Problem 8.116 if the true difference is 2 mm Hg.

Pulmonary Disease

A possible important environmental determinant of lung function in children is the level of cigarette smoke in the home. Suppose this question is studied by selecting two groups: Group 1 consists of 23 nonsmoking children 5–9 years of age, *both* of whose parents smoke, who have a mean FEV of 2.1 L and sd of 0.7 L; group 2 consists of 20 nonsmoking children of comparable age, *neither* of whose parents smoke, who have a mean FEV of 2.3 L and sd of 0.4 L.

8.118 What are the appropriate null and alternative hypotheses in this situation?

8.119 What is the appropriate test procedure for the hypotheses in Problem 8.118?

8.120 Carry out the test in Problem 8.119 using the critical value method.

8.121 Provide a 95% confidence interval for the true mean difference in FEV between 5–9-year-old children whose parents smoke and comparable children whose parents do not smoke.

8.122 If this is regarded as a pilot study, then how many children are needed in each group (assuming equal numbers in each group) to have a 95% chance of

detecting a significant difference using a two-sided test with $\alpha = .05$?

8.123 Answer the question in Problem 8.122 if the investigators intend to use a one-sided rather than a two-sided test.

Suppose 40 children, both of whose parents smoke, and 50 children, neither of whose parents smoke, are recruited for the study.

8.124 How much power would such a study have using a two-sided test with significance level = .05 and assuming that the estimates of the population parameters in the pilot study are correct?

8.125 Answer Problem 8.124 if a one-sided rather than a two-sided test is to be used.

Hypertension

The effect of sodium restriction on blood pressure remains a controversial subject. To test this hypothesis a group of 82 individuals participated in a study of restricted sodium intake (≤ 75 mEq/24 hrs) for a period of 12 weeks. The effect on diastolic blood pressure (DBP) was reported as in Table 8.21 [9].

8.126 What is the appropriate procedure to test for whether sodium restriction has had an impact on DBP?

8.127 Implement the procedure in Problem 8.126 using the critical value method for people age < 40.

One of the interesting findings is the difference in response to dietary therapy between people in the two age groups.

8.128 Test the hypothesis that the response to sodium restriction is different in the two groups and report a *p*-value.

8.129 Develop a 95% confidence interval for the response to dietary therapy in each age group separately.

TABLE 8.21 Effect of sodium restriction on diastolic blood pressure

	Baseline period			Diet period			Change from control (diet-baseline)		
	Mean	sd	n	Mean	sd	n	Mean	sd	n
Age < 40	69.7	8.5	61	69.1	8.5	61	−0.7	6.2	61
Age ⩾ 40	77.0	8.0	22	71.9	7.5	22	−5.0	4.7	22
Total							−1.8	6.1	83

8.130 Develop a 95% confidence interval for the difference in response between the 2 age groups.

Suppose the results of this study are to be used to plan a larger study on the effects of sodium restriction on DBP.

8.131 How many subjects need to be enrolled in the larger study to test if sodium restriction results in lower DBP if the mean and sd of decline in DBP over the total group of 83 subjects are used for planning purposes and a 90% chance of detecting a significant difference using a two-sided test with a 5% level of significance is desired?

8.132 Suppose 200 patients are enrolled in the larger study. How much power would the larger study have if a two-sided test with significance level = .05 is used?

Infectious Disease

The degree of clinical agreement among different physicians on the presence or absence of generalized lymphadenopathy was assessed in 32 randomly selected participants from a prospective study of male sexual contacts of men with AIDS or an AIDS-related condition (ARC) [10]. The total number of palpable lymph nodes was assessed by each of three physicians. Results

from two of the three physicians are presented in Table 8.22.

8.133 What is the appropriate test procedure to test if there is a systematic difference between the assessment of Doctor A versus Doctor B?

8.134 Should a one-sided or a two-sided test be performed? Why?

8.135 Perform the test in Problem 8.133 and report a p-value.

8.136 Compute a 95% confidence interval for the true mean difference between observers. How does it relate to your answer to Problem 8.135?

8.137 Suppose the result in Problem 8.135 shows no significant difference. Does this mean that this type of assessment is highly reproducible? Why or why not?

Environmental Health

A study was conducted relating lead level in umbilical-cord blood and cognitive development [11]. Three groups were identified at birth (high/medium/low) cord blood lead. One issue is the consistency of the differences in blood-lead levels over time between these three groups. The data in Table 8.23 were presented.

TABLE 8.22 Reproducibility of assessment of lymph nodes among sexual contacts of AIDS or ARC patients

| Patient | Number of palpable lymph nodes | | | Patient | Number of palpable lymph nodes | | |
	Doctor A	Doctor B	Difference		Doctor A	Doctor B	Difference
1	4	1	3	17	4	1	3
2	17	9	8	18	12	9	3
3	3	2	1	19	10	7	3
4	11	13	−2	20	9	11	−2
5	12	9	3	21	5	0	5
6	5	2	3	22	3	0	3
7	5	6	−1	23	12	12	0
8	6	3	3	24	5	1	4
9	3	0	3	25	13	9	4
10	5	0	5	26	12	6	6
11	9	6	3	27	6	9	−3
12	1	1	0	28	19	9	10
13	5	4	1	29	8	4	4
14	8	4	4	30	15	9	6
15	7	7	0	31	6	1	5
16	8	6	2	32	5	4	1
				Mean	7.91	5.16	2.75
				sd	4.35	3.93	2.83
				n	32	32	32

TABLE 8.23 Reproducibility of blood-lead level in the first two years of life

Cord blood-lead group		Blood-lead level (mean ± sd)				
		Birth	6 months	12 months	18 months	24 months
Low	Mean ± sd	1.8 ± 0.6	4.6 ± 3.9	5.8 ± 5.1	6.7 ± 5.5	5.4 ± 4.8
	n	85	70	69	65	61
Medium	Mean ± sd	6.5 ± 0.3	7.0 ± 7.8	8.5 ± 7.6	8.3 ± 5.8	7.2 ± 5.0
	n	88	70	70	65	63
High	Mean ± sd	14.6 ± 3.0	7.0 ± 8.7	8.8 ± 6.4	7.6 ± 5.8	7.7 ± 8.5
	n	76	61	60	57	58

8.138 What test procedure can be used to test if there are differences between the observed blood-lead levels at 24 months between the low and high groups (as defined at baseline)?

8.139 Implement the test procedure in Problem 8.138 and report a p-value.

8.140 Provide a 95% confidence interval for the difference in blood-lead levels between the low and high groups at 24 months.

Renal Disease

Ten patients with advanced diabetic nephropathy (kidney problems) were treated with captopril over an 8-week period [12]. Urinary protein was measured before and after drug therapy, with results listed in Table 8.24 in both the raw and log scale.

TABLE 8.24 Changes in urinary protein after treatment with captopril

Patient	Raw scale Urinary protein (g/24 hr)		Log scale Urinary protein (g/24 hr)	
	Before	After	Before	After
1	25.6	10.1	3.24	2.31
2	17.0	5.7	2.83	1.74
3	16.0	5.6	2.77	1.72
4	10.4	3.4	2.34	1.22
5	8.2	6.5	2.10	1.87
6	7.9	0.7	2.07	−0.36
7	5.8	6.1	1.76	1.81
8	5.4	4.7	1.69	1.55
9	5.1	2.0	1.63	0.69
10	4.7	2.9	1.55	1.06

8.141 What is the appropriate test procedure to test if the urinary protein has changed over the 8-week period?

8.142 Perform the test in Problem 8.141 using both the raw and log scale and report a p-value. Are there any advantages in using the raw or the log scale?

8.143 What is your best estimate of the percentage change in urinary protein based on the data in Table 8.24?

8.144 Provide a 95% confidence interval associated with your estimate in Problem 8.143.

Ophthalmology

Retinitis pigmentosa (RP) is the name given to a family of inherited retinal degenerative diseases that may be transmitted through various modes of inheritance. The most common features include a history of night blindness, loss of visual fields, and pigment clumping in the retina. Several reports of lipid abnormalities have been reported in RP patients. However, a consistent trend as regards either excesses or deficiencies in lipid levels has not been apparent. In one study, fatty-acid levels were measured in a group of RP patients and normal controls [13]. The data in Table 8.25 were reported on one particular fatty acid (docosahexaenoic acid) (labeled 22:6w3) in individuals with dominant disease and normal controls.

TABLE 8.25 Mean levels of plasma 22.6w3 (adjusted for age) (units are nmol/mL plasma)

	Mean	sd	n
Dominant affected individuals	34.8	20.8	36
Normal controls	47.8	30.3	68

8.145 What is an appropriate procedure to test if the mean levels of 22:6w3 differ between dominant affected individuals and normal controls? State the hypothesis being tested. Is a one-sided or a two-sided test appropriate here?

8.146 Perform the hypothesis test in Problem 8.145 and report a p-value.

8.147 Provide a 95% confidence interval for the difference in means between the two groups. What does it mean in words?

Nutrition

An important hypothesis in hypertension research is that sodium restriction may lower blood pressure. However, it is difficult to achieve sodium restriction over the long term and dietary counseling in a group setting is sometimes used to achieve this goal. The data on urinary sodium in Table 8.26 were obtained on 8 individuals enrolled in a sodium-restriction group. Data were collected at baseline and after 1 week of dietary counseling.

8.148 What are appropriate hypotheses to test if dietary counseling is effective in reducing sodium intake

over a 1-week period (as measured by overnight urinary sodium excretion)?

8.149 Conduct the test mentioned in Problem 8.148 and report a p-value.

8.150 Provide a 95% confidence interval for the true mean change in overnight Na excretion over a 1-week period.

8.151 How many subjects would be needed to have a 90% chance of detecting a significant change in urinary sodium excretion if a one-sided test is used with $\alpha = .05$ and the estimates from the data in Table 8.26 are used as the true population parameters?

Neurology

An article published in 1986 describes physical, social, and psychological problems in patients with multiple sclerosis [14]. Patients were classified as to mild, moderate, and severe disease and were graded on physical health using the McMaster physical health index, and mental health using the Rand mental health index scale. The data are shown in Table 8.27.

8.152 What test procedure should be used to test if there are differences in physical health between the two groups?

8.153 Perform the test in Problem 8.152 with the critical value method using a two-sided test with an α level of .05.

8.154 What test procedure should be used to test if there are differences in mental health between the two groups?

8.155 Perform the test in Problem 8.154 with the critical value method using an α level of .05.

8.156 What is the most accurate p-value corresponding to your answer for Problem 8.155?

Refer to Data Set 18, NIFED.DAT, in Appendix 2.

8.157 Assess if there is any difference between the nifedipine and placebo groups regarding their effects on blood pressure and heart rate.

TABLE 8.26 Overnight Na excretion (mEq/8 hr) before and after dietary counseling

Person	Week 0 (baseline)	Week 1	Difference
1	7.85	9.59	−1.74
2	12.03	34.50	−22.47
3	21.84	4.55	17.29
4	13.94	20.78	−6.84
5	16.68	11.69	4.99
6	41.78	32.51	9.27
7	14.97	5.46	9.51
8	12.07	12.95	−0.88
$\bar{x}$	17.65	16.50	1.14
s	10.56	11.63	12.22

TABLE 8.27 Physical and mental health indices for patients with multiple sclerosis

	Mild			Severe		
	Mean	sd	n	Mean	sd	n
McMaster physical health index	0.76	0.24	65	0.23	0.12	82
Rand mental health index	157.3	26.2	65	149.1	34.4	82

Hepatic Disease

An experiment was conducted to examine the influence of avian pancreatic polypeptide (aPP), cholecystokinin (CCK), vasoactive intestinal peptide (VIP), and secretin on pancreatic and biliary secretions in laying hens. In particular, researchers were concerned with the extent to which these hormones increase or decrease biliary and pancreatic flows and their pH values.

White leghorn hens, 14–29 weeks old, were surgically fitted with cannulas for collecting pancreatic and biliary secretions and a jugular cannula for continuous infusion of aPP, CCK, VIP, or secretin. One trial per day was conducted on a hen, as long as her implanted cannulas remained functional. Thus, there were varying numbers of trials per hen.

Each trial began with infusions of physiological saline for 20 minutes. At the end of this period, pancreatic and biliary secretions were collected and the cannulas were attached to new vials. The biliary and pancreatic flow rates (in micro-liters per minute) and pH values (if possible) were measured. Infusion of a hormone was then begun, and continued for 40 minutes. Measurements were then repeated.

Data Set 12, HORMONE.DAT contains data for the four hormones and saline, where saline indicates trials where physiological saline was infused in place of an active hormone during the second period. Each trial is one line in the file. There are 11 variables associated with each trial, as shown in Table 8.28.

8.158 Assess if there are significant changes in secretion rates or pH levels with any of the hormones or with saline.

8.159 Compare the changes in secretion rates or pH levels for each active hormone versus the placebo (saline) group. Use methods of hypothesis testing and/ or confidence intervals to express these comparisons statistically.

8.160 For each active hormone group, categorize dosage by high dose (above the median) versus low dose (at or below the median) and assess if there is any dose-response relationship (i.e., any differences in secretion or pH between the high- and low-dose groups).

Refer to Data Set 10, FEV.DAT, Appendix 2.

8.161 Compare the level of FEV between males and females separately in three distinct age groups (5–9, 10–14, 15–19).

8.162 Compare the level of FEV between smokers and nonsmokers separately for 10–14-year-old boys, 10–14-year-old girls, 15–19-year-old boys, and 15–19-year-old girls.

TABLE 8.28 Format of HORMONE.DAT

Column	Record Number	Format of HORMONE.DAT	Code
1–8	1	Unique identification number for each chicken	xx.x
10–17	1	Biliary secretion rate (pre)	xx.x
19–26	1	Biliary pH (pre)	xx.x
28–35	1	Pancreatic secretion rate (pre)	xx.x
37–44	1	Pancreatic pH (pre)	x.x
46–53	1	Dosage of hormone	xx.x
55–62	1	Biliary secretion rate (post)	xx.x
64–71	1	Biliary pH (post)	x.x
1–8	2	Pancreatic secretion rate (post)	xx.x
10–17	2	Pancreatic pH (post)	x.x
19–26	2	Hormone (1 = saline; 2 = APP; 3 = CCK; 4 = secretin; 5 = VIP)	xx.x

Zero values for pH indicate missing values. The units for dosages are nanograms per mL of plasma for aPP, and μg per kg per hour for CCK, VIP, and secretin.

References

[1] Landrigan, P. J., Whitworth, R. H., Baloh, R. W., Staehling, N. W., Barthel, W. F., & Rosenblum, B. F. (1975, March 29). Neuropsychological dysfunction in children with chronic low-level lead absorption. *The Lancet*, 708–715.

[2] Satterthwaite, F. W. (1946). An approximate distribution of estimates of variance components. *Biometrics Bulletin*, *2*, 110–114.

[3] Berson, E. L., Sandberg, M. A., Rosner, B., Birch, D. G., & Hanson, A. H. (1985). Progression of retinitis pigmentosa over a three-year interval. *American Journal of Ophthalmology*, *99*, 246–251.

[4] Yablonski, M. E., Maren, T. H., Hayashi, M., Naveh, N., Potash, S. D., & Pessah, N. (1988). Enhancement of the ocular hypertensive effect of acetazolamide by diflunisal. *American Journal of Ophthalmology*, *106*, 332–336.

[5] Alarcon, R. D., Jenkins, C. S., Heestand, D. E., Scott, L. K., & Cantor, L. (1982). The effectiveness of progressive relaxation in chronic hemodialysis patients. *Journal of Chronic Diseases*, *35*(10), 797–802.

[6] Schachter, J., Kuller, L. H., & Perfetti, C. (1984). Heart rate during the first five years of life: Relation to ethnic group (black or white) and to parental hypertension. *American Journal of Epidemiology*, *119*(4), 554–563.

[7] White, J. R., & Froeb, H. E., (1980). Small airway dysfunction in nonsmokers chronically exposed to tobacco smoke. *New England Journal of Medicine*, *302*(13), 720–723.

[8] Wallace, R. B., Barrett-Connor, E., Criqui, M., Wahl, P., Hoover, J., Hunninghake, D., & Heiss, G. (1982). Alteration in blood pressures associated with combined alcohol and oral contraceptive use—The Lipid Research Clinics prevalence study. *Journal of Chronic Diseases*, *35*(4), 251–257.

[9] Miller, J. Z., Weinberger, M. H., Daugherty, S. A., Fineberg, N. S., Christian, J. C., & Grim, C. E. (1987). Heterogeneity of blood pressure response to dietary sodium restriction in normotensive adults. *Journal of Chronic Disease*, *40*(3), 245–250.

[10] Coates, R. A., Fanning, M. M., Johnson, J. K., & Calzavara, L. (1988). Assessment of generalized lymphadenopathy in AIDS research: The degree of clinical agreement. *Journal of Clinical Epidemiology*, *41*(3), 267–273.

[11] Bellinger, D., Leviton, A., Waternaux, C., Needleman, H., & Rabinowitz, M. (1987). Longitudinal analyses of prenatal and postnatal lead exposure and early cognitive development. *New England Journal of Medicine*, *316*(17), 1037–1043.

[12] Taguma, Y., Kitamoto, Y., Futaki, G., Ueda, H., Monma, H., Ishizaki, M., Takahashi, H., Sekino, H., & Sasaki, Y. (1985). Effect of captopril on heavy proteinuria in azotemic diabetics. *New England Journal of Medicine*, *313*(26), 1617–1620.

[13] Anderson, R. E., Maude, M. B., Lewis, R. A., Newsome, D. A., & Fishman, G. A. (1987). Abnormal plasma levels of polyunsaturated fatty acid in autosomal dominant retinitis pigmentosa. *Experimental Eye Research*, *44*, 155–159.

[14] Harper, A. C., Harper, D. A., Chambers, L. W., Cino, P. M., & Singer, J. (1986). An epidemiological description of physical, social and psychological problems in multiple sclerosis. *Journal of Chronic Disease*, *39*(4), 305–310.

CHAPTER NINE

NONPARAMETRIC METHODS

Introduction

In the previous work in this text, the data were assumed to come from some underlying distribution, such as the normal or binomial, whose general form is known. Methods of estimation and hypothesis testing were developed, based on these assumptions. This procedure is usually referred to as **parametric statistical inference**, since the parametric form of the distribution is assumed known. If these assumptions about the shape of the distribution are not made, and if the central limit theorem also seems inapplicable, then **nonparametric statistical methods**, which make fewer assumptions about the shape of the distribution, must be used.

Another assumption made in the previous work in this text is that it is meaningful to measure the distance between possible data values. This assumption is characteristic of cardinal data.

DEFINITION 9.1 ■■■

Cardinal data are data that are on a scale where it is meaningful to measure the distance between possible data values. ■

EXAMPLE 9.1 Body weight is a cardinal variable because a difference of 6 lbs is actually twice as large as a difference of 3 lbs. ■

There are actually two types of cardinal data, interval-scale data and ratio-scale data.

DEFINITION 9.2 ■■■

For cardinal data, if the zero point is arbitrary, then the data are on an **interval scale**; if the zero point is fixed, then the data are on a **ratio scale**. ■

EXAMPLE 9.2 Body temperature is on an interval scale because the zero point is arbitrary. For example, the zero point has a different meaning for Fahrenheit and Celsius temperatures. ■

EXAMPLE 9.3 Blood pressure and body weight are on ratio scales because the zero point is well-defined in both instances. ■

It is meaningful to measure ratios between specific data values for data on a ratio scale (e.g., person A's weight is 10% higher than person B's), but it is not meaningful for data on an interval scale (e.g., the ratio of specific temperatures will be different in °F than °C). It is meaningful to use means and standard deviations for cardinal data of either type.

Another type of data that occurs frequently in medical and biological work but does not satisfy Definition 9.1 is ordinal data.

DEFINITION 9.3

▪▪

Ordinal data are data that can be ordered but do not have specific numerical values. Thus, common arithmetic *cannot* be performed on ordinal data in a meaningful way.   ▪

EXAMPLE 9.4   **Ophthalmology** Visual acuity can be measured on an ordinal scale, since we know that 20-20 vision is better than 20-30, which is better than 20-40,..., and so on. However, a numerical value cannot easily be assigned to each level of visual acuity that all ophthalmologists would agree upon.   ▪

EXAMPLE 9.5   In some clinical studies the major outcome variable is the change in a patient's condition after treatment. This variable is often measured on the following five-point scale: 1 = much improved, 2 = slightly improved, 3 = stays the same, 4 = slightly worse, 5 = much worse. This variable is ordinal because the different outcomes, 1, 2, 3, 4, 5, are ordered in the sense that condition 1 is better than condition 2, which is better than condition 3,..., and so on. However, we cannot say that the difference between categories 1 and 2 $(2 - 1)$ is the same as the difference between categories 2 and 3 $(3 - 2)$,..., and so on. If these categories were on a cardinal scale, the variable would have this property.   ▪

Because ordinal variables cannot be given a numerical scale that makes sense, computing means and standard deviations for such data is not meaningful. Therefore, none of the methods of estimation and hypothesis testing that were discussed in Chapters 6 through 8 can be used. However, we are still interested in making comparisons between groups for variables such as visual acuity and outcome of treatment, and nonparametric methods can be used for this purpose.

Another type of data scale, which has even less structure than an ordinal scale concerning relationships between data values, is a nominal scale.

DEFINITION 9.4

▪▪

Data are on a **nominal scale** if different data values can be classified into categories but the categories have no specific ordering.   ▪

EXAMPLE 9.6   **Renal Disease** In classifying cause of death among deceased patients with documented analgesic abuse, the following categories were used: (a) cardiovascular disease, (b) cancer, (c) renal or urogenital disease, (d) all other causes of death. Cause of death is a good example of a nominal scale, since the values (i.e., the categories of cause of death) have no specific ordering with respect to each other.   ▪

In this chapter the most commonly used nonparametric statistical tests, including the sign test, the Wilcoxon signed rank test, and the Wilcoxon rank sum test, are developed, assuming that the data are on either a cardinal or an ordinal scale. If they are on a cardinal scale, then the methods will be most useful if there is reason to question the normality of the underlying distribution of the data. For nominal (or categorical) data, discrete data methods, described in Chapter 10, are used.

SECTION 9.2   **The Sign Test**

9.2.1   **Normal Theory Method**

EXAMPLE 9.7   **Dermatology** Suppose we wish to compare the effectiveness of two ointments (A, B) in reducing excessive redness in people who cannot otherwise be exposed to sunlight. Ointment A is randomly applied either to the left or right arm and ointment B is applied to the corresponding area on the other arm. The person is then exposed to 1 hour of sunlight and the two arms are compared for degrees of redness. Suppose that only the following qualitative assessments can be made:

(1) The A arm is not as red as the B arm.

(2) The B arm is not as red as the A arm.

(3) The arms are equally red.

Of 45 people tested with the condition, 22 are better off on the A arm, 18 are better off on the B arm, and 5 are equally well off on both arms. How can we decide if this evidence is sufficient to conclude that ointment A is better than ointment B?   ∎

Suppose that the degree of redness could be measured on a quantitative scale. Let x_i = degree of redness on the A arm, y_i = degree of redness on the B arm for the ith person. We will focus on $d_i = x_i - y_i$ = difference in redness between the A and B arms and will test the hypothesis H_0: $\Delta = 0$ versus H_1: $\Delta \neq 0$, where Δ = the population median of the d_i or the 50th percentile of the underlying distribution of the d_i.

(1) If $\Delta = 0$, then the ointments are equally effective.

(2) If $\Delta < 0$, then ointment A is better, since arm A is less red than arm B.

(3) If $\Delta > 0$, then ointment B is better, since arm A is more red than arm B.

Notice that the actual d_i cannot be observed; we can only observe if $d_i > 0$, $d_i < 0$, or $d_i = 0$. The people for whom $d_i = 0$ will be excluded, since we cannot tell which ointment is better for them. The test will be based on the number of people C for whom $d_i > 0$ out of the total number of people n with nonzero d_i. This test makes sense, since if C is large, treatment B is preferred by most people over treatment A, whereas if C is small, treatment A is preferred over treatment B. We would expect under H_0 that $Pr(\text{nonzero } d_i > 0) = \frac{1}{2}$. We will assume that the normal approximation to the binomial is valid. This assumption will be true if

$$npq \geq 5 \quad \text{or} \quad n(\tfrac{1}{2})(\tfrac{1}{2}) \geq 5$$

or $$\frac{n}{4} \geq 5 \quad \text{or} \quad n \geq 20$$

where n = the number of nonzero d_i's.

The following test procedure for a two-sided level α test, referred to as the **sign test**, can then be used:

9.1 **The Sign Test**

To test the hypothesis H_0: $\Delta = 0$ versus H_1: $\Delta \neq 0$ where the number of nonzero d_i's $= n \geqslant 20$, if

$$C > \frac{n}{2} + \frac{1}{2} + z_{1-\alpha/2}\sqrt{n/4} \qquad \text{or} \qquad C < \frac{n}{2} - \frac{1}{2} - z_{1-\alpha/2}\sqrt{n/4}$$

then H_0 is rejected. Otherwise, H_0 is accepted.

The acceptance and rejection regions for this test are depicted in Figure 9.1.

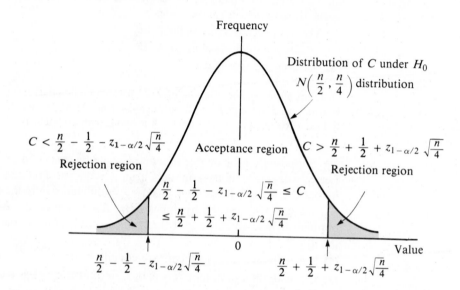

FIGURE 9.1
Acceptance and rejection regions for the sign test

Similarly, the exact p-value of the procedure is given by the following computation:

9.2 **Computation of the p-value for the Sign Test (Normal Theory Method)**

$$p = 2 \times \left[1 - \Phi\left(\frac{C - \frac{n}{2} - .5}{\sqrt{n/4}} \right) \right] \qquad \text{if } C \geqslant \frac{n}{2}$$

$$p = 2 \times \left[\Phi\left(\frac{C - \frac{n}{2} + .5}{\sqrt{n/4}} \right) \right] \qquad \text{if } C < \frac{n}{2}$$

This computation is illustrated in Figure 9.2.

This test is called the sign test because it depends only on the sign of the differences and not on their actual magnitude.

The sign test is actually a special case of the one-sample binomial test in Section 7.10, where the hypothesis H_0: $p = 1/2$ versus H_1: $p \neq 1/2$ was tested. In **(9.1)** and **(9.2)** a large sample test is being used and we are assuming that the normal approximation to the binomial distribution is valid. Under H_0, $p = 1/2$ and

If $C < n/2$, then $p = 2 \times$ area to the left of $\left(C - \dfrac{n}{2} + \dfrac{1}{2}\right)\bigg/\sqrt{\dfrac{n}{4}}$ under an $N(0, 1)$ distribution

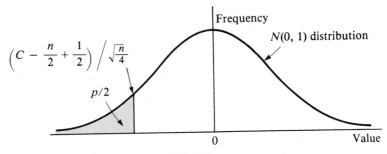

If $C \geqslant n/2$, then $p = 2 \times$ area to the right of $\left(C - \dfrac{n}{2} - \dfrac{1}{2}\right)\bigg/\sqrt{\dfrac{n}{4}}$ under an $N(0, 1)$ distribution

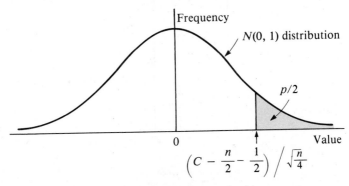

FIGURE 9.2
Computation of the p-value for the sign test

$E(C) = np = n/2$, $Var(C) = npq = n/4$, and $C \sim N(n/2, n/4)$. Furthermore, the .5 term in the computation of the critical region and p-value serves as a continuity correction and makes for a better approximation of the binomial distribution by the normal distribution.

EXAMPLE 9.8   **Dermatology** Assess the statistical significance of the skin-ointment data in Example 9.7.

SOLUTION   In this case there are 40 untied pairs, and $C = 22 \geqslant n/2 = 20$. From **(9.1)**, the critical values are given by

$$c_1 = n/2 + 1/2 + z_{1-\alpha/2}\sqrt{n/4}$$

$$= 40/2 + 1/2 + z_{.975}\sqrt{40/4} = 20.5 + 1.96(3.162) = 26.7$$

and $$c_2 = n/2 - 1/2 - z_{1-\alpha/2}\sqrt{n/4} = 19.5 - 1.96(3.162) = 13.3$$

Since $13.3 \leqslant C = 22 \leqslant 26.7$, H_0 is accepted using a two-sided test with $\alpha = .05$ and we conclude that the ointments are not significantly different in effectiveness. From **(9.2)**, since $C = 22 \geqslant n/2 = 20$, the exact p-value is given by

$$p = 2 \times \{1 - \Phi[(22 - 20 - \tfrac{1}{2})/\sqrt{40/4}]\} = 2 \times [1 - \Phi(0.47)] = 2 \times (1 - .6808) = .638$$

which is not statistically significant. Therefore, H_0, that the ointments are equally effective, is accepted.   ∎

9.2.2 **Exact Method**

If $n < 20$, then exact binomial probabilities rather than the normal approximation must be used to compute the p-value. H_0 should still be rejected if C is very large or very small. The expressions for the p-value based on exact binomial probabilities are given as follows:

9.3 | **Computation of the p-value for the Sign Test (Exact Test)**

If $C \geqslant n/2$,

$$p = 2 \times \sum_{k=C}^{n} \binom{n}{k}\left(\frac{1}{2}\right)^{n}$$

If $C < n/2$,

$$p = 2 \times \sum_{k=0}^{C} \binom{n}{k}\left(\frac{1}{2}\right)^{n}$$

This computation is depicted in Figure 9.3.

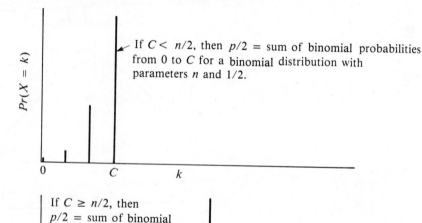

If $C < n/2$, then $p/2 =$ sum of binomial probabilities from 0 to C for a binomial distribution with parameters n and $1/2$.

If $C \geq n/2$, then $p/2 =$ sum of binomial probabilities from C to n for a binomial distribution with parameters n and $1/2$.

FIGURE 9.3
Computation of the p-value for the sign test (exact test)

This test is a special case of the small-sample, one-sample binomial test described in (7.36), where the hypothesis $H_0: p = \frac{1}{2}$ versus $H_1: p \neq \frac{1}{2}$ was tested.

EXAMPLE 9.9 | **Ophthalmology** Suppose we want to compare two different types of eye drops (A, B) that are intended to prevent redness in people with hay fever. Drug A is randomly given to one eye and drug B to the other eye. The redness is noted at baseline and after 10 minutes by an observer who is not aware of which drug was administered to which eye. We find that for 15 people with an equal amount of redness in each eye at baseline, after 10 minutes the drug-A eye is less red than the drug-B eye for 8 people; the drug-B eye is less red than the drug-A eye for 2 people; and the eyes are equally red for 5 people. Assess the statistical significance of the results.

SOLUTION The test is based on the 10 people who had a differential response to the two types of eye drops. Since $n = 10 < 20$, the normal theory method in **(9.2)** cannot be used; the exact method in **(9.3)** must be used instead. Since $C = 8 > \frac{10}{2} = 5$,

$$p = 2 \times \sum_{k=8}^{10} \binom{10}{k} \left(\frac{1}{2}\right)^{10}$$

Refer to the binomial tables (Table 1 in the Appendix) using $n = 10$, $p = .5$ and note that $Pr(X = 8) = .0439$, $Pr(X = 9) = .0098$, $Pr(X = 10) = .0010$. Thus, $p = 2 \times Pr(X \geqslant 8) = 2(.0439 + .0098 + .0010) = 2 \times .0547 = .109$, which is not statistically significant. Thus, H_0, that the two types of eye drops are equally effective in reducing redness in people with hay fever, is accepted. ∎

SECTION 9.3 **The Wilcoxon Signed Rank Test**

EXAMPLE 9.10 **Dermatology** Consider the data in Example 9.7 from a different perspective. We assumed that the only possible assessment was that the degree of sunburn with ointment A was either better or worse than that with ointment B. Suppose instead that the degree of burn can be quantified on a 10-point scale, with 10 being the worst burn and 1 being no burn at all. We can now compute $d_i = x_i - y_i$, where x_i = degree of burn for ointment A and y_i = degree of burn for ointment B. If d_i is positive, then ointment B is doing better than ointment A; if d_i is negative, then ointment A is doing better than ointment B. For example, if $d_i = +5$, then the degree of redness is 5 units greater with the ointment-A arm than with the ointment-B arm, whereas if $d_i = -3$, then the degree of redness is 3 units less with the ointment-A arm than with the ointment-B arm. How can this additional information be used to test if the ointments are equally effective? ∎

Suppose the sample data in Table 9.1 are obtained. The f_i represent the frequency or the number of people with difference in redness d_i between the ointment-A and ointment-B arms.

TABLE 9.1
Difference in degree of redness between ointment-A and ointment-B arms after 10 minutes of exposure to sunlight

$|d_i|$	Negative d_i	f_i	Positive d_i	f_i	Number of people with same absolute value	Range of ranks	Average rank
10	-10	0	10	0	0	—	—
9	-9	0	9	0	0	—	—
8	-8	1	8	0	1	40	40.0
7	-7	3	7	0	3	37–39	38.0
6	-6	2	6	0	2	35–36	35.5
5	-5	2	5	0	2	33–34	33.5
4	-4	1	4	0	1	32	32.0
3	-3	5	3	2	7	25–31	28.0
2	-2	4	2	6	10	15–24	19.5
1	-1	4	1	10	14	1–14	7.5
		22		18			
0	0	5					

Notice that there is only a slight excess of people with negative d_i, that is, who are better off with ointment A (22), than with positive d_i, that is, who are better off with ointment B (18). However, the extent to which the 22 people are better off appears to be far greater than that of the 18 people, since the negative d_i generally have a much greater absolute value than the positive d_i. This condition is illustrated in Figure 9.4.

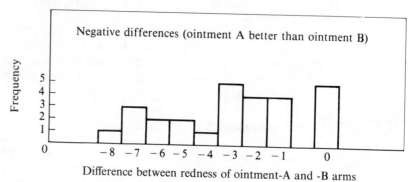

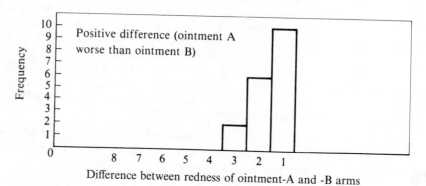

FIGURE 9.4
Bar graph of the differences in redness between the ointment-A and ointment-B arms for the data in Example 9.10

We wish to test the hypothesis $H_0: \Delta = 0$ versus $H_1: \Delta \neq 0$, where Δ = the median score difference between the ointment-A and ointment-B arms. If $\Delta < 0$, then ointment A is better; if $\Delta > 0$, then ointment B is better. We will assume that the d_i have an underlying continuous distribution.

Based on Figure 9.4, a seemingly reasonable test of this hypothesis would be to take into account both the magnitude and the sign of the differences d_i. A paired t test might be used here, but the problem is that the rating scale is ordinal. The measurement $d_i = -5$ does not mean that the difference in degree of burn is 5 times as great as $d_i = -1$, but simply that there is a relative ranking of differences in degree of burn, with -8 being most favorable to ointment A, -7 the next most favorable, ..., and so on. Thus a nonparametric test that is analogous to the paired t test is needed here. Such a test is the **Wilcoxon signed rank test**. It is nonparametric because it is based on the ranks of the observations rather than on their actual values, as is the paired t test.

The first step in performing this test is to compute ranks for each of the observations, as follows:

| **9.4** | **Ranking Procedure for the Wilcoxon Signed Rank Test** |

(1) Arrange the differences d_i in order of *absolute value* as has been done in Table 9.1.

(2) Count the number of differences with the same absolute value.

(3) Ignore the observations where $d_i = 0$ and rank the remaining observations from 1, for the observation with the lowest absolute value, up to n, for the observation with the highest absolute value.

(4) If there is a group of several observations with the same absolute value, then find the lowest rank in the range $= 1 + R$ and the highest rank in the range $= G + R$, where R = the highest rank used prior to considering this group and G = the number of differences in the *range of ranks* for the group. Assign the *average rank* = (lowest rank in the range + highest rank in the range)/2 as the rank for each difference in the group.

EXAMPLE 9.11   **Dermatology** Compute the ranks for the skin-ointment data in Table 9.1.

SOLUTION   First collect the differences with the same absolute value. Fourteen people have absolute value 1; this group has a rank range from 1–14 and an average rank of $(1 + 14)/2 = 7.5$. The group of 10 people with absolute value 2 has a rank range from $(1 + 14)$ to $(10 + 14) = 15 - 24$ and an average rank $= (15 + 24)/2 = 19.5, \ldots$, and so on.   ∎

The test is based on the sum of the ranks, or **rank sum** (R_1), for the group of people with positive d_i, that is, the rank sum for people for whom ointment A does worse than ointment B. A large rank sum indicates that differences in degree of burn in favor of treatment B tend to be larger than those for treatment A, whereas a small rank sum indicates that differences in degree of burn in favor of treatment A tend to be larger than those for treatment B. If the null hypothesis is true, then the expected value and variance of the rank sum are given by

$$E(R_1) = n(n + 1)/4, \ Var(R_1) = n(n + 1)(2n + 1)/24$$

where n is the number of nonzero differences.

If the number of nonzero d_i's is ≥ 16, then a normal approximation can be used for the test procedure. This test procedure, the **Wilcoxon signed rank test**, is given as follows:

| **9.5** | **Wilcoxon Signed Rank Test (Normal Approximation Method Two-Sided Level α Test)** |

(1) Rank the differences as shown in **(9.4)**.

(2) Compute the rank sum R_1 of the positive differences.

(3) (a) Compute

$$T = \left[\left| R_1 - \frac{n(n + 1)}{4} \right| - \frac{1}{2} \right] \Big/ \sqrt{n(n + 1)(2n + 1)/24}$$

if there *are no ties* (i.e., no groups of differences with the same absolute value).

(b) Compute

$$T = \left[\left| R_1 - \frac{n(n + 1)}{4} \right| - \frac{1}{2} \right] \Big/ \sqrt{n(n + 1)(2n + 1)/24 - \sum_{i=1}^{g} (t_i^3 - t_i)/2}$$

if there *are ties*, where t_i refers to the number of differences with the same absolute value in the ith tied group and g is the number of tied groups.

(4) If

$$T > z_{1-\alpha/2}$$

then reject H_0. Otherwise, accept H_0.

(5) The p-value for the test is given by

$$p = 2 \times [1 - \Phi(T)]$$

(6) This test should be used only if the number of nonzero differences ≥ 16 and if the difference scores have an underlying continuous symmetric distribution. The computation of the p-value is illustrated in Figure 9.5.

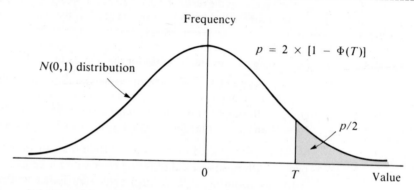

FIGURE 9.5
Computation of the p-value for the Wilcoxon signed rank test

The term $\frac{1}{2}$ in the computation of T serves as a continuity correction in the same manner as for the sign test in **(9.1)** and **(9.2)**. If there are no ties,

$$T = \left[\left| R_1 - \frac{n(n+1)}{4} \right| - \frac{1}{2} \right] \bigg/ \sqrt{n(n+1)(2n+1)/24}$$

If there are ties,

$$T = \left[\left| R_1 - \frac{n(n+1)}{4} \right| - \frac{1}{2} \right] \bigg/ \sqrt{n(n+1)(2n+1)/24 - \sum_{i=1}^{g} (t_i^3 - t_i)/2}$$

EXAMPLE 9.12 **Dermatology** Perform the Wilcoxon signed rank test for the data in Example 9.10.

SOLUTION Since the number of nonzero differences $(22 + 18 = 40) \geq 16$, the normal approximation method in **(9.5)** can be used. Compute the rank sum for the people with positive d_i, that is, where ointment B performs better than ointment A, as follows:

$$R_1 = 10(7.5) + 6(19.5) + 2(28.0) = 75 + 117 + 56 = 248$$

The expected rank sum is given by

$$E(R_1) = 40(41)/4 = 410$$

whereas the variance of the rank sum corrected for ties is given by

$$Var(R_1) = 40(41)(81)/24 - [(14^3 - 14) + (10^3 - 10) + (7^3 - 7) + (1^3 - 1)$$
$$+ (2^3 - 2) + (2^3 - 2) + (3^3 - 3) + (1^3 - 1)]/2$$
$$= 5535 - (2730 + 990 + 336 + 0 + 6 + 6 + 24 + 0)/2$$
$$= 5535 - 4092/2 = 3489$$

Thus, $sd(R_1) = \sqrt{3489} = 59.07$. Therefore, the test statistic T is given by

$$T = (|248 - 410| - \tfrac{1}{2})/59.07 = 161.5/59.07 = 2.73$$

The p-value of the test is given by

$$= 2[1 - \Phi(2.73)] = 2 \times (1 - .9968) = .006$$

We therefore can conclude that there is a significant difference between ointments, with ointment A doing better than ointment B, since the observed rank sum (248) is smaller than the expected rank sum (410). This conclusion is different from the conclusion based on the sign test in Example 9.8, where no significant difference between ointments was found. This result indicates that when the information is available, it is worthwhile to consider both the magnitude and the direction of the difference between treatments, as is done by the signed rank test, rather than just the direction of the difference, as is done by the sign test. ∎

In general, for a two-sided test, if the signed rank test is based on negative differences rather than positive differences, the same test statistic and p-value will always result. Thus, the rank sum can be arbitrarily computed based on either positive or negative differences.

EXAMPLE 9.13 **Dermatology** Perform the Wilcoxon signed rank test for the data in Example 9.10 based on negative rather than positive difference scores.

SOLUTION $R_2 = $ rank sum for negative differences

$$= 4(7.5) + 4(19.5) + 5(28.0) + 1(32.0) + 2(33.5) + 2(35.5) + 3(38.0) + 1(40.0)$$

$$= 572$$

Thus,

$$\left| R_2 - \frac{n(n+1)}{4} \right| - .5 = |572 - 410| - .5 = 161.5 = \left| R_1 - \frac{n(n+1)}{4} \right| - .5$$

Since $Var(R_1) = Var(R_2)$, the same test statistic $T = 2.73$ and p-value $= .006$ are obtained as when positive difference scores are used. ∎

The Wilcoxon signed rank test can also be used to test for one-sided alternatives:

9.6 | **Wilcoxon Signed Rank Test (Normal Approximation Method—One-Sided Level α Test**

Use the following procedure to test the hypothesis $H_0: \Delta = 0$ versus $H_1: \Delta < 0$, where $\Delta = $ the median score difference:

(1) Rank the differences as shown in **(9.4)**.

(2) Compute the rank sum R_2 of the negative differences.

(3) (a) Compute

$$T = \left[R_2 - \frac{n(n+1)}{2} - \frac{1}{2} \right] \Big/ \sqrt{\frac{n(n+1)(2n+1)}{24}}$$

if there *are no ties.*

(b) Compute

$$T = \left[R_2 - \frac{n(n+1)}{2} - \frac{1}{2} \right] \Big/ \sqrt{\frac{n(n+1)(2n+1)}{24} - \sum_{i=1}^{g} \frac{t_i^3 - t_i}{2}}$$

if there *are ties,* where t_i refers to the number of differences with the same absolute value in the ith tied group, and g is the number of tied groups.

(4) If $T > z_{1-\alpha}$, then reject H_0; otherwise, accept H_0.

(5) The p-value for the test is given by $p = 1 - \Phi(T)$.

(6) The test should only be used if the number of nonzero differences is ≥ 16, and if the difference scores have an underlying continuous symmetric distribution.

To test the hypothesis $H_0: \Delta = 0$ versus $H_1: \Delta > 0$, then follow the same test procedure except use R_1 = rank sum of positive differences instead of R_2 = rank sum of negative differences in computing the test statistic.

If the number of pairs with nonzero $d_i \leq 15$, then the normal approximation is no longer valid, and special small-sample tables giving significance levels for this test must be used. Such a table is Table 9 in Appendix 1, which gives upper and lower critical values for R_1 for a two-sided test with α levels of .10, .05, .02, and .01, respectively. In general, the results are statistically significant at a particular α level only if either $R_1 \leq$ the lower critical value or $R_1 \geq$ the upper critical value for that α level.

EXAMPLE 9.14   Suppose there are 9 untied pairs and a rank sum of 43. Evaluate the statistical significance of the results.

SOLUTION   Since $R_1 = 43 \geq 42$, it follows that $p < .02$. Since $R_1 = 43 < 44$, it follows that $p \geq .01$. Thus .01 $\leq p < .02$, and the results are statistically significant.   ∎

An example of the signed rank test with ordinal data has been presented. This test and the other nonparametric tests can be applied to cardinal data as well, particularly if the assumption of normality appears to be grossly violated. If the actual distribution turns out to be normal, then the signed rank test has less power than the paired t test, which is the penalty that is paid.

SECTION 9.4   **The Wilcoxon Rank Sum Test**

In the previous section a nonparametric analogue to the paired t test, namely, the Wilcoxon signed rank test, was presented. In this section a nonparametric analogue to the t test for two independent samples is described.

EXAMPLE 9.15   **Ophthalmology** Different genetic types of the disease retinitis pigmentosa are thought to have different rates of progression, with the dominant form of the disease progressing the slowest, the recessive form of the disease the next slowest, and the sex-linked form of the disease the quickest. This hypothesis can be tested by comparing the visual acuity of people aged 10–19 who have different genetic types. Suppose there are 25 people with dominant disease and 30 people with sex-linked disease. The best corrected visual acuities (i.e., with appropriate glasses) in the better eye of these people are presented in Table 9.2. How can these data be used to see if the median visual acuity is different in the two groups?   ∎

We wish to test the hypothesis $H_0: \text{median}_D = \text{median}_{SL}$ versus $H_1: \text{median}_D \neq \text{median}_{SL}$, where median_D and median_{SL} are the median visual acuities in the dominant and sex-linked groups, respectively. The two-sample t test for independent samples, discussed in Sections 8.4 and 8.7, would ordinarily be used for this type of problem. However, visual acuity cannot be given a specific numerical value that all ophthalmologists would agree on. Thus, the t test is

Visual acuity	Dominant	Sex-linked	Combined sample	Range of ranks	Average rank
20-20	5	1	6	1–6	3.5
20-25	9	5	14	7–20	13.5
20-30	6	4	10	21–30	25.5
20-40	3	4	7	31–37	34.0
20-50	2	8	10	38–47	42.5
20-60	0	5	5	48–52	50.0
20-70	0	2	2	53–54	53.5
20-80	0	1	1	55	55.0
	$\overline{25}$	$\overline{30}$			

TABLE 9.2
Comparison of visual acuity in people aged 10–19 with dominant and sex-linked retinitis pigmentosa

inapplicable, and a nonparametric analogue must be used. The nonparametric analogue to the independent samples t test is the **Wilcoxon rank sum test**. This test is nonparametric because it is based on the *ranks* of the individual observations rather than on their actual values, which would be used in the t test. The ranking procedure for this test is as follows:

9.7 **Ranking Procedure for the Wilcoxon Rank Sum Test**

(1) Combine the data from the two groups and order the values from the lowest to highest, or in the case of visual acuity, from best visual acuity (20-20) to worst visual acuity (20-80).

(2) Assign ranks to the individual values, with the best visual acuity (20-20) having the lowest rank and the worst visual acuity (20-80) having the highest rank, or vice versa.

(3) If a group of observations has the same value, then compute the *range of ranks* for the group, as was done for the signed rank test in **(9.4)**, and assign the *average rank* for each observation in the group.

EXAMPLE 9.16 Compute the ranks for the visual-acuity data in Table 9.2.

SOLUTION First collect all people with the same visual acuity over the two groups, as shown in Table 9.2. There are 6 people with visual acuity 20-20 who have a rank range of 1-6 and are assigned an average rank of $(1 + 6)/2 = 3.5$. There are 14 people over the two groups with visual acuity 20-25. The rank range for this group is from $(1 + 6)$ to $(14 + 6) = 7 - 20$. Thus, all people in this group are assigned the average rank $= (7 + 20)/2 = 13.5$, and similarly for the other groups. ∎

The test statistic used for this test is the sum of the ranks in the first sample (R_1). If this sum is large, then the dominant group has poorer visual acuity than the sex-linked group, whereas if it is small, the dominant group has better visual acuity. If the number of observations in the two groups are n_1 and n_2, respectively, then the average rank in the combined sample is $(1 + n_1 + n_2)/2$. Thus, under H_0, the expected rank sum in the first group $\equiv E(R_1) = n_1 \times$ average rank in the combined sample $= n_1(n_1 + n_2 + 1)/2$. It can be shown that the variance of R_1 under H_0 is given by $Var(R_1) = n_1 n_2(n_1 + n_2 + 1)/12$. Furthermore, we will assume that the smaller of the two groups is at least 10 and that the variable under study has an underlying continuous distribution. Under these assumptions, the distribution of the rank sum R_1 is approximately normal. Thus, the following test procedure is used:

9.8   **Wilcoxon Rank Sum Test (Normal Approximation Method for Two-Sided Level α Test)**

(1) Rank the observations as shown in **(9.7)**.

(2) Compute the rank sum R_1 in the first sample (the choice of sample is arbitrary).

(3) (a) Compute

$$T = \left[\left|R_1 - \frac{n_1(n_1 + n_2 + 1)}{2}\right| - \frac{1}{2}\right] \Bigg/ \sqrt{\left(\frac{n_1 n_2}{12}\right)(n_1 + n_2 + 1)}$$

if there *are no ties*.

(b) Compute

$$T = \left[\left|R_1 - \frac{n_1(n_1 + n_2 + 1)}{2}\right| - \frac{1}{2}\right] \Bigg/ \sqrt{\left(\frac{n_1 n_2}{12}\right)\left(n_1 + n_2 + 1 - \frac{\sum_{i=1}^{g} t_i(t_i^2 - 1)}{(n_1 + n_2)(n_1 + n_2 - 1)}\right)}$$

if there *are ties*, where t_i refers to the number of observations with the same value in the ith tied group, and g is the number of tied groups.

(4) If

$$T > z_{1-\alpha/2}$$

then reject H_0. Otherwise, accept H_0.

(5) Compute the exact p-value by

$$p = 2 \times [1 - \Phi(T)]$$

(6) This test should be used only if both n_1 and n_2 are at least 10, and if there is an underlying continuous distribution.

The computation of the p-value is illustrated in Figure 9.6.

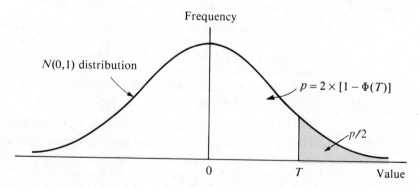

FIGURE 9.6
Computation of the p-value for the Wilcoxon rank sum test

If there are no ties,

$$T = \left[\left|R_1 - \frac{n_1(n_1 + n_2 + 1)}{2}\right| - \frac{1}{2}\right] \Bigg/ \sqrt{\frac{n_1 n_2(n_1 + n_2 + 1)}{12}}$$

If there are ties,

$$T = \left[\left| R_1 - \frac{n_1(n_1 + n_2 + 1)}{2} \right| - \frac{1}{2} \right] \Bigg/ \sqrt{ \left(\frac{n_1 n_2}{12} \right) \left[n_1 + n_2 + 1 - \frac{\sum_{i=1}^{g} t_i(t_i^2 - 1)}{(n_1 + n_2)(n_1 + n_2 - 1)} \right] }$$

EXAMPLE 9.17 Perform the Wilcoxon rank sum test for the data in Example 9.15.

SOLUTION Because the minimum sample size in the two samples is $25 \geq 10$, the normal approximation can be used. The rank sum in the dominant group is given by

$$R_1 = 5(3.5) + 9(13.5) + 6(25.5) + 3(34) + 2(42.5)$$

$$= 17.5 + 121.5 + 153 + 102 + 85 = 479$$

Furthermore,
$$E(R_1) = \frac{25(56)}{2} = 700$$

and $Var(R_1)$ corrected for ties is given by

$$[(25)(30)/12]\{56 - [6(6^2 - 1) + 14(14^2 - 1) + 10(10^2 - 1) + 7(7^2 - 1)$$
$$+ 10(10^2 - 1) + 5(5^2 - 1) + 2(2^2 - 1) + 1(1^2 - 1)]/[55(54)]\}$$
$$= 62.5(56 - 5382/2970) = 3386.74$$

Thus, the test statistic T is given by

$$T = \frac{(|479 - 700| - .5)}{\sqrt{3386.74}} = \frac{220.5}{58.2} = 3.79$$

which follows an $N(0, 1)$ distribution under H_0. The p-value of the test is

$$2 \times [1 - \Phi(3.79)] < .001$$

Since the observed rank sum in the dominant group (479) is lower than the expected rank sum (700), we conclude that the visual acuities of the two groups are significantly different; the dominant group has better visual acuity than the sex-linked group. ∎

The Wilcoxon rank sum test can also be used to test for one-sided alternatives:

9.9 **Wilcoxon Rank Sum Test (Normal Approximation Method—One-Sided Level α Test**

Use the following procedure to test the hypothesis H_0: median$_{\text{group 1}}$ = median$_{\text{group 2}}$ versus H_1: median$_{\text{group 1}}$ < median$_{\text{group 2}}$:

(1) Rank the observations as shown in **(9.7)**, with the lowest ranks corresponding to the lowest values.

(2) Compute the rank sum R_1 in the first group.

(3) (a) Compute

$$T = \left[R_1 - \frac{n_1(n_1 + n_2 + 1)}{2} - \frac{1}{2} \right] \Bigg/ \sqrt{(n_1 n_2/12)(n_1 + n_2 + 1)}$$

if there *are no ties.*

(b) Compute

$$T = \left[R_1 - \frac{n_1(n_1 + n_2 + 1)}{2} - \frac{1}{2} \right] \Bigg/ \sqrt{ \left(\frac{n_1 n_2}{12} \right) \left[n_1 + n_2 + 1 - \frac{\sum_{i=1}^{g} t_i^2(t_i - 1)}{(n_1 + n_2)(n_1 + n_2 - 1)} \right] }$$

if there *are ties,* where t_i refers to the number of observations with the same value in the ith tied group, and g is the number of tied groups.

(4) If $T < -z_{1-\alpha}$, then reject H_0; otherwise, accept H_0.

(5) The p-value for the test is given by $p = \Phi(T)$.

(6) The test should be used only if both n_1 and n_2 are at least 10 and there is an underlying continuous distribution.

To test the hypothesis H_0: median$_{\text{group 1}}$ = median$_{\text{group 2}}$ versus H_1: median$_{\text{group 1}}$ > median$_{\text{group 2}}$, compute the same test statistic as in steps 3(a) and 3(b) but use the following procedure instead of steps 4–6:

(4) If $T > z_{1-\alpha}$, then reject H_0; otherwise, accept H_0.

(5) The p-value for the test is given by $p = 1 - \Phi(T)$.

(6) The test should also be used only if both n_1 and n_2 are at least 10 and there is an underlying continuous distribution.

If either sample size is less than 10, the normal approximation is not valid, and a small-sample table of exact significance levels must be used. Table 10 in Appendix 1 gives upper and lower critical values for the rank sum in the first of two samples (T) for a two-sided test with α levels of .10, .05, .02, and .01, respectively. In general, the results are statistically significant at a particular α level only if either $T \leqslant T_l$ = the lower critical value or $T \geqslant T_r$ = the upper critical value for that α level.

EXAMPLE 9.18   Suppose there are two samples of size 8 and 15, with a rank sum of 73 in the sample size of 8. Evaluate the statistical significance of the results.

SOLUTION   Refer to $n_1 = 8$, $n_2 = 15$, $\alpha = .05$ and find that $T_l = 65$, $T_r = 127$. Since $T = 73 > 65$, $T < 127$, the results are not statistically significant using a two-sided test at the 5% level. ∎

The Wilcoxon rank sum test is sometimes referred to in the literature as the **Mann–Whitney U test.** The test statistic for the Mann–Whitney U test is based on the number of pairs of observations (x_i, y_j), one from each sample, such that $x_i < y_j$. The Mann–Whitney U test and the Wilcoxon rank sum test are completely equivalent, since the same p-value is obtained by applying either test. Therefore, the choice of which test to use is a matter of convenience.

Because ranking all the observations in a large sample is tedious, a computer program is useful in performing the Wilcoxon rank sum test. The SPSS^{X}/PC Wilcoxon rank sum test procedure was used on the data set in Table 9.2; the results are given in Table 9.3.

The average ranks in the dominant and sex-linked groups are listed first. Recall from Table 9.2 that a low rank corresponds to better visual acuity. Thus, the dominant group, with a lower average rank, has better visual acuity than the sex-linked group. Both the Mann–Whitney U statistic (labeled U) = 154 and the Wilcoxon rank sum test statistic (labeled W) = 479 are given in the output. The test statistic (labeled Z) = -3.7975 and the two-tailed p-value = .0001 < .001 are also given. The test statistic differs slightly from that computed in Example 9.17, since the program does not correct for continuity; that is, $|Z| = |479 - 700|/58.2 = 3.80$, whereas $T = (|479 - 700| - .5)/58.2 = 3.79$. The difference is minor in this case.

TABLE 9.3
SPSS^X/PC Wilcoxon rank
sum test program used
on the data in Table 9.2

TABLE 9.3
SPSS^X/PC Wilcoxon rank
sum test program used
on the data in Table 9.2

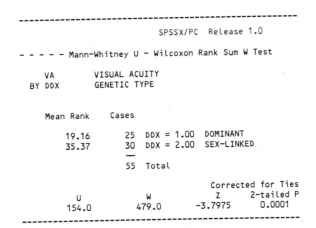

```
-----------------------------------------------------------

                          SPSSX/PC  Release 1.0

- - - - - Mann-Whitney U - Wilcoxon Rank Sum W Test

     VA          VISUAL ACUITY
   BY DDX        GENETIC TYPE


   Mean Rank      Cases

     19.16          25   DDX = 1.00   DOMINANT
     35.37          30   DDX = 2.00   SEX-LINKED
                    --
                    55   Total

                                  Corrected for Ties
                                    Z      2-tailed P
      U              W           -3.7975    0.0001
    154.0          479.0
-----------------------------------------------------------
```

Finally, a necessary condition for the strict validity of the rank sum test is that the underlying distributions being compared must be continuous. However, McNeil has investigated the use of this test in comparing discrete distributions and has found only small losses in power when applying this test to grouped data from normal distributions, as compared with the actual ungrouped observations from such distributions [1]. He concludes that the rank sum test is approximately valid in this case, with the appropriate provision for ties as given in (9.8).

The tests covered in this chapter, the log rank test in Chapter 10, the material on rank correlation in Chapter 11, and the Kruskal–Wallis test in Chapter 12 are among the most basic of nonparametric tests. Hollander and Wolfe provide a more comprehensive treatment of nonparametric statistics [2].

Summary

In this chapter some of the most widely used nonparametric statistical tests corresponding to the parametric procedures in Chapter 8 were presented. The key advantage of nonparametric methods is that the assumptions of normality made in previous chapters can be relaxed when such assumptions are unreasonable. One drawback of nonparametric procedures is that some power is lost relative to using a parametric procedure (such as a t test), if the data truly follow a normal distribution or if the central limit theorem is applicable. Also, the data typically have to be expressed in terms of ranks, a scale that some researchers find difficult to understand compared with maintaining the raw data in the original scale.

The specific procedures covered include the sign test, the Wilcoxon signed rank test, and the Wilcoxon rank sum test. Both the sign test and the signed rank test are nonparametric analogues to the paired t test. For the sign test it is necessary to determine only whether one member of a matched pair has a higher or lower score than the other member of the pair. For the signed rank test the magnitude of the difference score (expressed in the form of a rank), as well as its direction, is used in performing the significance test. Furthermore, the Wilcoxon rank sum test (also known as the Mann–Whitney U test) is an analogue to the two-sample t test for independent samples, in which the actual values are replaced by rank scores.

Other nonparametric procedures are introduced for life-table methods in Chapters 10 and 11, as analogues to correlation methods in Chapter 11, and the analysis of variance in Chapter 12.

PROBLEMS

Ophthalmology

Suppose an ophthalmologist reviews fundus photographs of 30 patients with macular degeneration both before and 3 months after receiving a laser treatment. To assess the efficacy of treatment, each patient is rated as improved, remained the same, or declined.

9.1 If 20 patients improved, 7 declined, and 3 remained the same, then assess whether or not patients undergoing this treatment are showing significant change from baseline to 3 months afterward. Report a *p*-value.

Suppose that the patients are divided into two groups according to initial visual acuity (VA). Of the 14 patients with VA 20-40 or better, 8 improved, 5 declined, and 1 stayed the same. Of the 16 patients with VA worse than 20-40, 12 improved, 2 declined, and 2 stayed the same.

9.2 Assess the results in the subgroup of patients with VA of 20-40 or better.

9.3 Assess the results in the subgroup of patients with VA worse than 20-40.

Dentistry

In a study, 28 adults with mild periodontal disease are assessed before and 6 months after the implementation of a dental education program intended to promote better oral hygiene. After 6 months, periodontal status improved in 15 patients, declined in 8, and remained the same in 5.

9.4 Assess the impact of the program statistically (use a two-sided test).

Diabetes

An experiment was conducted to study responses to different methods of taking insulin in patients with type I diabetes. The percentages of glycosolated hemoglobin initially and 3 months after taking insulin by nasal spray are given in Table 9.4 [3].

9.5 Perform a *t* test to compare the percentages of glycosolated hemoglobin before and 3 months after treatment.

9.6 Suppose normality is not assumed. Perform a nonparametric test corresponding to the *t* test in Problem 9.5.

TABLE 9.4 Percentages of glycosolated hemoglobin before and 3 months after taking insulin by nasal spray

Patient number	Before	3 months after
1	11.0	10.2
2	7.7	7.9
3	5.9	6.5
4	9.5	10.4
5	8.7	8.8
6	8.6	9.0
7	11.0	9.5
8	6.9	7.6

(Reprinted with permission of the *New England Journal of Medicine, 312*(17), 1078–1084, 1985.)

9.7 Compare your results in Problems 9.5 and 9.6.

Dentistry

Refer to Problem 9.4. Suppose that patients are graded on the degree of change in periodontal status on a 7-point scale, with +3 indicating the greatest improvement, 0 indicating no change, and −3 indicating the greatest decline. The data are given in Table 9.5.

TABLE 9.5 Degree of change in periodontal status

Change score	Number of patients
+3	4
+2	5
+1	6
0	5
−1	4
−2	2
−3	2

9.8 What nonparametric test can be used to determine whether or not a significant change in periodontal status has occurred over time?

9.9 Implement the procedure in Problem 9.8 and report a *p*-value.

9.10 Suppose we want to use the Wilcoxon signed rank test and have a rank sum of 27 based on 7 untied pairs. Evaluate the significance of the results.

9.11 Answer Problem 9.10 for a rank sum of 65 based on 13 untied pairs.

9.12 Answer Problem 9.10 for a rank sum of 90 with 15 untied pairs.

9.13 Suppose there are two samples of sizes 6 and 7, with a rank sum of 58 in the sample of size 6. Using the Wilcoxon rank sum test, evaluate the significance of the results.

9.14 Answer Problem 9.13 for two samples of sizes 7 and 10, with a rank sum of 47 in the sample of size 7.

9.15 Answer Problem 9.13 for two samples of sizes 12 and 15, with a rank sum of 220 in the sample of size 12 (assume that there are no ties).

Refer to Table 2.10 (p. 35).

9.16 Suppose normality is not assumed. What nonparametric test can be used to compare total heart weight of males with left heart disease with that of normal males?

9.17 Implement the test in Problem 9.16 and report a *p*-value.

Obstetrics

9.18 Reanalyze the data in Table 8.16 (p. 285) using nonparametric methods. Assume the samples are unpaired.

9.19 Would such methods be preferable to parametric methods in analyzing the data? Why or why not?

Cardiology

Propranolol is a standard drug given to ease the pain of patients with episodes of unstable angina. A new drug for the treatment of this disease is tested on 30 pairs of patients who are matched on a one-to-one basis according to age, sex, and clinical condition and are assessed as to the severity of their pain. Suppose that in 15 pairs of patients, the patient with the new drug has less pain; in 10 pairs of patients, the patient with propranolol has less pain; and in 5 pairs of patients, the pain is about the same with the two drugs.

9.20 What is the appropriate test to use here?

9.21 Perform the test in Problem 9.20 and report a *p*-value.

Health Services Administration

Suppose we wish to compare the length of stay in the hospital for patients with the same diagnosis at two different hospitals. The following results are found:

First hospital	21, 10, 32, 60, 8, 44, 29, 5, 13, 26, 33
Second hospital	86, 27, 10, 68, 87, 76, 125, 60, 35, 73, 96, 44, 238

9.22 Why might a *t* test not be very useful in this case?

9.23 Carry out a nonparametric procedure for testing the hypothesis that the lengths of stay are comparable in the two hospitals.

Ophthalmology

Table 8.14 presents data giving the median gray levels in the lens of the human eye for 6 cataractous and 6 normal people (p. 284).

9.24 What nonparametric test could be used to compare the median gray levels of normal and cataractous eyes?

9.25 Carry out the test in Problem 9.24 and report a *p*-value.

Infectious Disease

The distribution of white blood count is typically positively skewed, and assumptions of normality are usually not valid.

9.26 To compare the white blood counts of patients on the medical and surgical services in Table 2.11 (p. 36) when normality is not assumed, what test can be used?

9.27 Perform the test in Problem 9.26 and report a *p*-value.

Psychiatry

Suppose we are conducting a study of the effectiveness of lithium therapy for manic-depressive patients. The study is carried out at two different centers, and we want to determine if the patient populations are comparable at baseline. A self-rating questionnaire about their general psychological well-being is administered to the prospective patients at the two different centers. The outcome measure on the questionnaire is a four-category scale: (1) = feel good; (2) = usually feel good, once in a while feel nervous; (3) = feel nervous half the time; (4) = usually feel nervous. Suppose the data at the two different centers are as follows:

Center 1	3, 4, 1, 1, 3, 2, 3, 4, 4, 3, 2, 4, 4, 4
Center 2	1, 2, 1, 3, 2, 4, 1, 2, 1, 3, 1, 2, 2, 1, 3

9.28 What type of data does this type of scale represent?

9.29 Why might a parametric test not be useful with this type of data?

9.30 Assess if there is any significant difference in the responses of the two patient populations using a nonparametric test.

Sports Medicine

Many tennis players develop acute lateral epicondylitis whereby they experience acute elbow pain (tennis elbow). A variety of nonsurgical treatments are used for this condition, including rest, heat, and anti-inflammatory agents. A clinical trial was set up to compare the effects of the anti-inflammatory agent ibuprofen (Motrin) with those of a placebo in the treatment of tennis elbow. Patients were given 400 mg of ibuprofen orally or an identical looking placebo 4 times per day and were evaluated after 3 weeks of treatment. Patients subjectively rated their pain as (1) worse, (2) unchanged, (3) slightly improved, (4) moderately improved, (5) mostly improved, and (6) completely improved. The results are given in Table 9.6.

9.31 If all levels of improvement are regarded as equivalent, then what test can be used to assess whether or not patients on Motrin have improved over the 3-week period?

9.32 Perform the test in Problem 9.31 and report a *p*-value.

9.33 Perform the same test as in Problem 9.32 for patients on placebo and report a *p*-value.

Suppose the levels of improvement are differentiated using the scale in Table 9.6.

9.34 What test can be used to compare the degree of improvement between persons on Motrin and placebo?

9.35 Perform the test in Problem 9.34 and report a *p*-value.

9.36 What conclusions can you draw from your results in Problems 9.32, 9.33 and 9.35?

Otolaryngology, Pediatrics

A common symptom of otitis media in young children is the prolonged presence of fluid in the middle ear, known as *middle-ear effusion*. The presence of fluid may result in temporary hearing loss and interfere with normal learning skills in the first 2 years of life. One hypothesis is that babies who are breast-fed for at least 1 month build up some immunity against the effects of the disease and have less prolonged effusion than do bottle-fed babies. A small study of 24 pairs of babies is set up, where the babies are matched on a one-to-one basis according to age, sex, socioeconomic status, and type of medications taken. One member of the matched pair is a breast-fed baby whereas the other member is a bottle-fed baby. The outcome variable is the duration of middle-ear effusion after the first episode of otitis media. The results are given in Table 9.7.

9.37 What are the hypotheses being tested here?

9.38 Why might a nonparametric test be useful in testing the hypotheses?

9.39 Which nonparametric test should be used here?

9.40 Test the hypothesis that the duration of effusion is less prolonged among breast-fed babies than among bottle-fed babies using a nonparametric test.

Psychiatry

Much attention has been given in recent years to the role of transcendental meditation in improving health, particularly in lowering blood pressure. One hypothesis that emerges from this work is that transcendental meditation might also be useful in treating psychiatric patients with symptoms of anxiety. Suppose that a protocol of meditational therapy is administered once a day to 20 patients with anxiety. The patients are given a psychiatric exam at baseline and at a follow-up exam 2 months later. The degree of improvement is rated on a 10-point scale, with 1 indicating the most improvement and 10 the least improvement. Similarly, 26 comparably affected patients with anxiety are given standard psy-

TABLE 9.6 A comparison of pain during maximal activity after 3 weeks of therapy with Motrin and placebo as compared with baseline

Treatment group	Total	Worse	Unchanged	Slightly improved	Moderately improved	Mostly improved	Completely improved
Motrin	43	0	14	7	10	10	2
Placebo	44	5	20	11	3	5	0

TABLE 9.7 Duration of middle-ear effusion in breast-fed and bottle-fed babies

Pair number	Duration of effusion in breast-fed baby (days)	Duration of effusion in bottle-fed baby (days)	Pair number	Duration of effusion in breast-fed baby (days)	Duration of effusion in bottle-fed baby (days)
1	20	18	13	52	39
2	11	35	14	14	15
3	3	7	15	12	21
4	24	182	16	30	28
5	7	6	17	7	8
6	28	33	18	15	27
7	58	223	19	65	77
8	7	7	20	10	12
9	39	57	21	7	8
10	17	76	22	19	16
11	17	186	23	34	28
12	12	29	24	25	20

chotherapy and are asked to come back 2 months later for a follow-up exam. The results are given in Table 9.8.

TABLE 9.8 Degree of improvement in patients with anxiety who are treated with transcendental meditation or psychotherapy

Meditation		Psychotherapy	
d^*	$f†$	d^*	$f†$
1	3	1	0
2	4	2	2
3	7	3	5
4	3	4	3
5	2	5	8
6	1	6	4
7	0	7	2
8	0	8	1
9	0	9	1
10	0	10	0
	$\overline{20}$		$\overline{26}$

* d = degree of improvement
† f = frequency

9.41 Why might a parametric test not be useful here?

9.42 What nonparametric test should be used to analyze these data?

9.43 Compare the degree of improvement in the two groups using the test in Problem 9.42.

Hypertension

Polyunsaturated fatty acids in the diet favorably affect several risk factors for cardiovascular disease. The principal dietary polyunsaturated fat is linoleic acid. To test the effects of dietary supplements of linoleic acid on blood pressure, 17 adults consumed 23 g/day of safflower oil, high in linoleic acid, for 4 weeks. Blood-pressure measurements were taken at baseline (before ingestion of oil) and 1 month later, with the mean values over several readings at each visit given in Table 9.9.

9.44 What parametric test could be used to test for the effect of linoleic acid on blood pressure?

9.45 Perform the test in Problem 9.44 and report a p-value.

9.46 What nonparametric test could be used to test for the effect of linoleic acid on blood pressure?

9.47 Perform the test in Problem 9.46 and report a p-value.

9.48 Compare your results in Problems 9.45 and 9.47 and discuss which method you feel is appropriate here.

Ophthalmology

A new drug is developed to relieve the ocular symptoms of hay fever. The drug is composed of two components, A and B: A is supposed to relieve itching of the eye and

TABLE 9.9 Effect of linoleic acid on systolic blood pressure

Subject	Baseline blood pressure	1-month blood pressure	Baseline minus 1-month blood pressure
1	119.67	117.33	2.34
2	100.00	98.78	1.22
3	123.56	123.83	−0.27
4	109.89	107.67	2.22
5	96.22	95.67	0.55
6	133.33	128.89	4.44
7	115.78	113.22	2.56
8	126.39	121.56	4.83
9	122.78	126.33	−3.55
10	117.44	110.39	7.05
11	111.33	107.00	4.33
12	117.33	108.44	8.89
13	120.67	117.00	3.67
14	131.67	126.89	4.78
15	92.39	93.06	−0.67
16	134.44	126.67	7.77
17	108.67	108.67	0.00

B is supposed to prevent redness; the combination is supposed to relieve both itching and redness. Federal regulations require that each component be proven effective both separately and in combination. Three experiments are performed on a group of 25 patients with hay fever. In the first experiment, drug A is administered to a randomly selected eye and a placebo is administered to the other eye, and the change from baseline is noted for each eye. The data are given in Table 9.10. A plus sign represents more improvement in the drug-treated eye; a minus sign represents more improvement in the placebo-treated eye; 0 represents equal improvement in both eyes.

9.49 Why might a nonparametric statistical test be useful in comparing drug A with placebo for this experiment?

TABLE 9.10 Comparison of drug A versus placebo for the relief of redness and itching in hay-fever patients

Subject	Redness	Itching	Subject	Redness	Itching
1	0	+	13	+	+
2	0	+	14	0	0
3	−	−	15	0	0
4	+	−	16	−	0
5	0	+	17	−	+
6	+	+	18	−	0
7	+	+	19	0	0
8	−	0	20	−	+
9	−	+	21	−	0
10	0	−	22	−	+
11	+	+	23	0	0
12	0	+	24	−	0
			25	−	0

9.50 Why was it important to administer the placebo to the second eye of the same person rather than to a different group of people with hay fever?

9.51 What nonparametric statistical test would you use to compare drug A with placebo for redness or itching? Why?

9.52 Compare redness in the drug-A and placebo eyes and report a *p*-value.

9.53 Compare itching in the drug-A and placebo eyes and report a *p*-value.

In the second experiment, drug B is administered to a randomly selected eye and a placebo is administered to the other eye. The data are given in Table 9.11.

9.54 Compare redness in drug-B and placebo eyes and report a *p*-value.

9.55 Compare itching in drug-B and placebo eyes and report a *p*-value.

In the third experiment, the combination of drugs A and B is administered to a randomly selected eye and a placebo is administered to the other eye. The data are given in Table 9.12.

9.56 Compare redness in the combination and placebo eyes and report a *p*-value.

9.57 Compare itching in the combination and placebo eyes and report a *p*-value.

TABLE 9.11 Comparison of drug B versus placebo for the relief of redness and itching in hay-fever patients

Subject	Redness	Itching	Subject	Redness	Itching
1	+	−	13	+	0
2	+	−	14	+	0
3	+	+	15	+	0
4	+	+	16	0	0
5	+	+	17	+	−
6	0	−	18	+	−
7	+	0	19	+	+
8	+	+	20	+	+
9	+	0	21	+	0
10	0	0	22	+	0
11	+	−	23	+	0
12	+	+	24	+	0
			25	+	0

TABLE 9.12 Comparison of combination of drugs A and B versus placebo for the relief of redness and itching in hay-fever patients

Subject	Redness	Itching	Subject	Redness	Itching
1	+	+	13	+	+
2	+	0	14	+	0
3	+	+	15	+	0
4	+	+	16	+	+
5	+	+	17	+	+
6	+	−	18	+	+
7	+	+	19	0	+
8	+	0	20	0	+
9	+	+	21	+	0
10	+	+	22	+	0
11	+	−	23	+	0
12	+	−	24	+	0
			25	+	+

9.58 Summarize the results of the three experiments as concisely as possible.

Hypertension

An instrument that is in fairly common use in blood-pressure epidemiology is the random zero device, whereby the zero point of the machine is randomly set with each use and the observer is not aware of the actual level of blood pressure at the time of measurement. This instrument is intended to reduce observer bias. Before using such a machine, it is important to check that readings are, on the average, comparable to those of a standard cuff. For this purpose, two measurements were made on 20 children with both the standard cuff and the random zero machine. The mean blood

pressures for the two readings are given in Table 9.13. Suppose observers are reluctant to assume that the distribution of blood pressure is normal.

9.59 Which nonparametric test should be used to test the hypothesis that the two machines are comparable?

9.60 Conduct the test recommended in Problem 9.59.

Another aspect of the same study is to compare the variability of blood pressure with each method. This comparison is achieved by measuring $|x_1 - x_2|$ for each method (i.e., the absolute difference between first and second readings) and comparing the absolute differences between machines. The data are given in Table 9.14. The observers are reluctant to assume that the distributions are normal.

TABLE 9.13 Comparison of mean systolic blood pressure with the standard cuff and the random zero machine (mm Hg)

Person (*i*)	Mean systolic bp (standard cuff)	Mean systolic bp (random zero)	Person (*i*)	Mean systolic bp (standard cuff)	Mean systolic bp (random zero)
1	79	84	11	98	97
2	112	99	12	103	103
3	103	92	13	105	107
4	104	103	14	117	120
5	94	94	15	94	94
6	106	106	16	88	87
7	103	97	17	101	97
8	97	108	18	98	93
9	88	77	19	91	87
10	113	94	20	105	104

TABLE 9.14 Comparison of variability of systolic blood pressure with the standard cuff and the random zero machine

Person (*i*)	Absolute difference, standard cuff (a_s)	Absolute difference, random zero (a_r)	Person (*i*)	Absolute difference, standard cuff (a_s)	Absolute difference, random zero (a_r)
1	2	12	11	0	6
2	4	6	12	2	6
3	6	0	13	6	6
4	4	2	14	2	4
5	8	4	15	8	8
6	4	4	16	0	2
7	2	6	17	6	6
8	2	8	18	4	6
9	4	2	19	2	14
10	2	4	20	2	4

9.61 Which nonparametric test should be used to test the hypothesis that the variability of the two machines is comparable?

9.62 Conduct the test recommended in Problem 9.61.

Health Promotion

Refer to Data Set 24, SMOKE.DAT, Appendix 2.

9.63 Use nonparametric methods to test whether there is a difference between males and females regarding the number of days abstinent from smoking.

9.64 Divide the data set into age groups (perhaps above/below the median) and use nonparametric methods to test whether the number of days abstinent from smoking is related to age.

9.65 Use the same approach as in Problem 9.64 to test whether the amount previously smoked is related to the number of days abstinent from smoking.

9.66 Use the same approach as in Problem 9.64 to test whether the adjusted CO level is related to the number of days abstinent from smoking.

9.67 Why are nonparametric methods well-suited to a study of risk factors for smoking cessation?

Infectious Disease

Refer to the lymph-node data in Table 8.22 (p. 288).

9.68 Use nonparametric methods to assess whether there are systematic differences between the assessments of Doctor A versus Doctor B regarding the number of palpable lymph nodes sexual contacts of AIDS or in ARC patients.

9.69 Are there any differences between your conclusions in Problem 9.68 and your conclusions using parametric methods in Problem 8.135?

Refer to the urinary sodium data in Table 8.26 (p. 290).

9.70 Use nonparametric methods to assess whether dietary counseling is effective in reducing sodium intake as judged by urinary sodium excretion levels.

Refer to Data Set 12, HORMONE.DAT, Appendix 2.

9.71 Use nonparametric methods to answer Problem 8.158 (p. 291).

9.72 Use nonparametric methods to answer Problem 8.159 (p. 291).

9.73 Use nonparametric methods to answer Problem 8.160 (p. 291).

9.74 Compare your results in Problems 9.71–9.73 with the corresponding results using parametric methods in Problems 8.153–8.155.

References

[1] McNeil, D. R. (1967). Efficiency loss due to grouping in distribution free tests. *Journal of the American Statistical Association, 62,* 954–965.

[2] Hollander, M., & Wolfe, D. (1973). *Nonparametric Statistical Methods.* New York: Wiley.

[3] Salzman, R., Manson, J. E., Griffing, G. T., Kimmerle, R., Ruderman, N., McCall, A., Stoltz, E. I., Mullin, C., Small, D., Armstrong, J., & Melby, J. C. (1985). Intranasal aerosolized insulin: Mixed-meal studies and long term use in type I diabetes. *New England Journal of Medicine, 312*(17), 1078–1084.

HYPOTHESIS TESTING: CATEGORICAL DATA

Introduction

In Chapters 7 and 8 the basic methods of hypothesis testing for continuous data were presented. For each test, the data were assumed to come from an underlying normal distribution, and appropriate inference procedures were developed based on this assumption. In Chapter 9 the assumption of normality was relaxed and alternative nonparametric tests that did not depend as much on the underlying distribution of the data were developed.

If the variable under study is not continuous but is instead classified into categories, then the methods in Chapters 7 and 8 are not applicable, and different methods of inference should be used. Consider the problems in Examples 10.1 through 10.3.

EXAMPLE 10.1   **Cancer** Suppose we are interested in the association between the use of oral contraceptives (OC use) and the 1-year incidence of cervical cancer from January 1, 1988, to January 1, 1989. Women who are disease-free on January 1, 1988, are classified into two OC-use categories as of that date: ever users and never users. We are interested in whether or not the proportion of women who develop cervical cancer is different between ever users and never users. Hence, this is a two-sample problem comparing two binomial proportions, and the *t*-test methodology in Chapter 8 cannot be used because the outcome variable, the development of cervical cancer, is a discrete variable with two categories (yes/no), not a continuous variable. ∎

EXAMPLE 10.2   **Cancer** Suppose the OC users in Example 10.1 are subdivided into "heavy" users, who have used the pill for 5 years or more, and "light" users, who have used the pill for less than 5 years. We may be interested in comparing 1-year cervical cancer incidence rates among heavy users, light users, and nonusers. In this problem, *three* binomial proportions are being compared, and again, the methods in Chapter 8 are not applicable. ∎

EXAMPLE 10.3   **Infectious Disease** The fitting of a probability model based on the Poisson distribution to the random variable defined by the annual number of deaths due to polio in the United States during the period 1968–1976 has been discussed, as shown in Table 4.5 (p. 93). We want to develop a general procedure for testing the goodness of fit of this and other probability models based on actual sample data. ∎

In this chapter, methods of hypothesis testing for comparing two or more binomial proportions are developed. Methods for testing the goodness of fit of a previously specified probability model to actual data are also discussed.

Two-Sample Test for Binomial Proportions

EXAMPLE 10.4 **Cancer** A hypothesis has been proposed that breast cancer in women is caused in part by events that occur between the age at menarche (i.e., the age when menstruation begins) and the age at first childbirth. In particular, the hypothesis states that the risk of breast cancer increases as the length of this time interval increases. If this theory is correct, then an important risk factor for breast cancer is age at first birth. This theory would explain in part why breast cancer incidence seems to be higher for women in the upper socioeconomic groups, since they tend to have their children relatively late.

An international study was set up to test this hypothesis [1]. Breast cancer cases were identified among women in selected hospitals in the United States, Greece, Yugoslavia, Brazil, Taiwan, and Japan. Controls were chosen from women of comparable age who were in the hospital at the same time as the cases but who did *not* have breast cancer. All women were asked about their age at first birth.

Women with at least one birth are arbitrarily divided into two categories: (1) women whose age at first birth was $\leqslant 29$, and (2) women whose age at first birth was $\geqslant 30$. The following results are found among women with at least one birth: 683 out of 3220 (21.2%) women with breast cancer (case women) and 1498 out of 10,245 (14.6%) women without breast cancer (control women) had an age at first birth $\geqslant 30$. How can we assess whether this difference is significant or simply due to chance? ■

Let $p_1 =$ the probability that age at first birth is $\geqslant 30$ in case women with at least one birth and $p_2 =$ the probability that age at first birth is $\geqslant 30$ in control women with at least one birth. The question is whether or not the underlying probability of having an age at first birth of $\geqslant 30$ is different in the two groups. This problem is equivalent to testing the hypothesis $H_0: p_1 = p_2 = p$ versus $H_1: p_1 \neq p_2$ for some constant p.

Two approaches for testing the hypothesis will be presented. One approach uses normal theory methods similar to those developed in Chapter 8 and is discussed in Section 10.2.1. A second approach uses contingency table methods and is discussed in Section 10.2.2. These two approaches are *equivalent* in that they always yield the same p-values, and which one is used is a matter of convenience.

10.2.1 ### Normal Theory Method

It is reasonable to base the significance test on the difference between the sample proportions $(\hat{p}_1 - \hat{p}_2)$. If this difference is either very large or very small, then H_0 would be rejected; otherwise, H_0 would be accepted. The samples will be assumed large enough so that the *normal approximation to the binomial distribution is valid.* Then, under H_0, $\hat{p}_1$ is normally distributed with mean p and variance pq/n_1, and $\hat{p}_2$ is normally distributed with mean p and variance pq/n_2. Therefore, from **(5.8)**, since the samples are independent, $\hat{p}_1 - \hat{p}_2$ is normally distributed with mean 0 and variance

$$\frac{pq}{n_1} + \frac{pq}{n_2} = pq\left(\frac{1}{n_1} + \frac{1}{n_2}\right)$$

If we divide by the standard deviation

$$\sqrt{pq\left(\frac{1}{n_1} + \frac{1}{n_2}\right)}$$

then under H_0,

10.1
$$Z = (\hat{p}_1 - \hat{p}_2)/\sqrt{pq(1/n_1 + 1/n_2)} \sim N(0, 1)$$

The problem is that p and q are unknown, and thus the denominator of Z cannot be computed unless some estimate for p is found. The best estimator for p is based on a weighted average of the sample proportions $\hat{p}_1$, $\hat{p}_2$. This weighted average, referred to as $\hat{p}$, is given by

10.2
$$\hat{p} = \frac{n_1\hat{p}_1 + n_2\hat{p}_2}{n_1 + n_2} = \frac{x_1 + x_2}{n_1 + n_2}$$

where $x_1 =$ the observed number of events in the first sample and $x_2 =$ the observed number of events in the second sample. This estimate makes intuitive sense, since each of the sample proportions is weighted by the number of people in the sample. Thus, if the estimate $\hat{p}$ in **(10.2)** is substituted for p in **(10.1)**, then the following test procedure can be used:

10.3 **Two-Sample Test for Binomial Proportions (Normal Theory Test)**

To test the hypothesis $H_0: p_1 = p_2$ versus $H_1: p_1 \neq p_2$, where the proportions are obtained from two independent samples, use the following procedure:

(1) Compute the test statistic

$$Z = \frac{\hat{p}_1 - \hat{p}_2}{\sqrt{\hat{p}\hat{q}\left(\frac{1}{n_1} + \frac{1}{n_2}\right)}}$$

where
$$\hat{p} = \frac{(n_1\hat{p}_1 + n_2\hat{p}_2)}{(n_1 + n_2)} = \frac{(x_1 + x_2)}{(n_1 + n_2)}, \hat{q} = 1 - \hat{p}$$

and x_1, x_2 are the number of events in the first and second samples, respectively.

(2) For a two-sided level α test, if

$$Z > z_{1-\alpha/2} \quad \text{or} \quad Z < z_{\alpha/2}$$

then reject H_0; if
$$z_{\alpha/2} \leqslant Z \leqslant z_{1-\alpha/2}$$

then accept H_0.

(3) The exact p-value for this test is given by

$$p = 2[1 - \Phi(Z)] \quad \text{if } Z \geqslant 0$$
$$= 2\Phi(Z) \quad \text{if } Z < 0$$

(4) Use this test only when the normal approximation to the binomial distribution is valid for each of the two samples, that is, when $n_1\hat{p}\hat{q} \geqslant 5$ and $n_2\hat{p}\hat{q} \geqslant 5$.

The acceptance and rejection regions for this test are depicted in Figure 10.1. The computation of the exact p-value is illustrated in Figure 10.2.

$$Z = \frac{\hat{p}_1 - \hat{p}_2}{\sqrt{\hat{p}\hat{q}(1/n_1 + 1/n_2)}}, \quad \text{where} \quad \hat{p} = \frac{n_1\hat{p}_1 + n_2\hat{p}_2}{n_1 + n_2}, \hat{q} = 1 - \hat{p}$$

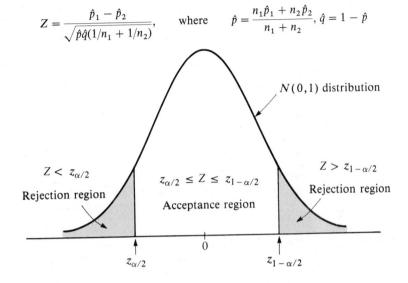

FIGURE 10.1
Acceptance and rejection regions for the two-sample test for binomial proportions (normal theory test)

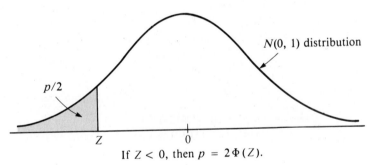

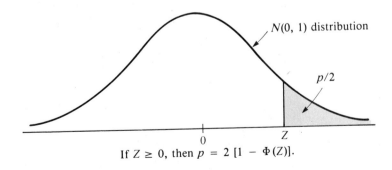

FIGURE 10.2
Computation of the exact p-value for the two-sample test for binomial proportions (normal theory test)

EXAMPLE 10.5   **Cancer** Assess the statistical significance of the results from the international study in Example 10.4.

SOLUTION   The sample proportion of case women whose age at first birth was ≥ 30 is $683/3220 = .212 = \hat{p}_1$, and the sample proportion of control women whose age at first birth was ≥ 30 is

$1498/10,245 = .146 = \hat{p}_2$. To compute the test statistic Z in **(10.3)**, the estimated common proportion $\hat{p}$, which is given by

$$\hat{p} = (683 + 1498)/(3220 + 10,245) = 2181/13,465 = .162$$

$$\hat{q} = 1 - 0.162 = 0.838$$

must be computed. Note that

$$n_1\hat{p}\hat{q} = 3220(.162)(.838) = 437.1 \geqslant 5$$

and

$$n_2\hat{p}\hat{q} = 10,245(.162)(.838) = 1390.8 \geqslant 5$$

Thus, the test in **(10.3)** can be used.

The test statistic is given by

$$Z = (.212 - .146) \Big/ \sqrt{(.162)(.838)\left(\frac{1}{3220} + \frac{1}{10,245}\right)} = .0660/.00744 = 8.9$$

Since $Z > 0$, the p-value $= 2 \times [1 - \Phi(8.9)] < .001$, and the results are extremely significant. Therefore, we can conclude that women with breast cancer are significantly more likely to have had their children after the age of 30 than are comparable women without breast cancer. ∎

EXAMPLE 10.6   **Cardiovascular Disease** A study was conducted to look at the effects of oral contraceptives (OC) on heart disease in women 40–44 years of age. It is found that among 5000 current OC users at baseline, 13 women develop a myocardial infarction (MI) over a 3-year period, while among 10,000 non-OC users, 7 develop an MI over a 3-year period. Assess the statistical significance of the results.

SOLUTION   Note that $n_1 = 5000$, $\hat{p}_1 = 13/5000 = .0026$, $n_2 = 10,000$, $\hat{p}_2 = 7/10,000 = .0007$. We wish to test the hypothesis $H_0: p_1 = p_2$ versus $H_1: p_1 \neq p_2$. The best estimate of the common proportion p is given by

$$\hat{p} = \frac{13 + 7}{15,000} = \frac{20}{15,000} = .00133$$

Since $n_1\hat{p}\hat{q} = 5000(.00133)(.99867) = 6.64$, $n_2\hat{p}\hat{q} = 10,000(.00133)(.99867) = 13.28$, the normal theory test in **(10.3)** can be used. The test statistic is given by

$$Z = \frac{.0026 - .0007}{\sqrt{(.00133)(.99867)(1/5000 + 1/10,000)}} = \frac{.0019}{.00063} = 3.02$$

The p-value is given by $2 \times [1 - \Phi(3.02)] = .003$. Thus, there is a highly significant difference between MI incidence rates for current OC users and non-OC users. In other words, OC use is significantly associated with MI incidence over a three-year period. ∎

10.2.2 Contingency Table Method

The same test posed in Section 10.2.1 is now approached from a different perspective.

EXAMPLE 10.7   **Cancer** Suppose all women with at least one birth in the international study in Example 10.4 are classified as either cases or controls and with age at first birth as either $\leqslant 29$ or $\geqslant 30$. The four possible combinations can be displayed as in Table 10.1.

The case/control status is displayed along the rows of the table and the age at first birth down the columns of the table. Hence, each woman falls into one of the four boxes, or *cells*, of the table. In particular, there are 683 women with breast cancer whose age at first birth is $\geqslant 30$; 2537 women with breast cancer whose age at first birth is $\leqslant 29$; 1498 control women

TABLE 10.1

Data for the international study in Example 10.4 comparing age at first birth in breast cancer cases with comparable controls

Status	Age at first birth		Total
	$\geqslant 30$	$\leqslant 29$	
Case	683	2537	3220
Control	1498	8747	10,245
Total	2181	11,284	13,465

(Reprinted with permission of *WHO Bulletin, 43*, 209–221, 1970.)

whose age at first birth is $\geqslant 30$; and 8747 control women whose age at first birth is $\leqslant 29$. Furthermore, the number of women in each row and column can be totaled and displayed in the margins of the table. Thus, there are 3220 case women (683 + 2537); 10,245 control women (1498 + 8747); 2181 women with age at first birth $\geqslant 30$ (683 + 1498); and 11,284 women with age at first birth $\leqslant 29$ (2537 + 8747). These sums are referred to as row margins and column margins, respectively. Finally, the total number of units = 13,465 as given in the lower-right-hand corner of the table, which can be obtained either by summing the four cells (683 + 2537 + 1498 + 8747) or by summing the row margins (3220 + 10,245) or the column margins (2181 + 11,284). This sum is sometimes referred to as the grand total. ∎

Table 10.1 is called a 2×2 contingency table because of the two groups for case/control status and the two groups for age-at-first-birth status.

DEFINITION 10.1 ▪▪▪

A **2 × 2 contingency table** is a table composed of two rows and two columns. It is an appropriate way to display data that can be classified by two different variables, *each* of which has only two possible outcomes. One variable is arbitrarily assigned to the rows and the other to the columns. Each of the four *cells* represents the number of units (women, in the previous example), with a specific value for each of the two variables. The cells are sometimes referred to by number, with the (1, 1) cell being the cell in the first row and first column, the (1, 2) cell being the cell in the first row and second column, the (2, 1) cell being the cell in the second row and first column, and the (2, 2) cell being the cell in the second row and second column. The observed number of units in the four cells are likewise referred to as O_{11}, O_{12}, O_{21}, and O_{22}, respectively. Furthermore, it is customary to total

(1) The number of units in each row and display them in the right margins, which are referred to as **row margins**.

(2) The number of units in each column and display them in the bottom margins, which are referred to as **column margins**.

(3) The total number of units in the four cells, which is displayed in the lower-right-hand corner of the table and is referred to as the **grand total**. ∎

EXAMPLE 10.8   **Cardiovascular Disease** Display the myocardial-infarction data in Example 10.6 in the form of a 2×2 contingency table.

SOLUTION   Let the rows of the table represent the OC-use group, with the first row representing current OC users and the second row representing non-OC users. Let the columns of the table represent MI, with the first column representing "yes" and the second column representing "no." We have studied 5000 current OC users, of whom 13 developed MI and 4987 did not. We have studied 10,000 non-OC users, of whom 7 developed MI and 9993 did not. Thus, the contingency table should look like Table 10.2. ∎

OC-use group	MI status over 3 years		Total
	Yes	No	
OC users	13	4987	5000
Non-OC users	7	9993	10,000
Total	20	14,980	15,000

Two different sampling designs lend themselves to a contingency table framework. In the breast cancer data in Example 10.4, there are two independent samples (i.e., case women and control women) and we want to compare the proportion of women in each group who have a first birth at a late age. Similarly, in the OC-MI data in Example 10.6, there are two independent samples of women with different contraceptive use patterns and we wish to compare the proportion of women in each group with an MI. In both instances we want to test whether or not the proportions are the same in the two independent samples. This test is referred to as a **test for homogeneity of binomial proportions**. In this situation, one set of margins is fixed (e.g., the rows) and the number of successes in each row is a random variable. For example, in Example 10.4, the total number of breast cancer cases and controls is fixed, and the number of women with age at first birth ≥ 30 is a binomial random variable conditional on the fixed-row margins (i.e., 3220 cases and 10,245 controls).

Another possible situation from which contingency tables arise is in testing for the independence of two characteristics in the same sample when neither characteristic is particularly appropriate as a denominator. In this setting, both sets of margins are assumed to be fixed. The number of units in one particular cell of the table [e.g., the (1, 1) cell] is a random variable and all other cells can be determined from the fixed margins and the (1, 1) cell. An example of this design is given in Example 10.9.

EXAMPLE 10.9   **Nutrition** The food frequency questionnaire is widely used to measure dietary intake. A person specifies the number of servings consumed per week of each of many different food items. The total nutrient composition is then calculated from the specific dietary components of each food item. One way to judge how well a questionnaire measures dietary intake is by its reproducibility. To assess reproducibility the questionnaire is distributed at two different times to 50 people and the reported nutrient intakes from the two questionnaires are compared. Suppose dietary cholesterol is quantified on each questionnaire as high if > 300 mg/day and normal otherwise. The contingency table in Table 10.3 is a natural way to

First food frequency questionnaire	Second food frequency questionnaire		Total
	High	Normal	
High	15	5	20
Normal	9	21	30
Total	24	26	50

compare the results of the two surveys. Notice that there is no natural denominator in this example. We simply want to test whether there is some relationship between the two reported measures of dietary cholesterol for the same person. This test is referred to as a **test of independence** or a **test of association** between the two characteristics.

Fortunately, the same test procedure is used whether a test of homogeneity or a test of independence is performed, and we will no longer distinguish between these two tests in this section.   ∎

10.2.3 Significance Testing Using the Contingency Table Approach

Table 10.1 is an **observed contingency table** or an **observed table**. In order to determine statistical significance, we need to develop an **expected table**, which is the contingency table that would be expected if there were no relationship between breast cancer and age at first birth, that is, if $H_0: p_1 = p_2 = p$ were true. In this example p_1 and p_2 are the probabilities (among women with at least one birth) of a breast cancer case and a control, respectively, having a first birth at an age ≥ 30. For this purpose a general observed table, if there were x_1 events out of n_1 women with breast cancer and x_2 events out of n_2 control women, is given in Table 10.4.

TABLE 10.4

General contingency table for the international study data in Example 10.4 if (1) of n_1 women in the case group, x_1 have events, and (2) of n_2 women in the control group, x_2 have events (An event here means having an age at first birth ≥ 30.)

Case/control status	Age at first birth		Total
	≥ 30	≤ 29	
Case	x_1	$n_1 - x_1$	n_1
Control	x_2	$n_2 - x_2$	n_2
Total	$x_1 + x_2$	$n_1 + n_2 - (x_1 + x_2)$	$n_1 + n_2$

If H_0 were true, then the best estimate of the common proportion p is $\hat{p}$, which is given in (10.2) as

$$(n_1 \hat{p}_1 + n_2 \hat{p}_2)/(n_1 + n_2)$$

or, alternatively, as

$$(x_1 + x_2)/(n_1 + n_2)$$

where x_1 and x_2 are the numbers of events in groups 1 and 2, respectively. Furthermore, under H_0 the expected number of units in the $(1, 1)$ cell equals the expected number of women with age at first birth ≥ 30 among women with breast cancer, which is given by

$$n_1 \hat{p} = n_1(x_1 + x_2)/(n_1 + n_2)$$

However, in Table 10.4 this number is simply the product of the first row margin (n_1) multiplied by the first column margin ($x_1 + x_2$), divided by the grand total ($n_1 + n_2$). Similarly, the expected number of units in the $(2, 1)$ cell equals the expected number of women with age at first birth ≥ 30 among control women:

$$n_2 \hat{p} = n_2(x_1 + x_2)/(n_1 + n_2)$$

which is equal to the product of the second row margin multiplied by the first column margin, divided by the grand total. In general, the following rule can be applied:

10.4 | **Computation of Expected Values for 2 × 2 Contingency Tables**

The **expected number of units** in the (i, j) cell, which is usually denoted by E_{ij}, is the product of the ith row margin multiplied by the jth column margin, divided by the grand total.

EXAMPLE 10.10   **Cancer** Compute the expected table for the breast cancer data in Example 10.4.

SOLUTION   Refer to Table 10.1, which gives the observed table for these data. The row totals are 3220 and 10,245; the column totals are 2181 and 11,284; and the grand total is 13,465. Thus,

$$E_{11} = \text{expected number of units in the } (1, 1) \text{ cell}$$
$$= 3220(2181)/13{,}465 = 521.6$$
$$E_{12} = \text{expected number of units in the } (1, 2) \text{ cell}$$
$$= 3220(11{,}284)/13{,}465 = 2698.4$$
$$E_{21} = \text{expected number of units in the } (2, 1) \text{ cell}$$
$$= 10{,}245(2181)/13{,}465 = 1659.4$$
$$E_{22} = \text{expected number of units in the } (2, 2) \text{ cell}$$
$$= 10{,}245(11{,}284)/13{,}465 = 8585.6$$

These expected values are displayed in Table 10.5. ∎

TABLE 10.5
Expected table for the breast cancer data in Example 10.4

Case/control status	Age at first birth		Total
	$\geqslant 30$	$\leqslant 29$	
Case	521.6	2698.4	3220
Control	1659.4	8585.6	10,245
Total	2181	11,284	13,465

EXAMPLE 10.11   **Cardiovascular Disease** Compute the expected table for the OC-MI data in Example 10.6.

SOLUTION   From Table 10.2, which gives the observed table for these data,

$$E_{11} = \frac{5000(20)}{15{,}000} = 6.7$$

$$E_{12} = \frac{5000(14{,}980)}{15{,}000} = 4993.3$$

$$E_{21} = \frac{10{,}000(20)}{15{,}000} = 13.3$$

$$E_{22} = \frac{10{,}000(14{,}980)}{15{,}000} = 9986.7$$

These expected values are displayed in Table 10.6. ∎

TABLE 10.6
Expected table for
the OC-MI data in
Example 10.6

OC-use group	MI status over 3 years		Total
	Yes	No	
OC users	6.7	4993.3	5000
Non-OC users	13.3	9986.7	10,000
Total	20	14,980	15,000

We can show from **(10.4)** that the *total* of the expected number of units in any row or column should be the same as the corresponding observed row or column total. This relationship provides a useful check that the expected values are computed correctly.

EXAMPLE 10.12   Check that the expected values in Table 10.5 are computed correctly.

SOLUTION   The following information is given:

(1) The total of the expected values in the first row $= E_{11} + E_{12} = 521.6 + 2698.4 = 3220 =$ first row total in the observed table.

(2) The total of the expected values in the second row $= E_{21} + E_{22} = 1659.4 + 8585.6 = 10{,}245 =$ second row total in the observed table.

(3) The total of the expected values in the first column $= E_{11} + E_{21} = 521.6 + 1659.4 = 2181 =$ first column total in the observed table.

(4) The total of the expected values in the second column $= E_{12} + E_{22} = 2698.4 + 8585.6 = 11{,}284 =$ second column total in the observed table.   ∎

We now wish to compare the observed table in Table 10.1 with the expected table in Table 10.5. If the corresponding cells in these two tables are close, then H_0 will be accepted; if they are sufficiently different, then H_0 will be rejected. How should we decide how different the cells should be for us to reject H_0? It can be shown that the best way of comparing the cells in the two tables is to use the statistic $(O - E)^2/E$, where O and E are the observed and expected number of units, respectively, in a particular cell. In particular, under H_0 it can be shown that the sum of $(O - E)^2/E$ over the four cells in the table approximately follows a chi-square distribution with 1 df. H_0 is rejected only if this sum is large and accepted otherwise, since small values of this sum correspond to good agreement between the two tables, whereas large values correspond to poor agreement. This test procedure will be used only when the normal approximation to the binomial distribution is valid. In this setting it can be shown to be approximately true if *no expected value in the table is less than 5.*

Furthermore, under certain circumstances a version of this test statistic with a *continuity correction* yields more accurate *p*-values than does the uncorrected version when approximated by a chi-square distribution. For the continuity corrected version, the statistic $(|O - E| - \frac{1}{2})^2/E$ rather than $(O - E)^2/E$ is computed for each cell and the preceding expression is summed over the four cells. This test procedure is referred to as the chi-square test using the Yates correction and is summarized as follows:

10.5 **Yates-Corrected Chi-Square Test for a 2 × 2 Contingency Table**

Suppose we wish to test the hypothesis $H_0: p_1 = p_2$ versus $H_1: p_1 \neq p_2$ using a contingency table format, where O_{ij} represents the observed number of units in the (i, j) cell and E_{ij} represents the expected number of units in the (i, j) cell.

(1) Compute the test statistic

$$X^2 = (|O_{11} - E_{11}| - .5)^2/E_{11} + (|O_{12} - E_{12}| - .5)^2/E_{12}$$
$$+ (|O_{21} - E_{21}| - .5)^2/E_{21} + (|O_{22} - E_{22}| - .5)^2/E_{22}$$

which under H_0 approximately follows a χ_1^2 distribution.

(2) For a level α test, reject H_0 if $X^2 > \chi_{1,1-\alpha}^2$ and accept H_0 if $X^2 \leq \chi_{1,1-\alpha}^2$.

(3) The exact p-value is given by the area to the right of X^2 under a χ_1^2 distribution.

(4) Use this test only if none of the four expected values is less than 5.

The acceptance and rejection regions for this test are depicted in Figure 10.3. The computation of the exact p-value is illustrated in Figure 10.4.

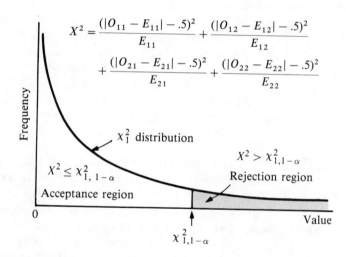

FIGURE 10.3
Acceptance and rejection regions for the Yates-corrected chi-square test for a 2 × 2 contingency table

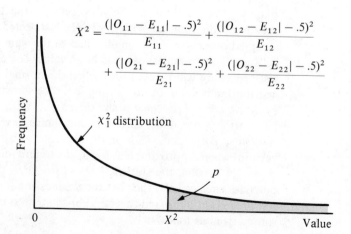

FIGURE 10.4
Computation of the p-value for the Yates-corrected chi-square test for a 2 × 2 contingency table

The Yates-corrected chi-square test is a *two-sided* test even though the critical region, based on the chi-square distribution, is one-sided. The rationale for this is that large values of $|O_{ij} - E_{ij}|$ and, correspondingly, of the test statistic X^2 will be obtained under H_1 regardless of whether $p_1 < p_2$ or $p_1 > p_2$. Small values of X^2 are evidence in favor of H_0.

EXAMPLE 10.13   **Cancer** Assess the breast cancer data in Example 10.4 (p. 319) for statistical significance.

SOLUTION   First compute the observed and expected tables as given in Tables 10.1 and 10.5, respectively. Check that all expected values in Table 10.5 are at least 5, which is clearly the case. Thus, following **(10.5)**,

$$X^2 = \frac{(|683 - 521.6| - .5)^2}{521.6} + \frac{(|2537 - 2698.4| - .5)^2}{2698.4}$$

$$+ \frac{(|1498 - 1659.4| - .5)^2}{1659.4} + \frac{(|8747 - 8585.6| - .5)^2}{8585.6}$$

$$= \frac{(160.9)^2}{521.6} + \frac{(160.9)^2}{2698.4} + \frac{(160.9)^2}{1659.4} + \frac{(160.9)^2}{8585.6}$$

$$= 49.633 + 9.594 + 15.601 + 3.015 = 77.84 \sim \chi_1^2 \text{ under } H_0$$

Since

$$\chi_{1,.999}^2 = 10.83 < 77.84 = X^2$$

we have

$$p < 1 - .999 = .001$$

and the results are extremely significant. Thus, breast cancer is significantly associated with having a first child after age 30.   ∎

EXAMPLE 10.14   **Cardiovascular Disease** Assess the OC-MI data in Example 10.6 (p. 322) for statistical significance.

First compute the observed and expected tables as given in Tables 10.2 and 10.6, respectively. Note that the minimum expected value in Table 10.6 is 6.7, which is $\geqslant 5$. Thus, the test procedure in **(10.5)** can be used:

$$X^2 = \frac{(|13 - 6.7| - .5)^2}{6.7} + \frac{(|4987 - 4993.3| - .5)^2}{4993.3}$$

$$+ \frac{(|7 - 13.3| - .5)^2}{13.3} + \frac{(|9993 - 9986.7| - .5)^2}{9986.7}$$

$$= \frac{(5.8)^2}{6.7} + \frac{(5.8)^2}{4993.3} + \frac{(5.8)^2}{13.3} + \frac{(5.8)^2}{9986.7}$$

$$= 5.021 + .007 + 2.529 + .003 = 7.56 \sim \chi_1^2 \text{ under } H_0$$

Since $\chi_{1,.99}^2 = 6.63$, $\chi_{1,.995}^2 = 7.88$, and $6.63 < 7.56 < 7.88$, it follows that $1 - .999 < p < 1 - .995$, or $.001 < p < .005$, and the results are highly significant. Thus, there is a significant difference between incidence rates of MI for OC users and non-OC users among 40–44-year-old women, with OC users having higher rates.   ∎

The test procedures in **(10.3)** and **(10.5)** used without continuity correction are equivalent in the sense that they always give the same *p*-values and always result in the same decisions about accepting or rejecting H_0. Which test procedure is used is a matter of convenience. Most research workers find the contingency table approach more understandable, and results are more frequently reported in this format in the scientific literature.

At this time statisticians disagree widely about whether or not a continuity correction is needed for the contingency table test in **(10.5)**. Generally, *p*-values obtained using the continuity correction are slightly larger. Thus, results obtained are slightly less significant than comparable results obtained without using a continuity correction. However, the difference in results obtained using these two methods should be small for tables based on large sample sizes. I believe that the Yates-corrected test statistic is slightly more widely used in the applied literature and therefore I use it in this section. Another possible approach for performing hypotheses tests based on 2 × 2 contingency tables is to use Fisher's exact test. This procedure will be discussed in Section 10.4.

10.2.4   **Short Computational Form for the Yates-Corrected Chi-Square Test for 2 × 2 Contingency Tables**

The test statistic X^2 in **(10.5)** has another computational version that is more convenient to use with a hand calculator and does not require the computation of an expected table:

10.6   **Short Computational Form for the Yates-Corrected Chi-Square Test for 2 × 2 Contingency Tables**

Suppose we have the 2 × 2 contingency table in Table 10.7. The test statistic X^2 in **(10.5)** can be written in the form

$$X^2 = n\left(|ad - bc| - \frac{n}{2}\right)^2 \Big/ [(a + b)(c + d)(a + c)(b + d)]$$

Thus, the test statistic X^2 depends only on (1) the grand total n, (2) the row and column margins $a + b$, $c + d$, $a + c$, $b + d$, and (3) the magnitude of the quantity $ad - bc$. To compute X^2, proceed as follows:

(1) Compute

$$\left(|ad - bc| - \frac{n}{2}\right)^2$$

Start with the first column margin and proceed counterclockwise.

(2) Divide by each of the two column margins.

(3) Multiply by the grand total.

(4) Divide by each of the two row margins.

This computation is particularly easy with a hand calculator, since previous products and quotients can be maintained in the display and used for further calculations.

TABLE 10.7
General contingency
table

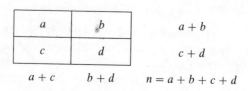

EXAMPLE 10.15   **Nutrition** Compute the chi-square statistic for the nutrition data in Example 10.9 using the short computational form in **(10.6)**.

SOLUTION   From Table 10.3,

$$a = 15 \quad b = 5 \quad c = 9 \quad d = 21 \quad n = 50$$

Furthermore, the smallest expected value $= (24 \times 20)/50 = 9.6 \geqslant 5$. Thus, it is valid to use the chi-square test. Use the approach in **(10.6)** as follows:

(1) Compute

$$\left(|ad - bc| - \frac{n}{2} \right)^2 = \left[|15 \times 21 - 5 \times 9| - \frac{50}{2} \right]^2$$
$$= (270 - 25)^2 = 245^2 = 60{,}025$$

(2) Divide the result in step 1 (60,025) by each of the two column margins (24 and 26), thus obtaining 96.194.

(3) Multiply the result in step 2 (96.194) by the grand total (50), thus obtaining 4809.70.

(4) Divide the result in step 3 (4809.70) by each of the two row margins (20 and 30), thus obtaining 8.02.

Since the critical value $= \chi^2_{1,.95} = 3.84$ and $X^2 = 8.02 > 3.84$, the results are statistically significant. To obtain a range for the p-value, note from the chi-square table that $\chi^2_{1,.995} = 7.88$, $\chi^2_{1,.999} = 10.83$, and thus, since $7.88 < 8.02 < 10.83$, $.001 < p < .005$.

These data have also been analyzed using the SPSS^X/PC CROSSTABS program, as shown in Table 10.8. The program prints out the cell counts, the row and column totals and percentages, and the grand total. Furthermore, it prints out the minimum expected frequency (min E.F. $= 9.6$), and it is noted that there are no expected frequencies < 5. Finally, the significance test is performed using both the Yates-corrected chi-square test (chi-square $= 8.02$, $df = 1$, p-value $= .0046$) and the uncorrected chi-square test (chi-square $= 9.74$, $df = 1$, p-value $= .0018$).

TABLE 10.8
Use of SPSS^X/PC
CROSSTABS program to
analyze the nutrition
data in Table 10.3

```
                                   SPSSX/PC   Release 1.0

          Crosstabulation:    CHOL1    1ST FOOD FREQUENCY QUESTIONNAIRE
                           By CHOL2    2ND FOOD FREQUENCY QUESTIONNAIRE


                   Count |HIGH   |NORMAL |
          CHOL2->         |        |        |  Row
                         |  1.00 |  2.00 | Total
          CHOL1          |--------|--------|
                 1.00    |   15   |   5    |   20
          HIGH           |        |        |  40.0
                         |--------|--------|
                 2.00    |   9    |   21   |   30
          NORMAL         |        |        |  60.0
                         |--------|--------|
                 Column      24       26      50
                 Total      48.0     52.0   100.0


          Chi-Square    D.F.     Significance      Min E.F.    Cells with E.F.< 5
          ----------    ----     ------------      --------    ------------------

            8.01616      1          0.0046           9.600          None
            9.73558      1         .0018                       ( Before Yates Correction )

          Number of Missing Observations =      0
```

The results show a highly significant association between dietary cholesterol intake reported by the same person at two different points in time, which gives us some confidence in the reproducibility of our instrument. The issue of reproducibility for discrete data is discussed in more detail in Section 11.15. ∎

Interval Estimates for Binomial Proportions

In Section 10.2 the focus was on methods for testing hypotheses concerning two binomial proportions. In many applications, interval estimates for either the difference or the ratio between two proportions, in addition to determining statistical significance, are important.

10.3.1 ### Interval Estimates for the Difference Between Two Proportions

If the normal approximation to the binomial distribution holds, then, from Chapter 6, $\hat{p}_1 \sim N(p_1, p_1 q_1/n_1)$, $\hat{p}_2 \sim N(p_2, p_2 q_2/n_2)$. Since there are two independent samples, from **(5.8)**,

$$\hat{p}_1 - \hat{p}_2 \sim N\left(p_1 - p_2, \frac{p_1 q_1}{n_1} + \frac{p_2 q_2}{n_2}\right)$$

Therefore, if $p_1 q_1/n_1 + p_2 q_2/n_2$ is approximated by $\hat{p}_1 \hat{q}_1/n_1 + \hat{p}_2 \hat{q}_2/n_2$, then

$$Pr\left(p_1 - p_2 - z_{1-\alpha/2}\sqrt{\frac{\hat{p}_1 \hat{q}_1}{n_1} + \frac{\hat{p}_2 \hat{q}_2}{n_2}} \leq \hat{p}_1 - \hat{p}_2 \leq p_1 - p_2\right.$$
$$\left. + z_{1-\alpha/2}\sqrt{\frac{\hat{p}_1 \hat{q}_1}{n_1} + \frac{\hat{p}_2 \hat{q}_2}{n_2}}\right) = 1 - \alpha$$

This can be rewritten as two inequalities:

$$p_1 - p_2 - z_{1-\alpha/2}\sqrt{\frac{\hat{p}_1 \hat{q}_1}{n_1} + \frac{\hat{p}_2 \hat{q}_2}{n_2}} \leq \hat{p}_1 - \hat{p}_2$$

and

$$\hat{p}_1 - \hat{p}_2 \leq p_1 - p_2 + z_{1-\alpha/2}\sqrt{\frac{\hat{p}_1 \hat{q}_1}{n_1} + \frac{\hat{p}_2 \hat{q}_2}{n_2}}$$

If $z_{1-\alpha/2}\sqrt{\hat{p}_1 \hat{q}_1/n_1 + \hat{p}_2 \hat{q}_2/n_2}$ is added to both sides of the first inequality and subtracted from both sides of the second inequality, the following two-sided $100\% \times (1 - \alpha)$ confidence interval for $p_1 - p_2$ is obtained:

10.7
$$Pr\left(\hat{p}_1 - \hat{p}_2 - z_{1-\alpha/2}\sqrt{\frac{\hat{p}_1 \hat{q}_1}{n_1} + \frac{\hat{p}_2 \hat{q}_2}{n_2}} \leq p_1 - p_2 \leq \hat{p}_1 - \hat{p}_2 + z_{1-\alpha/2}\sqrt{\frac{\hat{p}_1 \hat{q}_1}{n_1} + \frac{\hat{p}_2 \hat{q}_2}{n_2}}\right) = 1 - \alpha$$

Use this expression for the confidence interval only if $n_1 \hat{p}_1 \hat{q}_1 \geq 5$ and $n_2 \hat{p}_2 \hat{q}_2 \geq 5$.

EXAMPLE 10.16   **Cardiovascular Disease** Referring to the OC-MI data in Table 10.2, compute a 95% confidence interval for the difference between the proportion of women who develop MI among OC users and the comparable proportion among non-OC users.

SOLUTION   We have that $n_1 = 5000$, $\hat{p}_1 = 13/5000 = .0026$, $n_2 = 10,000$, $\hat{p}_2 = 7/10,000 = .0007$. Since $n_1\hat{p}_1\hat{q}_1 = 5000(.0026)(.9974) = 13.0 \geqslant 5$, $n_2\hat{p}_2\hat{q}_2 = 10,000(.0007)(.9993) = 7.0 \geqslant 5$, the large-sample confidence interval in **(10.7)** can be used. The 95% confidence interval is given by

$$\left[.0026 - .0007 - 1.96 \sqrt{\frac{(.0026)(.9974)}{5000} + \frac{(.0007)(.9993)}{10,000}}, \right.$$

$$\left. .0026 - .0007 + 1.96 \sqrt{\frac{(.0026)(.9974)}{5000} + \frac{(.0007)(.9993)}{10,000}} \right]$$

$$= (.0019 - .0015, .0019 + .0015) = (.0004, .0034)$$ ∎

10.3.2   The Odds Ratio

The relationship between proportions is often understood more easily in terms of a ratio than in terms of a difference. For this purpose the odds in favor of a success are defined as follows:

DEFINITION 10.2 ■■

If the probability of a success $= p$, then the **odds in favor of success** $= p/(1 - p)$. ∎

If two proportions p_1, p_2 are considered and the odds in favor of success are computed for each proportion, then the ratio of odds, or odds ratio, becomes a useful measure for relating the two proportions.

DEFINITION 10.3 ■■

Let p_1, p_2 be the probability of success for two populations. The **odds ratio** (OR) is defined as

$$OR = \frac{p_1/q_1}{p_2/q_2} = \frac{p_1 q_2}{p_2 q_1} \quad \text{and is estimated by} \quad \widehat{OR} = \frac{\hat{p}_1 \hat{q}_2}{\hat{p}_2 \hat{q}_1}$$

Equivalently, if the four cells of the 2×2 contingency table are labeled by a, b, c, d, as they are in Table 10.7, then

$$\widehat{OR} = \frac{[a/(a + b)] \times [d/(c + d)]}{[c/(c + d)] \times [b/(a + b)]} = \frac{ad}{bc}$$ ∎

EXAMPLE 10.17   **Cardiovascular Disease** Using the OC-MI data in Table 10.2, compute the odds ratio in favor of MI for an OC user compared with a non-OC user.

SOLUTION   We have $\hat{p}_1 = .0026$, $\hat{q}_1 = .9974$, $\hat{p}_2 = .0007$, $\hat{q}_2 = .9993$. Thus,

$$OR = \frac{(.0026)(.9993)}{(.0007)(.9974)} = 3.72$$

This means that the odds in favor of an MI for an OC user is 3.7 times the odds in favor of an MI for a non-OC user. The OR could also have been computed from the contingency table in Table 10.2, whereby

$$OR = \frac{13 \times 9993}{7 \times 4987} = 3.72$$ ∎

If the probability of success is the same in the two groups (i.e., H_0 is true), then $OR = 1$. Conversely, odds ratios greater than 1 indicate a greater likelihood of success in the first group than in the second group, whereas odds ratios less than 1 indicate a greater likelihood of success in the second group than in the first group.

In Chapter 3 a related measure of association, the **relative risk** (RR), was introduced. The relative risk can be expressed as the ratio of the success rates in the two groups or, in symbols, as p_1/p_2. Although easily understood the relative risk has the disadvantage of being constrained by the denominator probability (p_2). For example, if $p_2 = .5$, then the RR can be no larger than $1/.5 = 2$; if $p_2 = .8$, then the RR can be no larger than $1/.8 = 1.25$. The odds ratio, on the other hand, has no such restriction and can range from 0 to ∞, regardless of the denominator probability. This property is particularly advantageous when combining results over several 2×2 tables, as is discussed in Section 10.8. Finally, if the probabilities of success are low (i.e., p_1, p_2 are small), then $1 - p_1$ and $1 - p_2$ will each be close to 1, and the odds ratio will be approximately the same as the relative risk. Thus, the odds ratio is often used as an approximation to the relative risk for rare diseases.

10.3.3 Interval Estimates for the Odds Ratio

Several methods exist for obtaining interval estimates for the odds ratio. The **test-based method** of Miettinen has the advantages of easy implementation and the determination of the significance of results that are always consistent with the chi-square test in **(10.5)**. The method is given as follows:

10.8 | **Interval Estimates for the Odds Ratio (Test-Based Method)**

A two-sided $100\% \times (1 - \alpha)$ confidence interval for OR is given by

$$(\widehat{OR}^{1 - \sqrt{\chi^2_{1,1-\alpha}/X^2}}, \ \widehat{OR}^{1 + \sqrt{\chi^2_{1,1-\alpha}/X^2}}) \qquad \text{if } \widehat{OR} \geq 1$$

$$(\widehat{OR}^{1 + \sqrt{\chi^2_{1,1-\alpha}/X^2}}, \ \widehat{OR}^{1 - \sqrt{\chi^2_{1,1-\alpha}/X^2}}) \qquad \text{if } \widehat{OR} < 1$$

where $\hat{p}_1, \hat{p}_2$ are the estimated probabilities of success in the two samples and

$$\widehat{OR} = \frac{(\hat{p}_1 \hat{q}_2)}{(\hat{p}_2 \hat{q}_1)}$$

$$X^2 = \text{chi-square statistic in } \textbf{(10.5)}$$

If the chi-square statistic is statistically significant at level α, then $X^2 > \chi^2_{1,1-\alpha}$ and $1 - \sqrt{\chi^2_{1,1-\alpha}/X^2} > 0$. Therefore, if $\widehat{OR} > 1$, then the lower confidence limit $>$ 1.0, whereas if $\widehat{OR} < 1$, then the upper confidence limit < 1.0. In either case the confidence interval will not include 1 (the value under the null hypothesis). By similar logic, if the chi-square statistic is not statistically significant at level α, then the confidence interval will always include 1. This relationship is summarized as follows:

10.9 | **Relationship of the Test-Based Confidence Interval and the Chi-Square Test**

The two-sided $100\% \times (1 - \alpha)$ test-based confidence interval in **(10.8)** will contain 1 (the null value) if the chi-square test statistic in **(10.5)** is not statistically significant at level α.

Similarly, the $100\% \times (1 - \alpha)$ test-based confidence interval in **(10.8)** will exclude 1 if the chi-square test statistic in **(10.5)** is statistically significant at level α.

EXAMPLE 10.18   **Cardiovascular Disease** Using the OC-MI data in Table 10.2, compute a 95 % confidence interval for the odds ratio in favor of MI for OC users versus non-OC users.

SOLUTION   We have from Example 10.17 that $\widehat{OR} = 3.72$ and from Example 10.14 that $X^2 = 7.56$. Therefore, from **(10.8)**, since $\widehat{OR} > 1$,

$$c_1 = 3.72^{1-\sqrt{3.84/7.56}} = 3.72^{1-0.71} = 3.72^{0.29} = 1.46$$

$$c_2 = 3.72^{1+\sqrt{3.84/7.56}} = 3.72^{1+0.71} = 3.72^{1.71} = 9.45$$

Therefore, the 95 % confidence interval for $OR = (1.46, 9.45)$. ∎

EXAMPLE 10.19   **Cancer** Using the international study data in Table 10.1 (p. 323), compute the odds ratio for an age at first birth $\geqslant 30$ for the breast cancer cases compared with that for the controls.

SOLUTION   From Table 10.1,

$$\widehat{OR} = \frac{683 \times 8747}{2537 \times 1498} = 1.57$$

Furthermore, from Example 10.13 (p. 329), $X^2 = 77.84$. Therefore, from **(10.8)**,

$$c_1 = 1.57^{1-\sqrt{3.84/77.84}} = 1.57^{1-0.22} = 1.57^{0.78} = 1.42$$

$$c_2 = 1.57^{1+\sqrt{3.84/77.84}} = 1.57^{1+0.22} = 1.57^{1.22} = 1.73$$

Thus, the 95 % confidence interval for $OR = (1.42, 1.73)$. Note that the confidence interval does not contain 1 (the value under the null hypothesis), which is consistent with the highly significant chi-square statistic. ∎

EXAMPLE 10.20   **Cancer** Suppose the successful outcome in Example 10.19 is changed to age at first birth $\leqslant 29$. Compute the odds ratio and the associated 95 % confidence interval for breast cancer cases versus controls.

SOLUTION   If the first and second columns in Table 10.1 are interchanged, then $a = 2537$, $b = 683$, $c = 8747$, $d = 1498$. Therefore,

$$\widehat{OR} = \frac{2537(1498)}{683(8747)} = 0.64$$

Furthermore, we have the same chi-square statistic as in Example 10.13 (i.e., $X^2 = 77.84$). Since $\widehat{OR} < 1$, from **(10.8)**,

$$c_1 = 0.64^{1+\sqrt{3.84/77.84}} = 0.64^{1+0.22} = 0.64^{1.22} = 0.58$$

$$c_2 = 0.64^{1-\sqrt{3.84/77.84}} = 0.64^{1-0.22} = 0.64^{0.78} = 0.71$$

Therefore, the estimated odds ratio = 0.64, with the associated 95 % confidence interval = (0.58, 0.71). Once again, the confidence interval does not contain the value under the null hypothesis (1), since the chi-square test statistic was significant at the 5 % level. ∎

Some statisticians have found that the test-based method is inaccurate if the estimated odds ratio is too far from 1 in either direction. Specifically, this method should be used only if $0.2 \leqslant \widehat{OR} \leqslant 5.0$. If the $\widehat{OR}$ is outside this range, then more sophisticated methods should be used for confidence interval estimation. See Kleinbaum, Kupper, and Morgenstern for more details on this subject [2].

Fisher's Exact Test

EXAMPLE 10.21 **Cardiovascular Disease, Nutrition** Suppose we wish to investigate the relationship between high salt intake and the occurrence of death from cardiovascular disease (CVD). Groups of high- and low-salt users could be identified and followed over a long period of time to compare the relative frequency of death from CVD in the two groups. On the other hand, a much less expensive study would involve looking at death records, separating the CVD deaths from the non-CVD deaths, and then asking a close relative (such as a spouse) about the dietary habits of the deceased, and comparing salt intake between CVD deaths versus non-CVD deaths. ∎

The latter type of study, a retrospective study, may be impossible to perform for a number of reasons. But if it is possible, it will almost always be less expensive than the former type of study, a prospective study.

DEFINITION 10.4 ■■
A **prospective study** is a study in which a group of disease-free individuals are identified at one point in time and followed over a period of time until some of them develop the disease. The development of disease over time is then related to other variables measured at baseline. ∎

DEFINITION 10.5 ■■
A **retrospective study** is a study in which two groups of individuals are initially identified: (1) a group that has the disease under study (the cases) and (2) a group that does not have the disease under study (the controls). An attempt is then made to relate their *prior* health habits to their current disease status. ∎

What are the advantages of the two types of studies? A prospective study is usually more definitive because the patients' knowledge of their current health habits will be more precise than their (or related individuals') recall of their past health habits. Second, there is a greater chance of bias with a retrospective study because (1) it is much more difficult to obtain a representative sample of people who already have the disease in question, since, for example, some of the diseased individuals may have already died and only the mildest cases (or if it is a study of deceased cases, the most severe cases) may be included, and (2) the diseased individuals, if still alive, or their surrogates will tend to give biased answers about prior health habits if they *believe* there is a relationship between these prior health habits and the disease. For instance, in Example 10.21 it may be difficult to obtain dietary histories from the relatives. However, a retrospective study is much less expensive to perform and can be completed in much less time than a prospective study. Thus, an inexpensive retrospective study may initially be done as a justification for the ultimate, definitive, prospective study.

EXAMPLE 10.22 **Cardiovascular Disease, Nutrition** Suppose a retrospective study is done on the deaths of all men aged 50–54 in a specific county over a 1-month period. It is found that of 35 people who died from CVD, 5 were on a high-salt diet before they died, whereas of 25 people who died from other causes, 2 were on such a diet. These data, presented in Table 10.9, are in the form of a 2×2 contingency table, and thus the methods of Section 10.2.2 may be applicable.

TABLE 10.9

Data concerning the possible relationship between cause of death and high salt intake

| | Type of diet | | |
Cause of death	High salt	Low salt	Total
Non-CVD	2	23	25
CVD	5	30	35
Total	7	53	60

However, the expected values of this table are too small to validly use such methods. Indeed,

$$E_{11} = 7(25)/60 = 2.92$$
$$E_{21} = 7(35)/60 = 4.08$$

and thus two of the four cells have expected values less than 5. How should a relationship between cause of death and type of diet be tested for? ∎

In this case, **Fisher's exact test** can be used. This procedure gives exact results for any 2×2 table but is only necessary for tables with small expected values, where the standard chi-square test as given in **(10.5)** is not applicable. For tables in which the use of the chi-square test is appropriate, the two tests give very similar results. Suppose the probability that a person was on a high-salt diet given that his or her cause of death was noncardiovascular (non-CVD) $= p_1$, and the probability that a person was on a high-salt diet given that his or her cause of death was cardiovascular (CVD) $= p_2$. We wish to test the hypothesis $H_0: p_1 = p_2 = p$ versus $H_1: p_1 \neq p_2$. Table 10.10 gives the general layout of the data.

TABLE 10.10

General layout of data for Fisher's exact test example

| | Type of diet | | |
Cause of death	High salt	Low salt	Total
Non-CVD	a	b	$a + b$
CVD	c	d	$c + d$
Total	$a + c$	$b + d$	n

For mathematical convenience, we will assume that the margins of this table are *fixed*; that is, the numbers of non-CVD deaths and CVD deaths are fixed at $a + b$ and $c + d$, respectively, whereas the numbers of people on a high- and low-salt diet are fixed at $a + c$ and $b + d$, respectively. Indeed, it is difficult to compute exact probabilities unless one makes the assumption of fixed margins. The *exact* probability of observing the table with cells a, b, c, d is given as follows:

10.10 | **Exact Probability of Observing a Table with Cells a, b, c, d**

$$Pr(a, b, c, d) = \frac{(a + b)!(c + d)!(a + c)!(b + d)!}{n!a!b!c!d!}$$

The formula in **(10.10)** is easy to remember, since the numerator is the product of factorials of each of the row and column margins, and the denominator is the product of the factorial of the grand total and the factorials of the individual cells.

EXAMPLE 10.23   Suppose we have the 2×2 table as shown in Table 10.11. Compute the exact probability of obtaining this table assuming that the margins are fixed.

SOLUTION
$$Pr(2, 5, 3, 1) = \frac{7!4!5!6!}{11!2!5!3!1!} = \frac{(5040)(24)(120)(720)}{(39,916,800)(2)(120)(6)} = \frac{1.0450944 \times 10^{10}}{5.7480192 \times 10^{10}} = .182$$

Note that in some cases the factorials can be canceled, simplifying the computations:

$$Pr(2, 5, 3, 1) = \frac{7!4!5!6!}{11!2!5!3!1!} = \left(\frac{7!}{11!}\right) \times \left(\frac{4!}{2!}\right) \times \left(\frac{6!}{3!}\right)$$

$$= \frac{1}{11 \times 10 \times 9 \times 8} \times \frac{4 \times 3}{1} \times \frac{6 \times 5 \times 4}{1} = \frac{1440}{7920} = .182 \quad \blacksquare$$

TABLE 10.11
Hypothetical 2×2 contingency table in Example 10.23

2	5	7
3	1	4
5	6	11

The basic strategy in testing the hypothesis will be to enumerate all possible tables with the same margins as in the observed table and compute the exact probability for each such table. A method for accomplishing this task is given as follows:

10.11 | **Enumeration of All Possible Tables with the Same Margins as the Observed Table**

(1) Rearrange the rows and columns of the observed table so that the smaller row total is in the first row and the smaller column total is in the first column.

Suppose that after the rearrangement, the cells in the observed table are a, b, c, d, as depicted in Table 10.10.

(2) Start with the table with 0 in the (1, 1) cell. The other cells in this table are then determined from the row and column margins. Indeed, to maintain the same row and column margins as the observed table, the (1, 2) element must be $a + b$, the (2, 1) cell must be $a + c$, and the (2, 2) element must be $(c + d) - (a + c) = d - a$.

(3) Construct the next table by increasing the (1, 1) cell by 1 (i.e., from 0 to 1), decreasing the (1, 2) and (2, 1) cells by 1, and increasing the (2, 2) cell by 1.

(4) Continue increasing and decreasing the cells by 1, as in step 3, until one of the cells is 0, at which point all possible tables with the given row and column margins have been enumerated. Each table in the sequence of tables is referred to by the (1, 1) element. Thus, the first table is the 0 table, the next table is the 1 table, and so on.

EXAMPLE 10.24   **Cardiovascular Disease, Nutrition** Enumerate all possible tables with the same row and column margins as the observed table in Table 10.9.

SOLUTION The observed table has $a = 2, b = 23, c = 5, d = 30$. The rows or columns do not need to be rearranged, since the first row total is smaller than the second row total, and the first column total is smaller than the second column total. Start with the 0 table, which has 0 in the $(1, 1)$ cell, 25 in the $(1, 2)$ cell, 7 in the $(2, 1)$ cell, and $30 - 2$, or 28, in the $(2, 2)$ cell. The 1 table then has 1 in the $(1, 1)$ cell, $25 - 1 = 24$ in the $(1, 2)$ cell, $7 - 1 = 6$ in the $(2, 1)$ cell, and $28 + 1 = 29$ in the $(2, 2)$ cell. Continue in this fashion until the 7 table is reached, which has 0 in the $(2, 1)$ cell, at which point all possible tables with the given row and column margins have been enumerated. This collection of tables is shown in Table 10.12.

TABLE 10.12
Enumeration of all possible tables with fixed margins for Example 10.24

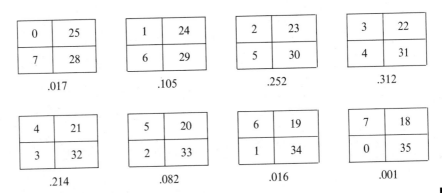

The exact probability of each table that has been enumerated must now be computed. This computation could be done directly by evaluating the probability of each table individually, using the formula in **(10.10)**. However, a **recursion rule** is often easier for this purpose. The advantage of the recursion rule is that once all the tables are generated, the probabilities of all tables can be calculated *at once* without computing a single factorial. The recursion rule is given as follows:

10.12 | **Recursion Rule for Computing Exact Probabilities**

(1) Enumerate all possible tables with the same margins as the observed table, starting with the 0 table as presented in **(10.11)**. The probability of the a table is referred to as $Pr(a)$. Suppose that the a table and the $a + 1$ table are among the tables that have been enumerated. We can show that:

(2)
$$Pr(a + 1) = Pr(a) \times \frac{bc}{(a + 1)(d + 1)}$$

Notice that $bc/[(a + 1)(d + 1)]$ can be computed by starting at the lower left of the a table, multiplying by the terms along the dotted diagonal, and dividing by the terms along the dotted diagonal of the $a + 1$ table, as shown in Table 10.13. Note that the $(1, 2)$, $(2, 1)$, and $(2, 2)$ cells of the $a + 1$ table must be $b - 1$, $c - 1$, and $d + 1$, provided that $b \geqslant 1$, $c \geqslant 1$, which is assumed true.

(3) This procedure can be used to calculate the probabilities of all possible tables. Since $Pr(1)$ can be expressed in terms of $Pr(0)$, $Pr(2)$ can also be expressed in terms of $Pr(1)$, which can in turn be expressed in terms of $Pr(0)$, and so on. Thus, the probability of each table can be expressed as a multiple of $Pr(0)$, and since all the probabilities add up to 1, we can solve first for $Pr(0)$ and then for each of the other probabilities. In particular, suppose that the k table is the last table enumerated. If

$$Pr(1) = \lambda_1 Pr(0), Pr(2) = \lambda_2 Pr(0), \ldots, Pr(k) = \lambda_k Pr(0)$$

then
$$1 = Pr(0) + Pr(1) + \cdots + Pr(k)$$
$$= Pr(0) + \lambda_1 Pr(0) + \lambda_2 Pr(0) + \cdots + \lambda_k Pr(0)$$
$$= Pr(0)(1 + \lambda_1 + \lambda_2 + \cdots + \lambda_k)$$

Therefore,
$$Pr(0) = \frac{1}{1 + \lambda_1 + \lambda_2 + \cdots + \lambda_k}$$

$$Pr(1) = \frac{\lambda_1}{1 + \lambda_1 + \lambda_2 + \cdots + \lambda_k}$$

$$Pr(2) = \frac{\lambda_2}{1 + \lambda_1 + \lambda_2 + \cdots + \lambda_k}$$

$$\vdots$$

$$Pr(k) = \frac{\lambda_k}{1 + \lambda_1 + \lambda_2 + \cdots + \lambda_k}$$

TABLE 10.13
Illustration of the
recursion rule for
computation of exact
probabilities

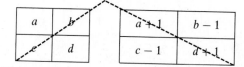

EXAMPLE 10.25   **Cardiovascular Disease, Nutrition** Evaluate the exact probabilities of all tables enumerated in Table 10.12.

SOLUTION   Refer to Table 10.12. Using the recursion rule,

$$Pr(0) = 1Pr(0)$$

$$Pr(1) = \frac{7 \times 25}{1 \times 29} Pr(0) = 6.034 Pr(0)$$

$$Pr(2) = \frac{6 \times 24}{2 \times 30} Pr(1) = 2.40 Pr(1) = 2.40 \times 6.034 Pr(0) = 14.482 Pr(0)$$

$$Pr(3) = \frac{5 \times 23}{3 \times 31} Pr(2) = 17.908 Pr(0)$$

$$Pr(4) = \frac{4 \times 22}{4 \times 32} Pr(3) = 12.312 Pr(0)$$

$$Pr(5) = \frac{3 \times 21}{5 \times 33} Pr(4) = 4.701 Pr(0)$$

$$Pr(6) = \frac{2 \times 20}{6 \times 34} Pr(5) = 0.922 Pr(0)$$

$$Pr(7) = \frac{1 \times 19}{7 \times 35} Pr(6) = 0.072 Pr(0)$$

All these probabilities must add up to 1. Hence,

$$Pr(0)(1 + 6.034 + 14.482 + 17.908 + 12.312 + 4.701 + 0.922 + 0.072) = 1$$

or $$Pr(0)(57.431) = 1 \quad \text{or} \quad Pr(0) = .0174$$

We can now solve for all the other probabilities in terms of $Pr(0)$. These probabilities are listed under the corresponding tables in Table 10.12. It is advisable to check that these probabilities add up to 1. They do in this case, except for roundoff error. ∎

The question now is what should be done with these probabilities to evaluate the significance of the results. The answer depends on whether a one-sided or a two-sided alternative is being used. In general, the following procedure can be used:

10.13 | **Fisher's Exact Test: General Procedure and Computation of p-value**

To test the hypothesis $H_0: p_1 = p_2$ versus $H_1: p_1 \neq p_2$, where the expected value of at least one cell is <5 when the data are analyzed in the form of a 2×2 contingency table, use the following procedure:

(1) Enumerate all possible tables with the same row and column margins as the observed table, as shown in **(10.11)**.

(2) Compute the exact probability of each table enumerated in step 1, using either the direct method in **(10.10)** or the recursion rule in **(10.12)**.

(3) Suppose that the observed table is the a table and that the last table enumerated is the k table.

    (a) To test the hypothesis $H_0: p_1 = p_2$ versus $H_1: p_1 \neq p_2$, the p-value $= 2 \times min[Pr(0) + Pr(1) + \cdots + Pr(a), Pr(a) + Pr(a + 1) + \cdots + Pr(k)]$.

    (b) To test the hypothesis $H_0: p_1 = p_2$ versus $H_1: p_1 < p_2$, the p-value $= Pr(0) + Pr(1) + \cdots + Pr(a)$.

    (c) To test the hypothesis $H_0: p_1 = p_2$ versus $H_1: p_1 > p_2$, the p-value $= Pr(a) + Pr(a + 1) + \cdots + Pr(k)$.

For each of these three alternative hypotheses, the p-value can be interpreted as the probability of obtaining a table as extreme as or more extreme than the observed table.

EXAMPLE 10.26

Cardiovascular Disease, Nutrition Evaluate the statistical significance of the data in Example 10.22.

SOLUTION

Suppose there is a two-sided alternative of the form $H_0: p_1 = p_2$ versus $H_1: p_1 \neq p_2$. Our table is the 2 table in Table 10.12. Thus, to compute the p-value, the smaller of the tail probabilities corresponding to the 2 table is computed and doubled. This strategy corresponds to the procedures for the various normal theory tests studied in Chapters 7 and 8. First compute the left-hand tail area,

$$Pr(0) + Pr(1) + Pr(2) = .017 + .105 + .252 = .374$$

and the right-hand tail area,

$$Pr(2) + Pr(3) + \cdots + Pr(7) = .252 + .312 + .214 + .082 + .016 + .001 = .877$$

Then
$$p = 2 \times min(.374, .877) = 2(.374) = .748$$

If a one-sided alternative of the form $H_0: p_1 = p_2$ versus $H_1: p_1 < p_2$ is used, then the p-value equals

$$Pr(0) + Pr(1) + Pr(2) = .017 + .105 + .252 = .374$$

Thus, the two proportions in this example are *not* significantly different with either a one-sided or two-sided alternative, and we *cannot* say, on the basis of this limited amount of data, that there is a significant association between salt intake and cause of death. ∎

If there are a large number of tables with the same margins as the observed table, then computer programs can be used to implement Fisher's exact test using statistical packages such as SAS.

SECTION 10.5

Two-Sample Test for Binomial Proportions for Matched-Pair Data (McNemar's Test)

EXAMPLE 10.27   **Cancer** Suppose we want to compare two different treatments for a rare form of cancer. Since relatively few cases of this disease are seen, the two treatment groups should be as comparable as possible. To accomplish this goal, a matched study is set up such that a random member of each matched pair gets treatment A (chemotherapy), whereas the other member gets treatment B (surgery). The patients are assigned to pairs matched on age (within 5 years), sex, and clinical condition. The patients are followed for 5 years, with survival as the outcome variable. The data are displayed in a 2 × 2 table, as shown in Table 10.14. Notice the small difference in survival between the two treatment groups: the 5-year survival rate for treatment $A = 106/621 = .171$ and for treatment $B = 95/621 = .153$. Indeed, the Yates-corrected chi-square statistic as given in **(10.5)** is 0.59 with 1 df, which is not significant. However, *the use of this test is valid only if the two samples are independent.* From the manner in which the samples were selected it is obvious that they are *not* independent, since each pair is similar in age, sex, and clinical condition. Thus, the Yates-corrected chi-square test *cannot* be used in this situation, since the p-values will not be correct. How can the two treatments be compared using a hypothesis test?   ■

TABLE 10.14
A 2 × 2 contingency table comparing treatments A and B for a rare form of cancer based on 1242 patients

	Outcome		
Treatment	**Survive for 5 years**	**Die within 5 years**	**Total**
A	106	515	621
B	95	526	621
Total	201	1041	1242

Suppose a different kind of 2 × 2 table is constructed to illustrate these data. In Table 10.14 the *person* was the basic unit, and the sample size was 1242 people. In Table 10.15 the *matched pair* is the basic unit, and *pairs* are classified according to whether or not each treatment worked for the members of that pair. Notice that Table 10.15 has 621 units rather than the 1242 in Table 10.14. Furthermore, there are 90 pairs in which both patients survived, 510 pairs in which both patients died, 16 pairs in which the treatment-A patient survived and the treatment-B patient died, and 5 pairs in which the treatment-B patient survived and the treatment-A patient died. The dependence of the two samples can be illustrated by noting that the probability that the treatment-B member of the pair survived given that the treatment-A member of the pair survived $= 90/106 = .849$, while the probability that the treatment-B member of the pair survived given that the treatment-A member of the pair died $= 5/515 = .010$. If the samples were independent, then

TABLE 10.15
A 2 × 2 table with the
matched pair as the
sampling unit based on
621 matched pairs

Outcome of treatment-A patient	Outcome of treatment-B patient		Total
	Survive	Die	
Survive	90	16	106
Die	5	510	515
Total	95	526	621

these two probabilities should be about the same. Thus, we conclude that the samples are highly dependent and that the chi-square test cannot be used.

In Table 10.15, for 600 pairs (90 + 510), the outcomes of the two treatments are the same, whereas for 21 pairs (16 + 5), the outcomes of the two treatments are different. The following special names are given to each of these types of pairs:

DEFINITION 10.6 ■■
A **concordant pair** is a matched pair in which the outcome is the same for each member of the pair. ■

DEFINITION 10.7 ■■
A **discordant pair** is a matched pair in which the outcomes are different for the members of the pair. ■

EXAMPLE 10.28   There are 600 concordant pairs and 21 discordant pairs for the data in Table 10.15.   ■

The concordant pairs provide no information about *differences between treatments* and will not be used in the assessment. Instead, the focus will be on the discordant pairs, which can be divided into two types:

DEFINITION 10.8 ■■
A **type A discordant pair** is a discordant pair in which the treatment-A member of the pair has the event and the treatment-B member does not. Similarly, a **type B discordant pair** is a discordant pair in which the treatment-B member of the pair has the event and the treatment-A member does not. ■

EXAMPLE 10.29   There are 16 type A discordant pairs and 5 type B discordant pairs from the data in Table 10.15.   ■

Let p = the probability that a discordant pair is of type A. If the treatments are equally effective, then about an equal number of type A and type B discordant pairs would be expected, and p should $= \frac{1}{2}$. If treatment A is more effective than treatment B, then more type A than type B discordant pairs would be expected, and p should be $> \frac{1}{2}$. Finally, if treatment B is more effective than treatment A, then more type B than type A discordant pairs would be expected, and p should be $< \frac{1}{2}$.

Thus we wish to test the hypothesis $H_0: p = \frac{1}{2}$ versus $H_1: p \neq \frac{1}{2}$.

10.5.1   Normal Theory Test

Suppose that of n_D discordant pairs, n_A are type A. Then under H_0, $E(n_A) = n_D/2$ and $Var(n_A) = n_D/4$, from the mean and variance of a binomial distribution, respectively. We will assume that the normal approximation to the binomial distribution holds, but will use a continuity correction for a better approximation. This approximation will be valid if $npq = n_D/4 \geqslant 5$ or $n_D \geqslant 20$. The following test procedure, referred to as McNemar's test, can then be used:

10.14 | **McNemar's Test for Correlated Proportions—Normal Theory Test**

(1) Form a 2×2 table of matched pairs, where the outcomes for the treatment-A members of the matched pairs are listed along the rows and the outcomes for the treatment-B members are listed along the columns.

(2) Count the total number of discordant pairs (n_D) and the number of type A discordant pairs (n_A).

(3) Compute the test statistic

$$X^2 = \left(\left| n_A - \frac{n_D}{2} \right| - \frac{1}{2} \right)^2 \Big/ \left(\frac{n_D}{4} \right)$$

(4) For a two-sided level α test, if

$$X^2 > \chi^2_{1,1-\alpha}$$

then reject H_0; if

$$X^2 \leqslant \chi^2_{1,1-\alpha}$$

then accept H_0.

(5) The exact p-value is given by $p = Pr(\chi^2_1 \geqslant X^2)$.

(6) Use this test only if $n_D \geqslant 20$.

The acceptance and rejection regions for this test are depicted in Figure 10.5. The computation of the p-value for McNemar's test is depicted in Figure 10.6.

This is a two-sided test despite the one-sided nature of the critical region in Figure 10.5. The rationale for this is that if either $p < \frac{1}{2}$ or $p > \frac{1}{2}$, $|n_A - n_D/2|$ will be large and, correspondingly, X^2 will be large. Thus, for alternatives on either side of the null hypothesis ($p = \frac{1}{2}$), H_0 is rejected if X^2 is large and accepted if X^2 is small.

$$X^2 = \left(\left| n_A - \frac{n_D}{2} \right| - \frac{1}{2} \right)^2 \Big/ \left(\frac{n_D}{4} \right)$$

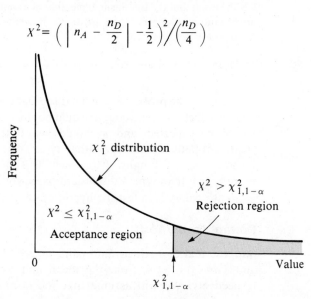

FIGURE 10.5

Acceptance and rejection regions for McNemar's test— normal theory method

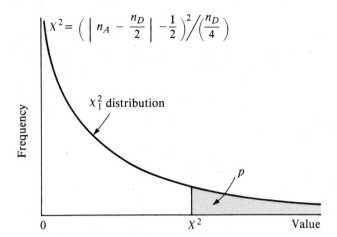

$$X^2 = \left(\left|\,n_A - \frac{n_D}{2}\,\right| - \frac{1}{2}\right)^2 \Big/ \left(\frac{n_D}{4}\right)$$

X_1^2 distribution

Frequency

p

0   X^2 Value

FIGURE 10.6
Computation of the p-value for McNemar's test—normal theory method

EXAMPLE 10.30   **Cancer** Assess the statistical significance of the data in Table 10.15.

SOLUTION   Note that $n_D = 21$. Since $n_D(\frac{1}{2})(\frac{1}{2}) = 5.25 \geqslant 5$, the normal approximation to the binomial distribution and the test in **(10.14)** can be used:

$$X^2 = \frac{(|16 - 10.5| - \frac{1}{2})^2}{(21/4)} = \frac{(5.5 - \frac{1}{2})^2}{5.25} = \frac{5^2}{5.25} = \frac{25}{5.25} = 4.76$$

From Table 6 in Appendix 1, note that

$$\chi^2_{1,.95} = 3.84$$

$$\chi^2_{1,.975} = 5.02$$

Thus, $.025 < p < .05$, and the results are statistically significant.   ∎

We conclude that *if the treatments give different results from each other* for the members of a matched pair, then the treatment-A member of the pair is significantly more likely to survive than the treatment-B member. Thus, all other things being equal (such as side effects, cost, etc.), treatment A would be the treatment of choice.

10.5.2   **Exact Test**

If $n_D/4 < 5$, that is, if $n_D < 20$, then the normal approximation to the binomial distribution cannot be used, and a test based on exact binomial probabilities is required. The details of the test procedure are similar to the one-sample binomial test in **(7.36)** and are summarized as follows:

10.15   **McNemar's Test for Correlated Proportions—Exact Test**

(1) Follow the procedure in step 1 in **(10.14)**.

(2) Follow the procedure in step 2 in **(10.14)**.

(3)
$$p = 2 \times \sum_{k=0}^{n_A} \binom{n_D}{k}\left(\frac{1}{2}\right)^{n_D} \quad \text{if } n_A < n_D/2$$

$$p = 2 \times \sum_{k=n_A}^{n_D} \binom{n_D}{k}\left(\frac{1}{2}\right)^{n_D} \quad \text{if } n_A \geqslant n_D/2$$

(4) This test is valid for any number of discordant pairs (n_D) but is particularly useful for $n_D < 20$, when the normal theory test in **(10.14)** cannot be used.

The computation of the p-value for this test is depicted in Figure 10.7.

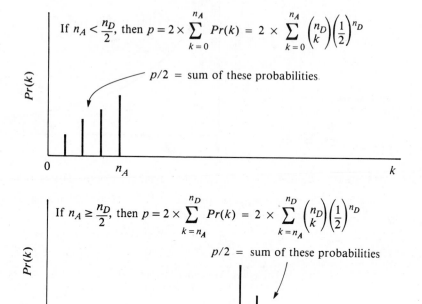

If $n_A < \dfrac{n_D}{2}$, then $p = 2 \times \displaystyle\sum_{k=0}^{n_A} Pr(k) = 2 \times \displaystyle\sum_{k=0}^{n_A} \binom{n_D}{k}\left(\frac{1}{2}\right)^{n_D}$

$p/2$ = sum of these probabilities

If $n_A \geq \dfrac{n_D}{2}$, then $p = 2 \times \displaystyle\sum_{k=n_A}^{n_D} Pr(k) = 2 \times \displaystyle\sum_{k=n_A}^{n_D} \binom{n_D}{k}\left(\frac{1}{2}\right)^{n_D}$

$p/2$ = sum of these probabilities

FIGURE 10.7
Computation of the p-value for McNemar's test—exact method

EXAMPLE 10.31

Hypertension A recent phenomenon in the recording of blood pressure is the development of the automated blood-pressure machine, where for a small fee a person can sit in a booth and have his or her blood pressure measured by a computer device. A study is conducted to compare the computer device with standard methods of measuring blood pressure. Twenty patients are recruited, and their hypertensive status is assessed by both the computer device and a trained observer. Hypertensive status is defined as either hypertensive $(+)$, if either systolic bp ≥ 160 or diastolic bp ≥ 95, or normotensive $(-)$ otherwise. The data are given in Table 10.16. Assess the statistical significance of these data.

SOLUTION

An ordinary Yates-corrected chi-square test cannot be used on these data, since each person is being used as his or her own control and there are *not* two independent samples. Instead, a 2×2 table of matched pairs is formed, as shown in Table 10.17. Note that 3 people are measured as hypertensive by both the computer device and the trained observer, 9 people are normotensive by both methods, 7 people are hypertensive by the computer device and normotensive by the trained observer, and 1 person is normotensive by the computer device and hypertensive by the trained observer. Therefore, there are 12 $(9 + 3)$ concordant pairs and 8 $(7 + 1)$ discordant pairs (n_D). Since $n_D < 20$, the exact method must be used. We see that $n_A = 7$, $n_D = 8$. Therefore, since $n_A \geq n_D/2 = 4$, it follows from **(10.15)** that

$$p = 2 \times \sum_{k=7}^{8} \binom{8}{k}\left(\frac{1}{2}\right)^{8}$$

TABLE 10.16
Hypertensive status of 20 patients as judged by a computer device and a trained observer

Person	Hypertensive status		Person	Hypertensive status	
	Computer device	Trained observer		Computer device	Trained observer
1	−	−	11	+	−
2	−	−	12	+	−
3	+	−	13	−	−
4	+	+	14	+	−
5	−	−	15	−	+
6	+	−	16	+	−
7	−	−	17	+	−
8	+	+	18	−	−
9	+	+	19	−	−
10	−	−	20	−	−

TABLE 10.17
Comparison of hypertensive status as judged by a computer device and a trained observer

Computer device	Trained observer	
	+	−
+	3	7
−	1	9

This expression can be evaluated using Table 1 in Appendix 1 by referring to $n = 8$, $p = .5$ and noting that $Pr(X \geqslant 7 | p = .5) = .0313 + .0039 = .0352$. Thus, the two-tailed p-value = $2 \times .0352 = .070$.

Alternatively, a computer program could be used to perform the computations, as shown in Table 10.18. Note that the first and second columns have been interchanged so that the discordant pairs appear in the diagonal elements (and are easier to identify). In summary, the results are not statistically significant, and we cannot conclude that there is a significant difference between the two methods, although a *trend* toward the computer device identifying more hypertensives than the trained observer can be detected.

TABLE 10.18
Use of SPSS^X/PC McNemar Test program to evaluate the significance of the data in Table 10.17

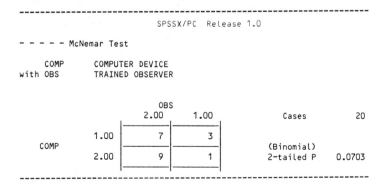

```
                           SPSSX/PC  Release 1.0

- - - - - McNemar Test

        COMP        COMPUTER DEVICE
     with OBS       TRAINED OBSERVER


                          OBS
                       2.00    1.00         Cases        20

                1.00     7       3
         COMP                              (Binomial)
                2.00     9       1         2-tailed P   0.0703
```

To use McNemar's test for correlated proportions, or the sign test in Section 9.2, the same test procedure is followed. For the sign test, the number of positive versus the number of negative difference scores is compared, while for McNemar's test the number of type A versus type B discordant pairs is compared. Similarly, for a two-sided one-sample binomial test, the hypothesis $H_0: p = p_0$ versus $H_1: p \neq p_0$ is tested. In the special case where $p_0 = \frac{1}{2}$, the same test procedure as for McNemar's test and the sign test in Section 9.2 is also followed.

Estimation of Sample Size and Power for Comparing Two Binomial Proportions

In Section 8.8, methods for estimating the sample size needed to compare means from two normally distributed populations were presented. In this section, similar methods for estimating the sample size required to compare two proportions are developed.

10.6.1 Independent Samples

EXAMPLE 10.32   **Cancer, Nutrition** Suppose we know from Connecticut tumor registry data that the incidence rate of breast cancer over a 1-year period for initially disease-free women ages 45–49 is 150 cases per 100,000 [3]. We wish to study whether or not the ingestion of large doses of vitamin A in tablet form will prevent breast cancer. The study is set up with (1) a control group of 45–49-year-old women who are given placebo pills by mail and are anticipated to have the same disease rate as indicated in the Connecticut tumor registry data and (2) a study group of similarly aged women who are given vitamin A pills by mail and are anticipated to have a 20% reduction in risk. How large a sample is needed to do a two-sided test with a significance level of .05 and a power of 80%? ∎

We wish to test the hypothesis $H_0: p_1 = p_2$ versus $H_1: p_1 \neq p_2$. Suppose that we wish to conduct a test with significance level α and power $1 - \beta$ and anticipate that there will be k times as many people in group 2 as in group 1; that is, $n_2 = kn_1$. The sample size required in each of the two groups to achieve these objectives is given as follows:

10.16 | **Sample Size Needed to Compare Two Binomial Proportions Using a Two-Sided Test with Significance Level α and Power $1 - \beta$, Where One Sample (n_2) Is k Times as Large as the Other Sample (n_1) (Independent-Sample Case)**

To test the hypothesis $H_0: p_1 = p_2$ versus $H_1: p_1 \neq p_2$ for the specific alternative $|p_1 - p_2| = \Delta$, with significance level α and power $1 - \beta$, the following sample size is required:

$$n_1 = \left[\sqrt{\bar{p}\bar{q}\left(1 + \frac{1}{k}\right)}\, z_{1-\alpha/2} + \sqrt{p_1 q_1 + \frac{p_2 q_2}{k}}\, z_{1-\beta} \right]^2 \Big/ \Delta^2$$

$$n_2 = kn_1$$

where

p_1, p_2 = projected true probabilities of success in the two groups

$q_1, q_2 = 1 - p_1, 1 - p_2$

$\Delta = p_2 - p_1$

$\bar{p} = \dfrac{p_1 + kp_2}{1 + k}$

$\bar{q} = 1 - \bar{p}$

EXAMPLE 10.33 **Cancer, Nutrition** Estimate the sample size required for the study proposed in Example 10.32 if an equal sample size is anticipated in each group.

SOLUTION
$$p_1 = 150 \text{ per } 100,000 \text{ or } 150/10^5 = .00150$$
$$q_1 = 1 - .00150 = .99850$$

If we wish to detect a 20% reduction in risk, then $p_2 = 0.8p_1$ or
$$p_2 = (150 \times .8)/10^5 = 120/10^5 = .00120$$
$$q_2 = 1 - .00120 = .99880$$
$$\alpha = .05$$
$$1 - \beta = .8$$
$$k = 1 \text{ (since } n_1 = n_2)$$
$$\bar{p} = \frac{.00150 + .00120}{2} = .00135$$
$$\bar{q} = 1 - .00135 = .99865$$
$$z_{1-\alpha/2} = z_{.975} = 1.96$$
$$z_{1-\beta} = z_{.80} = 0.84$$

Thus, referring to **(10.16)**,

$$n_1 = \frac{[\sqrt{(.00135)(.99865)(1+1)}(1.96) + \sqrt{(.00150)(.99850) + (.00120)(.99880)}(0.84)]^2}{(.00150 - .00120)^2}$$

$$= \frac{[(.05193)(1.96) + (.05193)(0.84)]^2}{(.00030)^2} = \frac{(.14540)^2}{(.00030)^2} = 234,902 = n_2$$

or about 235,000 women in each group. ∎

To perform a one-tailed rather than a two-tailed test, simply substitute α for $\alpha/2$ in the sample-size formula in **(10.16)** as follows:

10.17 | **Sample Size Needed to Compare Two Binomial Proportions Using a One-Sided Test with Significance Level α and Power $1 - \beta$, Where One Sample (n_2) Is k Times as Large as the Other Sample (n_1) (Independent-Sample Case)**

To test the hypothesis $H_0: p_1 = p_2$ versus $H_1: p_1 < p_2$ for the specific alternative $p_1 = p_2 - \Delta$, with significance level α and power $1 - \beta$, use the formula

$$n_1 = \left[\sqrt{\bar{p}\bar{q}\left(1 + \frac{1}{k}\right)} z_{1-\alpha} + \sqrt{\frac{p_1 q_1 + p_2 q_2}{k}} z_{1-\beta} \right]^2 \bigg/ \Delta^2$$

$$n_2 = kn_1$$

To test the hypothesis $H_0: p_1 = p_2$ versus $H_1: p_1 > p_2$ for the specific alternative $p_2 = p_1 + \Delta$, use the same formula, where

$$p_1, p_2 = \text{projected true probabilities of success in the two groups}$$
$$q_1, q_2 = 1 - p_1, 1 - p_2$$
$$\bar{p} = \frac{p_1 + kp_2}{1 + k}$$
$$\bar{q} = 1 - \bar{p}$$

EXAMPLE 10.34   **Cancer, Nutrition** Suppose that in considering the study design proposed in Example 10.32 more carefully, we decide that a one-sided test is really better, since we anticipate that vitamin A consumption will either lower the risk of breast cancer or will have no effect, but will not raise it. Estimate the required sample size for the study proposed in Example 10.32 if an equal sample size is anticipated for each group.

SOLUTION   We have the same parameters as in Example 10.33. Use $1 - \alpha = .95$ here rather than $1 - \alpha/2 = .975$, as was the case in Example 10.33. From **(10.17)**,

$$n_1 = \frac{[\sqrt{(.00135)(.99865)(2)}(1.645) + \sqrt{(.00150)(.99850) + (.00120)(.99880)}(0.84)]^2}{(.00030)^2}$$

$$= \frac{[(.05193)(1.645) + (.05193)(0.84)]^2}{(.00030)^2} = \frac{(0.12905)^2}{(.00030)^2} = 185,043 = n_2$$

or about 185,000 women in each group.   ∎

Clearly, from the results in Examples 10.33 and 10.34, it would not be feasible to conduct such a large study over a 1-year period. The sample size needed would be reduced considerably if the period of study was lengthened beyond 1 year, since the expected number of events would increase in a multiyear study.

In many instances the sample size available for investigation is fixed by practical constraints, and what is desired is an estimate of statistical power with the anticipated available sample size. In other instances, after a study is completed, we want to calculate the power using the sample sizes that were actually used in the study. For these purposes the following estimate of power is provided to test the hypothesis $H_0: p_1 = p_2$ versus $H_1: p_1 \neq p_2$, with significance level α and sample sizes of n_1 and n_2 in the two groups.

10.18   **Power Achieved in Comparing Two Binomial Proportions Using a Two-Sided Test with Significance Level α and Samples of Sizes n_1 and n_2 (Independent-Sample Case)**

To test the hypothesis $H_0: p_1 = p_2$ versus $H_1: p_1 \neq p_2$ for the specific alternative $|p_1 - p_2| = \Delta$, compute

$$\text{Power} = \Phi\left[\frac{\Delta}{\sqrt{p_1 q_1/n_1 + p_2 q_2/n_2}} - z_{1-\alpha/2}\frac{\sqrt{\bar{p}\bar{q}(1/n_1 + 1/n_2)}}{\sqrt{p_1 q_1/n_1 + p_2 q_2/n_2}}\right]$$

where

p_1, p_2 = projected true probabilities of success in groups 1, 2, respectively

$q_1, q_2 = 1 - p_1, 1 - p_2$

$\Delta = |p_2 - p_1|$

$\bar{p} = \dfrac{n_1 p_1 + n_2 p_2}{n_1 + n_2}$

$\bar{q} = 1 - \bar{p}$

EXAMPLE 10.35   **Otolaryngology** Suppose a study comparing a medical and a surgical treatment for children who have an excessive number of episodes of otitis media (OTM) during the first 3 years of life is planned. Success rates of 50% and 70% are assumed in the medical and surgical groups, respectively, and the recruitment of 100 patients for each group is realistically anticipated. Success is defined as $\leqslant 1$ episode of OTM in the first 12 months after treatment. How much power does such a study have of detecting a significant difference if a two-sided test with an α level of .05 is to be used?

SOLUTION   Note that $p_1 = .5$, $p_2 = .7$, $q_1 = .5$, $q_2 = .3$, $n_1 = n_2 = 100$, $\Delta = .2$, $\bar{p} = (.5 + .7)/2 = .6$, $\bar{q} = .4$, $\alpha = .05$, $z_{1-\alpha/2} = z_{.975} = 1.96$. Thus, from **(10.18)** the power can be computed as follows:

$$\text{Power} = \Phi\left[\frac{.2}{\sqrt{[(.5)(.5) + (.7)(.3)]/100}} - \frac{1.96\sqrt{(.6)(.4)(1/100 + 1/100)}}{\sqrt{[(.5)(.5) + (.7)(.3)]/100}}\right]$$

$$= \Phi\left(\frac{.2}{.0678} - 1.96\frac{.0693}{.0678}\right) = \Phi(2.950 - 2.003) = \Phi(0.947) = .83$$

Thus, there is an 83% chance of finding a significant difference using the anticipated sample sizes.   ∎

If a one-sided test is used, then **(10.18)** can be used after replacing $z_{1-\alpha/2}$ by $z_{1-\alpha}$:

10.19   **Power Achieved in Comparing Two Binomial Proportions Using a One-Sided Test with Significance Level α and Samples of Size n_1 and n_2 (Independent-Sample Case)**

To test the hypothesis $H_0: p_1 = p_2$ versus $H_1: p_1 < p_2$ for the specific alternative $p_1 = p_2 - \Delta$, use the following formula:

$$\text{Power} = \Phi\left[\frac{\Delta}{\sqrt{p_1 q_1/n_1 + p_2 q_2/n_2}} - z_{1-\alpha}\frac{\sqrt{\bar{p}\bar{q}(1/n_1 + 1/n_2)}}{\sqrt{p_1 q_1/n_1 + p_2 q_2/n_2}}\right]$$

where p_1, p_2 = projected true probabilities of success in groups 1, 2, respectively

$q_1, q_2 = 1 - p_1, 1 - p_2$

$\bar{p} = \dfrac{n_1 p_1 + n_2 p_2}{n_1 + n_2}$

$\bar{q} = 1 - \bar{p}$

To test the hypothesis $H_0: p_1 = p_2$ versus $H_1: p_1 > p_2$ for the specific alternative $p_1 = p_2 + \Delta$, use the same formula.

10.6.2   **Paired Samples**

In Section 10.5, McNemar's test for comparing binomial proportions in paired samples was introduced. As noted there, this test is a special case of the one-sample binomial test. Therefore, to estimate sample size and power, the more general formulas for the one-sample binomial test given in Section 7.10.3 can be used. Specifically, referring to **(7.38)** (p. 231), to test the hypothesis $H_0: p = p_0$ versus $H_1: p \neq p_0$ using a two-sided test with significance level α and power $1 - \beta$ for the specific alternative $p = p_1$, a sample size of

$$n = \frac{p_0 q_0[z_{1-\alpha/2} + z_{1-\beta}\sqrt{p_1 q_1/(p_0 q_0)}]^2}{(p_1 - p_0)^2}$$

is needed. To use this formula in the case of McNemar's test, set $p_0 = q_0 = \frac{1}{2}$, $p_1 = p_A =$ the proportion of discordant pairs that are of type A, and $n = n_D =$ the number of discordant pairs. Upon substitution,

$$n_D = \frac{(z_{1-\alpha/2} + 2z_{1-\beta}\sqrt{p_A q_A})^2}{4(p_A - .5)^2}$$

However, the number of discordant pairs $(n_D) =$ the total number of pairs $(n) \times$ the probability that a matched pair is discordant. If the latter probability is denoted by p_D, $n_D = np_D$, or $n = n_D/p_D$. Therefore, the following sample size formula can be used:

10.20 | **Sample Size Needed to Compare Two Binomial Proportions Using a Two-Sided Test with Significance Level α and Power $1 - \beta$ (Paired-Sample Case)**

If McNemar's test for correlated proportions is used to test the hypothesis $H_0: p = \frac{1}{2}$ versus $H_1: p \neq \frac{1}{2}$, for the specific alternative $p = p_A$, where $p =$ the probability that a discordant pair is of type A, then use

$$n = \frac{(z_{1-\alpha/2} + 2z_{1-\beta}\sqrt{p_A q_A})^2}{4(p_A - .5)^2 p_D} \text{ matched pairs}$$

or

$$2n = \frac{(z_{1-\alpha/2} + 2z_{1-\beta}\sqrt{p_A q_A})^2}{2(p_A - .5)^2 p_D} \text{ individuals}$$

where $p_D =$ projected proportion of discordant pairs among all pairs

   $p_A =$ projected proportion of discordant pairs of type A among discordant pairs

---

EXAMPLE 10.36   **Cancer** Suppose we wish to compare two different modes of chemotherapy (A, B) for the treatment of breast cancer. A matched-pair design is used, where patients are matched on age and clinical stage of disease, and one patient in a matched pair is assigned to treatment A, while the other is assigned to treatment B. Based on previous work, it is estimated that patients in a matched pair will respond similarly to the treatments in 85% of matched pairs. Furthermore, for matched pairs where there is a difference in response, it is estimated that in $\frac{2}{3}$ of the pairs the treatment-A patient will respond, while the treatment-B patient will not; in $\frac{1}{3}$ of the pairs the treatment-B patient will respond, while the treatment-A patient will not. How many subjects (or matched pairs) need to be enrolled in the study to have a 90% chance of finding a significant difference using a two-sided test with type I error = .05?

SOLUTION   We have that $\alpha = .05$, $\beta = .10$, $p_D = 1 - .85 = .15$, $p_A = \frac{2}{3}$, $q_A = \frac{1}{3}$. Therefore, from **(10.20)**,

$$n \text{ (pairs)} = \frac{[z_{.975} + 2z_{.90}\sqrt{(2/3)(1/3)}]^2}{4(2/3 - 1/2)^2(.15)}$$

$$= \frac{[1.96 + 2(1.28)(.4714)]^2}{4(1/6)^2(.15)} = \frac{(3.1668)^2}{.0167} = 601 \text{ matched pairs}$$

$$2n = 2 \times 601 = 1202 \text{ individuals}$$

Therefore, 1202 individuals in 601 matched pairs need to be enrolled. This will yield approximately $.15 \times 601 = 90$ discordant pairs.   ∎

In some instances, the sample size is fixed and we want to determine what power a study has (or had) to detect specific alternatives. For a two-sided one-sample binomial test with significance level α, to test the hypothesis $H_0: p = p_0$ versus $H_1: p \neq p_0$ for the specific alternative $p = p_1$, the power is given by [see (7.37)]

$$\text{Power} = \Phi\left\{\sqrt{p_0 q_0/(p_1 q_1)}\left[z_{\alpha/2} + \frac{|p_1 - p_0|\sqrt{n}}{\sqrt{p_0 q_0}}\right]\right\}$$

For McNemar's test, set $p_0 = q_0 = \frac{1}{2}$, $p_1 = p_A$, and $n = n_D$, yielding

$$\text{Power} = \Phi\left[\frac{1}{2\sqrt{p_A q_A}}(z_{\alpha/2} + 2|p_A - .5|\sqrt{n_D})\right]$$

Upon substituting $n_D = np_D$, the following power formula is obtained:

10.21 | **Power Achieved in Comparing Two Binomial Proportions Using a Two-Sided Test with Significance Level α (Paired-Sample Case)**

If McNemar's test for correlated proportions is used to test the hypothesis $H_0: p = \frac{1}{2}$ versus $H_1: p \neq \frac{1}{2}$, for the specific alternative $p = p_A$, where $p =$ the probability that a discordant pair is of type A,

$$\text{Power} = \Phi\left\{\frac{1}{2\sqrt{p_A q_A}}[z_{\alpha/2} + 2(p_A - .5)\sqrt{np_D}]\right\} \quad \text{if } p_A \geq .5$$

$$= \Phi\left\{\frac{1}{2\sqrt{p_A q_A}}[z_{\alpha/2} + 2(.5 - p_A)\sqrt{np_D}]\right\} \quad \text{if } p_A < .5$$

where

$n =$ number of matched pairs

$p_D =$ projected proportion of discordant pairs among all pairs

$p_A =$ projected proportion of discordant pairs of type A among discordant pairs

---

EXAMPLE 10.37   **Cancer**   Consider the study in Example 10.36. If 400 matched pairs are enrolled, how much power would such a study have?

SOLUTION   We have that $\alpha = .05$, $p_D = .15$, $p_A = \frac{2}{3}$, $n = 400$. Therefore, from **(10.21)**,

$$\text{Power} = \Phi\left\{\frac{1}{2\sqrt{(2/3)(1/3)}}[z_{.025} + 2(2/3 - .5)\sqrt{400(.15)}]\right\}$$

$$= \Phi\{1.0607[-1.96 + 2(1/6)(7.7460)]\}$$

$$= \Phi[1.0607(0.6220)] = \Phi(0.660) = .745$$

Therefore, the study would have 74.5% power, or a 74.5% chance of detecting a statistically significant difference. ∎

To compute sample size and power for a one-sided alternative, substitute α for $\alpha/2$ in the formulas in **(10.20)** and **(10.21)**, respectively.

$R \times C$ Contingency Tables

10.7.1. **Tests for Association for $R \times C$ Tables**

In the previous sections of this chapter, methods of analyzing data that can be organized in the form of a 2×2 contingency table, that is, where each of the variables under study has only two categories, were studied. Frequently, one or both variables under study have more than two categories.

DEFINITION 10.9 ∎∎

An $R \times C$ **contingency table** is a table with R rows and C columns. It displays the relationship between two variables, where the variable depicted in the rows has R categories and the variable depicted in the columns has C categories. ∎

EXAMPLE 10.38 **Cancer** Suppose we wish to study further the relationship between age at first birth and the development of breast cancer, as given in Example 10.4. In particular, we would like to know if the effect of age at first birth follows a consistent trend, that is, (1) more protection for women whose age at first birth is <20 than for women whose age at first birth is 25–29 and (2) higher risk for women whose age at first birth is ≥ 35 than for women whose age at first birth is 30–34. The data are presented in Table 10.19, where case/control status is indicated along the rows and age at first birth is indicated along the columns. The data are arranged in the form of a 2×5 contingency table, since case/control status has two categories and age at first birth has five categories. We wish to test for a relationship between age at first birth and case/control status. How should this be done? ∎

TABLE 10.19
Data from the international study in Example 10.4 investigating the possible association between age at first birth and case/control status

Case/control status	Age at first birth					Total
	<20	20–24	25–29	30–34	≥ 35	
Case	320	1206	1011	463	220	3220
Control	1422	4432	2893	1092	406	10,245
Total	1742	5638	3904	1555	626	13,465
% cases	.184	.214	.259	.298	.351	.239

(Reprinted with permission of *WHO Bulletin*, 43, 209–221, 1970.)

Generalizing our experience from the 2×2 situation, the expected table for an $R \times C$ table can be formed in the same way as for a 2×2 table.

10.22 **Computation of the Expected Table for an $R \times C$ Contingency Table**

The expected number of units that fall in the (i, j) cell $= E_{ij} =$ the product of the number of units in the ith row multiplied by the number of units in jth column, divided by the total number of units in the table.

EXAMPLE 10.39 **Cancer** Compute the expected table for the data in Table 10.19.

SOLUTION

$$\text{Expected value of the (1, 1) cell} = \frac{\text{first row total} \times \text{first column total}}{\text{grand total}} = \frac{3220(1742)}{13,465} = 416.6$$

$$\text{Expected value of the (1, 2) cell} = \frac{\text{first row total} \times \text{second column total}}{\text{grand total}} = \frac{(3220)(5638)}{13,465} = 1348.3$$

$$\vdots$$

$$\text{Expected value of the (2, 5) cell} = \frac{\text{second row total} \times \text{fifth column total}}{\text{grand total}} = \frac{(10,245)(626)}{13,465} = 476.3$$

All 10 expected values are given in Table 10.20. ∎

TABLE 10.20
Expected table for the international study data in Table 10.19

Case/control status	Age at first birth					Total
	<20	20–24	25–29	30–34	≥35	
Case	416.6	1348.3	933.6	371.9	149.7	3220
Control	1325.4	4289.7	2970.4	1183.1	476.3	10.245
Total	1742	5638	3904	1555	626	13,465

The sum of the expected values across any row or column must equal the corresponding row or column total, as was the case for 2×2 tables. This fact provides a good check that the expected values are computed correctly. The expected values in Table 10.20 fulfill this criterion except for roundoff error.

We again want to compare the observed table with the expected table. The more similar these tables are, the more willing we will be to accept the null hypothesis that there is no relationship between the two variables. The more different the tables are, the more willing we will be to reject H_0. Again the criterion $(O - E)^2/E$ is used to compare the observed and expected tables for a particular cell. Furthermore, $(O - E)^2/E$ is summed over all the cells in the table to get an overall measure of how close the observed and expected tables are. Under H_0, for an $R \times C$ contingency table, the sum of $(O - E)^2/E$ over the RC cells in the table will approximately follow a chi-square distribution with $(R - 1) \times (C - 1)$ df. H_0 will be rejected for large values of this sum and will be accepted for small values.

Generally speaking, the continuity correction is not used for contingency tables larger than 2×2, since it has been found empirically that it does not aid in the approximation of the test statistic by the chi-square distribution. As was the case for 2×2 tables, this test should not be used if the expected values of the cells are too small. Cochran has studied the validity of the approximation in this case and recommends its use under the following conditions [4]:

(1) No more than $\frac{1}{5}$ of the cells have expected values <5.

(2) No cells have expected values <1.

The test procedure can be summarized as follows:

10.23 | **Chi-Square Test for an $R \times C$ Contingency Table**

To test for the relationship between two discrete variables, where one variable has R categories and the other has C categories, use the following procedure:

(1) Analyze the data in the form of an $R \times C$ contingency table, where O_{ij} represents the observed number of units in the (i, j) cell.

(2) Compute the expected table as shown in **(10.22)**, where E_{ij} represents the expected number of units in the (i, j) cell.

(3) Compute the test statistic

$$X^2 = (O_{11} - E_{11})^2/E_{11} + (O_{12} - E_{12})^2/E_{12} + \cdots + (O_{RC} - E_{RC})^2/E_{RC}$$

which under H_0 approximately follows a chi-square distribution with $(R - 1) \times (C - 1)$ df.

(4) For a level α test, if

$$X^2 > \chi^2_{(R-1)\times(C-1),1-\alpha}$$

reject H_0. If

$$X^2 \leqslant \chi^2_{(R-1)\times(C-1),1-\alpha}$$

accept H_0.

(5) The exact p-value is given by the area to the right of X^2 under a $\chi^2_{(R-1)\times(C-1)}$ distribution.

(6) Use this test only if the following two conditions are satisfied:

(a) No more than $\frac{1}{5}$ of the cells should have expected values <5.

(b) No cell should have expected value <1.

The acceptance and rejection regions for this test are depicted in Figure 10.8. The computation of the p-value for this test is illustrated in Figure 10.9.

$$X^2 = (O_{11} - E_{11})^2/E_{11} + \cdots + (O_{RC} - E_{RC})^2/E_{RC}$$

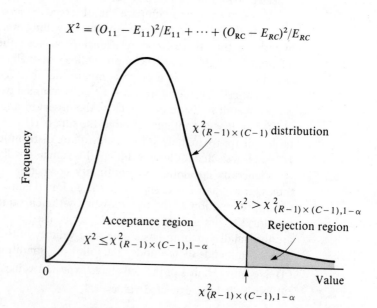

FIGURE 10.8
Acceptance and
rejection regions for the
chi-square test for an
$R \times C$ contingency
table

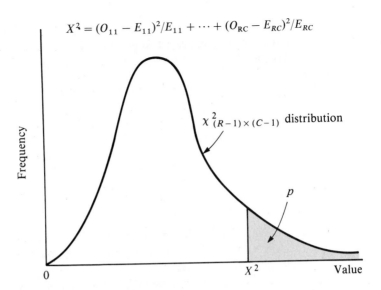

$$X^2 = (O_{11} - E_{11})^2/E_{11} + \cdots + (O_{RC} - E_{RC})^2/E_{RC}$$

$\chi^2_{(R-1) \times (C-1)}$ distribution

FIGURE 10.9
Computation of the p-value for the chi-square test for an $R \times C$ contingency table

EXAMPLE 10.40   **Cancer** Assess the statistical significance of the data in Example 10.38.

SOLUTION   From Table 10.20 we see that all expected values are ≥ 5; so the test procedure in **(10.22)** can be used. From Tables 10.19 and 10.20,

$$X^2 = \frac{(320 - 416.6)^2}{416.6} + \frac{(1206 - 1348.3)^2}{1348.3} + \cdots + \frac{(406 - 476.3)^2}{476.3} = 130.33$$

Under H_0, X^2 follows a chi-square distribution with $(2-1) \times (5-1)$, or 4, df. Since

$$\chi^2_{4, .999} = 18.47 < 130.33 = X^2$$

it follows that

$$p < 1 - .999 = .001$$

Therefore, the results are very highly significant, and we can conclude that there is a relationship between age at first birth and the development of breast cancer. ∎

10.7.2   Chi-Square Test for Trend in Binomial Proportions

Refer again to the international study data in Table 10.19. In Example 10.40 the test procedure in **(10.23)** was used to analyze these data. For the special case of a $2 \times k$ table, this test procedure enables us to test the hypothesis $H_0: p_1 = p_2 = \cdots = p_k$ versus H_1: at least two of the p_i are unequal, where p_i = probability of success for the ith group = probability that an observation from the ith column falls in the first row. When this test procedure was employed in Example 10.40, a chi-square statistic of 130.33 with 4 df was found, which was highly significant $(p < .001)$. As a result H_0 was rejected and we concluded that the proportion of breast cancer cases in at least two of the five age-at-first-birth groups were different. However, although this result shows that some relationship exists between breast cancer and age at first birth, it does not tell specifically about the nature of the relationship. In particular, from Table 10.19 we notice an increasing *trend* in the proportion of women with breast cancer in each succeeding column. We would like to employ a specific test to detect such trends. For this purpose a **score variable** S_i is introduced to correspond to the ith group. The score variable can represent some particular numerical attribute of the group. In other instances, for simplicity, 1 is assigned to the first group, 2 to the second group, ..., k to the kth (last) group.

EXAMPLE 10.41 **Cancer** Construct a score variable for the international study data in Table 10.19.

SOLUTION It is natural to use the average age at first birth within a group as the score variable for that group. This rule presents no problem for the second, third, and fourth groups, in which the average age is estimated as 22.5 $[(20 + 25)/2]$, 27.5, and 32.5 years, respectively. However, a similar calculation cannot be performed for the first and fifth groups, since they are defined as <20 and ≥ 35, respectively. By symmetry, a score of 17.5 years could be assigned to the first group and 37.5 years to the fifth group. However, since the scores are equally spaced, our purposes will be equally well served by assigning scores of 1, 2, 3, 4, and 5 to the five groups. For simplicity, this scoring method will be adopted. ∎

We wish to relate the proportion of breast cancer cases in a group with the score variable for that group. In other words, we wish to test whether the proportion of breast cancer cases increases or decreases as age at first birth increases. For this purpose the following test procedure is introduced:

10.24 **Chi-Square Test for Trend in Binomial Proportions (Two-Sided Test)**

Suppose there are k groups and we wish to test if there is an increasing (or decreasing) trend in the proportion of "successes" p_i (i.e., the proportion of units in the first row of the ith group) as i increases.

(1) Set up the data in the form of a $2 \times k$ contingency table, where success or failure is listed along the rows and the k groups are listed along the columns.

(2) Denote the number of successes in the ith group by x_i, the total number of units in the ith group by n_i, and the proportion of successes in the ith group by $\hat{p}_i = x_i/n_i$. Denote the total number of successes over all groups by x, the total number of units over all groups by N, the overall proportion of successes by $\bar{p} = x/N$, and the overall proportion of failures by $\bar{q} = 1 - \bar{p}$.

(3) Construct a score variable S_i to correspond to the ith group. This variable will usually either be 1, 2, ..., k for the k groups or be defined to correspond to some other numerical attribute of the group.

(4) To relate p_i and S_i, compute the test statistic $X_1^2 = A^2/B$, where

$$A = \sum_{i=1}^{k} n_i(\hat{p}_i - \bar{p})(S_i - \bar{S})$$

$$= \left(\sum_{i=1}^{k} x_i S_i\right) - x\bar{S} = \left(\sum_{i=1}^{k} x_i S_i\right) - x\left(\sum_{i=1}^{k} n_i S_i\right)\bigg/ N$$

$$B = \bar{p}\bar{q}\left[\left(\sum_{i=1}^{k} n_i S_i^2\right) - \left(\sum_{i=1}^{k} n_i S_i\right)^2 \bigg/ N\right]$$

which under H_0 approximately follows a chi-square distribution with 1 df.

(5) For a two-sided level α test, if

$$X_1^2 > \chi_{1,1-\alpha}^2$$

then reject H_0. If
$$X_1^2 \leq \chi_{1,1-\alpha}^2$$

then accept H_0.

(6) The exact *p*-value is given by the area to the right of X_1^2 under a χ_1^2 distribution.

(7) The direction of the trend in proportions is indicated by the sign of A. If $A > 0$, then the proportions increase with increasing score; if $A < 0$, then the proportions decrease with increasing score.

(8) Use this test only if $N\bar{p}\bar{q} \geqslant 5.0$.

The acceptance and rejection regions for this test are depicted in Figure 10.10. The computation of the *p*-value is illustrated in Figure 10.11.

$$X_1^2 = A^2/B, \text{ where } A = \sum_{i=1}^{k} x_i S_i - x\,\bar{S}$$

$$B = \bar{p}\,\bar{q}\left[\sum_{i=1}^{k} n_i S_i^2 - \left(\sum_{i=1}^{k} n_i S_i\right)^2/N\right]$$

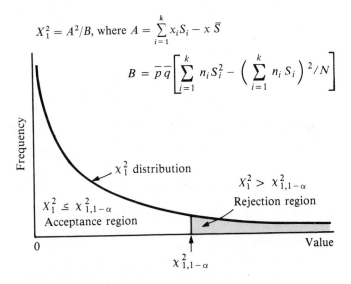

FIGURE 10.10
Acceptance and rejection regions for the chi-square test for trend in binomial proportions

$$X_1^2 = A^2/B, \text{ where } A = \sum_{i=1}^{k} x_i S_i - x\,\bar{S}$$

$$B = \bar{p}\,\bar{q}\left[\sum_{i=1}^{k} n_i S_i^2 - \left(\sum_{i=1}^{k} n_i S_i\right)^2/N\right]$$

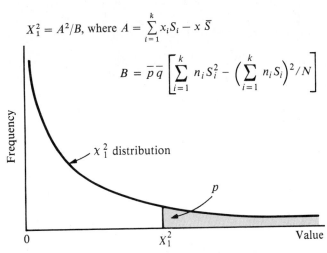

FIGURE 10.11
Computation of the *p*-value for the chi-square test for trend in binomial proportions

The test statistic in **(10.24)** is reasonable, since if $\hat{p}_i$ (or $\hat{p}_i - \bar{p}$) increases as S_i increases, then $A > 0$, whereas if $\hat{p}_i$ decreases as S_i increases, then $A < 0$. In either case A^2 and the test statistic, X_1^2 will be large. On the other hand, if $\hat{p}_i$ shows no particular trend regarding S_i, then A will be close to 0 and the test statistic X_1^2 will be small. This test can be used even if some of the groups have small sample size, since the test is based on the overall trend in the proportions. This property is in contrast to the overall chi-square test in (10.23), which tests for heterogeneity among proportions and requires that the expected number of units in individual cells not be too small.

EXAMPLE 10.42   **Cancer** Using the international study data in Table 10.19, assess whether or not there is an increasing trend in the proportion of breast cancer cases as age at first birth increases.

SOLUTION   Note that $S_i = 1, 2, 3, 4, 5$ in the five groups, respectively. Furthermore, from Table 10.19, $x_i = 320, 1206, 1011, 463, 220$, and $n_i = 1742, 5638, 3904, 1555, 626$ in the five respective groups, whereas $x = 3220, N = 13,465, \bar{p} = x/N = .239, \bar{q} = 1 - \bar{p} = .761$. From (10.24) it follows that

$$A = (320)(1) + (1206)(2) + \cdots + (220)(5)$$
$$- (3220)[1742(1) + 5638(2) + \cdots + 626(5)]/13,465$$
$$= 8717 - (3220)(34,080)/13,465 = 8717 - 8149.84 = 567.16$$
$$B = (.239)(.761)\{1742(1^2) + 5638(2^2) + \cdots + 626(5^2)$$
$$- [1742(1) + (5638)(2) + \cdots + 626(5)]^2/13,465\}$$
$$= (.239)(.761)[99,960 - (34,080)^2/13,465]$$
$$= (.239)(.761)(99,960 - 86,256.70) = 2492.34$$

Thus, $$X_1^2 = A^2/B = \frac{567.16^2}{2492.34} = 129.06 \sim \chi_1^2 \text{ under } H_0$$

Since $\chi_{1,.999}^2 = 10.83 < 129.06 = X_1^2$, H_0 can be rejected with $p < .001$ and we can conclude that there is a significant trend in the proportion of breast cancer cases among age-at-first-birth groups. Since $A > 0$, it follows that as age at first birth increases, the proportion of breast cancer cases rises. ∎

With a $2 \times k$ table, the chi-square test for trend in (10.24) is often more relevant to the hypotheses of interest than the chi-square test for heterogeneity in (10.23), since the former procedure tests for specific trends in the proportions, whereas the latter tests for any differences in the proportions, where the proportions may follow any pattern. Other, more advanced methods for assessing $R \times C$ contingency tables are given in Maxwell's *Analyzing Qualitative Data* [5].

SECTION 10.8   **Mantel–Haenszel Test**

When looking at the relationship between a disease and an exposure variable, it is often important to control for the effect of some other variable that may be associated with either the disease, the exposure, or both.

DEFINITION 10.10   ■■

A **confounding variable** is a variable that may be associated with either the disease, the exposure, or both. Such a variable must usually be controlled for before looking at the disease-exposure relationship. ■

EXAMPLE 10.43 **Cancer** A 1985 study identified a group of 518 cancer cases ages 15–59 and a group of 518 age- and sex-matched controls by mail questionnaire [6]. The main purpose of the study was to look at the effect of passive smoking on cancer risk. In the study passive smoking was defined as exposure to the cigarette smoke of a spouse who smoked at least one cigarette per day for at least 6 months. One potential confounding variable was smoking by the test subjects themselves (i.e., personal smoking), since personal smoking is related to both cancer risk and spouse smoking. Therefore, it was important to control for personal smoking before looking at the relationship between passive smoking and cancer risk. ■

To analyze the data, a 2×2 table relating case-control status to passive smoking can be constructed for both nonsmokers and smokers. The data are given in Table 10.21 for nonsmokers and Table 10.22 for smokers.

TABLE 10.21

Relationship of passive smoking to cancer risk among nonsmokers

	Passive smoker		
Case-control status	Yes	No	Total
Case	120	111	231
Control	80	155	235
Total	200	266	466

(Reprinted with permission of the *American Journal of Epidemiology*, *121*(1), 37–48, 1985.)

TABLE 10.22

Relationship of passive smoking to cancer risk among smokers

	Passive smoker		
Case-control status	Yes	No	Total
Case	161	117	278
Control	130	124	254
Total	291	241	532

(Reprinted with permission of the *American Journal of Epidemiology*, *121*(1), 37–48, 1985.)

The passive smoking effect can be assessed separately for nonsmokers and smokers. Indeed, we notice from Table 10.21 that the odds ratio in favor of a case being exposed to cigarette smoke from a spouse who smokes versus a control is $(120 \times 155)/(80 \times 111) = 2.1$ for nonsmokers, whereas the corresponding odds ratio for smokers is $(161 \times 124)/(130 \times 117) = 1.3$. Thus for both subgroups the trend is in the direction of more passive smoking among cases than controls. The key question is how to combine the results of the two tables to obtain an overall test of significance for the passive smoking effect.

In general, the data will be stratified into k subgroups according to one or more confounding variables to make the units within a stratum as homogeneous as possible. The data for each stratum consist of a 2×2 contingency table relating exposure to disease, as shown in Table 10.23 for the ith stratum.

TABLE 10.23
Relationship of disease
to exposure in the ith
stratum

Disease	Exposure		Total
	Yes	No	
Yes	a_i	b_i	$a_i + b_i$
No	c_i	d_i	$c_i + d_i$
Total	$a_i + c_i$	$b_i + d_i$	N_i

The test procedure will be based on a comparison of the observed number of units in the $(1, 1)$ cell of each stratum (denoted by $O_i = a_i$) with the expected number of units in that cell (denoted by E_i). The test procedure is the same regardless of the order of the rows and columns; that is, which row (or column) is designated as the first row (or column) is arbitrary. Based on the margins, the expected number of units in the $(1, 1)$ cell of the ith stratum is given by

10.25

$$E_i = \frac{(a_i + b_i)(a_i + c_i)}{N_i}$$

The observed and expected numbers of units in the $(1, 1)$ cell are then summed over all strata, obtaining $O = \Sigma_{i=1}^k O_i$, $E = \Sigma_{i=1}^k E_i$, and the test is based on $O - E$. It can be shown that the variance of $(O_i - E_i)$ is given by

10.26

$$V_i = \frac{(a_i + b_i)(c_i + d_i)(a_i + c_i)(b_i + d_i)}{N_i^2(N_i - 1)}$$

Furthermore, the variance of $O - E = V = \Sigma_{i=1}^k V_i$. The test statistic is given by $X_{MH}^2 = (|O - E| - .5)^2/V$, which should follow a chi-square distribution with 1 df under the null hypothesis of no association between disease and exposure. H_0 is rejected if X_{MH}^2 is large. This procedure is known as the Mantel–Haenszel test and is summarized as follows:

10.27 **Mantel–Haenszel Test**

To assess the association between a dichotomous disease and a dichotomous exposure variable after controlling for one or more confounding variables, use the following procedure:

(1) Form k strata, based on the level of the confounding variable(s), and construct a 2×2 table relating disease and exposure within each stratum, as shown in Table 10.23.

(2) Compute the total observed number of units (O) in the $(1, 1)$ cell over all strata, where

$$O = \sum_{i=1}^k O_i = \sum_{i=1}^k a_i$$

(3) Compute the total expected number of units (E) in the $(1, 1)$ cell over all strata, where

$$E = \sum_{i=1}^{k} E_i = \sum_{i=1}^{k} \frac{(a_i + b_i)(a_i + c_i)}{N_i}$$

(4) Compute the variance (V) of the difference $O - E$, where

$$V = \sum_{i=1}^{k} V_i = \sum_{i=1}^{k} \frac{(a_i + b_i)(c_i + d_i)(a_i + c_i)(b_i + d_i)}{N_i^2(N_i - 1)}$$

(5) The test statistic is then given by

$$X_{MH}^2 = \frac{(|O - E| - .5)^2}{V}$$

which under H_0 follows a chi-square distribution with 1 df (MH refers to Mantel–Haenszel).

(6) For a two-sided test with significance level α, if

$$X_{MH}^2 > \chi_{1,1-\alpha}^2$$

then reject H_0. If

$$X_{MH}^2 \leqslant \chi_{1,1-\alpha}^2$$

then accept H_0.

(7) The exact p-value for this test is given by

$$p = Pr(\chi_1^2 > X_{MH}^2)$$

(8) Use this test only if the variance $V \geqslant 5$.

(9) Which row or column is designated as first is arbitrary. The test statistic X_{MH}^2 and the assessment of significance are the same regardless of the order of the rows and columns.

The acceptance and rejection regions for the Mantel–Haenszel test are depicted in Figure 10.12. The computation of the p-value for the Mantel–Haenszel test is illustrated in Figure 10.13.

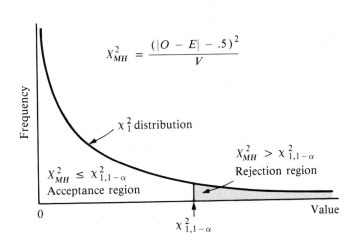

FIGURE 10.12

Acceptance and
rejection regions for the
Mantel–Haenszel test

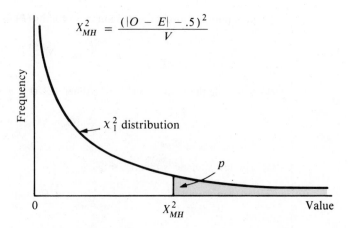

$$X^2_{MH} = \frac{(|O - E| - .5)^2}{V}$$

χ^2_1 distribution

p

0 X^2_{MH} Value

FIGURE 10.13
Computation of the p-value for the Mantel–Haenszel test

EXAMPLE 10.44   **Cancer** Assess the relationship between passive smoking and cancer risk using the data stratified by personal smoking status in Tables 10.21 and 10.22.

SOLUTION   Denote the nonsmokers as stratum 1 and the smokers as stratum 2.

O_1 = observed number of nonsmoking cases who are passive smokers = 120

O_2 = observed number of smoking cases who are passive smokers = 161

Furthermore,

$$E_1 = \frac{231 \times 200}{466} = 99.1$$

$$E_2 = \frac{278 \times 291}{532} = 152.1$$

Thus, the total observed and expected numbers of cases who are passive smokers are, respectively,

$$O = O_1 + O_2 = 120 + 161 = 281$$

$$E = E_1 + E_2 = 99.1 + 152.1 = 251.2$$

Therefore, there are more cases who are passive smokers than would be expected based on their personal smoking habits. Now compute the variance to assess if this difference is statistically significant.

$$V_1 = \frac{231 \times 235 \times 200 \times 266}{466^2 \times 465} = 28.60$$

$$V_2 = \frac{278 \times 254 \times 291 \times 241}{532^2 \times 531} = 32.95$$

Therefore, $$V = V_1 + V_2 = 28.60 + 32.95 = 61.55$$

Thus, the test statistic X^2_{MH} is given by

$$X^2_{MH} = \frac{(|281 - 251.2| - .5)^2}{61.55} = \frac{858.49}{61.55} = 13.95 \sim \chi^2_1 \text{ under } H_0$$

Since $\chi^2_{1,.999} = 10.83 < 13.95 = X^2_{MH}$, it follows that $p < .001$. Thus, there is a highly significant positive association between case-control status and passive smoking exposure, even after controlling for personal cigarette smoking habit.   ∎

10.8.1 **Estimation of the Odds Ratio for Stratified Data**

The Mantel–Haenszel test tells about the statistical significance of the relationship between disease and exposure. However, it does not give a measure of the strength of the association. Ideally, we would like a measure similar to the odds ratio presented for a single 2×2 contingency table in Definition 10.3 (p. 333). Assuming that the true odds ratio is the same for each stratum, an estimate of the common odds ratio is provided by the Mantel–Haenszel estimator as follows:

| 10.28 | **Mantel–Haenszel Estimator of the Common Odds Ratio for Stratified Data** |

In a collection of k 2×2 contingency tables, where the ith table corresponding to the ith stratum is denoted as in Table 10.23, the Mantel–Haenszel estimator of the common odds ratio is given by

$$\widehat{OR}_{MH} = \frac{\sum\limits_{i=1}^{k} (a_i d_i / N_i)}{\sum\limits_{i=1}^{k} (b_i c_i / N_i)}$$

EXAMPLE 10.45 **Cancer** Estimate the odds ratio in favor of having experienced passive smoking for cancer cases versus controls after controlling for personal smoking habit.

SOLUTION From **(10.28)**, Table 10.21, and Table 10.22,

$$\widehat{OR}_{MH} = \frac{(120 \times 155/466) + (161 \times 124/532)}{(80 \times 111/466) + (130 \times 117/532)} = \frac{77.44}{47.65} = 1.63$$

Thus, the odds in favor of having experienced passive smoking for a cancer case is 1.6 times as large as that for a control. ∎

We are also interested in estimating confidence limits for the odds ratio in **(10.28)**. A test-based method similar to that presented for a single 2×2 contingency table in **(10.8)** (p. 334) can be used:

| 10.29 | **Interval Estimates for the Common Odds Ratio from a Collection of k 2×2 Contingency Tables (Test-Based Method)** |

A two-sided $100\% \times (1 - \alpha)$ confidence interval for the common odds ratio from a collection of k 2×2 contingency tables is given by

$$(\widehat{OR}_{MH}^{1 - \sqrt{\chi_{1,1-\alpha}^2 / X_{MH}^2}}, \; \widehat{OR}_{MH}^{1 + \sqrt{\chi_{1,1-\alpha}^2 / X_{MH}^2}}) \quad \text{if } \widehat{OR}_{MH} \geq 1$$

$$(\widehat{OR}_{MH}^{1 + \sqrt{\chi_{1,1-\alpha}^2 / X_{MH}^2}}, \; \widehat{OR}_{MH}^{1 - \sqrt{\chi_{1,1-\alpha}^2 / X_{MH}^2}}) \quad \text{if } \widehat{OR}_{MH} < 1$$

where $\widehat{OR}_{MH}$ is the Mantel–Haenszel estimator of the common odds ratio given in **(10.28)**, and X_{MH}^2 is the Mantel–Haenszel test statistic given in **(10.27)**. This method should be used only if $0.2 \leq \widehat{OR}_{MH} \leq 5.0$.

EXAMPLE 10.46 **Cancer** Estimate 95% confidence limits for the common odds ratio using the data in Tables 10.21 and 10.22.

SOLUTION Note from Example 10.45 that the point estimate of the odds ratio $= \widehat{OR}_{MH} = 1.63$. Furthermore, from Example 10.44 the Mantel–Haenszel test statistic is 13.95. Finally, since a

95% confidence interval is desired, it follows that $\chi^2_{1,1-\alpha} = \chi^2_{1,.95} = 3.84$. Therefore, using **(10.29)**, the 95% confidence interval for OR is

$$(1.63^{1-\sqrt{3.84/13.95}}, 1.63^{1+\sqrt{3.84/13.95}})$$

$$= (1.63^{1-0.52}, 1.63^{1+0.52}) = (1.63^{0.48}, 1.63^{1.52}) = (1.26, 2.10)$$

Notice that the 95% confidence interval for OR does not include 1, which agrees with the level of significance of the test statistic ($X^2_{MH} = 13.95$, $p < .001$). This relationship is one advantage of the test-based method: if the Mantel–Haenszel test statistic is significant at the 5% level, then the test-based 95% confidence interval for OR will always exclude 1. If the Mantel–Haenszel test statistic is not significant at the 5% level, then the 95% confidence interval for OR will always include 1. This relationship was also true for the test-based method for a single 2×2 contingency table in **(10.8)**.   ∎

As was the case for a single 2×2 contingency table, the test-based method should be used only if the estimated odds ratio is not too different from 1, that is, if $0.2 \leqslant \widehat{OR}_{MH} \leqslant 5.0$. If $\widehat{OR}_{MH}$ is outside this range, then more sophisticated methods should be used (see [2]).

Finally, one assumption made in the estimation of a common odds ratio in **(10.28)** is that the strength of association is the same in each stratum. If the underlying odds ratio is different in the various strata, then it makes little sense to estimate a common odds ratio. A method for testing for the homogeneity of the odds ratio over multiple strata is given in Fleiss [7].

SECTION 10.9

Two-Sample Test for Incidence-Density Measures

We often want to study incidence rates, where each person is not followed for the same length of time. In this case, the methods developed earlier in this chapter are not applicable.

EXAMPLE 10.47   **Cancer**   A hypothesis of much recent interest is the possible association between the use of oral contraceptives (OC) and the development of breast cancer. To address this issue, data were collected in the Nurses' Health Study whereby disease-free women were classified in 1976 according to OC status (current user/past user/never user). A mail questionnaire was sent out every 2 years in which OC status was updated and breast cancer status was ascertained over the next two years. For each woman, an amount of time that the woman is a current user or a never user of OC's (ignoring past use) can be calculated and this *person-time* can be accumulated over the entire cohort of nurses. Thus, each nurse contributes a different amount of person-time to the analysis. The data are presented in Table 10.24 for current and never users of OC's among women initially 45–49 years of age. How should these data be used to assess any differences in incidence rate of breast cancer by OC-use group?   ∎

TABLE 10.24
Relationship between breast cancer incidence and OC use among 45–49-year-old women in the Nurses' Health Study

OC-use group	Number of cases	Number of person-years
Current users	9	2935
Never users	239	135,130

The first issue to consider is the appropriate unit of analysis for each group. If the woman is used as the unit of analysis, then the problem is that different women may contribute different amounts of person-time to the analysis, and the assumption of a constant probability of an event for each woman would then be violated. If a person-year is used as the unit of analysis (i.e., 1 person followed for 1 year), then since each woman can contribute more than 1 person-year to the analysis, the important assumption of independence for the binomial distribution would be violated.

The approach we will take is to use a *conditional* test. Specifically, suppose we consider only the case of two exposure groups and have the general table in Table 10.25.

TABLE 10.25

General observed table for comparing incidence rates in two groups

Exposure group	Number of events	Person-time
1	a_1	t_1
2	a_2	t_2
Total	$a_1 + a_2$	$t_1 + t_2$

DEFINITION 10.11　▪▪

The **incidence density** in a group is defined by the number of events in that group divided by the total person-time accumulated during the study in that group.　▪

EXAMPLE 10.48　　**Cancer** Compute the estimated incidence density among current and never OC users in Table 10.24.

SOLUTION　　The incidence density among current users $= 9/2935 = .00307$ events per person-year $= 307$ events per 100,000 person-years. The incidence density among never users $= 239/135,130 = .00177$ events per person-year $= 177$ events per 100,000 person-years.　▪

We wish to test the hypothesis $H_0: ID_1 = ID_2$ versus $H_1: ID_1 \neq ID_2$, where $ID_1 =$ true incidence density in group 1 = the number of events per unit of person-time in group 1, and ID_2 is the comparable rate in group 2. Under the null hypothesis, the fraction $t_1/(t_1 + t_2)$ of the total number of events would be expected to occur in group 1 and $t_2/(t_1 + t_2)$ of the total number of events to occur in group 2. Furthermore, conditional on the observed total number of events $= a_1 + a_2$, it follows under H_0 that

10.30

$$\text{Expected number of events in group 1} = E_1 = (a_1 + a_2)t_1/(t_1 + t_2)$$

$$\text{Expected number of events in group 2} = E_2 = (a_1 + a_2)t_2/(t_1 + t_2)$$

EXAMPLE 10.49　　**Cancer** Compute the expected number of events among current and never users for the OC–breast cancer data in Table 10.24.

SOLUTION   We have that $a_1 = 9$, $a_2 = 239$, $t_1 = 2935$ person-years, $t_2 = 135{,}130$ person-years. There-fore, under H_0, from (10.30), $2935/(2935 + 135{,}130) = .0213$ of the cases would be expected to occur among current OC users and $135{,}130/(2935 + 135{,}130) = .9787$ of the cases to occur among never OC users. Thus,

$$E_1 = .0213(248) = 5.28$$
$$E_2 = .9787(248) = 242.72$$

∎

10.9.1 Normal Theory Test

To assess statistical significance, the number of events in group 1 under H_0 is treated as a binomial random variable with parameters $n = a_1 + a_2$ and $p_0 = t_1/(t_1 + t_2)$. Under this assumption, the hypotheses can be stated as $H_0: p = p_0$ versus $H_1: p \neq p_0$, where $p =$ the true probability that an event will occur in group 1, given that it has occurred. We will also assume that the normal approximation to the binomial distribution is valid. Using the normal approximation to the binomial distribution, the observed number of events in group $1 = a_1$ is normally distributed with mean $= np_0 = (a_1 + a_2)t_1/(t_1 + t_2) = E_1$, and variance $= np_0q_0 = (a_1 + a_2)t_1t_2/(t_1 + t_2)^2 = V_1$. H_0 will be rejected if a_1 is much smaller or larger than E_1. This is an application of the large-sample one-sample binomial test, given by the following:

10.31

Comparison of Incidence Rates (Large-Sample Test)

To test the hypothesis $H_0: ID_1 = ID_2$ versus $H_1: ID_1 \neq ID_2$, where ID_1 and ID_2 is true incidence densities in groups 1 and 2, use the following procedure:

(1) Compute the test statistic

$$Z = \frac{|a_1 - E_1| - .5}{\sqrt{V_1}} \sim N(0, 1) \text{ under } H_0$$

where
$$E_1 = (a_1 + a_2)t_1/(t_1 + t_2)$$
$$V_1 = (a_1 + a_2)t_1t_2/(t_1 + t_2)^2$$

$a_1, a_2 =$ number of events in groups 1 and 2

$t_1, t_2 =$ amount of person-time in groups 1 and 2

(2) For a two-sided level α test, if

$$Z > z_{1-\alpha/2} \quad \text{or} \quad Z < z_{\alpha/2}$$

then reject H_0. If

$$z_{\alpha/2} \leqslant Z \leqslant z_{1-\alpha/2}$$

then accept H_0.

(3) Use this test only if $V_1 \geqslant 5$.

(4) The p-value for this test is given by

$$2 \times [1 - \Phi(Z)] \quad \text{if } Z \geqslant 0$$
$$2 \times \Phi(Z) \quad \text{if } Z < 0$$

The critical region and p-value are illustrated in Figures 10.14 and 10.15, respectively.

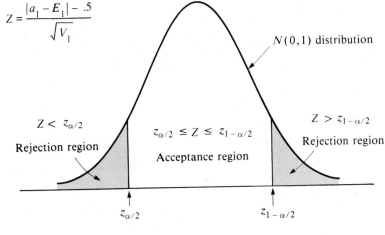

$$Z = \frac{|a_1 - E_1| - .5}{\sqrt{V_1}}$$

$N(0,1)$ distribution

$Z < z_{\alpha/2}$

Rejection region

$z_{\alpha/2} \leq Z \leq z_{1-\alpha/2}$

Acceptance region

$Z > z_{1-\alpha/2}$

Rejection region

$z_{\alpha/2}$ $z_{1-\alpha/2}$

FIGURE 10.14
Acceptance and rejection regions for the two-sided test for incidence rates (normal theory method)

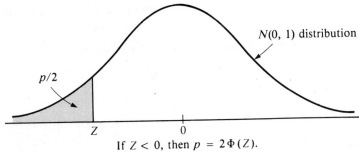

$N(0, 1)$ distribution

$p/2$

Z 0

If $Z < 0$, then $p = 2\Phi(Z)$.

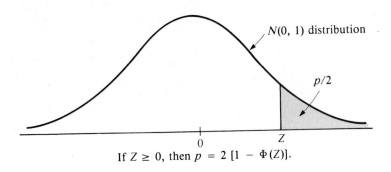

$N(0, 1)$ distribution

$p/2$

0 Z

If $Z \geq 0$, then $p = 2[1 - \Phi(Z)]$.

FIGURE 10.15
Computation of the p-value for the two-sided test for incidence rates (normal theory test)

EXAMPLE 10.50   **Cancer** Assess the statistical significance of the OC–breast cancer data in Table 10.24.

SOLUTION   From Example 10.49, $a_1 = 9$, $a_2 = 239$, $t_1 = 2935$, $t_2 = 135,130$, $E_1 = 5.28$, $E_2 = 242.72$. Furthermore,

$$V_1 = \frac{(a_1 + a_2)t_1 t_2}{(t_1 + t_2)^2}$$

$$= \frac{(9 + 239)(2935)(135,130)}{(2935 + 135,130)^2} = \frac{9.8358 \times 10^{10}}{(138,065)^2} = 5.16$$

Since $V_1 \geq 5$, we can use the large-sample test in **(10.31)**.

Therefore,
$$Z = \frac{|9 - 5.28| - .5}{\sqrt{5.16}} = \frac{3.22}{2.27} = 1.42 \sim N(0, 1)$$

The p-value $= 2 \times [1 - \Phi(1.42)] = 2 \times (1 - .9222) = .156$. Thus, the results are not statistically significant and there is no significant difference in incidence rates of breast cancer between current OC users and never OC users in this age group. ∎

10.9.2 Exact Test

Suppose that the number of events is too small to apply the normal theory test (i.e., $V_1 < 5$). In this case, an exact test based on the binomial distribution must be used. From Section 10.9.1, under H_0, the number of events in group 1 (a_1) will follow a binomial distribution with parameters $n = a_1 + a_2$ and $p_0 = t_1/(t_1 + t_2)$. We wish to test the hypothesis $H_0: p = p_0$ versus $H_1: p \neq p_0$, where p is the conditional probability that a case occurs in group 1, given that it has occurred at all. This is an application of the exact one-sample binomial test. H_0 will be rejected if the observed number of events a_1 is much smaller or much larger than the expected number of events $= E_1 = np_0$. The following test procedure is used:

| **10.32** | **Comparison of Incidence Rates—Exact Test** |

Let a_1, a_2 be the observed number of events and t_1, t_2 the amount of person-time in groups 1 and 2, respectively. To test the hypothesis $H_0: ID_1 = ID_2$ versus $H_1: ID_1 \neq ID_2$, where

$$ID_1 = \text{true incidence density in group 1}$$

$$ID_2 = \text{true incidence density in group 2}$$

$$p_0 = t_1/(t_1 + t_2), q_0 = 1 - p_0$$

using a two-sided test with significance level α, use the following procedure:

(1) If $a_1 < (a_1 + a_2)p_0$, then

$$p\text{-value} = 2 \times \sum_{k=0}^{a_1} \binom{a_1 + a_2}{k} p_0^k q_0^{a_1 + a_2 - k}$$

(2) If $a_1 \geq (a_1 + a_2)p_0$, then

$$p\text{-value} = 2 \times \sum_{k=a_1}^{a_1 + a_2} \binom{a_1 + a_2}{k} p_0^k q_0^{a_1 + a_2 - k}$$

(3) This test is valid in general for comparing two incidence densities but is particularly useful when $V_1 < 5$, in which case the normal theory test in **(10.31)** should not be used. The computation of the p-value is illustrated in Figure 10.16.

EXAMPLE 10.51 **Cancer** Suppose we have the data in Table 10.26 relating OC use and incidence of breast cancer among women aged 30–34. Assess the statistical significance of these data.

TABLE 10.26
Relationship between breast cancer incidence and OC use among 30–34-year-old women in the Nurses' Health Study

OC-use group	Number of cases	Number of person-years
Current users	3	8250
Never users	9	17,430

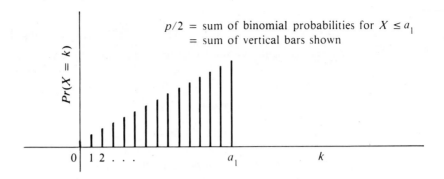

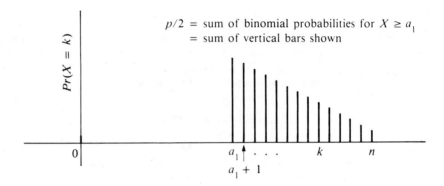

FIGURE 10.16
Illustration of the p-value
for the two-sample test
for incidence rates
exact method (two-
sided alternative)

SOLUTION   Note that $a_1 = 3$, $a_2 = 9$, $t_1 = 8250$, $t_2 = 17{,}430$. Thus,

$$V_1 = \frac{12(8250)(17{,}430)}{(8250 + 17{,}430)^2} = 2.62 < 5$$

Since $V_1 < 5$, the small-sample test must be used. From **(10.32)**, $p_0 = 8250/25{,}680 = .321$, $n = a_1 + a_2 = 12$. Since $a_1 = 3 < 12(0.321) = 3.9$,

$$p\text{-value} = 2 \times \sum_{k=0}^{3} \binom{12}{k}(.321)^k(.679)^{12-k}$$

Let X be a random variable representing the number of events in group 1 and use the recursion rule to evaluate the p-value:

$$Pr(X = 0) = (.679)^{12} = .0096$$

$$Pr(X = 1) = \frac{12}{1} \times \frac{.321}{.679} \times .0096 = .0545$$

$$Pr(X = 2) = \frac{11}{2} \times \frac{.321}{.679} \times .0545 = .1417$$

$$Pr(X = 3) = \frac{10}{3} \times \frac{.321}{.679} \times .1417 = .2233$$

Thus,

$$p\text{-value} = 2 \times (.0096 + .0545 + .1417 + .2233) = 2 \times .4291 = .858$$

Therefore, there is no significant effect of current OC use on breast cancer incidence in this age group as well. ■

SECTION 10.10

The Log-Rank Test

In the previous section, methods for comparing incidence rates in two groups, where the period of follow-up may be different for the two groups considered, were discussed. One assumption made in performing these analyses is that incidence remains *constant* over time. In many instances this assumption is not warranted and the number of disease events in two groups where the incidence of disease varies over time must be compared.

EXAMPLE 10.52

Health Promotion Consider Data Set 24, SMOKE.DAT in Appendix 2. In this data set, 234 smokers who expressed a willingness to quit smoking were followed for 1 year to estimate the proportion of recidivism, that is, the proportion of smokers who quit for a time but who start smoking again. One hypothesis is that older smokers are less likely to be successful quitters (and more likely to be recidivists). How can this hypothesis be tested? ■

The data in Table 10.27 were computed after subdividing the study population by age ($>40/\leqslant 40$).

TABLE 10.27 Number of days quit smoking by age

	Number of days quit smoking					
Age	**≤90**	**91–180**	**181–270**	**271–364**	**365**	**Total**
>40	92	4	4	1	19	120
≤40	88	7	3	2	14	114
Total	180	11	7	3	33	234
Percent	76.9	4.7	3.0	1.3	14.1	

Seventy-seven percent of the population resumed smoking within 90 days; an additional 9% resumed smoking between 91 days and 1 year; 14% succeeded in quitting smoking for the entire year of follow-up. The incidence rate of recidivism is clearly much higher immediately after the quit attempt, and it would be invalid to use the methods of Section 10.9, where a constant incidence rate is assumed.

These data could be analyzed in terms of cumulative incidence over 1 year; that is, the percentage of older versus younger ex-smokers who were successful quitters for 1 year could be compared. However, if incidence changes greatly over the year this will not be as powerful as the log-rank test described in **(10.33)**. Using this procedure, *when* an event occurs (in this case the event is recidivism) rather than simply *whether* it occurs is taken into account.

To implement this procedure, the total period of follow-up is subdivided into shorter time periods over which incidence is relatively constant. In Example 10.52, the ideal situation would be to subdivide the 1-year interval into 365 daily time intervals. However, for the purpose of illustrating the method, time has been subdivided into 3-month intervals. For each time interval, the number of people who have been successful quitters up to the beginning of the interval are identified. These are the people who are at risk for recidivism during this time interval. This group is then categorized according to whether they remained successful quitters or became recidivists during the time interval. For each time interval, the data are displayed in the form of a 2 × 2 contingency table relating age to incidence of recidivism over the time interval.

EXAMPLE 10.53   **Health Promotion** Display the smoking-cessation data in Table 10.27 in the form of incidence data by age for each of the four time intervals, 0–90 days, 91–180 days, 181–270 days, and 271–365 days.

SOLUTION   For the first time interval, (0–90 days), there are 120 older smokers who were successful quitters at time 0, of whom 92 became recidivists during the 0–90 day period; similarly, there were 114 younger smokers, of whom 88 became recidivists during this time period. These data are shown in a 2 × 2 contingency table in Table 10.28. For the second time period, (91–180 days), there are 28 older smokers who were successful quitters at day 91, of whom 4 became recidivists during the period from day 91 to day 180; similarly, there were 26 younger smokers who were successful quitters at 91 days, of whom 7 became recidivists from day 91 to day 180. Thus, the second 2 × 2 contingency table would look like Table 10.29. Similarly, 2 × 2 contingency tables for the time periods 181–270 days and 271–365 days can be developed, as shown in Tables 10.30 and 10.31, respectively.

TABLE 10.28
Incidence data by age for the 0–90-day period

	Outcome		
Age	Recidivist	Successful quitter	Total
>40	92	28	120
≤40	88	26	114
Total	180	54	234

TABLE 10.29
Incidence data by age for the 91–180-day period

	Outcome		
Age	Recidivist	Successful quitter	Total
>40	4	24	28
≤40	7	19	26
Total	11	43	54

TABLE 10.30
Incidence data by age for the 181–270-day period

	Outcome		
Age	Recidivist	Successful quitter	Total
>40	4	20	24
≤40	3	16	19
Total	7	36	43

TABLE 10.31

Incidence data by age for the 271–365-day period

	Outcome		
Age	**Recidivist**	**Successful quitter**	**Total**
>40	1	19	20
≤40	2	14	16
Total	3	33	36

If age has no association with recidivism, then the incidence rate for recidivism for older and younger smokers within each of the four time intervals should be the same. Conversely, if it is harder for older smokers than younger smokers to remain quitters, then the incidence rate of recidivism should be consistently higher for older smokers within each of the four time intervals considered. Note that incidence is allowed to vary over different time intervals under either hypothesis. To accumulate evidence over the entire period of follow-up, the Mantel–Haenszel procedure in **(10.27)**, based on the 2 × 2 tables in Tables 10.28–10.31, is used. This procedure is referred to as the log-rank test and is summarized as follows:

10.33 | **The Log-Rank Test**

To compare incidence rates for an event between two exposure groups, where incidence varies over the period of follow-up (T), use the following procedure:

(1) Subdivide T into k smaller time intervals, where incidence is homogeneous over the shorter time intervals.

(2) Compute a 2 × 2 contingency table corresponding to each time interval relating incidence over the time interval to exposure status ($+/-$). The ith table is of the form of Table 10.32,

where n_{i1} = the number of exposed people who have not yet had the event at the beginning of the ith time interval

n_{i2} = the number of unexposed people who have not yet had the event at the beginning of the ith time interval

a_i = the number of exposed people who had an event during the ith time interval

b_i = the number of exposed people who did not have an event during the ith time interval

and c_i, d_i are defined similarly for unexposed people.

(3) Perform the Mantel–Haenszel test over the collection of 2 × 2 tables defined in step 2. Specifically, compute the test statistic

$$X_{LR}^2 = \frac{(|O - E| - .5)^2}{V}$$

where

$$E = \sum_{i=1}^{k} E_i = \sum_{i=1}^{k} \frac{(a_i + b_i)(a_i + c_i)}{N_i}$$

$$V = \sum_{i=1}^{k} V_i = \sum_{i=1}^{k} \frac{(a_i + b_i)(c_i + d_i)(a_i + c_i)(b_i + d_i)}{N_i^2(N_i - 1)}$$

which under H_0 follows a chi-square distribution with one degree of freedom.

(4) For a two-sided test with significance level α, if $X^2_{LR} > \chi^2_{1,1-\alpha}$ then reject H_0. If $X^2_{LR} \leqslant \chi^2_{1,1-\alpha}$ then accept H_0.

(5) The exact p-value for this test is given by

$$p\text{-value} = Pr(\chi^2_1 > X^2_{LR})$$

(6) This test should be used only if $V \geqslant 5$.

The acceptance and rejection regions for the log-rank test are depicted in Figure 10.17. The computation of the exact p-value is depicted in Figure 10.18.

TABLE 10.32
Relationship at incidence to exposure status over the ith time interval

Exposure	Event +	Event −	Total
+	a_i	b_i	n_{i1}
−	c_i	d_i	n_{i2}
Total	$a_i + c_i$	$b_i + d_i$	N_i

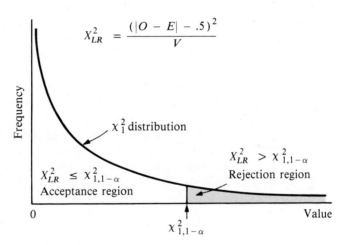

FIGURE 10.17
Acceptance and rejection regions for the log-rank test

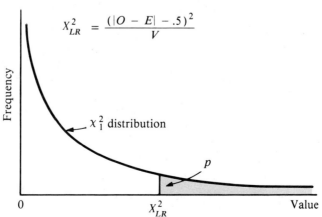

FIGURE 10.18
Computation of the p-value for the log-rank test

EXAMPLE 10.54

Health Promotion Evaluate the statistical significance of the association between age and the incidence of recidivism based on the smoking-cessation data in Table 10.27.

SOLUTION

Refer to the four 2×2 tables (Tables 10.28–10.31) developed in Example 10.53. We have that

$$O = 92 + 4 + 4 + 1 = 101$$

$$E = \frac{120 \times 180}{234} + \frac{28 \times 11}{54} + \frac{24 \times 7}{43} + \frac{20 \times 3}{36}$$

$$= 92.308 + 5.704 + 3.907 + 1.667 = 103.586$$

$$V = \frac{120 \times 114 \times 180 \times 54}{234^2 \times 233} + \frac{28 \times 26 \times 11 \times 43}{54^2 \times 53}$$

$$+ \frac{24 \times 19 \times 7 \times 36}{43^2 \times 42} + \frac{20 \times 16 \times 3 \times 33}{36^2 \times 35}$$

$$= 10.422 + 2.228 + 1.480 + 0.698 = 14.828$$

Since $V \geqslant 5$, the Mantel–Haenszel test can be used. The test statistic is given by

$$X_{LR}^2 = \frac{(|101 - 103.586| - .5)^2}{14.828} = \frac{2.086^2}{14.828} = 0.29 \sim \chi_1^2 \text{ under } H_0.$$

Since $\chi_{1,.95}^2 = 3.84 > 0.29$, it follows that the p-value $> .05$, and there is no significant difference in recidivism rates between younger and older smokers. ■

The log-rank test is a very powerful method for analyzing data where the time to an event is important rather than simply whether or not the event occurs. The methodology can be extended to allow for variable periods of follow-up for each individual and can be used to provide tests of trend. It can also be extended to allow for looking at effects of several risk factors on incidence at the same time. A popular technique used to perform this type of analysis is the Cox regression model, which is discussed in Section 11.11. The special field of statistics that is concerned with data where the outcome is time to an event is called *survival analysis*. It is covered in much greater detail in Lee [8] and Miller [9].

SECTION 10.11

Chi-Square Goodness-of-Fit Test

In our previous work on estimation and hypothesis testing, we usually assumed that the data came from a specific underlying probability model and then proceeded either to estimate the parameters of the model or test hypotheses concerning different possible values of the parameters. In this section a general method of testing for the *goodness of fit of a probability model* is presented. Consider the problem in Example 10.55.

EXAMPLE 10.55

Hypertension Suppose diastolic blood-pressure measurements were collected at home in a community-wide screening program of 14,736 adults ages 30–69 in East Boston, Massachusetts, as part of a nationwide study to detect and treat hypertensive people [10]. The people in the study were each screened in the home with two measurements taken at one visit. A frequency distribution of the mean blood pressure is given in Table 10.33 in 10-mm Hg intervals.

We would like to assume that these measurements came from an underlying normal distribution, since standard methods of statistical inference could then be applied on these data as presented in this text. How can the validity of this assumption be tested? ∎

TABLE 10.33
Frequency distribution of mean diastolic blood pressure for adults 30–69 years old in a community-wide screening program in East Boston, Massachusetts

Group	Observed frequency	Expected frequency	Group	Observed frequency	Expected frequency
<50	57	77	$\geq 80, <90$	4604	4511
$\geq 50, <60$	330	553	$\geq 90, <100$	2119	2417
$\geq 60, <70$	2132	2122	$\geq 100, <110$	659	684
$\geq 70, <80$	4584	4265	$\geq 110, <120$	251	107
			Total	14,736	14,736

This assumption can be tested by first computing what the expected frequencies would be in each group if the data did come from an underlying normal distribution and then comparing these expected frequencies with the corresponding observed frequencies.

EXAMPLE 10.56   **Hypertension** Compute the expected frequencies for the data in Table 10.33 assuming an underlying normal distribution.

SOLUTION   Assume that the mean and standard deviation of this hypothetical normal distribution are given by the sample mean and standard deviation, respectively ($\bar{x} = 80.68$, $s = 12.00$). The expected frequency within a group interval from a to b would then be given by

$$14,736\{\Phi[(b - \mu)/\sigma] - \Phi[(a - \mu)/\sigma]\}$$

Thus, the expected frequency within the (≥ 50, <60) group would be

$$14,736 \times \{\Phi[(60 - 80.68)/12] - \Phi[(50 - 80.68)/12]\}$$
$$= 14,736 \times [\Phi(-1.72) - \Phi(-2.56)]$$
$$= 14,736 \times (.0427 - .0052) = 14,736(.0375) = 552.6$$

The expected frequencies for all the groups are given in Table 10.33. ∎

The same measure of agreement between the observed and expected frequencies in a group will be used as was used in our work on contingency tables, namely, $(O - E)^2/E$. Furthermore, the agreement between observed and expected frequencies can be summarized over the whole table by summing $(O - E)^2/E$ over all the groups. If we have the correct underlying model, then this sum will approximately follow a chi-square distribution with $g - 1 - k$ df, where $g =$ the number of groups and $k =$ the number of parameters estimated from the data to compute the expected frequencies. This approximation will again be valid only if the expected values in the groups are not too small. In particular, the requirement is that no expected value can be <1 and no more than $\frac{1}{5}$ of the expected values can be <5. If there are too many groups with small expected frequencies, then some of them should be combined with other groups so that the preceding rule is not violated. The test procedure can be summarized as follows:

10.34   **Chi-Square Goodness-of-Fit Test**

To test for the goodness of fit of a probability model, use the following procedure:

(1) Divide the raw data into groups. The considerations for grouping data are similar to those given in Section 2.7 (p. 24). In particular, the groups must not be too small, so that step 7 is not violated.

(2) Estimate the k parameters of the probability model from the data using the methods of Chapter 6.

(3) Use the estimates in step 2 to compute the probability $\hat{p}$ of obtaining a value within a particular group and the corresponding expected frequency within that group $(n\hat{p})$, where n is the total number of data points.

(4) If O and E are, respectively, the observed and expected number of units within a particular group, then compute

$$X^2 = (O_1 - E_1)^2/E_1 + (O_2 - E_2)^2/E_2 + \cdots + (O_g - E_g)^2/E_g$$

where g = the number of groups.

(5) For a test with significance level α, if

$$X^2 > \chi^2_{g-k-1, 1-\alpha}$$

then reject H_0; if

$$X^2 \leq \chi^2_{g-k-1, 1-\alpha}$$

then accept H_0.

(6) The exact p-value for this test is given by

$$Pr(\chi^2_{g-k-1} > X^2)$$

(7) Use this test only if

(a) No more than $\frac{1}{5}$ of the expected values are <5.

(b) No expected values are <1.

The acceptance and rejection regions for this test are depicted in Figure 10.19. The computation of the p-value for this test is illustrated in Figure 10.20.

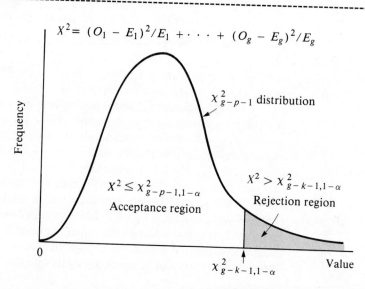

$$X^2 = (O_1 - E_1)^2/E_1 + \cdots + (O_g - E_g)^2/E_g$$

χ^2_{g-p-1} distribution

$X^2 \leq \chi^2_{g-p-1, 1-\alpha}$
Acceptance region

$X^2 > \chi^2_{g-k-1, 1-\alpha}$
Rejection region

Frequency

0

$\chi^2_{g-k-1, 1-\alpha}$ Value

FIGURE 10.19
Acceptance and
rejection regions for the
chi-square goodness-
of-fit test

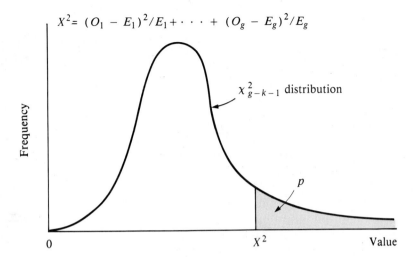

FIGURE 10.20
Computation of the p-value for the chi-square goodness-of-fit test

EXAMPLE 10.57 **Hypertension** Test for the goodness of fit of the normal probability model using the data in Table 10.33.

SOLUTION Two parameters have been estimated from the data (μ, σ^2) and there are eight groups. Therefore, $k = 2$, $g = 8$. Under H_0, X^2 follows a chi-square distribution with $8 - 2 - 1 = 5 \, df$.

$$X^2 = (O_1 - E_1)^2/E_1 + \cdots + (O_8 - E_8)^2/E_8$$
$$= (57 - 77)^2/77 + \cdots + (251 - 107)^2/107 = 352.6 \sim \chi_5^2 \text{ under } H_0$$

Since $\chi_{5,.999}^2 = 20.52 < 352.6 = X^2$, the p-value $< 1 - .999 = .001$ and the results are very highly significant.

Thus, the adequacy of the normal model is *not* accepted. The normal model appears to fit fairly well in the middle of the distribution (between 60 and 110 mm Hg) but fails badly in the tails, predicting too many blood pressures below 60 mm Hg and too few over 110 mm Hg. ∎

The test procedure in **(10.34)** can be used to assess the goodness of fit of any probability model, not just the normal model. The expected frequencies would be computed from the probability distribution of the proposed model and then the same goodness-of-fit test statistic as given in **(10.34)** would be used.

SECTION 10.12 **Summary**

In this chapter the most widely used techniques for analyzing qualitative (or categorical) data were discussed. First, the problem of how to compare binomial proportions from two independent samples was studied. For the large-sample case, this problem was solved in two different (but equivalent) ways: using either the two-sample test for binomial proportions or the chi-square test for 2×2 contingency tables. The former method is similar to the t test methodology introduced in Chapter 8, whereas the contingency table approach can be easily generalized to more complex problems involving qualitative data. For the small-sample case,

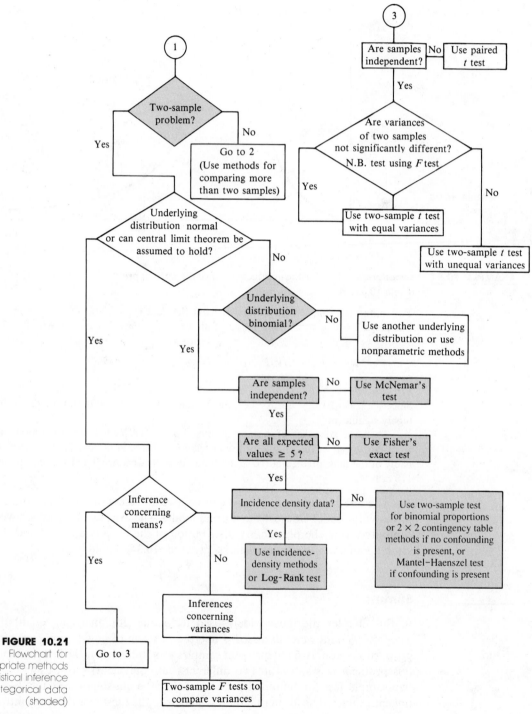

FIGURE 10.21

Flowchart for appropriate methods of statistical inference for categorical data (shaded)

Fisher's exact test is used to compare binomial proportions in two independent samples. To compare binomial proportions in paired samples, such as when a person is used as his or her own control, McNemar's test for correlated proportions should be used.

The 2×2 contingency table problem was then extended to the investigation of the relationship between two qualitative variables, in which one or both variables have more than two possible categories of response. A chi-square test for $R \times C$ contingency tables was developed, which is a direct generalization of the 2×2 contingency table test. The question of how to compare two binomial proportions when confounding is present was also considered. Our approach was to partition the data into k strata according to the values of one or more confounding variables. The Mantel–Haenszel test was then used to combine evidence over more than one stratum. The statistical treatment of incidence data, where the unit of analysis is person-time rather than person was then considered. The Log-Rank test was also introduced to treat data where incidence varied over the follow-up period. Finally, the problem of how to assess the goodness of fit of the probability models proposed in earlier chapter was studied. The chi-square goodness-of-fit test was used to address this problem. These strategies are illustrated in the shaded boxes of the flowchart in Figure 10.21.

In this chapter the focus was on how to assess the relationship between two qualitative variables. In the next two chapters, a similar problem, in which the variables under study are either all continuous or some are continuous and others are categorical, is discussed.

PROBLEMS

Cardiovascular Disease

In a 1985 study of the effectiveness of streptokinase in the treatment of patients who have been hospitalized after myocardial infarction, 9 of 199 males receiving streptokinase and 13 of 97 males in the control group died within 12 months [11]

10.1 Use the normal theory method to test for significant differences in 12-month mortality between the two groups.

10.2 Construct the observed and expected contingency tables for these data.

10.3 Perform the test in Problem 10.1 using the contingency table method.

10.4 Compare your results in Problems 10.1 and 10.3.

Gynecology

In a 1985 study of the relationship between contraceptive use and infertility, 89 out of 283 infertile women, compared with 640 out of 3833 control women, had used an IUD at some time in their lives [12].

10.5 Use the normal theory method to test for significant differences in contraceptive-use patterns between the two groups.

10.6 Use the contingency table method to perform the test in Problem 10.5.

10.7 Compare your results in Problems 10.5 and 10.6.

10.8 Compute a 95% confidence interval for the difference in 12-month mortality rates between the streptokinase and control groups in Problem 10.1.

10.9 Compute a 95% confidence interval for the difference in the proportion of women who have ever used IUD's between the case and control groups in Problem 10.5.

10.10 Compute the odds ratio in favor of death within 12 months for streptokinase therapy versus control therapy using the data in Problem 10.1.

10.11 Provide a 95% confidence interval for the true odds ratio corresponding to your answer to Problem 10.10.

10.12 What is the relationship between your answers to Problems 10.3 and 10.11?

10.13 Compute the odds ratio in favor of ever using an IUD for infertile women versus control women.

10.14 Provide a 95% confidence interval for the true odds ratio corresponding to your answer to Problem 10.13.

10.15 What is the relationship between your answers to Problems 10.6 and 10.14?

Cardiovascular Disease

Suppose a 5-year clinical trial comparing aspirin versus placebo for the prevention of myocardial infarction (MI) in men is planned. A 2.5% incidence rate of MI over 5 years among 40–64-year-old placebo males and a 2.0% rate over 5 years among comparably aged males taking aspirin is anticipated.

10.16 How many subjects need to be enrolled in each group to have an 80% chance of detecting a significant difference using a two-sided test with $\alpha = .05$?

10.17 Answer Problem 10.16 if a two-sided test with power = .9 is used.

10.18 Answer Problem 10.16 if a one-sided test with power = .8 is used.

10.19 Suppose 5000 men are actually enrolled in each treatment group. What would be the power of such a study if a two-sided test with $\alpha = .05$ were used?

10.20 Answer Problem 10.19 if 7000 men are enrolled in each group.

Cardiovascular Disease

In the streptokinase study in Problem 10.1, 2 of 15 females receiving streptokinase and 4 of 19 females in the control group died within 12 months.

10.21 Why is Fisher's exact test the appropriate procedure to test for differences in 12-month mortality rates between these two groups?

10.22 Write down all possible tables with the same row and column margins as given in the observed data.

10.23 Calculate the probability of each of the tables enumerated in Problem 10.22.

10.24 Evaluate whether or not there is a significant difference between the mortality rates for streptokinase and control group females using a two-sided test based on your results in Problem 10.23.

Refer to Table 2.11 (p. 36).

10.25 What significance test can be used to detect a relationship between receiving an antibiotic and receiving a bacterial culture, while in the hospital?

10.26 Perform the test in Problem 10.25 and report a p-value.

Gastroenterology

Two drugs (A, B) are compared for the medical treatment of duodenal ulcer. For this purpose patients are carefully matched on age, sex, and clinical condition. The treatment results based on 200 matched pairs show that for 89 matched pairs both treatments are effective; for 90 matched pairs both treatments are ineffective; for 5 matched pairs drug A is effective, whereas drug B is ineffective; for 16 matched pairs drug B is effective, whereas drug A is ineffective.

10.27 What test procedure can be used to assess the results?

10.28 Perform the test in Problem 10.27 and report a p-value.

In the same study, if the focus is on the 100 matched pairs consisting of male patients, then the following results are obtained: for 52 matched pairs both drugs are effective; for 35 matched pairs both drugs are ineffective; for 4 matched pairs drug A is effective, whereas drug B is ineffective; for 9 matched pairs drug B is effective, whereas drug A is ineffective.

10.29 How many concordant pairs are among the male matched pairs?

10.30 How many discordant pairs are among the male matched pairs?

10.31 Perform a significance test to assess any differences in effectiveness between the drugs among males. Report a p-value.

Gynecology

Women were subdivided by duration of IUD use in the study presented in Problem 10.5. The data are given in Table 10.34.

10.32 Perform a test for heterogeneity of the proportions of cases in the four groups.

TABLE 10.34 Relationship between duration of IUD use and infertility among IUD users

	Duration of IUD use (months)			
	<3	≥3, <18	≥18, ≤36	>36
Cases	10	23	20	36
Controls	53	200	168	219

(Reprinted with permission of the *New England Journal of Medicine, 312*(15), 941–947, 1985.)

10.33 Suppose the score variable 1, 2, 3, 4 is assigned to the four duration groups. Perform a test for trend on these data; that is, does the proportion of cases increase or decrease as the duration of IUD use increases?

10.34 Interpret your results in Problems 10.32 and 10.33.

Refer to the streptokinase data presented for males in Problem 10.1 and for females in Problem 10.21.

10.35 After stratifying the data by sex, perform a significance test for association between treatment group and 12-month mortality.

10.36 After stratifying the data by sex, estimate the odds ratio in favor of 12-month mortality in the streptokinase group versus the control group.

10.37 Provide a 95% confidence interval for the true odds ratio corresponding to your answer to Problem 10.36.

10.38 Test for the goodness of fit of the normal model for the distribution of survival times of mice given in Table 6.13 (p. 183).

10.39 Test for the adequacy of the normal model for the distribution of duration of hospitalization given in Table 2.11 (p. 36).

10.40 Answer Problem 10.39 for the distribution of $\log_e$ (duration of hospitalization).

Pulmonary Disease

Suppose we wish to investigate the familial aggregation of respiratory disease on a disease-specific basis. One hundred families in which the head of household or the spouse has asthma, referred to as type A families, and 200 families in which neither the head of household nor the spouse has asthma, referred to as type B families, are identified. Suppose that in 15 of the type A families the first-born child has asthma, whereas in 3 other type A families the first-born child has some nonasthmatic respiratory disease. Furthermore, in 4 of the type B households the first-born child has asthma, whereas in 2 other type B households the first-born child has some nonasthmatic respiratory disease.

10.41 Compare the prevalence rates of asthma in the two types of families. State all hypotheses being tested.

10.42 Compare the prevalence rates of nonasthmatic respiratory disease in the two types of families. State all hypotheses being tested.

Venereal Disease

Suppose an epidemiologic investigation of people entering a VD clinic is performed. It is found that 160 of 200 patients who are diagnosed as having gonorrhea and 50 of 105 patients who are diagnosed as having nongonococcal urethritis (NGU) have had previous episodes of urethritis.

10.43 Is there an association between the present diagnosis and prior episodes of urethritis?

Cardiovascular Disease

A 1979 study investigated the relationship between cigarette smoking and subsequent mortality in men with a prior history of coronary disease [13]. It was found that 264 out of 1731 nonsmokers and 208 out of 1058 smokers had died in the 5-year period after the study began.

10.44 Assuming that the age distributions of the two groups are comparable, compare the mortality rates in the two groups.

Cancer

10.45 A 1980 study investigated the relationship between the use of oral contraceptives and the development of endometrial cancer [14]. It was found that of 117 endometrial cancer patients, 6 had used the oral contraceptive Oracon at some time in their lives, whereas of 395 controls, 8 had used this agent. Test for an association between the use of Oracon and the incidence of endometrial cancer using a two-tailed test.

Obstetrics

Suppose there are 500 pairs of pregnant women who participate in a prematurity study and are paired in such a way that the body weight of the 2 women in a pair are within 5 lb of each other. One of the 2 women is given a placebo and the other drug A to see if drug A has an effect in preventing prematurity. Suppose that in 30 pairs of women, *both* women in a pair have a premature child; in 420 pairs of women, *both* women have a normal child; in 35 pairs of women, the woman taking drug A has a normal child and the woman taking the placebo has a premature child; in 15 pairs of women, the woman taking drug A has a premature child and the woman taking the placebo has a normal child.

10.46 Assess the statistical significance of these results.

Ophthalmology

Retinitis pigmentosa is a disease that manifests itself via different genetic modes of inheritance. Cases have been documented with a dominant, recessive, and sex-linked mode of inheritance. It has been conjectured that the mode of inheritance is related to the ethnic origin of the individual. Cases of the disease have been surveyed in an English and a Swiss population with the following results: Out of 125 English cases, 46 had sex-linked disease, 25 had recessive disease, and 54 had dominant disease. Out of 110 Swiss cases, 1 had sex-linked disease, 99 had recessive disease, and 10 had dominant disease.

10.47 Do these data show a significant association between ethnic origin and genetic type?

Cancer

Suppose we wish to compare the following two treatments for breast cancer: simple mastectomy (S) and radical mastectomy (R). Matched pairs of women who are within the same decade of age and with the same clinical condition are formed. They receive the two treatments, and their subsequent 5-year survival is monitored. The results are given in Table 10.35. We wish to test for significant differences between the treatments.

10.48 What test should be used to analyze these data? State the hypothesis being tested.

10.49 Conduct the test mentioned in Problem 10.48.

Venereal Disease

Suppose we are interested in comparing the effectiveness of two different antibiotics, A and B, in treating gonorrhea. Each person receiving antibiotic A is matched with an equivalent person (age within 5 years, same sex), to whom antibiotic B is given. These people are asked to return to the clinic within 1 week to see if the gonorrhea has been eliminated. Suppose the results are as follows:

(a) For 40 pairs of people, both antibiotics are successful.

(b) For 20 pairs of people, antibiotic A is effective whereas antibiotic B is not.

(c) For 16 pairs of people, antibiotic B is effective whereas antibiotic A is not.

(d) For 3 pairs of people, neither antibiotic is effective.

10.50 Test for the relative effectiveness of the two antibiotics.

10.51 How many matched pairs should be enrolled in a future study if we wish to conduct a two-sided test with $\alpha = .05$ and power $= .80$? Use the results in (a), (b), (c), and (d) for planning purposes.

The annual incidence in 1973 of a rare disease in two communities, A and B, was compiled and it was found that of 100,000 people in community A, 5 have the disease, whereas of 200,000 people in community B, only 1 has the disease.

10.52 Test whether the underlying incidence rates are significantly different.

Pulmonary Disease

One important aspect of medical diagnosis is its reproducibility. Suppose that two different doctors examine 100 patients for dyspnea in a respiratory disease clinic and that 15 patients are diagnosed as having dyspnea by Doctor A, 10 patients are diagnosed as having dyspnea by Doctor B, and 7 patients are diagnosed as having dyspnea by *both* Doctor A and Doctor B.

TABLE 10.35 Comparison of simple and radical mastectomy in treating breast cancer

Pair	Treatment S woman	Treatment R woman	Pair	Treatment S woman	Treatment R woman
1	L*	L	11	D	D
2	L	D	12	L	D
3	L	L	13	L	L
4	L	L	14	L	L
5	L	L	15	L	D
6	D†	L	16	L	L
7	L	L	17	L	D
8	L	D	18	L	D
9	L	D	19	L	L
10	L	L	20	L	D

* L = lived at least 5 years
† D = died within 5 years

10.53 Test whether or not the diagnoses of the two doctors are comparable.

Cardiovascular Disease

Much controversy has arisen recently on the possible association of myocardial infarction (MI) and coffee drinking. Suppose the information in Table 10.36 on coffee drinking and prior MI status is obtained from 200 60–64-year-old males in the general population.

TABLE 10.36 Coffee drinking and prior MI status

Coffee drinking (cups/day)	MI in last 5 years	Number of people
0	Yes	3
0	No	57
1	Yes	7
1	No	43
2	Yes	8
2	No	42
3 or more	Yes	12
3 or more	No	28
	Total yes	30
	Total no	170

10.54 Test for the association between history of MI and coffee drinking status, which is categorized as follows: 0 cups, 1 or more cups.

10.55 Suppose coffee drinking is categorized as follows: 0 cups, 1 cup, 2 cups, 3 or more cups. Perform a test to investigate whether or not there is a consistent association between these two variables using this categorization.

Infectious Disease

Suppose there is a computerized data bank consisting of all charts of patients at nine hospitals in Cleveland, Ohio. One concern of the group conducting the study is the possibility that the attending physician under- or overreports various diagnoses that seem consistent with a patient's chart. An investigator notes that 50 out of the 10,000 people in the data bank are reported as having a particular viral infection by their attending physician. A computer using an automated method of diagnosis claims that 68 out of the 10,000 people have the infection, 48 of them from the attending physician's 50 positives and 20 from the attending physician's 9950 negatives.

10.56 Test the hypothesis that the computer's and the attending physician's diagnoses are comparable.

Obstetrics

10.57 Test for the adequacy of the goodness of fit of the normal distribution when applied to the distribution of birthweights in Figure 2.8 (p. 30). The sample mean and standard deviation for these data are 111.26 oz and 20.95 oz respectively.

Cardiovascular Disease

An investigator wishes to study the effect of cigarette smoking on the development of myocardial infarction (MI) in women. In particular, there is some question in the literature as to the relationship of the timing of cigarette smoking to the development of disease. One school of thought says that current smokers are at much higher risk than ex-smokers. Another school of thought says that a considerable latent period of non-smoking is needed before the risk of ex-smokers becomes less than that of current smokers. To test this hypothesis, 2000 disease-free currently smoking women and 1000 disease-free ex-smoking women, aged 50–59, are identified in 1976, and the incidence of MI between 1976 and 1978 is noted at follow-up visits 2 years later. Investigators find that 40 currently smoking women and 10 ex-smoking women have developed the disease.

10.58 Is a one-sample or two-sample test needed here?

10.59 Is a one-sided or two-sided test needed here?

10.60 Which of the following test procedures should be used to test this hypothesis? (More than one may be necessary.)

(a) χ^2 test for 2×2 contingency tables

(b) Fisher's exact test

(c) McNemar's test

(d) One-sample binomial test

(e) One-sample t test

(f) Two-sample t test with equal variances

10.61 Carry out the test procedure(s) mentioned in Problem 10.60 and report a p value.

Cardiovascular Disease

A hypothesis has been suggested that a principal benefit of physical activity is to prevent sudden death from heart attack. The following study was designed to test this hypothesis: 100 men who died from a first heart attack and 100 men who survived a first heart attack in the age group 50–59 were identified and their wives were each given a detailed questionnaire concerning their husband's physical activity in the year preceding their heart attacks. The men were then classified as active or inactive. Suppose that 30 of the 100 who survived and 10 of the 100 who died were physically active. If we wish to test the hypothesis, then:

10.62 Is a one-sample or two-sample test needed here?

10.63 Which one of the following test procedures should be used to test the hypothesis?

(a) Paired t test

(b) Two-sample t test with independent samples

(c) χ^2 test for 2×2 contingency tables

(d) Fisher's exact test

(e) McNemar's test

10.64 Carry out the test procedure(s) in Problem 10.63 and report a p-value.

10.65 Compute the odds ratio in favor of prior physical activity for MI survivors versus MI deceased.

10.66 Compute a 95% confidence interval for the odds ratio referred to in Problem 10.65.

Cardiovascular Disease
A longitudinal study in apparently normal men is organized to relate *changes* in cardiovascular risk parameters to subsequent mortality. The hypothesis being tested is that men whose cholesterol level rises have a different subsequent mortality than those whose cholesterol level drops. In particular, two groups of 50–59-year-old men with initially normal cholesterol levels are identified: (1) group $A = 25$ men whose cholesterol level rises by 50 mg% over a 5-year period, and (2) group $B = 25$ men whose cholesterol level drops by 50 mg% over a 5-year period. The groups are then followed for mortality over the next 5 years. The results are given in Table 10.37.

10.67 Is a one-sample or two-sample test needed here?

10.68 Is a one-sided or two-sided test needed here?

10.69 Which of the following test procedures should be used? (More than one may be necessary.)

(a) Paired t test

(b) Two-sample t test with equal variances

(c) χ^2 test for 2×2 tables

(d) Fisher's exact test

(e) McNemar's test

(f) One-sample binomial test

10.70 Carry out the test procedure in Problem 10.69 and report a p-value.

10.71 How many people should be enrolled in a future study if 90% power is required and a two-sided test with $\alpha = .05$ is used? (Use the results given in Table 10.37 for planning purposes.)

10.72 Suppose 200 people are enrolled in the study. How much power would such a study have? (Use the results given in Table 10.37 for planning purposes.)

TABLE 10.37 Association between cardiovascular mortality and cholesterol change (+ = dead within the next 5 years; − = alive after 5 years)

Number of pairs	Mortality outcome, group A	Mortality outcome, group B	Number of pairs	Mortality outcome, group A	Mortality outcome, group B
1	−	−	14	−	−
2	+	−	15	−	−
3	−	−	16	−	−
4	−	+	17	+	−
5	−	−	18	−	−
6	−	−	19	−	−
7	−	−	20	−	−
8	+	−	21	+	−
9	−	−	22	−	−
10	−	−	23	−	−
11	+	−	24	−	−
12	−	−	25	−	−
13	−	−			

Mental Health

An observational study is set up to assess the effects of lithium in treating manic-depressive patients. New patients in an out-patient service are matched according to age, sex, and clinical condition, with one patient receiving lithium and the other a placebo. Suppose the outcome variable is whether or not the patient has any manic-depressive episodes in the next 3 months. The results are as follows: In 20 cases both the lithium and placebo members of the pair have manic-depressive episodes; in 10 cases only the placebo member has an episode (the lithium member does not); in 2 cases only the lithium member has an episode (the placebo member does not); in 36 cases neither member has an episode.

10.73 State an appropriate hypothesis to test whether lithium has any effect in treating manic-depressive patients.

10.74 Test the hypothesis mentioned in Problem 10.73.

Cancer

The following data are survival rates for cancer of the pancreas for the years 1955–1964, published by the U.S. Department of Health, Education and Welfare [15]: there were 256 reported cases of disease in the under-45 age group, of whom 8% survived for at least 3 years; 710 reported cases of disease in the 45–54 age group, of whom 2% survived for at least 3 years; 1348 reported cases of disease in the 55–64 age group, of whom 2% survived for at least 3 years; 1768 cases of disease in the 65–74 age group, of whom 1% survived for at least 3 years; 1292 cases of disease in the 75+ age group, of whom 1% survived for at least 3 years.

10.75 What significance test can be used to test if there is an age trend in the 3-year survival rates?

10.76 Perform the test in Problem 10.75 and report a *p*-value.

Cardiovascular Disease

In some studies heart disease has been associated with being overweight. Suppose this association is examined in a large-scale epidemiological study and it is found that of 2000 men in the age group 55–59, 200 have myocardial infarctions in the next 5 years. Suppose the men are grouped by body weight as given in Table 10.38.

TABLE 10.38 Association between body weight and myocardial infarction

Body weight (lb)	Number of myocardial infarctions	Total number of men
120–139	10	300
140–159	20	700
160–179	50	600
180–199	95	300
200+	25	100
Total	200	2000

10.77 Comment in detail on these data.

Venereal Disease

Suppose a study to examine the relative efficacy of penicillin and spectinomycin in the treatment of gonorrhea is conducted. Three treatments are looked at: (1) penicillin, (2) spectinomycin, low dose, and (3) spectinomycin, high dose. Three possible responses are recorded: (1) positive smear, (2) negative smear, positive culture, (3) negative smear, negative culture. The data in Table 10.39 are obtained.

10.78 Is there any relationship between type of treatment and response? What form does the relationship take?

TABLE 10.39 Efficacy of different treatments for gonorrhea

Treatment	Response			
	+ Smear	− Smear + Culture	− Smear − Culture	Total
Penicillin	40	30	130	200
Spectinomycin (low dose)	10	20	70	100
Spectinomycin (high dose)	15	40	45	100
Total	65	90	245	400

10.79 Suppose either a positive smear or a positive culture is regarded as a positive reponse and distinguished from the negative smear, negative culture response. Is there an association between the type of treatment and this measure of response?

Cerebrovascular Disease

Atrial fibrillation (AF) is widely recognized to predispose patients to embolic stroke. Although oral anticoagulant therapy has been suggested to decrease the number of embolic events, its benefits have not been proven. A study is proposed in which patients with AF are randomly divided into two groups: one receives the anticoagulant Warfarin, the other a placebo. The groups are then followed for the incidence of embolic stroke.

10.80 Suppose that 5% of treated patients and 22% of control patients are anticipated to experience an embolic stroke over 3 years. If 100 patients are to be randomized to each group, then how much power would such a study have of detecting a significant difference if a two-sided test with $\alpha = .05$ is used?

10.81 How large should such a study be to have an 80% chance of finding a significant difference given the same assumptions as in Problem 10.80?

10.82 Answer Problem 10.81 for a power of 90% rather than 80%.

Diabetes

Improvement in control of blood glucose levels is an important motivation for the use of insulin pumps for diabetic patients. However, certain side effects have been reported with pump therapy. Table 10.40 provides data on the occurrence of diabetic ketoacidosis (DKA) in patients before and after the onset of pump therapy [16].

TABLE 10.40 Occurrence of DKA in patients before and after the onset of insulin pump therapy

After pump therapy	Before pump therapy	
	No DKA	DKA
No DKA	128	7
DKA	19	7

(Reprinted with permission of *JAMA*, *252*(23), 3265–3269, 1984.)

10.83 What is the appropriate procedure to test if the rate of DKA is different before and after the onset of pump therapy?

10.84 Perform the significance test in Problem 10.83 and report a *p*-value.

Pulmonary Disease

Each year approximately 4% of current smokers attempt to quit smoking, and about 50% of those who try to quit are successful; that is, they are able to abstain from smoking for at least 1 year from the date they quit. Investigators have attempted to identify risk factors that might influence these two probabilities. One such variable is the number of cigarettes currently smoked per day. In particular, the investigators found that among 75 current smokers who smoked ⩽1 pack/day, 5 attempted to quit, whereas among 50 current smokers who smoked more than 1 pack/day, 1 attempted to quit.

10.85 Assess the statistical significance of these results and report a *p*-value.

Similarly, a different study reported that out of 311 people who had attempted to quit smoking, 16 out of 33 with less than a high school education were successful quitters; 47 out of 76 who had finished high school but had not gone to college were successful quitters; 69 out of 125 who attended college but did not finish 4 years of college were successful quitters; and 52 out of 77 who had completed college were successful quitters.

10.86 Do these data show an association between the number of years of education and the rate of successful quitting?

Renal Disease

A study group of 586 working women 30–49 years old who took phenacetin-containing analgesics and a control group of 559 comparably aged women without such intake were identified in 1968 and followed for mortality and morbidity outcomes. One hypothesis to be tested was that phenacetin intake may influence renal (kidney) function and hence have an effect on specific indices of renal morbidity and mortality. The mortality data of these women were traced from 1968 to 1979. Ten of the women in the study group and two of the women in the control group died, where at least one of the causes of death was deemed to be renal [17].

10.87 To test for differences in renal mortality between the two groups in either direction, what statistical test should be used?

10.88 Implement the test in Problem 10.87 and report a *p*-value.

10.89 Provide a 95% confidence interval for the difference in renal mortality rates between the two groups.

From the study group a subgroup of 309 women with a very high intake of phenacetin at baseline was identified, and 8 of them died during 1968–1979.

10.90 What statistical test should be used to compare the renal mortality experience of the high-intake group with that of the control group?

10.91 Implement the test in Problem 10.90 and report a *p*-value.

Infectious Disease, Hepatic Disease

Read "Foodborne Hepatitis A Infection: A Report of Two Urban Restaurant-Associated Out-breaks" by Denes et al., in the *American Journal of Epidemiology*, *105*(2) (1977), pages 156–162, and answer the following questions based on it.

10.92 The authors analyzed the results of Table 1 using a chi-square statistic. Is this method of analysis reasonable for this table? If not, suggest an alternative method.

10.93 Analyze the results in Table 1 using the method suggested in Problem 10.92. Do your results agree with the authors'?

10.94 Student's *t* test with 40 *df* was used to analyze the results in Table 2. Is this method of analysis reasonable for this table? If not, suggest an alternative method.

10.95 The authors claim that there is a significant difference ($p = .01$) between the rate of consumption of salad among those who did and did not feel well. Check this result using the method of analysis suggested in Problem 10.94.

Cancer

Read "Smoking and Carcinoma of the Lung" by R. Doll and A. B. Hill in the *British Medical Journal*, September 30, 1950, pages 739–748. Refer to Table IV in this paper and answer the following questions based on it.

10.96 Test for the association between cigarette smoking and disease status among males only.

10.97 Compute the odds ratio in favor of cigarette smoking for male lung cancer cases versus controls.

10.98 Compute a 95% confidence interval for the odds ratio computed in Problem 10.97.

10.99 Test for the association between cigarette smoking and disease status among females.

10.100 Answer Problem 10.97 for females.

10.101 Answer Problem 10.98 for females.

The people in the study were also classified in Table V according to the number of cigarettes per day smoked regularly just prior to the onset of their present illness.

10.102 Is there a consistent trend between the number of cigarettes smoked and disease status among males? Perform the appropriate significance test.

10.103 Answer Problem 10.102 for females.

Cancer

Suppose a clinical trial is performed to assess the effect of a new treatment for cancer of the esophagus. No attempt is made to match the patients in any way because the number of cases is too small. It is found that of 100 patients who were given the standard treatment, 6 lived for 3 years and 5 lived for 5 years. Correspondingly, of 47 patients who were given the new treatment, 10 survived for 3 years, whereas 2 survived for 5 years.

10.104 Is there any evidence that the new treatment is helpful for the 3-year prognosis of the patient?

10.105 Is there any evidence that the new treatment is helpful for the 5-year prognosis?

10.106 Suppose a person has survived for 3 years. Is there any evidence for a treatment effect on the prognosis for the next 2 years?

Infectious Disease

The presence of bacteria in the urine (bacteriuria) has been associated with kidney disease. Conflicting results have been reported from several studies concerning the possible role of oral contraceptives (OC) in bacteriuria. The following data were collected in a population-based group of nonpregnant premenopausal women below the age of 50 [18]. The data are presented on an age-specific basis in Table 10.41.

10.107 Why is controlling for age important in looking at the relationship between bacteriuria and OC use?

10.108 Perform a significance test to examine the association between OC use and bacteriuria after controlling for age.

10.109 Estimate the odds ratio in favor of bacteriuria for OC users versus non-OC users after controlling for age.

10.110 Provide a 95% confidence interval for the odds ratio in Problem 10.109.

TABLE 10.41 Rates of bacteriuria among oral contraceptive users and nonusers

Age group	OC users %	OC users n	Non-OC users %	Non-OC users n
16–19	1.2	84	3.2	281
20–29	5.6	284	4.0	552
30–39	6.3	96	5.5	623
40–49	22.2	18	2.7	482

(Reprinted with permission of the *New England Journal of Medicine*, 299, 536–537, 1978.)

10.111 How do your answers in Problems 10.108 and 10.110 relate to each other?

10.112 Suppose you did not control for age in the preceding analyses. Calculate the crude (unadjusted for age) odds ratio in favor of bacteriuria for OC users versus non-OC users.

10.113 How do your answers to Problems 10.112 and 10.109 relate to each other? Try to explain any differences found.

Diabetes, Ophthalmology

Diabetic retinopathy is an ocular condition that can result in significant visual loss and hence is of major concern to diabetic patients. A recent hypothesis is that use of aldose reductase inhibitors may lead to a major breakthrough in the prevention of diabetic retinopathy. To test this hypothesis, a treatment trial is planned for juvenile-onset, insulin-dependent diabetics, comparing the aldose reductase inhibitor Sorbinil with placebo. Suppose that, based on previous population estimates, 26% of patients on placebo with no signs of retinopathy at baseline are expected to show signs of diabetic retinopathy over a 30-month period.

10.114 If the true rate of disease in the Sorbinil group is 50% of that in the placebo group, then how much power would a study based on 100 randomized patients in each group have if a two-sided test with $\alpha = .05$ is used?

10.115 Answer Problem 10.114 if the true rate of disease in the Sorbinil group is assumed to be 30%, rather than 50%, less than that of placebo.

10.116 How large a sample is needed in each group to achieve 80% power under the assumptions stated in Problem 10.114?

10.117 How large a sample is needed in each group to achieve 80% power under the assumptions stated in Problem 10.115?

Infectious Disease, Cardiology

Kawasaki's syndrome is an acute illness of unknown cause that occurs predominantly in children under the age of 5. It is characterized by persistent high fever and other clinical signs and can result in death and/or coronary artery aneurysms. In the early 1980s, standard therapy for this condition was aspirin to prevent blood clotting. A Japanese group began experimentally treating children with intravenous gamma globulin in addition to aspirin to prevent cardiac symptoms in these patients [19].

A clinical trial is planned in the United States to compare the combined therapy of gamma globulin and aspirin versus aspirin therapy alone. Suppose the rate of coronary artery aneurysms is 15% in the aspirin-treated group, based on previous experience, and the investigators intend to use a two-sided significance test with $\alpha = .05$.

10.118 If the rate of coronary aneurysms in the combined therapy group is 5%, then how much statistical power will such a study have if 125 patients are to be recruited in each treatment group?

10.119 Answer Problem 10.118 if 150 patients are recruited in each group.

10.120 How many patients would have to be recruited in each group to have a 95% chance of finding a significant difference?

Emergency Medicine

Mannitol and Decadron are drugs that are often administered to patients with severe head injury when they are admitted to the emergency room of a hospital. One hypothesis is that this type of treatment would be more beneficial if administered by paramedics to patients in the field before they are transported to the hospital. To plan such a study, a pilot study is performed, whereby 4 of 10 patients with field treatment and 6 of 10 patients with no field treatment die before discharge from the hospital.

10.121 If a clinical trial is planned based on a two-tailed test with $\alpha = .05$, assuming that the pilot study results are valid, how much power would such a study have if 20 patients are randomized to each of the two groups?

10.122 How many patients are needed in each group to achieve an 80% power with the preceding study design if the differences found in the pilot study are assumed to be valid?

10.123 If the true mortality rate in the field-treated group were actually .50 rather than .40, how would these new data affect the power estimate in Problem 10.121 and the sample size estimate in Problem 10.122? (Provide only a qualitative answer.)

Pulmonary Disease

Read "Influence of Passive Smoking and Parental Phlegm on Pneumonia and Bronchitis in Early Childhood" by J. R. T. Colley, W. W. Holland, and R. T. Corkhill in *The Lancet*, November, 2, 1974, pages 1031–1034, and answer the following questions based on it.

10.124 Perform a statistical test comparing the incidence rates of pneumonia and bronchitis for children in their first year of life in families in which both parents are nonsmokers versus families in which both parents are smokers.

10.125 Compute an odds ratio to compare the incidence rates of pneumonia and bronchitis in families in which both parents are smokers versus families in which both parents are nonsmokers.

10.126 Compute a 95% confidence interval corresponding to the odds ratio computed in Problem 10.125.

10.127 Compare the incidence rates of pneumonia and bronchitis for children in their first year of life in families in which both parents are nonsmokers versus families in which one parent is a smoker.

10.128 Answer Problem 10.125, comparing families in which one parent is a smoker with families in which both parents are nonsmokers.

10.129 Compute a 95% confidence interval corresponding to the odds ratio computed in Problem 10.128.

10.130 Is there a significant trend in the percentage of children with pneumonia and bronchitis in the first year of life according to the number of smoking parents? Report a *p*-value.

10.131 Compare the incidence rates of pneumonia and bronchitis for children in their third year of life in families in which both parents are nonsmokers as compared with the rates for families in which both parents are smokers.

10.132 Suppose we wish to compare the incidence rates of disease for children in their first and second years of life in families in which both parents are

nonsmokers. Rates of 7.8% and 8.1% based on samples of size 372 and 358, respectively, are presented in Table II. Would it be reasonable to use a chi-square test to compare these rates?

10.133 Perform a statistical test comparing the incidence rates of pneumonia and bronchitis for children in the first year of life when stratified by number of cigarettes per day. (Use the groupings in Table IV in the appendix.) Restrict your analysis to families in which one or both parents are current smokers and have normal respiratory function.

10.134 Is there a consistent trend in the incidence rates referred to in Problem 10.133 as the number of cigarettes smoked increases?

10.135 Does the number of siblings in the family affect the incidence rate of pneumonia and bronchitis for children in the first year of life in families in which one or both parents have respiratory disease and both parents are nonsmokers? (Specifically, in Table VI, compare children with 0 siblings to children with 1 or more siblings.)

Obstetrics

An issue of current interest is the effect of delayed childbearing on pregnancy outcome. In a recent paper a population of first deliveries was assessed for low-birthweight deliveries (<2500 g) according to the woman's age and prior pregnancy history [20]. The data in Table 10.42 were presented.

TABLE 10.42 Relationship of age and pregnancy history to low-birthweight deliveries

Age	History*	n	Percentage low birthweight
≥30	No	225	3.56
≥30	Yes	88	6.82
<30	No	906	3.31
<30	Yes	153	1.31

* History = yes if the women had a prior history of spontaneous abortion or infertility
　　　　= no otherwise

(Reprinted with permission of the *American Journal of Epidemiology*, *125*(1), 101–109, 1987.)

10.136 What test can be used to assess the effect of age on low-birthweight deliveries among women with a negative history?

10.137 Perform the test in Problem 10.136 and report a *p*-value.

10.138 What test can be used to assess the effect of age on low-birthweight deliveries among women with a positive history?

10.139 Perform the test in Problem 10.138 and report a p-value.

10.140 Can you think of a way of assessing whether the effect of age is different among women with a positive versus a negative history?

Mental Health

A study was performed in Lebanon looking at the effect of widowhood on mortality [21]. Each of 151 widowers and 544 widows were matched to a person married at the time of widowhood and of the same age (± 2 years) and sex. The people in the matched pairs were followed until one member of the matched pair died. The results in Table 10.43 were obtained for those matched pairs in which at least one member had died by 1980.

TABLE 10.43 Effect of widowhood on mortality

Age (years)	Males		Females	
	n_1*	n_2**	n_1	n_2
36–45	4	8	3	2
46–55	20	17	17	10
56–65	42	26	16	15
66–75	21	10	18	11
Unknown	0	2	3	2
Total	87	63	57	40

* n_1 = number of pairs in which the widowed subject is deceased and the married subject is alive

** n_2 = number of pairs in which the widowed subject is alive and the married subject is deceased

(Reprinted with the permission of the *American Journal of Epidemiology*, *125*(1), 127–132, 1987.)

10.141 Suppose all of the matched pairs in Table 10.43 are considered. What method of analysis can be used to test if there is an association between widowhood and mortality?

10.142 Implement the test in Problem 10.141 and report a p-value.

10.143 Answer the same question as Problem 10.141 considering 36–45-year-old males only.

10.144 Implement the test in Problem 10.143 and report a p-value.

10.145 How much power did the study mentioned above have versus the alternative hypothesis that a widower is twice as likely to die before a married person of the same age and sex, assuming that all age groups are considered?

10.146 Can you think of any confounding variables that might affect your conclusions in Problems 10.141–10.144?

Hepatic Disease

Refer to Data Set 12, HORMONE.DAT, in Appendix 2.

10.147 What test procedure can be used to compare the percentage of hens whose pancreatic secretions increased among the 5 treatment regimens?

10.148 Implement the test procedure in Problem 10.147 and report a p-value.

Cancer

A recent study looked at the association between breast cancer incidence and alcohol consumption [22]. The data in Table 10.44 were presented for 50–54-year-old women.

10.149 What test procedure can be used to test if there is an association between breast cancer incidence and alcohol consumption, where alcohol consumption is coded as (drinker/nondrinker)?

10.150 Perform the test mentioned in Problem 10.149 and report a p-value.

10.151 Perform a test for linear trend based on the data in Table 10.44.

Cardiovascular Disease

A clinical trial is planned to look at the effect of aspirin in reducing total mortality in middle-aged men (ages 40–84) with no prior coronary disease. It is anticipated that 20,000 men will be enrolled in the study, of whom $\frac{1}{2}$ will be randomized to aspirin and $\frac{1}{2}$ to placebo.

10.152 If investigators expect 700 deaths over a 5-year period in those randomized to placebo and project a 10% reduction in mortality from aspirin, then how much power does the study have versus this alternative if a two-sided test with a type I error of .05 is used?

10.153 The *dropout* rate is defined as the percentage of subjects randomized to the active drug who do not take their study medications; the *drop-in* rate is defined as the percentage of subjects randomized to placebo who take the active drug on their own. If it is anticipated that the dropout rate will be a 10% and the drop-in rate will be 5%, then what is the expected mortality rate in

TABLE 10.44 Association between alcohol consumption and breast cancer in 50–54-year-old women

Group	Alcohol consumption (g/day)				
	None	<1.5	1.5–4.9	5.0–14.9	≥15.0
Breast cancer case	43	15	22	42	24
Total number of women	5944	2069	3449	3570	2917

(Reprinted with permission of the *New England Journal of Medicine, 316*(19), 1174–1180, 1987.)

the active drug and placebo group under these assumptions?

10.154 What is the true power of the study under the assumptions in Problem 10.153?

Cardiovascular Disease

A study was performed to look at the association between a parental history of myocardial infarction (MI) and coronary heart disease in women. In the subgroup of 50–55-year-old women, it was found that 4 of 900 women with a parental history of MI at age >60 and 1 of 1700 women without any parental history of MI had died from coronary heart disease over a 4-year period. All women were disease-free at the beginning of the period.

10.155 What test procedure can be used to test if there is an association between a parental history of MI at age >60 and the occurrence of fatal coronary heart disease?

10.156 Implement the procedure in Problem 10.155 and report a *p*-value.

Cancer

A case-control study was performed to look at the effect of cigarette smoking on the risk of cervical cancer [23]. There were 230 cervical cancer cases and 230 controls consisting of women ages 22–74 matched to the cases for age within 5-year intervals and admitted to the

hospital for acute conditions other than malignancy, hormonal, or gynecologic disorders. The smoking habits of these women are given in Table 10.45.

10.157 What is the appropriate test procedure to test if there is an association between smoking habit and cervical cancer?

10.158 Perform the test mentioned in Problem 10.157 and report a *p*-value.

Suppose we wish to directly compare the current smokers as a group with the nonsmokers (and exclude the ex-smokers).

10.159 What is the appropriate test procedure to test if there is an association between current smoking and cervical cancer risk?

10.160 Perform the test mentioned in Problem 10.159 and report a *p*-value.

10.161 Are there other ways of analyzing the data from this study if the data were displayed in a different manner?

Cardiovascular Disease

A secondary prevention trial of lipid lowering is planned in patients with previous myocardial infarction (MI). Patients are to be randomized to either a treatment group getting diet therapy and cholesterol-lowering drugs or a control group getting diet therapy and

TABLE 10.45

	Non-smokers	Ex-smokers	Current smokers <15 cigarettes/day	Current smokers ≥15 cigarettes/day
Cervical cancer cases	155	13	33	29
Controls	169	21	22	18

(Reprinted with permission by the *American Journal of Epidemiology, 123*(1), 22–29, 1986.)

placebo pills. The study endpoint is to be a combined endpoint consisting of either definite fatal coronary heart disease or nonfatal MI (i.e., a new nonfatal MI distinct from previous events). Suppose it is projected that the incidence of combined events among controls is 7% per year.

10.162 What proportion of controls will have events over 5 years?

Suppose the treatment benefit is projected to be a reduction in the event rate by 30%.

10.163 What is the expected event rate in the treated group?

10.164 How many subjects will be needed in each group if a one-sided test with $\alpha = .05$ is to be used and an 80% chance of finding a significant difference is desired if the rates in Problems 10.162 and 10.163 are the true rates?

An expectation of the investigators is that not all subjects will comply. In particular, it is projected that 5% of the treatment group will not comply with drug therapy, while 10% of the control group will start taking cholesterol-lowering drugs outside the study.

10.165 What will be the expected rates in Problems 10.162 and 10.163 if this level of lack of compliance is realized?

10.166 What will be the revised sample size estimate in Problem 10.164 if the lack of compliance is taken into account?

Cancer

A study was conducted to look at the relationship between self-reports of mole count and prevalence of malignant melanoma [24]. 98 cases of malignant melanoma occurring between 1976 and 1982 and 190 age-matched controls among members of the Nurses' Health Study (age 30–54) were selected for analysis. The total number of moles in the left and right arms combined was compared in cases and controls. The data are given in Table 10.46.

10.167 What method can be used to test if the distribution of mole count is different between cases and controls?

10.168 Implement the method and report a *p*-value.

Hypertension

A study was conducted in Wales relating blood pressure and blood-lead levels [25]. It was reported that 4 out of 455 men with blood-lead levels $\leqslant 11 \, \mu g/100 \, mL$ had elevated systolic blood pressure (SBP $\geqslant 160$ mm Hg), while 16 out of 410 men with blood-lead levels $\geqslant 12 \, \mu g/100 \, mL$ also had elevated SBP.

10.169 What is an appropriate procedure to test the hypothesis that there is an association between blood pressure and blood lead in men?

10.170 Implement the procedure in Problem 10.169 and report a *p*-value.

It was also reported that 6 out of 663 women with blood-lead levels $\leqslant 11 \, \mu g/100 \, mL$ had elevated SBP, while 1 out of 192 women with blood-lead levels $\geqslant 12 \, \mu g/100 \, mL$ had elevated SBP.

10.171 What is an appropriate procedure to test the hypothesis that there is an association between blood pressure and blood lead in women?

10.172 Implement the procedure in Problem 10.171 and report a *p*-value.

Cancer

An investigation was conducted looking at the incidence of leukemia after treatment among people with Hodgkin's disease according to the type of treatment received [26]. The data in Table 10.47 were reported.

Suppose we assume under the null hypothesis that the type of treatment has no effect on the incidence of leukemia.

10.173 What is the expected proportion of leukemias in the no-treatment group under the null hypothesis?

TABLE 10.46 Number of moles in left and right arm (combined)

	0	1–5	6–15	16–30	31+	Total
Cases	16	20	20	12	30	98
Controls	57	40	45	20	28	190

(Reprinted with permission of the *American Journal of Epidemiology*, *127*(4), 703–712, 1988.)

TABLE 10.47 Relationship of type of treatment for Hodgkin's disease and incidence of leukemia

Treatment procedure	Number of leukemias	Number of person-years of follow-up after treatment
(A) No intensive therapy	2	3322.8
(B) Either intensive radiotherapy or chemotherapy	4	1858.2

(Reprinted with permission of the *Journal of the National Cancer Institute, 672,* 751–760, 1981.)

10.174 What test procedure can be used to test if there is a different incidence density of leukemia in group A than expected under the null hypothesis if the total number of leukemias are considered fixed ($=6$)?

10.175 Implement the test procedure in Problem 10.174 and report a *p*-value.

Cancer

The data relating oral contraceptive use and the incidence of breast cancer in the age group 40–44 in the Nurses' Health Study are given in Table 10.48.

TABLE 10.48 Relationship between breast cancer incidence and OC use among 40–44-year-old women in the Nurses' Health Study

OC-use group	Number of cases	Number of person-years
Current users	13	4,761
Past users	164	121,091
Never users	113	98,091

10.176 Compare the incidence density of breast cancer in current users versus never users and report a *p*-value.

10.177 Compare the incidence density of breast cancer in past users versus never users and report a *p*-value.

Infectious Disease

Aminoglycoside antibiotics are particularly useful clinically in the treatment of serious gram-negative bacillary infections among hospitalized patients. Despite their potential for toxicity, as well as the continued development of newer antimicrobial agents of other classes, it seems likely that the clinical use of aminoglycosides will continue to be widespread. The choice of a particular aminoglycoside antibiotic for a given patient depends on several factors, including the specific clinical situation, differences in antimicrobial spectrum and cost, and risks of side effects, particularly nephrotoxicity and auditory toxicity. Many randomized, controlled trials have been published that compare the various aminoglycoside antibiotics with respect to efficacy, nephrotoxicity, and, to a lesser extent, auditory toxicity. These individual trials have varied widely with respect to their design features and their conclusions. A major limitation to their interpretability is that the majority of the individual trials have lacked an adequate sample size to detect the small to moderate differences between treatment groups that are most plausible. As a result, the individual trials published to date have generally not permitted firm conclusions, especially concerning the relative potential for toxicity of aminoglycosides.

In these circumstances, one method to estimate the true effects of these agents more precisely is to conduct an overview, or meta-analysis, of the data from all randomized trials. In this way, a true increase in risk could emerge that otherwise would not be apparent in any single trial due to small sample size. Therefore, a quantitative overview of the results of all published randomized controlled trials that assessed the efficacy and toxicity of individual aminoglycoside antibiotics was undertaken.

Forty-five randomized clinical trials, published between 1975 and September 1985, were identified that compared two or more of five aminoglycoside antibiotics: amikacin, gentamicin, netilmicin, sisomicin, and tobramycin. Thirty-seven of these trials could provide data suitable for comparative purposes.

The specific endpoints of interest were efficacy, nephrotoxicity, and auditory toxicity. Efficacy was defined as bacterial or clinical response to treatment as reported in each individual trial. Nephrotoxicity was defined as the percent of toxic events to the kidney reported, whether or not the published paper suggested some explanation other than use of the study drug, such as use of another potentially nephrotoxic agent, or the presence of an underlying disease affecting kidney function. Auditory toxicity was defined as reported differences between pre- and post treatment audiograms.

The data are organized into three data files: EFF.DAT, NEPHRO.DAT, and OTO.DAT (Data Sets 8, 16, and 20, in Appendix 2 respectively). A separate record is presented for each antibiotic studied for each endpoint. The format is given in the files EFF, DOC, NEPHRO.DOC in Appendix 2.

Columns 1–8: Study name
  10–11: Study number (number on reference list)
     13: Endpoint (1 = efficacy; 2 = nephrotoxicity; 3 = ototoxicity)
     15: antibiotic (1 = Amikacin; 2 = Gentamicin; 3 = Netilmicin; 4 = Sisomicin; 5 = Tobramycin
  17–19: Sample size
  21–23: Number cured (for efficacy) or number with side effect (for nephrotoxicity or ototoxicity)

10.178 Present an analysis comparing pairs of antibiotics as regards efficacy, nephrotoxicity, and ototoxicity. Are there specific antibiotics you would recommend or not recommend if you were the statistical advisor to a hospital infection-control committee?

Refer to the data set SMOKE.DAT, Appendix 2.

10.179 Compare the incidence of recidivism between males and females. Use 3 month intervals of time.

10.180 Divide the total study population into two approximately equal groups according to number of cigarettes smoked and compare the incidence of recidivism in these two subgroups.

10.181 Divide the total study population into two approximately equal groups according to adjusted log (Co) concentration and compare the incidence of recidivism in these two subgroups.

10.182 Write a computer program to implement the procedure used in Problems 10.179–10.181. Run this program using a time interval of 1 day rather than 3 months, comparing the subgroups identified in Problems 10.179–10.181.

References

[1] MacMahon, B., Cole, P., Lin, T. M., Lowe, C. R., Mirra, A. P., Ravnihar, B., Salber, E. J., Valaoras, V. G., & Yuasa, S. (1970). Age at first birth and breast cancer risk. *Bulletin of the World Health Organization, 43,* 209–221.

[2] Kleinbaum, D. G., Kupper, L. L., & Morgenstern, H. (1982). *Epidemiologic research: Principles and quantitative methods.* Belmont, CA: Wadsworth.

[3] Doll, R., Muir, C., & Waterhouse, J. (Eds.). (1970). *Cancer in five continents* (Vol. II). Berlin: Springer-Verlag.

[4] Cochran, W. G. (1954). Some methods for strengthening the common χ^2 test. *Biometrics, 10,* 417–451.

[5] Maxwell, A. W. (1961). *Analyzing qualitative data.* London: Methuen.

[6] Sandler, D. P., Everson, R. B., & Wilcox, A. J. (1985). Passive smoking in adulthood and cancer risk. *American Journal of Epidemiology, 121*(1) 37–48.

[7] Fleiss, J. (1981). *Statistical methods for rates and proportions.* New York: Wiley.

[8] Lee, E. T. (1986). *Statistical methods for survival data analysis.* Belmont, CA: Wadsworth.

[9] Miller, R. G., Jr. (1981). *Survival Analysis.* New York: Wiley.

[10] Hypertension Detection and Follow-up Program Cooperative Group. (1977). Blood pressure studies in 14 communities—A two-stage screen for hypertension. *Journal of the American Medical Association, 237*(22), 2385–2391.

[11] Kennedy, J. W., Ritchie, J. L., Davis, K. B., Stadius, M. L., Maynard, C., & Fritz, J. K. (1985). The western Washington randomized trial of intracoronary streptokinase in acute myocardial infarction: A 12-month follow-up report. *New England Journal of Medicine, 312*(17), 1073–1078.

[12] Cramer, D. W., Schiff, I., Schoenbaum, S. C., Gibson, M., Belisle, J., Albrecht, B., Stillman, R. J., Berger, M. J., Wilson, E., Stadel, B. V., & Seibel, M. (1985). Tubal infertility and the intrauterine device. *New England Journal of Medicine, 312*(15), 941–947.

[13] The Coronary Drug Project Research Group. (1979). Cigarette smoking as a risk factor in men with a history of myocardial infarction, *Journal of Chronic Diseases, 32*(6), 415–425.

[14] Weiss, N. S., & Sayetz, T. A. (1980). Incidence of endometrial cancer in relation to the use of oral contraceptives, *New England Journal of Medicine, 302*(10), 551–554.

[15] U.S. Department of Health, Education, and Welfare. (1972). *End results in cancer* (Report No. 4).

[16] Mecklenburg, R. S., Benson, E. A., Benson, J. W., Fredlung, P. N., Guinn, T., Metz, R. J., Nielsen, R. L., & Sannar, C. A. (1984). Acute complications associated with insulin pump therapy: Report of experience with 161 patients, *JAMA, 252*(23), 3265–3269.

[17] Dubach, U. C., Rosner, B., & Pfister, E. (1983). Epidemiological study of abuse of analgesics containing phenacetin: 1968–1979. Renal morbidity and mortality. *New England Journal of Medicine, 308*, 357–362.

[18] Evans, D. A., Hennekens, C. H., Miao, L., Laughlin, L. W., Chapman, W. G., Rosner, B., Taylor, J. O., & Kass, E. H. (1978). Oral contraceptives and bacteriuria in a community-based study. *New England Journal of Medicine, 299*, 536–537.

[19] Furusko, K., Sato, K., Socda, T., et al. (1983, December 10). High dose intravenous gamma globulin for Kawasaki's syndrome [Letter]. *Lancet*, 1359.

[20] Barkan, S. E., & Bracken, M. (1987). Delayed childbearing: No evidence for increased risk of low birth weight and preterm delivery. *American Journal of Epidemiology, 125*(1), 101–109.

[21] Armenian, H., Saadeh, F. M., & Armenian, S. L. (1987). Widowhood and mortality in an Armenian church parish in Lebanon. *American Journal of Epidemiology, 125*(1), 127–132.

[22] Willett, W., Stampfer, M. J., Colditz, G. A., Rosner, B. A., Hennekens, C. H., & Speizer, F. E. (1987). Moderate alcohol consumption and the risk of breast cancer. *New England Journal of Medicine, 316*(19), 1174–1180.

[23] La Vecchia, C., Franceschi, S., Decarli, A., Fasoli, M., Gentile, A., & Tognoni, G. (1986). Cigarette smoking and the risk of cervical neoplasia. *American Journal of Epidemiology, 123*(1), 22–29.

[24] Bain, C., Colditz, G. A., Willett, W. C., Stampfer, M. J., Green, A., Bronstein, B. R., Mihm, M. C., Rosner, B., Hennekens, C. H., & Speizer, F. E. (1988). Self-reports of mole counts and cutaneous malignant melanoma in women: Methodological issues and risk of disease. *American Journal of Epidemiology, 127*(4), 703–712.

[25] Elwood, P. C., Yarnell, J. W. G., Oldham, P. D., Catford, J. C., Nutbeam, D., Davey-Smith, G., & Toothill, C. (1988). Blood pressure and blood lead in surveys in Wales. *American Journal of Epidemiology, 127*(5), 942–945.

[26] Boivin, J. F., & Hutchison, G. B. (1981). Leukemia and other cancers after radiotherapy and chemotherapy for Hodgkin's Disease. *Journal of the National Cancer Institute, 67*(4), 751–760.

CHAPTER ELEVEN

REGRESSION AND CORRELATION METHODS

Introduction

In the previous chapters methods of estimation and hypothesis testing where only one variable was of interest were covered. Frequently, it is important to study the relationship between two or more variables in a particular sample.

EXAMPLE 11.1   **Obstetrics** Obstetricians sometimes order tests for estriol levels from 24-hour urine specimens taken from pregnant women who are near term, since the level of estriol has been found to be related to the birthweight of the infant. The test can provide indirect evidence of an abnormally small fetus. The relationship between estriol level and birthweight can be quantified by fitting a *regression line* that relates the two variables. ∎

EXAMPLE 11.2   **Hypertension** Much discussion has taken place in recent years on the familial aggregation of blood pressure. In general, children whose parents have high blood pressure tend to have higher blood pressure than their peers. One way of expressing this relationship is to compute a *correlation coefficient* relating the blood pressure of parents and children over a large collection of families. ∎

In this chapter methods of regression and correlation analysis in which *two* different variables in the same sample are related are studied. The extension of these methods to the case of multiple regression analysis, where the relationship between more than two variables at a time is considered, is also discussed briefly. Multiple logistic regression and Cox regression methods are also considered as extensions of regression methods for binary outcome variables.

SECTION 11.2 ## General Concepts

EXAMPLE 11.3   **Obstetrics** Greene and Touchstone conducted a study to relate birthweight to the estriol level of pregnant women [1]. Figure 11.1 is a plot of the data from the study, and the actual data points are listed in Table 11.1. As can be seen from the figure, there appears to be a linear relationship between estriol level and birthweight, although this relationship is not consistent and considerable scatter exists throughout the plot. How can this relationship be quantified? ∎

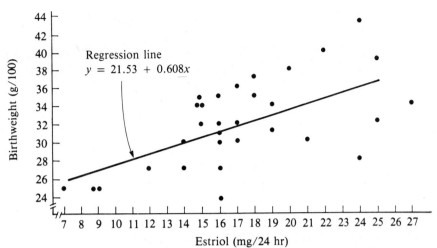

(Reprinted with permission of the *American Journal of Obstetrics and Gynecology*, 85(1), 1–9, 1963.)

FIGURE 11.1
Data from Greene–Touchstone study relating birthweight and estriol level in pregnant women near term

TABLE 11.1
Sample data from the Greene–Touchstone study relating birthweight and estriol level in pregnant women near term

i	Estriol (mg/24 hr) x_i	Birthweight (g/100) y_i	i	Estriol (mg/24 hr) x_i	Birthweight (g/100) y_i
1	7	25	17	17	32
2	9	25	18	25	32
3	9	25	19	27	34
4	12	27	20	15	34
5	14	27	21	15	34
6	16	27	22	15	35
7	16	24	23	16	35
8	14	30	24	19	34
9	16	30	25	18	35
10	16	31	26	17	36
11	17	30	27	18	37
12	19	31	28	20	38
13	21	30	29	22	40
14	24	28	30	25	39
15	15	32	31	24	43
16	16	32			

(Reprinted with permission of the *American Journal of Obstetrics and Gynecology*, 85(1), 1–9, 1963.)

If x = estriol level and y = birthweight, then a relationship between y and x that is of the following form may be postulated:

11.1

$$E(y|x) = \alpha + \beta x$$

That is, for a given estriol level x, the expected birthweight $E(y|x)$ is $\alpha + \beta x$.

DEFINITION 11.1 ■■■

The line $y = \alpha + \beta x$ is the **regression line**, where α is the intercept and β is the slope of the line.   ■

The relationship $y = \alpha + \beta x$ is not expected to hold exactly for every woman. Thus, an error term e, which represents the variance of birthweight among all babies of women with a given estriol level x, is introduced into the model. We will assume that e follows a normal distribution with mean 0 and variance σ^2. The full linear regression model then takes the following form:

11.2

$$y = \alpha + \beta x + e$$

where e is normally distributed with mean 0 and variance σ^2.

DEFINITION 11.2 ■■■

For any linear regression equation of the form $y = \alpha + \beta x + e$, y is referred to as the **dependent variable** and x as the **independent variable**, since we are trying to predict y from x.

■

EXAMPLE 11.4   **Obstetrics** Birthweight is the dependent variable and estriol is the independent variable for the problem posed in Example 11.3, since estriol levels are being used to try to predict birthweight.

■

One interpretation of the regression line is that for a woman with estriol level x, the corresponding birthweight will be normally distributed with mean $\alpha + \beta x$ and variance σ^2. If σ^2 were 0, then every point would fall exactly on the regression line, whereas the larger σ^2 is, the more scatter occurs about the regression line. This effect is illustrated in Figure 11.2.

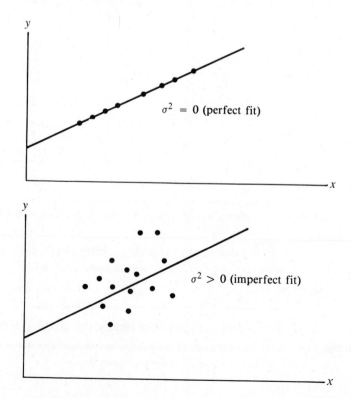

FIGURE 11.2
The effect of σ^2 on the goodness of fit of a regression line

How can β be interpreted? If β is greater than 0, then as x increases, the expected value of $y = \alpha + \beta x$ will increase.

EXAMPLE 11.5   **Obstetrics** This situation appears to be the case in Figure 11.3(a) for birthweight (y) and estriol (x), since as estriol increases, birthweight correspondingly increases.   ∎

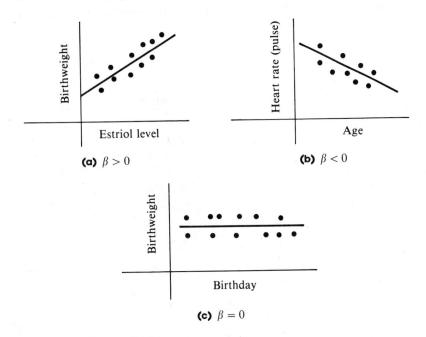

FIGURE 11.3
The interpretation of the regression line for different values of β

If β is less than 0, then as x increases, the expected value of y will decrease.

EXAMPLE 11.6   **Pediatrics** This situation might occur in a plot of pulse rate (y) versus age (x), as illustrated in Figure 11.3(b), since infants are born with rapid pulse rates that gradually decline with age.   ∎

If β is equal to 0, then there is no relationship between x and y.

EXAMPLE 11.7   This situation might occur in a plot of birthweight versus birthday, as shown in Figure 11.3(c), since there is no relationship between birthweight and birthday.   ∎

SECTION 11.3   **Fitting Regression Lines—The Method of Least Squares**

The question remains as to how to fit a regression line (or, equivalently, to obtain estimates of α and β, denoted by a and b, respectively) when data appear in the form of Figure 11.1. We could eyeball the data and draw a line that is not too distant from any of the points, but this approach is difficult in practice and can be quite imprecise with either a large number of points or a lot of scatter. A better method is to set up a specific criterion that defines the closeness of a line to a set of points and to find the line closest to the sample data according to this criterion.

Consider the data in Figure 11.4 and the estimated regression line $y = a + bx$. The distance d_i of a typical sample point (x_i, y_i) from the line could be measured along a direction parallel to the y-axis. If we let $(x_i, \hat{y}_i) = (x_i, a + bx_i)$ be the point on the estimated regression line at x_i, then this distance is given by $d_i = y_i - \hat{y}_i = y_i - a - bx_i$. A good-fitting line would make these distances as small as possible. Since the d_i cannot all be 0, the criterion $S_1 = $ sum of the absolute deviations of the sample points from the line $= \sum_{i=1}^{n} |d_i|$ can be used and the line that minimizes S_1 can be found. This strategy has proven to be analytically difficult. Instead, for both theoretical reasons and ease of derivation, the following least-squares criterion is commonly used:

$$S = \text{sum of the squared distances of the points from the line}$$

$$= \sum_{i=1}^{n} d_i^2 = \sum_{i=1}^{n} (y_i - a - bx_i)^2$$

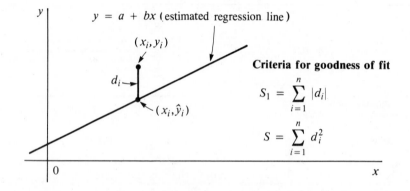

FIGURE 11.4
Possible criteria for measuring the goodness of fit of a regression line

DEFINITION 11.3

The **least-squares line**, or **estimated regression line**, is the line $y = a + bx$ that minimizes the sum of squared distances of the sample points from the line given by

$$S = \sum_{i=1}^{n} d_i^2$$

This method of estimating the parameters of the regression line is known as the **principle of least squares.** ∎

The following notation is needed to define the slope and intercept of a regression line:

DEFINITION 11.4

The **raw sum of squares for x** is defined by

$$\sum_{i=1}^{n} x_i^2$$

The **corrected sum of squares for x** is defined by

$$\sum_{i=1}^{n} x_i^2 - \left(\sum_{i=1}^{n} x_i\right)^2 \Big/ n$$

which is denoted by L_{xx}. Similarly, the **raw sum of squares for** y is defined by

$$\sum_{i=1}^{n} y_i^2$$

The **corrected sum of squares for** y is defined by

$$\sum_{i=1}^{n} y_i^2 - \left(\sum_{i=1}^{n} y_i\right)^2 \Big/ n$$

which is denoted by L_{yy}.   ∎

Notice that L_{xx} and L_{yy} are simply the numerators of the expressions for the sample variances of x(i.e., s_x^2) and y(i.e., s_y^2), respectively, since

$$s_x^2 = \sum_{i=1}^{n} (x_i - \bar{x})^2/(n-1) \qquad \text{and} \qquad s_y^2 = \sum_{i=1}^{n} (y_i - \bar{y})^2/(n-1)$$

DEFINITION 11.5 ■■
The **raw sum of cross products** is defined by

$$\sum_{i=1}^{n} x_i y_i$$

The **corrected sum of cross products** is defined by

$$\sum_{i=1}^{n} x_i y_i - \left(\sum_{i=1}^{n} x_i\right)\left(\sum_{i=1}^{n} y_i\right) \Big/ n$$

which is denoted by L_{xy}.   ∎

The coefficients of the line that satisfy the least-squares criterion in Definition 11.3 are given as follows:

11.3 | **Estimation of the Least-Squares Line**

The coefficients of the least-squares line $y = a + bx$ are given by

$$b = L_{xy}/L_{xx} \qquad \text{and} \qquad a = \bar{y} - b\bar{x} = \left(\sum_{i=1}^{n} y_i - b\sum_{i=1}^{n} x_i\right)\Big/ n$$

EXAMPLE 11.8   **Obstetrics** Derive the regression line for the data in Table 11.1.

SOLUTION   First

$$\sum_{i=1}^{31} x_i \qquad \sum_{i=1}^{31} x_i^2 \qquad \sum_{i=1}^{31} y_i \qquad \sum_{i=1}^{31} x_i y_i$$

must be obtained so as to compute the corrected sums of squares (L_{xx}) and cross products (L_{xy}). These quantities are given as follows:

$$\sum_{i=1}^{31} x_i = 534 \qquad \sum_{i=1}^{31} x_i^2 = 9876 \qquad \sum_{i=1}^{31} y_i = 992 \qquad \sum_{i=1}^{31} x_i y_i = 17{,}500$$

Then compute L_{xy} and L_{xx} as follows:

$$L_{xy} = \sum_{i=1}^{31} x_i y_i - \left(\sum_{i=1}^{31} x_i\right)\left(\sum_{i=1}^{31} y_i\right)\bigg/ 31 = 17{,}500 - (534)(992)/31 = 412$$

$$L_{xx} = \sum_{i=1}^{31} x_i^2 - \left(\sum_{i=1}^{31} x_i\right)^2\bigg/ 31 = 9876 - (534)^2/31 = 677.42$$

Finally, compute the slope of the regression line as follows:

$$b = L_{xy}/L_{xx} = 412/677.42 = 0.608$$

The intercept of the regression line can also be computed. Note from **(11.3)** that

$$a = \left(\sum_{i=1}^{31} y_i - 0.608 \sum_{i=1}^{31} x_i\right)\bigg/ 31 = [992 - 0.608(534)]/31 = 21.53$$

Thus the regression line is given by $y = 21.53 + 0.608x$. This regression line is depicted in Figure 11.1. ∎

How can the regression line be used? One of its uses is to *predict* values of y for given values of x.

DEFINITION 11.6 ▪▪▪

The **predicted**, or **expected**, **value of y** for a given value of x, as obtained from the regression line, is denoted by $\hat{y} = a + bx$. Thus the point $(x, a + bx)$ is always on the regression line. ∎

EXAMPLE 11.9 **Obstetrics** What is the expected birthweight if a pregnant woman has an estriol level of 15 mg/24 hr?

SOLUTION If the estriol level were 15 mg/24 hr, then the best prediction of birthweight would be

$$\hat{y} = 21.53 + 0.608(15) = 30.65 \times 100 \text{ g} = 3065 \text{ g}$$ ∎

One possible use of estriol levels is to identify women who are carrying a low-birthweight fetus. If such women can be identified, then drugs might be used to prolong the pregnancy until the fetus grows larger, since low-birthweight infants are at greater risk than normal infants for infant mortality in the first year of life and poor growth and development in childhood.

EXAMPLE 11.10 **Obstetrics** Low birthweight is defined here as $\leqslant 2500$ g. For what estriol level would the expected birthweight be 2500 g?

SOLUTION Note that the expected birthweight

$$\hat{y} = 21.53 + 0.608x$$

If $\hat{y} = 2500/100 = 25$, then x can be solved from the equation

$$25 = 21.53 + 0.608x \quad \text{or} \quad x = (25 - 21.53)/0.608 = 3.47/0.608 = 5.71$$

Thus if a woman has an estriol level of 5.71 mg/24 hr, then the expected birthweight would be 2500 g. Furthermore, the expected birthweight for all women with estriol levels of $\leqslant 5.71$ mg/24 hr would be $\leqslant 2500$ g. This level could serve as a critical value for identifying high-risk women and attempting to prolong their pregnancies. ∎

How can the slope of the regression line be interpreted? The slope of the regression line tells the amount that y increases per unit increase in x.

EXAMPLE 11.11   **Obstetrics** Interpret the slope of the regression line for the birthweight-estriol data in Example 11.1.

SOLUTION   The slope of 0.608 tells that the expected birthweight increases by about 0.6×100 g per 1 mg/24 hr, or 60 g per 1 mg/24 hr, increase in estriol.   ∎

SECTION 11.4   **Testing the Goodness of Fit of Regression Lines (*F* Test)**

In Section 11.3 the fitting of regression lines using the method of least squares was discussed. Since this method can be used with any set of points, criteria for goodness of fit to distinguish regression lines that fit the data well from those that do not must be established. Consider the typical situation depicted in Figure 11.5.

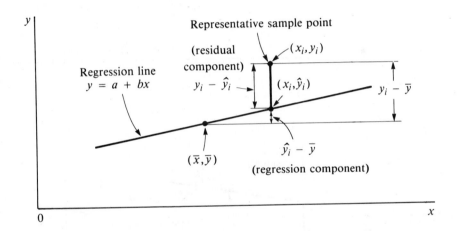

FIGURE 11.5
Goodness of fit of
a regression line

A hypothetical regression line and a representative sample point have been drawn. First, notice that the point $(\bar{x}, \bar{y})$ falls on the regression line. This feature is common to all regression lines, since a regression line can be represented as

$$y = a + bx = \bar{y} - b\bar{x} + bx = \bar{y} + b(x - \bar{x})$$

or, equivalently,

11.4
$$y - \bar{y} = b(x - \bar{x})$$

If $\bar{x}$ is substituted for x and $\bar{y}$ for y in **(11.4)**, then 0 is obtained on both sides of the equation, which shows that *the point $(\bar{x}, \bar{y})$ must always fall on the regression line.* If a typical sample point (x_i, y_i) is selected and a line is drawn through this point parallel to the y-axis, then the representation in Figure 11.5 is obtained.

DEFINITION 11.7   ■■
For any sample point (x_i, y_i), the **residual**, or **residual component**, of that point about the regression line is defined by $y_i - \hat{y}_i$.   ∎

DEFINITION 11.8 ■■

For any sample point (x_i, y_i), the **regression component** of that point about the regression line is defined by $\hat{y}_i - \bar{y}_i$.

■

In Figure 11.5 the deviation $y_i - \bar{y}$ can be separated into residual $(y_i - \hat{y}_i)$ and regression $(\hat{y}_i - \bar{y})$ components. Note that if the point (x_i, y_i) fell exactly on the regression line, then $y_i = \hat{y}_i$ and the residual component $y_i - \hat{y}_i$ would be 0 and $y_i - \bar{y} = \hat{y}_i - \bar{y}$. Generally speaking, a good-fitting regression line will have regression components large in absolute value relative to the residual components, whereas the opposite is true for poor-fitting regression lines. Some typical situations are depicted in Figure 11.6.

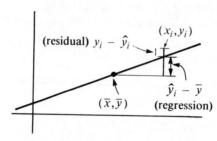

(a) Large regression, small residual components

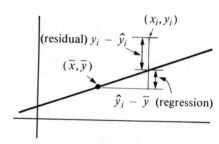

(b) Large regression, large residual components

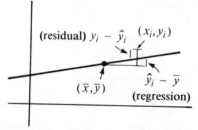

(c) Small regression, small residual components

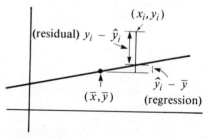

(d) Small regression, large residual components

FIGURE 11.6
Regression lines with varying goodness-of-fit relationships

The best-fitting regression line is depicted in (a), with large regression components and small residual components. The worst-fitting regression line is depicted in (d), which has small regression components and large residual components. Intermediate situations for goodness of fit are depicted in (b) and (c).

How can the plots in Figure 11.6 be quantified? One strategy is to square the deviations about the mean $y_i - \bar{y}$, sum them up over all points, and decompose this sum of squares into regression and residual components.

DEFINITION 11.9 ■■

The **total sum of squares**, or Total SS, is the sum of squares of the deviations of the individual sample points from the sample mean:

$$\sum_{i=1}^{n} (y_i - \bar{y})^2$$

■

DEFINITION 11.10 ▪▪
The **regression sum of squares**, or Reg SS, is the sum of squares of the regression components:

$$\sum_{i=1}^{n} (\hat{y}_i - \bar{y})^2$$ ▪

DEFINITION 11.11 ▪▪
The **residual sum of squares**, or Res SS, is the sum of squares of the residual components:

$$\sum_{i=1}^{n} (y_i - \hat{y}_i)^2$$ ▪

It can be shown that the following relationship is true:

11.5 **Decomposition of the Total Sum of Squares into Regression and Residual Components**

$$\sum_{i=1}^{n} (y_i - \bar{y})^2 = \sum_{i=1}^{n} (\hat{y}_i - \bar{y})^2 + \sum_{i=1}^{n} (y_i - \hat{y}_i)^2$$

Total SS = Reg SS + Res SS

or

The criterion for goodness of fit that will be used is the ratio of the regression sum of squares to the residual sum of squares. A large ratio indicates a good fit, whereas a small ratio indicates a poor fit. In hypothesis-testing terms we wish to test the hypothesis $H_0: \beta = 0$ versus $H_1: \beta \neq 0$, where β is the underlying slope of the regression line in **(11.2)**.

The following terms are introduced for ease of notation in describing the hypothesis test:

DEFINITION 11.12 ▪▪▪
The **regression mean square**, or Reg MS, is the Reg SS divided by the number of predictor variables (k) in the model. Thus Reg MS = Reg SS/k. For simple linear regression, which we have been discussing, $k = 1$ and thus Reg MS = Reg SS. For multiple regression in Section 11.8, k will be >1. We will refer to k as the degrees of freedom for the regression sum of squares, or Reg df. ▪

DEFINITION 11.13 ▪▪▪
The **residual mean square**, or Res MS, is the ratio of the Res SS divided by $(n - k - 1)$, or Res MS = Res SS/$(n - k - 1)$. For simple linear regression, $k = 1$ and Res MS = Res SS/$(n - 2)$. We will refer to $n - k - 1$ as the degrees of freedom for the residual sum of squares, or Res df. Res MS is also sometimes denoted by $s_{y \cdot x}^2$ in the literature. ▪

Under H_0, F = Reg MS/Res MS follows an F distribution with 1 and $n - 2$ df, respectively. H_0 should be rejected for large values of F. Thus, for a level α test, H_0 will be rejected if $F > F_{1, n-2, 1-\alpha}$ and accepted otherwise.

The expressions for the regression and residual sums of squares in **(11.5)** simplify for computational purposes as follows:

11.6 **Short Computational Form for Regression and Residual SS**

Regression SS = $bL_{xy} = b^2 L_{xx} = L_{xy}^2/L_{xx}$

Residual SS = Total SS − Regression SS = $L_{yy} - L_{xy}^2/L_{xx}$

Thus the test procedure can be summarized as follows:

EQUATION 11.7

F Test for Simple Linear Regression

To test $H_0: \beta = 0$ versus $H_1: \beta \neq 0$, use the following procedure:

(1) Compute the test statistic

$$F = \text{Reg MS/Res MS} = (L_{xy}^2/L_{xx})/[(L_{yy} - L_{xy}^2/L_{xx})/(n-2)]$$

that follows an $F_{1,n-2}$ distribution under H_0.

(2) For a two-sided test with significance level α, if

$$F > F_{1,n-2,1-\alpha} \qquad \text{then reject } H_0; \text{ if}$$

$$F \leqslant F_{1,n-2,1-\alpha} \qquad \text{then accept } H_0.$$

(3) The exact p-value is given by $Pr(F_{1,n-2} > F)$.

The acceptance and rejection regions for the regression F test are illustrated in Figure 11.7. The computation of the exact p-value for the regression F test is depicted in Figure 11.8. These results are typically summarized in an analysis of variance (ANOVA) table, as shown in Table 11.2.

FIGURE 11.7
Acceptance and rejection regions for the simple linear regression F test

$$F = \text{Reg MS/Res MS}$$
$$= (L_{xy}^2/L_{xx})/[(L_{yy} - L_{xy}^2/L_{xx})/(n-2)]$$

$F_{1,n-2}$ distribution

$F > F_{1,n-2,1-\alpha}$
Rejection region

$F \leq F_{1,n-2,1-\alpha}$
Acceptance region

$F_{1,n-2,1-\alpha}$

FIGURE 11.8
Computation of the p-value for the simple linear regression F test

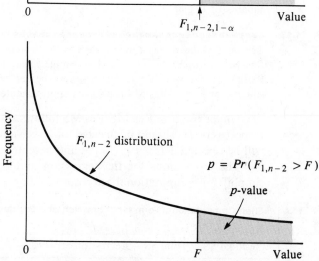

$F_{1,n-2}$ distribution

$$p = Pr(F_{1,n-2} > F)$$

p-value

F

Value

TABLE 11.2
ANOVA table for
displaying regression
results

	SS	df	MS	F Statistic	p-value
Regression	(a)*	1	$(a)/1$	$F = [(a)/1]/[(b)/(n-2)]$	$Pr(F_{1,n-2} > F)$
Residual	(b)†	$n-2$	$(b)/(n-2)$		
Total	$(a) + (b)$				

* (a) = Regression SS
† (b) = Residual SS

EXAMPLE 11.12   **Obstetrics** Test for the significance of the regression line derived for the birthweight-estriol data in Example 11.8 (p. 403).

SOLUTION   From Example 11.8,

$$L_{xy} = 412, L_{xx} = 677.42$$

Furthermore,

$$\sum_{i=1}^{31} y_i^2 = 32,418 \qquad L_{yy} = \sum_{i=1}^{31} y_i^2 - \left(\sum_{i=1}^{31} y_i\right)^2 \bigg/ 31 = 32,418 - (992)^2/31 = 674$$

Therefore,

$$\text{Reg SS} = L_{xy}^2/L_{xx} = \text{Reg MS} = (412)^2/677.42 = 250.57$$

$$\text{Total SS} = L_{yy} = 674$$

$$\text{Res SS} = \text{Total SS} - \text{Reg SS} = 674 - 250.57 = 423.43$$

$$\text{Res MS} = \text{Res SS}/(31-2) = \text{Res SS}/29 = 423.43/29 = 14.60$$

$$F = \text{Reg MS}/\text{Res MS} = 250.57/14.60 = 17.16 \sim F_{1,29} \text{ under } H_0$$

From Table 8 in Appendix 1,

$$F_{1,29,.999} < F_{1,20,.999} = 14.82 < 17.16 = F$$

Therefore,
$$p < .001$$

and H_0 is rejected and the alternative hypothesis, namely, that the slope of the regression line is significantly different from 0, is accepted, implying a *significant linear relationship* between birthweight and estriol level. These results are summarized in the ANOVA table (Table 11.3) using the SPSS^X/PC REGRESSION program.

TABLE 11.3
ANOVA results for the
birthweight-estriol data
in Example 11.12

```
Analysis of Variance
                   DF      Sum of Squares      Mean Square
Regression          1         250.57448         250.57448
Residual           29         423.42552          14.60088

F =     17.16160      Signif F = 0.0003
```

A summary measure of goodness of fit frequently referred to in the literature in R^2.

DEFINITION 11.14   ■■
R^2 is defined as Reg SS/Total SS.   ■

R^2 can be thought of as the proportion of the variance of y that can be explained by the variable x. If $R^2 = 1$, then all the variation in y can be explained by the variation in x, and all the data points fall on the regression line. In other words, once x is known, y can be predicted exactly, with no error or variability in the prediction. If $R^2 = 0$, then x gives no information about y, and the variance of y is the same with or without knowing x. If R^2 is between 0 and 1, then for a given value of x, the variance of y is lower than it would be if x were unknown but is still greater than 0. In particular, the best estimate of the variance of y given x [or σ^2 in the regression model in **(11.2)**] is given by Res MS (or $s_{y \cdot x}^2$). For large n, $s_{y \cdot x}^2 \approx s_y^2(1 - R^2)$. Thus R^2 represents the proportion of the variance of y that is explained by x.

EXAMPLE 11.13   **Obstetrics** Compute and interpret R^2 and $s_{y \cdot x}^2$ for the birthweight-estriol data in Example 11.12.

SOLUTION   From Table 11.3, the R^2 for the birthweight-estriol regression line is given by $250.57/674 = .372$. Thus about 37% of the variance of birthweight can be explained by estriol level. Furthermore, $s_{y \cdot x}^2 = 14.60$, as compared with

$$s_y^2 = \sum_{i=1}^{n} (y_i - \bar{y})^2/(n - 1) = 674/30 = 22.47$$

Thus, for the subgroup of women with a specific estriol level, such as 10 mg/24 hr, the variance of birthweight is 14.60, whereas for *all* women with any estriol level, the variance of birthweight is 22.47. Note that

$$s_{y \cdot x}^2/s_y^2 = 14.60/22.47 = .650 \approx 1 - R^2 = 1 - .372 = .628 \qquad \blacksquare$$

EXAMPLE 11.14   **Pulmonary Function** Forced expiratory volume (FEV) is a standard measure of pulmonary function. To identify people with abnormal pulmonary function, standards of FEV for normal people must be established. One problem here is that FEV is related to both age and height. Let us focus on boys who are ages 10–15 and postulate a regression model of the form FEV = $\alpha + \beta$(height) + e. Data were collected on FEV and height for 655 boys in this age group residing in Tecumseh, Michigan [2]. The mean FEV in liters is presented for each of the twelve 4-cm height groups in Table 11.4. Find the best-fitting regression line and test it for statistical significance. What proportion of the variance of FEV can be explained by height?

TABLE 11.4
Mean FEV and height for boys ages 10–15 in Tecumseh, Michigan

Height (cm)	Mean FEV (L)	Height (cm)	Mean FEV (L)
134*	1.7	158	2.7
138	1.9	162	3.0
142	2.0	166	3.1
146	2.1	170	3.4
150	2.2	174	3.8
154	2.5	178	3.9

* The middle value of each 4-cm height group is given here.
(Reprinted with permission of the *American Review of Respiratory Disease*, 108, 258–272, 1973.)

SOLUTION   A linear regression line is fitted to the points in Table 11.4:

$$\sum_{i=1}^{12} x_i = 1872 \qquad \sum_{i=1}^{12} x_i^2 = 294{,}320 \qquad \sum_{i=1}^{12} y_i = 32.3$$

$$\sum_{i=1}^{12} y_i^2 = 93.11 \qquad \sum_{i=1}^{12} x_i y_i = 5156.20$$

Therefore,

$$L_{xy} = 5156.20 - \frac{(1872)(32.3)}{12} = 117.4$$

$$L_{xx} = 294{,}320 - \frac{(1872)^2}{12} = 2288$$

$$b = L_{xy}/L_{xx} = 0.051$$

$$a = \left(\sum_{i=1}^{12} y_i - b\sum_{i=1}^{12} x_i\right)\Big/ 12 = [32.3 - 0.051(1872)]/12 = -5.264$$

Thus, the fitted regression line is

$$\text{FEV} = -5.264 + 0.051 \times \text{height}.$$

Statistical significance is tested for by computing the F statistic in **(11.7)** as follows:

$$\text{Reg SS} = L_{xy}^2/L_{xx} = (117.4)^2/2288 = 6.024 = \text{Reg MS}$$

$$\text{Total SS} = L_{yy} = 93.11 - (32.3)^2/12 = 6.169$$

$$\text{Res SS} = 6.169 - 6.024 = 0.145$$

$$\text{Res MS} = \text{Res SS}/(n-2) = 0.145/10 = 0.0145$$

$$F = \text{Reg MS}/\text{Res MS} = 415.4 \sim F_{1,10} \text{ under } H_0$$

Clearly, the fitted line is statistically significant, because from Table 8 in Appendix 1, $F_{1,10,.999} = 21.04$, so $p < .001$. These results can be displayed in an ANOVA table (Table 11.5).

TABLE 11.5
ANOVA table for the
FEV-height regression
results in Example 11.14

```
Analysis of Variance
                    DF      Sum of Squares     Mean Square
Regression           1         6.02393           6.02393
Residual            10         0.14523           0.01452

F =    414.77690      Signif F = 0.0000
```

Finally, the proportion of the variance of FEV that is explained by height is given by $R^2 = 6.024/6.169 = .976$. Thus, differences in height explain almost all the variability in FEV among boys in this age group. A scatter plot of the raw data and the fitted regression line is given in Figure 11.9. ∎

The scatter plot is useful for looking at the goodness of fit of the regression line to the observed data. In particular, we can look for patterns in the residuals, that is, patterns in deviations of the observed data points from the fitted regression line. These patterns can give clues to an underlying curvilinear relationship between y and x, which would be missed by fitting a simple linear model. Also, the residuals can alert us to the presence of outlying values, which might exert undue influence

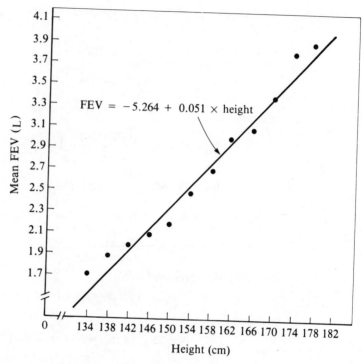

FIGURE 11.9
Scatter plot and
estimated regression
line of mean FEV by
height in boys ages
10–15 in Tecumseh,
Michigan

on the fitting of the regression line. In Figure 11.9 there is no particular pattern in the residuals nor are any outlying values evident. Thus we can conclude that the linear model seems to fit adequately in this age range.

However, the relationship of mean systolic blood pressure to age among girls 8–18 years old is curvilinear, with blood pressure increasing up to age 13–14 and then flattening out and remaining virtually constant throughout adolescence. If a linear regression line is fitted to the data, then the residual plot shown in Figure 11.10 is obtained. Notice how the residuals are virtually always positive for young

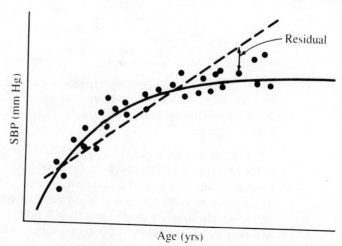

FIGURE 11.10
Relationship of systolic
blood pressure to age
among girls 8–18
years old

ages and virtually always negative for older ages. Fitting a curvilinear regression model to these data would provide a more random distribution of residuals with no particular pattern. How to fit curvilinear regression models in the context of multiple regression is discussed in Section 11.8.

To illustrate the influence of outliers on the fit of a regression line, suppose a mean FEV of 2.5 L instead of 3.8 L is substituted for boys of height 174 cm in Table 11.4. This might arise due to a data-entry or coding error. The fitted regression line is then $y = -3.826 + 0.041x$ and is displayed in Figure 11.11 together with the regression line fitted in the absence of outliers. Notice the large residual corresponding to the outlier and the generally poorer fit of the regression line in the presence of the outlier (solid line) compared with the fit when the correct value for mean FEV (3.8 L) is substituted for the outlying value (2.5 L).

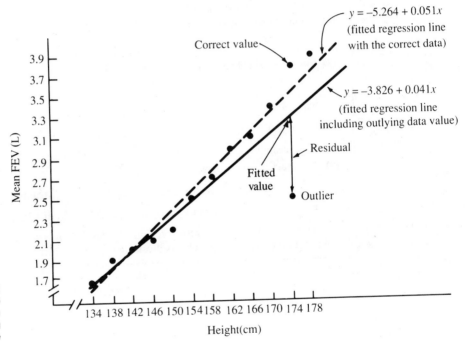

FIGURE 11.11
Relationship of FEV (L) to
height (cm) fitted with
and without an outlying
data value

Testing the Goodness of Fit of Regression Lines (*t* Test)

In this section an alternative method for testing the hypothesis $H_0: \beta = 0$ versus $H_1: \beta \neq 0$ is presented. This method is based on the *t* test and is equivalent to the *F* test presented in Section 11.4. The procedure is widely used and also provides interval estimates for β.

The hypothesis test here is based on the sample regression coefficient *b*, or, more specifically, on $b/se(b)$, and H_0 will be rejected if $|b|/se(b) > c$ for some constant *c* and will be accepted otherwise.

The sample regression coefficient b is an **unbiased estimator** of the population regression coefficient β and, in particular, under H_0, $E(b) = 0$. Furthermore, the variance of b is given by

$$\sigma^2 \Big/ \sum_{i=1}^{n} (x_i - \bar{x})^2 = \sigma^2/L_{xx}$$

In general, σ^2 is unknown. However, the best estimate of σ^2 is given by $s_{y \cdot x}^2$. Hence

$$se(b) \approx s_{y \cdot x}/(L_{xx})^{1/2}$$

Finally, under H_0, $t = b/se(b)$ follows a t distribution with $n - 2$ df. Therefore, the following test procedure for a two-sided test with significance level α is used:

11.8   **t Test for Simple Linear Regression**

To test the hypothesis $H_0 : \beta = 0$ versus $H_1 : \beta \neq 0$, use the following procedure:

(1) Compute the test statistic

$$t = b/(s_{y \cdot x}^2/L_{xx})^{1/2}$$

(2) For a two-sided test with significance level α, if

$$t > t_{n-2,1-\alpha/2}$$

or

$$t < t_{n-2,\alpha/2} = -t_{n-2,1-\alpha/2}$$

then reject H_0; if

$$-t_{n-2,1-\alpha/2} \leqslant t \leqslant t_{n-2,1-\alpha/2}$$

then accept H_0.

(3) The exact p-value is given by

$$p = 2 \times (\text{area to the left of } t \text{ under a } t_{n-2} \text{ distribution}) \quad \text{if } t < 0$$

$$p = 2 \times (\text{area to the right of } t \text{ under a } t_{n-2} \text{ distribution}) \quad \text{if } t \geqslant 0$$

The acceptance and rejection regions for this test are depicted in Figure 11.12. The computation of the exact p-value is illustrated in Figure 11.13.

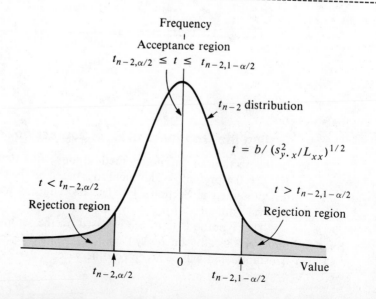

FIGURE 11.12
Acceptance and rejection regions for the t test for simple linear regression

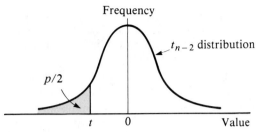

Frequency

t_{n-2} distribution

$p/2$

t 0 Value

(a) If $t < 0$, then $p = 2 \times$ (area to the left of t under a t_{n-2} distribution).

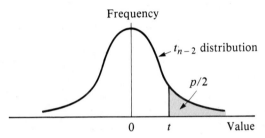

Frequency

t_{n-2} distribution

$p/2$

0 t Value

FIGURE 11.13
Computation of
the exact p-value for
the t test for simple linear
regression

(b) If $t \geqslant 0$, then $p = 2 \times$ (area to the right of t under a t_{n-2} distribution).

The t test in this section and the F test in Section 11.4 are equivalent in that they always provide the same p-values. Which test is used is a matter of personal preference; both appear in the literature.

EXAMPLE 11.15   **Obstetrics** Assess the statistical significance for the birthweight-estriol data using the t test in **(11.8)**.

SOLUTION   From Example 11.8 (p. 403), $b = L_{xy}/L_{xx} = 0.608$. Furthermore, from Table 11.3 (p. 409) and Example 11.12 (p. 409),

$$se(b) = (s_{y \cdot x}^2/L_{xx})^{1/2} = (14.60/677.42)^{1/2} = 0.147$$

Thus, $t = b/se(b) = 0.608/0.147 = 4.14 \sim t_{29}$ under H_0

Since $t_{29,.9995} = 3.659 < 4.14 = t$

we have $p < 2 \times (1 - .9995) = .001$

This information is summarized in Table 11.6. Note that the p-values based on the F test in Table 11.3 and the t test in Table 11.6 are the same ($p = .0003$).   ∎

TABLE 11.6
The t test approach for
the birthweight-estriol
example

```
--------------------------------------------------------------
                    SPSSX/PC  Release 1.0

      * * * *   M U L T I P L E   R E G R E S S I O N   * * * *

Equation Number 1   Dependent Variable..   BWT   BIRTHWEIGHT

------------------ Variables in the Equation ------------------

Variable              B         SE B       Beta        T  Sig T

ESTRIOL          0.60819    0.14681    0.60973    4.143 0.0003
(Constant)      21.52343    2.62042               8.214 0.0000
--------------------------------------------------------------
```

Interval Estimation for the Parameters of a Regression Line

Standard errors and interval estimates for the parameters of a regression line are often computed to obtain some idea of the precision of the estimates. Furthermore, if we want to compare our regression line with previously published regression coefficients β_0 and α_0, where these estimates are based on much larger samples than ours, then, based on our data, we can check whether β_0 and α_0 fall within the 95% confidence intervals for β and α, respectively, to decide whether the two sets of results are comparable.

The standard errors of the estimated regression parameters are given as follows:

11.9 | **Standard Errors of Estimated Parameters in Simple Linear Regression**

$$se(b) = \sqrt{\frac{s_{y \cdot x}^2}{L_{xx}}}$$

$$se(a) = \sqrt{s_{y \cdot x}^2 \left(\frac{1}{n} + \frac{\bar{x}^2}{L_{xx}} \right)}$$

Furthermore, the two-sided $100\% \times (1 - \alpha)$ confidence intervals for β and α are given by:

11.10 | **Two-Sided $100\% \times (1 - \alpha)$ Confidence Intervals for the Parameters of a Regression Line**

If b and a are, respectively, the estimated slope and intercept of a regression line as given in **(11.3)** and $se(b)$, $se(a)$ are the estimated standard errors as given in **(11.9)**, then the two-sided $100\% \times (1 - \alpha)$ confidence intervals for β and α are given by

$$b \pm t_{n-2, 1-\alpha/2}\, se(b) \quad \text{and} \quad a \pm t_{n-2, 1-\alpha/2}\, se(a) \quad \text{respectively.}$$

EXAMPLE 11.16   **Obstetrics** Provide standard errors and 95% confidence intervals for the regression parameters of the birthweight-estriol data in Table 11.1.

SOLUTION   The standard error of b is given by

$$\sqrt{14.60/677.42} = 0.147$$

Thus, a 95% confidence interval for β is obtained from

$$0.608 \pm t_{29, .975}(0.147) = 0.608 \pm 2.045(0.147) = 0.608 \pm 0.301 = (0.307, 0.909)$$

Compute $\bar{x}$ to obtain the standard error of a. From Example 11.8 (p. 403),

$$\bar{x} = \frac{\sum_{i=1}^{31} x_i}{31} = \frac{534}{31} = 17.23$$

Thus, the standard error of a is given by

$$\sqrt{14.60 \left[\frac{1}{31} + \frac{(17.23)^2}{677.42} \right]} = 2.62$$

It follows that a 95% confidence interval for α is provided by

$$21.53 \pm t_{29,.975}(2.62) = 21.53 \pm 2.045(2.62) = 21.53 \pm 5.36 = (16.17, 26.89)$$

These intervals are rather wide, which is not surprising due to the small sample size.

Suppose another data set based on 500 pregnancies, where the birthweight-estriol regression line is estimated as $y = 25.04 + 0.52x$ is found in the literature. Since 0.52 is within the 95% confidence interval for the slope and 25.04 is within the 95% confidence interval for the intercept, our results are compatible with the earlier study. ∎

SECTION 11.7 Interval Estimation for Predictions Made from Regression Lines

One important use for regression lines is in making predictions. Frequently, the accuracy of these predictions must be assessed.

EXAMPLE 11.17 **Pulmonary Function** Suppose we wish to use the FEV-height regression line computed in Example 11.14 to develop normal ranges for 10–15-year-old boys of particular heights. In particular, consider John H., who is 12 years old and 160 cm tall and whose FEV is 2.5 L. Can his FEV be considered abnormal for his age and height? ∎

In general, if all boys of height x are considered, then the average FEV for such boys can be best estimated from the regression equation by $\hat{y} = a + bx$. How accurate is this estimate? The answer to this question depends on whether we are predicting for *one specific boy* or for the *mean value of all boys of a given height*. The first estimate would be useful to a pediatrician interested in assessing the lung function of a particular patient, whereas the second estimate would be useful to a researcher interested in relationships between pulmonary function and height over large populations of boys. The standard error (se_1) of the first type of estimate and the resulting confidence interval are given as follows:

11.11 **Standard Error and Confidence Interval of Predictions Made from Regression Lines for a New Sample Point**

Suppose we wish to make predictions from a regression line for a new sample point with independent variable x that was not used in constructing the regression line. The observed value of y will be normally distributed with mean $= \hat{y} = a + bx$ and standard error given by

$$se_1(\hat{y}) = \sqrt{s_{y \cdot x}^2 \left[1 + \frac{1}{n} + \frac{(x - \bar{x})^2}{L_{xx}} \right]}$$

Furthermore, a two-sided $100\% \times (1 - \alpha)$ confidence interval for y is given by

$$\hat{y} \pm t_{n-2, 1-\alpha/2} se_1(\hat{y})$$

EXAMPLE 11.18 **Pulmonary Function** Construct a 95% confidence interval for the FEV of John H. in Example 11.17.

SOLUTION John's observed FEV is 2.5 L. The regression equation relating FEV and height was computed in Example 11.14 and is given by $y = -5.264 + 0.051 \times$ height (p. 410). Thus, his expected FEV is

$$\hat{y} = -5.264 + 160 \times 0.051 = 2.90 \, \text{L}$$

We need to obtain $\bar{x}$ before computing the $se(\hat{y})$. From Example 11.14,

$$\bar{x} = \frac{\sum\limits_{i=1}^{12} x_i}{12} = \frac{1872}{12} = 156.0$$

Thus, $se_1(\hat{y})$ is given by

$$se_1(\hat{y}) = \sqrt{(0.0145)\left[1 + \frac{1}{12} + \frac{(160 - 156)^2}{2288}\right]} = \sqrt{(0.0145)(1.090)} = 0.126$$

Finally, a 95% confidence interval for the FEV of John H. is given by

$$2.90 \pm t_{10,.975}(0.126) = 2.90 \pm 2.228(0.126) = 2.90 \pm 0.28 = (2.62, 3.18)$$

How can this confidence interval be used? Since the observed FEV (2.5 L) does not fall within the confidence interval, we can say that John's lung function is abnormally low for a boy of his age and height, and that, if possible, further exploration is needed to find a reason for this abnormality. ∎

The magnitude of the standard error in **(11.11)** depends on how far the observed value of x for the new sample point is from the mean value of x for the data points used in computing the regression line ($\bar{x}$). The standard error is smaller when x is close to $\bar{x}$ than when x is far from $\bar{x}$. In general, making predictions from a regression line for values of x that are very far from $\bar{x}$ is dangerous, since the predictions are likely to be very inaccurate.

EXAMPLE 11.19   **Pulmonary Function** Suppose that Bill Y. has a height of 190 cm with an FEV of 3.5 L. Compare the standard error of his predicted value with that for John H. given in Example 11.18.

SOLUTION   From **(11.11)**,

$$se_1(\hat{y}) = \sqrt{(0.0145)\left[1 + \frac{1}{12} + \frac{(190 - 156)^2}{2288}\right]}$$

$$= \sqrt{(0.0145)(1.589)} = 0.152 > 0.126 = se_1 \text{ (John H.)}$$

This result is expected, since 190 cm is further than 160 cm from $\bar{x} = 156$ cm. ∎

Suppose we want to assess the mean value of FEV for a large number of boys of a particular height rather than for one particular boy. This parameter might be of interest to a researcher interested in growth curves of pulmonary function in children. How can the mean FEV and the standard error of the estimate be obtained? The procedure is given as follows:

11.12   **Standard Error and Confidence Interval Predictions Made from Regression Lines for the Expected Value of _y_ for a Given _x_**

The best estimate of the expected value of y for a given x is $\hat{y} = a + bx$. Its standard error, denoted by $se_2(\hat{y})$, is given by

$$se_2(\hat{y}) = \sqrt{s_{y \cdot x}^2 \left[\frac{1}{n} + \frac{(x - \bar{x})^2}{L_{xx}}\right]}$$

Furthermore, a two-sided $100\% \times (1 - \alpha)$ confidence interval for the expected value of y is

$$\hat{y} \pm t_{n-2, 1-\alpha/2} se_2(\hat{y})$$

EXAMPLE 11.20   **Pulmonary Function** Compute the standard error and 95% confidence interval for the mean value of FEV over a large number of boys with height of 160 cm.

SOLUTION   Refer to the results of Example 11.18 for the necessary raw data to perform the computations. The best estimate of the mean value of FEV is the same as the estimate for one boy (John H.), which was 2.90 L as computed in Example 11.18. However, the standard error is computed differently. From **(11.12)**,

$$se_2(\hat{y}) = \sqrt{(0.0145)\left[\frac{1}{12} + \frac{(160 - 156)^2}{2288}\right]} = \sqrt{(0.0145)(0.090)} = 0.036$$

Therefore, a 95% confidence interval for the mean value of FEV over a large number of boys with height 160 cm is given by

$$2.90 \pm t_{10,.975}(0.036) = 2.90 \pm 2.228(0.036) = 2.90 \pm 0.08 = (2.82, 2.98)$$

Notice that this standard error $[se_2(\hat{y}) = 0.036]$ is much smaller than the corresponding standard error $[se_1(y) = 0.126]$ computed in Example 11.18 for the FEV of one particular boy. Similarly, the 95% confidence interval is much narrower here (2.82, 2.98) than the corresponding confidence interval in Example 11.18 (2.62, 3.18). This disparity reflects the intuitive idea that there is much more precision in estimating the mean value of y for a large number of boys with the same height x than in estimating y for one particular boy with height x.   ■

Note again that the standard error for the estimated expected value of y for a given value of x is not the same for all values of x, but gets larger the further x is from the mean value of x ($\bar{x}$) used to estimate the regression line.

EXAMPLE 11.21   **Pulmonary Function** Compare the standard error of the expected FEV for boys of height 190 cm with that for boys of 160 cm.

SOLUTION   From **(11.12)**,

$$se_2(\hat{y}) = \sqrt{(0.0145)\left[\frac{1}{12} + \frac{(190 - 156)^2}{2288}\right]} = \sqrt{(0.0145)(0.589)}$$

$$= 0.092 > 0.036 = se_2(\hat{y}) \text{ for } x = 160 \text{ cm}$$

This result is expected, since 190 cm is further than 160 cm from $\bar{x} = 156$ cm.   ■

SECTION 11.8   ## Multiple Regression

In Sections 11.2 through 11.7 problems in linear regression analysis in which there is one independent variable (x), one dependent variable (y), and a linear relationship between x and y, were discussed. In practice there is often more than one independent variable and we would like to look at the relationship between each of the independent variables ($x_1, \ldots, x_k$) and the dependent variable (y) after taking into account the remaining independent variables. This type of problem is the subject matter of **multiple regression analysis**.

EXAMPLE 11.22   **Hypertension, Pediatrics** A topic of current interest in hypertension research is how the relationship between blood-pressure levels of newborns and blood-pressure levels of infants relates to the etiology of hypertension. One problem that arises is that the blood pressure of a newborn is affected by several extraneous factors that make this relationship difficult to study. In particular, newborn blood pressures are affected by (1) birthweight and (2) the day of life on which blood pressure is measured. We would like to be able to adjust the observed blood pressure for these two factors before looking at this relationship.   ∎

11.8.1 Estimation of the Regression Equation

Suppose a relationship between systolic blood pressure (y), birthweight (x_1), and age in days (x_2) that is of the form

11.13

$$y = \alpha + \beta_1 x_1 + \beta_2 x_2 + e$$

where e is an error term that is normally distributed with mean 0 and variance σ^2 is postulated. We would like to estimate the parameters of this model and test various hypotheses concerning it. The same principle of least squares that was introduced in Section 11.3 for simple linear regression will be used to fit the parameters of this multiple regression model. In particular, α, β_1, β_2 will be estimated by a, b_1, and b_2, where we choose a, b_1, and b_2 to minimize the sum of

$$[y - (a + b_1 x_1 + b_2 x_2)]^2$$

over all the data points. For multiple regression, computing the least-squares estimators of the regression parameters involves matrix inversion and thus is sufficiently complex to almost always require a computer.

EXAMPLE 11.23   **Hypertension, Pediatrics** Suppose systolic blood pressure, birthweight (oz), and age (days) are measured for 16 infants and the data are as shown in Table 11.7. Estimate the parameters of the multiple-regression equation in **(11.13)**.

TABLE 11.7
Sample data for infant blood pressure, age, and birthweight for 16 infants

i	Birthweight in oz (x_1)	Age in days (x_2)	Systolic blood pressure (mm Hg) (y)
1	135	3	89
2	120	4	90
3	100	3	83
4	105	2	77
5	130	4	92
6	125	5	98
7	125	2	82
8	105	3	85
9	120	5	96
10	90	4	95
11	120	2	80
12	95	3	79
13	120	3	86
14	150	4	97
15	160	3	92
16	125	3	88

SOLUTION   Use the SPSS^X/PC REGRESSION program to obtain the least-squares estimators. The results are given in Table 11.8.

According to the **B** column, the regression equation is given by

$$y = 53.45 + 0.126x_1 + 5.89x_2$$ ■

TABLE 11.8
Least-squares
estimators of the
regression parameters
for the newborn
blood-pressure data in
Table 11.7 using the
SPSS^X/PC REGRESSION
program

```
-------------------------------------------------------------------
                      SPSSX/PC   Release 1.0

          * * * *   M U L T I P L E     R E G R E S S I O N   * * * *

     Equation Number 1    Dependent Variable..   SYSBP   SYSTOLIC BLOOD PRESSURE


     ------------------- Variables in the Equation -------------------

     Variable          B        SE B      Beta       T  Sig T

     BWT            0.12558   0.03434   0.35208    3.657 0.0029
     AGE            5.88772   0.68021   0.83323    8.656 0.0000
     (Constant)    53.45019   4.53189             11.794 0.0000
-------------------------------------------------------------------
```

The regression equation tells us that for an average newborn the expected blood pressure increases by 0.126 mm Hg per ounce of birthweight and 5.89 mm Hg per day of age.

EXAMPLE 11.24   **Hypertension, Pediatrics** Calculate the expected systolic blood pressure of a baby with birthweight 8 lb (128 oz) measured at 3 days of life.

SOLUTION   The expected systolic blood pressure is given by

$$53.45 + 0.126(128) + 5.89(3) = 87.2 \text{ mm Hg}$$ ■

We are often interested in ranking the independent variables according to their predictive relationship with the dependent variable y. It is difficult to rank the variables based on the ordinary regression coefficients, since the independent variables are often in different units. Specifically, from the multiple-regression model in **(11.13)**, we see that b estimates the increase in y per unit increase in x. If x is increased by 1 standard deviation unit (i.e., s_x) to $x + s_x$, then y would be expected to increase by $b \times s_x$ raw units or $(b \times s_x)/s_y$ standard deviation units of y (s_y).

DEFINITION 11.15   ■■

The **standardized regression coefficient** (b_s) is given by $b \times (s_x/s_y)$. It represents the predicted increase in y (expressed in standard deviation units of y) that would be expected per standard deviation increase in x.   ■

Thus the standardized regression coefficient is a useful measure for comparing the predictive value of several independent variables, since it tells us the predicted increase in standard deviation units of y that would be expected per standard deviation increase in x. By expressing change in standard deviation units of x, we can control for differences in the units of measurement for different independent variables.

EXAMPLE 11.25 Compute the standardized regression coefficients for birthweight and age in days using the data in Table 11.7.

SOLUTION From Table 11.7, $s_y = 6.69$, $s_{x_1} = 18.75$, $s_{x_2} = 0.946$. Therefore, referring to the regression coefficients (B) in Table 11.8,

$$b_s \text{ (birthweight)} = \frac{0.1256 \times 18.75}{6.69} = 0.352$$

$$b_s \text{ (age in days)} = \frac{5.888 \times 0.946}{6.69} = 0.833$$

These quantities are given under the BETA column in Table 11.8. Thus, the expected increase in systolic blood pressure is 0.352 standard deviation units of blood pressure per standard deviation increase in birthweight and 0.833 standard deviation units of blood pressure per standard deviation increase in age in days. Thus, age in days appears to be the more important variable after controlling for both variances simultaneously in the multiple regression model. ∎

11.8.2 Hypothesis Testing (Two-Independent-Variables Case)

EXAMPLE 11.26 **Hypertension, Pediatrics** We would like to test various hypotheses concerning the data in Table 11.7. First, we would like to test the overall hypothesis that birthweight and age in days when taken together are significant predictors of blood pressure. How can this be done? ∎

Specifically, we will test the hypothesis $H_0: \beta_1 = \beta_2 = 0$ versus H_1: either $\beta_1 \neq 0$ or $\beta_2 \neq 0$. The test of significance is similar to the F test in Section 11.4. The test procedure for a level α test is given as follows:

11.14 | **F Test for Testing the Hypothesis $H_0: \beta_1 = \beta_2 = 0$ versus H_1: Either $\beta_1 \neq 0$ or $\beta_2 \neq 0$ in Multiple Linear Regression**

(1) Compute

$$\text{Reg SS} \quad \text{and} \quad \text{Res SS}$$

(2) Compute

$$\text{Reg MS} = \frac{\text{Reg SS}}{2} \quad \text{and} \quad \text{Res MS} = \frac{\text{Res SS}}{n-3}$$

(3) Compute the test statistic

$$F = \text{Reg MS/Res MS}$$

which follows an $F_{2,n-3}$ distribution under H_0.

(4) For a level α test, if

$$F > F_{2,n-3,1-\alpha}$$

then reject H_0; if

$$F \leq F_{2,n-3,1-\alpha}$$

then accept H_0.

(5) The exact p-value is given by the area to the right of F under an $F_{2,n-3}$ distribution $= Pr(F_{2,n-3} > F)$.

The acceptance and rejection regions for this test procedure are depicted in Figure 11.14. The computation of the exact p-value is illustrated in Figure 11.15.

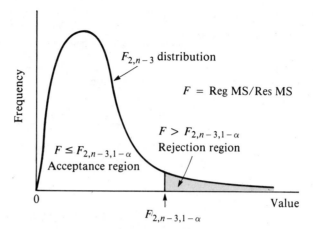

FIGURE 11.14
Acceptance and rejection regions for testing the hypothesis $H_0: \beta_1 = \beta_2 = 0$ versus H_1: either $\beta_1 \neq 0$ or $\beta_2 \neq 0$ in multiple linear regression

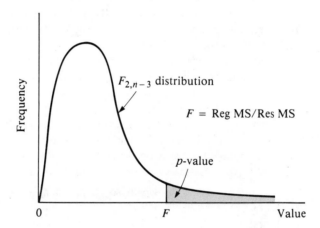

FIGURE 11.15
Computation of the exact p-value for testing the hypothesis $H_0: \beta_1 = \beta_2 = 0$ versus H_1: either $\beta_1 \neq 0$ or $\beta_2 \neq 0$ in multiple linear regression

EXAMPLE 11.27 **Hypertension, Pediatrics** Test the hypothesis $H_0: \beta_1 = \beta_2 = 0$ versus H_1: either $\beta_1 \neq 0$ or $\beta_2 \neq 0$ using the data in Table 11.7.

SOLUTION Refer to Table 11.9 and note that

$$\text{Reg SS} = 591.04$$

$$\text{Reg MS} = 591.04/2 = 295.52$$

$$\text{Res SS} = 79.90$$

$$\text{Res MS} = 79.90/13 = 6.146$$

$$F = \text{Reg MS/Res MS} = 48.08 \sim F_{2.13} \text{ under } H_0$$

Since
$$F_{2,13,.999} < F_{2,12,.999} = 12.97 < 48.08 = F$$

$p < .001$. Thus, we can conclude that the two variables when considered together are significant predictors of blood pressure. ∎

TABLE 11.9
ANOVA table for newborn blood-pressure data in Example 11.27

```
--------------------------------------------------------------------
                      SPSSX/PC  Release 1.0

        * * * *   M U L T I P L E   R E G R E S S I O N   * * * *

Equation Number 1    Dependent Variable..   SYSBP   SYSTOLIC BLOOD PRESSURE


Analysis of Variance
                    DF      Sum of Squares      Mean Square
Regression           2          591.03564        295.51782
Residual            13           79.90186          6.14630

F =     48.08063      Signif F = 0.0000
--------------------------------------------------------------------
```

The significant p-value for this test could be attributed to either variable. We would like to perform significance tests to identify the independent contributions of each variable. How can this be done?

In particular, to assess the independent contribution of birthweight, we will assume that age in days is making a contribution under either hypothesis, and we will test the hypothesis $H_0: \beta_1 = 0, \beta_2 \neq 0$ versus $H_1: \beta_1 \neq 0, \beta_2 \neq 0$. Similarly, to assess the independent contribution of age in days, we will assume that birthweight is making a contribution under either hypothesis and will test the hypothesis $H_0: \beta_2 = 0, \beta_1 \neq 0$ versus $H_1: \beta_2 \neq 0, \beta_1 \neq 0$. We will focus on assessing the independent contribution of birthweight. Our approach will be to compute the standard error of the regression coefficient for birthweight and base our test on $t = b/se(b)$, which will follow a t distribution with $n - 3$ df under H_0. Specifically, the following test procedure for a level α test is used:

11.15 | **t Test for Testing the Hypothesis $H_0: \beta_1 = 0, \beta_2 \neq 0$ versus H_1: $\beta_1 \neq 0, \beta_2 \neq 0$ in Multiple Linear Regression**

(1) Compute

$$t = b_1/se(b_1)$$

which should follow a t distribution with $n - 3$ df under H_0.

(2) If

$$t < t_{n-3, \alpha/2} \qquad \text{or} \qquad t > t_{n-3, 1-\alpha/2}$$

then reject H_0; if

$$t_{n-3, \alpha/2} \leqslant t \leqslant t_{n-3, 1-\alpha/2}$$

then accept H_0.

(3) The exact p-value is given by

$$2 \times Pr(t_{n-3} > t) \qquad \text{if } t \geqslant 0$$
$$2 \times Pr(t_{n-3} \leqslant t) \qquad \text{if } t < 0$$

The acceptance and rejection regions for this test are depicted in Figure 11.16. The computation of the exact p-value is illustrated in Figure 11.17.

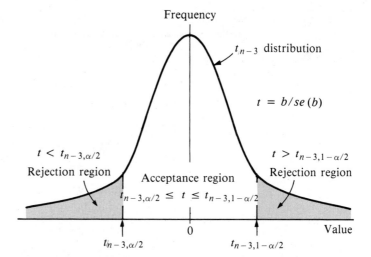

FIGURE 11.16
Acceptance and rejection regions for the *t* test for multiple linear regression (two-variable case)

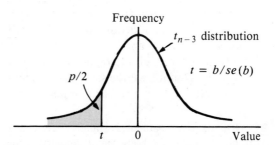

(a) If $t < 0$, then $p = 2 \times$ area to the left of t under a t_{n-3} distribution.

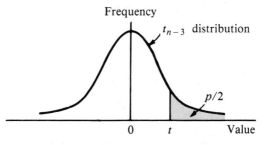

FIGURE 11.17
Computation of the exact *p*-value for the *t* test for multiple linear regression (two-variable case)

(b) If $t \geqslant 0$, then $p = 2 \times$ area to the right of t under a t_{n-3} distribution.

EXAMPLE 11.28   **Hypertension, Pediatrics** Test for the independent contributions of birthweight and age in days in predicting systolic blood pressure in infants using the data in Table 11.8.

SOLUTION   From Table 11.8,

$$b_1 = 0.1256$$

$$se(b_1) = 0.0343$$

$$t(\text{birthweight}) = b_1/se(b_1) = 3.66$$

$$p = 2 \times Pr(t_{13} > 3.66) = .003$$

$$b_2 = 5.888$$

$$se(b_2) = 0.6802$$

$$t(\text{age in days}) = b_2/se(b_2) = 8.66$$

$$p = 2 \times Pr(t_{13} > 8.66) < .001$$

Therefore, both birthweight and age in days have highly significant associations with systolic blood pressure even after controlling for the other variable.   ∎

It is possible that an independent variable (x_1) will seem to have an important effect on a dependent variable (y) when considered by itself but will not be significant after adjusting for another independent variable (x_2). This case usually occurs when x_1 and x_2 are strongly related to each other. Indeed, one of the advantages of multiple-regression analysis is that it allows us to identify which few variables among a large set of independent variables have a significant relationship to the dependent variable *after adjusting for other important independent variables.*

EXAMPLE 11.29   **Hypertension, Pediatrics** Suppose we consider the two independent variables $x_1 =$ birthweight, $x_2 =$ body length and try to use these variables to predict systolic blood pressure in newborns (y). Perhaps both x_1 and x_2, *when considered separately* in a simple linear regression model as given in **(11.2)**, have a significant relationship to blood pressure. However, since birthweight and body length are closely related to each other, after adjusting for birthweight, body length may not be significantly related to blood pressure based on the test procedure in **(11.15)**. One possible interpretation of this result is that the effect of body length on blood pressure can be explained by its strong relationship to birthweight.   ∎

11.8.3   **Extension to k Independent Variables**

The methods of this section can be extended to the general case of k independent variables. The general linear regression model is of the form

| **11.16** |

$$y = \alpha + \beta_1 x_1 + \beta_2 x_2 + \cdots + \beta_k x_k + e$$

The methods of estimation and hypothesis testing are similar to the procedures developed in Section 11.8.2. Specifically, to test the overall hypothesis $H_0: \beta_1 = \beta_2 = \cdots = \beta_k = 0$ versus $H_1:$ at least one of the $\beta_i \neq 0$, that is, at least some of the independent variables have an association with y, proceed as follows:

11.17 | *F* **Test for Testing the Hypothesis** $H_0: \beta_1 = \beta_2 = \cdots = \beta_k = 0$ **versus** H_1: **At Least One of the** $\beta_i \neq 0$ **in Multiple Linear Regression**

(1) Fit the regression parameters using the method of least squares and compute Reg SS and Res SS.

(2) Compute Reg MS = Reg SS/k, Res MS = Res SS/$(n - k - 1)$.

(3) Compute the test statistic

$$F = \text{Reg MS}/\text{Res MS}$$

which follows an $F_{k,n-k-1}$ distribution under H_0.

(4) For a level α test, if

$$F > F_{k,n-k-1,1-\alpha}$$

then reject H_0; if

$$F \leqslant F_{k,n-k-1,1-\alpha}$$

then accept H_0.

(5) The exact *p*-value is given by the area to the right of *F* under an $F_{k,n-k-1}$ distribution = $Pr(F_{k,n-k-1} > F)$.

The acceptance and rejection regions for this test procedure are depicted in Figure 11.18. The computation of the exact *p*-value is illustrated in Figure 11.19.

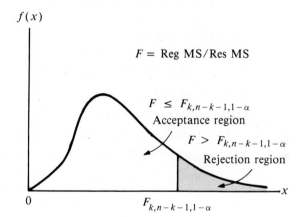

FIGURE 11.18
Acceptance and rejection regions for testing the hypothesis $H_0: \beta_1 = \beta_2 = \cdots = \beta_k = 0$ versus H_1: at least one of the $\beta_i \neq 0$ in multiple linear regression

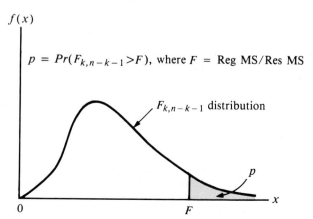

FIGURE 11.19
Computation of the *p*-value for testing the hypothesis $H_0: \beta_1 = \beta_2 = \cdots = \beta_k = 0$ versus H_1: at least one of the $\beta_i \neq 0$ in multiple linear regression

This test will not identify which specific independent variables are associated with the dependent variable. The following t test is performed to investigate the specific association of the ith independent variable with the dependent variable, after controlling for the effects of the other independent variables.

11.18   **t Test for Testing the Hypothesis H_0: $\beta_i = 0$, All Other $\beta_j \neq 0$ versus H_1: $\beta_i \neq 0$, All Other $\beta_j \neq 0$ in Multiple Linear Regression**

(1) Compute
$$t = b_i/se(b_i)$$
which should follow a t distribution with $n - k - 1$ df under H_0.

(2) If
$$t < t_{n-k-1,\alpha/2} \quad \text{or} \quad t > t_{n-k-1,1-\alpha/2}$$
then reject H_0; if $t_{n-k-1,\alpha/2} \leqslant t \leqslant t_{n-k-1,1-\alpha/2}$

then accept H_0.

(3) The exact p-value is given by
$$2 \times Pr(t_{n-k-1} > t) \quad \text{if } t \geqslant 0$$
$$2 \times Pr(t_{n-k-1} \leqslant t) \quad \text{if } t < 0$$

The acceptance and rejection regions for this test are depicted in Figure 11.20. The computation of the exact p-value is illustrated in Figure 11.21.

EXAMPLE 11.30   **Hypertension, Pediatrics** An example of the multiple-regression methods in **(11.17)** and **(11.18)** is given in Table 11.10. The sample consists of 650 infants whose blood pressure was measured in the hospital shortly after birth. The dependent variable in this regression equation is MSYSHP = mean of three systolic blood-pressure readings taken in the hospital. The independent variables are

ARMCIRHP = arm circumference (cm)

ARMLENHP = arm length (cm)

AGEHP = age (in days)

CONDHP = condition of infant at time of measurement (1 = awake, 0 = asleep)

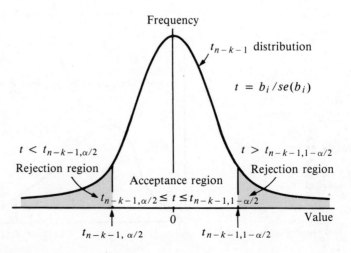

FIGURE 11.20
Acceptance and rejection regions for the t test for multiple linear regression (general case)

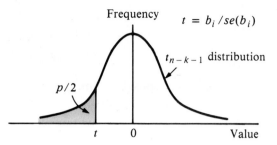

(a) If $t < 0$, then $p = 2 \times$ (area to the left of t under a t_{n-k-1} distribution).

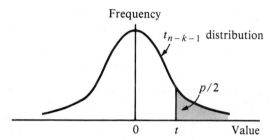

(b) If $t \geqslant 0$, then $p = 2 \times$ (area to the right of t under a t_{n-k-1} distribution).

FIGURE 11.21
Computation of the exact p-value for the t test for multiple linear regression (general case)

TABLE 11.10
An example of multiple-regression methods used to predict mean systolic blood pressure (MSYSHP) as a function of four other independent variables in a sample of 650 newborns

	df	SS	MS	F	p
Regression	4	10,580.44	2645.11	36.45	<.001
Error	645	46,806.44	72.57		
Total	649	57,386.88			

	b	se	t	p
INTERCEPT	38.132			
ARMCIRHP	2.758	0.386	7.15	<.001
ARMLENHP	−0.796	0.385	−2.07	.039
AGEHP	3.252	0.356	9.13	<.001
CONDHP	2.236	0.677	3.30	.001

Note that the overall F statistic $= F = 36.45 \sim F_{4, .645}$ and is statistically significant, with $p < .001$. This finding indicates that at least some of the variables are associated with systolic blood pressure. Furthermore, to test for the effects of specific variables, t tests as in **(11.18)**, which are given in the bottom of Table 11.10, are performed. For arm circumference, $t = 7.15$, $p < .001$; for arm length, $t = -2.07$, $p = .039$; for age, $t = 9.13$, $p < .001$; for condition at the time of measurement, $t = 3.30$, $p = .001$. Thus, each of the variables is associated with systolic blood pressure *even after controlling for the other three variables*.

The final regression equation consists of

$$\text{MSYSHP} = 38.13 + 2.76 \times \text{ARMCIRHP} - 0.80 \times \text{ARMLENHP}$$
$$+ 3.25 \times \text{AGEHP} + 2.24 \times \text{CONDHP} \qquad \blacksquare$$

A more detailed treatment of multiple-regression methods, including model-selection strategies, is given in Draper and Smith [3] and Kleinbaum and Kupper [4].

SECTION 11.9 ## Limitations on the Use of Linear Regression

A number of assumptions were made in using the methods of simple and multiple linear regression in the previous sections of this chapter. What are some of these assumptions and what possible situations could be encountered that would make these assumptions not viable?

11.19 | **Assumptions Made in Linear-Regression Models**

(1) For any given value of x, the corresponding value of y has an expected value $\alpha + \beta x$, which is a linear function of x.

(2) For any given value of x, the corresponding value of y is normally distributed about $\alpha + \beta x$ with the same variance σ^2 for any x.

(3) For any two data points (x_1, y_1), (x_2, y_2), the error terms e_1, e_2 are independent of each other.

EXAMPLE 11.31 **Pulmonary Disease** Cigarette smoking is one of the leading causes of pulmonary disease. One measure of cigarette consumption is the total number of *pack-years* of cigarettes consumed in a lifetime. This measure is frequently estimated by multiplying the number of packs of cigarettes per day currently smoked (or last smoked, if the person is not a current smoker) by the number of years smoked. Thus, if a person has smoked 2 packs per day for 20 years, then he or she has smoked 40 pack-years. Suppose we wish to relate current level of FEV (y) to total number of pack-years of cigarettes consumed (x) using a regression equation. How should this be done? $\qquad \blacksquare$

A model of the form $y = \alpha + \beta x + e$ could be used, as in **(11.2)**. This model is not likely to work well because the distribution of the number of pack-years is very skewed, and FEV is unlikely to be linearly related to this variable. Instead, pack-years are typically transformed to the scale of $\log(\text{pack-years} + 1) = z$ and a regression model of the following form is fitted:

11.20 |

$$y = \alpha + \beta z + e$$

This transformation makes it much more likely that assumptions 1 and 2 in **(11.19)** will hold.

EXAMPLE 11.32 **Hypertension, Pediatrics** A topic of current research interest is to look for risk factors in childhood for future cardiovascular disease. One such risk factor is blood pressure. One problem with quantifying blood pressure in children is that it is related to age. Another problem is that blood pressure tends to be more variable for older children than for younger children. Suppose we fit a linear-regression model of the form $y = \alpha + \beta x + e$, where y = systolic blood pressure and x = age. A typical plot of blood pressure versus age and the associated regression line is given in Figure 11.22.

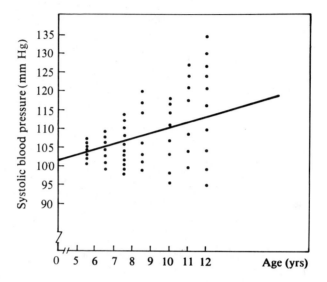

FIGURE 11.22
Scatter plot and regression line of systolic blood pressure vs. age in children

The problem with this type of regression line is that there is more scatter about the regression line for older children than for younger children. This violates assumption 2 in **(11.19)**, which requires that the variance of blood pressure about the regression line for any given age be the same. One strategy that is frequently used to overcome this problem is to express blood pressure in the form of a z score rather than in the form of raw blood pressure. For a child of age x with blood pressure y, the z score is defined as

$$z = \frac{(y - \bar{y})}{s}$$

where $\bar{y}$ and s are the sample mean and standard deviation, respectively, of blood pressure for children of the same age group. Thus, if a 10-year-old child has a systolic blood pressure of 120 mm Hg and the mean and standard deviation of systolic blood pressure for 10-year-olds is 110 mm Hg and 10 mm Hg, respectively, then the z score = $(120 - 110)/10 = 1.0$. The advantage of the z score method is that regardless of age, all blood-pressure z scores are approximately normally distributed with mean 0 and variance 1. ∎

EXAMPLE 11.33   **Cerebrovascular Disease** Cerebral blood flow is sometimes measured to confirm or, in some cases, to predict the occurrence of stroke. Its advantage over an angiogram, which is the standard test used for this purpose, is that it is noninvasive and can be performed by placing several leads at various positions on the head to measure blood flow at different parts of the brain. Before this test can have wide clinical applicability, standards for blood flow in normal people must be established. For this purpose, we would like to fit a regression line predicting blood flow (y) as a linear function of age (x) in the form $y = \alpha + \beta x$, since blood flow is known to decline with age. Suppose 10 measurements of blood flow have been made at different parts of the brain on 50 people of different ages. How should these data be used? ∎

Specifically, should 500 points (50 people × 10 measurements per person) be entered and a simple linear regression model of the type in **(11.2)** be fit? Probably not, since this model would violate assumption 3 in **(11.19)**, namely, that each sample point used to fit the regression line is independent of any other sample point. In particular, the 10 readings from one person are likely to be very similar to each other, and if the sample point corresponding to one of the readings is above the regression line, then most of the other readings for that person will also be above the regression line. A simple method for dealing with this problem is to compute the average blood flow over all readings for a particular person and use one sample point for that person $(x, \bar{y})$ in fitting the regression line consisting of age and mean blood flow. This method would be used for each of the 50 people, and thus 50 sample points rather than 500 would be used to fit the regression line. This method is valid although probably not the most efficient for handling this problem. (See Draper and Smith for a more advanced treatment of this subject [3].) The general issue here is that each sample point used in fitting a regression line should be independent of all other points, which is not likely to occur if several measurements are made for the same person, and each is considered as a separate data point in fitting a regression line.

SECTION 11.10   **Multiple Logistic Regression**

In the previous sections of this chapter, methods for relating one or more independent variables to a normally distributed outcome variable were discussed. In many instances we want to perform similar analyses in which the outcome variable follows a binomial rather than a normal distribution.

EXAMPLE 11.34   **Infectious Disease** *Chlamydia trachomatis* is a microorganism that has been established as an important cause of nongonococcal urethritis, pelvic inflammatory disease, and other infectious diseases. A study of risk factors for *C. trachomatis* was conducted in a population of 431 female college students [5]. Since multiple risk factors may be involved, several risk factors must be controlled for simultaneously in the analysis of variables associated with *C. trachomatis*.   ■

A model of the following form might be considered:

11.21
$$p = \alpha + \beta_1 x_1 + \cdots + \beta_k x_k$$

where p = probability of disease. However, since the right-hand side of **(11.21)** could be less than 0 or greater than 1 for certain values of $x_1, \ldots, x_k$, predicted probabilities that are either less than 0 or greater than 1 could be obtained, which is impossible. Instead, the logit (logistic) transformation of p is used as the dependent variable.

DEFINITION 11.16   ■■
The logit transformation logit(p) is defined as

$$\text{logit}(p) = \ln[p/(1 - p)]$$

Unlike p, the logit transformation can take on any value from $-\infty$ to $+\infty$.   ■

EXAMPLE 11.35 Compute logit(.1), logit(.95).

SOLUTION
$$\text{logit}(.1) = \ln(.1/.9) = \ln(1/9) = -\ln(9) = -2.20$$
$$\text{logit}(.95) = \ln(.95/.05) = \ln(19) = 2.94$$
∎

If logit(p) is modeled as a linear function of the independent variables $x_1, \ldots, x_k$, then the following multiple logistic regression model is obtained:

11.22 | **Multiple Logistic Regression Model**

If $x_1, \ldots, x_k$ are a collection of independent variables and y is a binomial outcome variable with probability of success $= p$, then the multiple logistic regression model is given by

$$\text{logit}(p) = \ln\left(\frac{p}{1-p}\right) = \alpha + \beta_1 x_1 + \cdots + \beta_k x_k$$

or, equivalently, if we solve for p, then the model can be expressed in the form

$$p = \frac{e^{\alpha + \beta_1 x_1 + \cdots + \beta_k x_k}}{1 + e^{\alpha + \beta_1 x_1 + \cdots + \beta_k x_k}}$$

In the second form of the model, we see that p must always lie between 0 and 1 regardless of the values of $x_1, \ldots, x_k$. Complex numerical algorithms are generally required to fit the parameters of the model in **(11.22)**. The best-fitting model relating the prevalence of *C. trachomatis* to the risk factors (1) race and (2) the lifetime number of sexual partners is presented in Table 11.11.

TABLE 11.11
Multiple logistic regression model relating prevalence of *C. trachomatis* to race and number of lifetime sexual partners

Risk factor	Regression coefficient ($\hat{\beta}_i$)	Standard error se($\hat{\beta}_i$)	z ($\hat{\beta}_i / se(\hat{\beta}_i)$)
Constant	−1.637		
Black race	+2.242	0.529	+4.24
Lifetime number of sexual partners among users of nonbarrier* methods of contraception†	+0.102	0.040	+2.55

* Barrier methods of contraception include diaphragm, diaphragm and foam, and condom; nonbarrier methods include all other forms of contraception or no contraception.
† This variable is defined as 0 for users of barrier methods of contraception.
(Reprinted with permission of the *American Journal of Epidemiology, 121*(1), 107–115, 1985.)

How can the results in Table 11.11 be interpreted? The significance of each of the independent variables after controlling for all other independent variables in the model should be assessed. This task can be accomplished by first computing the test statistic $z = \hat{\beta}_i / se(\hat{\beta}_i)$, which should follow an $N(0, 1)$ distribution under the null hypothesis that the ith independent variable has no association with the dependent variable after controlling for the other variables. H_0 will be rejected for either large positive or large negative values of z. This procedure is summarized as follows:

11.23	**Hypothesis Testing in Multiple Logistic Regression**

To test the hypothesis H_0: $\beta_i = 0$, all other $\beta_j \neq 0$, versus H_1: all $\beta_j \neq 0$ for the multiple logistic regression model in **(11.22)**, use the following procedure:

(1) Compute the test statistic $z = \hat{\beta}_i / se(\hat{\beta}_i) \sim N(0, 1)$ under H_0.

(2) To conduct a two-sided test with significance level α, if

$$z < z_{\alpha/2} \quad \text{or} \quad z > z_{1-\alpha/2}$$

then reject H_0; if

$$z_{\alpha/2} \leqslant z \leqslant z_{1-\alpha/2}$$

then accept H_0.

(3) The exact p-value is given by

$$2 \times [1 - \Phi(z)] \qquad \text{if } z \geqslant 0$$

$$2 \times \Phi(z) \qquad\quad \text{if } z < 0$$

The acceptance and rejection regions for this test are depicted in Figure 11.23. The computation of the exact p-value is illustrated in Figure 11.24.

FIGURE 11.23
Acceptance and rejection regions for the test of the hypothesis H_0: $\beta_i = 0$, all other $\beta_j \neq 0$, vs. H_1: all $\beta_j \neq 0$ in multiple logistic regression

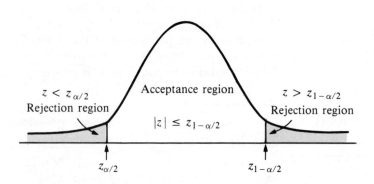

EXAMPLE 11.36	**Infectious Disease** Assess the significance of the independent variables in the multiple-logistic regression model presented in Table 11.11.

SOLUTION   First compute the test statistic $z = \hat{\beta}_i / se(\hat{\beta}_i)$ for each of the independent variables, as shown in Table 11.11. For an α level of .05, compare $|z|$ with $z_{.975} = 1.96$ to assess statistical significance. Since both of the independent variables satisfy this criterion, they are both significant at the 5% level. The exact p-values are given by

$$p(\text{race}) = 2 \times [1 - \Phi(4.24)] < .001$$

$$p(\text{number of sexual partners}) = 2 \times [1 - \Phi(2.55)] = .011$$

Thus, both variables are significantly associated with *C. trachomatis*. Specifically, after controlling for the other variable in the model, there is an increased probability of infection for black women versus white women and for women with greater previous sexual experience versus women with lesser previous sexual experience. ∎

In addition to assessing statistical significance, an odds ratio can also be used to measure the strength of the association between each dichotomous independent

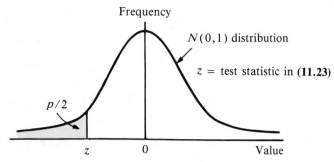

(a) If $z = \hat{\beta}_i/se(\hat{\beta}_i) < 0$, then $p = 2 \times$ (area to the left of z under an $N(0, 1)$ distribution).

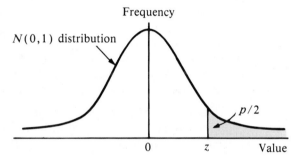

FIGURE 11.24
Computation of the p-value for the test of the hypothesis H_0: $\beta_j = 0$, all other $\beta_j \neq 0$, versus H_1: all $\beta_j \neq 0$ in multiple logistic regression

(b) If $z = \hat{\beta}_i/se(\hat{\beta}_i) \geqslant 0$, then $p = 2 \times$ (area to the right of z under an $N(0, 1)$ distribution).

variable and the dependent variable after controlling for the other variables in the model. This procedure is summarized as follows:

11.24   **Estimation of Odds Ratios in Multiple Logistic Regression for Dichotomous Independent Variables**

Suppose there is a dichotomous independent variable (x_i), which is coded as 1 if present and 0 if absent. For the multiple-logistic regression model in **(11.22)**, the odds ratio relating this independent variable to the dependent variable is estimated by

$$\widehat{OR} = e^{\hat{\beta}_i}$$

This relationship expresses the odds in favor of success if $x = 1$ divided by the odds in favor of success if $x = 0$ *after controlling for all other variables in the logistic regression model.* Furthermore, a two-sided $100\% \times (1 - \alpha)$ confidence interval for the true odds ratio is given by

$$[e^{\hat{\beta}_i - z_{1-\alpha/2}se(\hat{\beta}_i)}, e^{\hat{\beta}_i + z_{1-\alpha/2}se(\hat{\beta}_i)}]$$

EXAMPLE 11.37   **Infectious Disease** Estimate the odds in favor of infection with *C. trachomatis* for black women compared with white women after controlling for previous sexual experience and provide a 95% confidence interval about this estimate.

SOLUTION   From Table 11.11,

$$\widehat{OR} = e^{2.242} = 9.41$$

Thus, the odds in favor of infection for black women are 9 times as large as that for white women after controlling for previous sexual experience. Furthermore, since $z_{1-\alpha/2} = z_{.975} = 1.96$ and $se(\hat{\beta}_i) = 0.529$, a 95% confidence interval for OR is given by

$$[e^{2.242 - 1.96(0.529)}, e^{2.242 + 1.96(0.529)}] = (e^{1.205}, e^{3.279}) = (3.34, 26.55) \quad \blacksquare$$

We are also interested in expressing the strength of association between a continuous independent variable and the dependent variable in terms of an odds ratio after controlling for the other independent variables in the model.

11.25 | **Estimation of Odds Ratios in Multiple Logistic Regression for Continuous Independent Variables**

Suppose there is a continuous independent variable (x_i). Consider two individuals who have values of $x + \Delta$ and x for x_i, respectively, and have the same values for all other independent variables in the model. The odds ratio in favor of success for the first individual versus the second individual is estimated by

$$\widehat{OR} = e^{\hat{\beta}_i \Delta}$$

Furthermore, a two-sided $100\% \times (1 - \alpha)$ confidence interval for OR is given by

$$\left\{ e^{[\hat{\beta}_i - z_{1-\alpha/2} se(\hat{\beta}_i)]\Delta}, \; e^{[\hat{\beta}_i + z_{1-\alpha/2} se(\hat{\beta}_i)]\Delta} \right\}$$

Thus, OR represents the odds in favor of success for an individual with level $x + \Delta$ for x_i versus an individual with level x for x_i, after controlling for all other variables in the model.

EXAMPLE 11.38 | **Infectious Disease** Based on the data in Table 11.11, what is the extra risk of infection for each additional sexual partner for women of a particular race who use nonbarrier methods of contraception? Provide a 95% confidence interval associated with this estimate.

SOLUTION | We have that $\Delta = 1$. From Table 11.11, $\hat{\beta}_i = 0.102$, $se(\hat{\beta}_i) = 0.040$. Thus,

$$\widehat{OR} = e^{0.102 \times 1} = e^{0.102} = 1.11$$

Thus, the odds in favor of infection increase an estimated 11% for each additional sexual partner for women of a particular race who use nonbarrier methods of contraception. A 95% confidence interval for OR is given by

$$\left\{ e^{[0.102 - 1.96(0.040)]}, e^{[0.102 + 1.96(0.040)]} \right\} = (e^{0.0236}, e^{0.1804}) = (1.02, 1.20) \quad \blacksquare$$

Finally, the 95% confidence intervals in **(11.24)** and **(11.25)** will contain 1 only if there is a nonsignificant association between x_i and the dependent variable. Similarly, these intervals will not contain 1 only if there is a significant association between x_i and the dependent variable. Thus, since both independent variables in Table 11.11 are statistically significant, the confidence intervals in Example 11.37 and 11.38 both exclude 1.

SECTION 11.11 | **Survival Analysis**

In the previous section, the use of multiple-logistic regression analysis to predict a binary outcome variable (y) as a function of one or more independent variables ($x_1, \ldots, x_k$) was discussed. This model is useful for identifying risk factors related to the presence or absence of a disease or condition, such as the relationship of sexual history variables to the presence or absence of *C. trachomatis* among female college students in Example 11.34.

In some instances, the concern is not only with whether or not a particular condition has occurred, but *when* it occurs. For example, in mortality studies, if a population is followed for a sufficiently long period of time, then all people in the population will die regardless of the characteristics of the subjects involved. Thus, it is useful to identify the risk factors for differentiating people who die early from those who die late. In other instances, not all people in the population will experience an event, but it is still useful to know which risk factors identify people who experience early events versus late events.

EXAMPLE 11.39   **Health Promotion** Data Set 24, SMOKE.DAT, was introduced in Chapter 4 (p. 104). This data set is concerned with the experience of 234 individuals who tried to quit smoking. For each person, the number of days that they were able to abstain from smoking during the first year after the quit attempt is given as well as various risk factors that might be related to success in quitting (e.g., number of cigarettes smoked and serum CO concentration just prior to quitting). It is of great public health importance to identify risk factors that favorably impact the ability of former smokers to remain abstinent. Specifically, we are interested in relating the *time to relapse* to these risk factors.   ∎

One might try to regress time to relapse on the risk factors. However, survival times do not usually follow a normal distribution. Also, not all subjects may have relapsed by the end of the study. This is one example of *censored* data (see Lee [6] and Miller [7] for details). These problems are overcome by the set of techniques for the treatment of time-to-failure data commonly called *survival analysis*.

DEFINITION 11.17   ■■■
The **survival curve** $S(t)$ gives the probability of survival up to time t for each $t > 0$.

EXAMPLE 11.40   **Health Promotion** In the context of the smoking-cessation example mentioned in Example 11.39, the survival curve might look like Figure 11.25. The curve gives the probability of abstaining from smoking up to day t ($0 \leqslant t \leqslant 365$). In this context, a person "survives" to

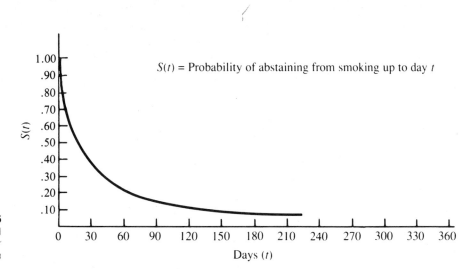

FIGURE 11.25
Hypothetical
survival curve for
smoking-cessation data

day t if she has abstained from smoking up to that time. Note that $S(0) = 1$, since all people initially abstained from smoking. For $t > 0$, $S(t) < 1$, since some people have resumed smoking by time t. By 1 year, $S(t) = S(365)$ is likely to be around 0.1–0.2, indicating that 10–20% of people have succeeded in abstaining from smoking for 1 year. ■

As was the case for logistic regression, we are often interested in studying the relationship between a collection of independent variables $x_1, \ldots, x_k$ and survival time. This will provide a generalization of the log-rank test as presented in Section 10.10, which was useful for comparing survival time between two groups. For this purpose, it is useful to characterize survival curves in terms of their hazard.

DEFINITION 11.18 ■■■

The **hazard function** $h(t)$ is the *instantaneous* probability of having an event at time t (per unit time) given that one has survived up to time t. In particular,

$$h(t) = \left[\frac{S(t) - S(t + \Delta t)}{\Delta t} \right] \bigg/ S(t) \quad \text{as } \Delta t \text{ approaches } 0$$

■

EXAMPLE 11.41 **Demography** Use the life table data in Table 4.8 (p. 101) to compute the approximate mortality hazard at age 60 and 80, respectively, for U.S. males in 1960.

SOLUTION Table 4.10 shows that there were 73,887 men who survived up to age 60 and 72,151 men who survived up to age 61 among the original 100,000 men at time 0 (birth). Therefore, the hazard at age 60 is approximately

$$h(60) = \frac{73,887 - 72,151}{73,887} = \frac{1736}{73,887} = .023$$

Similarly, since there were 25,300 men who survived to age 80 and 22,619 men who survived to age 81, the hazard at age 80 is approximately given by

$$h(80) = \frac{25,300 - 22,619}{25,300} = \frac{2681}{25,300} = .106$$

Thus, in words, the probability of dying in the next year is 2.3% given that one has survived to age 60 and 10.6% given that one has survived to age 80. The percentages 2.3% and 10.6% represent the approximate hazard at ages 60 and 80, respectively. To improve the approximation, shorter time intervals than 1 year would need to be considered. ■

Many different models can be used to relate survival to a collection of other risk factors. One of the most frequently used models was first proposed by D. R. Cox [8] and is called a proportional hazards model.

11.26 **Proportional Hazards Model**

Under a **proportional hazards model**, the hazard $h(t)$ is modeled as

$$h(t) = h_0(t)\exp(\beta_1 x_1 + \cdots + \beta_k x_k)$$

where $x_1, \ldots, x_k$ are a collection of independent variables, and $h_0(t)$ is the baseline hazard at time t, representing the hazard for a person with the value 0 for all the independent variables. The hypothesis $H_0: \beta_i = 0$ versus $H_1: \beta_i \neq 0$ can be tested as follows:

(1) Compute the test statistic $z = \hat{\beta}_i / se(\hat{\beta}_i)$.

(2) To conduct a two-sided level α significance test, if

$$z < z_{\alpha/2} \quad \text{or} \quad z > z_{1-\alpha/2}$$

reject H_0; if

$$z_{\alpha/2} \leqslant z \leqslant z_{1-\alpha/2}$$

accept H_0.

(3) The exact p-value is given by

$$2 \times [1 - \Phi(z)] \quad \text{if } z \geqslant 0$$
$$2 \times \Phi(z) \qquad \text{if } z < 0$$

By dividing both sides of **(11.26)** by $h_0(t)$ and taking logarithms, a proportional hazards model can be written in the form

$$\ln\left[\frac{h(t)}{h_0(t)}\right] = \beta_1 x_1 + \cdots + \beta_k x_k$$

This representation allows us to interpret the coefficients of a proportional hazards model in a similar manner to that of a multiple-logistic regression model. In particular, if x_i is a dichotomous independent variable, then the following principle applies:

11.27 | **Estimation of Hazard Ratio for Proportional Hazards Models for Dichotomous Independent Variables**

Suppose there is a dichotomous independent variable (x_i) that is coded as 1 if present and 0 if absent. For the proportional hazards model in **(11.26)**, the quantity $\exp(\beta_i)$ represents the ratio of hazards for two individuals, one with the risk factor present and the other with the risk factor absent, given that the individuals have the same values for all other covariates. This hazard ratio or relative hazard can be interpreted as the *instantaneous* relative risk of an event per unit time for an individual with the risk factor present compared with an individual with the risk factor absent, given that both individuals have survived to time t.

Similarly, if x_i is a continuous independent variable, then the following interpretation of the regression coefficient β_i is used:

11.28 | **Estimation of Hazard Ratio for Proportional Hazards Models for Continuous Independent Variables**

Suppose there is a continuous independent variable (x_i). Consider two individuals who differ by the quantity Δ on the ith independent variable and are the same for all other independent variables. The quantity $\exp(\beta_i \Delta)$ represents the ratio of hazards between the two individuals. The hazard ratio can also be interpreted as the *instantaneous* relative risk of an event per unit time for an individual with risk factor level $x_i + \Delta$ compared to an individual with risk factor level x_i, given that both individuals have survived to time t.

EXAMPLE 11.42 | **Health Promotion** Fit a proportional hazards model to the smoking-cessation data in Example 11.39 using the risk factors sex (coded as 1 if male and 2 if female) and adjusted log(CO concentration), which is an index of inhalation of smoke prior to quitting. Assess the statistical significance of the results and interpret the regression coefficients.

SOLUTION   The SAS PHGLM (Proportional Hazards General Linear Model) Procedure has been used to fit the Cox model to the smoking-cessation data. For ease of interpretation, sex was recoded as (1 = male/0 = female) from the original coding of (1 = male/2 = female). The results are given in Table 11.12.

TABLE 11.12
Proportional hazards model fitted to the smoking-cessation data in Example 11.39

Risk factor	Regression coefficient $(\hat{\beta}_i)$	Standard error $se(\hat{\beta}_i)$	z $(\hat{\beta}_i/se(\hat{\beta}_i))$
$\log_{10}CO$ (adjusted)*	0.833	0.350	2.380
Sex(1 = M/0 = F)	−0.117	0.135	−0.867

* This variable represents CO values adjusted for minutes elapsed since last cigarette smoked prior to quitting.

To assess significance, compute the test statistic given in **(11.26)** as follows:

$$z(\log_{10}CO) = 0.833/0.350 = 2.380$$

$$p(\log_{10}CO) = 2 \times [1 - \Phi(2.380)] = 2 \times (1 - .9913) = .017$$

$$z(\text{sex}) = -0.117/0.135 = -0.867$$

$$p(\text{sex}) = 2 \times \Phi(-0.867) = 2 \times [1 - \Phi(0.867)] = 2 \times (1 - .8070) = .386$$

Thus, there is a significant effect of CO concentration on the hazard or risk of recidivism (i.e., propensity to start smoking again), with the higher the CO concentration, the higher the hazard (risk). There is no significant effect of sex on risk of recidivism based on these data.

The effect of CO can be quantified in terms of relative risk. Specifically, if two individuals of the same sex who differ by one unit on adjusted $\log_{10}CO$ are considered (i.e., who differ by 10-fold in CO concentration), then the instantaneous relative risk of recidivism for a person with adjusted $\log_{10}CO = x_i + 1$ (person A) compared to a person with adjusted $\log_{10}CO = x_i$ (person B) is given by

$$RR = \exp(0.833) = 2.30$$

Thus, given that person A and person B have not started smoking up to time t, person A is 2.3 times as likely to quit over a short period of time than person B. ∎

Although the proportional hazards model is perhaps the most frequently used model for survival data in the medical literature, many other models have been introduced for this type of data. An extensive survey of the field is given in Lee [6] and Miller [7].

SECTION 11.12   **The Correlation Coefficient**

The primary focus of the discussion of linear regression analysis in Sections 11.2–11.9 was on methods of predicting one dependent variable (y) from one or more independent variables ($x_1, \ldots, x_k$). These methods were referred to as simple linear regression methods if there was only one independent variable and as multiple linear regression methods if there was more than one independent variable. Often we are interested not in predicting one variable from another but rather in investigating whether or not there is a relationship between two variables. The **correlation coefficient** is a useful tool for quantifying the relationship between variables and is better suited for this purpose than the regression coefficient.

EXAMPLE 11.43 **Cardiovascular Disease** Serum cholesterol is an important risk factor in the etiology of cardiovascular disease. Much research has been devoted to understanding the environmental factors that cause elevated cholesterol levels. For this purpose, cholesterol levels were measured on 100 genetically unrelated spouse pairs. We are not interested in predicting the cholesterol level of a husband from that of his wife but rather would like some quantitative measure of the relationship between their levels. What measure should be used? ■

DEFINITION 11.19 ■■■

The **sample (Pearson) correlation coefficient** (r) is defined by

$$L_{xy}/\sqrt{L_{xx}L_{yy}}$$ ■

The correlation is not affected by changes in location or scale in either variable and must lie between -1 and $+1$. The correlation can be interpreted as follows:

11.29 **Interpretation of the Correlation Coefficient**

(1) If the correlation is greater than 0, such as for birthweight and estriol, then the variables are said to be **positively correlated**. Two variables (x, y) are positively correlated if as x increases, y tends to increase, whereas as x decreases, y tends to decrease.

(2) If the correlation is less than 0, such as for pulse rate and age, then the variables are said to be **negatively correlated**. Two variables (x, y) are negatively correlated if as x increases, y tends to decrease, whereas as x decreases, y tends to increase.

(3) If the correlation is exactly 0, such as for birthweight and birthday, then the variables are said to be **uncorrelated**. Two variables (x, y) are uncorrelated if there is no relationship between x and y.

Thus the correlation coefficient provides a *quantitative* measure of the dependence between two variables: the closer $|r|$ is to 1, the more closely related the variables are; if $|r| = 1$, then one variable can be predicted exactly from the other. These interpretations are illustrated in Figure 11.26

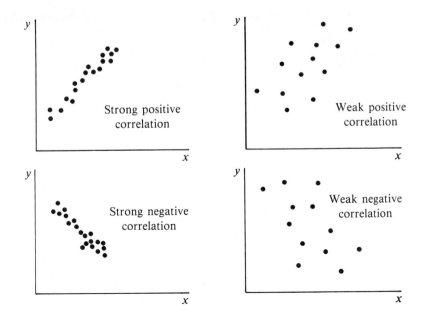

FIGURE 11.26
Interpretation of various
degrees of correlation

EXAMPLE 11.44   Suppose the two variables under study are temperature in °F (y) and temperature in °C (x). The correlation between these two variables must be 1, since one variable can be predicted exactly from the other $(y = \frac{9}{5}x + 32)$. ∎

EXAMPLE 11.45   **Obstetrics** Compute the sample correlation coefficient for the birthweight-estriol data discussed in Examples 11.3, 11.8, and 11.12 (p. 398, p. 403, and p. 409).

SOLUTION   From Examples 11.8 and 11.12,

$$L_{xy} = 412 \qquad L_{xx} = 677.42 \qquad L_{yy} = 674$$

Therefore, $r = L_{xy}/\sqrt{L_{xx}L_{xy}} = 412/\sqrt{(677.42)(674)} = 412/675.71 = .61$ ∎

What is the relationship between the sample regression coefficient (b) and the sample correlation coefficient (r)? Note from **(11.3)** that $b = L_{xy}/L_{xx}$ and from Definition 11.19 that $r = L_{xy}/\sqrt{L_{xx}L_{yy}}$. Therefore, if r is multiplied by $\sqrt{L_{yy}/L_{xx}}$,

11.30
$$r\sqrt{\frac{L_{yy}}{L_{xx}}} = \frac{L_{xy}}{\sqrt{L_{xx}L_{yy}}} \times \frac{\sqrt{L_{yy}}}{\sqrt{L_{xx}}} = \frac{L_{xy}}{L_{xx}} = b$$

Furthermore, from Definition 11.4,

$$s_y^2 = \frac{L_{yy}}{n-1}$$

$$s_x^2 = \frac{L_{xx}}{n-1}$$

$$s_y^2/s_x^2 = L_{yy}/L_{xx}$$

or $$s_y/s_x = \sqrt{L_{yy}/L_{xx}}$$

Substituting s_y/s_x for $\sqrt{L_{yy}/L_{xx}}$ on the left-hand side of **(11.30)** yields the following relationship:

11.31
$$b = \frac{rs_y}{s_x}$$

How can **(11.31)** be interpreted? The regression coefficient (b) can be interpreted as a rescaled version of the correlation coefficient (r), where the scale factor is the ratio of the standard deviation of y to that of x. Note that r will be unchanged by a change in the units of x or y (or even by which variable is designated as x and which is designated as y), whereas b is in the units of y/x.

EXAMPLE 11.46   **Pulmonary Function** Compute the correlation coefficient between FEV and height for the pulmonary-function data in Example 11.14 (p. 410).

SOLUTION   From Example 11.14,

$$L_{xy} = 117.4 \qquad L_{xx} = 2288 \qquad L_{yy} = 6.169$$

Therefore, $$r = \frac{117.4}{\sqrt{(2288)(6.169)}} = \frac{117.4}{118.81} = .988$$

Thus, a very strong positive correlation exists between FEV and height. The sample regression coefficient b was calculated as 0.051 in Example 11.14. Furthermore, the sample standard deviation of x and y can be computed as follows:

$$s_x = \sqrt{\frac{\sum_{i=1}^{n}(x_i - \bar{x})^2}{n-1}} = \sqrt{\frac{L_{xx}}{n-1}} = \sqrt{\frac{2288}{11}} = \sqrt{208} = 14.42$$

$$s_y = \sqrt{\frac{\sum_{i=1}^{n}(y_i - \bar{y})^2}{n-1}} = \sqrt{\frac{L_{yy}}{n-1}} = \sqrt{\frac{6.169}{11}} = \sqrt{0.561} = 0.749$$

and their ratio is thus given by

$$s_y/s_x = 0.749/14.42 = .052$$

Finally, b can be expressed as a rescaled version of r as

$$b = r(s_y/s_x) \quad \text{or} \quad .051 = (.988)(.052)$$

Notice that if height is re-expressed in inches rather than centimeters (1 in. = 2.54 cm), then s_x is divided by 2.54, and b is multiplied by 2.54; that is,

$$b_{\text{in.}} = b_{\text{cm}} \times 2.54 = .051 \times 2.54 = .130$$

However, the correlation coefficient remains the same at .988. ■

When should the regression coefficient be used and when should the correlation coefficient be used? The regression coefficient is used when we specifically want to predict one variable from another. The correlation coefficient is used when we simply want to describe the relationship between two variables but do not want to make predictions. In cases when it is not clear which of these two aims is primary, both a regression and a correlation coefficient can be reported.

EXAMPLE 11.47   **Obstetrics, Pulmonary Disease, Cardiovascular Disease** For the birthweight-estriol data in Example 11.3, the obstetrician is interested in using a regression equation to predict birthweight from estriol levels. Thus the regression coefficient is more appropriate. Similarly, for the FEV-height data in Example 11.14, the pediatrician is interested in using a growth curve relating a child's pulmonary function to height, and again the regression coefficient is more appropriate. However, in collecting data on cholesterol levels in spouse pairs in Example 11.43, the geneticist is interested simply in describing if there is a relationship between cholesterol levels of spouse pairs and is not interested in prediction. Thus the correlation coefficient is more appropriate here. ■

SECTION 11.13   **Hypothesis Testing for Correlation Coefficients**

In the previous section the sample correlation coefficient was defined. If every unit in the reference population could be sampled, then the sample correlation coefficient (r) would be the same as the population correlation coefficient, which is denoted by ρ.

DEFINITION 11.20   ■■
The **population correlation coefficient** (ρ) is the population analogue to the sample correlation coefficient; that is, if every member of the reference population could be enumerated, then the computed sample correlation (r) would be the same as the population correlation (ρ). For finite samples, r is used to estimate ρ. ■

11.13.1 One-Sample *t* Test for a Correlation Coefficient

EXAMPLE 11.48 **Cardiovascular Disease** We would like to test hypotheses concerning ρ. In particular, suppose serum cholesterol levels in spouse pairs are measured to determine whether or not there is a correlation between cholesterol levels in spouses. Specifically, suppose that $r = .25$ based on 100 spouse pairs. Is this evidence sufficient to warrant rejecting H_0?   ∎

We wish to test the hypothesis $H_0: \rho = 0$ versus $H_1: \rho \neq 0$. In this instance, the hypothesis test would naturally be based on the sample correlation coefficient r and H_0 would be rejected if $|r|$ is sufficiently far from 0. Assuming that each of the random variables $x =$ serum cholesterol level for the husband and $y =$ serum cholesterol level for the wife is normally distributed, then the best procedure for testing the hypothesis is given as follows:

11.32 **One-Sample *t* Test for a Correlation Coefficient**

To test the hypothesis $H_0: \rho = 0$ versus $H_1: \rho \neq 0$, use the following procedure:

(1) Compute the sample correlation coefficient r.

(2) Compute the test statistic

$$t = r(n-2)^{1/2}/(1-r^2)^{1/2}$$

which under H_0 follows a t distribution with $n - 2$ df.

(3) For a two-sided level α test, if

$$t > t_{n-2,1-\alpha/2} \qquad \text{or} \qquad t < -t_{n-2,1-\alpha/2}$$

then reject H_0. If

$$-t_{n-2,1-\alpha/2} \leqslant t \leqslant t_{n-2,1-\alpha/2}$$

then accept H_0.

(4) The exact p-value is given by

$$p = 2 \times (\text{area to the left of } t \text{ under a } t_{n-2} \text{ distribution}) \qquad \text{if } t < 0$$
$$p = 2 \times (\text{area to the right of } t \text{ under a } t_{n-2} \text{ distribution}) \qquad \text{if } t \geqslant 0$$

(5) We assume an underlying normal distribution for each of the random variables used to compute r.

The acceptance and rejection regions for this test are depicted in Figure 11.27. The computation of the exact p-value is illustrated in Figure 11.28.

EXAMPLE 11.49 Perform a test of significance for the data in Example 11.48.

SOLUTION We have $n = 100$, $r = .25$. Thus in this case,

$$t = (.25)\sqrt{98}/\sqrt{1 - .25^2} = 2.475/.968 = 2.56$$

From Table 5 in Appendix 1,

$$t_{60,.99} = 2.39 \qquad t_{60,.995} = 2.66 \qquad t_{120,.99} = 2.358 \qquad t_{120,.995} = 2.617$$

Therefore, since $60 < 98 < 120$,

$$.005 < p/2 < .01 \qquad \text{or} \qquad .01 < p < .02$$

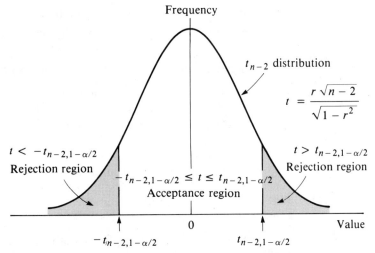

FIGURE 11.27
Acceptance and rejection regions for the one-sample t test for a correlation coefficient

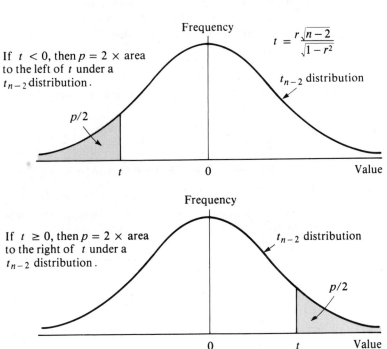

FIGURE 11.28
Computation of the exact p-value for the one-sample t test for a correlation coefficient

and H_0 is rejected. We conclude that there is a significant aggregation of cholesterol among spouse pairs. This result is possibly due to common environmental factors such as diet. But it could also be due to the tendency for people of similar body build to marry each other, and their cholesterol levels may have been correlated at the time of marriage.   ■

Interestingly, the test procedure for testing correlations in **(11.32)** is mathematically equivalent to the F test in **(11.7)** and the t test in **(11.8)** for testing regression coefficients, in that they always yield the same p-values. The question as to which test is more appropriate is best answered by whether a regression or a correlation coefficient is the parameter of primary interest.

11.13.2 One-Sample z Test for a Correlation Coefficient

In Section 11.13.1 a test of the hypothesis H_0: $\rho = 0$ versus H_1: $\rho \neq 0$ was considered. Sometimes the correlation between two random variables is expected to be some quantity ρ_0 other than 0 and we wish to test the hypothesis H_0: $\rho = \rho_0$ versus H_1: $\rho \neq \rho_0$.

EXAMPLE 11.50   Suppose the body weights of 100 fathers (x) and first-born sons (y) are measured and a sample correlation coefficient r of .38 is found. We might ask whether or not this sample correlation is compatible with an underlying correlation of .5 that might be expected on genetic grounds. How can this hypothesis be tested? ∎

In this case we want to test the hypothesis H_0: $\rho = .5$ versus H_1: $\rho \neq .5$. The problem with using the t test formulation in **(11.32)** is that the sample correlation coefficient r has a skewed distribution for nonzero ρ that cannot be easily approximated by a normal distribution. Fisher considered this problem and proposed the following transformation to better approximate a normal distribution:

11.33 | **Fisher's z Transformation of the Sample Correlation Coefficient r**

The z transformation of r given by

$$z = \frac{1}{2} \ln\left[\frac{(1 + r)}{(1 - r)}\right]$$

is approximately normally distributed under H_0 with mean

$$z_0 = \tfrac{1}{2} \ln[(1 + \rho_0)/(1 - \rho_0)]$$

and variance $1/(n - 3)$. The z transformation is very close to r for small values of r but tends to deviate substantially from r for larger values of r. A table of the z transformation is given in Table 11 in Appendix 1.

EXAMPLE 11.51   Compute the z transformation of $r = .38$.

SOLUTION   The z transformation can be computed from **(11.33)** as follows:

$$z = \frac{1}{2} \ln\left[\frac{(1 + 0.38)}{(1 - 0.38)}\right] = \frac{1}{2} \ln\left(\frac{1.38}{0.62}\right) = \frac{1}{2} \ln(2.226) = \frac{1}{2}(0.800) = 0.400$$

Alternatively, we could refer to Table 11 in Appendix 1 with $r = .38$ to obtain $z = 0.400$. ∎

Fisher's z transformation can be used for the hypothesis test as follows: Under H_0, z is normally distributed with mean z_0 and variance $1/(n - 3)$ or, equivalently,

$$\lambda = (z - z_0)\sqrt{n - 3} \sim N(0, 1)$$

H_0 will be rejected if z is far from z_0. Thus, the following test procedure for a two-sided level α test is used:

11.34   **One-Sample z Test for a Correlation Coefficient**

To test the hypothesis $H_0: \rho = \rho_0$ versus $H_1: \rho \neq \rho_0$, use the following procedure:

(1) Compute the sample correlation coefficient r and the z transformation of r.

(2) Compute the test statistic

$$\lambda = (z - z_0)\sqrt{n - 3}$$

(3) If $\lambda > z_{1-\alpha/2}$ or $\lambda < -z_{1-\alpha/2}$

reject H_0. If $-z_{1-\alpha/2} \leqslant \lambda \leqslant z_{1-\alpha/2}$

accept H_0.

(4) The exact p-value is given by

$$p = 2 \times \Phi(\lambda) \qquad \text{if } \lambda \leqslant 0$$
$$p = 2 \times [1 - \Phi(\lambda)] \qquad \text{if } \lambda > 0$$

(5) Assume an underlying normal distribution for each of the random variables used to compute r and z.

The acceptance and rejection regions for this test are depicted in Figure 11.29. The computation of the exact p-value is illustrated in Figure 11.30.

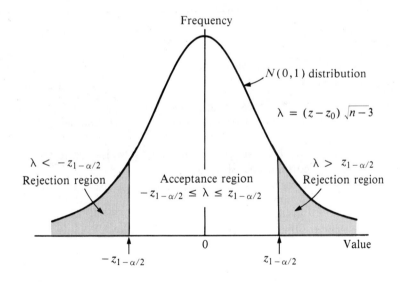

FIGURE 11.29
Acceptance and rejection regions for the one-sample z test for a correlation coefficient

EXAMPLE 11.52   Perform a test of significance for the data in Example 11.50.

SOLUTION   In this case $r = .38$, $n = 100$, $\rho_0 = .50$. From Table 11 in Appendix 1,

$$z_0 = \frac{1}{2} \ln\left[\frac{(1 + .5)}{(1 - .5)}\right] = .549 \qquad z = \frac{1}{2} \ln\left[\frac{(1 + .38)}{(1 - .38)}\right] = .400$$

Hence,

$$\lambda = (0.400 - 0.549)\sqrt{97} = (-0.149)(9.849) = -1.47 \sim N(0, 1)$$

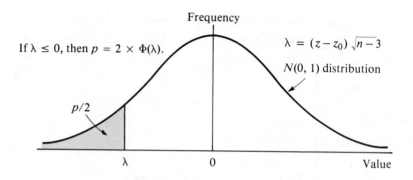

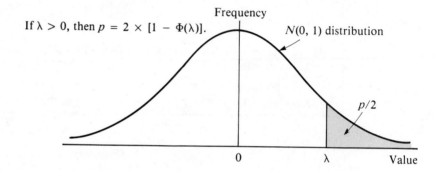

FIGURE 11.30
Computation of the
exact p-value for the
one-sample z test for a
correlation coefficient

Thus, the p-value is given by

$$2 \times [1 - \Phi(1.47)] = 2 \times (1 - .9292) = .142$$

Therefore, we accept H_0 that the sample estimate of .38 is compatible with an underlying correlation of 0.50; this would be expected on purely genetic groups. ∎

To sum up, the z test in **(11.34)** is used to test hypotheses about nonzero null correlations, whereas the t test in **(11.32)** is used to test hypotheses about null correlations of zero. The z test can also be used to test correlations of zero under the null hypothesis, but the t test is slightly more powerful in this case and is preferred.

11.13.3   **Two-Sample Tests for Correlations**

The use of Fisher's z transformation can be extended to two-sample problems.

EXAMPLE 11.53   **Hypertension**   Suppose there are two groups of children. Children in one group live with their natural parents, whereas children in the other group live with adopted parents. One question that arises is whether or not the correlation between the blood pressure of mother and child is different in these two groups. A different correlation would suggest a genetic effect on blood pressure. Suppose there are 1000 mother-child pairs in the first group, with correlation 0.35, and 100 children in the second group, with correlation 0.06. How can this question be answered? ∎

We wish to test the hypothesis $H_0: \rho_1 = \rho_2$ versus $H_1: \rho_1 \neq \rho_2$. It is reasonable to base the test on the difference between the z's in the two samples. If this difference is large or small, then H_0 will be rejected; otherwise, H_0 will be accepted. This principle suggests the following test procedure for a two-sided level α test:

11.35 **Fisher's z Test for Comparing Two Correlation Coefficients**

To test the hypothesis $H_0: \rho_1 = \rho_2$ versus $H_1: \rho_1 \neq \rho_2$, use the following procedure:

(1) Compute the sample correlation coefficients (r_1, r_2) and Fisher's z transformation (z_1, z_2) for each of the two samples.

(2) Compute the test statistic

$$\lambda = \frac{z_1 - z_2}{\sqrt{\dfrac{1}{n_1 - 3} + \dfrac{1}{n_2 - 3}}} \sim N(0, 1) \text{ under } H_0$$

(3) If

$$\lambda > z_{1 - \alpha/2} \quad \text{or} \quad \lambda < -z_{1 - \alpha/2}$$

reject H_0. If

$$-z_{1 - \alpha/2} \leqslant \lambda \leqslant z_{1 - \alpha/2}$$

accept H_0.

(4) The exact p-value is given by

$$p = 2\Phi(\lambda) \qquad \text{if } \lambda \leqslant 0$$
$$p = 2 \times [1 - \Phi(\lambda)] \qquad \text{if } \lambda > 0$$

(5) Assume an underlying normal distribution for each of the random variables used to compute r_1, r_2 and z_1, z_2.

The acceptance and rejection regions for this test are depicted in Figure 11.31. The computation of the exact p-value is illustrated in Figure 11.32.

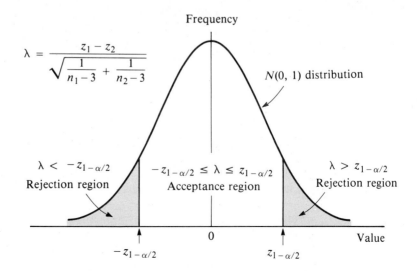

FIGURE 11.31
Acceptance and
rejection regions for
Fisher's z test for
comparing two
correlation coefficients

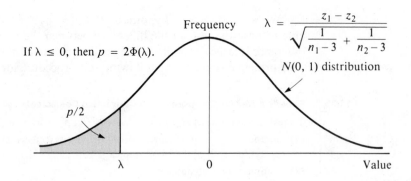

If $\lambda \leq 0$, then $p = 2\Phi(\lambda)$.

$$\lambda = \frac{z_1 - z_2}{\sqrt{\dfrac{1}{n_1 - 3} + \dfrac{1}{n_2 - 3}}}$$

$N(0, 1)$ distribution

$p/2$

λ 0   Value

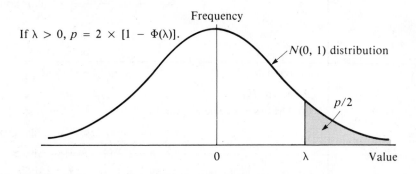

If $\lambda > 0$, $p = 2 \times [1 - \Phi(\lambda)]$.

$N(0, 1)$ distribution

$p/2$

0   λ Value

FIGURE 11.32
Computation of the exact p-value for Fisher's z test for comparing two correlation coefficients

EXAMPLE 11.54   Perform a significance test for the data in Example 11.53.

SOLUTION
$$r_1 = .35 \quad n_1 = 1000 \quad r_2 = .06 \quad n_2 = 100$$

Thus, from Table 11 in Appendix 1,

$$z_1 = 0.365 \quad z_2 = 0.060$$

and
$$\lambda = \frac{(0.365 - 0.060)}{\sqrt{\dfrac{1}{997} + \dfrac{1}{97}}} = 9.402(0.305) = 2.87 \sim N(0, 1) \text{ under } H_0$$

Hence the p-value is given by

$$2 \times [1 - \Phi(2.87)] = .004$$

Therefore, there is a significant difference between the mother-child correlations in the two groups, implying a significant genetic effect on blood pressure. ■

SECTION 11.14   **Rank Correlation**

Sometimes we may want to look at the relationship between two variables, but one or both of the variables are either ordinal or have a distribution that is far from normal. The significance tests in Section 11.13 then will no longer be valid, and nonparametric analogues to these tests are needed.

EXAMPLE 11.55   **Obstetrics**   The Apgar score was developed in 1952 as a measure of the physical condition of an infant at 1 and 5 minutes after birth [9]. The score is obtained by summing five components, each of which is rated as 0, 1, or 2 and represents different aspects of the condition of an infant at birth [10]. The method of scoring is indicated in Table 11.13. The score is routinely calculated for most newborn infants in U.S. hospitals. Suppose we are given the data in Table 11.14. We wish to relate the Apgar scores at 1 and 5 minutes and assess the significance of this relationship. How should this be done?   ∎

TABLE 11.13
Method of Apgar
scoring

	Score		
Sign	**0**	**1**	**2**
Heart rate	Absent	Slow (<100)	$\geqslant 100$
Respiratory effort	Absent	Weak cry; hypoventilation	Good; strong cry
Muscle tone	Limp	Some flexion of extremities	Well flexed
Reflex irritability	No response	Some motion	Cry
Color	Blue; pale	Body pink; extremities blue	Completely pink

(Reprinted with permission of *JAMA*, *168*(15), 1985–88, 1958.)

TABLE 11.14
Apgar scores at 1 and
5 minutes for 24
newborns

Infant	Apgar score, 1 min	Rank	Apgar score, 5 min	Rank	Rank difference (d_i)
1	10	23.5	10	19.5	$+4.0$
2	3	1.5	6	2.0	-0.5
3	8	10.0	9	8.5	$+1.5$
4	9	18.5	10	19.5	-1.0
5	8	10.0	9	8.5	$+1.5$
6	9	18.5	10	19.5	-1.0
7	8	10.0	9	8.5	$+1.5$
8	8	10.0	9	8.5	$+1.5$
9	8	10.0	9	8.5	$+1.5$
10	8	10.0	9	8.5	$+1.5$
11	7	4.5	9	8.5	-4.0
12	8	10.0	9	8.5	$+1.5$
13	6	3.0	9	8.5	-5.5
14	8	10.0	10	19.5	-9.5
15	9	18.5	10	19.5	-1.0
16	9	18.5	10	19.5	-1.0
17	9	18.5	10	19.5	-1.0
18	9	18.5	9	8.5	$+10.0$
19	8	10.0	10	19.5	-9.5
20	9	18.5	9	8.5	$+10.0$
21	3	1.5	3	1.0	$+0.5$
22	9	18.5	9	8.5	$+10.0$
23	7	4.5	10	19.5	-15.0
24	10	23.5	10	19.5	$+4.0$

The ordinary correlation coefficient developed in Section 11.12 should not be used, since the significance of this measure can be assessed only if the distribution of each Apgar score is assumed to be normally distributed. Instead, a nonparametric analogue to the correlation coefficient based on ranks is used.

DEFINITION 11.21 ■■■

The **Spearman rank correlation coefficient** (r_s) is an ordinary correlation coefficient based on ranks.

Thus

$$r_s = \frac{L_{xy}}{\sqrt{L_{xx} \times L_{yy}}}$$

where the L's are computed from the ranks rather than from the actual scores. ■

However, a simpler method for the computation of r_s is given as follows:

11.36 **Method for Computing the Spearman Rank Correlation Coefficient**

If there are no ties,

$$r_s = 1 - \frac{6 \times \Sigma d_i^2}{n^3 - n}$$

If there are ties,

$$r_s = \frac{\dfrac{(n^3 - n)}{6} - \displaystyle\sum_{i=1}^{g_x} \frac{(t_i^3 - t_i)}{12} - \sum_{j=1}^{g_y} \frac{(t_j^3 - t_j)}{12} - \Sigma d_i^2}{2\sqrt{\left[\dfrac{(n^3 - n)}{12} - \displaystyle\sum_{i=1}^{g_x} \frac{(t_i^3 - t_i)}{12}\right]\left[\dfrac{(n^3 - n)}{12} - \displaystyle\sum_{j=1}^{g_y} \frac{(t_j^3 - t_j)}{12}\right]}}$$

where

t_i refers to the ith group of tied observations for the first variable (of which there are g_x tied groups in total)

t_j refers to the jth group of tied observations for the second variable (of which there are g_y tied groups in total)

d_i equals the rank for the ith observation for the first variable minus the rank for the ith observation for the second variable

The rationale for this estimator is that if there were a perfect correlation between the two variables, then the ranks for each person on each variable would be the same. Thus, d_i would equal 0 for each person, and in the absence of ties $r_s = 1$. The less perfect the correlation, the larger Σd_i^2 would be and the smaller r_s would be.

EXAMPLE 11.56 **Obstetrics** Compute the Spearman rank correlation coefficient for the Apgar score data in Table 11.14.

SOLUTION Rank the collection of 24 Apgar 1-minute scores and 24 Apgar 5-minute scores in the usual way, assigning the average rank to tied values, as indicated in Table 11.15. The ranks and rank differences have been entered in Table 11.14. Now compute Σd_i^2 as follows:

$$\Sigma d_i^2 = (4.0)^2 + (-0.5)^2 + \cdots + (4.0)^2 = 805.0$$

TABLE 11.15	Apgar score, 1 min	Frequency	Range of ranks	Average rank	Apgar score, 5 min	Frequency	Range of ranks	Average rank
Computation of average ranks for Apgar scores in Table 11.14	3	2	1–2	1.5	3	1	1	1.0
	6	1	3	3.0	6	1	2	2.0
	7	2	4–5	4.5	9	12	3–14	8.5
	8	9	6–14	10.0	10	$\overline{10}$	15–24	19.5
	9	8	15–22	18.5		24		
	10	$\overline{2}$	23–24	23.5				
		24						

Then compute the correction terms for ties for the two scores. For Apgar 1-minute scores,

$$\frac{\Sigma (t_i^3 - t_i)}{12} = \frac{(2^3 - 2) + (2^3 - 2) + (9^3 - 9) + (8^3 - 8) + (2^3 - 2)}{12} = \frac{1242}{12} = 103.5$$

For Apgar 5-minute scores,

$$\frac{\Sigma (t_j^3 - t_j)}{12} = \frac{(12^3 - 12) + (10^3 - 10)}{12} = \frac{2706}{12} = 225.5$$

Thus, from (11.36),

$$r_s = \frac{\dfrac{(24^3 - 24)}{6} - 103.5 - 225.5 - 805.0}{2\sqrt{\left[\dfrac{(24^3 - 24)}{12} - 103.5\right] \times \left[\dfrac{(24^3 - 24)}{12} - 225.5\right]}}$$

$$= \frac{1166.0}{2\sqrt{(1046.5)(924.5)}} = \frac{1166.0}{1967.2} = .593 \qquad \blacksquare$$

We would now like to test the rank correlation for statistical significance. A similar test to that given in (11.32) can be performed, as follows:

11.37 | **t Test for Spearman Rank Correlation**

(1) Compute the test statistic

$$t_s = \frac{r_s \sqrt{n - 2}}{\sqrt{1 - r_s^2}}$$

which under the null hypothesis of no correlation follows a t distribution with $n - 2$ degrees of freedom.

(2) For a two-sided level α test, if

$$t_s > t_{n-2, 1-\alpha/2} \qquad \text{or} \qquad t_s < t_{n-2, \alpha/2}$$

then reject H_0; otherwise, accept H_0.

(3) The exact p-value is given by

$$p = 2 \times (\text{area to the left of } t_s \text{ under a } t_{n-2} \text{ distribution}) \qquad \text{if } t_s < 0$$
$$p = 2 \times (\text{area to the right of } t_s \text{ under a } t_{n-2} \text{ distribution}) \qquad \text{if } t_s \geq 0$$

(4) This test is valid only if $n \geq 10$.

The acceptance and rejection regions for this test are given in Figure 11.33. The computation of the exact p-value is illustrated in Figure 11.34.

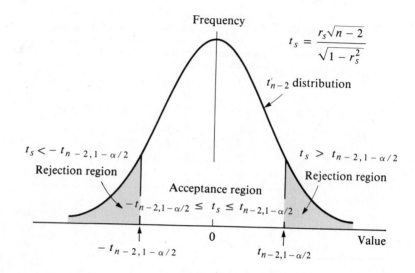

FIGURE 11.33
Acceptance and
rejection regions for the
t test for a Spearman
rank correlation
coefficient

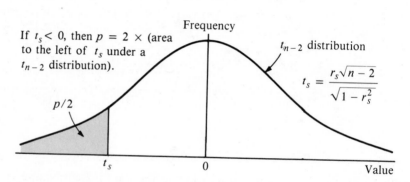

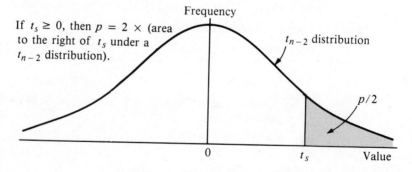

FIGURE 11.34
Computation of the
exact p-value for the t
test for a Spearman
rank correlation
coefficient

EXAMPLE 11.57   **Obstetrics** Perform a significance test for the Spearman rank correlation coefficient based on the Apgar score data in Table 11.14.

SOLUTION   Note that $r_s = .593$ from Example 11.56. The test statistic is given by

$$t_s = \frac{r_s\sqrt{n-2}}{\sqrt{1-r_s^2}} = \frac{.593\sqrt{22}}{\sqrt{1-.593^2}} = \frac{2.781}{.805} = 3.45$$

which follows a t_{22} distribution under H_0. Note that

$$t_{22,.995} = 2.819 \qquad t_{22,.9995} = 3.792$$

Thus, the two-tailed p-value is given by

$$2 \times (1 - .9995) < p < 2 \times (1 - .995) \qquad \text{or} \qquad .001 < p < .01$$

Thus, there is a significant rank correlation between the two scores. ∎

Note that the test procedure given in **(11.37)** is valid only for $n \geqslant 10$. If $n < 10$, then the t distribution is not a good approximation to the distribution of t_s, and a table giving exact significance levels must be used. For this purpose, exact two-sided critical values for r_s when $n \leqslant 9$ are presented in Table 12 in the Appendix. This table can be used in the following way:

(1) Suppose the critical value in the table for significance level α is c.

(2) Reject H_0 using a two-sided test with significance level α if $r_s \geqslant c$ or $r_s \leqslant -c$ and accept H_0 otherwise.

EXAMPLE 11.58   Suppose that $r_s = .750$ based on a sample of size 9. Assess the statistical significance of the results.

SOLUTION   From Table 12 in Appendix 1, the critical value for $\alpha = .05$, $n = 9$ is .683 and for $\alpha = .02$, $n = 9$ is .783. Since $.683 < .750 < .783$, it follows that the two-tailed p-value is given by $.02 \leqslant p < .05$. Similarly, the one-tailed p-value is given by $.01 \leqslant p < .025$. ∎

SECTION 11.15   **The Kappa Statistic**

In the previous sections the notion of correlation for continuous data and of rank correlation for ordinal data was discussed. A measure of reproducibility for categorical data is also needed.

EXAMPLE 11.59   **Nutrition** A diet questionnaire was administered by mail to 537 female American nurses on two separate occasions several months apart. The questions asked included the quantities eaten of over 100 separate food items. The data obtained from the two surveys for beef consumption are presented in Table 11.16. Notice that the responses on the two surveys are the same only for $136 + 240 = 376$ out of 537 (70.0%) women. How can the reproducibility of response for these data be quantified? ∎

A chi-square test for association between the survey-1 and survey-2 responses could be performed. However, this test would not give a quantitative measure of reproducibility between the responses at the two surveys. Instead, we will focus on the percentage of women with concordant responses in the two surveys. We noted

TABLE 11.16
Beef consumption reported by 537 female American nurses at two different surveys

	Survey 2		
Survey 1	≤1 serving/week	>1 serving/week	Total
≤1 serving/week	136	92	228
>1 serving/week	69	240	309
Total	205	332	537

in Example 11.59 that 70.0% of the women gave concordant responses. We would like to compare the observed concordance rate (p_o) with the expected concordance rate (p_e) if the responses of the women in the two surveys were statistically independent. The motivation behind this definition is that the questionnaire would be virtually worthless if the frequency of consumption reported at one survey had no relationship to the frequency of consumption reported at a second survey. Suppose there are c response categories and the probability of response in the ith category is a_i for the first survey and b_i for the second survey. These probabilities can be estimated from the row and column margins of the contingency table. The expected concordance rate (p_e) if the survey responses are independent is given by $\Sigma\,(a_i b_i)$.

EXAMPLE 11.60   **Nutrition** Compute the expected concordance rate using the beef-consumption data in Table 11.16.

SOLUTION   From Table 11.16,

$$a_1 = \frac{228}{537} = .425$$

$$a_2 = \frac{309}{537} = .575$$

$$b_1 = \frac{205}{537} = .382$$

$$b_2 = \frac{332}{537} = .618$$

Thus, $$p_e = (.425 \times .382) + (.575 \times .618) = .518$$

Therefore, 51.8% concordance would be expected if the subjects were responding independently at the two surveys. ∎

We could use $p_o - p_e$ as the measure of reproducibility. However, it is preferable to use a measure that equals $+1.0$ in the case of perfect agreement and 0.0 if the responses on the two surveys are completely independent. Indeed, the maximum possible value for $p_o - p_e$ is $1 - p_e$, which is achieved when $p_o = 1$. Therefore, the Kappa statistic, which is defined as $(p_o - p_e)/(1 - p_e)$, is used as the measure of reproducibility:

11.38   **The Kappa Statistic**

(1) If a categorical variable is measured at two surveys, then the Kappa statistic (κ) is used to measure reproducibility between surveys, where

$$\kappa = \frac{p_o - p_e}{1 - p_e}$$

and p_o = observed probability of concordance between the two surveys

p_e = expected probability of concordance between the two surveys

$= \Sigma\,(a_i b_i)$

where a_i, b_i are the marginal probabilities for the ith category in the $c \times c$ contingency table relating response at the two surveys.

(2) Furthermore,

$$se(\kappa) = \sqrt{\frac{1}{N(1 - p_e)^2} \times \left\{ p_e + p_e^2 - \sum_{i=1}^{c} [a_i b_i (a_i + b_i)] \right\}}$$

To test the one-sided hypothesis $H_0: \kappa = 0$ versus $H_1: \kappa > 0$, use the test statistic

$$z = \frac{\kappa}{se(\kappa)}$$

which follows an $N(0, 1)$ distribution under H_0.

(3) Reject H_0 at level α if $z > z_{1-\alpha}$ and accept H_0 otherwise.

(4) The exact p-value is given by $p = 1 - \Phi(z)$.

The acceptance and rejection regions for this test are depicted in Figure 11.35. The computation of the p-value is shown in Figure 11.36.

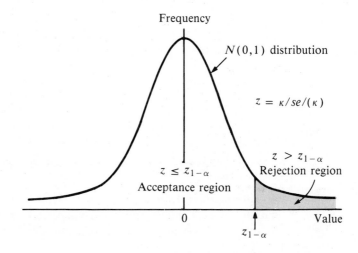

FIGURE 11.35
Acceptance and rejection regions for the significance test for Kappa

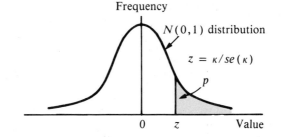

FIGURE 11.36
Computation of the exact p-value for the significance test for Kappa

Note that we are customarily interested in one-tailed tests in **(11.38)**, since negative values for Kappa usually have no biological significance.

EXAMPLE 11.61 **Nutrition** Compute the Kappa statistic and assess its statistical significance using the beef-consumption data in Table 11.16.

SOLUTION   From Examples 11.59 and 11.60,

$$p_o = .700$$

$$p_e = .518$$

Therefore, the Kappa statistic is given by

$$\kappa = \frac{.700 - .518}{1 - .518} = \frac{.182}{.482} = .378$$

Furthermore, from **(11.38)** and the results of Example 11.60, the standard error of κ is given by

$$se(\kappa) = \sqrt{\frac{1}{537(1 - .518)^2} \times \{.518 + .518^2 - \sum [a_i b_i(a_i + b_i)]\}}$$

where

$$\sum [a_i b_i(a_i + b_i)] = .425 \times .382 \times (.425 + .382) + .575 \times .618 \times (.575 + .618)$$

$$= .555$$

Thus, $$se(\kappa) = \sqrt{\frac{1}{537 \times .232} \times (.518 + .268 - .555)} = \sqrt{\frac{1}{124.6} \times .231} = .0431$$

The test statistic is given by

$$z = \frac{.378}{.0431} = 8.8 \sim N(0, 1) \text{ under } H_0$$

The p-value is $$p = 1 - \Phi(8.8) < .001$$

Thus, the Kappa statistic indicates highly significant reproducibility between the first and second surveys.   ■

Although the Kappa statistic was significant in Example 11.61, it still shows that the reproducibility was far from perfect. Indeed, Landis and Koch (1977) provide the following guidelines for the evaluation of Kappa [11]:

11.39 | **Guidelines for the Evaluation of Kappa**

$\kappa > .75$ denotes *excellent* reproducibility.

$.4 \leqslant \kappa \leqslant .75$ denotes *good* reproducibility.

$0 \leqslant \kappa < .4$ denotes *marginal* reproducibility.

In general, reproducibility is not good for many items on dietary surveys, indicating the need for multiple dietary assessments to reduce variability. See Fleiss (1981) for further information about the Kappa statistic, including assessments of reproducibility for more than two surveys [12].

SECTION 11.16   **Summary**

In this chapter methods of statistical inference that are appropriate for investigating the relationship between two or more variables were studied. If only two variables, both of which are continuous, are being studied, and we wish to predict

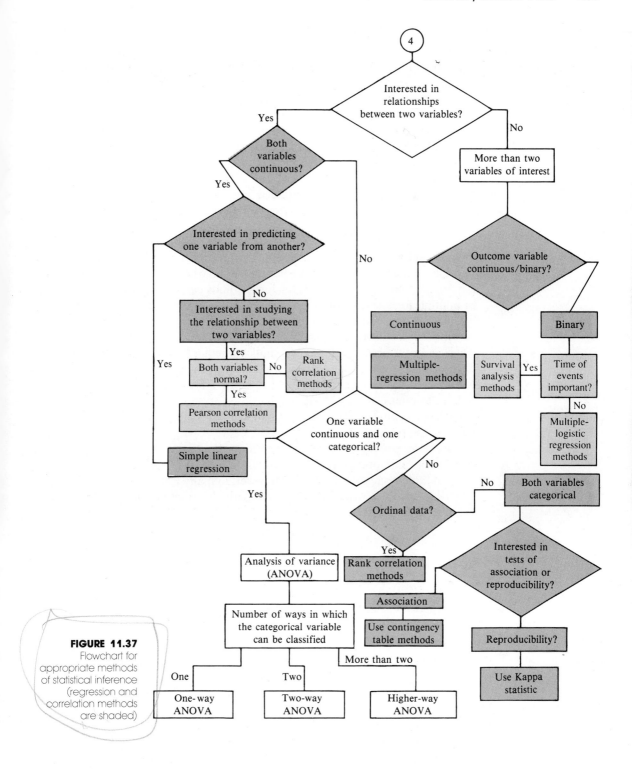

FIGURE 11.37
Flowchart for appropriate methods of statistical inference (regression and correlation methods are shaded)

one variable (the dependent variable) as a function of the other variable (the independent variable), then simple linear regression analysis is used. If we simply wish to look at the association between the two variables without distinguishing between dependent and independent variables, then Pearson correlation methods are more appropriate. If the scale of measurement is ordinal rather than cardinal or the distributions are very far from being normal, then rank correlation methods are more appropriate than Pearson correlation methods for quantifying association. If both variables of interest are categorical and we are interested in the association between the two variables, then the contingency table methods of Chapter 10 can be used. If, on the other hand, we are almost certain that there will be some association between the two variables and we wish to quantify the degree of association, then the Kappa statistic can be used.

In many instances we are interested in more than two variables and we wish to predict the value of one variable (the dependent variable) as a function of several independent variables. If the dependent variable is normally distributed, then multiple regression methods can be used; similarly, if the dependent variable is binary, then multiple logistic regression methods can be used. Multiple regression or multiple logistic regression methods can be very powerful, since the independent variables can be either continuous or categorical, or a combination of both.

In some instances, the response variable is binary, but we are also interested in when an event occurs, rather than simply whether or not it occurs. Under these circumstances, methods of survival analysis (e.g., the proportional hazards model) can be used to relate survival time to one or more risk factors in an analogous fashion to multiple-logistic regression.

The preceding methods are summarized in the flowchart in Figure 11.37 and again in the back of the book.

In many situations we have a continuous outcome variable that we wish to relate to one or more categorical variables. In general, this situation can be handled with multiple-regression methods. However, in many instances, the formulation is easier if analysis of variance (ANOVA) methods are used, which are discussed in detail in Chapter 12.

PROBLEMS

Cardiology

The data in Table 11.17 are given for 27 patients with acute dilated cardiomyopathy [13].

11.1 Fit a regression line relating age (x) to LVEF (y).

11.2 What is the expected LVEF for a 45-year-old patient with this condition?

11.3 Test for the significance of the regression line in Problem 11.1 using the F test.

11.4 What is the R^2 for the regression line in Problem 11.1?

11.5 What does R^2 mean in Problem 11.4?

11.6 Test for the significance of the regression line in Problem 11.1 using the t test.

11.7 What are the standard errors of the slope and intercept for the regression line in Problem 11.1?

11.8 Provide 95% confidence limits for the slope and intercept.

Hematology

The data in Table 11.18 are given for 9 patients with aplastic anemia [14].

11.9 Fit a regression line relating the percentage of reticulytes (x) to the number of lymphocytes (y).

TABLE 11.17 Data for patients with acute dilated cardiomyopathy

Patient number	Age, x	Left ventricular ejection fraction (LVEF), y
1	35	0.19
2	28	0.24
3	25	0.17
4	75	0.40
5	42	0.40
6	19	0.23
7	54	0.20
8	35	0.20
9	30	0.30
10	65	0.19
11	26	0.24
12	56	0.32
13	60	0.32
14	47	0.28
15	50	0.24
16	43	0.18
17	30	0.22
18	56	0.23
19	23	0.14
20	26	0.14
21	58	0.30
22	65	0.07
23	34	0.12
24	63	0.13
25	23	0.17
26	23	0.24
27	46	0.19

$\Sigma x_i = 1137$ $\Sigma x_i^2 = 54{,}749$ $\Sigma y_i = 6.05$
$\Sigma y_i^2 = 1.522$ $\Sigma x_i y_i = 262.93$
(Reprinted with permission of the *New England Journal of Medicine*, *312*(14), 885–890, 1985.)

11.10 Test for the statistical significance of this regression line using the F test.

11.11 What is R^2 for the regression line in Problem 11.9?

11.12 What does R^2 mean in Problem 11.11?

11.13 What is $s_{y \cdot x}^2$?

11.14 Test for the statistical significance of the regression line using the t test.

11.15 What are the standard errors of the slope and intercept for the regression line in Problem 11.9?

TABLE 11.18 Hematologic data for patients with aplastic anemia

Patient number	% reticulytes	Lymphocytes (per mm³)
1	3.6	1700
2	2.0	3078
3	0.3	1820
4	0.3	2706
5	0.2	2086
6	3.0	2299
7	0.0	676
8	1.0	2088
9	2.2	2013

(Reprinted with permission of the *New England Journal of Medicine*, *312*(16), 1015–1022, 1985.)

Endocrinology, Bone and Joint Disease
A study was conducted to relate serum estrogens to bone metabolism.

11.16 Suppose that a correlation of .52 is found between serum estradiol and the bone density of the lumbar region of the spine, as computed by CT scan among 23 postmenopausal women. Test for the statistical significance of these results and report a *p*-value.

11.17 Similarly, a correlation of .34 is found between serum estradiol and the bone density of the distal radius of the nondominant arm among the same 23 postmenopausal women. Test for the statistical significance of these results and report a *p*-value.

11.18 Finally, a correlation coefficient of .395 was computed between the two preceding measures of bone metabolism in Problems 11.16 and 11.17 based on the same group of women. Test for the statistical significance of these results and report a *p*-value.

11.19 Suppose in a previous large study, other investigators found a correlation of .5 between the two measures of bone metabolism in Problems 11.16 and 11.17. Test for whether or not the sample results in Problem 11.18 are compatible with these previous results. Report a *p*-value.

11.20 What is the z transformation of .34?

11.21 What is the z transformation of .435?

Pulmonary Function
Suppose the correlation coefficient between FEV for 100 sets of identical twins is .7, whereas the comparable correlation for 120 sets of fraternal twins is .38.

11.22 What test procedure can be used to compare the two correlation coefficients?

11.23 Perform the procedure in Problem 11.22 using the critical value method.

11.24 What is the *p*-value of the test?

Suppose the correlation coefficient between weight is .78 for the 100 sets of identical twins and .50 for the 120 sets of fraternal twins.

11.25 Test for whether or not the true correlation coefficients are different between these groups. Report a *p*-value.

Psychiatry

Suppose as a reliability check the same psychiatric questionnaire is administered by two different observers to each of 30 psychiatric outpatients. An anxiety score is computed from each questionnaire. Since the investigators do not wish to assume normality for this score, the Spearman rank correlation coefficient is to be used as the index of reliability.

11.26 If the estimated rank correlation coefficient is .4, then test whether or not the rank correlation is significantly *greater* than 0 using the critical value method.

11.27 What is the *p*-value of the test?

Suppose the same procedure is employed for 8 psychiatric inpatients.

11.28 If the estimated rank correlation is .7 for this group, then test whether or not the rank correlation is significantly *greater* than 0 for this subgroup. Report a *p*-value.

Pathology

Two pathologists independently review biopsy specimens from 40 people with self-reported malignant melanoma as ascertained by mail questionnaire. The results are given in Table 11.19.

TABLE 11.19 Pathology review for patients with self-reported malignant melanoma

Pathologist 1 diagnosis	Pathologist 2 diagnosis		
	+	−	Total
+	24	8	32
−	4	4	8
Total	28	12	40

11.29 What index can be used to measure the reproducibility of the diagnoses made by the two pathologists?

11.30 Compute the index in Problem 11.29.

11.31 Test for whether or not the true index of reproducibility is significantly greater than 0 and report a *p*-value.

Hypertension

Many hypertension studies involve more than one observer, and it is important to verify that the different observers are in fact comparable. A new observer and an experienced observer simultaneously take bp readings on 50 people (x_i, y_i) $i = 1, \ldots, 50$, and the sample correlation coefficient is .75. Suppose the minimum acceptable population correlation is .9.

11.32 Are the sample data compatible with this population correlation?

Hospital Epidemiology

Refer to the data on hospital stays given in Table 2.11 (p. 36).

11.33 Find the best-fitting linear relationship between duration of hospitalization and age.

11.34 Test for the significance of this relationship. State any underlying assumptions you have used.

11.35 What is R^2 for this regression?

Hypertension

A frequently encountered phenomenon in Western society is the positive correlation of blood pressure and age. One finding of recent interest is the apparent absence of this correlation in many underdeveloped countries. Suppose that in a sample of 903 American males, the observed correlation is .402, whereas in a sample of 444 Polynesian males, the observed correlation is .053.

11.36 Test for whether or not a significant difference exists between the two underlying correlations.

Nutrition

In Table 10.3 (p. 324), data were presented relating reported cholesterol intake for two different food frequency questionnaires. A test of association was performed on the data in Example 10.15 (p. 331). Suppose we wish to quantify the degree of reproducibility of reported cholesterol intake.

11.37 What statistical index can be used to measure reproducibility in this case?

11.38 Compute the index mentioned in Problem 11.37 and perform an appropriate significance test concerning this index.

11.39 What information does this index provide that was not provided in the test for association in Example 10.15?

Pathology

11.40 Assuming normality, use the data in Table 2.10 (p. 35), to test if there is any relation between total heart weight and body weight within each of the two groups.

11.41 Suppose normality is not assumed. Perform a nonparametric procedure to address the question posed in Problem 11.40.

11.42 Which method of analysis do you feel is appropriate here? Plot the data to answer this question.

Hypertension

Suppose that 20 married couples, each in the age group 25–34, have their systolic blood pressures taken, with the data listed in Table 11.20.

TABLE 11.20 Systolic bp measurements from 20 married couples

	Male	Female		Male	Female
1	136	110	11	156	135
2	121	112	12	98	115
3	128	128	13	132	125
4	100	106	14	142	130
5	110	127	15	138	132
6	116	100	16	126	146
7	127	98	17	124	127
8	150	142	18	137	128
9	180	143	19	160	135
10	172	150	20	125	110

11.43 Test the hypothesis that there is no correlation between the male and female scores.

11.44 Suppose a study in the literature has found $\rho = .1$ based on a sample of size 100. Are the data consistent with this finding?

Environmental Health

Suppose we are interested in the relation between carbon monoxide concentrations and the density of cars in some geographical area. The number of cars per hour to the nearest 500 cars per hour and the concentration of carbon monoxide (CO) in parts per million at a particular street corner are measured and the data are grouped by cars per hour. The data are given in Table 11.21.

TABLE 11.21 CO concentration and car density at a particular street corner

Cars/hour ($\times 10^3$)	CO concentrations				Number of samples
1.0	9.0	6.8	7.7		3
1.5	9.6	6.8	11.3		3
2.0	12.3	11.8			2
3.0	20.7	19.2	21.6	20.6	4

11.45 Is there an association between cars per hour and CO concentration?

11.46 What is the expected CO concentration if 2500 cars per hour are on the road?

11.47 What is the standard error for the expected CO concentration over a large number of days when 2500 cars per hour are on the road?

Hypertension

It has been observed that blood-pressure levels tend to be similar when measured in the same person over time. However, because of physiologic variations and measurement error, this similarity is far from being exact. Furthermore, a hypothesis that older people "track" better than younger people in the sense that their blood-pressure levels at two points in time tend to be more similar than those for younger people has been proposed. Suppose systolic blood pressure is measured at two points in time, 4 years apart, on a group of 300 50–54-year-old males and a tracking correlation (i.e., a correlation between successive readings) of .6 is found; whereas for a group of 200 30–34-year-old males, the estimated tracking correlation is .4.

11.48 Is there a significant difference between the two underlying tracking correlations for these age groups?

Obstetrics

The data in Table 11.22 give the infant mortality rates per 1000 live births in the United States for the period 1960–1979 [15].

TABLE 11.22 U.S. infant mortality rates per 1000 live births, 1960–1979

x	y	x	y
1960	26.0	1974	16.7
1965	24.7	1975	16.1
1970	20.0	1976	15.2
1971	19.1	1977	14.1
1972	18.5	1978	13.8
1973	17.7	1979	13.0

Suppose the following information is given:

$$\sum_{i=1}^{12} x_i = 23{,}670$$

$$\sum_{i=1}^{12} x_i^2 = 46{,}689{,}410$$

$$\sum_{i=1}^{12} y_i = 214.9$$

$$\sum_{i=1}^{12} y_i^2 = 4033.83$$

$$\sum_{i=1}^{12} x_i y_i = 423{,}643.3$$

11.49 Fit a linear regression line relating infant mortality rate to chronological year using these data.

11.50 Test for the significance of the linear relationship developed in Problem 11.49.

11.51 If the present trends continue for the next 10 years, then what would be the expected infant mortality rate in 1989?

11.52 Provide a standard error for the estimate in Problem 11.51.

11.53 Can the linear relationship developed in Problem 11.49 be expected to continue indefinitely? Why or why not?

Hypertension

The level of a catecholamine called kallikrein in the urine is a variable that has been associated in some studies with level of blood pressure in adults. Generally, people with low levels of kallikrein tend to have high levels of blood pressure. A study was undertaken in a group of 18 infants to see if this relationship persisted. The data in Table 11.23 were obtained.

TABLE 11.23 The relationship of kallikrein to blood pressure

Observation	$\log_e$* (kallikrein), x	Systolic blood pressure z score,† y
1	2.773	1.929
2	5.545	−1.372
3	3.434	−0.620
4	3.434	1.738
5	2.639	0.302
6	3.091	0.679
7	4.836	0.999
8	3.611	0.656
9	4.554	0.027
10	3.807	−0.057
11	4.500	1.083
12	2.639	−2.265
13	3.555	0.963
14	3.258	−1.062
15	4.605	2.771
16	3.296	−0.160
17	4.787	−0.217
18	3.401	−1.290

* The log transformation was used to better normalize the underlying distribution.
† Instead of raw blood pressure a z score was used, which is defined as the raw blood pressure standardized for body weight and expressed in standard deviation units, with positive scores indicating high blood pressure and negative scores indicating low blood pressure.

Suppose we are given the following statistics:

$$\sum_{i=1}^{18} x_i = 67.765 \qquad \sum_{i=1}^{18} x_i^2 = 267.217 \qquad \sum_{i=1}^{18} y_i = 4.104$$

$$\sum_{i=1}^{18} y_i^2 = 28.767 \qquad \sum_{i=1}^{18} x_i y_i = 17.249$$

11.54 Assuming that a linear relationship exists between systolic blood pressure z score and $\log_e$ kallikrein, derive the best-fitting linear relationship between these two variables.

11.55 Is there a significant relationship between these two variables based on your answer to Problem 11.54? Report a p-value.

11.56 In words, what does a p-value mean in the context of Problem 11.55?

Nutrition

The assessment of the relationship between dietary intake and disease is one of the more prominent areas of current medical research. One of the problems is the difficulty in accurately assessing a person's diet, since the reported dietary intake varies over different surveys because of both (a) faulty memory of the actual diet and (b) true changes in diet over time. To assess reproducibility, a food frequency questionnaire with over 100 food items was administered to 537 American nurses at two different points in time 6 months apart. A number of food nutrients, such as total protein and fat, were calculated on the basis of the individual items, and correlations were calculated over the responses to the two questionnaires. Suppose the correlation coefficient for total protein intake as assessed at two points in time is .362. We wish to test for a significant relationship between reported total protein intake at two points in time.

11.57 Is a one-sided or two-sided test appropriate here?

11.58 Perform a significance test for this relationship and report a p-value.

An alternative method for assessing dietary intake is the 7-day diet record, in which a person writes down each food item eaten over a 1-week period, and the total nutrient intake is computed from these data. Suppose that two 7-day diet records 6 months apart are completed for a group of 50 nurses (different from the first group of 537 nurses), and it is found that the correlation between total protein intake is .45 using this method. We wish to test if the two methods are equally reproducible.

11.59 Is a one-sided or two-sided test needed here?

11.60 Perform a significance test for this hypothesis and report a p-value.

Hypertension

The variability of blood pressure is important in planning screening programs for detecting people with high blood pressure. There are two schools of thought concerning this variability: Some researchers feel that a subgroup of people have extremely variable blood pressures, whereas most people have relatively stable blood pressures; other researchers feel that this variability is common to all people. A study was set up to answer this question. A group of 15 people had their blood pressures measured on three separate visits in each of 2 years, and the between-visit variance of blood

pressure was measured each year. The results for systolic blood pressure are given in Table 11.24.

TABLE 11.24 Blood-pressure variability measured at two points in time

Person	Between-visit variance, year 1	Between-visit variance, year 2
1	8.8	19.0
2	10.0	25.7
3	6.7	17.9
4	13.3	95.7
5	10.0	16.7
6	6.2	15.2
7	35.0	3.9
8	51.2	11.9
9	30.9	14.6
10	61.0	21.6
11	5.4	31.3
12	46.6	21.4
13	37.0	10.1
14	2.6	18.7
15	2.0	22.7

We wish to assess if there is any relationship between the year-1 and year-2 variances.

11.61 Why might a rank correlation be a useful method for expressing such a relationship?

11.62 Test for the significance of the rank correlation based on the previous data.

11.63 What are your conclusions based on these data?

Pediatrics

It is often mentioned anecdotally that very young children have a higher metabolism that gives them more energy than older children and adults. We decide to test this hypothesis by measuring the pulse rates on a selected group of children. The children are randomly chosen from a community census so that two male children are selected from each 2-year age group starting with age 0 and ending with age 21 (i.e., 0–1/2–3/4–5/.../20–21). The data are given in Table 11.25. We are given the following basic statistics:

$$n = 22 \qquad \sum_{i=1}^{22} y_i = 1725 \qquad \sum_{i=1}^{22} y_i^2 = 140{,}933$$

$$\sum_{i=1}^{22} x_i = 233 \qquad \sum_{i=1}^{22} x_i^2 = 3345 \qquad \sum_{i=1}^{22} x_i y_i = 16{,}748$$

TABLE 11.25 Pulse rate and age in children aged 0–21

Age group	Age, x_i	Pulse rate, y_i
0–1	1 / 0	103 / 125
2–3	3 / 3	102 / 86
4–5	5 / 5	88 / 78
6–7	6 / 6	77 / 68
8–9	9 / 8	90 / 75
10–11	11 / 11	78 / 66
12–13	12 / 13	76 / 82
14–15	14 / 14	58 / 56
16–17	16 / 17	72 / 70
18–19	19 / 18	56 / 64
20–21	21 / 21	81 / 74

11.64 Suppose we hypothesize that there is a linear regression model relating pulse rate and age. What are the assumptions for such a model?

11.65 Fit the parameters for the model in Problem 11.64.

11.66 Test the model fitted in Problem 11.65 for statistical significance.

11.67 What is the predicted pulse for an average 12-year-old child?

11.68 What is the standard error of the estimate in Problem 11.67?

11.69 Suppose John Smith is 12 years old. What is his estimated pulse rate from the regression line?

11.70 What is the standard error of the estimate in Problem 11.69?

11.71 How does John compare with other children in his age group if his actual pulse rate is 75?

11.72 What is the difference between the two predictions and standard errors in Problems 11.67, 11.68, 11.69, and 11.70?

11.73 Plot the data and assess whether the assumptions of the model are satisfied.

Hypertension

Adults show a strong relationship between blood pressure and age. Data were collected from schoolchildren in Muscatine, Iowa, to see if this relationship continued to hold in the age group 5–18 [16]. The data in Table 11.26 were obtained from boys in this age group.

TABLE 11.26 The relationship between blood pressure and age

Age, x	Mean systolic blood pressure (mm Hg), y
5	94.4
6	97.7
7	101.9
8	104.5
9	106.3
10	109.3
11	112.6
12	113.8
13	117.7
14	121.6
15	122.3
16	123.6
17	124.9
18	131.0

$\Sigma x_i = 161 \quad \Sigma x_i^2 = 2079$
$\Sigma y_i = 1581.6 \quad \Sigma y_i^2 = 180,271$
$\Sigma x_i y_i = 18,787.3$

11.74 Fit a linear model relating blood pressure to age using the method of least squares.

11.75 What are the standard errors of the regression parameters obtained in Problem 11.74?

11.76 Test for the significance of the regression line in Problem 11.74 and report a p-value.

11.77 What is the expected blood pressure for an average 13-year-old boy as predicted from the regression line?

11.78 What is the standard error of the estimate in Problem 11.77?

11.79 What is the expected value and standard error for the change in blood pressure over the next 5 years for an average 13-year-old boy (i.e., from 13–18 years of age)?

Hypertension, Genetics

Much research has been devoted to the etiology of hypertension. One general problem is to determine to what extent hypertension is a genetic phenomenon. This issue can be examined in 20 families by measuring the systolic bp of the mother, father, and first-born child in the family. The data are given in Table 11.27.

11.80 Test the assumption that the mother's bp and father's bp are uncorrelated. (We would expect a lack of correlation, since the mother and father are genetically unrelated.)

11.81 We would expect from genetic principles that the correlation between the mother's bp and the child's bp is .5. Can this expectation be tested?

11.82 Suppose a nonzero correlation has been found in Problem 11.81. Is there some explanation other than a genetic one for the existence of this correlation?

We would like to be able to predict the child's bp on the basis of the parents' bp.

11.83 Find the best-fitting linear relationship between the child's bp and the mother's bp.

11.84 Test for the significance of this relationship.

11.85 What would be the expected average child's bp if the mother's bp is 130 mm Hg?

11.86 What would be the expected average child's bp if the mother's bp is 150 mm Hg?

11.87 What would be the expected average child's bp if the mother's bp is 170 mm Hg?

11.88 Find the standard errors for the estimates in Problems 11.85, 11.86, and 11.87.

TABLE 11.27 Familial blood-pressure relationships

Family	Systolic bp mother (mm Hg), y	Systolic bp father (mm Hg), x	Systolic bp child (mm Hg), t	
1	130	140	90	$\sum_{i=1}^{20} x_i = 2980$
2	125	120	85	
3	140	180	120	
4	110	150	100	$\sum_{i=1}^{20} x_i^2 = 451{,}350$
5	145	175	105	
6	160	120	100	$\sum_{i=1}^{20} y_i = 2620$
7	120	145	110	
8	180	160	140	
9	120	190	115	$\sum_{i=1}^{20} y_i^2 = 351{,}350$
10	130	135	105	
11	125	150	100	$\sum_{i=1}^{20} x_i y_i = 390{,}825$
12	110	125	80	
13	90	140	70	
14	120	170	115	$\sum_{i=1}^{20} t_i = 2030$
15	150	150	90	
16	145	155	90	$\sum_{i=1}^{20} t_i^2 = 210{,}850$
17	130	160	115	
18	155	125	110	
19	110	140	90	$\sum_{i=1}^{20} x_i t_i = 305{,}700$
20	125	150	100	
				$\sum_{i=1}^{20} y_i t_i = 269{,}550$

11.89 Why are the three standard errors calculated in Problem 11.88 not the same?

The two-variable regression model with the father's bp and mother's bp as independent variables and the child's bp as the dependent variable is fitted in Table 11.28.

TABLE 11.28 Two-variable regression model predicting child's blood pressure as a function of the parental blood pressures

```
               SPSSX/PC  Release 1.0

* * * *   M U L T I P L E   R E G R E S S I O N   * * * *

Equation Number 1    Dependent Variable..  CBP   CHILD BP

Variable(s) Entered on Step Number

1..   FBP      FATHER BP
2..   MBP      MOTHER BP

Analysis of Variance

               DF     Sum of Squares    Mean Square
Regression      2        2870.08458      1435.04229
Residual       17        1934.91542       113.81855

* * * *   M U L T I P L E   R E G R E S S I O N   * * * *

Equation Number 1    Dependent Variable..  CBP   CHILD BP

Variable              B             SE B

FBP              0.41500         0.12482
MBP              0.42255         0.11852
(Constant)     -15.68925        23.65025
```

Suppose we wish to assess the independent effects of the mother's bp and father's bp.

11.90 What hypotheses should be tested for this assessment?

11.91 Test the hypotheses proposed in Problem 11.90.

11.92 What are the standardized regression coefficients in Table 11.28 and what do they mean?

Cancer

The following statistics are taken from an article by P. Burch relating cigarette smoking to lung cancer [17]. The article presents some data relating mortality from lung cancer to average cigarette consumption (lb/person) for females in England and Wales over a 40-year period. These data are given in Table 11.29.

$$\left(\sum_{i=1}^{8} x_i = 2.38 \quad \sum_{i=1}^{8} x_i^2 = 1.31 \quad \sum_{i=1}^{8} y_i = -15.55\right.$$

$$\left. \sum_{i=1}^{8} y_i^2 = 30.71 \quad \sum_{i=1}^{8} x_i y_i = -4.12\right)$$

TABLE 11.29 Cigarette consumption and lung cancer mortality in England and Wales, 1930–1969

Period	$\log_{10}$ mortality (over 5 years), y	$\log_{10}$ annual cigarette consumption (lb/person), x
1930–1934	−2.35	−0.26
1935–1939	−2.20	−0.03
1940–1944	−2.12	0.30
1945–1949	−1.95	0.37
1950–1954	−1.85	0.40
1955–1959	−1.80	0.50
1960–1964	−1.70	0.55
1965–1969	−1.58	0.55

(Reprinted with permission of the *Journal of the Royal Statistical Society, A.*, *141*, 437–477, 1978.)

11.93 Compute the correlation between 5-year mortality and annual cigarette consumption when expressed in the $\log_{10}$ scale.

11.94 Test this correlation for statistical significance and report a *p*-value.

11.95 Fit a regression line relating 5-year mortality to annual cigarette consumption.

11.96 To test the significance of this regression line, is it necessary to perform any additional tests other than those in Problem 11.94? If so, perform them.

11.97 What is the expected mortality rate with an annual cigarette consumption of 1 lb/person?

11.98 Why are the variables mortality rate and annual cigarette consumption expressed in the log scale?

Cardiovascular Disease

Many studies have demonstrated that the level of HDL cholesterol (high-density lipoprotein cholesterol) is positively related to alcohol consumption. This relationship has intrigued researchers because the level of HDL cholesterol is inversely correlated with the incidence of heart disease.

One possible mechanism was explored by Kuller et al. in an analysis of participants in the MRFIT study relating the level of HDL cholesterol to the level of SGOT, a parameter commonly used to assess liver function [18]. The data in Table 11.30 were presented.

11.99 Fit a regression line predicting mean HDL cholesterol as a function of SGOT. For this purpose assume that all levels of SGOT ≤ 9 and ≥21 are 9.5 and 20.5, respectively, and that levels of SGOT within

TABLE 11.30 The relationship of HDL cholesterol to SGOT in the MRFIT population

SGOT	Mean HDL cholesterol
⩽9	40.0
10–12	41.2
13–14	42.3
15–16	42.8
17–18	43.8
19–20	43.6
⩾21	46.5

(Reprinted with permission of the *American Journal of Epidemiology*, *117*(4), 406–418, 1983.)

all other groups occur at the midpoint of the group. Assume also that group assignments are made after rounding; for example, 10–12 represents an actual range from 9.5 to 12.5. Using this convention, if $x =$ SGOT and $y =$ mean HDL cholesterol, then

$$\sum_{i=1}^{7} x_i = 107 \qquad \sum_{i=1}^{7} x_i^2 = 1740.5 \qquad \sum_{i=1}^{7} y_i = 300.2$$

$$\sum_{i=1}^{7} y_i^2 = 12{,}900.2 \qquad \sum_{i=1}^{7} x_i y_i = 4637.6$$

11.100 Test the line fitted in Problem 11.99 for statistical significance and report a p-value.

11.101 What is the best estimate of the HDL cholesterol for an average person with SGOT level of 11, and what is the standard error of this estimate?

11.102 Suppose that a mistake was made in Problem 11.99 and the SGOT groups were assigned by truncation rather than rounding (e.g., 10–12 would represent 10.0–12.99 rather than 9.5–12.5). It would follow that average levels of SGOT within each group would increase by 0.5 (assume this is also the case for the groups ⩽9 and ⩾21, respectively). What effect does the truncation assignment rule have on the estimates of the regression parameters in Problem 11.99? (No further calculation should be necessary.)

As part of the same study, a multiple regression analysis was performed relating the change in HDL cholesterol to baseline levels and changes in other risk factors. The results are presented in Table 11.31.

11.103 Interpret the coefficients for change in SGOT and drinks/week in the multiple regression model.

TABLE 11.31 Multiple regression of change in HDL cholesterol from the second screening visit to the 48-month follow-up visit (follow-up minus baseline) among participants receiving standard medical care ($n = 5,112$)

Variable	Regression coefficient	Regression coefficient/ standard error
Age (years)	−0.0054	−0.3
Screen 2 HDL cholesterol (mg/dL)	−0.3109	−29.3
Screen 1 diastolic blood pressure (mm Hg)	−0.0322	−1.7
Screen 1 cigarettes/day	−0.0324	−4.6
Change in diastolic blood pressure (mm Hg)	0.1233	9.1
Change in thiocyanate* (mg/L)	0.0106	−4.4
Change in body mass index (kg/m²)	−1.2702	−18.0
Change in drinks/week	0.1053	9.5
Change in SGOT (IU/L)	0.1160	8.4

* A biochemical marker used to measure the actual number of cigarettes smoked recently.

Cardiovascular Disease

Sudden death is an important, lethal cardiovascular end point. Most previous studies of risk factors for sudden death have focused on men. Looking at this issue for women is important as well. For this purpose, data were used from the Framingham Heart Study [19]. Several potential risk factors, such as age, blood pressure, and cigarette smoking, are of interest and need to be controlled for simultaneously. Therefore, a multiple logistic regression model was fitted to these data, as shown in Table 11.32.

11.104 Assess the statistical significance of the individual risk factors.

11.105 What do these statistical tests mean in this instance?

11.106 Compute the odds ratio relating the additional risk of sudden death per 100 centiliter decrease in vital capacity after adjustment for the other risk factors.

11.107 Provide a 95% confidence interval for the estimate in Problem 11.106.

TABLE 11.32 Multiple logistic regression model relating 2-year incidence of sudden death in females without prior coronary heart disease (data taken from the Framingham Heart Study) to several risk factors

Risk factor	Regression coefficient, $\hat{\beta}_l$	$se(\hat{\beta}_l)$
Constant	−15.3	
Systolic blood pressure (mm Hg)	0.0019	0.0070
Framingham relative weight (%)	−0.0060	0.0100
Cholesterol (mg/100 ml)	0.0056	0.0029
Glucose (mg/100 ml)	0.0066	0.0038
Cigarette smoking (cigarettes/day)	0.0069	0.0199
Hematocrit (%)	0.111	0.049
Vital capacity (centiliters)	−0.0098	0.0036
Age (years)	0.0686	0.0225

(Reprinted with permission of the *American Journal of Epidemiology*, *120*(6), 888–899, 1984.)

Ophthalmology

A study was conducted using 225 lenses to assess the interobserver reproducibility of the classification of cataracts using the Cooperative Cataract Research Group classification system [20]. One aspect of the study involved the rating of cataracts by two different observers as either immature, mature, or hypermature. The results are given in Table 11.33.

11.108 Compute a measure of reproducibility between observers for this index.

11.109 Assess the statistical significance of this measure.

Obstetrics

Interest has increased in possible relationships between the health habits of the mother during pregnancy and adverse delivery outcomes. An issue that is receiving attention is the effect of marijuana usage on the occurrence of congenital malformations. One potential problem is that many other maternal factors are associated with adverse pregnancy outcomes, including age, race, socioeconomic status (SES), smoking, and so forth. Since these factors may also be related to marijuana usage, they need to be simultaneously controlled for in the analysis. Therefore, data from the Delivery Interview Program (DIP) were utilized, and a multiple logistic regression was performed relating the occurrence of major congenital malformations to several maternal risk factors [21]. The results are given in Table 11.34.

11.110 Perform a test to assess the significance of each of the risk factors after controlling for the other risk factors.

11.111 Compute an odds ratio and an associated 95% confidence interval relating the presence or absence of each independent variable to the risk of malformations.

11.112 How do you interpret the effect of marijuana usage on congenital malformations based on your findings in Problems 11.110 and 11.111?

Hypertension

Refer to the data in Table 3.12 (p. 68). Another method for relating measures of reactivity for the automated and manual blood pressures is the correlation coefficient. Suppose the correlation coefficient relating these

TABLE 11.33 Reproducibility of the classification of cataracts

Observer 1	Observer 2			Total
	Hypermature	Mature	Immature	
Hypermature	8	2	0	10
Mature	1	8	3	12
Immature	0	1	202	203
Total	9	11	205	225

TABLE 11.34 Multiple logistic regression analysis relating the occurrence of major congenital malformations to several risk factors, using 12,424 women in the DIP study

Variable*	Regression coefficient	Standard error
Marijuana usage (any frequency)	0.307	0.173
Any previous miscarriage	0.239	0.141
White race	0.191	0.161
Alcohol use in pregnancy	0.174	0.145
Age 35+	0.178	0.186
Any previous stillbirth	0.049	0.328
On welfare	0.030	0.182
Smoking 3+ cigarettes per day at delivery	−0.174	0.144
Any previous induced abortion	−0.198	0.168
Parity > 1†	−0.301	0.113

* All variables are coded as 1 if yes and 0 if no.
† Women with at least one previous pregnancy.
(Reprinted with permission of the *American Journal of Public Health*, 73(10), 1161–1164, 1983.)

two measures of reactivity is .19, based on 79 individuals having reactivity measured by each type of bp monitor.

11.113 What is the appropriate procedure to test if there is a relationship between reactivity as measured by the automated and manual monitors?

11.114 Conduct the test procedure in Problem 11.113 and report a *p*-value. What do the results mean in words?

Bioavailability

The intake of high doses of beta carotene in food substances has been associated in many observational studies with a decreased incidence of cancer. A clinical trial was planned comparing the incidence of cancer in a group getting beta carotene in capsule form compared with a group getting beta carotene placebo capsules. One issue in planning such a study is which preparation to use for the beta carotene capsules. Four preparations were considered: (1) Solatene (30 mg capsules), (2) Roche (60 mg capsules), (3) BASF (30 mg capsules), and (4) BASF (60 mg capsules). To test the efficacy of the four agents in raising plasma carotene levels, a small bioavailability study was conducted. After two consecutive-day fasting blood samples, 23 volunteers were randomized to one of the four preparations, taking 1 pill every other day for 12 weeks: (1) Solatene 30 mg, (2) Roche 60 mg, (3) BASF 30 mg, and (4) BASF 60 mg. The primary endpoint was the level of plasma carotene attained after a moderately prolonged steady ingestion. For this purpose, blood samples were drawn at 6, 8, 10, and 12 weeks, with results given in Data Set 2, BETA-CAR.DAT in Appendix 2. The format of the data is given in Table 11.35.

11.115 How do the three alternative preparations compare with Solatene as to bioavailability? Perform pairwise comparisons.

11.116 How do the three alternative preparations compare among themselves as to bioavailability? Perform pairwise comparisons.

11.117 Which preparation do you recommend be used in the large clinical trial?

Nutrition

Refer to Data Set 26, VALID.DAT in Appendix 2.

11.118 Assess the agreement between the food frequency questionnaire and the dietary record with regard to total fat intake, saturated fat intake, alcohol intake, and total caloric intake. Quantify the level of

TABLE 11.35 Format of BETACAR.DAT

Variable	Column	Code
Preparation	1	1 = SOL; 2 = ROCHE; 3 = BASF-30; 4 = BASF-60
Subject number	3–4	
First baseline level	6–8	
Second baseline level	10–12	
Week 6 level	14–16	
Week 8 level	18–20	
Week 10 level	22–24	
Week 12 level	26–28	

agreement by representing dietary intake both in the original continuous scale and on a quintile scale.

Pulmonary Disease
Refer to Data Set 10, FEV.DAT in Appendix 2.

11.119 Use regression methods to look at the relationship between level of pulmonary function and other factors (e.g., age, sex, height, and personal smoking) when considered separately and simultaneously.

Health Promotion
Refer to Data Set 24, SMOKE.DAT in Appendix 2. A proportional hazards model was fit to these data to assess the relationship of age, sex, number of cigarettes smoked, and log CO concentration, when considered simultaneously, on the ability to remain abstinent from smoking. The results are given in Table 11.36.

11.120 Assess the significance of the variables.

11.121 Estimate the effects of each variable in terms of hazard ratios.

Heptic Disease
Refer to Data Set 12, HORMONE.DAT in Appendix 2.

TABLE 11.36 Proportional hazards model relating the hazard of recidivism to age, sex, number of cigarettes smoked prior to quitting, and log CO concentration

Risk factor	Regression coefficient $(\hat{\beta}_i)$	Standard error $se(\hat{\beta}_i)$
Age	0.0023	0.0058
Sex(1 = M/0 = F)	−0.127	0.143
Number of cigarettes smoked	−0.0038	0.0050
$\log_{10}CO$ (adjusted)*	0.912	0.366

* This variable represents CO values adjusted for minutes elapsed since last cigarette smoked prior to quitting.

11.122 Use methods of regression analysis to assess whether there are dose-response relationships with regard to biliary and pancreatic secretion levels. Perform separate analyses for each of the four active hormones tested.

11.123 Use methods of regression analysis to assess whether there are dose-response relationships with regard to biliary and pancreatic pH levels. Perform separate analyses for each of the four active hormones tested.

References

[1] Greene, J., & Touchstone, J. (1963). Urinary tract estriol: An index of placental function. *American Journal of Obstetrics and Gynecology, 85*(1), 1–9.

[2] Higgins, M., & Keller, J. (1973). Seven measures of ventilatory lung function. *American Review of Respiratory Disease, 108,* 258–272.

[3] Draper, N., & Smith, H. (1981). *Applied regression analysis* (2nd edition). New York: Wiley.

[4] Kleinbaum, D. G., Kupper, L. L., & Muller, K. E. (1988). *Applied regression analysis and other multivariable methods* (2nd edition). Boston: Duxbury.

[5] McCormack, W. M., Rosner, B., McComb, D. E., Evrard, J. R., & Zinner, S. H. (1985). Infection with *Chlamydia trachomatis* in female college students. *American Journal of Epidemiology, 121*(1), 107–115.

[6] Lee, E. T. (1986). *Statistical methods for survival data analysis.* Belmont, CA: Wadsworth.

[7] Miller, R. G., Jr. (1981). *Survival analysis.* New York: Wiley.

[8] Cox, D. R. (1972). Regression models and life tables (with Discussion). *Journal of the Royal Statistical Society, Ser. B, 34,* 187–220.

[9] Apgar, V. (1953). A proposal for a new method of evaluation of the newborn infant. *Current Researches in Anesthesia and Analgesia,* 260–267.

[10] Apgar, V. et al. (1958). Evaluation of the newborn infant—second report. *Journal of American Medical Association, 168*(15), 1985–1988.

[11] Landis, J. R., & Koch, G. G. (1977). The measurement of observer agreement for categorical data. *Biometrics, 33,* 159–174.

[12] Fleiss, J. L. (1981). *Statistical methods for rates and proportions.* New York: Wiley.

[13] Dec, G. W., Palacios, I. F., Fallon, J. T., Aretz, H. T., Mills, J., Lee, D. C. S., & Johnson, R. A. (1985). Active myocarditis in the spectrum of acute dilated cardiomyopathies: Clinical features, histologic correlates and clinical outcome. *New England Journal of Medicine, 312*(14), 885–890.

[14] Torok-Storb, B., Doney, K., Sale, G., Thomas, E. D., & Storb, R. (1985). Subsets of patients with aplastic anemia identified by flow microfluorometry. *New England Journal of Medicine, 312*(16), 1015–1022.

[15] National Center for Health Statistics. (1979). *Monthly vital statistics report, annual summary.*

[16] Report of the Task Force on Blood Pressure Control in Children. (1977). *Pediatrics, 59*(5, part 2), 797–820.

[17] Burch, P. R. B. (1978). Smoking and lung cancer: The problem of inferring cause. *Journal of the Royal Statistical Society, 141*, 437–477.

[18] Kuller, L. H., Hulley, S. B., Laporte, R. E., Neaton, J., & Dai, W. S. (1983). Environmental determinants, liver function, and high density lipoprotein cholesterol levels. *American Journal of Epidemiology, 117*(4), 406–418.

[19] Schatzkin, A., Cupples, L. A., Heeren, T., Morelock, S., & Kannel, W. S. (1984). Sudden death in the Framingham Heart Study: Differences in incidence and risk factors by sex and coronary disease status. *American Journal of Epidemiology, 120*(6), 888–899.

[20] Chylack, L. T., Jr., White, O., & Tung, W. H. (1984). Classification of human senile cataractous change by the American Cooperative Cataract Research Group (CCRG) method: II. Staged simplification of cataract classification. *Investigative Ophthalmology and Visual Science, 25*, 166–173.

[21] Linn, S., Schoenbaum, S. C., Monson, R. R., Rosner, B., Stubblefield, P. C., & Ryan, K. J. (1983). The association of marijuana use with outcome of pregnancy. *American Journal of Public Health, 73*(10), 1161–1164.

CHAPTER TWELVE

ANALYSIS OF VARIANCE

Introduction

In Chapter 8 we were concerned with comparing the means of two normal distributions using the two-sample t test for independent samples. Frequently, the means of more than two distributions need to be compared.

EXAMPLE 12.1 **Pulmonary Disease** A topic of ongoing public health interest is whether or not *passive smoking* (i.e., exposure to cigarette smoke in the atmosphere among nonsmokers) has a measurable effect on pulmonary health. White and Froeb studied this question by measuring pulmonary function in several ways in the following six groups [1]:

(1) Nonsmokers (NS) People who themselves did not smoke and were not exposed to cigarette smoke either at home or on the job.

(2) Passive Smokers (PS) People who themselves did not smoke and were not exposed to cigarette smoke in the home but were employed for 20 or more years in an enclosed working area that routinely contained tobacco smoke.

(3) Noninhaling Smokers (NI) People who smoked pipes, cigars, or cigarettes but who did not inhale.

(4) Light Smokers (LS) People who smoked and inhaled 1–10 cigarettes per day for 20 or more years. (*Note:* There are 20 cigarettes in a pack.)

(5) Moderate Smokers (MS) People who smoked and inhaled 11–39 cigarettes per day for 20 or more years.

(6) Heavy Smokers (HS) People who smoked and inhaled 40 or more cigarettes per day for 20 or more years.

A principal measure used by the authors to assess pulmonary function was forced mid-expiratory flow (FEF). The authors were interested in comparing FEF in the six groups. ■

The t test methodology generalizes nicely in this case to a procedure referred to as the **one-way analysis of variance**.

In some instances the comparison groups can be classified in two different ways.

EXAMPLE 12.2 **Cardiovascular Disease** Suppose we wish to look at the relationship between serum cholesterol level as measured in 1976 and oral contraceptive (OC) use as assessed at two different points in time (in 1973 and 1976). Women can be categorized as OC users or non-OC users in 1976 and also as OC users or non-OC users in 1973. This classification might be useful if the effect of both current and past use of OCs on serum cholesterol level is being studied. ■

In this type of example, when the groups can be classified in the form of a 2×2 contingency table, but the response of interest has a continuous distribution, specifically normal, the data can be analyzed by a technique known as the **two-way analysis of variance**.

SECTION 12.2 **One-Way Analysis of Variance—General Model**

EXAMPLE 12.3 **Pulmonary Disease** Refer to Example 12.1. The authors were able to identify 200 males and 200 females in each of the six groups except for the NI group, which, because of the small number of such people available, was limited to 50 males and 50 females. The mean and standard deviation of FEF for each of the six groups for males are presented in Table 12.1. How can the means of these six groups be compared? ∎

TABLE 12.1
FEF data for smoking
and nonsmoking males

Group number, i	Group name	Mean FEF (L/s)	sd FEF (L/s)	n_i
1	NS	3.78	0.79	200
2	PS	3.30	0.77	200
3	NI	3.32	0.86	50
4	LS	3.23	0.78	200
5	MS	2.73	0.81	200
6	HS	2.59	0.82	200

(Reprinted by permission of *The New England Journal of Medicine, 302*(13), 720–723, 1980.)

Suppose that there are k groups with n_i observations in the ith group. The jth observation in the ith group will be denoted by y_{ij}. We will assume that the following model holds:

12.1
$$y_{ij} = \mu + \alpha_i + e_{ij}$$

where μ is a constant, α_i is a constant that is different for each group, and e_{ij} is an error term, which is normally distributed with mean 0 and variance σ^2. The α_i are typically constrained in this model, so that the sum of the α_i's over all groups is 0. Thus, a typical observation from the ith group is normally distributed with mean $\mu + \alpha_i$ and variance σ^2.

DEFINITION 12.1 ■■■
The model in **(12.1)** is a **one-way analysis of variance**, or a **one-way ANOVA** model. With this model, the means of an arbitrary number of groups, each of which follows a normal distribution, can be compared. Whether the variability in the data comes mostly from variability within groups or can truly be attributed to variability between groups can also be determined. ∎

The parameters in **(12.1)** can be interpreted as follows:

| 12.2 | **Interpretation of the Parameters of a One-Way Analysis of Variance Model** |

(1) μ represents the underlying mean of all groups taken together.

(2) α_i represents the difference between the mean of the ith group and the overall mean.

(3) e_{ij} represents random error about the mean $\mu + \alpha_i$ for an individual observation from the ith group.

Intuitively, in Table 12.1, FEF will be predicted by an overall mean FEF plus the effect of each smoking group plus random variability within each smoking group. Group means will be compared within the context of this model.

SECTION 12.3   **Hypothesis Testing in One-Way ANOVA**

The null hypothesis (H_0) in this case is that the underlying mean FEF of each of the six groups is the same. This hypothesis is equivalent to stating that each $\alpha_i = 0$, since the α_i sum up to 0. The alternative hypothesis (H_1) is that at least two of the group means are not the same. This hypothesis is equivalent to stating that at least one $\alpha_i \neq 0$. Thus, we wish to test the hypothesis H_0: all $\alpha_i = 0$ versus H_1: at least one $\alpha_i \neq 0$.

12.3.1   ***F* Test for Overall Comparison of Group Means**

The mean FEF for the ith group will be denoted by $\bar{y}_i$, and the mean FEF over all groups by $\bar{y}$. The deviation of an individual observation from the overall mean can be represented as

| 12.3 |

$$y_{ij} - \bar{y} = (y_{ij} - \bar{y}_i) + (\bar{y}_i - \bar{y})$$

The first term on the right-hand side $(y_{ij} - \bar{y}_i)$ represents the deviation of an individual observation from the group mean for that observation and is an indication of *within-group variability*. The second term on the right-hand side $(\bar{y}_i - \bar{y})$ represents the deviation of a group mean from the overall mean and is an indication of *between-group variability*. These terms are depicted in Figure 12.1.

Generally speaking, if the between-group variability is large and the within-group variability is small, as in Figure 12.1(a), then H_0 will be rejected and the underlying group means will be declared significantly different. Conversely, if the between-group variability is small and the within-group variability is large, as in Figure 12.1(b), then H_0, that the underlying group means are the same, will be accepted.

If both sides of **(12.3)** are squared and the squared deviations are summed over all observations over all groups, then the following relationship is obtained

| 12.4 |

$$\sum_{i=1}^{k} \sum_{j=1}^{n_i} (y_{ij} - \bar{y})^2 = \sum_{i=1}^{k} \sum_{j=1}^{n_i} (y_{ij} - \bar{y}_i)^2 + \sum_{i=1}^{k} \sum_{j=1}^{n_i} (\bar{y}_i - \bar{y})^2$$

since the cross-product term can be shown to be zero.

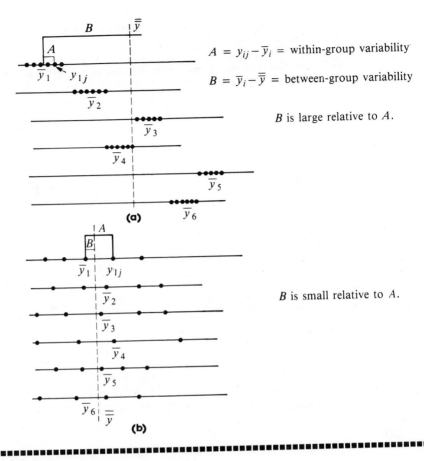

$A = y_{ij} - \bar{y}_i$ = within-group variability

$B = \bar{y}_i - \bar{\bar{y}}$ = between-group variability

B is large relative to A.

B is small relative to A.

FIGURE 12.1
Comparison of between-group and within-group variability

DEFINITION 12.2 ■■■

The term

$$\sum_{i=1}^{k} \sum_{j=1}^{n_i} (y_{ij} - \bar{\bar{y}})^2$$

is denoted as the **Total Sum of Squares** (Total SS).   ■

DEFINITION 12.3 ■■■

The term

$$\sum_{i=1}^{k} \sum_{j=1}^{n_i} (y_{ij} - \bar{y}_i)^2$$

is denoted as the **Within Sum of Squares** (Within SS).   ■

DEFINITION 12.4 ■■■

The term

$$\sum_{i=1}^{k} \sum_{j=1}^{n_i} (\bar{y}_i - \bar{\bar{y}})^2$$

is denoted as the **Between Sum of Squares** (Between SS).   ■

Thus, the relationship in **(12.4)** can be written as Total SS = Within SS + Between SS. The Within SS and Between SS play an analogous role to the Res SS and Reg SS in linear regression analysis in Chapter 11.

It will be easier to use the short computational forms for the Within SS and Between SS in **(12.5)** for performing the hypothesis test.

12.5 | **Short Computational Forms for the Between SS and Within SS Based on Raw Data**

Let

$$y_{i.} = \sum_{j=1}^{n_i} y_{ij} = \text{sum of the observations in the } i\text{th group}$$

(handwritten: n_i (not n))

$$y_{..} = \sum_{i=1}^{k} \sum_{j=1}^{n_i} y_{ij} = \text{sum of the observations across all groups}$$

$$n = \sum_{i=1}^{k} n_i = \text{total number of observations over all groups}$$

The short computational forms are given as follows:

$$\text{Between SS} = \sum_{i=1}^{k} \frac{y_{i.}^2}{n_i} - \frac{y_{..}^2}{n}$$

$$\text{Total SS} = \sum_{i=1}^{k} \sum_{j=1}^{n_i} y_{ij}^2 - \frac{y_{..}^2}{n}$$

$$\text{Within SS} = \text{Total SS} - \text{Between SS}$$

This computational method is preferable if the raw data are available, since it minimizes roundoff error. This procedure is also used in most computer programs. If the raw data are not available, but rather the data are presented in terms of group means and variances, then the following computational formula can be used:

12.6 | **Short Computational Forms for the Between SS and Within SS Based on Grouped Data**

$$\text{Between SS} = \sum_{i=1}^{k} n_i \bar{y}_i^2 - \frac{\left(\sum_{i=1}^{k} n_i \bar{y}_i\right)^2}{n}$$

$$\text{Within SS} = \sum_{i=1}^{k} (n_i - 1)s_i^2$$

EXAMPLE 12.4 **Pulmonary Disease** Compute the Within SS and Between SS for the FEF data in Table 12.1.

SOLUTION Since grouped data are presented, the computational form in **(12.6)** is used:

$$\text{Between SS} = [200(3.78)^2 + 200(3.30)^2 + \cdots + 200(2.59)^2]$$

$$- \frac{[200(3.78) + 200(3.30) + \cdots + 200(2.59)]^2}{1050}$$

$$= 10{,}505.58 - (3292)^2/1050 = 10{,}505.58 - 10{,}321.20 = 184.38.$$

$$\text{Within SS} = 199(0.79)^2 + 199(0.77)^2 + 49(0.86)^2 + 199(0.78)^2$$

$$+ 199(0.81)^2 + 199(0.82)^2$$

$$= 124.20 + 117.99 + 36.24 + 121.07 + 130.56 + 133.81 = 663.87 \quad \blacksquare$$

Finally, the following definitions are important:

DEFINITION 12.5 ■■
Between Mean Square = Between MS = Between SS/$(k - 1)$ ■

DEFINITION 12.6 ■■
Within Mean Square = Within MS = Within SS/$(n - k)$ ■

The significance test will be based on the ratio of the Between MS to the Within MS. If this ratio is large, then we will reject H_0; if it is small, we will accept (or fail to reject) H_0. Furthermore, under H_0, the ratio of Between MS to Within MS follows an F distribution with $k - 1$ and $n - k$ degrees of freedom. Thus, the following test procedure for a level α test is used:

12.7 | **Overall F Test for One-Way ANOVA**

To test the hypothesis H_0: $\alpha_i = 0$ for all i versus H_1: at least one $\alpha_i \neq 0$, use the following procedure:

(1) Compute the Between SS, Between MS, Within SS, and Within MS using either **(12.5)** or **(12.6)** and Definitions 12.5 and 12.6.

(2) Compute the test statistic F = Between MS/Within MS, which follows an F distribution with $k - 1$ and $n - k$ df under H_0.

(3) If

$$F > F_{k-1, n-k, 1-\alpha}$$

then reject H_0. If

$$F \leq F_{k-1, n-k, 1-\alpha}$$

then accept H_0.

(4) The exact p-value is given by the area to the right of F under an $F_{k-1, n-k}$ distribution = $Pr(F_{k-1, n-k} > F)$.

The acceptance and rejection regions for this test are depicted in Figure 12.2. The computation of the exact p-value is illustrated in Figure 12.3. The results from the analysis of variance are typically displayed in an ANOVA table, as in Table 12.2.

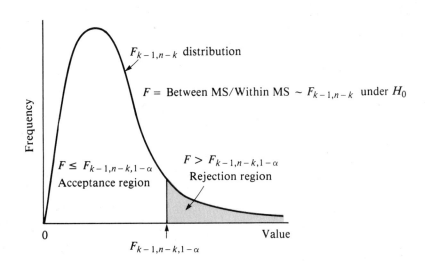

FIGURE 12.2
Acceptance and
rejection regions for the
overall F test for one-
way ANOVA

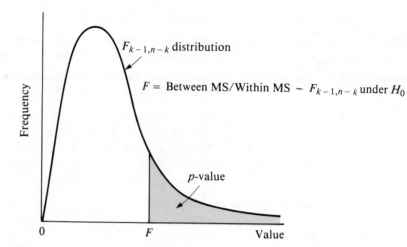

FIGURE 12.3
Computation of the
exact p-value for the
overall F test for one-
way ANOVA

TABLE 12.2
Display of one-way
ANOVA results

Source of variation	SS	df	MS	F statistic	p-value
Between	$\sum_{i=1}^{k} \dfrac{y_{i.}^2}{n_i} - \dfrac{y_{..}^2}{n} = A$	$k-1$	$\dfrac{A}{k-1}$	$\dfrac{A/(k-1)}{B/(n-k)} = F$	$Pr(F_{k-1,n-k} > F)$
Within	Total SS − Between SS $= B$	$n-k$	$\dfrac{B}{n-k}$		
Total	$\sum_{i=1}^{k} \sum_{j=1}^{n_i} y_{ij}^2 - \dfrac{y_{..}^2}{n}$				

EXAMPLE 12.5   **Pulmonary Disease** Test if the mean FEF scores are significantly different in the six groups in Table 12.1.

SOLUTION   From Example 12.4, Between SS = 184.38 and Within SS = 663.87. Therefore, since there are 1050 observations combined over all groups, it follows that

$$\text{Between MS} = 184.38/5 = 36.876$$

$$\text{Within MS} = 663.87/(1050 - 6) = 663.87/1044 = 0.636$$

$$F = \text{Between MS/Within MS} = 36.876/0.636 = 58.0 \sim F_{5,1044} \text{ under } H_0$$

Refer to Table 8 in Appendix 1 and find that

$$F_{5,120,.999} = 4.42$$

Since $$F_{5,1044,.999} < F_{5,120,.999} = 4.42 < 58.0 = F$$

it follows that $p < .001$. Therefore, H_0, that all the means are equal, can be rejected and we can conclude that at least two of the means are not the same. These results are displayed in an ANOVA table (Table 12.3).   ∎

TABLE 12.3
ANOVA table for FEF
data in Table 12.1

	SS	df	MS	F statistic	p-value
Between	184.38	5	36.876	58.0	$p < .001$
Within	663.87	1044	0.636		
Total	848.25				

SECTION 12.4

Comparisons of Specific Groups in One-Way ANOVA

In the previous section a test of the hypothesis H_0: all group means are equal versus H_1: at least two group means are different was presented. This test enables us to detect when at least two groups have different underlying means, but it does not allow us to state which of the groups have means that are different from each other. The usual practice is to perform the overall F test just discussed. If H_0 is rejected, then specific groups are compared, as discussed in this section.

12.4.1 *t* Test for Comparison of Pairs of Groups

Suppose at this point we want to test if groups 1 and 2 have means that are significantly different from each other. From the underlying model in **(12.1)**, under either hypothesis,

12.8

$\bar{y}_1$ is normally distributed with mean $\mu + \alpha_1$ and variance σ^2/n_1

and $\bar{y}_2$ is normally distributed with mean $\mu + \alpha_2$ and variance σ^2/n_2

The difference of the sample means $(\bar{y}_1 - \bar{y}_2)$ will be used as a test criterion. Thus, from **(12.8)**, it follows that

12.9

$$\bar{y}_1 - \bar{y}_2 \sim N\left[\alpha_1 - \alpha_2,\ \sigma^2\left(\frac{1}{n_1} + \frac{1}{n_2}\right)\right]$$

However, under H_0, $\alpha_1 = \alpha_2$ and **(12.9)** reduce to

12.10

$$\bar{y}_1 - \bar{y}_2 \sim N\left[0,\ \sigma^2\left(\frac{1}{n_1} + \frac{1}{n_2}\right)\right]$$

If σ^2 were known, then we could divide by the standard error

$$\sigma\sqrt{\frac{1}{n_1} + \frac{1}{n_2}} \quad \text{and obtain the test statistic:}$$

12.11

$$t^* = \frac{\bar{y}_1 - \bar{y}_2}{\sqrt{\sigma^2\left(\frac{1}{n_1} + \frac{1}{n_2}\right)}}$$

The test statistic t^* would follow an $N(0, 1)$ distribution under H_0. Since σ^2 is in general unknown, the best estimate of it, denoted by s^2, is substituted, and the test statistic is revised accordingly.

How should σ^2 be estimated? Recall that when a pooled estimate of the variance from two independent samples had to be obtained in Chapter 8, a weighted average of the sample variances from the individual samples, where the weights were the number of degrees of freedom in each sample, was used. In particular, from **(8.10)**,

$$s^2 = [(n_1 - 1)s_1^2 + (n_2 - 1)s_2^2]/(n_1 + n_2 - 2)$$

For the one-way ANOVA, there are k sample variances and a similar approach is used to estimate σ^2 by computing a weighted average of k individual sample variances, where the weights are the number of degrees of freedom in each of the k samples. This formula is given as follows:

12.12 | **Pooled Estimate of the Variance for One-Way ANOVA**

$$s^2 = \sum_{i=1}^{k} (n_i - 1)s_i^2 \Big/ \sum_{i=1}^{k} (n_i - 1) = \left[\sum_{i=1}^{k} (n_i - 1)s_i^2 \right] \Big/ (n - k) = \text{Within MS}$$

However, note from **(12.6)**, **(12.12)**, and Definition 12.6 that this weighted average is the same as the Within MS. Thus, the Within MS is used to estimate σ^2. Note that s^2 had $(n_1 - 1) + (n_2 - 1) = n_1 + n_2 - 2$ df in the two-sample case. Similarly, for the one-way ANOVA, s^2 has

$$(n_1 - 1) + (n_2 - 1) + \cdots + (n_k - 1) \; df = (n_1 + n_2 + \cdots + n_k) - k = n - k \; df$$

EXAMPLE 12.6   **Pulmonary Disease** What is the best estimate of σ^2 for the FEF data in Table 12.1? How many df does it have?

SOLUTION   From Table 12.3, the best estimate of the variance is the Within MS = 0.636. It has $n - k$ $df = 1044 \; df$. ∎

Hence, the test statistic in **(12.11)** will be revised, substituting s^2 for σ^2, with the new test statistic distributed as t_{n-k} rather than $N(0, 1)$. The test procedure is given as follows:

12.13 | ***t* Test for the Comparison of Pairs of Groups in One-Way ANOVA**

Suppose we wish to compare two specific groups, arbitrarily labeled as group 1 and group 2, among k groups. To test the hypothesis $H_0: \alpha_1 = \alpha_2$ versus $H_1: \alpha_1 \neq \alpha_2$, use the following procedure:

(1) Compute the pooled estimate of the variance $s^2 = $ Within MS from the one-way ANOVA.

(2) Compute the test statistic

$$t = \frac{\bar{y}_1 - \bar{y}_2}{\sqrt{s^2 \left(\dfrac{1}{n_1} + \dfrac{1}{n_2} \right)}}$$

which follows a t_{n-k} distribution under H_0.

(3) For a two-sided level α test, if

$$t > t_{n-k, 1-\alpha/2} \quad \text{or} \quad t < t_{n-k, \alpha/2}$$

then reject H_0; if

$$t_{n-k, \alpha/2} \leqslant t \leqslant t_{n-k, 1-\alpha/2}$$

then accept H_0.

(4) The exact p-value is given by

$p = 2 \times$ the area to the left of t under a t_{n-k} distribution if $t < 0$

$\qquad = 2 \times Pr(t_{n-k} < t)$

$p = 2 \times$ the area to the right of t under a t_{n-k} distribution if $t \geqslant 0$

$\qquad = 2 \times Pr(t_{n-k} > t)$

The acceptance and rejection regions for this test are given in Figure 12.4. The computation of the exact p-value is illustrated in Figure 12.5. This test is often referred to as the **Least Significant Difference method**.

EXAMPLE 12.7   **Pulmonary Disease** Compare each pair of groups for the FEF data in Table 12.1 and report any significant differences.

SOLUTION   First plot the mean $\pm se$ of the FEF values for each of the six groups in Figure 12.6 to obtain some idea of the magnitude of the differences between groups. The standard error for an individual group mean is estimated by $s/\sqrt{n_i}$, where $s^2 =$ Within MS. Notice that the nonsmokers have the best pulmonary function; the passive smokers, noninhaling smokers, and light smokers have about the same pulmonary function and are worse off than the nonsmokers; and the moderate and heavy smokers have the poorest pulmonary function. Note also that the standard error bars are wider for the noninhaling smokers than for the other groups, since this group has only 50 people compared with 200 for all other groups. Are the observed differences in the figure statistically significant as assessed by the t test procedure in **(12.13)**? The results are presented in Table 12.4.

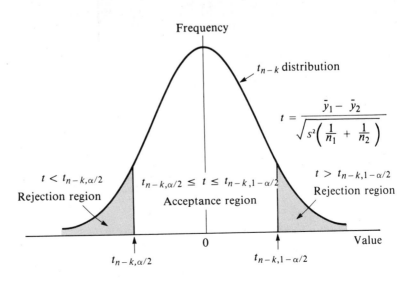

FIGURE 12.4
Acceptance and rejection regions for the t test for the comparison of pairs of groups in one-way ANOVA

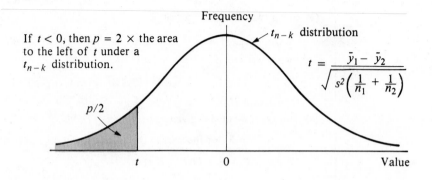

If $t < 0$, then $p = 2 \times$ the area to the left of t under a t_{n-k} distribution.

$$t = \frac{\bar{y}_1 - \bar{y}_2}{\sqrt{s^2\left(\dfrac{1}{n_1} + \dfrac{1}{n_2}\right)}}$$

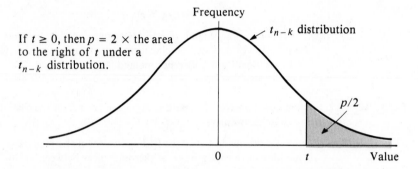

If $t \geq 0$, then $p = 2 \times$ the area to the right of t under a t_{n-k} distribution.

FIGURE 12.5

Computation of the exact p-value for the t test for the comparison of pairs of groups in one-way ANOVA

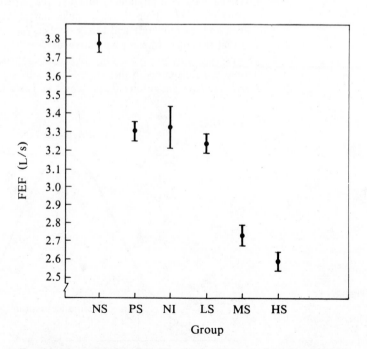

FIGURE 12.6

Mean $\pm$ *se* for FEF for each of six smoking groups

(Reprinted by permission of the *New England Journal of Medicine, 302*(13), 720–723, 1980.)

Groups compared	Test statistic	p-value
NS, PS	$t = \dfrac{3.78 - 3.30}{\sqrt{0.636\left(\dfrac{1}{200} + \dfrac{1}{200}\right)}} = \dfrac{0.48}{0.08} = 6.00 \sim t_{1044}$	$<.001$
NS, NI	$t = \dfrac{3.78 - 3.32}{\sqrt{0.636\left(\dfrac{1}{200} + \dfrac{1}{50}\right)}} = \dfrac{0.46}{0.126} = 3.65 \sim t_{1044}$	$<.001$
NS, LS	$t = \dfrac{3.78 - 3.23}{\sqrt{0.636\left(\dfrac{1}{200} + \dfrac{1}{200}\right)}} = \dfrac{0.55}{0.08} = 6.88 \sim t_{1044}$	$<.001$
NS, MS	$t = \dfrac{3.78 - 2.73}{0.080} = \dfrac{1.05}{0.08} = 13.13 \sim t_{1044}$	$<.001$
NS, HS	$t = \dfrac{3.78 - 2.59}{0.080} = \dfrac{1.19}{0.08} = 14.88 \sim t_{1044}$	$<.001$
PS, NI	$t = \dfrac{3.30 - 3.32}{0.126} = \dfrac{-0.02}{0.126} = -0.16 \sim t_{1044}$	NS
PS, LS	$t = \dfrac{3.30 - 3.23}{0.080} = \dfrac{0.07}{0.08} = 0.88 \sim t_{1044}$	NS
PS, MS	$t = \dfrac{3.30 - 2.73}{0.080} = \dfrac{0.57}{0.08} = 7.13 \sim t_{1044}$	$<.001$
PS, HS	$t = \dfrac{3.30 - 2.59}{0.080} = \dfrac{0.71}{0.08} = 8.88 \sim t_{1044}$	$<.001$
NI, LS	$t = \dfrac{3.32 - 3.23}{0.126} = \dfrac{0.09}{0.126} = 0.71 \sim t_{1044}$	NS
NI, MS	$t = \dfrac{3.32 - 2.73}{0.126} = \dfrac{0.59}{0.126} = 4.68 \sim t_{1044}$	$<.001$
NI, HS	$t = \dfrac{3.32 - 2.59}{0.126} = \dfrac{0.73}{0.126} = 5.79 \sim t_{1044}$	$<.001$
LS, MS	$t = \dfrac{3.23 - 2.73}{0.08} = \dfrac{0.50}{0.08} = 6.25 \sim t_{1044}$	$<.001$
LS, HS	$t = \dfrac{3.23 - 2.59}{0.08} = \dfrac{0.64}{0.08} = 8.00 \sim t_{1044}$	$<.001$
MS, HS	$t = \dfrac{2.73 - 2.59}{0.08} = \dfrac{0.14}{0.08} = 1.75 \sim t_{1044}$	NS

There are very highly significant differences (1) between the nonsmokers and all other groups, (2) between the passive smokers and the moderate and heavy smokers, (3) between the noninhalers and the moderate and heavy smokers, and (4) between the light smokers and the moderate and heavy smokers. There are no significant differences between the passive smokers, noninhalers, and light smokers and no significant differences between the moderate and heavy smokers, although there is a trend towards significance with the latter comparison. Thus, these results tend to confirm what Figure 12.6 showed. They are very interesting because they show that the pulmonary function of passive smokers is significantly worse than that of nonsmokers and is essentially the same as that of noninhaling and light inhaling smokers ($\leqslant \frac{1}{2}$ pack cigarettes per day). In this case the $N(0, 1)$ distribution has been used to evaluate the p-values in Table 12.4, since the t_{1044} and the $N(0, 1)$ distributions are essentially identical, and a table of $N(0, 1)$ percentiles is available in Table 3 in the Appendix.

∎

A frequent error in performing the t test in **(12.13)** when comparing groups 1 and 2 is to use only the sample variances from *these two groups* rather than from *all k groups* to estimate σ^2. If the sample variances from only two groups are used, then different estimates of σ^2 are obtained for each pair of groups considered, which is not reasonable because *all* the groups have the same underlying variance σ^2. Furthermore, the estimate of σ^2 obtained by using all k groups will be more accurate than that obtained from using any two groups, since the estimate of the variance will be based on more information. This is the principal advantage of performing the t tests in the framework of a one-way ANOVA rather than by considering each pair of groups separately and performing t tests for two independent samples as given in **(8.11)** for each pair of samples. If, on the other hand, there is reason to believe that not all groups have the same underlying variance (σ^2), then one-way ANOVA should not be performed, and t tests based on pairs of groups can be used instead.

12.4.2 Linear Contrasts

In Section 12.4.1 methods for comparing specific groups within the context of the analysis of variance were developed. More general comparisons, such as the comparison of a collection of ℓ_1 groups with another collection of ℓ_2 groups, are frequently desired.

EXAMPLE 12.8 **Pulmonary Disease** Suppose we want to compare the pulmonary function of the group of smokers who inhale cigarettes with the group of nonsmokers. The three groups of inhaling smokers in Table 12.1 could just be combined to form one group of 600 inhaling smokers. However, these three groups were selected so as to be of the same size, whereas in the general population the proportions of light, moderate, and heavy smokers are not likely to be the same. Suppose large population surveys report that 70 % of inhaling smokers are moderate smokers, 20 % are heavy smokers, and 10 % are light smokers. How can inhaling smokers as a group be compared with nonsmokers?

∎

It is for this type of question that the estimation and testing of hypotheses for linear contrasts is used.

DEFINITION 12.7 ■■

A **linear contrast** (L) is any linear combination of the individual group means such that the linear coefficients add up to 0. Specifically,

$$L = \sum_{i=1}^{k} c_i \bar{y}_i$$

where

$$\sum_{i=1}^{k} c_i = 0$$ ■

Notice that the comparison of two means that was considered in the previous section is a special case of a linear contrast.

EXAMPLE 12.9   **Pulmonary Disease** Suppose we wish to compare the pulmonary function of the non-smokers and passive smokers. Represent this comparison as a linear contrast.

SOLUTION   Since the nonsmokers are the first group and the passive smokers are the second group, this comparison can be represented by the linear contrast

$$L = \bar{y}_1 - \bar{y}_2 \qquad \text{that is,} \qquad c_1 = +1 \qquad c_2 = -1$$ ■

EXAMPLE 12.10   **Pulmonary Disease** Suppose we wish to compare the pulmonary function of nonsmokers with that of the total group of inhaling smokers, assuming that 10% of inhaling smokers are light smokers, 70% are moderate smokers, and 20% are heavy smokers. Represent this comparison as a linear contrast.

SOLUTION   This comparison can be represented by the linear contrast

$$\bar{y}_1 - 0.1\bar{y}_4 - 0.7\bar{y}_5 - 0.2\bar{y}_6$$

since the nonsmokers are group 1, the light smokers group 4, the moderate smokers group 5, and the heavy smokers group 6. ■

How can we test if a linear contrast is different from 0? In general, for any linear contrast,

$$L = c_1\bar{y}_1 + c_2\bar{y}_2 + \cdots + c_k\bar{y}_k$$

we wish to test the hypothesis $H_0: \mu_L = 0$ versus $H_1: \mu_L \neq 0$, where μ_L is the mean of the linear contrast L:

$$c_1\alpha_1 + c_2\alpha_2 + \cdots + c_k\alpha_k$$

The following test procedure, which is analogous to the t test for pairs of groups in **(12.13)**, can be used:

12.14   ***t* Test for Linear Contrasts in One-Way ANOVA**

Suppose we want to test if the mean of a specific linear contrast L is significantly different from 0, using a two-sided test with significance level α.

(1) Compute the pooled estimate of the variance $= s^2 =$ Within MS from the one-way ANOVA.

(2) Compute the linear contrast

$$L = \sum_{i=1}^{k} c_i \bar{x}_i$$

(3) Compute the test statistic

$$t = \frac{L}{\sqrt{s^2 \sum_{i=1}^{k} \frac{c_i^2}{n_i}}}$$

(4) If $\qquad t > t_{n-k, 1-\alpha/2} \qquad$ or $\qquad t < t_{n-k, \alpha/2}$

then reject H_0. If $\qquad t_{n-k, \alpha/2} \leqslant t \leqslant t_{n-k, 1-\alpha/2}$

then accept H_0.

(5) The exact p-value is given by

$\qquad p = 2 \times$ the area to the left of t under a t_{n-k} distribution if $t < 0 = 2 \times Pr(t_{n-k} < t)$

$\qquad p = 2 \times$ the area to the right of t under a t_{n-k} distribution if $t \geqslant 0 = 2 \times Pr(t_{n-k} > t)$

EXAMPLE 12.11

Pulmonary Disease   Test the hypothesis that the linear contrast defined in Example 12.10 is significantly different from 0.

SOLUTION   From Table 12.3, $s^2 = 0.636$. Furthermore, the linear contrast L is given by

$$L = \bar{y}_1 - 0.1\bar{y}_4 - 0.7\bar{y}_5 - 0.2\bar{y}_6 = 3.78 - 0.1(3.23) - 0.7(2.73) - 0.2(2.59) = 1.03$$

The standard error of this linear contrast is given by

$$se(L) = \sqrt{s^2 \sum_{i=1}^{k} \frac{c_i^2}{n_i}} = \sqrt{(0.636)\left[\frac{(1)^2}{200} + \frac{(-0.1)^2}{200} + \frac{(-0.7)^2}{200} + \frac{(-0.2)^2}{200}\right]} = 0.070$$

Thus, $\qquad t = L/se(L) = 1.03/0.070 = 14.71 \sim t_{1044}$ under H_0

Clearly, this linear contrast is very highly significant ($p < .001$), and the inhaling smokers as a group have strikingly poorer pulmonary function than the nonsmokers.   ■

Another useful application of linear contrasts is when the different groups correspond to different dose levels of a particular quantity, and the coefficients of the contrast are chosen to reflect a particular dose-response relationship. This application is particularly useful if the sample sizes of the individual groups are small and a comparison of any pair of groups does not show a significant difference, but the overall trend is consistent in one direction.

EXAMPLE 12.12

Pulmonary Disease   Suppose we wish to study whether or not the amount of smoke inhaled is related to level of FEF among those smokers who inhale cigarettes. Perform a test of significance for this trend.

SOLUTION   Focus on the light smokers, moderate smokers, and heavy smokers in this analysis. We know the light smokers smoke from 1 to 10 cigarettes per day, and we will assume they smoke an average of $(1 + 10)/2 = 5.5$ cigarettes per day. The moderate smokers smoke from 11 to 39 cigarettes per day, and we will assume they smoke an average of $(11 + 39)/2 = 25$ cigarettes per day. The heavy smokers smoke at least 40 cigarettes per day. We will assume they smoke exactly 40 cigarettes per day, which will underestimate the trend but is the best we can do with the information presented. We wish to test the contrast

$$L = 5.5\bar{y}_4 + 25\bar{y}_5 + 40\bar{y}_6$$

for statistical significance. The problem is that the coefficients of this contrast do not add up to 0; indeed, they add up to $5.5 + 25 + 40 = 70.5$. However, if $70.5/3 = 23.5$ is subtracted from each coefficient, then they will add up to 0. Thus, we wish to test the contrast

$$L = (5.5 - 23.5)\bar{y}_4 + (25 - 23.5)\bar{y}_5 + (40 - 23.5)\bar{y}_6 = -18\bar{y}_4 + 1.5\bar{y}_5 + 16.5\bar{y}_6$$

for statistical significance. This contrast represents the increasing number of cigarettes per day smoked in the three groups. From **(12.14)**,

$$L = -18(3.23) + 1.5(2.73) + 16.5(2.59) = -58.14 + 4.10 + 42.74 = -11.30$$

$$se(L) = \sqrt{0.636\left[\frac{(-18)^2}{200} + \frac{(1.5)^2}{200} + \frac{(16.5)^2}{200}\right]} = \sqrt{0.636(2.99)} = \sqrt{1.902} = 1.38$$

Thus, $\qquad t = L/se(L) = -11.30/1.38 = -8.19 \sim t_{1044}$ under H_0

Clearly, this trend is very highly significant ($p < .001$), and we can say that among smokers who inhale, the greater the number of cigarettes smoked per day, the worse the pulmonary function. ∎

12.4.3 Multiple Comparisons

In many studies the comparisons of interest are specified before looking at the actual data, in which case the t test procedure in **(12.13)** and the linear contrast procedure in **(12.14)** are appropriate. In other instances the comparisons of interest will only be specified after looking at the data. In this case a large number of potential comparisons are often possible. Specifically, if there are a large number of groups and every pair of groups is compared using the t test procedure in **(12.13)**, then some significant differences are likely to be found just by chance.

EXAMPLE 12.13   Suppose there are 10 groups. Thus, there are $\binom{10}{2} = 45$ possible pairs of groups to be compared. Using a 5% level of significance would imply that .05(45), or about two comparisons, are likely to be significant by chance alone. How can we protect ourselves against the detection of falsely significant differences resulting from making too many comparisons? ∎

Several procedures, referred to as **multiple comparisons procedures**, ensure that too many falsely significant differences are not declared. The basic idea of these procedures is to ensure that the *overall probability of declaring any significant differences between all possible pairs of groups* is maintained at some fixed significance level (say α). One such procedure, discussed here, is the **Newman–Keuls procedure**, which is in common use today. (See Kleinbaum, Kupper and Muller for more details on other multiple comparisons procedures for one-way ANOVA [2].)

To use this procedure, the studentized range statistic must be used.

DEFINITION 12.8   ■■

Suppose there is a group of c means, the largest of which is $\bar{y}_1$ with sample size n_1 and the smallest of which is $\bar{y}_2$ with sample size n_2. The c means can be either all the means or a subset of the means identified in a one-way ANOVA problem. Let $s^2 = $ Within MS from the one-way ANOVA. The **studentized range statistic** q for this group of means is defined by

$$q = \frac{\bar{y}_1 - \bar{y}_2}{\sqrt{\frac{s^2}{2}\left(\frac{1}{n_1} + \frac{1}{n_2}\right)}}$$ ■

The basic idea for using this statistic is that if q is sufficiently small, then all of the means in the group are considered equal; otherwise, some of the means are considered significantly different. Table 13 in Appendix 1 gives the 5th and 1st percentiles of the studentized range statistic for varying values of $c = $ the number of means in the group and $d =$ the number of degrees of freedom for s^2. The appropriate critical value in the table is found in the cth column and the dth row of the table.

DEFINITION 12.9 ■■

The **upper α percentile of the studentized range statistic** based on c means, where the Within MS has d degrees of freedom, is denoted by $q_{c,d,1-\alpha}$. ■

Thus, $q_{c,d,1-\alpha}$ is found in the entry at the cth column and dth row of the table for either $\alpha = .05$ or $\alpha = .01$.

EXAMPLE 12.14 Find the upper 5% point of the studentized range statistic based on four means, where the Within MS is based on 10 df.

SOLUTION Refer to the 4th column and the 10th row in the $\alpha = .05$ table. We find that $q_{4,10,.95} = 4.33$. ■

The basic strategy for declaring whether a group of c means is significantly different or not is summarized as follows:

12.15 | **Use of the Studentized Range Statistic to Decide Whether a Group of Means Is Significantly Different**

If there is a group of c means, the largest of which is $\bar{y}_1$ with sample size n_1 and the smallest of which is $\bar{y}_2$ with sample size n_2, and a Within MS $= s^2$ with d degrees of freedom, then:

(1) Compute
$$q = \frac{\bar{y}_1 - \bar{y}_2}{\sqrt{\frac{s^2}{2}\left(\frac{1}{n_1} + \frac{1}{n_2}\right)}}$$

(2) For a level α test, if
$$q \leq q_{c,d,1-\alpha}$$
then declare *all* the means in the group as *not significantly different*; if
$$q > q_{c,d,1-\alpha}$$
then declare *some* of the means in the group as *significantly different*.

EXAMPLE 12.15 Suppose there is a group of three means, the largest of which is 28.0 based on a sample of size 10 and the smallest of which is 15.0 based on a sample of size 5. Suppose also that the Within MS $= 72$ based on 20 df. Test if the means, when considered as a group, are significantly different at the 5% level.

SOLUTION Compute the studentized range statistic as follows:
$$q = \frac{28 - 15}{\sqrt{\frac{72}{2}\left(\frac{1}{10} + \frac{1}{5}\right)}} = \frac{13.0}{3.286} = 3.96$$

We find $q_{3,20,.95}$ from the 3 column and 20 row of the 5% table $= 3.58$. Since $q = 3.96 > 3.58$, some of the three means in this group are declared significantly different. ■

Notice that from Table 13 the larger the number of means in the group, the more difficult it is to declare significant differences among means.

EXAMPLE 12.16 Suppose the group in Example 12.15 consisted of five rather than three means. Test if the means, when considered as a group, are significantly different.

SOLUTION   We find $q_{5, 20, .95}$ from the 5 column and 20 row of the table $= 4.23$. Since $q = 3.96 < 4.23$, the group of five means is *not* declared significantly different. ∎

The problem remains that if we decide from **(12.15)** that a *group* of means is significantly different, then we still do not know which means are actually different. A unified strategy for dealing with the problem of deciding which of k means in a one-way ANOVA setting are significantly different is needed. This strategy is given by the following Newman–Keuls multiple comparisons procedure:

| **12.16** | **Newman–Keuls Multiple Comparisons Procedure** |

(1) Rank order the k group means from smallest to largest and renumber the groups so that group 1 has the smallest mean, group 2 has the next smallest mean, ..., and group k has the largest mean.

(2) For a level α test, compare the entire group of k means using the studentized range procedure in **(12.15)** with significance level α. If all means in the group are not significantly different, then stop; otherwise, go on to step 3.

(3) Compare all possible subgroups of $(k - 1)$ means using **(12.15)**. If *all* subgroups of $(k - 1)$ means are not significantly different, then stop; otherwise, go on to step 4.

(4) Compare all possible subgroups of $(k - 2)$ means within the groups of $(k - 1)$ means that were declared significantly different in step 3. If all such subgroups of $(k - 2)$ means are not significantly different, then stop; otherwise, go on to step 5.

(5) Continue this process until subgroups of two means are being compared or until you have stopped at an earlier step. Notice that because of this procedure, any two means will be considered significantly different only if all subgroups of means that contain these two means are also declared significantly different at an earlier step of the procedure.

EXAMPLE 12.17   **Pulmonary Disease** Perform a significance test at the 5% level for all pairs of means in the FEF data of Table 12.1, using the Newman–Keuls multiple comparisons procedure.

SOLUTION   First, renumber the groups from smallest to largest. Thus, we have the renumbered group means as shown in Table 12.5. Furthermore, the Within MS $= 0.636$ with 1044 df from Table 12.3. The row marked ∞ df in Table 13 and $\alpha = .05$ will be used for simplicity, since the actual percentiles for 1044 df will be close to these percentiles. (Harmonic interpolation,

TABLE 12.5
Renumbered group means of FEF data for use with the Newman–Keuls procedure

Group number	Group name	Mean $(\bar{y}_i)$	Number in group (n_i)
1	HS	2.59	200
2	MS	2.73	200
3	LS	3.23	200
4	PS	3.30	200
5	NI	3.32	50
6	NS	3.78	200

(Reprinted with permission of *The New England Journal of Medicine, 302*(13), 720–723, 1980.)

as was done for the F distribution in **(8.15)**, could be used to obtain the exact percentiles.) First compare the entire group of six means:

$$q = \frac{3.78 - 2.59}{\sqrt{\frac{0.636}{2}\left(\frac{1}{200} + \frac{1}{200}\right)}} = \frac{1.19}{0.0564} = 21.1$$

Refer to $q_{6,\infty,.95} = 4.03$. Since $21.1 > 4.03$, some of the six means are declared significantly different. Next compare subgroups of five means. Start by comparing groups 1–5:

$$q = \frac{3.32 - 2.59}{\sqrt{\frac{0.636}{2}\left(\frac{1}{200} + \frac{1}{50}\right)}} = \frac{0.73}{0.0892} = 8.18$$

Compare this value with $q_{5,\infty,.95} = 3.86$. Since $8.18 > 3.86$, some of these five means are declared significantly different. Similarly, compare groups 2–6:

$$q = \frac{3.78 - 2.73}{0.0564} = 18.6 > q_{5,\infty,.95} = 3.86$$

Thus, some of the means in groups 2–6 are significantly different as well. Now focus on subgroups of four means. Test groups 1–4, 2–5, and 3–6. The following test statistics are obtained:

$$\text{Groups 1–4: } q = \frac{3.30 - 2.59}{0.0564} = 12.6 > q_{4,\infty,.95} = 3.63$$

$$\text{Groups 2–5: } q = \frac{3.32 - 2.73}{0.0892} = 6.61 > 3.63$$

$$\text{Groups 3–6: } q = \frac{3.78 - 3.23}{0.0564} = 9.75 > 3.63$$

Thus, each of these subgroups of means contains means that are significantly different.

Now focus on subgroups of three means. Test groups 1–3, 2–4, 3–5, and 4–6. The following test statistics are obtained:

$$\text{Groups 1–3: } q = \frac{3.23 - 2.59}{0.0564} = 11.3 > q_{3,\infty,.95} = 3.31$$

$$\text{Groups 2–4: } q = \frac{3.30 - 2.73}{0.0564} = 10.1 > 3.31$$

$$\text{Groups 3–5: } q = \frac{3.32 - 3.23}{0.0892} = 1.01 < 3.31$$

$$\text{Groups 4–6: } q = \frac{3.78 - 3.30}{0.0564} = 8.51 > 3.31$$

Thus, some means in groups 1–3, 2–4, and 4–6 are significantly different but none in groups 3–5. Finally, look at subgroups of two means within the groups 1–3, 2–4, or 4–6 but *not* within the groups 3–5. Therefore, test groups 1–2, 2–3, and 5–6 but not 3–4 or 4–5, since these groups are contained in groups 3–5, which have already been declared not significantly different. The following test statistics are obtained:

$$\text{Groups 1--2: } q = \frac{2.73 - 2.59}{0.0564} = 2.48 < q_{2,\,\infty,\,.95} = 2.77$$

$$\text{Groups 2--3: } q = \frac{3.23 - 2.73}{0.0564} = 8.87 > 2.77$$

$$\text{Groups 5--6: } q = \frac{3.78 - 3.32}{0.0892} = 5.16 > 2.77$$

Thus, groups 2–3 and 5–6 are significantly different, whereas groups 1–2 are not. Thus, we conclude that there are three distinct groups of means, groups 1–2, 3–5, and 6. The means are not significantly different within groups but are significantly different between groups. The results of the multiple comparisons procedure are typically displayed as in Figure 12.7. A line is drawn between the names or numbers of each pair of means that are *not* significantly different. This line enables us to visually summarize the results of many comparisons of pairs of means in one concise display. ∎

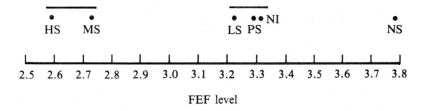

FIGURE 12.7
Display of results of
Newman–Keuls
multiple comparisons
procedure on FEF data
in Table 12.1

Note that the results of the *t* tests in Table 12.4 and the multiple comparisons procedures in Example 12.17 are the same. Namely, there are three distinct groups: heavy and moderate smokers; light smokers, passive smokers, and noninhaling smokers; and nonsmokers. In general, the multiple comparison procedures are more strict than the ordinary *t* tests if more than two means are being compared. That is, there are comparisons between pairs of groups for which the *t* test would declare a significant difference but the multiple comparisons procedure would not. This is the price that is paid for trying to fix the α level of finding *any* significant difference among pairs of groups in using the multiple comparisons procedure rather than for *particular* pairs of groups in using the *t* test. If only two means are being compared, then the *p*-values obtained from using the two procedures are identical.

Also note from Example 12.17 that the further apart two means are in the rank ordering of means, the larger the critical value for declaring significance. Thus, if there are two means between the pair of means considered, then the critical value is 2.77, whereas if there are three means, the critical value is 3.31, and so forth. This property illustrates another difference between the procedures, since for the *t* test in **(12.13)**, the same critical value is used $(t_{n-k,\,1-\alpha/2})$ regardless of how many means are between the pair of means considered.

When should the more conservative multiple comparisons procedure in **(12.16)** rather than the *t* test procedure in **(12.13)** be used to identify specific differences between groups? This area is controversial. Some research workers routinely use multiple comparisons procedures for all one-way ANOVA problems; others never use them. My opinion is that multiple comparisons procedures should be used if there are many groups and not all comparisons between individual groups have been thought out in advance. On the other hand, if there are relatively few groups and only specific comparisons of interest are intended, which have been thought out in advance, then I prefer to use ordinary t tests rather than multiple comparison procedures.

12.4.4   Multiple Comparisons Procedures for Linear Contrasts

The multiple comparisons procedures in Section 12.4.3 are applicable if pairs of means are being compared. In some situations, linear contrasts involving more complex comparisons than simple contrasts based on pairs of means are of interest. In this context, if linear contrasts, which have not been planned in advance, are suggested by looking at the data, then a multiple comparisons procedure might be used to ensure that under H_0, the probability that any linear contrast will be significant is no larger than α. Scheffé's multiple comparisons procedure is applicable in this situation and is summarized as follows:

12.17	**Scheffé's Multiple Comparisons Procedure**

Suppose we wish to test the hypothesis $H_0: \mu_L = 0$ versus $H_1: \mu_L \neq 0$, at significance level α, where

$$L = \sum_{i=1}^{k} c_i \bar{x}_i \quad \text{and} \quad \mu_L = \sum_{i=1}^{k} c_i \mu_i$$

To use Scheffé's multiple comparisons procedure in this situation, take the following steps:

(1) Compute the test statistic

$$t = \frac{L}{\sqrt{s^2 \sum_{i=1}^{k} \frac{c_i^2}{n_i}}}$$

as given in **(12.14)**.

(2) If

$$t > c_2 = \sqrt{(k-1)F_{k-1,n-k,1-\alpha}} \quad \text{or} \quad t < c_1 = -\sqrt{(k-1)F_{k-1,n-k,1-\alpha}}$$

then reject H_0; if

$$c_1 \leqslant t \leqslant c_2$$

then accept H_0.

EXAMPLE 12.18   **Pulmonary Disease**   Test the hypothesis that the linear contrast defined in Example 12.12 (p. 488) is significantly different from 0 using Scheffé's multiple comparisons procedure.

SOLUTION   From Example 12.12, $t = L/se(L) = -8.19$. There are six groups and 1050 subjects. Thus, since t is negative, the critical value is given by $c_1 = -\sqrt{(k-1)F_{k-1,n-k,1-\alpha}} = -\sqrt{5F_{5,1044,0.95}}$. $F_{5,1044,.95}$ will be approximated by $F_{5,\infty,.95} = 2.21$. (We could interpolate from the F table (Table 8 in Appendix 1) for more precise results.) We have $c_1 = -\sqrt{5(2.21)} = -3.32$. Since $t = -8.19 < c_1 = -3.32$, H_0 is rejected at the 5% level and a significant trend among inhaling smokers, with pulmonary function decreasing as the number of cigarettes smoked per day increases, is declared. ∎

Scheffé's multiple comparisons procedure could also have been used when pairs of means were being compared, since a difference between means is a special case of a linear contrast. However, the Newman–Keuls procedure introduced in Section 12.4.3 is preferable in this instance, since, if only pairs of means are being compared, then significant differences can appropriately be declared more often than with Scheffé's procedure (which is designed for a broader set of alternative hypotheses) when true differences exist in this situation.

Once again, if a few linear contrasts, which have been specified in advance, are to be tested, then it may not be necessary to use a multiple comparisons procedure, since if such procedures are used, there will be less power to detect differences for linear contrasts whose means are truly different from zero than the t tests introduced in **(12.14)**. Conversely, if many contrasts are to be tested, or if the contrasts have not been specified before looking at the data, then the multiple comparisons procedures in this section may be useful in protecting against declaring too many significant differences.

SECTION 12.5

Bartlett's Test for Homogeneity of Variance

In our previous work on the analysis of variance, we assumed in the underlying model in **(12.1)** that the population variances of the k groups are the same. This assumption should be tested before using the analysis of variance, and in this section a test for the homogeneity of variance over k groups is presented.

The F test for the equality of two variances given in Section 8.6.3 has already been studied. The test presented here is a generalization of this test to the k group situation. We wish to test the hypothesis $H_0: \sigma_1^2 = \sigma_2^2 = \cdots = \sigma_k^2$ versus H_1: at least two of the σ_i^2 are unequal. Let s_i^2 = sample variance of the ith group and let s^2 = Within Mean Square from the one-way ANOVA table, which represents the best estimate of the common variance σ^2 under H_0. The test will be based on deviations of the individual sample variances s_i^2 from s^2. Specifically, Bartlett devised a test based on the deviations of each of the $\log(s_i^2)$ from $\log(s^2)$. If these deviations are large, then H_0 is rejected; otherwise, H_0 is accepted. In particular, Bartlett has shown that the following test procedure can be used:

12.18 | **Bartlett's Test for the Homogeneity of Variances**

To test the hypothesis $H_0: \sigma_1^2 = \sigma_2^2 = \cdots = \sigma_k^2$ versus H_1: at least two of the σ_i^2 are unequal, use the following procedure:

(1) Compute the test statistic $X^2 = \lambda/c$, which follows a χ^2_{k-1} distribution under H_0, where

$$\lambda = 2.326 \sum_{i=1}^{k} (n_i - 1) \log_{10}\left(\frac{s^2}{s_i^2}\right) = \sum_{i=1}^{k} (n_i - 1) \ln\left(\frac{s^2}{s_i^2}\right)$$

$$c = 1 + \frac{1}{3(k-1)}\left[\left(\sum_{i=1}^{k} \frac{1}{n_i - 1}\right) - \frac{1}{n-k}\right]$$

s_i^2 = sample variance of the ith group

s^2 = Within MS from the one-way ANOVA

(2) If

$$X^2 > \chi^2_{k-1,1-\alpha}$$

then reject H_0; if

$$X^2 \leqslant \chi^2_{k-1,1-\alpha}$$

then accept H_0.

(3) The exact p-value is given by $p =$ the area to the right of X^2 under a χ^2_{k-1} distribution $= Pr(\chi^2_{k-1} > X^2)$.

The acceptance and rejection regions for this test are depicted in Figure 12.8. The computation of the exact p-value is given in Figure 12.9.

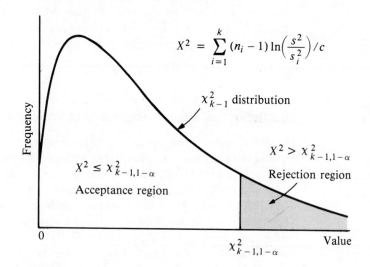

$$X^2 = \sum_{i=1}^{k} (n_i - 1) \ln\left(\frac{s^2}{s_i^2}\right) / c$$

χ^2_{k-1} distribution

$X^2 > \chi^2_{k-1,1-\alpha}$

Rejection region

$X^2 \leq \chi^2_{k-1,1-\alpha}$

Acceptance region

FIGURE 12.8
Acceptance and rejection regions for Bartlett's test for the homogeneity of variances

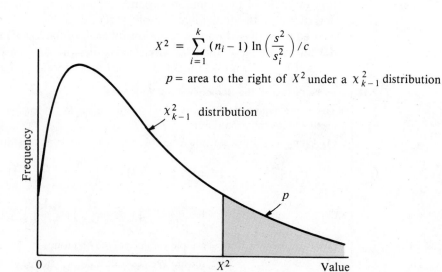

$$X^2 = \sum_{i=1}^{k} (n_i - 1) \ln\left(\frac{s^2}{s_i^2}\right) / c$$

$p =$ area to the right of X^2 under a χ^2_{k-1} distribution

χ^2_{k-1} distribution

FIGURE 12.9
Computation of the exact p-value for Bartlett's test for the homogeneity of variances

EXAMPLE 12.19   **Pulmonary Disease** Test for the homogeneity of variances for the FEF data in Table 12.1 (p. 475).

SOLUTION   From Table 12.1,

$$s_1 = 0.79, \; s_2 = 0.77, \; s_3 = 0.86, \; s_4 = 0.78, \; s_5 = 0.81, \; s_6 = 0.82$$

$$n_1 = n_2 = n_4 = n_5 = n_6 = 200 \quad \text{and} \quad n_3 = 50$$

Furthermore, from Table 12.3 (p. 481),

$$s^2 = 0.636 = (0.797)^2$$

Thus,

$$\lambda = 2.326 \left[199 \log_{10}\left(\frac{0.797}{0.79}\right)^2 + 199 \log_{10}\left(\frac{0.797}{0.77}\right)^2 + 49 \log_{10}\left(\frac{0.797}{0.86}\right)^2 \right.$$

$$\left. + 199 \log_{10}\left(\frac{0.797}{0.78}\right)^2 + 199 \log_{10}\left(\frac{0.797}{0.81}\right)^2 + 199 \log_{10}\left(\frac{0.797}{0.82}\right)^2 \right]$$

$$= 2.326[199(0.008) + 199(0.030) + 49(-0.066) + 199(0.019) + 199(-0.014)$$

$$+ 199(-0.025)]$$

$$= 2.326(1.592 + 5.970 - 3.234 + 3.781 - 2.786 - 4.975)$$

$$= 2.326(0.348) = 0.809$$

Furthermore,

$$c = 1 + \frac{1}{3(5)} \left[\frac{1}{199} + \frac{1}{199} + \frac{1}{49} + \frac{1}{199} + \frac{1}{199} + \frac{1}{199} - \frac{1}{1044} \right] = 1 + \frac{1}{15}(0.045)$$

$$= 1.003$$

Thus,

$$X^2 = 0.809/1.003 = 0.807 \sim \chi_5^2 \text{ under } H_0$$

From Table 6 in Appendix 1,

$$\chi_{5,.95}^2 = 11.07 > 0.807 = X^2$$

Thus, H_0, that the variances are *not* significantly different, is accepted. This result justifies the use of the one-way ANOVA procedure on these data. ∎

     The question arises as to what can be done in comparing the means of k groups if the variances *are* found to be significantly different using Bartlett's test. One possibility is to rescale the data using either a log or another transformation in an attempt to make the variances of the groups more homogeneous. Also, such a transformation often has the effect of stabilizing the underlying distribution and making the data appear more bell-shaped or at least more symmetric. If there is still a demonstrated heterogeneity of variances after transforming the data or if transforming the data does not seem advisable, then the groups can be compared in a pairwise fashion using t tests with either equal or unequal variances, as given in

Chapter 8, since there is no obvious advantage in considering the k groups together. The general procedure for comparing the means of k independent samples is summarized in Figure 12.10.

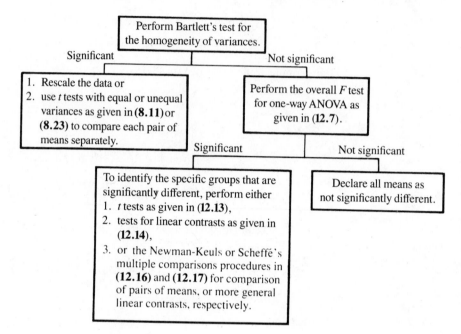

FIGURE 12.10

General procedure for comparing the means of k independent normally distributed samples

SECTION 12.6 **The Kruskal–Wallis Test**

In some instances we wish to compare means among more than two samples, but either the underlying distribution is far from being normal or we have ordinal data. In these situations, a nonparametric alternative to the one-way ANOVA described earlier in this chapter must be developed.

EXAMPLE 12.20   **Ophthalmology** Arachidonic acid is well known to have an effect on ocular metabolism. In particular, topical application of arachidonic acid has caused lid closure, itching, and ocular discharge, among other effects. A study was conducted to compare the anti-inflammatory effects of four different drugs in albino rabbits after administration of arachidonic acid [3]. For each group, one of the four drugs was administered to one eye and a saline solution was administered to the other eye. Ten minutes later arachidonic acid (sodium arachidonate) was delivered to both eyes. Both eyes were evaluated every 15 minutes thereafter for lid closure. At each assessment the lids of both eyes were examined and a lid closure score from 0–3 was determined, where 0 = eye completely open, 3 = eye completely closed, and 1, 2 = intermediate states. The measure of effectiveness (x) is the change in lid-closure scores (from baseline to follow-up) in the treated eye minus the change in lid-closure scores in the saline eye. A high value for x is indicative of an effective drug. The results, after 15 minutes of follow-up, are presented in Table 12.6. Since the scale of measurement was ordinal (0, 1, 2, 3), the use of a nonparametric technique to compare the four treatment groups is appropriate.

∎

TABLE 12.6
Ocular anti-inflammatory effects of four drugs on lid closure after administration of arachidonic acid

Rabbit number	Indomethicin		Aspirin		Piroxicam		BW755C	
	Score*	Rank	Score	Rank	Score	Rank	Score	Rank
1	+2	13.5	+1	9.0	+3	20.0	+1	9.0
2	+3	20.0	+3	20.0	+1	9.0	0	4.0
3	+3	20.0	+1	9.0	+2	13.5	0	4.0
4	+3	20.0	+2	13.5	+1	9.0	0	4.0
5	+3	20.0	+2	13.5	+3	20.0	0	4.0
6	0	4.0	+3	20.0	+3	20.0	−1	1.0

* (Lid-closure score at baseline − lid-closure score at 15 minutes)$_{\text{treated eye}}$ − (lid-closure score at baseline − lid-closure score at 15 minutes)$_{\text{saline eye}}$

We would like to generalize the Wilcoxon rank sum test to enable us to compare more than two samples. To accomplish this aim, the observations in all treatment groups are pooled and ranks are assigned to each observation in the combined sample. The average ranks ($\bar{R}_i$) in the individual treatment groups are then compared. If the average ranks are close to each other, then H_0, that the treatments are equally effective, will be accepted. If the average ranks are far apart, then H_0 will be rejected and we will conclude that at least some of the treatments are different. The test procedure for accomplishing this goal is known as the Kruskal–Wallis test.

12.19 The Kruskal–Wallis Test

To compare the means of k samples ($k > 2$) using nonparametric methods, use the following procedure:

(1) Pool the observations over all samples, thus constructing a combined sample of size $N = \Sigma n_i$.

(2) Assign ranks to the individual observations, using the average rank in the case of tied observations.

(3) Compute the rank sum R_i for each of the k samples.

(4) If there are no ties, compute the test statistic

$$H = H^* = \frac{12}{N(N + 1)} \times \sum_{i=1}^{k} \frac{R_i^2}{n_i} - 3(N + 1)$$

If there are ties, compute the test statistic

$$H = \frac{H^*}{1 - \dfrac{\sum_{j=1}^{g} (t_j^3 - t_j)}{N^3 - N}}$$

where t_j refers to the jth cluster of tied observations.

(5) For a level α test, if

$$H > \chi^2_{k-1, 1-\alpha}$$

then reject H_0; if

$$H \leqslant \chi^2_{k-1, 1-\alpha}$$

then accept H_0.

(6) To assess statistical significance, the *p*-value is given by

$$p = Pr(\chi^2_{k-1} > H)$$

(7) This test procedure should be used only if minimum $n_i \geq 5$ (i.e., if the smallest sample size for an individual group is at least 5).

The acceptance and rejection regions for this test are depicted in Figure 12.11. The computation of the exact *p*-value is given in Figure 12.12.

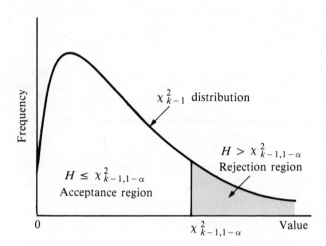

FIGURE 12.11
Acceptance and rejection regions for the Kruskal–Wallis test

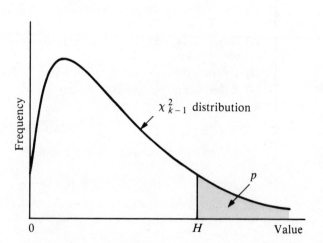

FIGURE 12.12
Computation of the exact *p*-value for the Kruskal–Wallis test

EXAMPLE 12.21   **Ophthalmology** Apply the Kruskal–Wallis test procedure to the ocular data in Table 12.6 and assess the statistical significance of the results.

SOLUTION   First pool the samples together and assign ranks to the individual observations. This procedure is performed in Table 12.7, with ranks given in Table 12.6.

Lid-closure score	Frequency	Range of ranks	Average rank
-1	1	1	1.0
0	5	2–6	4.0
$+1$	5	7–11	9.0
$+2$	4	12–15	13.5
$+3$	9	16–24	20.0

Then compute the rank sum in the four treatment groups:

$$R_1 = 13.5 + 20.0 + \cdots + 4.0 = 97.5$$

$$R_2 = 9.0 + 20.0 + \cdots + 20.0 = 85.0$$

$$R_3 = 20.0 + 9.0 + \cdots + 20.0 = 91.5$$

$$R_4 = 9.0 + 4.0 + \cdots + 1.0 = 26.0$$

Since there are ties, compute the Kruskal–Wallis test statistic H as follows:

$$H = \frac{\dfrac{12}{24 \times 25} \times \left[\dfrac{97.5^2}{6} + \dfrac{85.0^2}{6} + \dfrac{91.5^2}{6} + \dfrac{26.0^2}{6}\right] - 3(25)}{1 - \dfrac{(1^3 - 1) + (5^3 - 5) + (5^3 - 5) + (4^3 - 4) + (9^3 - 9)}{(24^3 - 24)}}$$

$$= \frac{0.020 \times 4296.583 - 75}{1 - \dfrac{1020}{13,800}} = \frac{10.932}{0.926} = 11.806$$

To assess statistical significance, compare H with a chi-square distribution with $k - 1 = 4 - 1 = 3$ df. Note from Table 6 of Appendix 1 that $\chi^2_{3,.99} = 11.34$, $\chi^2_{3,.995} = 12.84$. Since $11.34 < H < 12.84$, it follows that $.005 < p < .01$. Thus, there is a significant difference in the anti-inflammatory potency of the four drugs. ∎

Note that although the sample sizes in the individual treatment groups were the same in Table 12.6, the Kruskal–Wallis test procedure can, in fact, be used for samples of unequal size. Also, if there are no ties, the Kruskal–Wallis test statistic H in **(12.19)** can be written in the form

12.20

$$H = \frac{12}{N(N + 1)} \sum_{i=1}^{k} n_i (\bar{R}_i - \bar{\bar{R}})^2$$

where $\bar{R}_i$ = average rank in the ith sample and $\bar{\bar{R}}$ = average rank over all samples combined. Thus, if the average rank is about the same in all samples, then $|\bar{R}_i - \bar{\bar{R}}|$ will tend to be small and H_0 will be accepted. On the contrary, if the average rank is very different across samples, then $|\bar{R}_i - \bar{\bar{R}}|$ will tend to be large and H_0 will be rejected.

The test procedure in **(12.19)** is only applicable if minimum $n_i \geq 5$. If one of the sample sizes is smaller than 5, then either the sample should be combined with another sample, or special small-sample tables should be utilized. Table 14 in the Appendix provides critical values for selected sample sizes for the case of three samples (i.e., $k = 3$). The procedure for using this table is as follows:

(1) Reorder the samples so that $n_1 \leqslant n_2 \leqslant n_3$, that is, so that the first sample has the smallest sample size, whereas the third sample has the largest sample size.

(2) For a level α test, refer to the α column and the row corresponding to the sample sizes n_1, n_2, n_3 to find the critical value c.

(3) If $H \geqslant c$, then reject H_0 at level α (i.e., $p < \alpha$); if $H < c$, then accept H_0 at level α (i.e., $p \geqslant \alpha$).

EXAMPLE 12.22   Suppose there are three samples of sizes 2, 4, and 5 and $H = 6.141$. Assess the statistical significance of the results.

SOLUTION   Refer to the $n_1 = 2$, $n_2 = 4$, $n_3 = 5$ row. The critical values for $\alpha = .05$ and $\alpha = .02$ are 5.273 and 6.541, respectively. Since $H \geqslant 5.273$, it follows that the results are statistically significant $(p < .05)$. Since $H < 6.541$, it follows that $p \geqslant .02$. Thus, $.02 \leqslant p < .05$.   ■

12.6.1   Comparison of Specific Groups Under the Kruskal–Wallis Test

In Example 12.21 we determined that the treatments were not all equally effective. To determine which pairs of treatment groups are different, use the following procedure:

12.21   **Comparison of Specific Groups Under the Kruskal–Wallis Test (Dunn Procedure)**

To compare the i_1th and i_2th treatment groups under the Kruskal–Wallis test, use the following procedure:

(1) Compute

$$z = \frac{\bar{R}_{i_1} - \bar{R}_{i_2}}{\sqrt{\dfrac{N(N+1)}{12} \times \left(\dfrac{1}{n_{i_1}} + \dfrac{1}{n_{i_2}}\right)}}$$

(2) For a two-sided level α test, if

$$|z| > z_{1-\alpha^*}$$

then reject H_0; if

$$|z| \leqslant z_{1-\alpha^*}$$

then accept H_0, where

$$\alpha^* = \frac{\alpha}{k(k-1)}$$

The acceptance and rejection regions for this test are depicted in Figure 12.13.

$N(0, 1)$ distribution = distribution of z in **(12.21)** under H_0

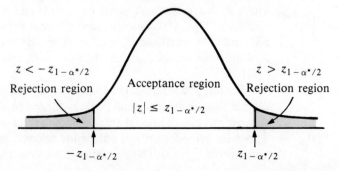

$z < -z_{1-\alpha^*/2}$
Rejection region

Acceptance region

$z > z_{1-\alpha^*/2}$
Rejection region

$|z| \leq z_{1-\alpha^*/2}$

$-z_{1-\alpha^*/2}$ $z_{1-\alpha^*/2}$

FIGURE 12.13
Acceptance and
rejection regions for the
Dunn procedure

EXAMPLE 12.23   **Ophthalmology** Determine which specific groups are different using the ocular data in Table 12.6.

SOLUTION   From Example 12.21,

$$\bar{R}_1 = \frac{97.5}{6} = 16.25$$

$$\bar{R}_2 = \frac{85.0}{6} = 14.17$$

$$\bar{R}_3 = \frac{91.5}{6} = 15.25$$

$$\bar{R}_4 = \frac{26.0}{6} = 4.33$$

Therefore, the following test statistics are used to compare each pair of groups:

$$\text{Groups 1 and 2: } z_{12} = \frac{16.25 - 14.17}{\sqrt{\frac{24 \times 25}{12} \times \left(\frac{1}{6} + \frac{1}{6}\right)}} = \frac{2.08}{4.082} = 0.51$$

$$\text{Groups 1 and 3: } z_{13} = \frac{16.25 - 15.25}{4.082} = \frac{1.0}{4.082} = 0.24$$

$$\text{Groups 1 and 4: } z_{14} = \frac{16.25 - 4.33}{4.082} = \frac{11.92}{4.082} = 2.92$$

$$\text{Groups 2 and 3: } z_{23} = \frac{14.17 - 15.25}{4.082} = \frac{-1.08}{4.082} = -0.26$$

$$\text{Groups 2 and 4: } z_{24} = \frac{14.17 - 4.33}{4.082} = \frac{9.84}{4.082} = 2.41$$

$$\text{Groups 3 and 4: } z_{34} = \frac{15.25 - 4.33}{4.082} = \frac{10.92}{4.082} = 2.68$$

The critical value for $\alpha = .05$ is given by $z_{1-\alpha^*}$, where

$$\alpha^* = \frac{.05}{4 \times 3} = .0042$$

From Table 3 in Appendix 1, $\Phi(2.635) = .9958 = 1 - .0042$. Thus $z_{1-.0042} = z_{.9958} = 2.635$ is the critical value. Since z_{14} and z_{34} are greater than the critical value, it follows that Indomethicin (group 1) and Piroxicam (group 3) have significantly better anti-inflammatory properties than BW755C (group 4), whereas the other treatment comparisons are not statistically significant.   ∎

SECTION 12.7   **Two-Way Analysis of Variance—General Model**

In Sections 12.1 through 12.5, the relationship between pulmonary function and cigarette smoking was used to illustrate the one-way analysis of variance. In this example, groups were defined by only one variable, cigarette smoking habit. In some instances, the groups being considered can be classified by two different variables and thus can be arranged in the form of an $r \times c$ contingency table. We would like to be able to look at the effects of each variable after controlling for the effects of the other variable. The latter type of data is usually analyzed using a technique known as the **two-way analysis of variance**.

EXAMPLE 12.24   **Cardiovascular Disease** Consider the data in Table 12.8. We are interested in the effects of OC use and cigarette smoking on serum cholesterol level. The effects of OC use and cigarette smoking may be independent or they may be related or "interact" with each other. One approach to the problem would be to construct a two-way ANOVA model predicting serum cholesterol level as a function of OC use and cigarette smoking.   ∎

TABLE 12.8
Data illustrating the effects of OC use and cigarette smoking in 1976 on serum cholesterol level in 1976

Cigarette smoking status		OC use	
		Current OC user	Non-OC user
Current smokers	Mean	206.15	188.58
	sd	38.32	29.62
	n	34	33
Noncurrent smokers	Mean	207.54	185.34
	sd	35.86	35.52
	n	28	38

DEFINITION 12.10 ▪▪
An **interaction effect** between two variables is defined as one in which the effect of one variable depends on the level of the other variable.   ∎

EXAMPLE 12.25   **Cardiovascular Disease** Suppose hypothetically that 1976 OC users have serum cholesterol levels that are 20 mg% higher than those of 1976 non-OC users. Suppose, however, that among the subgroup of women who are cigarette smokers, there is a 40 mg% difference in serum cholesterol levels between those women who are and who are not OC users. Correspondingly, among the subgroup of women who are nonsmokers, there is a 10 mg% difference in serum cholesterol levels between those who are and who are not OC users. This relationship would be an example of an interaction effect between OC use and cigarette smoking on serum cholesterol levels.   ∎

The general model for the two-way analysis of variance is given as follows:

12.22   **Two-Way Analysis of Variance—General Model**

$$y_{ijk} = \mu + \alpha_i + \beta_j + \gamma_{ij} + e_{ijk}$$

where
y_{ijk} is the serum cholesterol of the kth woman in the ith cigarette smoking group and the jth OC use group

μ is a constant

α_i is a constant representing the effect of cigarette smoking

β_j is a constant representing the effect of OC use

γ_{ij} is a constant representing the interaction effect between cigarette smoking and OC use

e_{ijk} is an error term, which is assumed to be normally distributed with mean 0 and variance σ^2

By convention,

$$\sum_{i=1}^{r} \alpha_i = \sum_{j=1}^{c} \beta_j = 0, \qquad \sum_{j=1}^{c} \gamma_{ij} = 0 \qquad \text{for all } i$$

$$\sum_{i=1}^{r} \gamma_{ij} = 0 \qquad \text{for all } j$$

Thus, from (**12.22**), y_{ijk} is normally distributed with mean $\mu + \alpha_i + \beta_j + \gamma_{ij}$ and variance σ^2.

<u>SECTION 12.8</u> **Hypothesis Testing in Two-Way ANOVA**

Let us denote the mean serum cholesterol for the ith row and jth column by $\bar{y}_{ij.}$, the mean cholesterol for the ith row by $\bar{y}_{i..}$, the mean cholesterol for the jth column by $\bar{y}_{.j.}$, and the overall mean by $\bar{y}_{...}$. The deviation of an individual observation from the overall mean can be represented as follows:

12.23
$$y_{ijk} - \bar{y}_{...} = (y_{ijk} - \bar{y}_{ij.}) + (\bar{y}_{i..} - \bar{y}_{...}) + (\bar{y}_{.j.} - \bar{y}_{...}) + (\bar{y}_{ij.} - \bar{y}_{i..} - \bar{y}_{.j.} + \bar{y}_{...})$$

DEFINITION 12.11 ■■
The first term on the right-hand side $(y_{ijk} - \bar{y}_{ij.})$ represents the deviation of an individual observation from the group mean for that observation. This expression is an indication of *within-group variability* and is the **error term**. ■

DEFINITION 12.12 ■■
The second term on the right-hand side $(\bar{y}_{i..} - \bar{y}_{...})$ represents the deviation of the mean of the ith row from the overall mean and is the **row effect**. ■

DEFINITION 12.13 ■■
The third term on the right-hand side $(\bar{y}_{.j.} - \bar{y}_{...})$ represents the deviation of the mean of the jth column from the overall mean and is the **column effect**. ■

DEFINITION 12.14 ■■
The fourth term on the right-hand side

$$(\bar{y}_{ij.} - \bar{y}_{i..} - \bar{y}_{.j.} + \bar{y}_{...}) = (\bar{y}_{ij.} - \bar{y}_{i..}) - (\bar{y}_{.j.} - \bar{y}_{...})$$

represents the deviation of the column effect in the ith row $(\bar{y}_{ij.} - \bar{y}_{i..})$ from the overall column effect $(\bar{y}_{.j.} - \bar{y}_{...})$ and is the **interaction effect**. ■

We would like to test the following hypotheses concerning these data:

(1) Test for the presence of row effects: H_0: all $\alpha_i = 0$ versus H_1: at least one $\alpha_i \neq 0$. This is a test for the effect of cigarette smoking on serum cholesterol levels after controlling for the effect of OC use.

(2) Test for the presence of column effects: H_0: all $\beta_j = 0$ versus H_1: at least one $\beta_j \neq 0$. This is a test for the effect of OC use on serum cholesterol levels after controlling for the effect of cigarette smoking.

(3) Test for the presence of interaction effects: H_0: all $\gamma_{ij} = 0$ versus H_1: at least one $\gamma_{ij} \neq 0$. This is a test of whether or not there is a differential effect of OC use among different cigarette smoking groups. For example, OC use may have an effect on serum cholesterol only among women who are smokers.

12.8.1   **The Method of Unweighted Means**

DEFINITION 12.15

The **total sum of squares** (total SS) for two-way ANOVA is the sum of $(y_{ijk} - \bar{y}_{..})^2$ over all observations over all samples. In symbols,

$$\text{Total SS} = \sum_{i=1}^{r} \sum_{j=1}^{c} \sum_{k=1}^{n_{ij}} (y_{ijk} - \bar{y}_{..})^2$$ ∎

We would like to decompose the total sum of squares into separate components that represent row, column, and interaction effects and perform exact significance tests based on these components. This strategy is only possible here under special circumstances, either when the number of observations is the same in each row and column combination or when the number of observations in each column within any particular row is proportional to the total number of observations in that column.

EXAMPLE 12.26   In Table 12.8 the number of observations is different for each row and column combination. Furthermore, the proportionality assumption does not hold, since there are 62 observations in the first column (34 + 28) out of a total of 133 observations (34 + 33 + 28 + 38), or 46.6%, whereas in the first row there are 34 out of 67 observations in the first column, or 50.7%. ∎

Since these conditions rarely occur in practice, except in designed experiments, we will not make this assumption. Instead, an approximate test procedure known as the **method of unweighted means,** which does not make such strong assumptions for the number of observations in specific rows and columns, will be used. However, this test procedure is only valid if the number of observations in different row-column combinations is not too different. To be more precise, there should not be more than a *twofold variation* in the number of observations in the rc row-column combinations.

12.24 | **Method of Unweighted Means for Two-Way ANOVA**

Let n_{ij} = the number of units in the ith row and jth column. To test for row, column, and interaction effects for two-way ANOVA take the following steps:

(1) Compute the Row Sum of Squares (Row SS)

$$\frac{\sum_{i=1}^{r} y_{i.}^{*2}}{c} - \frac{y_{..}^{*2}}{rc}$$

where

$$y_{i.}^{*} = \sum_{j=1}^{c} \bar{y}_{ij}$$

$$y_{..}^{*} = \sum_{i=1}^{r} \sum_{j=1}^{c} \bar{y}_{ij}$$

(2) Compute the Row Mean Square (Row MS) = Row SS/$(r - 1)$.

(3) Compute the Column Sum of Squares (Column SS)

$$\frac{\sum_{j=1}^{c} y_{.j}^{*2}}{r} - \frac{y_{..}^{*2}}{rc} \qquad \text{where} \qquad y_{.j}^{*} = \sum_{i=1}^{r} \bar{y}_{ij}$$

(4) Compute the Column Mean Square (Column MS) = Column SS/$(c - 1)$.

(5) Compute the Interaction Sum of Squares (Interaction SS)

$$\sum_{i=1}^{r} \sum_{j=1}^{c} \bar{y}_{ij}^2 - \frac{y_{..}^{*2}}{rc} - \text{Row SS} - \text{Column SS}$$

(6) Compute the Interaction Mean Square (Interaction MS)

$$\text{Interaction SS}/[(r - 1)(c - 1)]$$

(7) Compute the Error Mean Square (Error MS)

$$\sum_{i=1}^{r} \sum_{j=1}^{c} (n_{ij} - 1)s_{ij}^2/[(n - rc)n_h] = \left[\sum_{i=1}^{r} \sum_{j=1}^{c} \sum_{k=1}^{n_{ij}} y_{ijk}^2 - \sum_{i=1}^{r} \sum_{j=1}^{c} \frac{y_{ij.}^{*2}}{n_{ij}} \right] \bigg/ [(n - rc)n_h]$$

where

$$y_{ij.}^* = \sum_{k=1}^{n_{ij}} y_{ijk}, \quad n = \sum_{i=1}^{r} \sum_{j=1}^{c} n_{ij}$$

and n_h is defined by

$$\frac{1}{n_h} = \left(\sum_{i=1}^{r} \sum_{j=1}^{c} 1/n_{ij} \right) \bigg/ rc$$

(8) Perform a test for **row effects** (H_0: all $\alpha_i = 0$ versus H_1: at least one $\alpha_i \neq 0$) as follows:

(a) Compute the test statistic

$$F_{\text{ROW}} = \text{Row MS/Error MS}$$

which follows an $F_{r-1, n-rc}$ distribution under H_0.

(b) For a level α test, if

$$F_{\text{ROW}} > F_{r-1, n-rc, 1-\alpha}$$

then reject H_0: if $\qquad\qquad F_{\text{ROW}} \leqslant F_{r-1, n-rc, 1-\alpha}$

then accept H_0.

(c) The exact p-value = p_{ROW} is given by the area to the right of F_{ROW} under an $F_{r-1, n-rc}$ distribution = $Pr(F_{r-1, n-rc} > F_{\text{ROW}})$.

(9) Perform a test for **column effects** (H_0: all $\beta_j = 0$ versus H_1: at least one $\beta_j \neq 0$) as follows:

(a) Compute the test statistic

$$F_{\text{COLUMN}} = \text{Column MS/Error MS}$$

which follows an $F_{c-1, n-rc}$ distribution under H_0.

(b) If $\qquad\qquad\qquad F_{\text{COLUMN}} > F_{c-1, n-rc, 1-\alpha}$

then reject H_0; if $\qquad\qquad F_{\text{COLUMN}} \leqslant F_{c-1, n-rc, 1-\alpha}$

then accept H_0.

(c) The exact p-value = p_{COLUMN} is given by the area to the right of F_{COLUMN} under an $F_{c-1, n-rc}$ distribution = $Pr(F_{c-1, n-rc} > F_{\text{COLUMN}})$.

(10) Perform a test for **interaction effects** (H_0: all $\gamma_{ij} = 0$ versus H_1: at least one $\gamma_{ij} \neq 0$) as follows:

(a) Compute the test statistic

$$F_{\text{INT}} = \text{Interaction MS/Error MS}$$

which follows an $F_{(r-1) \times (c-1), n-rc}$ distribution under H_0.

(b) If
$$F_{INT} > F_{(r-1) \times (c-1), n-rc, 1-\alpha}$$

then reject H_0: if
$$F_{INT} \leq F_{(r-1) \times (c-1), n-rc, 1-\alpha}$$

then accept H_0.

(c) The exact p-value $= p_{INT}$ is given by the area to the right of F_{INT} under an $F_{(r-1) \times (c-1), n-rc}$ distribution $= Pr(F_{(r-1) \times (c-1), n-rc} > F_{INT})$.

(11) This test should be used only if there is no more than a twofold variation in the number of observations in the rc row-column combinations.

The results can be displayed in the form of an ANOVA table, as shown in Table 12.9.

TABLE 12.9 ANOVA table for method of unweighted means

Source of variation	SS	df	MS	F Statistic	p-value
Row	$\dfrac{\sum_{i=1}^{r} y_{i..}^{*2}}{c} - \dfrac{y_{...}^{*2}}{rc}$	$r-1$	SS/$(r-1)$	$F_{ROW} = \dfrac{\text{Row MS}}{\text{Error MS}}$	$Pr(F_{r-1, n-rc} > F_{ROW})$
Column	$\dfrac{\sum_{j=1}^{c} y_{.j.}^{*2}}{r} - \dfrac{y_{...}^{*2}}{rc}$	$c-1$	SS/$(c-1)$	$F_{COLUMN} = \dfrac{\text{Column MS}}{\text{Error MS}}$	$Pr(F_{c-1, n-rc} > F_{COLUMN})$
Interaction	$\left(\sum_{i=1}^{r} \sum_{j=1}^{c} \bar{y}_{ij}^2 - \dfrac{y_{...}^{*2}}{rc} \right.$ $\left. - \text{Row SS} - \text{Column SS} \right)$	$(r-1)(c-1)$	SS/$[(r-1)(c-1)]$	$F_{INT} = \dfrac{\text{Interaction MS}}{\text{Error MS}}$	$Pr[F_{(r-1) \times (c-1), n-rc} > F_{INT}]$
Error		$n-rc$	$\displaystyle\sum_{i=1}^{r} \sum_{j=1}^{c} \dfrac{(n_{ij}-1)s_{ij}^2}{(n-rc)n_h}$ $= \dfrac{\sum_{i=1}^{r} \sum_{j=1}^{c} \sum_{k=1}^{n_{ij}} y_{ijk}^2 - \sum_{i=1}^{r} \sum_{j=1}^{c} y_{ij.}^{*2}/n_{ij}}{(n-rc)n_h}$		

EXAMPLE 12.27   **Cardiovascular Disease** Apply the method of unweighted means to the cholesterol data in Table 12.8 to assess the effects of cigarette smoking, OC use, and the interaction of these effects on cholesterol levels.

SOLUTION   First check the validity of using the method of unweighted means on these data. The maximum variation in the number of observations in particular cells is from 28 to 38. Since $38/28 = 1.36 < 2$, it follows from rule 11 in **(12.24)** that this method is valid for these data. Now compute

$$y_{1.}^* = 206.15 + 188.58 = 394.73$$

$$y_{2.}^* = 207.54 + 185.34 = 392.88$$

$$y_{.1}^* = 206.15 + 207.54 = 413.69$$

$$y_{.2}^* = 188.58 + 185.34 = 373.92$$

$$y_{..}^* = 206.15 + 188.58 + 207.54 + 185.34 = 787.61$$

We then have

(1) Row SS $= \dfrac{(394.73)^2 + (392.88)^2}{2} - \dfrac{(787.61)^2}{4}$

$= 155{,}083.234 - 155{,}082.378 = 0.856$

(2) Row MS = Row SS/1 = 0.856

(3) Column SS = $\dfrac{(413.69)^2 + (373.92)^2}{2} - \dfrac{(787.61)^2}{4}$

$$= 155{,}477.791 - 155{,}082.378 = 395.413$$

(4) Column MS = Column SS/1 = 395.413

(5) Interaction SS = $(206.15)^2 + (188.58)^2 + (207.54)^2 + (185.34)^2$

$$-\dfrac{(787.61)^2}{4} - \text{Row SS} - \text{Column SS}$$

$$= 155{,}484.006 - 155{,}082.378 - 0.856 - 395.413 = 5.359$$

(6) Interaction MS = Interaction SS/(1 × 1) = 5.359

(7) Now compute the Error MS. We first need to compute n_h.

$$1/n_h = \left[\left(\frac{1}{34} + \frac{1}{33} + \frac{1}{28} + \frac{1}{38}\right)\middle/ 4\right] = 0.03044 \quad \text{and} \quad n_h = \frac{1}{0.03044} = 32.85$$

Thus,

$$\text{Error MS} = [33(38.32)^2 + 32(29.62)^2 + 27(35.86)^2 + 37(35.52)^2]/[(133 - 4)32.85]$$

$$= 157{,}935.134/[(129)(32.85)] = 37.27$$

(8) $F_{\text{ROW}} = 0.856/37.27 = 0.02 \sim F_{1,129}, \quad p > .05$

(9) $F_{\text{COLUMN}} = 395.413/37.27 = 10.61 \sim F_{1,129}, \quad .001 < p < .005$

(10) $F_{\text{INT}} = 5.359/37.27 = 0.14 \sim F_{1,129}, \quad p > .05$

These results are displayed in Table 12.10. Thus, there are significant column effects but nonsignificant row and interaction effects. This result implies that there are significant differences in cholesterol levels by OC-use groups but no significant differences in cholesterol by cigarette smoking groups. Furthermore, there are no significant interaction effects, which implies that the differences in cholesterol levels by OC use is the same for those who are or who are not current cigarette smokers. Thus, a reasonable interpretation of the data might be that current OC use is associated with elevated cholesterol levels, but current cigarette smoking is not. ∎

Source of variation	SS	df	MS	F Statistic	p-value
Cigarette smoking	0.856	1	0.856	0.02	NS
OC use	395.413	1	395.413	10.61	$.001 < p < .005$
Cigarette smoking × OC use	5.359	1	5.359	0.14	NS
Error		129	37.27		

TABLE 12.10
ANOVA table for cholesterol data in Table 12.8

The data in Table 12.8 could have been analyzed separately for the effects of cigarette smoking and OC use on serum cholesterol levels using either several two-sample t tests or several one-way ANOVAs. The advantage of using the two-way ANOVA for this and other similar problems is that we can look at the effect of one variable *after controlling for the effect of the other variable*. Thus, a two-way ANOVA can be thought of as a type of multiple regression in which both variables are controlled simultaneously, whereas performing several one-way ANOVAs can be thought of as doing several simple linear regressions. If the two variables under study are closely related to each other, then the results from these two approaches can be very different, and the two-way ANOVA is preferred.

12.8.2 **Unbalanced Designs in Two-Way ANOVA**

A key assumption with the method of unweighted means is that there is not too much variation in the number of observations for specific row-column combinations. If this criterion is not met, then the method of unweighted means is not valid, and a more exact test based on multiple regression methods must be used.

EXAMPLE 12.28

Hypertension, Nutrition A study was performed to look at the level of blood pressure in two different vegetarian groups, both compared with each other and with normals. A group of 226 strict vegetarians (SV), who eat no animal products of any kind, 63 lactovegetarians (LV), who eat dairy products but no other animal foods, and 460 normals (NOR), who eat a standard American diet, provided data for the study. Mean systolic blood pressure by dietary group and sex is given in Table 12.11. Note that the largest cell (normal males) has 240 subjects and the smallest cell (LV males) has 26 subjects, yielding a ratio of largest to smallest cell of $240/26 = 9.2 > 2$. Thus, the method of unweighted means cannot be used, and multiple regression methods must be used to analyze these data. ∎

TABLE 12.11
Mean systolic blood
pressure by dietary
group and sex

Dietary group		Sex Male	Female
SV	mean	109.9	102.6
	n	138	88
LV	mean	115.5	105.2
	n	26	37
NOR	mean	128.3	119.6
	n	240	220

The SAS General Linear Model procedure (PROC GLM) has been used to analyze the data. In particular, two "indicator" or "dummy" variables were set up to represent study group (x_1, x_2), whereby

$x_1 = 1$ if a person is in the first (SV) group

$= 0$ otherwise

$x_2 = 1$ if a person is in the second (LV) group

$= 0$ otherwise

Note that each group has a unique representation in terms of the variables x_1, x_2. Specifically, for people in the SV group, $x_1 = 1$ and $x_2 = 0$; for people in the LV group, $x_1 = 0$ and $x_2 = 1$; for people in the NOR group, $x_1 = 0$ and $x_2 = 0$. A variable x_3 is also included to represent sex, whereby

$x_3 = 1$ if male

$= 0$ if female

The multiple regression model can then be written as

| **12.25** |

$$y = \alpha + \beta_1 x_1 + \beta_2 x_2 + \beta_3 x_3 + e$$

The results from using the SAS procedure are shown in Table 12.12. The program first provides a test of the overall hypothesis $H_0: \beta_1 = \beta_2 = \beta_3 = 0$ versus H_1: at least one of the $\beta_i \neq 0$, as given in **(11.17)** (p. 427). The F statistic corresponding to this test is $105.85 \sim F_{3,745}$ under H_0, with p-value $<.001$. Thus, at least one of the effects (study group or sex) is significant. In the second part of the display, the program lists the type III SS and the corresponding F statistic (F value) and p-value ($Pr > F$). The type III SS provides an estimate of the effects of specific risk factors after controlling for the effects of all other variables in the model. Thus, to test the effect of study group after controlling for sex, we wish to test the hypothesis $H_0: \beta_1 = \beta_2 = 0$, $\beta_3 \neq 0$, versus H_1: at least one of $\beta_1, \beta_2 \neq 0$, $\beta_3 \neq 0$. The F statistic for this comparison is obtained by dividing the study MS $= (51,806.42/2) = 25,903.21$ by the error MS $= 195.89$, yielding 132.24 (except for roundoff error) $\sim F_{2,745}$ under H_0, and a p-value ($Pr > F$) of $<.001$. Thus, there are highly significant effects of dietary group on systolic blood pressure even after

TABLE 12.12 SAS GLM procedure output illustrating the effects of study group and sex on systolic blood pressure using the data set in Example 12.28

```
                                        SAS

                            GENERAL LINEAR MODELS PROCEDURE

DEPENDENT VARIABLE: MSYS

SOURCE            DF      SUM OF SQUARES      MEAN SQUARE     F VALUE       PR > F      R-SQUARE            C.V.

MODEL             3       62202.79213079     20734.26404360   105.85       0.0001      0.298854          11.8858

ERROR            745     145934.76850283      195.88559531                 ROOT MSE                     MSYS MEAN

CORRECTED TOTAL  748     208137.56063362                                  13.99591352                 117.75303516


SOURCE            DF         TYPE I SS      F VALUE    PR > F      DF         TYPE III SS     F VALUE    PR > F

STUDY             2       49146.49426085    125.45     0.0001      2       51806.42069945    132.24      0.0001
SEX               1       13056.29786994     66.65     0.0001      1       13056.29786994     66.65      0.0001


                       STUDY    PROB > |T|
                                I/J     1       2      3

                          1     1   .        0.0425  0.0001
                          2     2  0.0425   .        0.0001
                          3     3  0.0001  0.0001   .


                       SEX      PROB > |T|
                                   1       2

                          1        .     0.0001
                          2      0.0001    .


                                                    T FOR HO:      PR > |T|     STD ERROR OF
                 PARAMETER            ESTIMATE     PARAMETER=0                    ESTIMATE

                 INTERCEPT         119.75747587      141.53        0.0001       0.84614985
                 STUDY      1       -17.86546724     -15.66        0.0001       1.14061756
                            2       -13.79147908      -7.32        0.0001       1.88356205
                            3         0.00000000        .            .              .
                 SEX        1         8.42854624       8.16        0.0001       1.03239026
                            2         0.00000000        .            .              .
```

controlling for the effect of sex. Similarly, to test the effect of sex, we test the hypothesis $H_0: \beta_3 = 0$, at least one $\beta_1, \beta_2 \neq 0$, versus $H_1: \beta_3 \neq 0$, at least one $\beta_1, \beta_2 \neq 0$. The F statistic for the sex effect is given by $(13{,}056/1)/195.89 = 66.65 \sim F_{1,745}$ under H_0, $p < .001$. Thus, there are highly significant effects of sex as well after controlling for the effect of dietary group, with males having higher blood pressure than females. SAS also displays a type I SS as well as an associated F statistic and p-value. The purpose of the type I SS is to enter and test the variables in the order specified by the user. In this case, study group was specified first, and sex was specified second. Thus, the effect of study group is entered first (without controlling for sex), yielding an F statistic of $125.45 \sim F_{2,745}$ under H_0, $p < .001$. Second, the effect of sex is assessed after controlling for study group. This is the same hypothesis as was tested above using the type III SS. In general, except for the last user-specified risk factor, results from type I SS (where all variables above the current variable on the user-specified variable list are controlled for) and type III SS (where *all* other variables in the model are controlled for) will not necessarily be the same. Usually, unless we are interested in entering the variables in a pre-specified order, the hypotheses tested using the type III SS will be of greater interest.

Although there was a significant effect of study group after controlling for sex, this does not identify which specific dietary groups differ from each other on systolic blood pressure. For this purpose, t tests are provided comparing specific dietary groups ($1 = SV/2 = LV/3 = NOR$) after controlling for sex. Refer to the 3×3 table listed for STUDY under PROB > $|T|$. The (two-tailed) p-value comparing dietary group i versus dietary group j is given in the (i, j) cell of the table [as well as the (j, i) cell]. Thus, referring to the $(1, 2)$ cell, we see that the mean blood pressure of people in group 1 (SV) is significantly different from people in group 2 (LV) after controlling for sex ($p = .0425$). Similarly, referring to the $(1, 3)$ or $(3, 1)$ cells, we see that the mean blood pressure of people in group 1 is significantly different from the mean blood pressure of people in group 3 ($p = .0001$). Similar results are obtained from a comparison of people in groups 2 (LV) and 3 (NOR). Furthermore, a 2×2 table is listed for the sex effect, yielding a p-value for comparisons of the two sexes (refer to the $(1, 2)$ or $(2, 1)$ cell of the table) after controlling for the effect of study group ($p = .0001$). This test is actually superfluous in this instance, since there were only two groups under sex and the F test under type III SS for the sex effect is equivalent to the sex effect t test.

Finally, note that at the bottom of the display there are estimates of the regression parameters as well as their standard errors and associated t statistics. These have a similar interpretation to that of the multiple regression parameters in **(11.16)**. In particular, the regression coefficient $\beta_1 = -17.9$ mm Hg is an estimate of the difference in systolic blood pressure between the SV and NOR groups after controlling for the effect of sex. Similarly, the regression coefficient $\beta_2 = -13.8$ mm Hg is an estimate of the difference in systolic blood pressure between the LV and NOR groups after controlling for the effect of sex. Also, the estimated difference in systolic blood pressure between the SV and LV groups is given by $(-17.9 - (-13.8)) = -4.1$ mm Hg; thus, the strict vegetarians on average have systolic blood pressure 4.1 mm Hg lower than lactovegetarians after controlling for the effect of sex. Since there was no explicit parameter entered for the third study group, the program lists the default value of 0. Finally, the regression coefficient $\beta_3 = 8.4$ mm Hg, tells us that males have systolic blood pressure 8.4 mm Hg higher than females, even after controlling for the effect of study group.

It is possible to assess interaction effects for unbalanced designs (e.g., using the SAS PROC GLM program), but these were not included in this example for the sake of simplicity. A more detailed discussion of two-way and higher-way ANOVA in the context of unbalanced designs is given in Kleinbaum, Kupper, and Muller [2].

SECTION 12.9 Analysis of Covariance

We often will want to look at the relationship between one or more categorical variables and a continuous outcome variable. If there is one categorical variable, then one-way ANOVA can be used; if there are two (or more) categorical variables, then two-way (higher-way) ANOVA can be used. However, confounding variables are sometimes present, which makes it difficult to interpret these analyses.

EXAMPLE 12.29   **Hypertension, Nutrition** In Example 12.28, differences in systolic blood pressure (SBP) by dietary group and sex were presented using an unbalanced two-way ANOVA model. Highly significant differences were found among dietary groups, with mean SBP of SV < mean SBP of LV < mean SBP of NOR. However, there are other important differences between these groups, such as differences in weight, and possibly age, which might explain all or part of the apparent blood-pressure differences. How can we examine if these blood-pressure differences persist, after accounting for the confounding variables?   ■

The multiple regression model in **(12.25)** can be extended to allow for the effects of other covariates using the **analysis of covariance**. If weight is denoted by x_4 and age by x_5, then we have the multiple regression model

12.26
$$y = \alpha + \beta_1 x_1 + \beta_2 x_2 + \beta_3 x_3 + \beta_4 x_4 + \beta_5 x_5 + e$$

where $e \sim N(0, \sigma^2)$. We have fit this model using the SAS PROC GLM program as shown in Table 12.13.

Note from the top of Table 12.13 that the overall model is highly significant (F value = 103.16, $p = .0001$), indicating that some of the variables are having a significant effect on SBP. To identify the effects of specific variables, refer to the type III SS. Note that each of the risk factors has a significant effect on SBP after controlling for the effects of all other variables in the model ($p = .0001$). Finally, of key importance is whether there are differences in blood pressure by dietary group after controlling for the effects of age, sex, and weight. In this regard, different conclusions are reached than in Table 12.12. Referring to the t statistics for STUDY, we see that there are no significant differences in SBP between the strict vegetarians (group 1) and the lactovegetarians (group 2) after controlling for the other variables ($p = .7012$). There are still highly significant differences between each of the vegetarian groups and normals ($p = .0001$). Thus, there must have been differences in either age and/or weight between the SV and LV groups, which accounted for the significant blood-pressure difference between these groups in Table 12.12 ($p = .0425$). Finally, the estimates of specific regression parameters are given in the bottom of Table 12.13. Note that after controlling for age, sex, and weight, the estimated differences in mean SBP between the SV and NOR groups = $\beta_1 = -8.2$ mm Hg, between the LV and NOR groups = $\beta_2 = -9.0$ mm Hg, and

TABLE 12.13 SAS GLM procedure output illustrating the effects of study group, age, sex, and weight on systolic blood pressure using the data set in Example 12.28

SAS

GENERAL LINEAR MODELS PROCEDURE

DEPENDENT VARIABLE: MSYS

SOURCE	DF	SUM OF SQUARES	MEAN SQUARE	F VALUE	PR > F	R-SQUARE	C.V.
MODEL	5	85358.44910498	17071.68982100	103.16	0.0001	0.410402	10.9264
ERROR	741	122628.85342226	165.49103026		ROOT MSE		MSYS MEAN
CORRECTED TOTAL	746	207987.30252724			12.86433171		117.73630968

SOURCE	DF	TYPE I SS	F VALUE	PR > F	DF	TYPE III SS	F VALUE	PR > F
STUDY	2	49068.28440076	148.25	0.0001	2	8257.21427825	24.95	0.0001
SEX	1	13092.51273176	79.11	0.0001	1	4250.57708379	25.68	0.0001
AGE	1	12978.84918739	78.43	0.0001	1	10524.41438768	63.60	0.0001
WGT	1	10218.80278507	61.75	0.0001	1	10218.80278507	61.75	0.0001

```
                    STUDY    PROB > |T|
                             I/J    1       2       3

                             1   1   .     0.7012  0.0001
                             2   2  0.7012   .      0.0001
                             3   3  0.0001  0.0001   .


                    SEX      PROB > |T|
                             1       2

                             1   .      0.0001
                             2  0.0001    .
```

PARAMETER		ESTIMATE	T FOR HO: PARAMETER=0	PR > |T|	STD ERROR OF ESTIMATE
INTERCEPT		82.74987242	25.69	0.0001	3.22121552
STUDY	1	-8.22799340	-6.20	0.0001	1.32786689
	2	-8.95389632	-5.03	0.0001	1.78082376
	3	0.00000000	.	.	.
SEX	1	5.50352855	5.07	0.0001	1.08593669
	2	0.00000000	.	.	.
AGE		0.47488301	7.97	0.0001	0.05954906
WGT		0.13011703	7.86	0.0001	0.01655851

----

between the SV and LV groups $= \beta_1 - \beta_2 = -8.23 - (-8.95) = 0.7$ mm Hg. These differences are all much smaller than the estimated differences in Table 12.12, of -17.9 mm Hg, -13.8 mm Hg, and -4.1 mm Hg, respectively, where age and weight were not controlled for. The difference in SBP between males and females is also much smaller in Table 12.13 after controlling for age and weight (5.5 mm Hg) than in Table 12.12 (8.4 mm Hg), where these factors were not controlled for. Also, we see from Table 12.13 that the estimated effects of age and weight on SBP are 0.47 mm Hg per year and 0.13 mm Hg per lb, respectively. Thus, it is important to control for the effects of possibly confounding variables in performing regression analyses.

SECTION 12.10 **Summary**

In this chapter analysis of variance methods were studied. These methods enable us to relate outcome variables that are either continuous or ordinal to the levels of one or more categorical independent variables. If only a single categorical independent

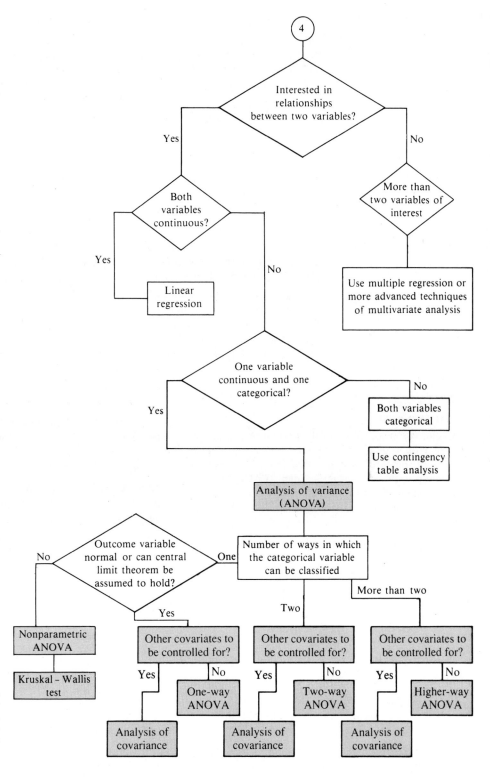

FIGURE 12.14
Flowchart for
appropriate methods
of statistical inference
(analysis of variance
methods are shaded)

variable is being considered and the dependent variable is normally distributed, then one-way analysis of variance (ANOVA) methods are appropriate. Using these methods, the hypothesis that the mean level of the dependent variable is different for different groups defined by the categorical variable can be tested. Which specific groups are different can also be identified, using t tests if the comparisons have been planned in advance or multiple comparison methods if they have not. If the dependent variable is ordinal or if it is far from being normally distributed, then a nonparametric analogue of one-way ANOVA is appropriate, namely, the Kruskal–Wallis test.

If the data are simultaneously stratified by two categorical variables, each of which we wish to simultaneously relate to a normally distributed outcome variable, then two-way ANOVA methods are appropriate. In this regard the method of unweighted means was introduced as a suitable method of analysis if the numbers of observations in the strata defined by the categorical variables are not that different. In other situations, in which there is an unbalanced design, the two-way ANOVA can be implemented using multiple regression methods. This procedure was illustrated using PROC GLM of SAS. If the data are stratified by more than two categorical variables, then higher-way ANOVA models, which are extensions of the methods in this chapter, can be used. Finally, to control for one or more confounding variables while looking at the effects of several categorical variables on a normally distributed outcome variable, the analysis of covariance can be used. The flowchart in Figure 12.14 summarizes these methods.

PROBLEMS

Nutrition, Arthritis

A comparison was made of the protein intake among three groups of premenopausal women: (1) women eating a standard American diet (STD), (2) women eating a lacto-ovo-vegetarian diet (LAC), and (3) women eating a strict vegetarian diet (VEG). The mean ± 1 sd for protein intake (mg) is presented in Table 12.14.

TABLE 12.14 Protein intake (mg) among three dietary groups of premenopausal women

Group	Mean	sd	n
STD	74	16	10
LAC	56	16	10
VEG	55	9	10

12.1 What parametric procedure can be used to compare the underlying variances of the three groups?

12.2 Implement the procedure in Problem 12.1 and report a p-value.

12.3 What parametric procedure can be used to compare the underlying means of the three groups?

12.4 Implement the procedure in Problem 12.3 using the critical value method.

12.5 Compare the underlying means of each specific pair of groups using the t test methodology.

12.6 Suppose that in the general population 70% of vegetarians are lactovegetarians, whereas 30% are strict vegetarians. Perform a statistical procedure to test if the contrast $L = 0.7\bar{y}_2 + 0.3\bar{y}_3 - \bar{y}_1$ is significantly different from 0. What does this contrast mean?

Similar data were collected for postmenopausal women (Table 12.15).

TABLE 12.15 Protein intake (mg) among three dietary groups of postmenopausal women

Group	Mean	sd	n
STD	75	9	10
LAC	57	13	10
VEG	47	17	6

12.7 Perform a statistical procedure to compare the variances of the three groups using the critical value method.

12.8 Perform a statistical procedure to compare the means of the three groups using the critical value method.

12.9 What is the p-value from the test performed in Problem 12.8?

12.10 Compare the means of each specific pair of groups using the t test methodology.

12.11 Perform the test indicated in Problem 12.6 for the postmenopausal women in Table 12.15.

12.12 Find the upper 5% point of the studentized range statistic based on three means, where the Within MS is based on 18 degrees of freedom.

12.13 Find the upper 1% point of the studentized range statistic based on five means, where the Within MS is based on 30 degrees of freedom.

12.14 Calculate the studentized range statistic for the group of three means in Table 12.14.

12.15 Using the data in Table 12.14, perform a multiple comparisons procedure to identify which specific underlying means are different.

12.16 Calculate the studentized range statistic for the group of three means in Table 12.15.

12.17 Using the data in Table 12.15, perform a multiple comparisons procedure to identify which specific underlying means are different.

12.18 Perform a multiple comparisons procedure to test if the linear contrasts in Problems 12.6 and 12.11 are significantly different from 0.

Pulmonary Disease

Twenty-two young asthmatic volunteers were studied to assess the short-term effects of sulfur dioxide (SO_2) exposure under various conditions [4]. The baseline data in Table 12.16 were presented regarding bronchial reactivity to SO_2 stratified by lung function (as defined by FEV_1/FVC) at screening.

12.19 Suppose we do not wish to assume normality. What nonparametric test can be used to compare the three groups?

12.20 Implement the test in Problem 12.19 and report a p-value.

Refer to Tables 12.14 and 12.15.

12.21 Suppose we wish to simultaneously look at the effects of dietary group and menopausal status on protein intake. What statistical procedure can be used?

TABLE 12.16 Relationship of bronchial reactivity to SO_2 (cm H_2O/s) grouped by lung function at screening among 22 asthmatic volunteers

Lung function group		
Group A FEV$_1$/FVC $\leqslant 74\%$	Group B FEV$_1$/FVC 75–84%	Group C FEV$_1$/FVC $\geqslant 85\%$
20.8	7.5	9.2
4.1	7.5	2.0
30.0	11.9	2.5
24.7	4.5	6.1
13.8	3.1	7.5
	8.0	
	4.7	
	28.1	
	10.3	
	10.0	
	5.1	
	2.2	

(Reprinted with permission of the *American Review of Respiratory Disease, 131*(2), 221–225, 1985.)

12.22 Implement the procedure in Problem 12.21. Discuss the significance of the effects of dietary group and menopausal status.

12.23 Is there a significant interaction effect between menopausal status and dietary group?

12.24 What does an interaction effect mean in Problem 12.23?

Cardiovascular Disease

Some reports in the literature have noted that there is an inverse relationship between moderate alcohol consumption and cholesterol levels. To test this hypothesis, a questionnaire is administered to a group of workers at a particular company as to their alcohol consumption, and blood samples are taken to measure their cholesterol levels. The workers are subdivided into those who report no alcohol consumption, those who drink $\leqslant 2$ oz of alcohol, and those who drink more than 2 oz of alcohol on an average day. The 23 nondrinkers had average cholesterol levels of 205.6 mg/dL with a standard deviation of 25.3. The 15 light drinkers had average cholesterol levels of 182.7 mg/dL with a standard deviation of 21.9. The 12 heavy drinkers had average cholesterol levels of 199.8 mg/dL with a standard deviation of 30.3.

12.25 In a one-way or two-way ANOVA appropriate here?

12.26 Test the hypothesis that there is an overall difference in underlying mean cholesterol levels among these three groups.

12.27 Use the method of multiple comparisons to test for significant differences in cholesterol levels between each pair of groups.

Hypertension

A recent phenomenon is the emergence of the automated blood-pressure device, which has appeared in many banks, drug stores, and other public places. A study was conducted to assess the comparability of the machine readings with those of the standard cuff [5]. Readings were taken using both the machine and the standard cuff at four separate locations. The results are given in Table 12.17. Suppose we wish to test if the mean difference between the machine and the standard cuff is consistent over the four locations (i.e., if the bias is comparable over all four locations).

12.28 Is a one-way or two-way ANOVA appropriate here?

12.29 Perform Bartlett's test for homogeneity of variance.

12.30 Test if the mean difference is consistent over all four locations.

12.31 Why was it necessary to perform Bartlett's test in Problem 12.29 before performing the test in Problem 12.30?

Cardiovascular Disease

Physical activity has been shown to have beneficial effects on cardiovascular disease outcomes. As part of a study of physical activity assessment methodology, the number of hours of sleep, light activity, moderate activity, hard activity, and very hard activity was computed for each person in the study [6]. These categories were defined in terms of levels of energy, or MET, where MET = ratio of working metabolic rate to resting metabolic rate. Using this classification, sleep = 1 MET; light activity = 1.1–2.9 MET; moderate activity = 3.0–5.0 MET; hard activity = 5.1–6.9 MET; very hard activity $\geq$ 7.0 MET. In Table 12.18 data relating the number of hours of moderate activity per week to age for males are presented.

TABLE 12.17 Mean systolic bp and difference between machine and human readings at four locations

Location	Systolic bp machine (mm Hg)			Systolic bp standard cuff (mm Hg)			Systolic bp machine minus systolic bp standard cuff (mm Hg)		
	Mean	sd	n	Mean	sd	n	Mean	sd	n
A	142.5	21.0	98	142.0	18.1	98	0.5	11.2	98
B	134.1	22.5	84	133.6	23.2	84	0.5	12.1	84
C	147.9	20.3	98	133.9	18.3	98	14.0	11.7	98
D	135.4	16.7	62	128.5	19.0	62	6.9	13.6	62

(Reprinted with permission of the American Heart Association. *Hypertension, 2*(2), 221–227, 1980.)

TABLE 12.18 Hours per week of moderate activity for men by age

Age group											
20–34			35–49			50–64			65–74		
Mean	sd	n	Mean	sd	n	Mean	sd	n	Mean	sd	n
8.1	10.4	487	9.7	10.2	233	7.9	10.2	191	5.8	10.3	82

(Reprinted with permission of the *American Journal of Epidemiology, 121*(1), 91–106, 1985.)

12.32 What is the appropriate method of analysis to test for the effect of age on the number of hours of moderate activity per week?

12.33 Perform the test mentioned in Problem 12.32 and report a *p*-value.

12.34 Comment on which specific age groups are different based on the data in Table 12.18.

12.35 Perform a test for whether or not there is a general trend for increasing or decreasing amounts of moderate activity by age based on the data in Table 12.18.

Psychiatry

For the purpose of identifying older nondemented people with early signs of senile dementia, a Mental Function Index was constructed based on three short tests of cognitive function. In Table 12.19, data relating the Mental Function Index at baseline to clinical status determined independently at baseline and follow-up, with a median follow-up period of 959 days, are presented [7].

12.36 What test procedure can be used to test for significant differences among groups?

12.37 Perform the test mentioned in Problem 12.36 and report appropriate *p*-values identifying differences between specific groups.

TABLE 12.19 Relationship between clinical status at baseline and follow-up (median follow-up period of 959 days) to mean Mental Function Index at baseline

Clinical status				
Baseline	Follow-up	Mean	sd	*n*
Normal	Unchanged	0.04	0.11	27
Normal	Questionably or mildly affected	0.22	0.17	9
Questionably affected	Progressed	0.43	0.35	7
Definitely affected	Progressed	0.76	0.58	10

(Reprinted with permission of the *American Journal of Epidemiology*, *121*(1), 91–106, 1985.)

Gastroenterology

A 1985 study was performed focusing on the protein concentration of duodenal secretions from patients with cystic fibrosis [8]. Table 12.20 provides data relating protein concentration to pancreatic function as measured by trypsin secretion.

12.38 If we do not wish to assume normality for these distributions, then what statistical procedure can be used to compare the three groups?

TABLE 12.20 Relationship between protein concentration (mg/mL) of duodenal secretions to pancreatic function as measured by trypsin secretion (u/(kg/hr))

	Trypsin secretion [*u*/(kg/hr)]				
⩽50		51–1000		>1000	
Subject number	Protein concentration	Subject number	Protein concentration	Subject number	Protein concentration
1	1.7	1	1.4	1	2.9
2	2.0	2	2.4	2	3.8
3	2.0	3	2.4	3	4.4
4	2.2	4	3.3	4	4.7
5	4.0	5	4.4	5	5.0
6	4.0	6	4.7	6	5.6
7	5.0	7	6.7	7	7.4
8	6.7	8	7.6	8	9.4
9	7.8	9	9.5	9	10.3
		10	11.7		

(Reprinted with permission of the *New England Journal of Medicine*, *312*(6), 329–334, 1985.)

12.39 Perform the test mentioned in Problem 12.38 and report a *p*-value.

Obstetrics

The birthweight of an infant has been hypothesized to be associated with the smoking status of the mother during the first trimester of pregnancy. This hypothesis is tested by recording the birthweights of infants and the smoking status of the mother during pregnancy for all mothers who register at the prenatal clinic at a particular hospital within a 1-month period. The mothers are divided into four groups according to smoking habit, and the sample of birthweights in pounds within each group is given as follows:

Group 1: Mother is a nonsmoker (NON):

> 7.5   6.2   6.9   7.4   9.2   8.3   7.6

Group 2: Mother is an exsmoker (smoked at some time prior to pregnancy but not during pregnancy) (EX):

> 5.8   7.3   8.2   7.1   7.8

Group 3: Mother is a current smoker and smokes less than 1 pack per day (CUR < 1):

> 5.9   6.2   5.8   4.7   8.3   7.2   6.2

Group 4: Mother is a current smoker and smokes greater than or equal to 1 pack per day (CUR ⩾ 1):

> 6.2   6.8   5.7   4.9   6.2   7.1   5.8   5.4

12.40 Should a one-way or two-way ANOVA be used on these data?

12.41 Test for the homogeneity of the variances in the four samples.

12.42 Are the mean birthweights different overall in the four groups?

12.43 Test for all differences between each pair of groups and summarize your results using a *t* test procedure.

12.44 Perform the same tests as in Problem 12.43 using the method of multiple comparisons.

12.45 Are there any differences between the results in Problems 12.43 and 12.44?

12.46 Suppose we assume that smokers of ⩾1 pack per day smoke an average of 1.3 packs per day, whereas smokers of <1 pack per day smoke an average of 0.5 pack per day. Use the method of linear contrasts to test if the amount of current cigarette consumption among *ever* smokers is significantly related to birthweight.

12.47 Use a multiple comparisons approach to answer the question posed in Problem 12.46.

Pharmacology

Suppose we wish to test the relative effects of three drugs (A, B, C) on the reduction of fever. Drug A is 100% aspirin; drug B is 50% aspirin and 50% other compounds; drug C is 25% aspirin and 75% other compounds. The drugs are prescribed to children aged 5–14 entering the outpatient ward of a hospital complaining of the "flu," with fever of 100.0°F to 100.9°F. The drugs are assigned in time-sequence order; that is, the first patient gets drug A, the second drug B, the third drug C, the fourth drug A, and so forth, until there is a set of 15 patients. The parents are then telephoned 4 hours after administration of the drug and the reduction in fever is noted. The results are given in Table 12.21. We assume that the timing of prescribing the drugs (i.e., when it is prescribed during the day) is irrelevant to the reduction in fever.

12.48 What are the appropriate null and alternative hypotheses to test whether or not all three drugs are equally effective?

12.49 Test for the homogeneity of variances in the three groups.

12.50 Perform the significance test in Problem 12.48.

12.51 Summarize the differences in treatment effects among the three drugs using the method of multiple comparisons.

TABLE 12.21 Reduction in fever for patients getting different doses of aspirin

Drug		Mean (°F)	sd (°F)	*n*
Drug A	2.0, 1.6, 2.1, 0.6, 1.3	1.52	0.61	5
Drug B	0.5, 1.2, 0.3, 0.2, −0.4	0.36	0.58	5
Drug C	1.1, −1.0, −0.2, +0.2, +0.3	0.08	0.77	5
Overall		0.65		15

12.52 Use nonparametric methods to test for significant differences among groups. Identify which groups are significantly different.

12.53 Compare your results for this problem using parametric and nonparametric methods. Which do you think is the more appropriate method here?

Pulmonary Function

In the same study referred to in Example 12.1, the authors also obtained other measures of pulmonary function on the 1050 men. In particular, the FEV data for these men are presented in Table 12.22.

TABLE 12.22 FEV data for smoking and nonsmoking males in the White and Froeb study [1]

Group number (i)	Group name	Mean FEV (L)	sd FEV (L)	n_i
1	NS	3.72	0.65	200
2	PS	3.54	0.61	200
3	NI	3.56	0.76	50
4	LS	3.49	0.62	200
5	MS	3.08	0.61	200
6	HS	2.77	0.60	200
Overall		3.331		1050

(Reprinted with permission of the *New England Journal of Medicine*, *302*(13), 720–723, 1980.)

12.54 Test for the homogeneity of variances in the six groups.

12.55 Are the mean FEVs different overall in the six groups?

12.56 Analyze the data for between-group differences using the conventional t test procedure.

12.57 Analyze the data for between-group differences using a multiple comparisons procedure.

Pulmonary Disease

To study the effects of parental smoking on children, 218 children 10–15 years of age are assessed. One problem is that lung function is believed to be influenced by age and height. Thus, these factors may have to be corrected for before looking at the effect of parental smoking. This question is explored by subdividing the children in each 2-year age group into height quartiles, with the first quartile of children having the smallest height and the fourth quartile the largest height. The data are given for boys in Table 12.23.

12.58 Is a one-way or two-way ANOVA appropriate here?

12.59 Are there significant differences in FEV by age? Report a p-value.

12.60 Are there significant differences in FEV by height quartile? Report a p-value.

12.61 Are the differences by height found in Problem 12.60 consistent over all age groups?

Hypertension

Some common strategies for treating hypertensive patients by nonpharmacologic methods include (1) weight reduction and (2) trying to get the patient to relax more by meditational or other techniques. Suppose these strategies are evaluated by establishing four

TABLE 12.23 FEV (L) by age and height for boys ages 10–15

Age		1st quartile	2nd quartile	3rd quartile	4th quartile
10–11	Mean	3.30	3.47	3.60	3.66
	sd	0.83	0.59	0.64	0.75
	n	17	18	12	21
12–13	Mean	3.38	3.52	3.79	3.80
	sd	0.50	0.92	0.73	0.64
	n	16	18	20	22
14–15	Mean	3.52	3.83	3.81	4.06
	sd	0.74	0.68	0.78	0.63
	n	15	21	20	18

groups of hypertensive patients who receive the following types of nonpharmacologic therapy:

Group 1: Patients receive counseling for both weight reduction and meditation.

Group 2: Patients receive counseling for weight reduction but not for meditation.

Group 3: Patients receive counseling for meditation but not for weight reduction.

Group 4: Patients receive no counseling at all.

Suppose that 20 hypertensive patients are assigned at random to each of the four groups, and the change in diastolic blood pressure is noted in these patients after a 1-month period. The results are given in Table 12.24.

TABLE 12.24 Change in diastolic blood pressure for four groups of hypertensive patients who receive different kinds of nonpharmacologic therapy

Group	Mean change in diastolic bp (baseline − follow-up) (mm Hg)	sd change	n
1	8.6	6.2	20
2	5.3	5.4	20
3	4.9	7.0	20
4	1.1	6.5	20

12.62 Is a one-way or two-way ANOVA appropriate here?

12.63 Analyze if counseling for weight reduction has having a significant effect in reducing blood pressure.

12.64 Analyze if meditation instruction has a significant effect in reducing blood pressure.

12.65 Is there any relationship between the effects of weight-reduction counseling and meditation counseling on blood-pressure reduction? That is, does weight-reduction counseling work better for people who receive meditational counseling or for people who do not receive meditational counseling, or is there no difference in effect between these two subgroups?

12.66 Reanalyze the data in Table 8.15 (p. 284) using analysis of variance methods to test if there are significant differences overall among the three groups.

12.67 Test for which specific groups are different using the t test methodology in this chapter (if appropriate).

12.68 What is the difference in your conclusions between the methods used in Problems 12.66 and 12.67 and the t test procedures in Chapter 8?

Hypertension

An instructor in health education wants to familiarize her students with the measurement of blood pressure. Each student in the class is given a portable blood-pressure machine to take home. Readings are to be taken for 10 consecutive days, with one reading on each arm. Two of the goals of this study are to investigate (1) if there is a difference in blood pressure between the first and second readings and (2) if there is a difference in blood pressure between the left and right arms. For this purpose, on day 1 the first reading is taken on the left arm and the second reading on the right arm; on day 2 the first reading is taken on the right arm and the second reading on the left arm; on day 3 the day 1 schedule is used; on day 4 the day 2 schedule is used; and so forth. This protocol is followed for 10 consecutive days. The data are given in Table 12.25 for one student from the class.

TABLE 12.25 Systolic blood pressure recordings on 10 consecutive days by arm and order of readings

Day	Left arm Systolic bp (mm Hg)	Reading order	Right arm Systolic bp (mm Hg)	Reading order
1	98	1	99	2
2	93	2	102	1
3	100	1	98	2
4	100	2	99	1
5	96	1	100	2
6	100	2	95	1
7	90	1	98	2
8	93	2	102	1
9	91	1	92	2
10	94	2	90	1

12.69 Why was it important to change the arm measured first on alternate days?

12.70 Test for whether or not there are significant differences between the first and second readings. (Assume that there are no day effects, i.e., that blood pressures are comparable on different days.)

12.71 Test for whether or not there are significant differences between the left and right arms. (Assume that there are no day effects i.e., that blood pressures are comparable on different days.)

12.72 Some previous literature reports that blood pressure tends to decline the more times it is measured.

Average the results over the left and right arms within a given day. Using these data, perform a test to study this phenomenon by relating the average blood pressure to the day of the study.

Bioavailability
Refer to Data Set 2, BETACAR.DAT, Appendix 2.

12.73 Use analysis of variance methods to test if baseline serum beta carotene levels are comparable in the four groups. Identify and test for any specific group differences.

12.74 Use regression and analysis of variance methods to test if the rate of change in serum beta carotene over time is comparable in the four groups. Identify and test for any specific group differences.

Hepatic Disease
Refer to Data Set 12, HORMONE.DAT, Appendix 2.

12.75 Use analysis of variance methods to test if the change in biliary secretion levels is comparable for the five hormone groups. Identify and test for any specific group differences.

12.76 Answer Problem 12.75 for changes in pancreatic secretion levels.

12.77 Answer Problem 12.75 for changes in biliary pH levels.

12.78 Answer Problem 12.75 for changes in pancreatic pH levels.

Hypertension
Refer to Example 12.28 (p. 510). A similar two-way ANOVA was run using PROC GLM of SAS comparing mean diastolic blood pressure by study group and sex. The results are given in Table 12.26.

12.79 Summarize the findings in a few sentences.

TABLE 12.26 SAS GLM procedure output illustrating the effects of study groups and sex on diastolic blood pressure using the data set in Example 12.28

```
------------------------------------------------------------------------------------------------
                                            SAS

                          GENERAL LINEAR MODELS PROCEDURE

DEPENDENT VARIABLE: MDIAS

SOURCE            DF      SUM OF SQUARES      MEAN SQUARE     F VALUE       PR > F     R-SQUARE        C.V.

MODEL              3      48186.99270094    16062.33090031    134.15       0.0001     0.350741      15.0906

ERROR            745      89199.44205496     119.73079470                  ROOT MSE              MDIAS MEAN

CORRECTED TOTAL  748     137386.43475590                                 10.94215677            72.50972853


SOURCE            DF        TYPE I SS      F VALUE    PR > F     DF        TYPE III SS      F VALUE    PR > F

STUDY              2      45269.88509153    189.05    0.0001      2      46573.92818903    194.49     0.0001
SEX                1       2917.10760942     24.36    0.0001      1       2917.10760942     24.36     0.0001


                         STUDY    PROB > |T|
                                  I/J     1        2        3

                           1      1   .       0.0001   0.0001
                           2      2  0.0001    .       0.0001
                           3      3  0.0001   0.0001    .


                         SEX      PROB > |T|
                                   1        2

                           1      .       0.0001
                           2     0.0001    .


                                                    T FOR H0:      PR > |T|     STD ERROR OF
              PARAMETER                ESTIMATE     PARAMETER=0                  ESTIMATE

              INTERCEPT               76.47708914     115.61        0.0001       0.66152912
              STUDY      1           -17.30065001     -19.40        0.0001       0.89174716
                         2           -10.65302392      -7.23        0.0001       1.47258921
                         3             0.00000000       .             .             .
              SEX        1             3.98399582       4.94        0.0001       0.80713389
                         2             0.00000000       .             .             .

------------------------------------------------------------------------------------------------
```

TABLE 12.27 Collagen-linked fluorescence in relation to the type and severity of diabetic complications

Retinopathy grade	n	mean ± sd	Nephropathy grade	n	mean ± sd
0	11	447 ± 17	0	28	487 ± 24
1	16	493 ± 30	1	6	481 ± 16
2	14	551 ± 35	2	7	567 ± 24

(Reprinted with permission from the *New England Journal of Medicine, 314*(7), 403–408, 1986.)

Diabetes

Collagen-linked fluorescence was measured in skin biopsy specimens from 41 subjects with longstanding type I diabetes [9]. Diabetics were subdivided by the severity of diabetic complications. In particular, diabetics were graded according to (1) level of retinopathy (ocular abnormalities), where grade 0 = normal, grade 1 = background retinopathy, grade 2 = extensive (proliferative) retinopathy, and (2) level of nephropathy (kidney abnormalities), where grade 0 = urinary protein $<0.5 \,\mathrm{g}/24\,\mathrm{hrs}$, grade 1 = urinary protein ≥ 0.5, $<1\,\mathrm{g}/24\,\mathrm{hrs}$, grade 2 = urinary protein $\geq 1\,\mathrm{g}/24\,\mathrm{hrs}$. The results are presented in Table 12.27.

12.80 Assess whether there is any overall difference in mean fluorescence level by retinopathy grade.

12.81 Identify which specific retinopathy grade levels are different using both ordinary *t* tests and the method of multiple comparisons.

12.82 Answer Problem 12.80 when patients are grouped by nephropathy grade.

12.83 Answer Problem 12.81 when patients are grouped by nephropathy grade.

Endocrinology

A study was conducted [10] concerning the effect of calcium supplementation on bone loss among post-menopausal women. Women were randomized to either (1) estrogen cream and calcium placebo ($n = 15$), (2) placebo estrogen cream and 2000 mg/day of calcium ($n = 15$), or (3) placebo estrogen cream and calcium placebo ($n = 13$). Subjects were seen every 3 months for a 2-year period. The rate of bone loss was computed for each woman by regression analysis and expressed as a percentage of the initial bone mass. The results are shown in Table 12.28.

TABLE 12.28 Mean (±1 sd) slope of total body bone mass (percentage per year) in the three treatment groups

Treatment group		
(1) Estrogen ($n = 15$)	**(2) Calcium** ($n = 15$)	**(3) Placebo** ($n = 13$)
-0.43 ± 1.60	-2.62 ± 2.68	-3.98 ± 1.63

(Reprinted with permission from the *New England Journal of Medicine, 316*(4), 173–177, 1987.

12.84 What test can be used to compare the rate of bone loss in the three groups?

12.85 Implement the test in Problem 12.84 and report a *p*-value.

12.86 Identify which pairs of groups are different from each other using both *t* tests and the method of multiple comparisons. Report a *p*-value for each comparison of treatment groups.

12.87 Which methodology do you think is more appropriate in Problem 12.86?

Hypertension

Refer to Example 12.29 (p. 513). An analysis of covariance was performed using PROC GLM of SAS comparing mean diastolic blood pressure by study group and sex after controlling for the effects of age and weight. The results are given in Table 12.29.

12.88 Summarize the findings in a few sentences and compare them with the results in Problem 12.79.

TABLE 12.29 SAS GLM procedure output illustrating the effects of study group and sex on diastolic blood pressure after controlling for age and weight, using the data set in Example 12.28

```
------------------------------------------------------------------------------------------------
                                           SAS

                            GENERAL LINEAR MODELS PROCEDURE

DEPENDENT VARIABLE: MDIAS

SOURCE            DF      SUM OF SQUARES       MEAN SQUARE    F VALUE     PR > F    R-SQUARE        C.V.

MODEL              5      57521.19225928    11504.23845186    107.08     0.0001    0.419457     14.2973

ERROR            741      79611.39165542      107.43777551               ROOT MSE           MDIAS MEAN

CORRECTED TOTAL  746     137132.58391470                               10.36521951          72.49770638


SOURCE            DF         TYPE I SS    F VALUE    PR > F     DF      TYPE III SS    F VALUE    PR > F

STUDY              2      45234.31356527    210.51    0.0001      2    12349.74237359     57.47    0.0001
SEX                1       2912.79699312     27.11    0.0001      1      624.47946004      5.81    0.0162
AGE                1       5237.52247659     48.75    0.0001      1     4245.67574526     39.52    0.0001
WGT                1       4136.55922429     38.50    0.0001      1     4136.55922429     38.50    0.0001


                          STUDY    PROB > |T|
                                   I/J     1        2        3

                            1      1    .       0.0186   0.0001
                            2      2  0.0186      .      0.0001
                            3      3  0.0001   0.0001      .


                          SEX      PROB > |T|
                                        1        2

                            1         .       0.0162
                            2      0.0162       .


                                                         T FOR HO:      PR > |T|    STD ERROR OF
                          PARAMETER             ESTIMATE  PARAMETER=0                  ESTIMATE

                          INTERCEPT          52.96724415     20.41       0.0001      2.59544038
                          STUDY       1      -11.18295628    -10.45       0.0001      1.06990647
                                      2       -7.58825363     -5.29       0.0001      1.43486888
                                      3        0.00000000       .           .
                          SEX         1        2.10948458      2.41       0.0162      0.87497527
                                      2        0.00000000       .           .
                          AGE                  0.30162065      6.29       0.0001      0.04798065
                          WGT                  0.08278540      6.20       0.0001      0.01334174
------------------------------------------------------------------------------------------------
```

References

[1] White, J. R., & Froeb, H. F. (1980). Small-airways dysfunction in nonsmokers chronically exposed to tobacco smoke. *New England Journal of Medicine, 302*(13), 720–723.

[2] Kleinbaum, D. G., Kupper, L. L., & Muller, K. E. (1988). *Applied regression analysis and other multivariable methods* (2nd edition). Boston: Duxbury.

[3] Abelson, M. B., Kliman, G. H., Butrus, S. I., & Weston, J. H. (1983). Modulation of arachidonic acid in the rabbit conjunctiva: Predominance of the cyclo-oxygenase pathway. Presented at the Annual Spring Meeting of the Association for Research in Vision and Ophthalmology, Sarasota, Florida, May 2–6, 1983.

[4] Linn, W. S., Shamoo, D. A., Anderson, K. R., Whynot, J. D., Avol, E. L., & Hackney, J. D. (1985). Effects of heat and humidity on the responses of exercising asthmatics to sulfur dioxide exposure. *American Review of Respiratory Disease, 131*(2), 221–225.

[5] Polk, B. F., Rosner, B., Feudo, R., & Van Denburgh, M. (1980). An evaluation of the Vita Stat automatic blood pressure measuring device. *Hypertension*, *2*(2), 221–227.

[6] Sallis, J. F., Haskell, W. L., Wood, P. D., Fortmann, S. P., Rogers, T., Blair, S. N., & Paffenbarger, R. S., Jr. (1985). Physical activity assessment methodology in the five-city project. *American Journal of Epidemiology*, *121*(1), 91–106.

[7] Pfeffer, R. I., Kurosaki, T. T., Chance, J. M., Filos, S., & Bates, D. (1984). Use of the mental function index in older adults: Reliability, validity and measurement of change over time. *American Journal of Epidemiology*, *120*(6), 922–935.

[8] Kopelman, H., Durie, P., Gaskin, K., Weizman, Z., & Forstner, G. (1985). Pancreatic fluid secretion and protein hyperconcentration in cystic fibrosis. *New England Journal of Medicine*, *312*(6), 329–334.

[9] Monnier, V. M., Vishwanath, V., Frank, K. E., Elmets, C. A., Dauchot, P., & Kohn, R. R. (1986). Relation between complications of type I diabetes mellitus and collagen-linked fluorescence. *New England Journal of Medicine*, *314*(7), 403–408.

[10] Riis, B., Thomsen, K., & Christiansen, C. (1987). Does calcium supplementation prevent post-menopausal bone loss? A double-blind controlled study. *New England Journal of Medicine*, *316*(4), 173–177.

TABLES

TABLE 1 Exact binomial probabilities $Pr(X = k) = \binom{n}{k} p^k q^{n-k}$

n	k	.05	.10	.15	.20	.25	.30	.35	.40	.45	.50
2	0	.9025	.8100	.7225	.6400	.5625	.4900	.4225	.3600	.3025	.2500
	1	.0950	.1800	.2550	.3200	.3750	.4200	.4550	.4800	.4950	.5000
	2	.0025	.0100	.0225	.0400	.0625	.0900	.1225	.1600	.2025	.2500
3	0	.8574	.7290	.6141	.5120	.4219	.3430	.2746	.2160	.1664	.1250
	1	.1354	.2430	.3251	.3840	.4219	.4410	.4436	.4320	.4084	.3750
	2	.0071	.0270	.0574	.0960	.1406	.1890	.2389	.2880	.3341	.3750
	3	.0001	.0010	.0034	.0080	.0156	.0270	.0429	.0640	.0911	.1250
4	0	.8145	.6561	.5220	.4096	.3164	.2401	.1785	.1296	.0915	.0625
	1	.1715	.2916	.3685	.4096	.4219	.4116	.3845	.3456	.2995	.2500
	2	.0135	.0486	.0975	.1536	.2109	.2646	.3105	.3456	.3675	.3750
	3	.0005	.0036	.0115	.0256	.0469	.0756	.1115	.1536	.2005	.2500
	4	.0000	.0001	.0005	.0016	.0039	.0081	.0150	.0256	.0410	.0625
5	0	.7738	.5905	.4437	.3277	.2373	.1681	.1160	.0778	.0503	.0313
	1	.2036	.3280	.3915	.4096	.3955	.3602	.3124	.2592	.2059	.1563
	2	.0214	.0729	.1382	.2048	.2637	.3087	.3364	.3456	.3369	.3125
	3	.0011	.0081	.0244	.0512	.0879	.1323	.1811	.2304	.2757	.3125
	4	.0000	.0004	.0022	.0064	.0146	.0283	.0488	.0768	.1128	.1563
	5	.0000	.0000	.0001	.0003	.0010	.0024	.0053	.0102	.0185	.0313
6	0	.7351	.5314	.3771	.2621	.1780	.1176	.0754	.0467	.0277	.0156
	1	.2321	.3543	.3993	.3932	.3560	.3025	.2437	.1866	.1359	.0938
	2	.0305	.0984	.1762	.2458	.2966	.3241	.3280	.3110	.2780	.2344
	3	.0021	.0146	.0415	.0819	.1318	.1852	.2355	.2765	.3032	.3125
	4	.0001	.0012	.0055	.0154	.0330	.0595	.0951	.1382	.1861	.2344
	5	.0000	.0001	.0004	.0015	.0044	.0102	.0205	.0369	.0609	.0938
	6	.0000	.0000	.0000	.0001	.0002	.0007	.0018	.0041	.0083	.0156
7	0	.6983	.4783	.3206	.2097	.1335	.0824	.0490	.0280	.0152	.0078
	1	.2573	.3720	.3960	.3670	.3115	.2471	.1848	.1306	.0872	.0547
	2	.0406	.1240	.2097	.2753	.3115	.3177	.2985	.2613	.2140	.1641
	3	.0036	.0230	.0617	.1147	.1730	.2269	.2679	.2903	.2918	.2734
	4	.0002	.0026	.0109	.0287	.0577	.0972	.1442	.1935	.2388	.2734
	5	.0000	.0002	.0012	.0043	.0115	.0250	.0466	.0774	.1172	.1641
	6	.0000	.0000	.0001	.0004	.0013	.0036	.0084	.0172	.0320	.0547
	7	.0000	.0000	.0000	.0000	.0001	.0002	.0006	.0016	.0037	.0078
8	0	.6634	.4305	.2725	.1678	.1001	.0576	.0319	.0168	.0084	.0039
	1	.2793	.3826	.3847	.3355	.2670	.1977	.1373	.0896	.0548	.0313
	2	.0515	.1488	.2376	.2936	.3115	.2965	.2587	.2090	.1569	.1094
	3	.0054	.0331	.0839	.1468	.2076	.2541	.2786	.2787	.2568	.2188

TABLE 1 (Continued)

n	k	.05	.10	.15	.20	.25	.30	.35	.40	.45	.50
	4	.0004	.0046	.0185	.0459	.0865	.1361	.1875	.2322	.2627	.2734
	5	.0000	.0004	.0026	.0092	.0231	.0467	.0808	.1239	.1719	.2188
	6	.0000	.0000	.0002	.0011	.0038	.0100	.0217	.0413	.0703	.1094
	7	.0000	.0000	.0000	.0001	.0004	.0012	.0033	.0079	.0164	.0313
	8	.0000	.0000	.0000	.0000	.0000	.0001	.0002	.0007	.0017	.0039
9	0	.6302	.3874	.2316	.1342	.0751	.0404	.0207	.0101	.0046	.0020
	1	.2985	.3874	.3679	.3020	.2253	.1556	.1004	.0605	.0339	.0176
	2	.0629	.1722	.2597	.3020	.3003	.2668	.2162	.1612	.1110	.0703
	3	.0077	.0446	.1069	.1762	.2336	.2668	.2716	.2508	.2119	.1641
	4	.0006	.0074	.0283	.0661	.1168	.1715	.2194	.2508	.2600	.2461
	5	.0000	.0008	.0050	.0165	.0389	.0735	.1181	.1672	.2128	.2461
	6	.0000	.0001	.0006	.0028	.0087	.0210	.0424	.0743	.1160	.1641
	7	.0000	.0000	.0000	.0003	.0012	.0039	.0098	.0212	.0407	.0703
	8	.0000	.0000	.0000	.0000	.0001	.0004	.0013	.0035	.0083	.0176
	9	.0000	.0000	.0000	.0000	.0000	.0000	.0001	.0003	.0008	.0020
10	0	.5987	.3487	.1969	.1074	.0563	.0282	.0135	.0060	.0025	.0010
	1	.3151	.3874	.3474	.2684	.1877	.1211	.0725	.0403	.0207	.0098
	2	.0746	.1937	.2759	.3020	.2816	.2335	.1757	.1209	.0763	.0439
	3	.0105	.0574	.1298	.2013	.2503	.2668	.2522	.2150	.1665	.1172
	4	.0010	.0112	.0401	.0881	.1460	.2001	.2377	.2508	.2384	.2051
	5	.0001	.0015	.0085	.0264	.0584	.1029	.1536	.2007	.2340	.2461
	6	.0000	.0001	.0012	.0055	.0162	.0368	.0689	.1115	.1596	.2051
	7	.0000	.0000	.0001	.0008	.0031	.0090	.0212	.0425	.0746	.1172
	8	.0000	.0000	.0000	.0001	.0004	.0014	.0043	.0106	.0229	.0439
	9	.0000	.0000	.0000	.0000	.0000	.0001	.0005	.0016	.0042	.0098
	10	.0000	.0000	.0000	.0000	.0000	.0000	.0000	.0001	.0003	.0010
11	0	.5688	.3138	.1673	.0859	.0422	.0198	.0088	.0036	.0014	.0005
	1	.3293	.3835	.3248	.2362	.1549	.0932	.0518	.0266	.0125	.0054
	2	.0867	.2131	.2866	.2953	.2581	.1998	.1395	.0887	.0513	.0269
	3	.0137	.0710	.1517	.2215	.2581	.2568	.2254	.1774	.1259	.0806
	4	.0014	.0158	.0536	.1107	.1721	.2201	.2428	.2365	.2060	.1611
	5	.0001	.0025	.0132	.0388	.0803	.1321	.1830	.2207	.2360	.2256
	6	.0000	.0003	.0023	.0097	.0268	.0566	.0985	.1471	.1931	.2256
	7	.0000	.0000	.0003	.0017	.0064	.0173	.0379	.0701	.1128	.1611
	8	.0000	.0000	.0000	.0002	.0011	.0037	.0102	.0234	.0462	.0806
	9	.0000	.0000	.0000	.0000	.0001	.0005	.0018	.0052	.0126	.0269
	10	.0000	.0000	.0000	.0000	.0000	.0000	.0002	.0007	.0021	.0054
	11	.0000	.0000	.0000	.0000	.0000	.0000	.0000	.0000	.0002	.0005
12	0	.5404	.2824	.1422	.0687	.0317	.0138	.0057	.0022	.0008	.0002
	1	.3413	.3766	.3012	.2062	.1267	.0712	.0368	.0174	.0075	.0029
	2	.0988	.2301	.2924	.2835	.2323	.1678	.1088	.0639	.0339	.0161
	3	.0173	.0852	.1720	.2362	.2581	.2397	.1954	.1419	.0923	.0537
	4	.0021	.0213	.0683	.1329	.1936	.2311	.2367	.2128	.1700	.1208
	5	.0002	.0038	.0193	.0532	.1032	.1585	.2039	.2270	.2225	.1934
	6	.0000	.0005	.0040	.0155	.0401	.0792	.1281	.1766	.2124	.2256
	7	.0000	.0000	.0006	.0033	.0115	.0291	.0591	.1009	.1489	.1934
	8	.0000	.0000	.0001	.0005	.0024	.0078	.0199	.0420	.0762	.1208
	9	.0000	.0000	.0000	.0001	.0004	.0015	.0048	.0125	.0277	.0537

TABLE 1 (Continued)

n	k	.05	.10	.15	.20	.25	.30	.35	.40	.45	.50
	10	.0000	.0000	.0000	.0000	.0000	.0002	.0008	.0025	.0068	.0161
	11	.0000	.0000	.0000	.0000	.0000	.0000	.0001	.0003	.0010	.0029
	12	.0000	.0000	.0000	.0000	.0000	.0000	.0000	.0000	.0001	.0002
13	0	.5133	.2542	.1209	.0550	.0238	.0097	.0037	.0013	.0004	.0001
	1	.3512	.3672	.2774	.1787	.1029	.0540	.0259	.0113	.0045	.0016
	2	.1109	.2448	.2937	.2680	.2059	.1388	.0836	.0453	.0220	.0095
	3	.0214	.0997	.1900	.2457	.2517	.2181	.1651	.1107	.0660	.0349
	4	.0028	.0277	.0838	.1535	.2097	.2337	.2222	.1845	.1350	.0873
	5	.0003	.0055	.0266	.0691	.1258	.1803	.2154	.2214	.1989	.1571
	6	.0000	.0008	.0063	.0230	.0559	.1030	.1546	.1968	.2169	.2095
	7	.0000	.0001	.0011	.0058	.0186	.0442	.0833	.1312	.1775	.2095
	8	.0000	.0000	.0001	.0011	.0047	.0142	.0336	.0656	.1089	.1571
	9	.0000	.0000	.0000	.0001	.0009	.0034	.0101	.0243	.0495	.0873
	10	.0000	.0000	.0000	.0000	.0001	.0006	.0022	.0065	.0162	.0349
	11	.0000	.0000	.0000	.0000	.0000	.0001	.0003	.0012	.0036	.0095
	12	.0000	.0000	.0000	.0000	.0000	.0000	.0000	.0001	.0005	.0016
	13	.0000	.0000	.0000	.0000	.0000	.0000	.0000	.0000	.0000	.0001
14	0	.4877	.2288	.1028	.0440	.0178	.0068	.0024	.0008	.0002	.0001
	1	.3593	.3559	.2539	.1539	.0832	.0407	.0181	.0073	.0027	.0009
	2	.1229	.2570	.2912	.2501	.1802	.1134	.0634	.0317	.0141	.0056
	3	.0259	.1142	.2056	.2501	.2402	.1943	.1366	.0845	.0462	.0222
	4	.0037	.0349	.0998	.1720	.2202	.2290	.2022	.1549	.1040	.0611
	5	.0004	.0078	.0352	.0860	.1468	.1963	.2178	.2066	.1701	.1222
	6	.0000	.0013	.0093	.0322	.0734	.1262	.1759	.2066	.2088	.1833
	7	.0000	.0002	.0019	.0092	.0280	.0618	.1082	.1574	.1952	.2095
	8	.0000	.0000	.0003	.0020	.0082	.0232	.0510	.0918	.1398	.1833
	9	.0000	.0000	.0000	.0003	.0018	.0066	.0183	.0408	.0762	.1222
	10	.0000	.0000	.0000	.0000	.0003	.0014	.0049	.0136	.0312	.0611
	11	.0000	.0000	.0000	.0000	.0000	.0002	.0010	.0033	.0093	.0222
	12	.0000	.0000	.0000	.0000	.0000	.0000	.0001	.0005	.0019	.0056
	13	.0000	.0000	.0000	.0000	.0000	.0000	.0000	.0001	.0002	.0009
	14	.0000	.0000	.0000	.0000	.0000	.0000	.0000	.0000	.0000	.0001
15	0	.4633	.2059	.0874	.0352	.0134	.0047	.0016	.0005	.0001	.0000
	1	.3658	.3432	.2312	.1319	.0668	.0305	.0126	.0047	.0016	.0005
	2	.1348	.2669	.2856	.2309	.1559	.0916	.0476	.0219	.0090	.0032
	3	.0307	.1285	.2184	.2501	.2252	.1700	.1110	.0634	.0318	.0139
	4	.0049	.0428	.1156	.1876	.2252	.2186	.1792	.1268	.0780	.0417
	5	.0006	.0105	.0449	.1032	.1651	.2061	.2123	.1859	.1404	.0916
	6	.0000	.0019	.0132	.0430	.0917	.1472	.1906	.2066	.1914	.1527
	7	.0000	.0003	.0030	.0138	.0393	.0811	.1319	.1771	.2013	.1964
	8	.0000	.0000	.0005	.0035	.0131	.0348	.0710	.1181	.1647	.1964
	9	.0000	.0000	.0001	.0007	.0034	.0116	.0298	.0612	.1048	.1527
	10	.0000	.0000	.0000	.0001	.0007	.0030	.0096	.0245	.0515	.0916
	11	.0000	.0000	.0000	.0000	.0001	.0006	.0024	.0074	.0191	.0417
	12	.0000	.0000	.0000	.0000	.0000	.0001	.0004	.0016	.0052	.0139
	13	.0000	.0000	.0000	.0000	.0000	.0000	.0001	.0003	.0010	.0032
	14	.0000	.0000	.0000	.0000	.0000	.0000	.0000	.0000	.0001	.0005
	15	.0000	.0000	.0000	.0000	.0000	.0000	.0000	.0000	.0000	.0000

TABLE 1 (*Continued*)

n	k	.05	.10	.15	.20	.25	.30	.35	.40	.45	.50
16	0	.4401	.1853	.0743	.0281	.0100	.0033	.0010	.0003	.0001	.0000
	1	.3706	.3294	.2097	.1126	.0535	.0228	.0087	.0030	.0009	.0002
	2	.1463	.2745	.2775	.2111	.1336	.0732	.0353	.0150	.0056	.0018
	3	.0359	.1423	.2285	.2463	.2079	.1465	.0888	.0468	.0215	.0085
	4	.0061	.0514	.1311	.2001	.2252	.2040	.1553	.1014	.0572	.0278
	5	.0008	.0137	.0555	.1201	.1802	.2099	.2008	.1623	.1123	.0667
	6	.0001	.0028	.0180	.0550	.1101	.1649	.1982	.1983	.1684	.1222
	7	.0000	.0004	.0045	.0197	.0524	.1010	.1524	.1889	.1969	.1746
	8	.0000	.0001	.0009	.0055	.0197	.0487	.0923	.1417	.1812	.1964
	9	.0000	.0000	.0001	.0012	.0058	.0185	.0442	.0840	.1318	.1746
	10	.0000	.0000	.0000	.0002	.0014	.0056	.0167	.0392	.0755	.1222
	11	.0000	.0000	.0000	.0000	.0002	.0013	.0049	.0142	.0337	.0667
	12	.0000	.0000	.0000	.0000	.0000	.0002	.0011	.0040	.0115	.0278
	13	.0000	.0000	.0000	.0000	.0000	.0000	.0002	.0008	.0029	.0085
	14	.0000	.0000	.0000	.0000	.0000	.0000	.0000	.0001	.0005	.0018
	15	.0000	.0000	.0000	.0000	.0000	.0000	.0000	.0000	.0001	.0002
	16	.0000	.0000	.0000	.0000	.0000	.0000	.0000	.0000	.0000	.0000
17	0	.4181	.1668	.0631	.0225	.0075	.0023	.0007	.0002	.0000	.0000
	1	.3741	.3150	.1893	.0957	.0426	.0169	.0060	.0019	.0005	.0001
	2	.1575	.2800	.2673	.1914	.1136	.0581	.0260	.0102	.0035	.0010
	3	.0415	.1556	.2359	.2393	.1893	.1245	.0701	.0341	.0144	.0052
	4	.0076	.0605	.1457	.2093	.2209	.1868	.1320	.0796	.0411	.0182
	5	.0010	.0175	.0668	.1361	.1914	.2081	.1849	.1379	.0875	.0472
	6	.0001	.0039	.0236	.0680	.1276	.1784	.1991	.1839	.1432	.0944
	7	.0000	.0007	.0065	.0267	.0668	.1201	.1685	.1927	.1841	.1484
	8	.0000	.0001	.0014	.0084	.0279	.0644	.1134	.1606	.1883	.1855
	9	.0000	.0000	.0003	.0021	.0093	.0276	.0611	.1070	.1540	.1855
	10	.0000	.0000	.0000	.0004	.0025	.0095	.0263	.0571	.1008	.1484
	11	.0000	.0000	.0000	.0001	.0005	.0026	.0090	.0242	.0525	.0944
	12	.0000	.0000	.0000	.0000	.0001	.0006	.0024	.0081	.0215	.0472
	13	.0000	.0000	.0000	.0000	.0000	.0001	.0005	.0021	.0068	.0182
	14	.0000	.0000	.0000	.0000	.0000	.0000	.0001	.0004	.0016	.0052
	15	.0000	.0000	.0000	.0000	.0000	.0000	.0000	.0001	.0003	.0010
	16	.0000	.0000	.0000	.0000	.0000	.0000	.0000	.0000	.0000	.0001
	17	.0000	.0000	.0000	.0000	.0000	.0000	.0000	.0000	.0000	.0000
18	0	.3972	.1501	.0536	.0180	.0056	.0016	.0004	.0001	.0000	.0000
	1	.3763	.3002	.1704	.0811	.0338	.0126	.0042	.0012	.0003	.0001
	2	.1683	.2835	.2556	.1723	.0958	.0458	.0190	.0069	.0022	.0006
	3	.0473	.1680	.2406	.2297	.1704	.1046	.0547	.0246	.0095	.0031
	4	.0093	.0700	.1592	.2153	.2130	.1681	.1104	.0614	.0291	.0117
	5	.0014	.0218	.0787	.1507	.1988	.2017	.1664	.1146	.0666	.0327
	6	.0002	.0052	.0301	.0816	.1436	.1873	.1941	.1655	.1181	.0708
	7	.0000	.0010	.0091	.0350	.0820	.1376	.1792	.1892	.1657	.1214
	8	.0000	.0002	.0022	.0120	.0376	.0811	.1327	.1734	.1864	.1669
	9	.0000	.0000	.0004	.0033	.0139	.0386	.0794	.1284	.1694	.1855
	10	.0000	.0000	.0001	.0008	.0042	.0149	.0385	.0771	.1248	.1669
	11	.0000	.0000	.0000	.0001	.0010	.0046	.0151	.0374	.0742	.1214
	12	.0000	.0000	.0000	.0000	.0002	.0012	.0047	.0145	.0354	.0708

TABLE 1 (*Continued*)

n	k	.05	.10	.15	.20	.25	.30	.35	.40	.45	.50
	13	.0000	.0000	.0000	.0000	.0000	.0002	.0012	.0045	.0134	.0327
	14	.0000	.0000	.0000	.0000	.0000	.0000	.0002	.0011	.0039	.0117
	15	.0000	.0000	.0000	.0000	.0000	.0000	.0000	.0002	.0009	.0031
	16	.0000	.0000	.0000	.0000	.0000	.0000	.0000	.0000	.0001	.0006
	17	.0000	.0000	.0000	.0000	.0000	.0000	.0000	.0000	.0000	.0001
	18	.0000	.0000	.0000	.0000	.0000	.0000	.0000	.0000	.0000	.0000
19	0	.3774	.1351	.0456	.0144	.0042	.0011	.0003	.0001	.0000	.0000
	1	.3774	.2852	.1529	.0685	.0268	.0093	.0029	.0008	.0002	.0000
	2	.1787	.2852	.2428	.1540	.0803	.0358	.0138	.0046	.0013	.0003
	3	.0533	.1796	.2428	.2182	.1517	.0869	.0422	.0175	.0062	.0018
	4	.0112	.0798	.1714	.2182	.2023	.1491	.0909	.0467	.0203	.0074
	5	.0018	.0266	.0907	.1636	.2023	.1916	.1468	.0933	.0497	.0222
	6	.0002	.0069	.0374	.0955	.1574	.1916	.1844	.1451	.0949	.0518
	7	.0000	.0014	.0122	.0443	.0974	.1525	.1844	.1797	.1443	.0961
	8	.0000	.0002	.0032	.0166	.0487	.0981	.1489	.1797	.1771	.1442
	9	.0000	.0000	.0007	.0051	.0198	.0514	.0980	.1464	.1771	.1762
	10	.0000	.0000	.0001	.0013	.0066	.0220	.0528	.0976	.1449	.1762
	11	.0000	.0000	.0000	.0003	.0018	.0077	.0233	.0532	.0970	.1442
	12	.0000	.0000	.0000	.0000	.0004	.0022	.0083	.0237	.0529	.0961
	13	.0000	.0000	.0000	.0000	.0001	.0005	.0024	.0085	.0233	.0518
	14	.0000	.0000	.0000	.0000	.0000	.0001	.0006	.0024	.0082	.0222
	15	.0000	.0000	.0000	.0000	.0000	.0000	.0001	.0005	.0022	.0074
	16	.0000	.0000	.0000	.0000	.0000	.0000	.0000	.0001	.0005	.0018
	17	.0000	.0000	.0000	.0000	.0000	.0000	.0000	.0000	.0001	.0003
	18	.0000	.0000	.0000	.0000	.0000	.0000	.0000	.0000	.0000	.0000
	19	.0000	.0000	.0000	.0000	.0000	.0000	.0000	.0000	.0000	.0000
20	0	.3585	.1216	.0388	.0115	.0032	.0008	.0002	.0000	.0000	.0000
	1	.3774	.2702	.1368	.0576	.0211	.0068	.0020	.0005	.0001	.0000
	2	.1887	.2852	.2293	.1369	.0669	.0278	.0100	.0031	.0008	.0002
	3	.0596	.1901	.2428	.2054	.1339	.0716	.0323	.0123	.0040	.0011
	4	.0133	.0898	.1821	.2182	.1897	.1304	.0738	.0350	.0139	.0046
	5	.0022	.0319	.1028	.1746	.2023	.1789	.1272	.0746	.0365	.0148
	6	.0003	.0089	.0454	.1091	.1686	.1916	.1712	.1244	.0746	.0370
	7	.0000	.0020	.0160	.0546	.1124	.1643	.1844	.1659	.1221	.0739
	8	.0000	.0004	.0046	.0222	.0609	.1144	.1614	.1797	.1623	.1201
	9	.0000	.0001	.0011	.0074	.0271	.0654	.1158	.1597	.1771	.1602
	10	.0000	.0000	.0002	.0020	.0099	.0308	.0686	.1171	.1593	.1762
	11	.0000	.0000	.0000	.0005	.0030	.0120	.0336	.0710	.1185	.1602
	12	.0000	.0000	.0000	.0001	.0008	.0039	.0136	.0355	.0727	.1201
	13	.0000	.0000	.0000	.0000	.0002	.0010	.0045	.0146	.0366	.0739
	14	.0000	.0000	.0000	.0000	.0000	.0002	.0012	.0049	.0150	.0370
	15	.0000	.0000	.0000	.0000	.0000	.0000	.0003	.0013	.0049	.0148
	16	.0000	.0000	.0000	.0000	.0000	.0000	.0000	.0003	.0013	.0046
	17	.0000	.0000	.0000	.0000	.0000	.0000	.0000	.0000	.0002	.0011
	18	.0000	.0000	.0000	.0000	.0000	.0000	.0000	.0000	.0000	.0002
	19	.0000	.0000	.0000	.0000	.0000	.0000	.0000	.0000	.0000	.0000
	20	.0000	.0000	.0000	.0000	.0000	.0000	.0000	.0000	.0000	.0000

TABLE 2 Exact Poisson Probabilities $Pr(X = k) = \dfrac{e^{-\mu}\mu^{k}}{k!}$

					μ					
k	0.5	1.0	1.5	2.0	2.5	3.0	3.5	4.0	4.5	5.0
0	.6065	.3679	.2231	.1353	.0821	.0498	.0302	.0183	.0111	.0067
1	.3033	.3679	.3347	.2707	.2052	.1494	.1057	.0733	.0500	.0337
2	.0758	.1839	.2510	.2707	.2565	.2240	.1850	.1465	.1125	.0842
3	.0126	.0613	.1255	.1804	.2138	.2240	.2158	.1954	.1687	.1404
4	.0016	.0153	.0471	.0902	.1336	.1680	.1888	.1954	.1898	.1755
5	.0002	.0031	.0141	.0361	.0668	.1008	.1322	.1563	.1708	.1755
6	.0000	.0005	.0035	.0120	.0278	.0504	.0771	.1042	.1281	.1462
7	.0000	.0001	.0008	.0034	.0099	.0216	.0385	.0595	.0824	.1044
8	.0000	.0000	.0001	.0009	.0031	.0081	.0169	.0298	.0463	.0653
9	.0000	.0000	.0000	.0002	.0009	.0027	.0066	.0132	.0232	.0363
10	.0000	.0000	.0000	.0000	.0002	.0008	.0023	.0053	.0104	.0181
11	.0000	.0000	.0000	.0000	.0000	.0002	.0007	.0019	.0043	.0082
12	.0000	.0000	.0000	.0000	.0000	.0001	.0002	.0006	.0016	.0034
13	.0000	.0000	.0000	.0000	.0000	.0000	.0001	.0002	.0006	.0013
14	.0000	.0000	.0000	.0000	.0000	.0000	.0000	.0001	.0002	.0005
15	.0000	.0000	.0000	.0000	.0000	.0000	.0000	.0000	.0001	.0002
16	.0000	.0000	.0000	.0000	.0000	.0000	.0000	.0000	.0000	.0000

					μ					
k	5.5	6.0	6.5	7.0	7.5	8.0	8.5	9.0	9.5	10.0
0	.0041	.0025	.0015	.0009	.0006	.0003	.0002	.0001	.0001	.0000
1	.0225	.0149	.0098	.0064	.0041	.0027	.0017	.0011	.0007	.0005
2	.0618	.0446	.0318	.0223	.0156	.0107	.0074	.0050	.0034	.0023
3	.1133	.0892	.0688	.0521	.0389	.0286	.0208	.0150	.0107	.0076
4	.1558	.1339	.1118	.0912	.0729	.0573	.0443	.0337	.0254	.0189
5	.1714	.1606	.1454	.1277	.1094	.0916	.0752	.0607	.0483	.0378
6	.1571	.1606	.1575	.1490	.1367	.1221	.1066	.0911	.0764	.0631
7	.1234	.1377	.1462	.1490	.1465	.1396	.1294	.1171	.1037	.0901
8	.0849	.1033	.1188	.1304	.1373	.1396	.1375	.1318	.1232	.1126
9	.0519	.0688	.0858	.1014	.1144	.1241	.1299	.1318	.1300	.1251
10	.0285	.0413	.0558	.0710	.0858	.0993	.1104	.1186	.1235	.1251
11	.0143	.0225	.0330	.0452	.0585	.0722	.0853	.0970	.1067	.1137
12	.0065	.0113	.0179	.0263	.0366	.0481	.0604	.0728	.0844	.0948
13	.0028	.0052	.0089	.0142	.0211	.0296	.0395	.0504	.0617	.0729
14	.0011	.0022	.0041	.0071	.0113	.0169	.0240	.0324	.0419	.0521
15	.0004	.0009	.0018	.0033	.0057	.0090	.0136	.0194	.0265	.0347
16	.0001	.0003	.0007	.0014	.0026	.0045	.0072	.0109	.0157	.0217
17	.0000	.0001	.0003	.0006	.0012	.0021	.0036	.0058	.0088	.0128
18	.0000	.0000	.0001	.0002	.0005	.0009	.0017	.0029	.0046	.0071
19	.0000	.0000	.0000	.0001	.0002	.0004	.0008	.0014	.0023	.0037
20	.0000	.0000	.0000	.0000	.0001	.0002	.0003	.0006	.0011	.0019
21	.0000	.0000	.0000	.0000	.0000	.0001	.0001	.0003	.0005	.0009

TABLE 2 (Continued)

					μ					
k	5.5	6.0	6.5	7.0	7.5	8.0	8.5	9.0	9.5	10.0
22	.0000	.0000	.0000	.0000	.0000	.0000	.0001	.0001	.0002	.0004
23	.0000	.0000	.0000	.0000	.0000	.0000	.0000	.0000	.0001	.0002
24	.0000	.0000	.0000	.0000	.0000	.0000	.0000	.0000	.0000	.0001
25	.0000	.0000	.0000	.0000	.0000	.0000	.0000	.0000	.0000	.0000

					μ					
k	10.5	11.0	11.5	12.0	12.5	13.0	13.5	14.0	14.5	15.0
0	.0000	.0000	.0000	.0000	.0000	.0000	.0000	.0000	.0000	.0000
1	.0003	.0002	.0001	.0001	.0000	.0000	.0000	.0000	.0000	.0000
2	.0015	.0010	.0007	.0004	.0003	.0002	.0001	.0001	.0001	.0000
3	.0053	.0037	.0026	.0018	.0012	.0008	.0006	.0004	.0003	.0002
4	.0139	.0102	.0074	.0053	.0038	.0027	.0019	.0013	.0009	.0006
5	.0293	.0224	.0170	.0127	.0095	.0070	.0051	.0037	.0027	.0019
6	.0513	.0411	.0325	.0255	.0197	.0152	.0115	.0087	.0065	.0048
7	.0769	.0646	.0535	.0437	.0353	.0281	.0222	.0174	.0135	.0104
8	.1009	.0888	.0769	.0655	.0551	.0457	.0375	.0304	.0244	.0194
9	.1177	.1085	.0982	.0874	.0765	.0661	.0563	.0473	.0394	.0324
10	.1236	.1194	.1129	.1048	.0956	.0859	.0760	.0663	.0571	.0486
11	.1180	.1194	.1181	.1144	.1087	.1015	.0932	.0844	.0753	.0663
12	.1032	.1094	.1131	.1144	.1132	.1099	.1049	.0984	.0910	.0829
13	.0834	.0926	.1001	.1056	.1089	.1099	.1089	.1060	.1014	.0956
14	.0625	.0728	.0822	.0905	.0972	.1021	.1050	.1060	.1051	.1024
15	.0438	.0534	.0630	.0724	.0810	.0885	.0945	.0989	.1016	.1024
16	.0287	.0367	.0453	.0543	.0633	.0719	.0798	.0866	.0920	.0960
17	.0177	.0237	.0306	.0383	.0465	.0550	.0633	.0713	.0785	.0847
18	.0104	.0145	.0196	.0255	.0323	.0397	.0475	.0554	.0632	.0706
19	.0057	.0084	.0119	.0161	.0213	.0272	.0337	.0409	.0483	.0557
20	.0030	.0046	.0068	.0097	.0133	.0177	.0228	.0286	.0350	.0418
21	.0015	.0024	.0037	.0055	.0079	.0109	.0146	.0191	.0242	.0299
22	.0007	.0012	.0020	.0030	.0045	.0065	.0090	.0121	.0159	.0204
23	.0003	.0006	.0010	.0016	.0024	.0037	.0053	.0074	.0100	.0133
24	.0001	.0003	.0005	.0008	.0013	.0020	.0030	.0043	.0061	.0083
25	.0001	.0001	.0002	.0004	.0006	.0010	.0016	.0024	.0035	.0050
26	.0000	.0000	.0001	.0002	.0003	.0005	.0008	.0013	.0020	.0029
27	.0000	.0000	.0000	.0001	.0001	.0002	.0004	.0007	.0011	.0016
28	.0000	.0000	.0000	.0000	.0001	.0001	.0002	.0003	.0005	.0009
29	.0000	.0000	.0000	.0000	.0000	.0001	.0001	.0002	.0003	.0004
30	.0000	.0000	.0000	.0000	.0000	.0000	.0000	.0001	.0001	.0002
31	.0000	.0000	.0000	.0000	.0000	.0000	.0000	.0000	.0001	.0001
32	.0000	.0000	.0000	.0000	.0000	.0000	.0000	.0000	.0000	.0001
33	.0000	.0000	.0000	.0000	.0000	.0000	.0000	.0000	.0000	.0000

TABLE 2 (Continued)

					μ					
k	15.5	16.0	16.5	17.0	17.5	18.0	18.5	19.0	19.5	20.0
0	.0000	.0000	.0000	.0000	.0000	.0000	.0000	.0000	.0000	.0000
1	.0000	.0000	.0000	.0000	.0000	.0000	.0000	.0000	.0000	.0000
2	.0000	.0000	.0000	.0000	.0000	.0000	.0000	.0000	.0000	.0000
3	.0001	.0001	.0001	.0000	.0000	.0000	.0000	.0000	.0000	.0000
4	.0004	.0003	.0002	.0001	.0001	.0001	.0000	.0000	.0000	.0000
5	.0014	.0010	.0007	.0005	.0003	.0002	.0002	.0001	.0001	.0001
6	.0036	.0026	.0019	.0014	.0010	.0007	.0005	.0004	.0003	.0002
7	.0079	.0060	.0045	.0034	.0025	.0019	.0014	.0010	.0007	.0005
8	.0153	.0120	.0093	.0072	.0055	.0042	.0031	.0024	.0018	.0013
9	.0264	.0213	.0171	.0135	.0107	.0083	.0065	.0050	.0038	.0029
10	.0409	.0341	.0281	.0230	.0186	.0150	.0120	.0095	.0074	.0058
11	.0577	.0496	.0422	.0355	.0297	.0245	.0201	.0164	.0132	.0106
12	.0745	.0661	.0580	.0504	.0432	.0368	.0310	.0259	.0214	.0176
13	.0888	.0814	.0736	.0658	.0582	.0509	.0441	.0378	.0322	.0271
14	.0983	.0930	.0868	.0800	.0728	.0655	.0583	.0514	.0448	.0387
15	.1016	.0992	.0955	.0906	.0849	.0786	.0719	.0650	.0582	.0516
16	.0984	.0992	.0985	.0963	.0929	.0884	.0831	.0772	.0710	.0646
17	.0897	.0934	.0956	.0963	.0956	.0936	.0904	.0863	.0814	.0760
18	.0773	.0830	.0876	.0909	.0929	.0936	.0930	.0911	.0882	.0844
19	.0630	.0699	.0761	.0814	.0856	.0887	.0905	.0911	.0905	.0888
20	.0489	.0559	.0628	.0692	.0749	.0798	.0837	.0866	.0883	.0888
21	.0361	.0426	.0493	.0560	.0624	.0684	.0738	.0783	.0820	.0846
22	.0254	.0310	.0370	.0433	.0496	.0560	.0620	.0676	.0727	.0769
23	.0171	.0216	.0265	.0320	.0378	.0438	.0499	.0559	.0616	.0669
24	.0111	.0144	.0182	.0226	.0275	.0328	.0385	.0442	.0500	.0557
25	.0069	.0092	.0120	.0154	.0193	.0237	.0285	.0336	.0390	.0446
26	.0041	.0057	.0076	.0101	.0130	.0164	.0202	.0246	.0293	.0343
27	.0023	.0034	.0047	.0063	.0084	.0109	.0139	.0173	.0211	.0254
28	.0013	.0019	.0028	.0038	.0053	.0070	.0092	.0117	.0147	.0181
29	.0007	.0011	.0016	.0023	.0032	.0044	.0058	.0077	.0099	.0125
30	.0004	.0006	.0009	.0013	.0019	.0026	.0036	.0049	.0064	.0083
31	.0002	.0003	.0005	.0007	.0010	.0015	.0022	.0030	.0040	.0054
32	.0001	.0001	.0002	.0004	.0006	.0009	.0012	.0018	.0025	.0034
33	.0000	.0001	.0001	.0002	.0003	.0005	.0007	.0010	.0015	.0020
34	.0000	.0000	.0001	.0001	.0002	.0002	.0004	.0006	.0008	.0012
35	.0000	.0000	.0000	.0000	.0001	.0001	.0002	.0003	.0005	.0007
36	.0000	.0000	.0000	.0000	.0000	.0001	.0001	.0002	.0003	.0004
37	.0000	.0000	.0000	.0000	.0000	.0000	.0001	.0001	.0001	.0002
38	.0000	.0000	.0000	.0000	.0000	.0000	.0000	.0000	.0001	.0001
39	.0000	.0000	.0000	.0000	.0000	.0000	.0000	.0000	.0000	.0001
40	.0000	.0000	.0000	.0000	.0000	.0000	.0000	.0000	.0000	.0000

TABLE 3 The Normal Distribution

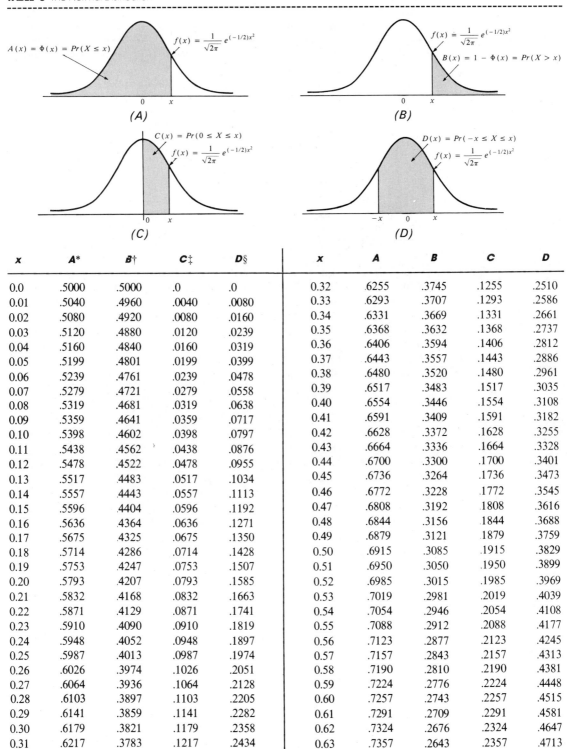

x	A*	B†	C‡	D§	x	A	B	C	D
0.0	.5000	.5000	.0	.0	0.32	.6255	.3745	.1255	.2510
0.01	.5040	.4960	.0040	.0080	0.33	.6293	.3707	.1293	.2586
0.02	.5080	.4920	.0080	.0160	0.34	.6331	.3669	.1331	.2661
0.03	.5120	.4880	.0120	.0239	0.35	.6368	.3632	.1368	.2737
0.04	.5160	.4840	.0160	.0319	0.36	.6406	.3594	.1406	.2812
0.05	.5199	.4801	.0199	.0399	0.37	.6443	.3557	.1443	.2886
0.06	.5239	.4761	.0239	.0478	0.38	.6480	.3520	.1480	.2961
0.07	.5279	.4721	.0279	.0558	0.39	.6517	.3483	.1517	.3035
0.08	.5319	.4681	.0319	.0638	0.40	.6554	.3446	.1554	.3108
0.09	.5359	.4641	.0359	.0717	0.41	.6591	.3409	.1591	.3182
0.10	.5398	.4602	.0398	.0797	0.42	.6628	.3372	.1628	.3255
0.11	.5438	.4562	.0438	.0876	0.43	.6664	.3336	.1664	.3328
0.12	.5478	.4522	.0478	.0955	0.44	.6700	.3300	.1700	.3401
0.13	.5517	.4483	.0517	.1034	0.45	.6736	.3264	.1736	.3473
0.14	.5557	.4443	.0557	.1113	0.46	.6772	.3228	.1772	.3545
0.15	.5596	.4404	.0596	.1192	0.47	.6808	.3192	.1808	.3616
0.16	.5636	.4364	.0636	.1271	0.48	.6844	.3156	.1844	.3688
0.17	.5675	.4325	.0675	.1350	0.49	.6879	.3121	.1879	.3759
0.18	.5714	.4286	.0714	.1428	0.50	.6915	.3085	.1915	.3829
0.19	.5753	.4247	.0753	.1507	0.51	.6950	.3050	.1950	.3899
0.20	.5793	.4207	.0793	.1585	0.52	.6985	.3015	.1985	.3969
0.21	.5832	.4168	.0832	.1663	0.53	.7019	.2981	.2019	.4039
0.22	.5871	.4129	.0871	.1741	0.54	.7054	.2946	.2054	.4108
0.23	.5910	.4090	.0910	.1819	0.55	.7088	.2912	.2088	.4177
0.24	.5948	.4052	.0948	.1897	0.56	.7123	.2877	.2123	.4245
0.25	.5987	.4013	.0987	.1974	0.57	.7157	.2843	.2157	.4313
0.26	.6026	.3974	.1026	.2051	0.58	.7190	.2810	.2190	.4381
0.27	.6064	.3936	.1064	.2128	0.59	.7224	.2776	.2224	.4448
0.28	.6103	.3897	.1103	.2205	0.60	.7257	.2743	.2257	.4515
0.29	.6141	.3859	.1141	.2282	0.61	.7291	.2709	.2291	.4581
0.30	.6179	.3821	.1179	.2358	0.62	.7324	.2676	.2324	.4647
0.31	.6217	.3783	.1217	.2434	0.63	.7357	.2643	.2357	.4713

TABLE 3 (*Continued*)

x	A*	B†	C‡	D§	x	A	B	C	D
0.64	.7389	.2611	.2389	.4778	1.10	.8643	.1357	.3643	.7287
0.65	.7422	.2578	.2422	.4843	1.11	.8665	.1335	.3665	.7330
0.66	.7454	.2546	.2454	.4907	1.12	.8686	.1314	.3686	.7373
0.67	.7486	.2514	.2486	.4971	1.13	.8708	.1292	.3708	.7415
0.68	.7517	.2483	.2517	.5035	1.14	.8729	.1271	.3729	.7457
0.69	.7549	.2451	.2549	.5098	1.15	.8749	.1251	.3749	.7499
0.70	.7580	.2420	.2580	.5161	1.16	.8770	.1230	.3770	.7540
0.71	.7611	.2389	.2611	.5223	1.17	.8790	.1210	.3790	.7580
0.72	.7642	.2358	.2642	.5285	1.18	.8810	.1190	.3810	.7620
0.73	.7673	.2327	.2673	.5346	1.19	.8830	.1170	.3830	.7660
0.74	.7703	.2297	.2703	.5407	1.20	.8849	.1151	.3849	.7699
0.75	.7734	.2266	.2734	.5467	1.21	.8869	.1131	.3869	.7737
0.76	.7764	.2236	.2764	.5527	1.22	.8888	.1112	.3888	.7775
0.77	.7793	.2207	.2793	.5587	1.23	.8907	.1093	.3907	.7813
0.78	.7823	.2177	.2823	.5646	1.24	.8925	.1075	.3925	.7850
0.79	.7852	.2148	.2852	.5705	1.25	.8944	.1056	.3944	.7887
0.80	.7881	.2119	.2881	.5763	1.26	.8962	.1038	.3962	.7923
0.81	.7910	.2090	.2910	.5821	1.27	.8980	.1020	.3980	.7959
0.82	.7939	.2061	.2939	.5878	1.28	.8997	.1003	.3997	.7995
0.83	.7967	.2033	.2967	.5935	1.29	.9015	.0985	.4015	.8029
0.84	.7995	.2005	.2995	.5991	1.30	.9032	.0968	.4032	.8064
0.85	.8023	.1977	.3023	.6047	1.31	.9049	.0951	.4049	.8098
0.86	.8051	.1949	.3051	.6102	1.32	.9066	.0934	.4066	.8132
0.87	.8078	.1922	.3078	.6157	1.33	.9082	.0918	.4082	.8165
0.88	.8106	.1894	.3106	.6211	1.34	.9099	.0901	.4099	.8198
0.89	.8133	.1867	.3133	.6265	1.35	.9115	.0885	.4115	.8230
0.90	.8159	.1841	.3159	.6319	1.36	.9131	.0869	.4131	.8262
0.91	.8186	.1814	.3186	.6372	1.37	.9147	.0853	.4147	.8293
0.92	.8212	.1788	.3212	.6424	1.38	.9162	.0838	.4162	.8324
0.93	.8238	.1762	.3238	.6476	1.39	.9177	.0823	.4177	.8355
0.94	.8264	.1736	.3264	.6528	1.40	.9192	.0808	.4192	.8385
0.95	.8289	.1711	.3289	.6579	1.41	.9207	.0793	.4207	.8415
0.96	.8315	.1685	.3315	.6629	1.42	.9222	.0778	.4222	.8444
0.97	.8340	.1660	.3340	.6680	1.43	.9236	.0764	.4236	.8473
0.98	.8365	.1635	.3365	.6729	1.44	.9251	.0749	.4251	.8501
0.99	.8389	.1611	.3389	.6778	1.45	.9265	.0735	.4265	.8529
1.00	.8413	.1587	.3413	.6827	1.46	.9279	.0721	.4279	.8557
1.01	.8438	.1562	.3438	.6875	1.47	.9292	.0708	.4292	.8584
1.02	.8461	.1539	.3461	.6923	1.48	.9306	.0694	.4306	.8611
1.03	.8485	.1515	.3485	.6970	1.49	.9319	.0681	.4319	.8638
1.04	.8508	.1492	.3508	.7017	1.50	.9332	.0668	.4332	.8664
1.05	.8531	.1469	.3531	.7063	1.51	.9345	.0655	.4345	.8690
1.06	.8554	.1446	.3554	.7109	1.52	.9357	.0643	.4357	.8715
1.07	.8577	.1423	.3577	.7154	1.53	.9370	.0630	.4370	.8740
1.08	.8599	.1401	.3599	.7199	1.54	.9382	.0618	.4382	.8764
1.09	.8621	.1379	.3621	.7243	1.55	.9394	.0606	.4394	.8789

TABLE 3 (Continued)

x	A*	B†	C‡	D§	x	A	B	C	D
1.56	.9406	.0594	.4406	.8812	2.03	.9788	.0212	.4788	.9576
1.57	.9418	.0582	.4418	.8836	2.04	.9793	.0207	.4793	.9586
1.58	.9429	.0571	.4429	.8859	2.05	.9798	.0202	.4798	.9596
1.59	.9441	.0559	.4441	.8882	2.06	.9803	.0197	.4803	.9606
1.60	.9452	.0548	.4452	.8904	2.07	.9808	.0192	.4808	.9615
1.61	.9463	.0537	.4463	.8926	2.08	.9812	.0188	.4812	.9625
1.62	.9474	.0526	.4474	.8948	2.09	.9817	.0183	.4817	.9634
1.63	.9484	.0516	.4484	.8969	2.10	.9821	.0179	.4821	.9643
1.64	.9495	.0505	.4495	.8990	2.11	.9826	.0174	.4826	.9651
1.65	.9505	.0495	.4505	.9011	2.12	.9830	.0170	.4830	.9660
1.66	.9515	.0485	.4515	.9031	2.13	.9834	.0166	.4834	.9668
1.67	.9525	.0475	.4525	.9051	2.14	.9838	.0162	.4838	.9676
1.68	.9535	.0465	.4535	.9070	2.15	.9842	.0158	.4842	.9684
1.69	.9545	.0455	.4545	.9090	2.16	.9846	.0154	.4846	.9692
1.70	.9554	.0446	.4554	.9109	2.17	.9850	.0150	.4850	.9700
1.71	.9564	.0436	.4564	.9127	2.18	.9854	.0146	.4854	.9707
1.72	.9573	.0427	.4573	.9146	2.19	.9857	.0143	.4857	.9715
1.73	.9582	.0418	.4582	.9164	2.20	.9861	.0139	.4861	.9722
1.74	.9591	.0409	.4591	.9181	2.21	.9864	.0136	.4864	.9729
1.75	.9599	.0401	.4599	.9199	2.22	.9868	.0132	.4868	.9736
1.76	.9608	.0392	.4608	.9216	2.23	.9871	.0129	.4871	.9743
1.77	.9616	.0384	.4616	.9233	2.24	.9875	.0125	.4875	.9749
1.78	.9625	.0375	.4625	.9249	2.25	.9878	.0122	.4878	.9756
1.79	.9633	.0367	.4633	.9265	2.26	.9881	.0119	.4881	.9762
1.80	.9641	.0359	.4641	.9281	2.27	.9884	.0116	.4884	.9768
1.81	.9649	.0351	.4649	.9297	2.28	.9887	.0113	.4887	.9774
1.82	.9656	.0344	.4656	.9312	2.29	.9890	.0110	.4890	.9780
1.83	.9664	.0336	.4664	.9327	2.30	.9893	.0107	.4893	.9786
1.84	.9671	.0329	.4671	.9342	2.31	.9896	.0104	.4896	.9791
1.85	.9678	.0322	.4678	.9357	2.32	.9898	.0102	.4898	.9797
1.86	.9686	.0314	.4686	.9371	2.33	.9901	.0099	.4901	.9802
1.87	.9693	.0307	.4693	.9385	2.34	.9904	.0096	.4904	.9807
1.88	.9699	.0301	.4699	.9399	2.35	.9906	.0094	.4906	.9812
1.89	.9706	.0294	.4706	.9412	2.36	.9909	.0091	.4909	.9817
1.90	.9713	.0287	.4713	.9426	2.37	.9911	.0089	.4911	.9822
1.91	.9719	.0281	.4719	.9439	2.38	.9913	.0087	.4913	.9827
1.92	.9726	.0274	.4726	.9451	2.39	.9916	.0084	.4916	.9832
1.93	.9732	.0268	.4732	.9464	2.40	.9918	.0082	.4918	.9836
1.94	.9738	.0262	.4738	.9476	2.41	.9920	.0080	.4920	.9840
1.95	.9744	.0256	.4744	.9488	2.42	.9922	.0078	.4922	.9845
1.96	.9750	.0250	.4750	.9500	2.43	.9925	.0075	.4925	.9849
1.97	.9756	.0244	.4756	.9512	2.44	.9927	.0073	.4927	.9853
1.98	.9761	.0239	.4761	.9523	2.45	.9929	.0071	.4929	.9857
1.99	.9767	.0233	.4767	.9534	2.46	.9931	.0069	.4931	.9861
2.00	.9772	.0228	.4772	.9545	2.47	.9932	.0068	.4932	.9865
2.01	.9778	.0222	.4778	.9556	2.48	.9934	.0066	.4934	.9869
2.02	.9783	.0217	.4783	.9566	2.49	.9936	.0064	.4936	.9872

TABLE 3 (Continued)

x	A*	B†	C‡	D	x	A	B	C	D
2.50	.9938	.0062	.4938	.9876	2.97	.9985	.0015	.4985	.9970
2.51	.9940	.0060	.4940	.9879	2.98	.9986	.0014	.4986	.9971
2.52	.9941	.0059	.4941	.9883	2.99	.9986	.0014	.4986	.9972
2.53	.9943	.0057	.4943	.9886	3.00	.9987	.0013	.4987	.9973
2.54	.9945	.0055	.4945	.9889	3.01	.9987	.0013	.4987	.9974
2.55	.9946	.0054	.4946	.9892	3.02	.9987	.0013	.4987	.9975
2.56	.9948	.0052	.4948	.9895	3.03	.9988	.0012	.4988	.9976
2.57	.9949	.0051	.4949	.9898	3.04	.9988	.0012	.4988	.9976
2.58	.9951	.0049	.4951	.9901	3.05	.9989	.0011	.4989	.9977
2.59	.9952	.0048	.4952	.9904	3.06	.9989	.0011	.4989	.9978
2.60	.9953	.0047	.4953	.9907	3.07	.9989	.0011	.4989	.9979
2.61	.9955	.0045	.4955	.9909	3.08	.9990	.0010	.4990	.9979
2.62	.9956	.0044	.4956	.9912	3.09	.9990	.0010	.4990	.9980
2.63	.9957	.0043	.4957	.9915	3.10	.9990	.0010	.4990	.9981
2.64	.9959	.0041	.4959	.9917	3.11	.9991	.0009	.4991	.9981
2.65	.9960	.0040	.4960	.9920	3.12	.9991	.0009	.4991	.9982
2.66	.9961	.0039	.4961	.9922	3.13	.9991	.0009	.4991	.9983
2.67	.9962	.0038	.4962	.9924	3.14	.9992	.0008	.4992	.9983
2.68	.9963	.0037	.4963	.9926	3.15	.9992	.0008	.4992	.9984
2.69	.9964	.0036	.4964	.9929	3.16	.9992	.0008	.4992	.9984
2.70	.9965	.0035	.4965	.9931	3.17	.9992	.0008	.4992	.9985
2.71	.9966	.0034	.4966	.9933	3.18	.9993	.0007	.4993	.9985
2.72	.9967	.0033	.4967	.9935	3.19	.9993	.0007	.4993	.9986
2.73	.9968	.0032	.4968	.9937	3.20	.9993	.0007	.4993	.9986
2.74	.9969	.0031	.4969	.9939	3.21	.9993	.0007	.4993	.9987
2.75	.9970	.0030	.4970	.9940	3.22	.9994	.0006	.4994	.9987
2.76	.9971	.0029	.4971	.9942	3.23	.9994	.0006	.4994	.9988
2.77	.9972	.0028	.4972	.9944	3.24	.9994	.0006	.4994	.9988
2.78	.9973	.0027	.4973	.9946	3.25	.9994	.0006	.4994	.9988
2.79	.9974	.0026	.4974	.9947	3.26	.9994	.0006	.4994	.9989
2.80	.9974	.0026	.4974	.9949	3.27	.9995	.0005	.4995	.9989
2.81	.9975	.0025	.4975	.9950	3.28	.9995	.0005	.4995	.9990
2.82	.9976	.0024	.4976	.9952	3.29	.9995	.0005	.4995	.9990
2.83	.9977	.0023	.4977	.9953	3.30	.9995	.0005	.4995	.9990
2.84	.9977	.0023	.4977	.9955	3.31	.9995	.0005	.4995	.9991
2.85	.9978	.0022	.4978	.9956	3.32	.9995	.0005	.4995	.9991
2.86	.9979	.0021	.4979	.9958	3.33	.9996	.0004	.4996	.9991
2.87	.9979	.0021	.4979	.9959	3.34	.9996	.0004	.4996	.9992
2.88	.9980	.0020	.4980	.9960	3.35	.9996	.0004	.4996	.9992
2.89	.9981	.0019	.4981	.9961	3.36	.9996	.0004	.4996	.9992
2.90	.9981	.0019	.4981	.9963	3.37	.9996	.0004	.4996	.9992
2.91	.9982	.0018	.4982	.9964	3.38	.9996	.0004	.4996	.9993
2.92	.9982	.0018	.4982	.9965	3.39	.9997	.0003	.4997	.9993
2.93	.9983	.0017	.4983	.9966	3.40	.9997	.0003	.4997	.9993
2.94	.9984	.0016	.4984	.9967	3.41	.9997	.0003	.4997	.9993
2.95	.9984	.0016	.4984	.9968	3.42	.9997	.0003	.4997	.9994
2.96	.9985	.0015	.4985	.9969	3.43	.9997	.0003	.4997	.9994

TABLE 3 (Continued)

x	A*	B†	C‡	D§	x	A	B	C	D
3.44	.9997	.0003	.4997	.9994	3.72	.9999	.0001	.4999	.9998
3.45	.9997	.0003	.4997	.9994	3.73	.9999	.0001	.4999	.9998
3.46	.9997	.0003	.4997	.9995	3.74	.9999	.0001	.4999	.9998
3.47	.9997	.0003	.4997	.9995	3.75	.9999	.0001	.4999	.9998
3.48	.9997	.0003	.4997	.9995	3.76	.9999	.0001	.4999	.9998
3.49	.9998	.0002	.4998	.9995	3.77	.9999	.0001	.4999	.9998
3.50	.9998	.0002	.4998	.9995	3.78	.9999	.0001	.4999	.9998
3.51	.9998	.0002	.4998	.9996	3.79	.9999	.0001	.4999	.9998
3.52	.9998	.0002	.4998	.9996	3.80	.9999	.0001	.4999	.9999
3.53	.9998	.0002	.4998	.9996	3.81	.9999	.0001	.4999	.9999
3.54	.9998	.0002	.4998	.9996	3.82	.9999	.0001	.4999	.9999
3.55	.9998	.0002	.4998	.9996	3.83	.9999	.0001	.4999	.9999
3.56	.9998	.0002	.4998	.9996	3.84	.9999	.0001	.4999	.9999
3.57	.9998	.0002	.4998	.9996	3.85	.9999	.0001	.4999	.9999
3.58	.9998	.0002	.4998	.9997	3.86	.9999	.0001	.4999	.9999
3.59	.9998	.0002	.4998	.9997	3.87	.9999	.0001	.4999	.9999
3.60	.9998	.0002	.4998	.9997	3.88	.9999	.0001	.4999	.9999
3.61	.9998	.0002	.4998	.9997	3.89	.9999	.0001	.4999	.9999
3.62	.9999	.0001	.4999	.9997	3.90	1.0000	.0000	.5000	.9999
3.63	.9999	.0001	.4999	.9997	3.91	1.0000	.0000	.5000	.9999
3.64	.9999	.0001	.4999	.9997	3.92	1.0000	.0000	.5000	.9999
3.65	.9999	.0001	.4999	.9997	3.93	1.0000	.0000	.5000	.9999
3.66	.9999	.0001	.4999	.9997	3.94	1.0000	.0000	.5000	.9999
3.67	.9999	.0001	.4999	.9998	3.95	1.0000	.0000	.5000	.9999
3.68	.9999	.0001	.4999	.9998	3.96	1.0000	.0000	.5000	.9999
3.69	.9999	.0001	.4999	.9998	3.97	1.0000	.0000	.5000	.9999
3.70	.9999	.0001	.4999	.9998	3.98	1.0000	.0000	.5000	.9999
3.71	.9999	.0001	.4999	.9998	3.99	1.0000	.0000	.5000	.9999

*$A(x) = \Phi(x) = Pr(X \leqslant x)$, where X is a standard normal distribution.
†$B(x) = 1 - \Phi(x) = Pr(X > x)$, where X is a standard normal distribution.
‡$C(x) = Pr(0 \leqslant X \leqslant x)$, where X is a standard normal distribution.
§$D(x) = Pr(-x \leqslant X \leqslant x)$, where X is a standard normal distribution.

TABLE 4 Table of 1000 random digits

01	32924	22324	18125	09077	26	96772	16443	39877	04653
02	54632	90374	94143	49295	27	52167	21038	14338	01395
03	88720	43035	97081	83373	28	69644	37198	00028	98195
04	21727	11904	41513	31653	29	71011	62004	81712	87536
05	80985	70799	57975	69282	30	31217	75877	85366	55500
06	40412	58826	94868	52632	31	64990	98735	02999	35521
07	43918	56807	75218	46077	32	48417	23569	59307	46550
08	26513	47480	77410	47741	33	07900	65059	48592	44087
09	18164	35784	44255	30124	34	74526	32601	24482	16981
10	39446	01375	75264	51173	35	51056	04402	58353	37332
11	16638	04680	98617	90298	36	39005	93458	63143	21817
12	16872	94749	44012	48884	37	67883	76343	78155	67733
13	65419	87092	78596	91512	38	06014	60999	87226	36071
14	05207	36702	56804	10498	39	93147	88766	04148	42471
15	78807	79243	13729	81222	40	01099	95731	47622	13294
16	69341	79028	64253	80447	41	89252	01201	58138	13809
17	41871	17566	61200	15994	42	41766	57239	50251	64675
18	25758	04625	43226	32986	43	92736	77800	81996	45646
19	06604	94486	40174	10742	44	45118	36600	68977	68831
20	82259	56512	48945	18183	45	73457	01579	00378	70197
21	07895	37090	50627	71320	46	49465	85251	42914	17277
22	59836	71148	42320	67816	47	15745	37285	23768	39302
23	57133	76610	89104	30481	48	28760	81331	78265	60690
24	76964	57126	87174	61025	49	82193	32787	70451	91141
25	27694	17145	32439	68245	50	89664	50242	12382	39379

TABLE 5 Percentage points of the t distribution $(t_{d,u})$ *

Degrees of freedom, d	.75	.80	.85	.90	.95	.975	.99	.995	.9995
1	1.000	1.376	1.963	3.078	6.314	12.706	31.821	63.657	636.619
2	0.816	1.061	1.386	1.886	2.920	4.303	6.965	9.925	31.598
3	0.765	0.978	1.250	1.638	2.353	3.182	4.541	5.841	12.924
4	0.741	0.941	1.190	1.533	2.132	2.776	3.747	4.604	8.610
5	0.727	0.920	1.156	1.476	2.015	2.571	3.365	4.032	6.869
6	0.718	0.906	1.134	1.440	1.943	2.447	3.143	3.707	5.959
7	0.711	0.896	1.119	1.415	1.895	2.365	2.998	3.499	5.408
8	0.706	0.889	1.108	1.397	1.860	2.306	2.896	3.355	5.041
9	0.703	0.883	1.100	1.383	1.833	2.262	2.821	3.250	4.781
10	0.700	0.879	1.093	1.372	1.812	2.228	2.764	3.169	4.587
11	0.697	0.876	1.088	1.363	1.796	2.201	2.718	3.106	4.437
12	0.695	0.873	1.083	1.356	1.782	2.179	2.681	3.055	4.318
13	0.694	0.870	1.079	1.350	1.771	2.160	2.650	3.012	4.221
14	0.692	0.868	1.076	1.345	1.761	2.145	2.624	2.977	4.140
15	0.691	0.866	1.074	1.341	1.753	2.131	2.602	2.947	4.073
16	0.690	0.865	1.071	1.337	1.746	2.120	2.583	2.921	4.015
17	0.689	0.863	1.069	1.333	1.740	2.110	2.567	2.898	3.965
18	0.688	0.862	1.067	1.330	1.734	2.101	2.552	2.878	3.922
19	0.688	0.861	1.066	1.328	1.729	2.093	2.539	2.861	3.883
20	0.687	0.860	1.064	1.325	1.725	2.086	2.528	2.845	3.850
21	0.686	0.859	1.063	1.323	1.721	2.080	2.518	2.831	3.819
22	0.686	0.858	1.061	1.321	1.717	2.074	2.508	2.819	3.792
23	0.685	0.858	1.060	1.319	1.714	2.069	2.500	2.807	3.767
24	0.685	0.857	1.059	1.318	1.711	2.064	2.492	2.797	3.745
25	0.684	0.856	1.058	1.316	1.708	2.060	2.485	2.787	3.725
26	0.684	0.856	1.058	1.315	1.706	2.056	2.479	2.779	3.707
27	0.684	0.855	1.057	1.314	1.703	2.052	2.473	2.771	3.690
28	0.683	0.855	1.056	1.313	1.701	2.048	2.467	2.763	3.674
29	0.683	0.854	1.055	1.311	1.699	2.045	2.462	2.756	3.659
30	0.683	0.854	1.055	1.310	1.697	2.042	2.457	2.750	3.646
40	0.681	0.851	1.050	1.303	1.684	2.021	2.423	2.704	3.551
60	0.679	0.848	1.046	1.296	1.671	2.000	2.390	2.660	3.460
120	0.677	0.845	1.041	1.289	1.658	1.980	2.358	2.617	3.373
∞	0.674	0.842	1.036	1.282	1.645	1.960	2.326	2.576	3.291

*The uth percentile of a t distribution with d degrees of freedom.
[Table 5 is taken from Table III of Fisher and Yates: "Statistical Tables for Biological, Agricultural and Medical Research," published by Longman Group Ltd., London (previously published by Oliver and Boyd Ltd., Edinburgh) and by permission of the authors and publishers.]

TABLE 6 Percentage points of the chi-square distribution $(\chi^2_{d,u})$§

d	.005	.01	.025	.05	.10	.25	.50	.75	.90	.95	.975	.99	.995	.999
1	0.0^4393*	0.0^3157†	0.0^3982‡	0.00393	0.02	0.10	0.45	1.32	2.71	3.84	5.02	6.63	7.88	10.83
2	0.0100	0.0201	0.0506	0.103	0.21	0.58	1.39	2.77	4.61	5.99	7.38	9.21	10.60	13.81
3	0.0717	0.115	0.216	0.352	0.58	1.21	2.37	4.11	6.25	7.81	9.35	11.34	12.84	16.27
4	0.207	0.297	0.484	0.711	1.06	1.92	3.36	5.39	7.78	9.49	11.14	13.28	14.86	18.47
5	0.412	0.554	0.831	1.15	1.61	2.67	4.35	6.63	9.24	11.07	12.83	15.09	16.75	20.52
6	0.676	0.872	1.24	1.64	2.20	3.45	5.35	7.84	10.64	12.59	14.45	16.81	18.55	22.46
7	0.989	1.24	1.69	2.17	2.83	4.25	6.35	9.04	12.02	14.07	16.01	18.48	20.28	24.32
8	1.34	1.65	2.18	2.73	3.49	5.07	7.34	10.22	13.36	15.51	17.53	20.09	21.95	26.12
9	1.73	2.09	2.70	3.33	4.17	5.90	8.34	11.39	14.68	16.92	19.02	21.67	23.59	27.88
10	2.16	2.56	3.25	3.94	4.87	6.74	9.34	12.55	15.99	18.31	20.48	23.21	25.19	29.59
11	2.60	3.05	3.82	4.57	5.58	7.58	10.34	13.70	17.28	19.68	21.92	24.72	26.76	31.26
12	3.07	3.57	4.40	5.23	6.30	8.44	11.34	14.85	18.55	21.03	23.34	26.22	28.30	32.91
13	3.57	4.11	5.01	5.89	7.04	9.30	12.34	15.98	19.81	22.36	24.74	27.69	29.82	34.53
14	4.07	4.66	5.63	6.57	7.79	10.17	13.34	17.12	21.06	23.68	26.12	29.14	31.32	36.12
15	4.60	5.23	6.27	7.26	8.55	11.04	14.34	18.25	22.31	25.00	27.49	30.58	32.80	37.70
16	5.14	5.81	6.91	7.96	9.31	11.91	15.34	19.37	23.54	26.30	28.85	32.00	34.27	39.25
17	5.70	6.41	7.56	8.67	10.09	12.79	16.34	20.49	24.77	27.59	30.19	33.41	35.72	40.79
18	6.26	7.01	8.23	9.39	10.86	13.68	17.34	21.60	25.99	28.87	31.53	34.81	37.16	42.31
19	6.84	7.63	8.91	10.12	11.65	14.56	18.34	22.72	27.20	30.14	32.85	36.19	38.58	43.82
20	7.43	8.26	9.59	10.85	12.44	15.45	19.34	23.83	28.41	31.41	34.17	37.57	40.00	45.32
21	8.03	8.90	10.28	11.59	13.24	16.34	20.34	24.93	29.62	32.67	35.48	38.93	41.40	46.80
22	8.64	9.54	10.98	12.34	14.04	17.24	21.34	26.04	30.81	33.92	36.78	40.29	42.80	48.27
23	9.26	10.20	11.69	13.09	14.85	18.14	22.34	27.14	32.01	35.17	38.08	41.64	44.18	49.73
24	9.89	10.86	12.40	13.85	15.66	19.04	23.34	28.24	33.20	36.42	39.36	42.98	45.56	51.18
25	10.52	11.52	13.12	14.61	16.47	19.94	24.34	29.34	34.38	37.65	40.65	44.31	46.93	52.62
26	11.16	12.20	13.84	15.38	17.29	20.84	25.34	30.43	35.56	38.89	41.92	45.64	48.29	54.05
27	11.81	12.88	14.57	16.15	18.11	21.75	26.34	31.53	36.74	40.11	43.19	46.96	49.64	55.48
28	12.46	13.56	15.31	16.93	18.94	22.66	27.34	32.62	37.92	41.34	44.46	48.28	50.99	56.89
29	13.12	14.26	16.05	17.71	19.77	23.57	28.34	33.71	39.09	42.56	45.72	49.59	52.34	58.30
30	13.79	14.95	16.79	18.49	20.60	24.48	29.34	34.80	40.26	43.77	46.98	50.89	53.67	59.70
40	20.71	22.16	24.43	26.51	29.05	33.66	39.34	45.62	51.81	55.76	59.34	63.69	66.77	73.40
50	27.99	29.71	32.36	34.76	37.69	42.94	49.33	56.33	63.17	67.50	71.42	76.15	79.49	86.66
60	35.53	37.48	40.48	43.19	46.46	52.29	59.33	66.98	74.40	79.08	83.30	88.38	91.95	99.61
70	43.28	45.44	48.76	51.74	55.33	61.70	69.33	77.58	85.53	90.53	95.02	100.42	104.22	112.32
80	51.17	53.54	57.15	60.39	64.28	71.14	79.33	88.13	96.58	101.88	106.63	112.33	116.32	124.84
90	59.20	61.75	65.65	69.13	73.29	80.62	89.33	98.64	107.56	113.14	118.14	124.12	128.30	137.21
100	67.33	70.06	74.22	77.93	82.36	90.13	99.33	109.14	118.50	124.34	129.56	135.81	140.17	149.45

* = 0.0000393   † = 0.000157   ‡ = 0.000982   § $\chi^2_{d,u}$ = uth percentile of a χ^2 distribution with d degrees of freedom.
(Reproduced in part with permission of the Biometrika Trustees, from Table 3 of "Biometrika Tables for Statisticians," Volume II, edited by E. S. Pearson and H. O. Hartley, published for the Biometrika Trustees, Cambridge University Press, Cambridge, England, 1972.)

TABLE 7a Exact two-sided 100% × (1 − α) confidence limits for binomial proportions (α = .05)

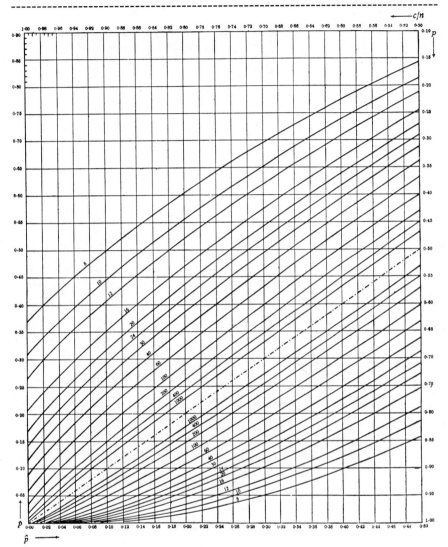

TABLE 7b Exact two-sided 100% × (1 − α) confidence limits for binomial proportions (α = .01)

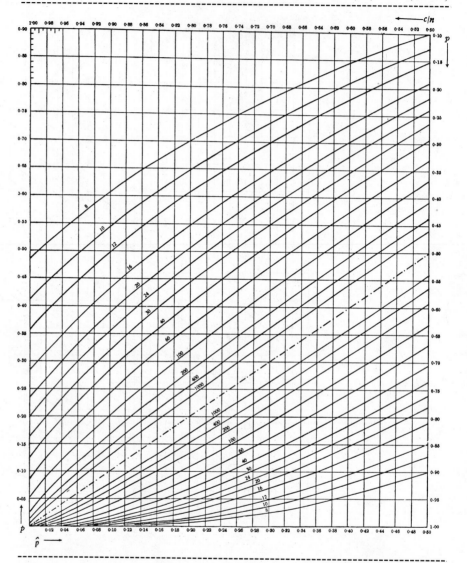

(Tables 7a and 7b have been reproduced with permission of the Biometrika Trustees, from Table 41 of "Biometrika Tables for Statisticians," 3rd Edition, Volume II, Published for the Biometrika Trustees, Cambridge University Press, Cambridge, England, 1966.)

TABLE 8 Percentage points of the F distribution ($F_{d_1, d_2, p}$)

df for denominator, d_2	p	1	2	3	4	5	6	7	8	12	24	∞
1	.90	39.86	49.50	53.59	55.83	57.24	58.20	58.91	59.44	60.71	62.00	63.33
	.95	161.4	199.5	215.7	224.6	230.2	234.0	236.8	238.9	243.9	249.1	254.3
	.975	647.8	799.5	864.2	899.6	921.8	937.1	948.2	956.7	976.7	997.2	1018.
	.99	4052.	5000.	5403.	5625.	5764.	5859.	5928.	5981.	6106.	6235.	6366.
	.995	16211.	20000.	21615.	22500.	23056.	23437.	23715.	23925.	24426.	24940.	25464.
	.999	405280.	500000.	540380.	562500.	576400.	585940.	592870.	598140.	610670.	623500.	636620.
2	.90	8.53	9.00	9.16	9.24	9.29	9.33	9.35	9.37	9.41	9.45	9.49
	.95	18.51	19.00	19.16	19.25	19.30	19.33	19.35	19.37	19.41	19.45	19.50
	.975	38.51	39.00	39.17	39.25	39.30	39.33	39.36	39.37	39.42	39.46	39.50
	.99	98.50	99.00	99.17	99.25	99.30	99.33	99.36	99.37	99.42	99.46	99.50
	.995	198.5	199.0	199.2	199.2	199.3	199.3	199.4	199.4	199.4	199.5	199.5
	.999	998.5	999.0	999.2	999.2	999.3	999.3	999.4	999.4	999.4	999.5	999.5
3	.90	5.54	5.46	5.39	5.34	5.31	5.28	5.27	5.25	5.22	5.18	5.13
	.95	10.13	9.55	9.28	9.12	9.01	8.94	8.89	8.85	8.74	8.64	8.53
	.975	17.44	16.04	15.44	15.10	14.88	14.74	14.62	14.54	14.34	14.12	13.90
	.99	34.12	30.82	29.46	28.71	28.24	27.91	27.67	27.49	27.05	26.60	26.13
	.995	55.55	49.80	47.47	46.20	45.39	44.84	44.43	44.13	43.39	42.62	41.83
	.999	167.00	148.5	141.1	137.1	134.6	132.8	131.6	130.6	128.3	125.9	123.5
4	90	4.54	4.32	4.19	4.11	4.05	4.01	3.98	3.95	3.90	3.83	3.76
	.95	7.71	6.94	6.59	6.39	6.26	6.16	6.09	6.04	5.91	5.77	5.63
	.975	12.22	10.65	9.98	9.60	9.36	9.20	9.07	8.98	8.75	8.51	8.26
	.99	21.20	18.00	16.69	15.98	15.52	15.21	14.98	14.80	14.37	13.93	13.46
	.995	31.33	26.28	24.26	23.16	22.46	21.98	21.62	21.35	20.70	20.03	19.32
	.999	74.14	61.25	56.18	53.44	51.71	50.53	49.66	49.00	47.41	45.77	44.05
5	.90	4.06	3.78	3.62	3.52	3.45	3.40	3.37	3.34	3.27	3.19	3.10
	.95	6.61	5.79	5.41	5.19	5.05	4.95	4.88	4.82	4.68	4.53	4.36
	.975	10.01	8.43	7.76	7.39	7.15	6.98	6.85	6.76	6.52	6.28	6.02
	.99	16.26	13.27	12.06	11.39	10.97	10.67	10.46	10.29	9.89	9.47	9.02
	.995	22.78	18.31	16.53	15.56	14.94	14.51	14.20	13.96	13.38	12.78	12.14
	.999	47.18	37.12	33.20	31.09	29.75	28.83	28.16	27.65	26.42	25.13	23.79
6	.90	3.78	3.46	3.29	3.18	3.11	3.05	3.01	2.98	2.90	2.82	2.72
	.95	5.99	5.14	4.76	4.53	4.39	4.28	4.21	4.15	4.00	3.84	3.67
	.975	8.81	7.26	6.60	6.23	5.99	5.82	5.70	5.60	5.37	5.12	4.85
	.99	13.75	10.92	9.78	9.15	8.75	8.47	8.26	8.10	7.72	7.31	6.88
	.995	18.64	14.54	12.92	12.03	11.46	11.07	10.79	10.57	10.03	9.47	8.88
	.999	35.51	27.00	23.70	21.92	20.80	20.03	19.46	19.03	17.99	16.90	15.75
7	.90	3.59	3.26	3.07	2.96	2.88	2.83	2.78	2.75	2.67	2.58	2.47
	.95	5.59	4.74	4.35	4.12	3.97	3.87	3.79	3.73	3.57	3.41	3.23
	.975	8.07	6.54	5.89	5.52	5.29	5.12	4.99	4.90	4.67	4.42	4.14
	.99	12.25	9.55	8.45	7.85	7.46	7.19	6.99	6.84	6.47	6.07	5.65
	.995	16.24	12.40	10.88	10.05	9.52	9.16	8.89	8.68	8.18	7.65	7.08
	.999	29.25	21.69	18.77	17.20	16.21	15.52	15.02	14.63	13.71	12.73	11.70
8	.90	3.46	3.11	2.92	2.81	2.73	2.67	2.62	2.59	2.50	2.40	2.29
	.95	5.32	4.46	4.07	3.84	3.69	3.58	3.50	3.44	3.28	3.12	2.93
	.975	7.57	6.06	5.42	5.05	4.82	4.65	4.53	4.43	4.20	3.95	3.67
	.99	11.26	8.65	7.59	7.01	6.63	6.37	6.18	6.03	5.67	5.28	4.86
	.995	14.69	11.04	9.60	8.81	8.30	7.95	7.69	7.50	7.01	6.50	5.95
	.999	25.42	18.49	15.83	14.39	13.49	12.86	12.40	12.04	11.19	10.30	9.33
9	.90	3.36	3.01	2.81	2.69	2.61	2.55	2.51	2.47	2.38	2.28	2.16
	.95	5.12	4.26	3.86	3.63	3.48	3.37	3.29	3.23	3.07	2.90	2.71
	.975	7.21	5.71	5.08	4.72	4.48	4.32	4.20	4.10	3.87	3.61	3.33
	.99	10.56	8.02	6.99	6.42	6.06	5.80	5.61	5.47	5.11	4.73	4.31
	.995	13.61	10.11	8.72	7.96	7.47	7.13	6.88	6.69	6.23	5.73	5.19
	.999	22.86	16.39	13.90	12.56	11.71	11.13	10.70	10.37	9.57	8.72	7.81

TABLE 8 (Continued)

df for denominator, d_2	p	1	2	3	4	5	6	7	8	12	24	∞
						df for numerator, d_1						
10	.90	3.29	2.92	2.73	2.61	2.52	2.46	2.41	2.38	2.28	2.18	2.06
	.95	4.96	4.10	3.71	3.48	3.33	3.22	3.14	3.07	2.91	2.74	2.54
	.975	6.94	5.46	4.83	4.47	4.24	4.07	3.95	3.85	3.62	3.37	3.08
	.99	10.04	7.56	6.55	5.99	5.64	5.39	5.20	5.06	4.71	4.33	3.91
	.995	12.83	9.43	8.08	7.34	6.87	6.54	6.30	6.12	5.66	5.17	4.64
	.999	21.04	14.91	12.55	11.28	10.48	9.93	9.52	9.20	8.45	7.64	6.76
12	.90	3.18	2.81	2.61	2.48	2.39	2.33	2.28	2.24	2.15	2.04	1.90
	.95	4.75	3.89	3.49	3.26	3.11	3.00	2.91	2.85	2.69	2.51	2.30
	.975	6.55	5.10	4.47	4.12	3.89	3.73	3.61	3.51	3.28	3.02	2.72
	.99	9.33	6.93	5.95	5.41	5.06	4.82	4.64	4.50	4.16	3.78	3.36
	.995	11.75	8.51	7.23	6.52	6.07	5.76	5.52	5.35	4.91	4.43	3.90
	.999	18.64	12.97	10.80	9.63	8.89	8.38	8.00	7.71	7.00	6.25	5.42
14	.90	3.10	2.73	2.52	2.39	2.31	2.24	2.19	2.15	2.05	1.94	1.80
	.95	4.60	3.74	3.34	3.11	2.96	2.85	2.76	2.70	2.53	2.35	2.13
	.975	6.30	4.86	4.24	3.89	3.66	3.50	3.38	3.29	3.05	2.79	2.49
	.99	8.86	6.51	5.56	5.04	4.69	4.46	4.28	4.14	3.80	3.43	3.00
	.995	11.06	7.92	6.68	6.00	5.56	5.26	5.03	4.86	4.43	3.96	3.44
	.999	17.14	11.78	9.73	8.62	7.92	7.44	7.08	6.80	6.13	5.41	4.60
16	.90	3.05	2.67	2.46	2.33	2.24	2.18	2.13	2.09	1.99	1.87	1.72
	.95	4.49	3.63	3.24	3.01	2.85	2.74	2.66	2.59	2.42	2.24	2.01
	.975	6.12	4.69	4.08	3.73	3.50	3.34	3.22	3.12	2.89	2.63	2.32
	.99	8.53	6.23	5.29	4.77	4.44	4.20	4.03	3.89	3.55	3.18	2.75
	.995	10.58	7.51	6.30	5.64	5.21	4.91	4.69	4.52	4.10	3.64	3.11
	.999	16.12	10.97	9.01	7.94	7.27	6.80	6.46	6.19	5.55	4.85	4.06
18	.90	3.01	2.62	2.42	2.29	2.20	2.13	2.08	2.04	1.93	1.81	1.66
	.95	4.41	3.55	3.16	2.93	2.77	2.66	2.58	2.51	2.34	2.15	1.92
	.975	5.98	4.56	3.95	3.61	3.38	3.22	3.10	3.01	2.77	2.50	2.19
	.99	8.29	6.01	5.09	4.58	4.25	4.01	3.84	3.71	3.37	3.00	2.57
	.995	10.22	7.21	6.03	5.37	4.96	4.66	4.44	4.28	3.86	3.40	2.87
	.999	15.38	10.39	8.49	7.46	6.81	6.35	6.02	5.76	5.13	4.45	3.67
20	.90	2.97	2.59	2.38	2.25	2.16	2.09	2.04	2.00	1.89	1.77	1.61
	.95	4.35	3.49	3.10	2.87	2.71	2.60	2.51	2.45	2.28	2.08	1.84
	.975	5.87	4.46	3.86	3.51	3.29	3.13	3.01	2.91	2.68	2.41	2.09
	.99	8.10	5.85	4.94	4.43	4.10	3.87	3.70	3.56	3.23	2.86	2.42
	.995	9.94	6.99	5.82	5.17	4.76	4.47	4.26	4.09	3.68	3.22	2.69
	.999	14.82	9.95	8.10	7.10	6.46	6.02	5.69	5.44	4.82	4.15	3.38
30	.90	2.88	2.49	2.28	2.14	2.05	1.98	1.93	1.88	1.77	1.64	1.46
	.95	4.17	3.32	2.92	2.69	2.53	2.42	2.33	2.27	2.09	1.89	1.62
	.975	5.57	4.18	3.59	3.25	3.03	2.87	2.75	2.65	2.41	2.14	1.79
	.99	7.56	5.39	4.51	4.02	3.70	3.47	3.30	3.17	2.84	2.47	2.01
	.995	9.18	6.35	5.24	4.62	4.23	3.95	3.74	3.58	3.18	2.73	2.18
	.999	13.29	8.77	7.05	6.12	5.53	5.12	4.82	4.58	4.00	3.36	2.59
40	.90	2.84	2.44	2.23	2.09	2.00	1.93	1.87	1.83	1.71	1.57	1.38
	.95	4.08	3.23	2.84	2.61	2.45	2.34	2.25	2.18	2.00	1.79	1.51
	.975	5.42	4.05	3.46	3.13	2.90	2.74	2.62	2.53	2.29	2.01	1.64
	.99	7.31	5.18	4.31	3.83	3.51	3.29	3.12	2.99	2.66	2.29	1.80
	.995	8.83	6.07	4.98	4.37	3.99	3.71	3.51	3.35	2.95	2.50	1.93
	.999	12.61	8.25	6.59	5.70	5.13	4.73	4.44	4.21	3.64	3.01	2.23
60	.90	2.79	2.39	2.18	2.04	1.95	1.87	1.82	1.77	1.66	1.51	1.29
	.95	4.00	3.15	2.76	2.53	2.37	2.25	2.17	2.10	1.92	1.70	1.39
	.975	5.29	3.93	3.34	3.01	2.79	2.63	2.51	2.41	2.17	1.88	1.48
	.99	7.08	4.98	4.13	3.65	3.34	3.12	2.95	2.82	2.50	2.12	1.60
	.995	8.49	5.80	4.73	4.14	3.76	3.49	3.29	3.13	2.74	2.29	1.69
	.999	11.97	7.77	6.17	5.31	4.76	4.37	4.09	3.86	3.32	2.69	1.89

TABLE 8 (*Continued*)

df for denominator, d_2	p	1	2	3	4	5	6	7	8	12	24	∞
						df for numerator, d_1						
120	.90	2.75	2.35	2.13	1.99	1.90	1.82	1.77	1.72	1.60	1.45	1.19
	.95	3.92	3.07	2.68	2.45	2.29	2.17	2.09	2.02	1.83	1.61	1.25
	.975	5.15	3.80	3.23	2.89	2.67	2.52	2.39	2.30	2.05	1.76	1.31
	.99	6.85	4.79	3.95	3.48	3.17	2.96	2.79	2.66	2.34	1.95	1.38
	.995	8.18	5.54	4.50	3.92	3.55	3.28	3.09	2.93	2.54	2.09	1.43
	.999	11.38	7.32	5.78	4.95	4.42	4.04	3.77	3.55	3.02	2.40	1.54
∞	.90	2.71	2.30	2.08	1.94	1.85	1.77	1.72	1.67	1.55	1.38	1.00
	.95	3.84	3.00	2.60	2.37	2.21	2.10	2.01	1.94	1.75	1.52	1.00
	.975	5.02	3.69	3.12	2.79	2.57	2.41	2.29	2.19	1.94	1.64	1.00
	.99	6.63	4.61	3.78	3.32	3.02	2.80	2.64	2.51	2.18	1.79	1.00
	.995	7.88	5.30	4.28	3.72	3.35	3.09	2.90	2.74	2.36	1.90	1.00
	.999	10.83	6.91	5.42	4.62	4.10	3.74	3.47	3.27	2.74	2.13	1.00

*$F_{d_1, d_2, p}$ = pth percentile of an F distribution with d_1 and d_2 degrees of freedom.
(This table has been reproduced in part with the permission of the Biometrika Trustees, from "Biometrika Tables for Statisticians," Volume II, edited by E. S. Pearson and H. O. Hartley, published for the Biometrika Trustees, Cambridge University Press, Cambridge, England, 1972.)

TABLE 9 Two-tailed critical values for the Wilcoxon sign rank test

n*	.10 Lower	.10 Upper	.05 Lower	.05 Upper	.02 Lower	.02 Upper	.01 Lower	.01 Upper
1	—		—		—		—	
2	—		—		—		—	
3	—		—		—		—	
4	—		—		—		—	
5	0	15	—		—		—	
6	2	19	0	21	—		—	
7	3	25	2	26	0	28	—	
8	5	31	3	33	1	35	0	36
9	8	37	5	40	3	42	1	44
10	10	45	8	47	5	50	3	52
11	13	53	10	56	7	59	5	61
12	17	61	13	65	9	69	7	71
13	21	70	17	74	12	79	9	82
14	25	80	21	84	15	90	12	93
15	30	90	25	95	19	101	15	105

*n = number of untied pairs.
(Figures from "Documenta Geigy Scientific Tables," 6th Edition. Reprinted with the kind permission of CIBA-GEIGY Limited, Basle, Switzerland.)

TABLE 10 Two-tailed critical values for the Wilcoxon rank sum test

Each cell shows $T_l - T_r$ for the given n_1.

n_2†	$\alpha = .10$, n_1* : 4	5	6	7	8	9	$\alpha = .05$, n_1 : 4	5	6	7	8	9
	T_l T_r	T_l T_r	T_l T_r	T_l T_r	T_l T_r	T_l T_r	T_l T_r	T_l T_r	T_l T_r	T_l T_r	T_l T_r	T_l T_r
4	11– 25	17– 33	24– 42	32– 52	41– 63	51– 75	10– 26	16– 34	23– 43	31– 53	40– 64	49– 77
5	12– 28	19– 36	26– 46	34– 57	44– 68	54– 81	11– 29	17– 38	24– 48	33– 58	42– 70	52– 83
6	13– 31	20– 40	28– 50	36– 62	46– 74	57– 87	12– 32	18– 42	26– 52	34– 64	44– 76	55– 89
7	14– 34	21– 44	29– 55	39– 66	49– 79	60– 93	13– 35	20– 45	27– 57	36– 69	46– 82	57– 96
8	15– 37	23– 47	31– 59	41– 71	51– 85	63– 99	14– 38	21– 49	29– 61	38– 74	49– 87	60–102
9	16– 40	24– 51	33– 63	43– 76	54– 90	66–105	14– 42	22– 53	31– 65	40– 79	51– 93	62–109
10	17– 43	26– 54	35– 67	45– 81	56– 96	69–111	15– 45	23– 57	32– 70	42– 84	53– 99	65–115
11	18– 46	27– 58	37– 71	47– 86	59–101	72–117	16– 48	24– 61	34– 74	44– 89	55–105	68–121
12	19– 49	28– 62	38– 76	49– 91	62–106	75–123	17– 51	26– 64	35– 79	46– 94	58–110	71–127
13	20– 52	30– 65	40– 80	52– 95	64–112	78–129	18– 54	27– 68	37– 83	48– 99	60–116	73–134
14	21– 55	31– 69	42– 84	54–100	67–117	81–135	19– 57	28– 72	38– 88	50–104	62–122	76–140
15	22– 58	33– 72	44– 88	56–105	69–123	84–141	20– 60	29– 76	40– 92	52–109	65–127	79–146
16	24– 60	34– 76	46– 92	58–110	72–128	87–147	21– 63	30– 80	42– 96	54–114	67–133	82–152
17	25– 63	35– 80	47– 97	61–114	75–133	90–153	21– 67	32– 83	43–101	56–119	70–138	84–159
18	26– 66	37– 83	49–101	63–119	77–139	93–159	22– 70	33– 87	45–105	58–124	72–144	87–165
19	27– 69	38– 87	51–105	65–124	80–144	96–165	23– 73	34– 91	46–110	60–129	74–150	90–171
20	28– 72	40– 90	53–109	67–129	83–149	99–171	24– 76	35– 95	48–114	62–134	77–155	93–177
21	29– 75	41– 94	55–113	69–134	85–155	102–177	25– 79	37– 98	50–118	64–139	79–161	95–184
22	30– 78	43– 97	57–117	72–138	88–160	105–183	26– 82	38–102	51–123	66–144	81–167	98–190
23	31– 81	44–101	58–122	74–143	90–166	108–189	27– 85	39–106	53–127	68–149	84–172	101–196
24	32– 84	45–105	60–126	76–148	93–171	111–195	27– 89	40–110	54–132	70–154	86–178	104–202
25	33– 87	47–108	62–130	78–153	96–176	114–201	28– 92	42–113	56–136	72–159	89–183	107–208
26	34– 90	48–112	64–134	81–157	98–182	117–207	29– 95	43–117	58–140	74–164	91–189	109–215
27	35– 93	50–115	66–138	83–162	101–187	120–213	30– 98	44–121	59–145	76–169	93–195	112–221
28	36– 96	51–119	67–143	85–167	103–193	123–219	31–101	45–125	61–149	78–174	96–200	115–227
29	37– 99	53–122	69–147	87–172	106–198	126–225	32–104	47–128	63–153	80–179	98–206	118–233
30	38–102	54–126	71–151	89–177	109–203	129–231	33–107	48–132	64–158	82–184	101–211	121–239
31	39–105	55–130	73–155	92–181	111–209	132–237	34–110	49–136	66–162	84–189	103–217	123–246
32	40–108	57–133	75–159	94–186	114–214	135–243	34–114	50–140	67–167	86–194	106–222	126–252
33	41–111	58–137	77–163	96–191	117–219	138–249	35–117	52–143	69–171	88–199	108–228	129–258
34	42–114	60–140	78–168	98–196	119–225	141–255	36–120	53–147	71–175	90–204	110–234	132–264
35	43–117	61–144	80–172	100–201	122–230	144–261	37–123	54–151	72–180	92–209	113–239	135–270
36	44–120	62–148	82–176	102–206	124–236	148–266	38–126	55–155	74–184	94–214	115–245	137–277
37	45–123	64–151	84–180	105–210	127–241	151–272	39–129	57–158	76–188	96–219	117–251	140–283
38	46–126	65–155	85–185	107–215	130–246	154–278	40–132	58–162	77–193	98–224	120–256	143–289
39	47–129	67–158	87–189	109–220	132–252	157–284	41–135	59–166	79–197	100–229	122–262	146–295
40	48–132	68–162	89–193	111–225	135–257	160–290	41–139	60–170	80–202	102–234	125–267	149–301
41	49–135	69–166	91–197	114–229	138–262	163–296	42–142	61–174	82–206	104–239	127–273	151–308
42	50–138	71–169	93–201	116–234	140–268	166–302	43–145	63–177	84–210	106–244	129–279	154–314
43	51–141	72–173	95–205	118–239	143–273	169–308	44–148	64–181	85–215	108–249	132–284	157–320
44	52–144	74–176	96–210	120–244	146–278	172–314	45–151	65–185	87–219	110–254	134–290	160–326
45	53–147	75–180	98–214	123–248	148–284	175–320	46–154	66–189	88–224	112–259	137–295	163–332
46	55–149	77–183	100–218	125–253	151–289	178–326	47–157	68–192	90–228	114–264	139–301	165–339
47	56–152	78–187	102–222	127–258	154–294	181–332	48–160	69–196	92–232	116–269	141–307	168–345
48	57–155	79–191	104–226	129–263	156–300	184–338	48–164	70–200	93–237	118–274	144–312	171–351
49	58–158	81–194	106–230	132–267	159–305	187–344	49–167	71–204	95–241	120–279	146–318	174–357
50	59–161	82–198	107–235	134–272	162–310	190–350	50–170	73–207	97–245	122–284	149–323	177–363

*n_1 = minimum of the two sample sizes.
†n_2 = maximum of the two sample sizes.

‡T_l = lower critical value for the rank sum in the first sample.
§T_r = upper critical value for the rank sum in the first sample.

TABLE 10 (Continued)

	$\alpha=.02$ n_1^*						$\alpha=.01$ n_1					
$n_2^\dagger$	4	5	6	7	8	9	4	5	6	7	8	9
	$T_l^\ddagger$ T_r	T_l T_r	T_l T_r	T_l T_r	T_l T_r	T_l T_r	T_l T_r	T_l T_r	T_l T_r	T_l T_r	T_l T_r	T_l T_r
4	— —	15– 35	22– 44	29– 55	38– 66	48– 78	— —	— —	21– 45	28– 56	37– 67	46– 80
5	10– 30	16– 39	23– 49	31– 60	40– 72	50– 85	— —	15– 40	22– 50	29– 62	38– 74	48– 87
6	11– 33	17– 43	24– 54	32– 66	42– 78	52– 92	10– 34	16– 44	23– 55	31– 67	40– 80	50– 94
7	11– 37	18– 47	25– 59	34– 71	43– 85	54– 99	10– 38	16– 49	24– 60	32– 73	42– 86	52–101
8	12– 40	19– 51	27– 63	35– 77	45– 91	56–106	11– 41	17– 53	25– 65	34– 78	43– 93	54–108
9	13– 43	20– 55	28– 68	37– 82	47– 97	59–112	11– 45	18– 57	26– 70	35– 84	45– 99	56–115
10	13– 47	21– 59	29– 73	39– 87	49–103	61–119	12– 48	19– 61	27– 75	37– 89	47–105	58–122
11	14– 50	22– 63	30– 78	40– 93	51–109	63–126	12– 52	20– 65	28– 80	38– 95	49–111	61–128
12	15– 53	23– 67	32– 82	42– 98	53–115	66–132	13– 55	21– 69	30– 84	40–100	51–117	63–135
13	15– 57	24– 71	33– 87	44–103	56–120	68–139	13– 59	22– 73	31– 89	41–106	53–123	65–142
14	16– 60	25– 75	34– 92	45–109	58–126	71–145	14– 62	22– 78	32– 94	43–111	54–130	67–149
15	17– 63	26– 79	36– 96	47–114	60–132	73–152	15– 65	23– 82	33– 99	44–117	56–136	69–156
16	17– 67	27– 83	37–101	49–119	62–138	76–158	15– 69	24– 86	34–104	46–122	58–142	72–162
17	18– 70	28– 87	39–105	51–124	64–144	78–165	16– 72	25– 90	36–108	47–128	60–148	74–169
18	19– 73	29– 91	40–110	52–130	66–150	81–171	16– 76	26– 94	37–113	49–133	62–154	76–176
19	19– 77	30– 95	41–115	54–135	68–156	83–178	17– 79	27– 98	38–118	50–139	64–160	78–183
20	20– 80	31– 99	43–119	56–140	70–162	85–185	18– 82	28–102	39–123	52–144	66–166	81–189
21	21– 83	32–103	44–124	58–145	72–168	88–191	18– 86	29–106	40–128	53–150	68–172	83 196
22	21– 87	33–107	45–129	59–151	74–174	90–198	19– 89	29–111	42–132	55–155	70–178	85–203
23	22– 90	34–111	47–133	61–156	76–180	93–204	19– 93	30–115	43–137	57–160	71–185	88–209
24	23– 93	35–115	48–138	63–161	78–186	95–211	20– 96	31–119	44–142	58–166	73–191	90–216
25	23– 97	36–119	50–142	64–167	81–191	98–217	20–100	32–123	45–147	60–171	75–197	92–223
26	24–100	37–123	51–147	66–172	83–197	100–224	21–103	33–127	46–152	61–177	77–203	94–230
27	25–103	38–127	52–152	68–177	85–203	103–230	22–106	34–131	48–156	63–182	79–209	97–236
28	26–106	39–131	54–156	70–182	87–209	105–237	22–110	35–135	49–161	64–188	81–215	99–243
29	26–110	40–135	55–161	71–188	89–215	108–243	23–113	36–139	50–166	66–193	83–221	101–250
30	27–113	41–139	56–166	73–193	91–221	110–250	23–117	37–143	51–171	68–198	85–227	103–257
31	28–116	42–143	58–170	75–198	93–227	112–257	24–120	37–148	53–175	68–204	87–233	106–263
32	28–120	43–147	59–175	77–203	95–233	115–263	24–124	38–152	54–180	71–209	89–239	108–270
33	29–123	44–151	61–179	78–209	97–239	117–270	25–127	39–156	55–185	72–215	90–246	110–277
34	30–126	45–155	62–184	79–215	99–245	120–276	26–130	40–160	56–190	73–221	92–252	112–284
35	30–130	46–159	63–189	81–220	101–251	122–283	26–134	41–164	57–195	75–226	94–258	114–291
36	31–133	47–163	65–193	83–225	103–257	125–289	27–137	42–168	58–200	76–232	96–264	117–297
37	32–136	48–167	66–198	84–231	105–263	127–296	28–140	43–172	60–204	78–237	98–270	119–304
38	32–140	49–171	67–203	86–236	107–269	129–303	28–144	44–176	61–209	79–243	100–276	121–311
39	33–143	50–175	69–207	88–241	109–275	132–309	29–147	45–180	62–214	81–248	102–282	123–318
40	34–146	51–179	70–212	90–246	111–281	134–316	29–151	46–184	63–219	82–254	103–289	126–324
41	34–150	52–183	72–216	91–252	113–287	137–322	30–154	46–189	65–223	84–259	105–295	128–331
42	35–153	53–187	73–221	93–257	116–292	139–329	31–157	47–193	66–228	85–265	107–301	130–338
43	35–157	54–191	74–226	95–262	118–298	142–335	31–161	48–197	67–233	87–270	109–307	133–344
44	36–160	55–195	76–230	97–267	120–304	144–342	32–164	49–201	68–238	88–276	111–313	135–351
45	37–163	56–199	77–235	98–273	122–310	147–348	32–168	50–205	69–243	90–281	113–319	137–358
46	37–167	57–203	78–240	100–278	124–316	149–355	33–171	51–209	71–247	91–287	115–325	139–365
47	38–170	58–207	80–244	102–283	126–322	152–361	34–174	52–213	72–252	93–292	117–331	142–371
48	39–173	59–211	81–249	103–289	128–328	154–368	34–178	53–217	73–257	95–297	118–338	144–378
49	39–177	60–215	82–254	105–294	130–334	157–374	35–181	54–221	74–262	96–303	120–344	146–385
50	40–180	61–219	84–258	107–299	132–340	159–381	36–184	55–225	76–266	98–308	122–350	148–392

(The data of this table are reproduced with permission from *Documenta Geigy Scientific Tables*, 6th Ed., pp. 124–127, Geigy Pharmaceuticals, Division of Geigy Chemical Corporation, Ardsley, N.Y. Figures from "Documenta Geigy Scientific Tables," 6th Edition. Reprinted with the kind permission of CIBA-GEIGY Limited, Basle, Switzerland.)

TABLE 11 Fisher's z transformation

r	z	r	z	r	z	r	z	r	z
.00	.000								
.01	.010	.21	.213	.41	.436	.61	.709	.81	1.127
.02	.020	.22	.224	.42	.448	.62	.725	.82	1.157
.03	.030	.23	.234	.43	.460	.63	.741	.83	1.188
.04	.040	.24	.245	.44	.472	.64	.758	.84	1.221
.05	.050	.25	.255	.45	.485	.65	.775	.85	1.256
.06	.060	.26	.266	.46	.497	.66	.793	.86	1.293
.07	.070	.27	.277	.47	.510	.67	.811	.87	1.333
.08	.080	.28	.288	.48	.523	.68	.829	.88	1.376
.09	.090	.29	.299	.49	.536	.69	.848	.89	1.422
.10	.100	.30	.310	.50	.549	.70	.867	.90	1.472
.11	.110	.31	.321	.51	.563	.71	.887	.91	1.528
.12	.121	.32	.332	.52	.576	.72	.908	.92	1.589
.13	.131	.33	.343	.53	.590	.73	.929	.93	1.658
.14	.141	.34	.354	.54	.604	.74	.950	.94	1.738
.15	.151	.35	.365	.55	.618	.75	.973	.95	1.832
.16	.161	.36	.377	.56	.633	.76	.996	.96	1.946
.17	.172	.37	.388	.57	.648	.77	1.020	.97	2.092
.18	.182	.38	.400	.58	.662	.78	1.045	.98	2.298
.19	.192	.39	.412	.59	.678	.79	1.071	.99	2.647
.20	.203	.40	.424	.60	.693	.80	1.099		

TABLE 12 Two-tailed upper critical values for the Spearman rank correlation coefficient (r_s)

n	α			
	.10	.05	.02	.01
1	—	—	—	—
2	—	—	—	—
3	—	—	—	—
4	1.0	—	—	—
5	.900	1.0	1.0	—
6	.829	.886	.943	1.0
7	.714	.786	.893	.929
8	.643	.738	.833	.881
9	.600	.683	.783	.833

(The data for this table have been adapted with permission from Olds, E. G. (1938) "Distributions of Sums of Squares of Rank Differences for Small Numbers of Individuals," *Ann. Math. Statist.*, 9, 133–148.)

TABLE 13 Upper α percentage points for the studentized range

α = .05   c*

d†	2	3	4	5	6	7	8	9	10	11	12	13	14	15	16	17	18	19	20
1	18.0	27.0	32.8	37.1	40.4	43.1	45.4	47.4	49.1	50.6	52.0	53.2	54.3	55.4	56.3	57.2	58.0	58.8	59.6
2	6.08	8.33	9.80	10.9	11.7	12.4	13.0	13.5	14.0	14.4	14.7	15.1	15.4	15.7	15.9	16.1	16.4	16.6	16.8
3	4.50	5.91	6.82	7.50	8.04	8.48	8.85	9.18	9.46	9.72	9.95	10.2	10.3	10.5	10.7	10.8	11.0	11.1	11.2
4	3.93	5.04	5.76	6.29	6.71	7.05	7.35	7.60	7.83	8.03	8.21	8.37	8.52	8.66	8.79	8.91	9.03	9.13	9.23
5	3.64	4.60	5.22	5.67	6.03	6.33	6.58	6.80	6.99	7.17	7.32	7.47	7.60	7.72	7.83	7.93	8.03	8.12	8.21
6	3.46	4.34	4.90	5.30	5.63	5.90	6.12	6.32	6.49	6.65	6.79	6.92	7.03	7.14	7.24	7.34	7.43	7.51	7.59
7	3.34	4.16	4.68	5.06	5.36	5.61	5.82	6.00	6.16	6.30	6.43	6.55	6.66	6.76	6.85	6.94	7.02	7.10	7.17
8	3.26	4.04	4.53	4.89	5.17	5.40	5.60	5.77	5.92	6.05	6.18	6.29	6.39	6.48	6.57	6.65	6.73	6.80	6.87
9	3.20	3.95	4.41	4.76	5.02	5.24	5.43	5.59	5.74	5.87	5.98	6.09	6.19	6.28	6.36	6.44	6.51	6.58	6.64
10	3.15	3.88	4.33	4.65	4.91	5.12	5.30	5.46	5.60	5.72	5.83	5.93	6.03	6.11	6.19	6.27	6.34	6.40	6.47
11	3.11	3.82	4.26	4.57	4.82	5.03	5.20	5.35	5.49	5.61	5.71	5.81	5.90	5.98	6.06	6.13	6.20	6.27	6.33
12	3.08	3.77	4.20	4.51	4.75	4.95	5.12	5.27	5.39	5.51	5.61	5.71	5.80	5.88	5.95	6.02	6.09	6.15	6.21
13	3.06	3.73	4.15	4.45	4.69	4.88	5.05	5.19	5.32	5.43	5.53	5.63	5.71	5.79	5.86	5.93	5.99	6.05	6.11
14	3.03	3.70	4.11	4.41	4.64	4.83	4.99	5.13	5.25	5.36	5.46	5.55	5.64	5.71	5.79	5.85	5.91	5.97	6.03
15	3.01	3.67	4.08	4.37	4.59	4.78	4.94	5.08	5.20	5.31	5.40	5.49	5.57	5.65	5.72	5.78	5.85	5.90	5.96
16	3.00	3.65	4.05	4.33	4.56	4.74	4.90	5.03	5.15	5.26	5.35	5.44	5.52	5.59	5.66	5.73	5.79	5.84	5.90
17	2.98	3.63	4.02	4.30	4.52	4.70	4.86	4.99	5.11	5.21	5.31	5.39	5.47	5.54	5.61	5.67	5.73	5.79	5.84
18	2.97	3.61	4.00	4.28	4.49	4.67	4.82	4.96	5.07	5.17	5.27	5.35	5.43	5.50	5.57	5.63	5.69	5.74	5.79
19	2.96	3.59	3.98	4.25	4.47	4.65	4.79	4.92	5.04	5.14	5.23	5.31	5.39	5.46	5.53	5.59	5.65	5.70	5.75
20	2.95	3.58	3.96	4.23	4.45	4.62	4.77	4.90	5.01	5.11	5.20	5.28	5.36	5.43	5.49	5.55	5.61	5.66	5.71
24	2.92	3.53	3.90	4.17	4.37	4.54	4.68	4.81	4.92	5.01	5.10	5.18	5.25	5.32	5.38	5.44	5.49	5.55	5.59
30	2.89	3.49	3.85	4.10	4.30	4.46	4.60	4.72	4.82	4.92	5.00	5.08	5.15	5.21	5.27	5.33	5.38	5.43	5.47
40	2.86	3.44	3.79	4.04	4.23	4.39	4.52	4.63	4.73	4.82	4.90	4.98	5.04	5.11	5.16	5.22	5.27	5.31	5.36
60	2.83	3.40	3.74	3.98	4.16	4.31	4.44	4.55	4.65	4.73	4.81	4.88	4.94	5.00	5.06	5.11	5.15	5.20	5.24
120	2.80	3.36	3.68	3.92	4.10	4.24	4.36	4.47	4.56	4.64	4.71	4.78	4.84	4.90	4.95	5.00	5.04	5.09	5.13
∞	2.77	3.31	3.63	3.86	4.03	4.17	4.29	4.39	4.47	4.55	4.62	4.68	4.74	4.80	4.85	4.89	4.93	4.97	5.01

α = .01

d†	2	3	4	5	6	7	8	9	10	11	12	13	14	15	16	17	18	19	20
1	90.0	135.	164.	186.	202.	216.	227.	237.	246.	253.	260.	266.	272.	277.	282.	286.	290.	294.	298.
2	14.0	19.0	22.3	24.7	26.6	28.2	29.5	30.7	31.7	32.6	33.4	34.1	34.8	35.4	36.0	36.5	37.0	37.5	37.9
3	8.26	10.6	12.2	13.3	14.2	15.0	15.6	16.2	16.7	17.1	17.5	17.9	18.2	18.5	18.8	19.1	19.3	19.5	19.8
4	6.51	8.12	9.17	9.96	10.6	11.1	11.5	11.9	12.3	12.6	12.8	13.1	13.3	13.5	13.7	13.9	14.1	14.2	14.4
5	5.70	6.98	7.80	8.42	8.91	9.32	9.67	9.97	10.2	10.5	10.7	10.9	11.1	11.2	11.4	11.6	11.7	11.8	11.9
6	5.24	6.33	7.03	7.56	7.97	8.32	8.61	8.87	9.10	9.30	9.48	9.65	9.81	9.95	10.1	10.2	10.3	10.4	10.5
7	4.95	5.92	6.54	7.01	7.37	7.68	7.94	8.17	8.37	8.55	8.71	8.86	9.00	9.12	9.24	9.35	9.46	9.55	9.65
8	4.75	5.64	6.20	6.62	6.96	7.24	7.47	7.68	7.86	8.03	8.18	8.31	8.44	8.55	8.66	8.76	8.85	8.94	9.03
9	4.60	5.43	5.96	6.35	6.66	6.91	7.13	7.33	7.49	7.65	7.78	7.91	8.03	8.13	8.23	8.33	8.41	8.49	8.57
10	4.48	5.27	5.77	6.14	6.43	6.67	6.87	7.05	7.21	7.36	7.49	7.60	7.71	7.81	7.91	7.99	8.08	8.15	8.23
11	4.39	5.15	5.62	5.97	6.25	6.48	6.67	6.84	6.99	7.13	7.25	7.36	7.46	7.56	7.65	7.73	7.81	7.88	7.95
12	4.32	5.05	5.50	5.84	6.10	6.32	6.51	6.67	6.81	6.94	7.06	7.17	7.26	7.36	7.44	7.52	7.59	7.66	7.73
13	4.26	4.96	5.40	5.73	5.98	6.19	6.37	6.53	6.67	6.79	6.90	7.01	7.10	7.19	7.27	7.35	7.42	7.48	7.55
14	4.21	4.89	5.32	5.63	5.88	6.08	6.26	6.41	6.54	6.66	6.77	6.87	6.96	7.05	7.13	7.20	7.27	7.33	7.39

TABLE 13 (Continued)

α = .01   c*

d†	2	3	4	5	6	7	8	9	10	11	12	13	14	15	16	17	18	19	20
15	4.17	4.84	5.25	5.56	5.80	5.99	6.16	6.31	6.44	6.55	6.66	6.76	6.84	6.93	7.00	7.07	7.14	7.20	7.26
16	4.13	4.79	5.19	5.49	5.72	5.92	6.08	6.22	6.35	6.46	6.56	6.66	6.74	6.82	6.90	6.97	7.03	7.09	7.15
17	4.10	4.74	5.14	5.43	5.66	5.85	6.01	6.15	6.27	6.38	6.48	6.57	6.66	6.73	6.81	6.87	6.94	7.00	7.05
18	4.07	4.70	5.09	5.38	5.60	5.79	5.94	6.08	6.20	6.31	6.41	6.50	6.58	6.65	6.73	6.79	6.85	6.91	6.96
19	4.05	4.67	5.05	5.33	5.55	5.73	5.89	6.02	6.14	6.25	6.34	6.43	6.51	6.58	6.65	6.72	6.78	6.84	6.89
20	4.02	4.64	5.02	5.29	5.51	5.69	5.84	5.97	6.09	6.19	6.28	6.37	6.45	6.52	6.59	6.65	6.71	6.77	6.82
24	3.96	4.55	4.91	5.17	5.37	5.54	5.69	5.81	5.92	6.02	6.11	6.19	6.26	6.33	6.39	6.45	6.51	6.56	6.61
30	3.89	4.45	4.80	5.05	5.24	5.40	5.54	5.65	5.76	5.85	5.93	6.01	6.08	6.14	6.20	6.26	6.31	6.36	6.41
40	3.82	4.37	4.70	4.93	5.11	5.26	5.39	5.50	5.60	5.69	5.76	5.83	5.90	5.96	6.02	6.07	6.12	6.16	6.21
60	3.76	4.28	4.59	4.82	4.99	5.13	5.25	5.36	5.45	5.53	5.60	5.67	5.73	5.78	5.84	5.89	5.93	5.97	6.01
120	3.70	4.20	4.50	4.71	4.87	5.01	5.12	5.21	5.30	5.37	5.44	5.50	5.56	5.61	5.66	5.71	5.75	5.79	5.83
∞	3.64	4.12	4.40	4.60	4.76	4.88	4.99	5.08	5.16	5.23	5.29	5.35	5.40	5.45	5.49	5.54	5.57	5.61	5.65

*c = number of means in the group of means considered.
†d = df for the Within MS.
(This table has been reproduced with permission of the Biometrika Trustees from "Biometrika Tables for Statisticians," Volume I, edited by E. S. Pearson and H. O. Hartley, published for the Biometrika Trustees, pp. 176–177. Cambridge University Press, Cambridge, England, 1959.)

TABLE 14 Critical values for the Kruskal–Wallis test statistic (H) for selected sample sizes for $k = 3$

| | | | α | | | |
n_1	n_2	n_3	.10	.05	.02	.01
1	1	2	—	—	—	—
1	1	3	—	—	—	—
1	1	4	—	—	—	—
1	1	5	—	—	—	—
1	2	2	—	—	—	—
1	2	3	4.286	—	—	—
1	2	4	4.500	—	—	—
1	2	5	4.200	5.000	—	—
1	3	3	4.571	5.143	—	—
1	3	4	4.056	5.389	—	—
1	3	5	4.018	4.960	6.400	—
1	4	4	4.167	4.967	6.667	—
1	4	5	3.987	4.986	6.431	6.954
1	5	5	4.109	5.127	6.146	7.309
2	2	2	4.571	—	—	—
2	2	3	4.500	4.714	—	—
2	2	4	4.500	5.333	6.000	—
2	2	5	4.373	5.160	6.000	6.533
2	3	3	4.694	5.361	6.250	—
2	3	4	4.511	5.444	6.144	6.444
2	3	5	4.651	5.251	6.294	6.909
2	4	4	4.554	5.454	6.600	7.036
2	4	5	4.541	5.273	6.541	7.204
2	5	5	4.623	5.338	6.469	7.392
3	3	3	5.067	5.689	6.489	7.200
3	3	4	4.709	5.791	6.564	7.000
3	3	5	4.533	5.648	6.533	7.079
3	4	4	4.546	5.598	6.712	7.212
3	4	5	4.549	5.656	6.703	7.477
3	5	5	4.571	5.706	6.866	7.622
4	4	4	4.654	5.692	6.962	7.654
4	4	5	4.668	5.657	6.976	7.760
4	5	5	4.523	5.666	7.000	7.903
5	5	5	4.580	5.780	7.220	8.000

(The data for this table have been adapted from Table F of *A Nonparametric Introduction to Statistics* by C. H. Kraft and C. Van Eeden, Macmillan, New York, 1968, with the permission of the publisher and the authors.)

INDEX OF DATA SETS

Directory of Data Sets Each of the following data sets has a documentation file and a data file. A page number in the text is provided, giving a complete description of the data set.*

----

* A list of the names of the 26 data sets is also provided in the data set DATASETS.DOC.

Disk Directory Volume in drive B has no label. Directory of B:

```
------------------------------------------------
FEV       DAT   17005   3-09-88    3:40a
VALID     DAT   14186   4-11-88   10:25p
EAR       DAT    3615  12-20-88    5:17a
SMOKE     DAT    5619   4-18-88    2:30p
SEXRAT    DAT     843   7-31-88    8:04a
FEV       DOC     482   7-26-89    3:45p
VALID     DOC     677   7-31-88    4:15p
EAR       DOC     493   7-31-88    4:19p
SMOKE     DOC     897   7-31-88    4:26p
SEXRAT    DOC     499   7-31-88    4:30p
NIFED     DAT    1179   8-06-88    7:12a
DISEASE   DAT    8119   7-26-89    3:43p
INFANTBP  DOC    1837   7-26-89    3:49p
BETACAR   DOC     532  10-07-88    4:34a
HORMONE   DAT   41078  10-07-88    4:50a
INFANTBP  DAT   13101   1-01-80   12:02a
EFF       DAT    1601   1-15-87    9:12a
NEPHRO    DAT    1801   1-15-87    9:13a
OTO       DAT    1251   1-15-87    9:13a
BETACAR   DAT     691  10-07-88    2:42a
NIFED     DOC    1102  10-07-88    4:23a
DISEASE   DOC    2785  10-07-88    4:26a
HORMONE   DOC     785  10-07-88    5:04a
EFF       DOC     541  10-08-88    3:35a
NEPHRO    DOC     557  10-08-88    3:38a
OTO       DOC     554  10-08-88    3:39a
DATASETS  DOC     657  10-08-88    3:53a
      27 File(s)    228352 bytes free
------------------------------------------------
```

Data Set 1 BETACAR.DOC

Variable	Column	Code
Preparation	1	$1 = SOL/2 = ROCHE/3 = BASF - 30/4 = BASF - 60$
Subject number	3–4	
1st baseline level	6–8	
2nd baseline level	10–12	
Week 6 level	14–16	
Week 8 level	18–20	
Week 10 level	22–24	
Week 12 level	26–28	

Data Set 2 BETACAR.DAT

```
-------------------------------------
1  71  298  116  174  178  218  190
1  73  124  146  294  278  244  262
1  80  176  200  276  286  308  334
1  83  116  180  164  238  308  226
1  90  152  142  290  300  270  268
1  92  106  106  246  206  304  356
2  78  114  110  280  220  178  210
2  82  106  114  114  176  100  104
2  84  100  100  144  114  154  142
2  85   92   92  164  116  140  112
2  87  212  212  354  430  352  382
2  89   92   94  160  200  150  170
3  72  180  162  432  336  440  472
3  79  186  198  242  252  240  336
3  88  202  208  476  408  414  416
3  94  192  160  264  252  320  350
3  95   80   88  160  152  208  226
4  74  174  182  206  268  202  232
4  75  252  234  590  594  522  566
4  76  210  230  474  500  472  444
4  77  162  152  202  204  140  180
4  86   68   64  262  216  214  216
4  93   74   72  218  164  218  184
-------------------------------------
```

Data Set 3 DISEASE.DOC

Variable	Record	Column	Format	Variable	Record	Column	Format
Disease	1	2–3		X26	2	14–15	.XX
Incidence	1	5–8	.XXX	X27	2	17–18	.XX
X1	1	11–12	.XX	X28	2	20–21	.XX
X2	1	14–15	.XX	X29	2	23–24	.XX
X3	1	17–18	.XX	X30	2	26–27	.XX
X4	1	20–21	.XX	X31	2	29–30	.XX
X5	1	23–24	.XX	X32	2	32–33	.XX
X6	1	26–27	.XX	X33	2	35–36	.XX
X7	1	29–30	.XX	X34	2	38–39	.XX
X8	1	32–33	.XX	X35	2	41–42	.XX
X9	1	35–36	.XX	X36	2	44–45	.XX
X10	1	38–39	.XX	X37	2	47–48	.XX
X11	1	41–42	.XX	X38	2	50–51	.XX
X12	1	44–45	.XX	X39	2	53–54	.XX
X13	1	47–48	.XX	X40	2	56–57	.XX
X14	1	50–51	.XX	X41	2	59–60	.XX
X15	1	53–54	.XX	X42	2	62–63	.XX
X16	1	56–57	.XX	X43	2	65–66	.XX
X17	1	59–60	.XX	X44	2	68–69	.XX
X18	1	62–63	.XX	X45	2	71–72	.XX
X19	1	65–66	.XX	X46	2	74–75	.XX
X20	1	68–69	.XX	Record #	2	80	
X21	1	71–72	.XX	Disease	3	2–3	.XXX
X22	1	74–75	.XX	X47	3	5–6	.XX
Record #	1	80		X48	3	8–9	.XX
Disease	2	2–3	.XXX	X49	3	11–12	.XX
X23	2	5–6	.XX	X50	3	14–15	.XX
X24	2	8–9	.XX				
X25	2	11–12	.XX	Record #	3	80	

Data Set 4 DISEASE.DAT

```
-------------------------------------------------------------------------------
Y01 .100  01 49 50 01 00 01 00 01 01 10 03 05 05 03 05 01 70 02 07 00 80 01    1
Y01 05 01 00 01 01 15 05 10 03 01 01 01 02 02 02 02 02 01 00 02 70 04 03 00    2
Y01 00 80 05 10                                                                3
Y02 .081  10 50 50 02 01 02 00 01 35 50 05 02 40 01 02 02 30 20 02 05 90 02    1
Y02 02 01 01 01 01 60 01 80 01 01 01 70 05 05 85 02 02 01 02 01 30 02 20 05    2
Y02 01 90 01 60                                                                3
Y03 .005  30 60 10 20 10 20 00 01 60 70 05 02 10 10 02 02 05 05 02 57 40 01    1
Y03 03 01 01 01 02 30 15 40 01 05 01 85 05 20 70 02 02 01 01 01 05 01 05 60    2
Y03 01 38 01 70                                                                3
Y04 .001  10 20 70 30 10 25 00 01 80 90 05 05 15 10 02 02 15 20 02 05 40 20    1
Y04 01 01 01 01 01 95 01 50 01 05 01 85 05 20 70 02 02 01 02 01 15 20 02 05    2
Y04 01 40 01 40                                                                3
Y05 .027  20 50 30 15 05 10 00 01 40 50 05 05 30 05 60 15 90 40 02 10 20 10    1
Y05 01 01 01 01 01 70 02 40 10 10 01 05 70 05 85 02 02 15 01 85 05 02 20 02    2
Y05 20 20 20 80                                                                3
Y06 .005  10 40 50 01 01 01 00 01 15 20 01 05 05 01 02 02 20 02 02 02 60 05    1
Y06 05 01 01 10 15 40 02 10 01 01 01 15 02 02 15 02 02 02 02 02 20 02 02 02    2
Y06 02 60 02 30                                                                3
Y07 .001  20 70 10 65 10 05 00 01 70 80 05 05 20 05 02 02 10 15 10 05 75 05    1
Y07 20 01 01 10 15 85 02 80 01 01 01 90 02 25 75 02 02 02 02 30 10 01 30 05    2
Y07 01 80 02 70                                                                3
Y08 .018  50 48 02 30 65 01 00 10 80 90 20 05 15 10 02 05 65 05 05 20 20 02    1
Y08 05 01 01 01 01 02 60 01 20 30 01 02 90 02 02 90 10 05 02 50 15 05 02 20    2
Y08 20 20 20 50                                                                3
Y09 .001  10 45 45 22 44 01 00 22 80 80 10 30 15 22 05 25 95 25 05 05 15 02    1
Y09 05 01 01 01 01 02 35 10 20 10 01 10 02 02 60 02 02 25 25 45 45 25 25 15    2
Y09 15 05 05 50                                                                3
Y10 .054  40 55 05 25 25 10 00 30 75 90 05 05 10 20 02 02 20 02 05 65 25 02    1
Y10 05 02 02 10 15 10 60 20 01 02 01 95 02 85 10 02 02 02 02 20 05 02 02 60    2
Y10 05 25 05 90                                                                3
Y11 .063  40 55 05 30 30 10 00 40 75 90 05 05 10 25 02 02 20 02 05 65 25 02    1
Y11 05 02 02 10 15 10 60 20 01 02 01 95 02 85 10 02 02 02 02 20 05 02 02 60    2
Y11 05 25 05 90                                                                3
Y12 .045  20 70 10 01 01 01 00 01 50 65 01 01 01 10 02 02 10 02 05 70 20 02    1
Y12 10 02 02 01 01 10 60 20 01 05 01 95 02 85 10 02 02 01 01 01 10 02 02 68    2
Y12 01 25 01 80                                                                3
Y13 .013  20 70 10 01 01 01 00 01 50 65 01 01 01 10 02 02 10 02 02 70 20 02    1
Y13 02 02 02 01 01 10 60 20 01 05 01 95 02 85 10 02 02 01 01 01 10 01 01 68    2
Y13 01 25 01 80                                                                3
Y14 .014  90 09 01 10 90 00 00 80 90 99 05 10 05 35 02 02 40 05 05 01 02 02    1
Y14 05 02 02 10 10 01 90 20 01 02 01 95 02 85 10 02 02 02 01 30 40 02 05 01    2
Y14 01 02 02 20                                                                3
Y15 .001  05 45 50 01 01 01 00 01 01 01 01 01 01 01 01 04 01 02 01 01 02 25 02 1
Y15 01 20 02 50 05 10 02 10 01 01 01 10 02 10 02 02 02 01 01 02 02 01 00 02    2
Y15 01 25 02 60                                                                3
Y16 .013  10 45 45 01 01 01 00 01 70 95 40 10 10 10 01 01 30 05 01 01 05 30    1
Y16 02 02 02 02 02 95 00 30 01 10 01 95 02 90 05 02 02 01 01 01 30 15 05 02    2
Y16 02 05 02 20                                                                3
Y17 .001  30 60 10 05 01 01 00 01 10 10 05 01 10 01 05 10 20 05 60 01 10 05    1
Y17 20 02 02 02 02 70 01 20 40 01 01 01 15 02 02 60 05 10 02 10 20 05 02 02    2
Y17 02 10 05 75                                                                3
```

Data Set 4 DISEASE.DAT (*Continued*)

```
------------------------------------------------------------------------------------------
Y18  .072   20 40 40 01 01 01 00 01 20 20 10 01 10 05 05 15 10 02 50 02 13 05    1
Y18  85 02 02 03 05 50 01 20 40 02 01 02 10 02 02 50 05 10 02 05 10 02 02 05     2
Y18  02 20 10 85                                                                 3
Y19  .002   20 30 50 45 45 01 00 01 10 20 05 01 01 10 05 02 10 02 20 02 10 02    1
Y19  05 01 01 05 70 05 05 20 01 01 01 05 05 02 02 02 02 02 02 10 10 02 02 02     2
Y19  02 10 10 30                                                                 3
Y20  .008   20 50 30 01 01 01 00 01 50 50 40 05 10 10 80 20 10 10 02 05 10 02    1
Y20  02 02 02 01 01 50 01 20 05 02 01 50 02 10 40 02 02 20 20 10 10 10 10 05     2
Y20  05 10 10 70                                                                 3
Y21  .013   70 29 01 01 01 01 00 01 40 50 20 01 05 05 15 02 05 02 02 02 05 02    1
Y21  02 10 02 01 01 20 02 10 50 02 01 05 10 05 05 40 90 02 02 10 10 02 02 02     2
Y21  02 05 05 10                                                                 3
Y22  .001   70 29 01 01 01 01 00 01 30 30 30 80 15 20 05 01 01 01 01 01 01 01    1
Y22  01 01 01 01 01 20 02 01 05 01 01 05 10 05 05 20 90 01 01 01 01 01 01 01     2
Y22  01 01 01 10                                                                 3
Y23  .036   10 80 10 01 01 01 00 01 20 30 20 15 01 35 20 02 20 10 02 05 05 01    1
Y23  01 95 05 01 01 20 10 01 40 01 05 05 15 02 02 70 15 02 02 02 20 10 02 05     2
Y23  01 05 01 90                                                                 3
Y24  .009   10 80 10 01 01 01 00 01 20 30 20 15 01 35 20 02 20 10 02 05 05 01    1
Y24  01 95 05 01 01 20 10 01 40 01 05 05 15 02 02 70 15 02 02 02 20 10 02 05     2
Y24  01 05 01 90                                                                 3
Y25  .054   10 70 20 01 01 01 00 01 20 30 20 01 01 05 05 01 20 10 02 02 10 01    1
Y25  05 15 10 80 15 10 10 01 30 01 99 05 05 02 02 40 04 01 01 05 20 10 02 02     2
Y25  02 10 05 65                                                                 3
Y26  .005   50 40 10 30 60 01 00 15 15 30 05 01 20 10 02 02 70 02 02 10 10 02    1
Y26  02 02 02 05 10 40 10 30 05 01 01 30 10 40 10 20 05 02 02 40 40 02 02 10     2
Y26  10 10 10 40                                                                 3
Y27  .063   90 10 00 20 60 05 10 05 60 70 20 01 05 10 05 02 50 02 02 03 10 02    1
Y27  02 05 02 01 01 20 10 20 20 02 02 40 20 30 05 20 05 02 02 30 30 02 02 03     2
Y27  03 10 10 50                                                                 3
Y28  .001   30 30 30 30 05 10 00 01 10 20 01 01 01 01 05 02 70 02 02 05 30 02    1
Y28  02 05 02 01 01 20 10 10 10 01 01 20 10 10 10 10 10 02 02 30 30 02 02 05     2
Y28  05 30 30 60                                                                 3
Y29  .001   60 39 01 01 01 01 80 30 10 50 05 20 01 20 05 02 50 02 02 10 30 02    1
Y29  02 05 02 01 01 90 02 40 05 01 10 70 05 80 05 10 05 02 02 30 30 02 02 10     2
Y29  10 30 30 20                                                                 3
Y30  .252   15 70 15 01 01 01 00 01 20 30 05 01 15 05 05 20 95 05 02 10 10 05    1
Y30  01 02 05 01 01 30 02 05 30 01 01 30 10 05 05 15 05 20 02 92 05 05 01 01     2
Y30  10 01 10 85                                                                 3
Y31  .081   30 60 10 30 50 10 00 05 60 70 20 10 20 10 05 01 50 10 02 05 05 25    1
Y31  01 02 05 01 01 90 02 30 05 05 01 70 05 75 15 10 05 01 01 30 30 10 02 01     2
Y31  05 01 05 50                                                                 3
Y32  .005   30 40 30 01 01 05 50 01 20 30 10 01 10 05 02 02 10 10 02 02 20 10    1
Y32  02 02 02 02 02 90 02 30 05 05 01 70 05 75 15 10 05 02 02 10 10 02 02 02     2
Y32  02 20 20 20                                                                 3
Y33  .009   40 55 05 50 20 10 00 01 80 90 20 01 30 05 05 10 70 05 02 10 30 10    1
Y33  02 02 01 01 01 30 10 01 20 30 01 02 90 02 02 90 10 10 02 30 30 05 05 10     2
Y33  10 30 30 50                                                                 3
------------------------------------------------------------------------------------------
```

Data Set 5 EAR.DOC

Column	Variable	Format or code
1–3	ID	
5	Clearance by 14 days	1 = yes/0 = no
7	Antibiotic	1 = CEF/2 = AMO
9	Age	1 = <2 yrs/2 = 2–5 yrs/3 = 6+ yrs
11	Ear	1 = 1st ear/2 = 2nd ear

Data Set 6 EAR.DAT

```
001 1 1 1 1    047 1 1 3 1    093 0 2 2 1    134 1 1 1 1    157 1 1 2 1    180 0 2 1 1
002 1 1 1 1    048 1 1 3 1    094 0 2 2 1    134 1 1 1 2    157 0 1 2 2    180 0 2 1 2
003 1 1 1 1    049 1 1 3 1    095 0 2 2 1    135 1 1 1 1    158 1 1 2 1    181 0 2 1 1
004 0 1 1 1    050 1 1 3 1    096 0 2 2 1    135 1 1 1 2    158 0 1 2 2    181 0 2 1 2
005 0 1 1 1    051 1 1 3 1    097 0 2 2 1    136 1 1 1 1    159 1 1 2 1    182 0 2 1 1
006 0 1 1 1    052 1 1 3 1    098 0 2 2 1    136 1 1 1 2    159 0 1 2 2    182 0 2 1 2
007 0 1 1 1    053 1 1 3 1    099 0 2 2 1    137 1 1 1 1    160 1 1 2 1    183 0 2 1 1
008 0 1 1 1    054 1 1 3 1    100 0 2 2 1    137 0 1 1 2    160 0 1 2 2    183 0 2 1 2
009 0 1 1 1    055 0 1 3 1    101 0 2 2 1    138 1 1 1 1    161 1 1 2 1    184 0 2 1 1
010 0 1 1 1    056 0 1 3 1    102 0 2 2 1    138 0 1 1 2    161 0 1 2 2    184 0 2 1 2
011 0 1 1 1    057 0 1 3 1    103 0 2 2 1    139 0 1 1 1    162 1 1 2 1    185 0 2 1 1
012 0 1 1 1    058 0 1 3 1    104 0 2 2 1    139 0 1 1 2    162 0 1 2 2    185 0 2 1 2
013 1 1 2 1    059 0 1 3 1    105 0 2 2 1    140 0 1 1 1    163 0 1 2 1    186 0 2 1 1
014 1 1 2 1    060 0 1 3 1    106 0 2 2 1    140 0 1 1 2    163 0 1 2 2    186 0 2 1 2
015 1 1 2 1    061 0 1 3 1    107 0 2 2 1    141 0 1 1 1    164 0 1 2 1    187 0 2 1 1
016 1 1 2 1    062 0 1 3 1    108 0 2 2 1    141 0 1 1 2    164 0 1 2 2    187 0 2 1 2
017 1 1 2 1    063 1 2 1 1    109 0 2 2 1    142 0 1 1 1    165 0 1 2 1    188 1 2 2 1
018 1 1 2 1    064 1 2 1 1    110 0 2 2 1    142 0 1 1 2    165 0 1 2 2    188 1 2 2 2
019 1 1 2 1    065 0 2 1 1    111 1 2 3 1    143 0 1 1 1    166 0 1 2 1    189 1 2 2 1
020 1 1 2 1    066 0 2 1 1    112 1 2 3 1    143 0 1 1 2    166 0 1 2 2    189 1 2 2 2
021 1 1 2 1    067 0 2 1 1    113 1 2 3 1    144 0 1 1 1    167 0 1 2 1    190 1 2 2 1
022 1 1 2 1    068 0 2 1 1    114 1 2 3 1    144 0 1 1 2    167 0 1 2 2    190 1 2 2 2
023 1 1 2 1    069 0 2 1 1    115 1 2 3 1    145 0 1 1 1    168 0 1 2 1    191 1 2 2 1
024 1 1 2 1    070 0 2 1 1    116 1 2 3 1    145 0 1 1 2    168 0 1 2 2    191 1 2 2 2
025 1 1 2 1    071 0 2 1 1    117 1 2 3 1    146 0 1 1 1    169 1 1 3 1    192 1 2 2 1
026 1 1 2 1    072 0 2 1 1    118 1 2 3 1    146 0 1 1 2    169 1 1 3 2    192 1 2 2 2
027 1 1 2 1    073 0 2 1 1    119 1 2 3 1    147 1 1 2 1    170 1 1 3 1    193 1 2 2 1
028 1 1 2 1    074 0 2 1 1    120 1 2 3 1    147 1 1 2 2    170 1 1 3 2    193 0 2 2 2
029 1 1 2 1    075 1 2 2 1    121 1 2 3 1    148 1 1 2 1    171 1 1 3 1    194 0 2 2 1
030 1 1 2 1    076 1 2 2 1    122 0 2 3 1    148 1 1 2 2    171 1 1 3 2    194 0 2 2 2
031 1 1 2 1    077 1 2 2 1    123 0 2 3 1    149 1 1 2 1    172 1 1 3 1    195 0 2 2 1
032 1 1 2 1    078 1 2 2 1    124 0 2 3 1    149 1 1 2 2    172 0 1 3 2    195 0 2 2 2
033 1 1 2 1    079 1 2 2 1    125 0 2 3 1    150 1 1 2 1    173 1 2 1 1    196 0 2 2 1
034 1 1 1 1    080 1 2 2 1    126 0 2 3 1    150 1 1 2 2    173 1 2 1 2    196 0 2 2 2
035 1 1 2 1    081 1 2 2 1    127 0 2 3 1    151 1 1 2 1    174 1 2 1 1    197 1 2 3 1
036 1 1 2 1    082 1 2 2 1    128 0 2 3 1    151 1 1 2 2    174 1 2 1 2    197 1 2 3 2
037 0 1 2 1    083 1 2 2 1    129 1 1 1 1    152 1 1 2 1    175 1 2 1 1    198 1 2 3 1
038 0 1 2 1    084 1 2 2 1    129 1 1 1 2    152 1 1 2 2    175 0 2 1 2    198 1 2 3 2
039 0 1 2 1    085 1 2 2 1    130 1 1 1 1    153 1 1 2 1    176 1 2 1 1    199 1 2 3 1
040 0 1 2 1    086 1 2 2 1    130 1 1 1 2    153 1 1 2 2    176 0 2 1 2    199 1 2 3 2
041 0 1 2 1    087 1 2 2 1    131 1 1 1 1    154 1 1 2 1    177 0 2 1 1    200 1 2 3 1
042 0 1 2 1    088 1 2 2 1    131 1 1 1 2    154 1 1 2 2    177 0 2 1 2    200 1 2 3 2
043 0 1 2 1    089 0 2 2 1    132 1 1 1 1    155 1 1 2 1    178 0 2 1 1    201 1 2 3 1
044 1 1 3 1    090 0 2 2 1    132 1 1 1 2    155 1 1 2 2    178 0 2 1 2    201 1 2 3 2
045 1 1 3 1    091 0 2 2 1    133 1 1 1 1    156 1 1 2 1    179 0 2 1 1    202 1 2 3 1
046 1 1 3 1    092 0 2 2 1    133 1 1 1 2    156 1 1 2 2    179 0 2 1 2    202 1 2 3 2
                                                                          203 0 2 3 1
                                                                          203 0 2 3 2
```

Data Set 7 EFF.DOC

Variable	Column	Code
Study name	1–8	
Study number	10–11	
Endpoint	13	1 = efficacy/2 = nephrotoxicity/3 = ototoxicity
Antibiotic	15	1 = Amikacin/2 = Gentamicin/3 = Netilmicin/ 4 = Sisomycin/5 = Tobramycin
Sample size	17–19	
Number cured	21–23	

Data Set 8 EFF.DAT

BODEY	03	1	2	5	3	KEATING	21	1 1	102	69
BODEY	03	1	4	5	5	KEATING	21	1 2	115	77
BODEY	03	1	5	1	0	KEATING	21	1 4	93	62
KLAST	04	1	2	21	13	LOVE	22	1 1	16	11
KLAST	04	1	4	25	22	LOVE	22	1 2	22	19
MADSDEN	05	1	2	37	24	LOVE	22	1 3	19	17
MADSDEN	05	1	5	38	24	LOVELESS	23	1 2	16	13
WALKER	06	1	2	40	32	LOVELESS	23	1 3	13	12
WALKER	06	1	5	40	32	MAKI	24	1 2	41	27
FELD	07	1	1	60	33	MAKI	24	1 4	40	23
FELD	07	1	5	51	29	SCHNIDER	25	1 2	33	8
GILBERT	08	1	1	15	10	SCHNIDER	25	1 3	42	6
GILBERT	08	1	2	15	9	BOCK	27	1 1	33	14
LAU	09	1	1	50	36	BOCK	27	1 3	34	17
LAU	09	1	2	47	39	LEAL	28	1 2	33	22
SMITH77	10	1	1	39	20	LEAL	28	1 4	33	28
SMITH77	10	1	2	32	14	SMITH80	29	1 2	59	46
EDEN	11	1	2	20	9	SMITH80	29	1 5	52	42
EDEN	11	1	3	20	10	FONG	31	1 2	102	65
HAHN	12	1	1	33	27	FONG	31	1 5	103	72
HAHN	12	1	2	33	30	HERTIG	32	1 2	212	193
HAHN	12	1	3	34	29	HERTIG	32	1 3	210	201
HOYME	13	1	3	30	17	LORBER	34	1 2	30	17
HOYME	13	1	5	31	20	LORBER	34	1 4	27	22
MAIGAARD	15	1	1	28	16	SCHIEKER	35	1 4	26	21
MAIGAARD	15	1	3	29	20	SCHIEKER	35	1 5	24	14
COX	19	1	2	24	22	WEISS	36	1 2	21	20
COX	19	1	3	26	21	WEISS	36	1 3	15	13
GREENE	20	1	1	10	7	HOLM	41	1 1	50	40
GREENE	20	1	2	9	6	HOLM	41	1 2	45	34
GREENE	20	1	3	9	5	LERNER83	42	1 3	81	67
GREENE	20	1	5	10	4	LERNER83	42	1 5	83	67

Data Set 9 FEV.DOC

Column	Variable	Format or code
1–5	ID number	
7–8	Age (yrs)	
10–15	FEV (L)	X.XXXX
17–20	Height (in.)	XX.X
22	Sex	0 = female/1 = male
24	Smoking status	0 = noncurrent smoker/1 = current smoker

Data Set 10 FEV.DAT

```
 301   9 1.7080 57.0 0 0        9801   8 2.4200 59.0 1 0       16501   8 2.2110 63.0 1 0
 451   8 1.7240 67.5 0 0       10041   5 1.7760 51.0 1 0       16601   8 1.7940 54.5 1 0
 501   7 1.7200 54.5 0 0       10401   8 1.9310 57.0 0 0       16651   7 1.9170 58.0 0 0
 642   9 1.5580 53.0 1 0       10451   5 1.3430 50.0 0 0       18101   8 2.1440 63.0 0 0
 901   9 1.8950 57.0 1 0       10601   9 2.0760 57.0 0 0       18501   7 1.2530 52.0 1 0
1701   8 2.3360 61.0 0 0       10701   7 1.6240 54.0 1 0       18843   9 2.6590 61.5 1 0
1752   6 1.9190 58.0 0 0       10751   8 1.3440 52.5 0 0       18844   5 1.5800 52.5 1 0
1753   6 1.4150 56.0 0 0       10841   6 1.6500 55.0 1 0       19401   9 2.1260 62.0 1 0
1901   8 1.9870 58.5 0 0       11201   8 2.7320 60.5 1 0       20152   9 3.0290 61.5 0 0
1951   9 1.9420 60.0 0 0       11241   5 2.0170 54.5 1 0       20801   9 2.9640 64.5 0 0
1952   6 1.6020 53.0 0 0       11501   9 2.7970 61.5 1 0       20841   7 1.6110 57.5 1 0
2001   8 1.7350 54.0 1 0       12001   9 3.5560 62.0 0 0       21552   8 2.2150 60.0 0 0
2101   8 2.1930 58.5 0 0       12201   8 1.7030 54.5 1 0       22201   8 2.3880 60.0 0 0
2401   8 2.1180 60.5 1 0       12241   6 1.6340 54.0 1 0       22901   9 2.1960 61.0 0 0
3102   8 2.2580 58.0 1 0       12402   9 2.5700 57.0 0 0       23101   9 1.7510 58.0 0 0
3142   7 1.9320 53.0 1 0       13351   9 3.0160 62.5 0 0       23441   9 2.1650 61.5 0 0
3541   5 1.4720 50.0 1 0       13701   7 2.4190 60.0 0 0       23442   7 1.6820 55.0 1 0
3551   6 1.8780 53.0 0 0       13751   4 1.5690 50.0 0 0       23641   8 1.5230 55.0 1 0
4201   9 2.3520 59.0 1 0       14051   8 1.6980 57.5 0 0       23653   8 1.2920 52.0 0 0
4301   9 2.6040 61.5 1 0       14101   8 2.1230 60.0 1 0       23801   7 1.6490 54.0 1 0
4351   5 1.4000 49.0 0 0       14201   8 2.4810 60.0 0 0       23802   9 2.5880 63.0 1 0
5151   5 1.2560 52.5 0 0       14251   6 1.4810 51.0 0 0       23841   4 0.7960 47.0 1 0
5152   4 0.8390 48.0 0 0       14252   4 1.5770 49.0 0 0       23902   9 2.5740 60.5 0 0
5201   7 2.5780 62.5 1 0       14501   8 1.9400 59.0 1 0       23941   6 1.9790 56.0 1 0
5601   9 2.9880 65.0 0 0       14541   6 1.7470 57.5 1 0       24401   8 2.3540 58.5 1 0
5642   3 1.4040 51.5 1 0       14601   9 2.0690 58.0 1 0       24843   6 1.7180 55.0 1 0
5702   9 2.3480 60.0 1 0       14651   7 1.6310 55.5 0 0       24851   7 1.7420 58.5 0 0
6042   5 1.7550 52.0 1 0       15153   5 1.5360 52.0 0 0       25001   7 1.6030 51.0 0 0
6101   8 2.9800 60.0 0 0       15201   9 2.5600 60.5 0 0       25052   8 2.6390 59.5 0 0
6801   9 2.1000 60.0 0 0       15241   8 1.9620 57.0 1 0       25201   7 1.8290 54.0 0 0
6851   5 1.2820 49.0 0 0       15301   8 2.5310 58.0 0 0       27001   7 2.0840 58.0 1 0
7201   9 3.0000 65.5 1 0       15401   9 2.7150 60.0 1 0       27401   7 2.2200 58.0 1 0
7251   8 2.6730 60.0 0 0       15443   9 2.4570 59.0 1 0       27952   7 1.4730 52.5 0 0
7252   7 2.0930 57.5 0 0       15601   9 2.0900 59.5 1 0       28452   8 2.3410 60.5 0 0
7253   5 1.6120 52.0 0 0       15701   7 1.7890 56.0 1 0       28501   7 1.6980 54.5 0 0
8501   8 2.1750 59.0 0 0       15741   5 1.8580 53.0 1 0       28551   5 1.1960 46.5 0 0
8801   9 2.7250 59.0 1 0       15842   5 1.4520 51.0 1 0       28801   8 1.8720 56.5 0 0
9101   8 2.0710 55.0 1 0       16101   9 3.8420 69.0 1 0       29001   7 2.2190 55.0 1 0
9201   8 1.5470 57.0 1 0       16151   6 1.7190 53.0 0 0       29101   9 2.4200 57.0 1 0
9301   8 2.0040 57.0 1 0       16251   7 2.1110 57.0 0 0       29151   7 1.8270 54.5 0 0
9501   9 3.1350 60.0 0 0       16252   6 1.6950 53.0 0 0       29401   7 1.4610 54.0 0 0
```

Data Set 10 FEV.DAT (*Continued*)

```
29443  6 1.3380 53.0 1 0      42552  9 2.0910 58.5 0 0      53651  6 1.3380 51.5 0 0
29601  8 2.0900 57.0 1 0      42901  9 2.3160 59.5 0 0      54201  8 2.0160 56.0 1 0
29701  8 1.6970 59.0 0 0      42941  5 1.7040 51.0 0 0      54701  9 2.6390 63.0 0 0
29741  8 1.5620 55.0 1 0      43201  9 1.6060 57.5 0 0      54751  4 1.3890 48.0 0 0
29901  9 2.0400 55.5 0 0      43242  7 1.1650 47.0 1 0      54941  7 1.6120 56.5 1 0
29954  7 1.6090 51.5 0 0      43651  6 2.1020 55.5 0 0      54952  8 2.1350 60.5 0 0
30001  8 2.4580 61.0 0 0      43901  9 2.3200 57.0 0 0      55645  8 2.6810 60.5 1 0
30043  9 2.6500 63.5 1 0      44201  9 2.2300 61.0 1 0      55652  9 3.2230 65.0 0 0
31001  8 1.4290 57.5 1 0      44301  9 1.7160 55.5 1 0      57651  6 1.7960 55.0 0 0
31501  8 1.6750 53.0 1 0      44501  7 1.7900 53.5 1 0      57901  8 2.0100 55.0 1 0
31551  9 1.9470 56.5 0 0      44551  5 1.1460 50.0 0 0      58341  6 1.5230 51.0 0 0
32501  8 2.0690 54.0 1 0      44701  8 2.1870 61.5 0 0      58601  8 1.7440 52.5 1 0
32542  6 1.5720 52.0 1 0      45001  9 2.7170 61.5 1 0      58602  9 2.4850 64.0 0 0
32742  6 1.3480 53.0 1 0      45041  7 1.7960 55.0 1 0      60051  8 2.3350 59.0 0 0
32751  8 2.2880 61.5 0 0      45241  9 1.9530 58.0 1 1      60251  7 1.4150 53.5 0 0
33301  9 1.7730 58.5 1 0      45251  8 1.3350 56.5 0 0      60801  9 2.0760 60.5 1 0
33351  5 0.7910 52.0 0 0      45301  9 2.1190 57.0 1 0      61101  8 2.4350 59.5 1 0
33501  7 1.9050 58.0 1 0      45641  6 1.6660 52.0 1 0      61601  7 1.7280 56.5 0 0
33601  9 2.4630 61.0 0 0      45642  6 1.8260 52.5 1 0      61801  9 2.8500 63.0 0 0
33641  6 1.4310 51.0 1 0      45653  8 2.7090 62.5 0 0      61901  8 1.8440 56.5 0 0
33801  9 2.6310 62.0 0 0      46101  9 2.8710 65.0 1 0      62301  9 1.7540 61.5 0 0
34101  9 3.1140 64.5 1 0      46151  5 1.0920 50.0 0 0      62351  6 1.3430 52.0 0 0
34301  9 2.1350 58.5 1 0      46641  6 2.2620 57.5 1 0      62641  8 2.3030 57.0 1 0
34341  6 1.5270 52.5 1 0      46642  6 2.1040 56.5 1 0      62702  9 2.2460 63.5 1 0
35102  8 2.2930 58.0 0 0      47001  9 2.1660 57.5 0 0      63102  8 2.4760 63.0 0 0
35103  9 3.0420 66.0 0 0      47053  7 1.6900 54.0 0 0      63201  9 3.2390 65.0 1 0
35601  8 2.9270 63.5 1 0      47242  9 2.9730 59.5 1 0      63941  9 2.4570 61.5 1 0
37201  8 2.6650 64.0 0 0      47301  8 2.1450 59.5 0 0      64101  8 2.3820 62.0 0 0
37301  9 2.3010 58.5 1 0      47341  5 1.9710 58.0 1 0      64151  7 1.6400 55.0 0 0
37901  9 2.4600 64.0 1 0      47552  7 2.0950 57.0 0 0      64152  5 1.5890 51.0 0 0
38051  9 2.5920 60.5 0 0      48052  6 1.6970 55.0 0 0      64201  7 2.0560 54.0 1 0
38152  7 1.7500 55.0 0 0      48101  9 2.4550 60.0 0 0      70001  8 2.2260 57.0 1 0
38241  8 1.7590 53.0 1 0      48401  7 1.9200 56.5 1 0      70401  9 1.8860 56.0 0 0
38242  6 1.5360 48.0 1 0      48402  9 2.1640 60.0 1 0      71101  9 2.8330 61.5 1 0
38801  9 2.2590 58.5 0 0      48901  9 2.1300 59.0 0 0      71241  6 1.7150 53.0 1 0
39001  9 2.0480 64.5 0 0      49301  8 2.9930 63.0 0 0      71441  8 2.6310 59.0 1 0
39101  9 2.5710 60.5 1 0      49401  9 2.5290 59.0 0 0      71444  7 2.5500 56.0 1 0
39141  7 2.0460 56.0 1 0      49501  7 1.7260 53.0 0 0      71851  9 1.9120 59.0 0 0
39201  8 1.7800 58.5 0 0      49542  9 2.4420 61.5 0 0      72001  7 1.8770 52.5 0 0
39251  5 1.5520 54.0 0 0      49551  4 1.1020 48.0 0 0      72051  7 1.9350 52.5 0 0
39301  8 1.9530 58.0 0 0      49701  9 2.0560 63.0 0 0      72052  5 1.5390 50.0 0 0
39401  9 2.8930 64.5 1 0      49741  5 1.8080 55.5 1 0      72501  9 2.8030 59.5 1 0
39741  6 1.7130 50.5 1 0      49751  8 2.3050 64.5 0 0      73001  9 2.9230 64.0 1 0
39801  9 2.8510 60.0 0 0      50501  9 1.9690 59.0 0 0      73151  8 2.3580 61.0 0 0
39841  6 1.6240 51.5 1 0      50701  8 1.5560 58.5 0 0      73342  8 2.0940 57.5 1 0
39901  8 2.6310 59.0 1 0      50951  3 1.0720 46.0 0 0      74201  9 1.8550 60.0 1 0
39941  5 1.8190 53.0 1 0      51241  9 2.0420 62.0 1 0      74241  6 1.5350 55.0 0 0
40001  7 1.6580 53.0 1 0      51301  8 1.5120 53.0 0 0      74401  7 2.1350 56.0 1 0
40501  7 2.1580 53.5 1 0      51341  6 1.4230 49.5 1 0      74441  5 1.9300 51.0 1 0
40541  4 1.7890 52.0 1 0      51501  9 3.6810 68.0 1 0      74601  9 2.1820 59.5 0 0
42101  9 3.0040 64.0 0 0      51542  8 1.9910 59.5 1 0      74641  5 1.3590 50.5 1 0
42201  8 2.5030 63.0 1 0      52101  8 1.8970 55.5 1 0      74652  7 2.0020 57.5 0 0
42501  9 1.9330 58.0 0 0      53601  7 1.3700 55.0 0 0      74941  6 1.6990 54.0 1 0
```

Data Set 10 FEV.DAT (*Continued*)

```
75701  8 2.5000 57.0 1 0      5351 13 3.1470 64.0 0 0     16901 11 3.1710 63.0 0 0
75751  7 2.3660 58.0 0 0      5352 10 2.5200 60.5 0 0     16941 13 3.8870 67.5 1 0
75901  8 2.0690 60.0 0 0      5641 10 2.2920 63.0 1 0     16951 13 2.6460 61.5 0 0
75951  4 1.4180 49.0 0 0      5701 12 2.8890 64.0 0 0     16952 10 2.5040 60.0 0 0
76501  8 2.3330 57.0 0 0      6001 10 2.2460 60.5 1 0     17301 11 3.5870 64.5 1 0
76541  5 1.5140 52.0 1 0      6041 10 1.9370 62.0 1 0     17401 11 3.8450 68.5 1 0
76751  8 1.7580 52.0 0 0      6145 10 2.6460 60.0 1 0     17801 12 2.9710 64.5 1 0
77141  7 2.5350 59.5 1 0      6201 11 2.9570 64.5 1 0     17851 10 2.8910 61.0 0 0
77901  7 2.5640 58.0 0 0      6401 11 4.0070 67.0 1 0     18502 10 1.8230 57.0 0 0
78301  9 2.4870 64.0 0 0      7101 11 2.3860 61.5 0 0     18541 11 2.4170 62.5 1 0
78352  9 1.5910 57.0 0 0      7241 10 3.2510 66.0 1 0     18551 10 2.1750 58.0 0 0
80001  8 1.6240 53.0 1 0      7901 11 2.7620 60.0 0 0     18801 11 2.7350 62.5 0 0
80301  9 2.7980 62.0 1 0      8301 11 3.0110 64.0 0 0     18841 14 4.2730 72.5 1 0
80341  6 1.6910 53.0 1 0      8541 13 4.3050 68.5 1 0     18842 13 2.9760 65.5 1 0
80601  8 1.9990 56.5 0 0      8542 13 3.9060 67.0 1 0     19601 12 3.8350 69.5 0 1
80801  9 1.8690 57.0 1 0      8842 11 3.5830 67.0 1 0     19901 11 4.0650 66.5 1 0
80841  4 1.0040 48.0 1 0      8901 11 3.2360 66.0 0 0     20101 11 2.3180 59.0 0 0
81241  6 1.4270 49.5 1 0      9141 14 3.4360 62.5 1 0     21301 11 3.5960 68.0 1 0
81401  7 1.8260 51.0 1 0      9142 11 3.0580 61.0 1 0     21352 14 3.3950 67.0 0 0
81451  9 2.6880 59.5 0 0      9502 10 3.0070 62.0 1 0     21353 12 2.7510 63.0 0 0
81501  8 1.6570 56.0 1 0      9802 10 3.4890 66.5 1 0     21501 10 2.6730 64.5 0 0
81751  6 1.6720 54.0 0 0     10001 10 2.8640 60.0 0 0     22253 12 2.5560 62.0 0 0
82701  8 2.0150 57.5 0 0     10053 14 3.4280 64.0 0 1     23151 11 2.5420 62.0 0 0
83801  7 2.3710 55.5 0 0     10054 13 2.8190 62.0 0 0     23401 10 2.6080 66.0 1 0
83841  5 2.1150 50.0 1 0     10501 10 2.2500 58.0 0 0     23601 11 2.3540 62.0 0 0
83901  8 2.3280 60.0 0 0     10642 14 4.6830 68.5 1 0     23651 13 2.5990 62.5 0 1
83952  7 1.4950 57.0 0 0     10801 10 2.3520 61.5 1 0     23652 10 1.4580 57.0 0 0
  201 11 2.8840 69.0 1 0     10901 11 3.1080 64.5 1 0     23901 10 3.7950 68.5 1 0
  202 10 2.3280 64.0 1 0     10942 13 3.9940 67.0 1 0     24201 11 2.4910 59.0 0 0
  341 14 3.3810 63.0 1 0     11101 12 4.3930 68.5 1 0     24251 13 3.0600 61.5 0 0
  351 11 2.1700 58.0 0 0     11151 13 3.2080 61.0 0 1     24501 10 2.5450 65.0 1 0
  401 11 3.4700 66.5 1 0     11301 10 2.5920 65.0 1 0     24543 11 2.9930 66.5 1 0
  551 12 3.0580 60.5 0 0     11341 13 3.1930 70.0 1 0     24601 10 3.3050 65.0 0 0
  601 10 1.8110 57.0 1 0     11601 11 1.6940 60.0 1 1     24642 13 4.7560 68.0 1 1
  641 11 2.5240 64.0 1 0     11642 14 3.9570 72.0 1 1     24701 11 3.7740 67.0 0 0
 1751 10 2.6420 61.0 0 0     11901 11 2.3460 59.0 0 0     24741 10 2.8550 64.5 1 0
 2041 14 3.7410 68.5 1 0     11942 13 4.7890 69.0 1 1     24801 11 2.9880 70.0 1 0
 2042 13 4.3360 69.5 1 0     12501 11 3.5150 67.5 1 0     25041 11 2.4980 60.0 1 0
 2142 14 4.8420 72.0 1 0     13301 11 2.7540 65.5 0 0     25051 14 3.1690 64.0 0 0
 2143 12 4.5500 71.0 1 0     14001 10 2.7200 65.5 1 0     25501 11 2.8870 62.5 1 0
 2451 12 2.8410 63.0 0 0     14143 11 2.4630 64.5 1 0     25551 13 2.7040 61.0 0 0
 2801 10 3.1660 61.5 0 0     14401 11 2.6330 62.0 0 0     25901 11 3.5150 64.0 0 0
 2851 13 3.8160 63.5 0 0     15101 10 3.0480 65.5 0 0     27441 11 3.4250 65.5 1 0
 3141 10 2.5610 62.0 1 0     15141 11 3.1110 67.5 1 0     27701 10 2.2870 61.0 0 0
 3501 11 3.6540 65.0 0 0     15152 13 3.7450 68.0 0 0     27751 13 2.4340 65.4 0 0
 3901 10 2.4810 61.0 1 0     15302 12 2.3840 63.5 0 1     27901 10 2.3650 63.5 0 0
 4001 11 2.6650 63.0 0 0     15342 10 2.0940 58.5 1 0     27951 13 3.0860 67.5 0 1
 4341 10 3.2030 66.0 1 0     15451 10 3.1830 65.5 0 0     28401 10 2.6960 66.0 1 0
 4901 13 3.5490 68.0 1 0     15751 14 3.0740 65.0 0 1     29952 12 2.8680 62.0 0 0
 4952 14 2.2360 66.0 0 1     15801 11 3.9770 70.5 1 0     29953 10 2.8130 61.5 0 0
 5001 11 3.2220 72.0 1 0     15841 10 3.3540 63.0 1 0     30042 14 4.3090 69.0 1 1
 5101 10 3.1110 66.0 1 0     16201 11 3.4110 63.5 0 0     30051 12 3.2550 66.0 0 0
 5251 11 3.4900 67.0 0 0     16551 10 2.3870 66.0 0 1     30052 10 3.4130 66.0 0 1
```

Data Set 10 FEV.DAT (Continued)

30401	11	4.5930	69.0	1	0
31042	14	4.1110	71.0	1	0
31201	12	1.9160	60.5	1	0
31242	10	1.8580	58.0	1	0
31502	10	2.9750	63.0	0	1
31901	10	3.3500	69.0	1	0
32541	10	2.9010	59.5	1	0
32701	12	2.2410	64.0	1	0
32741	13	4.2250	74.0	1	0
33001	11	3.2230	64.5	0	0
33041	12	5.2240	70.0	1	0
33502	11	4.0730	67.0	1	0
33541	12	4.0800	64.5	1	0
33701	11	2.6060	65.0	0	0
34201	11	3.1690	62.5	0	1
34243	12	4.4110	68.0	1	0
35101	12	3.7910	68.5	1	0
35901	13	3.0890	67.5	1	0
35941	11	2.4650	60.0	1	0
36001	12	3.3430	68.0	1	1
36041	10	3.2000	65.0	1	0
36101	12	2.9130	64.0	1	0
37241	13	4.8770	73.0	1	0
37351	10	2.3580	59.0	0	0
37401	12	3.2790	70.5	1	0
37701	10	2.5810	66.0	1	0
37702	12	2.3470	61.5	0	0
37951	10	2.6910	67.0	0	0
38001	11	2.8270	62.5	0	0
38201	10	1.8730	52.5	1	0
39041	12	3.7510	72.0	1	1
39052	14	2.5380	71.0	0	0
39601	10	2.7580	65.5	1	0
39701	10	3.0500	60.0	0	0
40051	12	3.0790	60.0	0	0
40301	10	2.2010	60.5	1	0
41601	10	1.8580	59.0	1	0
42151	13	2.2160	68.0	0	1
42251	12	3.4030	62.0	0	0
42301	12	3.5010	64.5	0	0
42502	11	2.5780	63.0	0	0
42551	13	3.0780	66.0	0	1
43002	12	3.1860	67.0	0	1
43241	10	1.6650	57.0	1	0
43251	12	2.0810	63.0	0	0
43301	11	2.9740	62.0	0	0
43351	13	3.2970	65.0	0	1
43601	12	4.0730	68.5	1	0
43641	13	4.4480	69.0	1	0
44242	13	3.9840	71.0	1	0
44302	10	2.2500	58.0	0	0
44351	12	2.7520	63.5	0	0
44601	12	2.3040	66.5	1	1
44742	14	3.6800	67.0	1	0
45201	11	3.1020	64.0	0	1
45601	10	2.8620	61.0	0	0
45651	13	2.6770	67.0	0	1
45652	11	3.0230	67.5	0	0
46301	11	3.6810	68.0	0	0
46353	13	3.2550	66.5	0	0
46601	12	3.6920	67.0	1	0
46651	10	2.3560	60.5	0	0
46901	10	4.5910	70.0	1	0
46951	12	3.0820	63.5	0	0
47051	13	3.2970	65.0	0	1
47052	11	3.2580	63.0	0	0
47801	10	2.2160	61.0	1	0
47841	11	3.2470	65.5	1	0
48001	11	4.3240	67.5	1	0
48142	11	2.3620	61.0	0	0
48951	11	2.5630	63.0	0	0
49201	11	3.2060	63.5	1	0
49341	14	3.5850	70.0	1	0
49342	12	4.7200	71.5	1	0
49451	13	3.3310	65.5	0	0
49541	13	5.0830	74.0	1	0
50301	10	3.4980	68.0	1	1
50551	12	2.4170	61.0	0	0
50702	10	2.3640	61.0	1	0
50901	10	2.3410	61.0	1	0
51101	12	2.7590	61.5	0	1
51201	11	2.9530	67.0	0	1
51302	12	3.2310	63.0	1	0
51541	11	3.0780	67.5	1	0
51601	11	3.3690	70.5	1	0
51641	12	3.5290	70.5	1	0
52401	12	2.8660	62.0	0	0
52452	14	2.8910	62.0	0	0
52801	11	3.0220	61.5	0	0
53101	10	3.1270	62.0	1	0
53151	11	2.8660	60.5	0	0
53301	12	2.6050	62.5	0	0
54251	13	3.0560	63.0	0	0
54901	12	2.5690	63.0	0	0
54951	11	2.5010	62.0	0	0
55601	11	3.3200	65.5	1	0
55901	11	2.1230	65.0	1	0
55941	14	3.7800	70.0	1	0
56601	11	3.8470	66.0	1	0
56651	13	3.7850	63.0	0	1
56901	12	3.9240	68.0	1	0
57001	10	2.1320	59.0	1	0
57042	12	2.7520	68.5	1	0
57053	13	2.4490	63.0	0	0
57601	10	3.4560	63.0	1	0
58301	10	3.0730	66.0	0	0
58603	10	2.6880	62.0	0	0
59901	10	3.3290	68.0	1	0
59941	14	4.2710	72.5	1	0
60001	12	3.5300	64.0	1	0
60201	11	2.9280	65.5	1	0
60851	11	2.6890	61.5	0	0
61301	12	2.3320	57.0	1	0
61351	14	2.9340	64.0	0	0
61941	14	2.2760	66.0	1	1
62601	10	3.1100	64.5	1	0
62701	11	2.8940	67.0	1	0
62801	11	4.6370	72.0	1	1
62802	10	2.4350	65.0	0	0
63101	10	2.8380	63.0	0	0
63252	12	3.0350	62.0	0	0
63901	12	4.8310	71.0	1	0
64241	11	2.8120	61.0	1	0
70051	12	2.7140	65.5	0	0
70052	10	3.0860	62.0	0	0
70201	12	3.5190	65.5	0	0
70441	13	4.2320	70.5	1	0
70442	10	2.7700	62.0	1	0
71151	12	3.3410	65.5	0	0
71201	10	3.0900	65.0	1	0
71443	13	2.5310	61.0	1	0
71801	12	2.8220	69.5	1	0
72552	10	3.0380	65.0	0	1
72901	12	2.9350	65.5	1	0
73101	10	2.5680	63.5	0	0
73301	11	2.3870	60.5	1	0
73341	12	2.4990	65.0	1	0
73701	11	4.1300	67.0	1	0
74651	12	3.0010	63.5	0	0
74701	10	3.1320	59.5	0	0
74751	13	3.5770	63.5	0	0
74752	12	3.2220	61.0	0	0
74901	11	3.2800	66.0	1	0
75801	11	2.6590	64.0	1	0
76601	11	2.8220	62.0	0	0
76701	11	2.1400	60.5	0	0
77101	12	4.2030	71.0	1	0
77152	14	2.9970	64.5	0	0
77501	11	3.1200	61.0	0	1
77601	11	2.5620	62.5	0	0
80051	12	3.0820	64.5	0	0
80641	14	3.8060	68.0	1	0
80642	11	3.3390	68.5	1	1
80651	13	3.1520	62.0	0	1
81201	11	2.4580	60.0	0	0
81541	10	2.3910	59.5	1	0
81551	13	3.1410	61.0	0	0
81552	12	2.5790	63.0	0	0
81701	11	3.1040	67.5	0	1

Data Set 10 FEV.DAT (*Continued*)

```
81741 13 4.0450 69.0 1 1     22251 18 3.0820 64.5 0 0     47241 15 3.7990 66.5 1 1
82743 14 4.7630 68.0 1 1     22252 16 3.3870 66.5 0 0     48141 18 4.0860 67.0 1 1
82744 10 2.1000 58.0 1 0     24541 17 3.0820 67.0 1 1     48152 15 2.8870 63.0 0 0
83951 11 3.0690 65.0 0 1     24552 16 2.9030 63.0 0 1     48441 16 4.0700 69.5 1 1
90001 11 2.7850 69.0 1 0     24553 15 3.0040 64.0 0 1     52841 17 3.9600 70.0 1 0
  441 15 4.2840 70.0 1 0     25941 15 5.7930 69.0 1 0     52842 16 4.2990 66.0 1 0
 4041 15 4.5060 71.0 1 1     30041 15 3.9850 71.0 1 0     55951 16 2.9810 66.0 0 0
 4051 18 2.9060 66.0 0 0     30441 18 4.2200 68.0 1 0     57052 15 2.2640 63.0 0 1
 6144 19 5.1020 72.0 1 0     30442 17 4.7240 70.5 1 0     59944 18 4.4040 70.5 1 1
 6252 19 3.5190 66.0 0 1     33741 15 3.7310 67.0 1 0     61951 15 2.2780 60.0 0 1
 6441 16 3.6880 68.0 1 1     34241 17 3.4060 69.0 1 1     63241 16 4.5040 72.0 1 0
 7142 17 4.4290 70.0 1 0     37251 17 3.5000 62.0 0 0     71141 17 5.6380 70.0 1 0
 8841 15 4.2790 67.5 1 0     37252 16 3.6740 67.5 0 0     71142 16 4.8720 72.0 1 1
10941 15 4.5000 70.0 1 0     37441 17 5.6330 73.0 1 0     73041 16 4.2700 67.0 1 1
15351 15 2.6350 64.0 0 0     37451 15 3.1220 64.0 0 1     73042 15 3.7270 68.0 1 1
15752 15 2.6790 66.0 0 1     39051 15 3.3300 68.5 0 1     73751 18 2.8530 60.0 0 0
20151 15 2.1980 62.0 0 1     40351 16 2.6080 62.0 0 1     75852 16 2.7950 63.0 0 1
21351 19 3.3450 65.5 0 1     44241 16 3.6450 73.5 1 0     77151 15 3.2110 66.5 0 0
```

Data Set 11 HORMONE.DOC

Variable	Record number	Column	Code
ID	1	1–8	
Biliary secretion-pre	1	10–17	
Biliary pH-pre	1	19–26	
Pancreatic secretion-pre	1	28–35	
Pancreatic pH-pre	1	37–44	
Dose	1	46–53	
Biliary secretion-post	1	55–62	
Biliary pH-post	1	64–71	
Pancreatic secretion-post	2	1–8	
Pancreatic pH-post	2	10–17	
Hormone	2	19–26	1 = SAL/2 = APP/3 = CCK/4 = SEC/5 = VIP

Data Set 12 HORMONE.DAT

```
 1.00      0.0       0.0      25.00     7.00     0.0      3.00      0.0
10.00      7.40      1.00
 1.00     14.30      0.0        .20     0.0      0.0      5.00      0.0
 6.00      0.0       1.00
 1.00      0.0       0.0       6.00     7.60     0.0      0.0       0.0
  .10      7.20      1.00
31.00     95.00      8.10       0.0     0.0      0.0    103.00      8.20
 0.0       0.0       1.00
31.00     40.00      8.40       0.0     0.0      0.0     40.00      8.40
 0.0       0.0       1.00
29.00     25.00      8.60       0.0     0.0      0.0     17.50      8.70
 0.0       0.0       1.00
25.00      0.0       0.0      20.00     8.20     0.0      1.20      0.0
10.00      8.40      1.00
25.00       .10      0.0      20.00     8.40     0.0     10.00      0.0
12.50      8.30      1.00
23.00     35.00      8.40       0.0     0.0      0.0     40.00      8.50
 0.0       0.0       1.00
23.00      0.0       0.0        0.0     0.0      0.0     30.00      8.40
 0.0       0.0       1.00
40.00      1.80      0.0        0.0     0.0      0.0      3.70      0.0
 0.0       0.0       1.00
35.00      7.40      0.0        0.0     0.0      0.0      5.00      0.0
 0.0       0.0       1.00
35.00      0.0       0.0       2.30     8.60     0.0      0.0       0.0
 1.40      8.60      1.00
35.00       .40      0.0        0.0     0.0      0.0     12.00      8.40
  .50      9.50      1.00
39.00     10.00      0.0        0.0     0.0      0.0     10.00      0.0
 1.40      8.10      1.00
39.00     13.50      0.0        0.0     0.0      0.0     12.50      8.60
 4.00      8.40      1.00
39.00     21.00      0.0       3.00     8.60     0.0     20.50      8.30
 3.00      8.80      1.00
```

Data Set 12 HORMONE.DAT (*Continued*)

```
41.00     4.80     0.0      0.0      0.0      0.0      2.20     0.0
  .20     8.50     1.00
41.00    18.50     0.0      0.0      0.0      0.0      5.00     0.0
  0.0     0.0      1.00
43.00     5.40     0.0      0.0      0.0      0.0      2.10     0.0
  0.0     0.0      1.00
42.00     4.50     0.0      2.70     8.70     0.0       .30     0.0
  .30     8.80     1.00
44.00     6.60     0.0      0.0      0.0      0.0      6.30     0.0
  0.0     0.0      1.00
44.00     7.20     0.0      0.0      0.0      0.0       .60     0.0
  0.0     0.0      1.00
34.00     2.60     0.0      0.0      0.0      0.0       .50     0.0
  0.0     0.0      1.00
49.00     0.0      0.0      0.0      0.0      0.0       .60     0.0
  0.0     0.0      1.00
49.00     9.60     0.0      0.0      0.0      0.0      2.30     0.0
  0.0     0.0      1.00
32.00    23.00     8.10     0.0      0.0      0.0     27.50     8.40
  0.0     0.0      1.00
32.00    20.00     0.0      0.0      0.0      0.0     16.30     0.0
  0.0     0.0      1.00
45.00     2.50     0.0      0.0      0.0      0.0      0.0      0.0
  0.0     0.0      1.00
47.00     0.0      0.0       .80     8.00     0.0      0.0      0.0
  0.0     0.0      1.00
 1.00     3.00     6.80    16.00     7.30    15.00     2.00     7.00
 5.00     7.20     2.00
 1.00     2.30     7.40    27.00     7.40    15.00     1.00     6.80
 5.00     7.50     2.00
 1.00    10.00     7.00    10.00     7.20    15.00     0.0      0.0
 5.00     7.40     2.00
 2.00     2.80     0.0      0.0      0.0     15.00     0.0      0.0
 0.0      0.0      2.00
 2.00     6.00     6.80     3.00     7.20    15.00     0.0      0.0
 0.0      0.0      2.00
 2.00     0.0      0.0       .10     7.00    15.00     0.0      0.0
 0.0      0.0      2.00
 2.00     3.40     6.50      .10     6.90    15.00     4.00     6.40
 0.0      0.0      2.00
 3.00     2.10     6.80     0.0      0.0     15.00     1.70     6.90
 0.0      0.0      2.00
 3.00     0.0      0.0      6.30     7.80    15.00     0.0      0.0
 1.20     7.60     2.00
 4.00      .70     7.40     0.0      0.0     15.00      .40     7.20
 0.0      0.0      2.00
 5.00    10.00     7.00     6.30     7.60    15.00    10.00     6.80
 5.00     8.00     2.00
 5.00     0.0      0.0       .60     7.80    15.00     0.0      0.0
 0.0      0.0      2.00
 6.00     2.30     6.80     1.90     8.00    15.00     3.00     6.80
 1.20     8.00     2.00
```

Data Set 12 HORMONE.DAT (*Continued*)

9.00	.90	7.40	0.0	0.0	15.00	2.00	6.80
0.0	0.0	2.00					
22.00	4.10	0.0	0.0	0.0	15.00	.50	0.0
0.0	0.0	2.00					
24.00	15.00	0.0	1.80	0.0	15.00	10.00	0.0
2.40	0.0	2.00					
24.00	9.70	0.0	7.70	8.80	15.00	5.00	0.0
0.0	0.0	2.00					
30.00	7.70	0.0	0.0	0.0	15.00	2.40	0.0
0.0	0.0	2.00					
23.00	5.70	0.0	0.0	0.0	7.50	5.00	0.0
0.0	0.0	2.00					
43.00	0.0	0.0	4.90	8.90	7.50	0.0	0.0
1.00	8.60	2.00					
39.00	3.00	0.0	0.0	0.0	7.50	2.20	0.0
.70	8.50	2.00					
36.00	0.0	0.0	7.60	9.00	7.50	0.0	0.0
1.20	9.00	2.00					
32.00	6.30	0.0	2.90	9.00	7.50	4.20	0.0
0.0	0.0	2.00					
32.00	9.00	0.0	0.0	0.0	15.00	9.00	0.0
0.0	0.0	2.00					
32.00	2.60	0.0	1.40	8.90	7.50	1.10	0.0
1.10	8.40	2.00					
32.00	0.0	0.0	3.80	8.40	7.50	0.0	0.0
0.0	0.0	2.00					
32.00	0.0	0.0	2.40	8.90	7.50	0.0	0.0
0.0	0.0	2.00					
34.00	3.40	0.0	1.60	9.10	15.00	6.30	0.0
0.0	0.0	2.00					
34.00	2.00	0.0	0.0	0.0	7.50	1.80	0.0
0.0	0.0	2.00					
44.00	0.0	0.0	0.0	0.0	15.00	0.0	0.0
.50	8.90	2.00					
62.00	0.0	0.0	1.50	9.20	7.50	0.0	0.0
.70	8.90	2.00					
63.00	0.0	0.0	1.80	8.80	7.50	0.0	0.0
1.20	9.00	2.00					
64.00	6.10	0.0	0.0	0.0	7.50	4.00	0.0
0.0	0.0	2.00					
64.00	2.90	0.0	0.0	0.0	7.50	2.50	0.0
0.0	0.0	2.00					
65.00	1.40	0.0	0.0	0.0	7.50	2.80	0.0
0.0	0.0	2.00					
9.00	.02	0.0	0.0	0.0	.40	.02	7.00
0.0	0.0	3.00					
15.00	.02	0.0	0.0	0.0	.40	.02	0.0
0.0	0.0	3.00					
12.00	8.00	0.0	0.0	0.0	.40	33.00	0.0
0.0	0.0	3.00					
56.00	0.0	0.0	2.30	0.0	.10	6.80	0.0
0.0	0.0	3.00					

Data Set 12 HORMONE.DAT (*Continued*)

19.00	23.50	0.0	0.0	0.0	.10	16.30	8.60
0.0	0.0	3.00					
19.00	0.0	0.0	0.0	0.0	.20	0.0	0.0
.20	0.0	3.00					
18.00	15.00	0.0	0.0	0.0	.20	12.50	8.60
0.0	0.0	3.00					
18.00	35.00	8.50	0.0	0.0	.80	37.50	8.40
0.0	0.0	3.00					
56.00	0.0	0.0	2.30	0.0	.10	6.80	0.0
0.0	0.0	3.00					
56.00	2.00	0.0	0.0	0.0	.80	6.30	0.0
0.0	0.0	3.00					
60.00	0.0	0.0	0.0	0.0	.10	3.40	0.0
1.90	8.60	3.00					
60.00	5.00	0.0	2.40	8.60	.10	5.00	0.0
1.30	8.60	3.00					
19.00	0.0	0.0	0.0	0.0	.20	0.0	0.0
.20	0.0	3.00					
20.00	5.00	0.0	0.0	0.0	.20	10.00	0.0
0.0	0.0	3.00					
20.00	13.00	0.0	0.0	0.0	.80	7.50	0.0
0.0	0.0	3.00					
20.00	10.30	0.0	7.50	0.0	.10	1.90	0.0
4.70	8.00	3.00					
20.00	0.0	0.0	0.0	0.0	.10	.80	0.0
0.0	0.0	3.00					
20.00	2.70	0.0	0.0	0.0	.05	5.00	0.0
0.0	0.0	3.00					
20.00	4.60	0.0	0.0	0.0	.05	1.30	0.0
0.0	0.0	3.00					
20.00	4.00	0.0	0.0	0.0	.05	.60	0.0
0.0	0.0	3.00					
20.00	8.70	0.0	0.0	0.0	.80	27.50	8.70
0.0	0.0	3.00					
21.00	0.0	0.0	0.0	0.0	.20	5.00	0.0
.30	0.0	3.00					
21.00	20.00	0.0	0.0	0.0	.80	20.00	0.0
0.0	0.0	3.00					
21.00	4.10	0.0	0.0	0.0	.10	1.60	0.0
0.0	0.0	3.00					
21.00	50.00	8.70	0.0	0.0	.10	37.50	8.20
0.0	0.0	3.00					
21.00	65.00	8.40	0.0	0.0	.05	55.00	8.40
3.10	9.10	3.00					
21.00	3.20	0.0	0.0	0.0	.05	0.0	0.0
0.0	0.0	3.00					
21.00	7.80	0.0	0.0	0.0	.05	2.70	0.0
.90	8.10	3.00					
21.00	0.0	0.0	0.0	0.0	.80	22.50	8.30
0.0	0.0	3.00					
22.00	3.50	0.0	0.0	0.0	.40	12.50	0.0
0.0	0.0	3.00					

Data Set 12 HORMONE.DAT (*Continued*)

22.00	25.00	0.0	0.0	0.0	.80	10.00	8.70
0.0	0.0	3.00					
22.00	28.00	8.40	0.0	0.0	.80	27.50	8.50
0.0	0.0	3.00					
24.00	27.50	0.0	2.40	9.40	.40	35.00	8.40
1.20	9.00	3.00					
24.00	34.00	8.40	0.0	0.0	.80	55.00	8.40
0.0	0.0	3.00					
30.00	2.50	0.0	1.60	8.90	.80	17.50	8.70
1.70	8.80	3.00					
30.00	3.70	0.0	0.0	0.0	.80	24.00	8.40
0.0	0.0	3.00					
30.00	0.0	0.0	0.0	0.0	.80	4.80	0.0
0.0	0.0	3.00					
30.00	0.0	0.0	0.0	0.0	.40	60.00	8.60
0.0	0.0	3.00					
30.00	0.0	0.0	2.60	0.0	.40	7.00	0.0
1.30	8.80	3.00					
59.00	7.00	0.0	2.70	8.50	.80	22.50	8.60
3.50	8.80	3.00					
59.00	1.90	0.0	0.0	0.0	.80	17.00	8.60
0.0	0.0	3.00					
26.00	10.00	0.0	0.0	0.0	.20	.80	0.0
0.0	0.0	3.00					
31.00	45.00	8.60	0.0	0.0	.20	32.50	8.50
0.0	0.0	3.00					
31.00	9.40	0.0	0.0	0.0	.20	5.00	0.0
0.0	0.0	3.00					
31.00	50.00	8.10	0.0	0.0	.20	30.00	8.10
0.0	0.0	3.00					
31.00	0.0	0.0	0.0	0.0	.80	150.00	8.10
0.0	0.0	3.00					
25.00	6.50	0.0	10.00	8.50	.80	15.00	8.60
12.50	8.20	3.00					
25.00	.80	0.0	10.00	8.10	.80	1.10	0.0
2.80	8.10	3.00					
23.00	35.00	8.40	0.0	0.0	.80	35.00	8.50
0.0	0.0	3.00					
23.00	45.00	8.50	0.0	0.0	.80	62.50	8.30
0.0	0.0	3.00					
23.00	30.00	8.80	0.0	0.0	.80	30.00	8.40
0.0	0.0	3.00					
40.00	2.20	0.0	0.0	0.0	.20	7.50	0.0
0.0	0.0	3.00					
35.00	1.70	0.0	0.0	0.0	.80	2.40	0.0
0.0	0.0	3.00					
41.00	10.00	0.0	0.0	0.0	.80	20.00	8.60
0.0	0.0	3.00					
41.00	17.50	0.0	0.0	0.0	3.20	29.80	8.50
0.0	0.0	3.00					
41.00	27.50	0.0	0.0	0.0	3.20	32.00	8.50
0.0	0.0	3.00					

Data Set 12 HORMONE.DAT (Continued)

41.00	25.00	8.70	0.0	0.0	3.20	42.50	8.30
0.0	0.0	3.00					
41.00	17.50	0.0	0.0	0.0	3.20	45.00	8.20
0.0	0.0	3.00					
41.00	10.00	0.0	0.0	0.0	.40	18.80	8.40
0.0	0.0	3.00					
43.00	3.50	0.0	0.0	0.0	.80	5.00	0.0
.30	8.30	3.00					
42.00	3.90	0.0	0.0	0.0	3.20	14.80	9.10
1.30	8.90	3.00					
42.00	13.50	0.0	0.0	0.0	3.20	30.00	8.50
0.0	0.0	3.00					
42.00	11.40	0.0	0.0	0.0	3.20	38.50	8.50
0.0	0.0	3.00					
42.00	0.0	0.0	0.0	0.0	3.20	22.50	8.50
0.0	0.0	3.00					
42.00	0.0	0.0	0.0	0.0	.40	.30	0.0
0.0	0.0	3.00					
39.00	17.00	0.0	0.0	0.0	3.20	36.30	8.60
0.0	0.0	3.00					
39.00	11.00	0.0	0.0	0.0	3.20	16.50	0.0
0.0	0.0	3.00					
39.00	0.0	0.0	0.0	0.0	3.20	35.00	8.50
.60	8.00	3.00					
39.00	30.00	8.60	0.0	0.0	3.20	37.00	8.50
0.0	0.0	3.00					
39.00	26.50	8.20	0.0	0.0	3.20	42.50	8.50
0.0	0.0	3.00					
38.00	4.00	0.0	0.0	0.0	3.20	25.00	0.0
2.70	8.10	3.00					
38.00	3.60	0.0	0.0	0.0	3.20	21.30	8.70
0.0	0.0	3.00					
38.00	0.0	0.0	1.00	9.00	.40	0.0	0.0
1.30	9.40	3.00					
38.00	0.0	0.0	0.0	0.0	.20	3.50	0.0
0.0	0.0	3.00					
38.00	0.0	0.0	0.0	0.0	.20	.90	0.0
0.0	0.0	3.00					
34.00	1.20	0.0	0.0	0.0	.40	7.00	0.0
0.0	0.0	3.00					
34.00	15.00	0.0	1.50	9.20	3.20	23.50	8.80
1.00	9.20	3.00					
32.00	11.00	0.0	0.0	0.0	3.20	30.00	8.50
0.0	0.0	3.00					
32.00	24.00	0.0	0.0	0.0	3.20	42.50	8.50
0.0	0.0	3.00					
32.00	8.00	0.0	0.0	0.0	.40	12.50	8.10
0.0	0.0	3.00					
32.00	52.50	8.40	6.70	8.80	.20	42.50	8.40
0.0	0.0	3.00					
32.00	34.00	0.0	0.0	0.0	.20	20.00	8.70
4.50	8.90	3.00					

Data Set 12 HORMONE.DAT (*Continued*)

```
-------------------------------------------------------------------------------
32.00    40.00     8.40     2.00     8.90     3.20     1.00     0.0
 1.10     8.70     3.00
32.00    35.00     8.40     0.0      0.0       .40    22.50     8.60
 0.0      0.0      3.00
32.00     7.80     0.0      0.0      0.0       .20     5.90     0.0
 0.0      0.0      3.00
32.00     0.0      0.0      4.90     8.70      .20     0.0      0.0
 9.50     8.60     3.00
32.00     0.0      0.0      8.10     8.40     3.20     0.0      0.0
 7.00     8.40     3.00
32.00     0.0      0.0       .30     9.20      .40     0.0      0.0
 0.0      0.0      3.00
32.00     0.0      0.0      4.50     8.80     3.20     0.0      0.0
  .40     9.20     3.00
32.00     0.0      0.0      4.80     8.40     3.20     0.0      0.0
  .60     8.80     3.00
44.00    20.00     0.0      0.0      0.0       .40    20.00     8.60
 0.0      0.0      3.00
44.00     0.0      0.0      0.0      0.0       .20    10.50     0.0
 1.10     9.50     3.00
44.00    24.00     0.0      0.0      0.0       .20    19.50     8.80
 0.0      0.0      3.00
44.00    25.00     8.80     0.0      0.0      3.20    40.00     8.50
 0.0      0.0      3.00
44.00     9.50     0.0      0.0      0.0       .40    11.50     0.0
  .10     8.80     3.00
44.00    19.50     0.0      0.0      0.0       .20     2.20     0.0
 0.0      0.0      3.00
36.00     2.30     0.0      3.20     8.10      .20     4.10     0.0
  .90     8.80     3.00
36.00     7.70     0.0      8.00     8.90      .20      .40     0.0
  .40     8.90     3.00
36.00     9.00     0.0      1.60     8.00     3.20    15.00     8.60
 0.0      0.0      3.00
36.00     5.30     0.0      4.40     8.90      .40    13.80     9.20
  .50     9.20     3.00
36.00     0.0      0.0       .60     9.10      .20     0.0      0.0
 0.0      0.0      3.00
54.00    28.00     7.90      .70     0.0      3.20    14.00     8.20
 0.0      0.0      3.00
54.00    22.50     8.10     3.00     9.20      .40     9.50     0.0
 0.0      0.0      3.00
54.00    10.00     0.0       .20     8.80     3.20     8.80     0.0
 0.0      0.0      3.00
54.00    15.00     0.0      0.0      0.0       .20    19.00     8.30
 0.0      0.0      3.00
55.00     5.00     0.0      0.0      0.0       .20     5.00     0.0
 0.0      0.0      3.00
54.00    25.00     8.30     0.0      0.0       .20    15.00     8.20
 0.0      0.0      3.00
55.00     0.0      0.0      0.0      0.0       .40     0.0      0.0
  .10     8.10     3.00
```

Data Set 12 HORMONE.DAT (*Continued*)

52.00	24.00	7.90	0.0	0.0	.40	22.30	8.50
0.0	0.0	3.00					
52.00	8.60	0.0	0.0	0.0	3.20	2.70	0.0
0.0	0.0	3.00					
52.00	7.10	0.0	0.0	0.0	3.20	15.00	8.40
0.0	0.0	3.00					
52.00	0.0	0.0	2.30	8.80	3.20	.20	0.0
.20	8.60	3.00					
52.00	0.0	0.0	3.00	9.30	.40	0.0	0.0
1.20	9.00	3.00					
51.00	9.30	0.0	.20	0.0	.40	6.30	0.0
1.10	9.00	3.00					
51.00	14.50	0.0	0.0	0.0	3.20	15.00	8.50
0.0	0.0	3.00					
51.00	4.50	0.0	0.0	0.0	3.20	0.0	0.0
0.0	0.0	3.00					
51.00	.80	0.0	1.00	8.90	3.20	1.50	0.0
0.0	0.0	3.00					
62.00	6.40	0.0	0.0	0.0	.40	6.50	0.0
.70	8.00	3.00					
62.00	0.0	0.0	.50	8.80	3.20	0.0	0.0
.60	8.80	3.00					
63.00	1.20	0.0	0.0	0.0	3.20	7.80	0.0
1.20	9.00	3.00					
63.00	12.50	0.0	0.0	0.0	3.20	.50	0.0
2.50	8.80	3.00					
63.00	7.40	0.0	2.50	8.80	3.20	0.0	0.0
2.10	8.80	3.00					
63.00	24.00	7.80	2.00	8.80	3.20	11.00	8.10
0.0	0.0	3.00					
64.00	4.40	0.0	0.0	0.0	3.20	3.00	0.0
0.0	0.0	3.00					
64.00	6.80	0.0	0.0	0.0	3.20	6.80	0.0
0.0	0.0	3.00					
64.00	9.10	0.0	0.0	0.0	3.20	9.00	0.0
0.0	0.0	3.00					
64.00	10.00	0.0	0.0	0.0	3.20	5.30	0.0
0.0	0.0	3.00					
64.00	20.00	0.0	0.0	0.0	3.20	.40	0.0
0.0	0.0	3.00					
64.00	19.00	0.0	0.0	0.0	3.20	0.0	0.0
0.0	0.0	3.00					
65.00	4.00	0.0	0.0	0.0	3.20	6.30	0.0
0.0	0.0	3.00					
65.00	8.40	0.0	0.0	0.0	3.20	.70	0.0
0.0	0.0	3.00					
65.00	11.00	0.0	0.0	0.0	3.20	.80	0.0
0.0	0.0	3.00					
65.00	17.50	0.0	0.0	0.0	3.20	9.50	0.0
0.0	0.0	3.00					
66.00	15.00	0.0	0.0	0.0	3.20	15.00	8.00
0.0	0.0	3.00					

Data Set 12 HORMONE.DAT (Continued)

66.00	7.50	0.0	0.0	0.0	3.20	8.00	0.0
0.0	0.0	3.00					
73.00	4.50	0.0	4.40	0.0	3.20	23.80	8.40
.20	7.80	3.00					
73.00	0.0	0.0	.30	9.00	3.20	0.0	0.0
.30	9.00	3.00					
74.00	10.50	0.0	0.0	0.0	3.20	13.80	7.90
0.0	0.0	3.00					
74.00	13.00	0.0	0.0	0.0	3.20	5.00	0.0
0.0	0.0	3.00					
74.00	5.70	0.0	0.0	0.0	3.20	6.30	0.0
0.0	0.0	3.00					
74.00	6.10	0.0	0.0	0.0	.40	9.00	0.0
0.0	0.0	3.00					
75.00	25.00	7.30	0.0	0.0	3.20	17.50	8.30
0.0	0.0	3.00					
75.00	29.50	8.30	0.0	0.0	3.20	7.80	0.0
0.0	0.0	3.00					
75.00	8.80	0.0	0.0	0.0	3.20	0.0	0.0
0.0	0.0	3.00					
75.00	2.90	0.0	0.0	0.0	.40	13.80	7.90
0.0	0.0	3.00					
76.00	27.50	8.20	0.0	0.0	.80	28.80	8.60
0.0	0.0	3.00					
76.00	6.70	0.0	0.0	0.0	.80	7.50	0.0
0.0	0.0	3.00					
68.00	0.0	0.0	0.0	0.0	.80	0.0	0.0
1.60	7.90	3.00					
77.00	7.20	0.0	0.0	0.0	.80	6.30	0.0
0.0	0.0	3.00					
77.00	7.20	0.0	0.0	0.0	.80	16.80	7.90
0.0	0.0	3.00					
77.00	15.00	0.0	0.0	0.0	.80	12.50	8.20
0.0	0.0	3.00					
78.00	9.00	0.0	0.0	0.0	.80	15.00	8.10
0.0	0.0	3.00					
78.00	1.30	0.0	0.0	0.0	.80	4.50	0.0
0.0	0.0	3.00					
78.00	25.00	8.30	0.0	0.0	.10	10.00	0.0
0.0	0.0	3.00					
78.00	30.00	8.90	0.0	0.0	.10	25.00	8.60
0.0	0.0	3.00					
78.00	1.30	0.0	0.0	0.0	.10	0.0	0.0
0.0	0.0	3.00					
70.00	0.0	0.0	1.70	8.00	.10	0.0	0.0
1.20	8.70	3.00					
70.00	15.00	0.0	0.0	0.0	.80	8.80	0.0
0.0	0.0	3.00					
79.00	15.00	0.0	2.90	7.70	.10	6.30	0.0
2.90	7.80	3.00					
79.00	0.0	0.0	0.0	0.0	.10	0.0	0.0
.20	8.40	3.00					

Data Set 12 HORMONE.DAT (*Continued*)

71.00	0.0	0.0	.40	8.00	.10	0.0	0.0
.80	8.40	3.00					
67.00	6.90	0.0	6.30	8.00	.80	7.50	0.0
4.00	8.10	3.00					
67.00	7.50	0.0	1.40	8.00	.80	11.30	0.0
2.20	8.00	3.00					
67.00	0.0	0.0	1.10	8.90	3.20	0.0	0.0
.70	9.10	3.00					
67.00	12.50	0.0	1.30	8.10	.10	12.50	8.40
1.20	8.10	3.00					
28.00	2.00	0.0	0.0	0.0	2.50	0.0	0.0
0.0	0.0	4.00					
28.00	0.0	0.0	0.0	0.0	5.00	4.90	0.0
2.60	9.40	4.00					
24.00	9.80	0.0	3.20	0.0	2.50	7.50	0.0
2.10	8.60	4.00					
24.00	25.00	8.70	0.0	0.0	5.00	30.00	8.50
0.0	0.0	4.00					
22.00	10.00	0.0	0.0	0.0	2.50	5.00	0.0
0.0	0.0	4.00					
30.00	0.0	0.0	25.00	8.90	5.00	0.0	0.0
1.10	9.40	4.00					
30.00	2.90	0.0	0.0	0.0	.50	4.10	0.0
0.0	0.0	4.00					
30.00	10.00	0.0	0.0	0.0	.50	.70	0.0
0.0	0.0	4.00					
59.00	15.00	0.0	0.0	0.0	.50	12.50	8.20
.80	8.10	4.00					
58.00	10.00	0.0	0.0	0.0	.50	5.00	0.0
0.0	0.0	4.00					
27.00	3.80	0.0	0.0	0.0	2.50	3.00	0.0
0.0	0.0	4.00					
27.00	5.50	0.0	0.0	0.0	5.00	1.80	0.0
0.0	0.0	4.00					
26.00	7.90	0.0	2.90	8.00	2.50	5.00	0.0
2.10	9.00	4.00					
31.00	15.00	0.0	0.0	0.0	.50	7.50	0.0
0.0	0.0	4.00					
31.00	55.00	8.40	0.0	0.0	.50	0.0	0.0
0.0	0.0	4.00					
31.00	30.00	8.40	0.0	0.0	.50	7.50	0.0
0.0	0.0	4.00					
23.00	15.00	0.0	5.00	8.30	.50	10.00	0.0
5.00	8.10	4.00					
23.00	30.00	8.90	15.00	8.50	.50	30.00	8.40
15.00	8.00	4.00					
23.00	35.00	8.40	0.0	0.0	.50	7.50	0.0
0.0	0.0	4.00					
23.00	10.00	0.0	0.0	0.0	.50	0.0	0.0
0.0	0.0	4.00					
29.00	10.00	0.0	0.0	0.0	.50	15.00	7.90
0.0	0.0	4.00					

Data Set 12 HORMONE.DAT (*Continued*)

```
-------------------------------------------------------------------------------
29.00     4.80      0.0       0.0       0.0        .50     15.00      8.20
 0.0      0.0       4.00
25.00    30.00      7.60      1.50      8.30       .50     37.50      8.50
 5.20     8.10      4.00
25.00     0.0       0.0       8.80      8.10       .50      4.00      0.0
 5.00     7.90      4.00
25.00     9.00      0.0      15.00      8.40       .50       .50      0.0
 5.00     8.20      4.00
40.00    30.00      8.90     15.00      8.50       .50     30.00      8.40
15.00     8.00      4.00
33.00    20.00      0.0       0.0       0.0        .50     12.50      8.10
 0.0      0.0       4.00
39.00    20.00      0.0       0.0       0.0        .50     15.00      8.30
 0.0      0.0       4.00
39.00    16.00      0.0       4.00      9.00     20.00     15.00      8.00
 0.0      0.0       4.00
41.00     3.10      0.0       2.00      8.60       .50      3.90      0.0
 3.00     8.90      4.00
41.00    14.00      0.0       0.0       0.0      20.00       .80      0.0
 0.0      0.0       4.00
43.00     0.0       0.0      12.50      8.00     20.00      0.0       0.0
 1.30     8.80      4.00
42.00    14.50      0.0        .90      8.90     20.00      0.0       0.0
 3.20     8.80      4.00
49.00     5.00      0.0       0.0       0.0      12.50      1.40      0.0
 0.0      0.0       4.00
45.00    17.50      0.0       3.20      7.00     12.50     15.50      8.70
 2.50     8.40      4.00
32.00    32.50      8.00      1.40      8.00     12.50     37.50      8.20
 9.30     8.40      4.00
66.00    26.00      8.00      0.0       0.0      12.50      1.30      0.0
 0.0      0.0       4.00
67.00     9.40      0.0       4.60      8.00     12.50      0.0       0.0
 5.00     8.50      4.00
10.00     4.50      0.0       0.0       0.0       8.00      3.40      0.0
 0.0      0.0       5.00
11.00     0.0       0.0       2.30      0.0       1.00      0.0       0.0
 1.70     0.0       5.00
10.00      .90      0.0       0.0       0.0      24.00       .80      0.0
  .70     0.0       5.00
14.00     0.0       0.0       0.0       0.0      12.00      0.0       0.0
  .10     0.0       5.00
14.00     0.0       0.0        .90      0.0      24.00      0.0       0.0
 1.60     0.0       5.00
11.00     0.0       0.0       1.00      0.0       8.00      0.0       0.0
  .40     0.0       5.00
10.00    10.00      0.0       0.0       0.0       1.00      3.90      0.0
 0.0      0.0       5.00
23.00    15.00      0.0       5.00      7.10       .80       .50      0.0
 0.0      0.0       5.00
23.00     0.0       0.0       0.0       0.0        .40      4.40      0.0
 5.10     8.40      5.00
```

Data Set 12 HORMONE.DAT (*Continued*)

35.00	.50	0.0	0.0	0.0	3.20	1.10	0.0
0.0	0.0	5.00					
17.00	4.80	0.0	.50	0.0	4.00	7.50	0.0
2.70	7.10	5.00					
17.00	1.40	0.0	.90	7.00	2.00	0.0	0.0
3.20	7.00	5.00					
17.00	0.0	0.0	1.70	7.80	1.00	0.0	0.0
1.30	7.80	5.00					
17.00	2.30	0.0	0.0	0.0	2.00	2.00	0.0
0.0	0.0	5.00					
17.00	0.0	0.0	0.0	0.0	.50	.10	0.0
0.0	0.0	5.00					
17.00	0.0	0.0	0.0	0.0	2.00	1.50	0.0
0.0	0.0	5.00					
17.00	.50	0.0	0.0	0.0	8.00	1.40	0.0
0.0	0.0	5.00					
16.00	0.0	0.0	.50	0.0	4.00	.70	0.0
0.0	0.0	5.00					
16.00	2.20	0.0	0.0	0.0	1.00	1.00	0.0
1.30	7.60	5.00					
16.00	0.0	0.0	0.0	0.0	4.00	1.10	0.0
0.0	0.0	5.00					
16.00	4.40	0.0	0.0	0.0	1.00	3.60	0.0
0.0	0.0	5.00					
16.00	0.0	0.0	0.0	0.0	2.00	5.00	0.0
0.0	0.0	5.00					
16.00	0.0	0.0	0.0	0.0	4.00	5.00	0.0
0.0	0.0	5.00					
18.00	2.80	0.0	0.0	0.0	4.00	7.50	0.0
0.0	0.0	5.00					
18.00	12.50	0.0	0.0	0.0	.50	11.30	0.0
0.0	0.0	5.00					
18.00	10.00	0.0	0.0	0.0	8.00	10.00	0.0
0.0	0.0	5.00					
18.00	45.00	8.40	0.0	0.0	.50	40.00	8.60
0.0	0.0	5.00					
18.00	25.00	8.40	0.0	0.0	1.00	25.00	8.50
0.0	0.0	5.00					
18.00	0.0	0.0	0.0	0.0	8.00	1.40	0.0
0.0	0.0	5.00					
21.00	27.50	0.0	0.0	0.0	2.00	25.00	0.0
0.0	0.0	5.00					
21.00	10.00	0.0	0.0	0.0	.50	6.30	0.0
0.0	0.0	5.00					
21.00	7.50	0.0	0.0	0.0	1.00	2.80	0.0
0.0	0.0	5.00					
21.00	4.50	0.0	0.0	0.0	8.00	6.20	0.0
.20	0.0	5.00					
19.00	3.00	0.0	.10	0.0	8.00	0.0	0.0
0.0	0.0	5.00					
19.00	40.00	8.40	6.80	8.40	.50	17.50	8.60
5.00	7.90	5.00					

Data Set 12 HORMONE.DAT (*Continued*)

```
------------------------------------------------------------------------------
19.00    27.50     8.60     8.50     8.60     1.00    22.50     8.50
 2.50     8.80     5.00
19.00    20.00     0.0      0.0      0.0      8.00     5.00     0.0
 1.50     8.50     5.00
20.00    10.50     0.0      0.0      0.0     16.00    12.00     0.0
 0.0      0.0      5.00
22.00    13.40     0.0      0.0      0.0       .80    25.00     8.60
 0.0      0.0      5.00
24.00    42.50     8.30     4.70     8.90      .80    32.50     8.50
 3.50     8.60     5.00
24.00    25.00     0.0      0.0      0.0      1.60    40.00     8.40
 0.0      0.0      5.00
24.00     2.10     0.0      0.0      0.0      1.60     3.00     0.0
 0.0      0.0      5.00
28.00     0.0      0.0      0.0      0.0      1.60     3.00     0.0
 0.0      0.0      5.00
30.00     0.0      0.0      0.0      0.0      1.60     1.10     0.0
 0.0      0.0      5.00
30.00     2.90     0.0      0.0      0.0      1.60     2.40     0.0
 2.10     9.40     5.00
30.00    10.00     0.0      0.0      0.0       .80     1.10     0.0
 0.0      0.0      5.00
26.00    11.10     0.0      0.0      0.0       .40      .50     0.0
 0.0      0.0      5.00
27.00     9.20     0.0      0.0      0.0       .40     0.0      0.0
 0.0      0.0      5.00
31.00    50.00     8.10     0.0      0.0       .40    45.00     8.40
 0.0      0.0      5.00
31.00    40.00     8.40     0.0      0.0       .40    25.00     8.40
 0.0      0.0      5.00
31.00     0.0      0.0      0.0      0.0      1.60    25.00     8.10
 0.0      0.0      5.00
31.00    35.00     8.30     0.0      0.0      3.20    17.50     8.40
 0.0      0.0      5.00
23.00    30.00     7.90     0.0      0.0      1.60    27.50     8.40
 0.0      0.0      5.00
23.00    10.00     0.0      0.0      0.0      3.20    25.00     8.60
 0.0      0.0      5.00
23.00     0.0      0.0      0.0      0.0      3.20      .80     0.0
 0.0      0.0      5.00
29.00    10.00     0.0      0.0      0.0       .40     5.00     0.0
 0.0      0.0      5.00
43.00    10.00     0.0      0.0      0.0      3.20     4.10     0.0
 4.30     8.90     5.00
43.00     0.0      0.0      4.00     8.20     9.60     0.0      0.0
  .80     8.60     5.00
39.00    30.00     8.10     0.0      0.0      3.20    37.50     8.30
 0.0      0.0      5.00
39.00    20.00     0.0      0.0      0.0      9.60    27.50     0.0
 0.0      0.0      5.00
39.00    26.00     8.00     0.0      0.0      9.60     0.0      0.0
 0.0      0.0      5.00
```

Data Set 12 HORMONE DAT (*Continued*)

39.00	19.00	0.0	0.0	0.0	8.00	25.00	8.60
0.0	0.0	5.00					
41.00	30.00	8.60	0.0	0.0	3.20	27.50	8.60
0.0	0.0	5.00					
41.00	15.00	0.0	0.0	0.0	9.60	31.30	0.0
0.0	0.0	5.00					
41.00	30.00	8.50	0.0	0.0	8.00	32.50	8.30
0.0	0.0	5.00					
42.00	7.00	0.0	0.0	0.0	9.60	2.70	0.0
.70	8.10	5.00					
42.00	9.00	0.0	0.0	0.0	9.60	1.30	0.0
0.0	0.0	5.00					
44.00	15.00	0.0	0.0	0.0	12.00	15.00	8.40
0.0	0.0	5.00					
44.00	10.00	0.0	0.0	0.0	8.00	11.00	0.0
0.0	0.0	5.00					
44.00	17.00	0.0	0.0	0.0	12.00	22.50	8.50
0.0	0.0	5.00					
44.00	24.00	8.00	0.0	0.0	12.00	3.80	0.0
0.0	0.0	5.00					
44.00	0.0	0.0	0.0	0.0	12.00	.20	0.0
0.0	0.0	5.00					
32.00	14.00	0.0	0.0	0.0	8.00	32.50	8.40
0.0	0.0	5.00					
32.00	40.00	8.40	0.0	0.0	12.00	72.50	8.30
0.0	0.0	5.00					
32.00	18.00	0.0	0.0	0.0	8.00	23.80	8.40
0.0	0.0	5.00					
32.00	23.00	8.40	7.90	8.40	12.00	.60	0.0
4.80	8.10	5.00					
32.00	19.00	0.0	0.0	0.0	24.00	0.0	0.0
0.0	0.0	5.00					
32.00	0.0	0.0	6.70	8.40	12.00	.10	0.0
2.60	8.50	5.00					
32.00	30.50	7.90	0.0	0.0	12.00	60.00	8.30
5.30	8.80	5.00					
32.00	27.50	7.90	0.0	0.0	24.00	85.00	8.20
0.0	0.0	5.00					
32.00	7.10	0.0	0.0	0.0	24.00	4.70	0.0
0.0	0.0	5.00					
32.00	7.30	0.0	0.0	0.0	8.00	20.00	8.50
0.0	0.0	5.00					
32.00	0.0	0.0	1.00	8.60	8.00	0.0	0.0
0.0	0.0	5.00					
32.00	0.0	0.0	.50	8.90	8.00	0.0	0.0
6.00	9.00	5.00					
38.00	0.0	0.0	0.0	0.0	8.00	11.50	0.0
0.0	0.0	5.00					
36.00	9.00	0.0	0.0	0.0	12.00	17.50	8.30
0.0	0.0	5.00					
36.00	12.50	0.0	0.0	0.0	8.00	16.50	0.0
0.0	0.0	5.00					

Data Set 12 HORMONE.DAT (*Continued*)

```
------------------------------------------------------------------------------------
37.00    12.50     0.0      2.30     8.90    12.00    24.50     8.70
 3.80     8.80     5.00
37.00    12.00     0.0      2.90     8.00     8.00    10.50     8.70
 2.30     8.60     5.00
37.00    40.00     8.40     7.90     8.40    12.00    32.50     8.20
 2.00     9.00     5.00
37.00    75.00     7.50     0.0      0.0     24.00    47.50     7.70
 1.30     9.30     5.00
37.00    67.50     7.00     0.0      0.0     12.00    17.50     8.30
 0.0      0.0      5.00
50.00     0.0      0.0      0.0      0.0     12.00     6.80     0.0
 0.0      0.0      5.00
34.00     9.30     0.0      0.0      0.0     12.00    10.00     8.70
 3.30     9.00     5.00
34.00     0.0      0.0      9.40     9.00    12.00      .30     0.0
 1.20     9.10     5.00
34.00     2.40     0.0      0.0      0.0     24.00     0.0      0.0
  .90     8.90     5.00
49.00    10.60     0.0      0.0      0.0     12.00    29.00     8.60
 0.0      0.0      5.00
45.00    12.50     0.0      3.60     7.90    12.00    13.50     8.80
16.30     8.00     5.00
45.00    32.50     8.60     0.0      0.0     24.00    24.00     8.70
 0.0      0.0      5.00
45.00    40.00     8.30     0.0      0.0     24.00    65.00     8.40
 0.0      0.0      5.00
45.00    20.00     0.0      0.0      0.0      8.00    45.00     8.40
 0.0      0.0      5.00
45.00    10.60     0.0      0.0      0.0      8.00    35.00     8.50
 3.20     8.80     5.00
46.00    15.00     0.0      0.0      0.0     12.00    52.50     8.60
 0.0      0.0      5.00
46.00     0.0      0.0      0.0      0.0     24.00      .90     0.0
 0.0      0.0      5.00
48.00    25.00     8.50     0.0      0.0      8.00    30.00     8.60
 0.0      0.0      5.00
47.00    30.00     8.40     0.0      0.0      8.00    32.50     8.60
  .50     0.0      5.00
47.00     0.0      0.0      0.0      0.0      8.00     1.30     0.0
 0.0      0.0      5.00
47.00     1.50     0.0      0.0      0.0     12.00     0.0      0.0
 0.0      0.0      5.00
47.00     0.0      0.0      0.0      0.0      8.00     0.0      0.0
 1.70     7.80     5.00
52.00     1.20     0.0      1.40     9.20    12.00     1.60     0.0
  .40     9.30     5.00
53.00    16.50     0.0      0.0      0.0     12.00    40.80     8.50
 0.0      0.0      5.00
62.00    12.50     0.0       .10     8.90    12.00     8.50     0.0
 2.40     8.70     5.00
62.00     0.0      0.0       .50     8.60    12.00     0.0      0.0
 1.60     8.90     5.00
```

Data Set 12 HORMONE.DAT (*Continued*)

65.00	14.00	0.0	1.50	8.70	24.00	11.30	8.60
2.50	8.80	5.00					
65.00	10.50	0.0	3.10	8.80	12.00	15.00	8.40
.20	8.40	5.00					
63.00	0.0	0.0	.60	8.30	24.00	12.00	8.20
.80	8.60	5.00					
63.00	9.20	0.0	1.80	8.00	12.00	32.00	8.20
10.00	8.60	5.00					
63.00	0.0	0.0	3.30	8.80	12.00	0.0	0.0
3.30	8.90	5.00					
64.00	10.10	0.0	0.0	0.0	24.00	33.80	8.10
1.30	0.0	5.00					
68.00	12.00	0.0	.60	8.80	4.00	27.80	8.60
.40	9.00	5.00					
68.00	0.0	0.0	.60	8.00	4.00	.50	0.0
.90	8.00	5.00					
69.00	15.00	0.0	4.80	8.90	2.00	17.50	8.40
5.00	8.40	5.00					
70.00	15.00	0.0	5.30	7.80	.50	7.50	0.0
2.50	7.90	5.00					
70.00	20.00	0.0	0.0	0.0	.50	17.50	8.10
.40	8.80	5.00					
71.00	0.0	0.0	0.0	0.0	4.00	1.80	0.0
1.00	8.60	5.00					
72.00	4.50	0.0	4.30	8.00	2.00	12.50	8.10
4.60	8.20	5.00					
72.00	10.00	0.0	5.20	8.00	4.00	10.00	0.0
5.00	8.20	5.00					
72.00	10.00	0.0	5.70	8.00	.50	12.50	8.70
6.30	8.00	5.00					
72.00	10.00	0.0	3.90	8.10	4.00	11.30	0.0
3.20	8.90	5.00					
72.00	10.00	0.0	2.40	8.90	.50	13.80	7.80
.80	9.30	5.00					

Data Set 13 INFANTBP.DOC

Variable	Record	Column		Comments
Salt taste variables				
ID	1	2–4		
Record number	1	5	=1	
Mean SBP	1	6–11	xxx.xx	
Mean DBP	1	12–17	xxx.xx	
MSB-trial 1*	1	18–24	xxxx.xx	water
MSB-trial 2	1	25–31	xxxx.xx	water
MSB-trial 3	1	32–38	xxxx.xx	0.1 molar salt + water
MSB-trial 4	1	39–45	xxxx.xx	0.1 molar salt + water
MSB-trial 5	1	46–52	xxxx.xx	water
MSB-trial 6	1	53–59	xxxx.xx	water
MSB-trial 7	1	60–66	xxxx.xx	0.3 molar salt + water
MSB-trial 8	1	67–73	xxxx.xx	0.3 molar salt + water
MSB-trial 9	1	74–80	xxxx.xx	water
ID	2	2–4		
Record number	2	5	=2	
MSB-trial 10	2	6–12	xxxx.xx	water

Variable	Record	Column		Comments
Sugar taste variables				
MSB-trial 1	2	13–19	xxxx.xx	nonnutritive sucking
MSB-trial 2	2	20–26	xxxx.xx	water
MSB-trial 3	2	27–33	xxxx.xx	5% sucrose + water
MSB-trial 4	2	34–40	xxxx.xx	15% sucrose + water
MSB-trial 5	2	41–47	xxxx.xx	nonnutritive sucking

----

* 999.99 is a missing value; 0 indicates the baby did not suck; MSB = mean number of sucks per burst.

Data Set 14 INFANTBP.DAT

```
 11 75.33 42.67    5.00    5.66    0.00    4.33    4.50    3.25    5.00    5.00    3.75
 12    0.00  12.85   40.00   39.00   68.00   10.28
 71 53.33 37.33   33.00   33.00   24.00   34.00   24.00   31.00   31.00   25.00   24.00
 72  27.00  13.85   26.75   16.00    1.18    0.00
 91 66.00 38.67    2.00    0.37    6.66    6.66   17.00   19.00    0.00    0.00    0.00
 92   6.00    3.90   14.25   20.00   44.33    5.25
101 70.67 43.33    0.00    0.00    0.00    0.00    0.00    0.00    0.00    0.00    0.00
102   0.00    0.00    0.00    0.00    0.00    0.00
111 84.67 53.33    0.00   15.00    6.33    6.33   10.00   17.00   16.50   15.00   10.66
112  13.00    0.00    0.00    0.00    0.00    0.00
171 72.67 42.00    0.00   23.00   17.00    6.50   24.00   17.00    1.00    0.00    0.00
172   0.00  15.00   25.25   27.75   37.33   12.16
191 63.33 48.00    6.33    8.50    5.33    4.00    8.00   12.00    2.00    3.50    7.00
192   0.00  11.60   22.79   49.00   85.50    5.40
```

Data Set 14 INFANTBP.DAT (*Continued*)

```
---------------------------------------------------------------------------
201 57.33 41.67    6.33   23.00    3.50    4.00    7.33    5.66   11.00    0.00   12.50
202  12.00    6.19   18.16   34.25   28.60   10.57
211 66.67 38.67    0.00    0.00    0.00    0.00   12.00   10.00    0.00    0.00    0.00
212   0.00    7.66   28.66   30.50   31.50    6.59
221 81.33 54.67    7.00   14.00    7.00    4.00    3.00    3.00    0.00    0.00    2.00
222  12.00   12.39   20.50   25.20  146.00   12.57
231 82.67 61.33    5.00    7.50    2.50    1.00    1.50    0.00    6.50    2.33   11.00
232  24.00    3.40   20.00   24.00   29.25   13.28
241 74.00 43.33    0.00    1.50    3.00    0.00    0.00    2.00    1.00    0.00    1.00
242   0.00    3.92   15.00   34.00   34.25    4.46
261 70.67 48.67    6.50    3.50    4.50    2.00    1.66   11.00    7.66   12.50   12.50
262  23.00    2.12    7.33   38.66   72.50    4.27
301 71.33 38.67    3.33    8.00    9.00    5.66    6.33    4.75    5.00    4.00    6.50
302   2.00    4.19   31.66   33.00   15.66   12.66
321 64.67 44.33   29.00   11.50   25.00   21.00    6.66   10.50   10.50    8.33   14.00
322   7.00    5.25   11.39   17.50   16.00    4.78
331 68.00 41.33    0.00    0.00    0.00    0.00    0.00    0.00    0.00    0.00    0.00
332   0.00    0.00    0.00    0.00    0.00    0.00
351 71.33 49.33    0.00    0.00    0.00    0.00    0.00    0.00    0.00    0.00    0.00
352   0.00    7.47   13.77   35.50   19.28    3.61
361 73.33 48.67    0.00    4.00    9.50    0.00    3.00    8.00   11.00    7.00    0.00
362   0.00   10.63    9.39   10.25   15.50    3.90
371 62.67 38.00   15.00   12.50   25.00   12.50    9.66    9.00   26.00    3.50   12.50
372   8.33    0.00    0.00    0.00    0.00    0.00
401 74.00 52.00    6.50   16.00    2.00   18.00    8.50    7.50   23.00    9.33    3.33
402   0.00   30.00  142.00  141.00  143.00   18.28
411 72.00 37.33    6.33   11.50    5.00   27.00    5.33    0.00   17.00   18.00    5.33
412   8.00    7.75   17.83   15.57   34.25   11.00
451 66.67 36.00   31.00   28.00   34.00   20.00   28.00   14.50   31.00    0.00    0.00
452   7.50    4.00   14.66   27.25   35.00    8.22
471 64.67 32.67    8.00   21.00    4.33    3.00    3.50    4.00    4.00    2.00    5.00
472  10.00   17.42   12.85   23.39   21.39   14.50
501 66.00 44.00    8.00    8.00   11.50   23.00   14.50    8.00    4.75    8.00   16.00
502   9.33    5.59   22.20   29.25   33.00    5.00
511 60.67 30.00    0.00   16.50    5.33    6.00    6.00    5.00    0.00    0.00    0.00
512   8.50    7.07   37.00   12.42   13.71    8.44
521 70.00 49.33    5.66    5.33    5.75   10.00    1.00    7.00    3.00    5.00    2.00
522   5.00    4.92   18.00   14.66   24.75    5.22
531 70.00 38.67    4.00    9.00    3.66   11.00    4.50    1.00    0.00    0.00    1.00
532   1.00    4.53    4.41   16.66   15.42    7.22
571 74.00 50.67    6.33    7.50    8.00   16.00   13.00    6.00   14.00    3.00   28.00
572  32.00   14.71   30.75   15.00    9.60    7.75
581 84.00 44.00    2.00   10.50    6.50    6.25    0.00   13.00   14.50   15.00   14.50
582   7.00    6.25   15.25   41.66   36.50    4.42
601 69.33 40.67    0.00    2.00    0.00    0.00    1.33    0.00    1.66    0.00    1.60
602   2.50    1.00    1.60    0.00   50.00    9.60
621 72.67 44.00    3.50    5.66   15.50   14.50   36.00   13.50   15.00   26.00   36.00
622  31.00   20.28   78.50   37.25   39.00   15.77
641 61.33 38.00    6.00    0.00    0.00    0.00    4.50    0.00    0.00    0.00    0.00
642   0.00    4.53   10.30   15.14   10.89    8.08
651 76.67 53.00   28.00   33.00   16.00   12.00   13.50   10.00    4.75    4.00    1.00
652  27.00   10.37    6.66    9.00   24.33    8.62
```

Data Set 14 INFANTBP.DAT (*Continued*)

```
----------------------------------------------------------------------------------------
661 68.00 44.00  12.00  12.00   6.66  12.00   0.00   7.66   9.00  10.33   8.66
662 20.00   2.25  16.87  36.00  31.00   7.72
691 70.00 40.67  11.00  16.00  13.00   7.50  24.00   9.50   2.66   4.00   5.50
692  5.50   5.00  21.00  23.79  28.50   9.54
701 73.33 48.67  34.00  16.50  15.00  10.00  24.00  17.50  16.00  15.00  30.00
702 12.50  15.25  66.00  70.50  25.20   7.40
711 64.67 41.33  17.50  27.00   6.66  12.50   7.00   1.75   4.00   2.00   0.00
712 10.00  12.12  19.83  32.50  49.00   8.85
731 63.33 34.67  31.00  21.00  14.00  14.00   5.00   0.00   8.00   0.00  10.00
732  0.00   6.00  11.57  23.39  14.85  20.00
741 94.67 62.67  16.00  12.50  14.00   9.50  11.50  10.00  16.50   7.66   9.00
742  8.50   8.18   6.62  11.37  21.66   6.22
751 82.67 57.33   2.50   0.00   0.00   0.00   0.00   4.66   5.00   0.00   0.00
752  2.00   9.50  17.00  19.39  20.79   7.00
761 74.67 50.00 999.99 999.99 999.99 999.99 999.99 999.99 999.99 999.99 999.99
762 999.99  11.12  23.75  20.20  16.00  16.00
771 63.33 43.33  25.00  27.00   8.33   9.66   7.33  27.00   4.50  13.50  14.50
772  9.00   7.30  15.16  18.66  34.00   8.55
781              22.00  17.00   5.33  10.00  22.00   9.00   8.00   6.50  11.00
782 12.00   6.33  15.50  22.79  30.25   8.10
791 95.33 62.67   8.00  22.00   7.00   0.00   5.50  13.00  14.00   3.66  19.00
792 13.00   3.00   8.80  11.00  23.50   4.50
801 60.00 40.00   9.50  25.00   6.50   0.00   0.00   0.00  29.00   5.33  21.00
802  9.66  10.33  31.75  65.00  15.28  15.00
811 60.00 32.67  12.00  17.50  17.00  38.00  20.00  20.50   5.00  15.50   7.00
812 12.00  25.33  27.60  51.00  24.16  16.00
821 59.33 40.67  14.00  12.00   0.00   0.00   0.00   0.00   0.00   0.00   0.00
822 11.00  11.50  12.11   9.88  14.42   6.88
831 90.67 53.33  12.00   5.00   4.66   2.50  12.00  22.00   6.00  15.00   8.33
832  8.50  12.66  24.39  29.25  42.00   9.12
841 73.33 38.67  13.50  14.50   8.66  13.50  28.00  14.50  13.50  30.00   7.25
842 14.50   7.36  14.14   4.19  10.22   7.44
851 68.67 34.00   0.00   1.00   2.00   0.00   1.00   3.00   0.00   0.00   0.00
852  1.00   3.62   2.38   2.00   1.75   1.66
861 77.67 53.67   8.25  13.00   9.50   4.00   2.50   9.33   6.00   3.66  26.00
862  6.66   6.91  26.00  42.33  68.00  12.42
871 67.00 45.00   3.50   5.00   2.00   0.00   0.00   0.00   0.00   0.00   0.00
872  0.00   9.85  23.60  42.66  83.50   6.66
881 70.67 33.33   9.00  26.00  27.00  15.00  28.00  28.00  10.50   2.00  28.00
882 17.00  28.75 155.00 139.00 146.00  15.16
901 70.67 37.33   5.50   4.50  24.50  36.00  25.00   4.66   3.00   2.00   3.00
902  7.00  11.33  23.39  28.75  19.42   4.38
921 78.00 42.67   9.00  10.00   8.00   1.00   4.50   7.00   3.50   3.00   4.75
922  7.33   5.33   6.54  18.83  29.60   4.59
931 76.67 52.00  14.00   7.00   5.00   2.00   4.50   9.00   5.00   0.00   4.66
932  2.25   8.12  10.62   5.90  11.00  11.83
941 84.67 42.00 999.99 999.99 999.99 999.99 999.99 999.99 999.99 999.99 999.99
942 999.99   7.11  16.00  12.16  15.50   9.55
961 70.33 34.33  10.50  22.00  12.50   9.00   8.00  18.00   7.50   5.00  14.00
962  6.33  21.25   7.11   6.50   8.50   7.16
991 89.33 34.33   4.00   1.00   0.00   0.00   0.00   0.00   0.00   0.00   0.00
992  0.00   1.79   1.75   1.62   4.19   7.00
```

Data Set 14 INFANTBP.DAT (*Continued*)

```
-----------------------------------------------------------------------------------------
1001 68.00 33.00  11.00   10.50    2.33    5.00    3.00    5.75    3.00    3.66   10.00
1002  25.00    8.69  18.83   46.33   30.50    5.00
1011 62.00 37.33   4.00    3.50    8.50    9.50    8.33    5.40    3.33   11.00    6.00
1012   5.33    5.18  13.62   32.75   40.66    8.77
1031 70.67 38.67   3.33    3.66    6.50    4.00    6.50    5.00    8.00    1.33    1.00
1032   4.00    3.77  14.28   13.62   18.33    5.41
1061 65.33 43.33  22.00   15.50   15.00   16.00   11.00   18.00   21.00   22.00   15.50
1062  13.50   11.50  28.25   49.00   27.25   13.19
1071 71.33 42.00  32.00   13.50    0.00    0.00    1.00    0.00    0.00    0.00    6.00
1072   1.00    6.63  11.50   26.79    5.19  999.99
1081 89.67 39.67  14.00   19.00    8.50    3.50   14.00   10.50    6.33    3.50   13.00
1082  22.00    7.90  31.00   23.50   11.25    3.59
1101 78.33 35.33   7.00    1.00    1.00    1.00   31.00    6.00    3.00    0.00    3.00
1102   1.00    6.50  15.33   14.80   14.22    8.85
1111 83.33 40.00   7.33    4.66    4.33   10.00    3.40    6.00    3.33    4.00   10.50
1112   9.00   11.30  21.83   22.33   12.50   13.39
1121 74.00 41.00  17.50   13.00   37.00    8.33   13.00   27.00   10.00    8.33   30.00
1122  35.00    3.25  18.00   13.33   39.00    5.71
1131 68.00 39.00  13.00    8.66    4.40   11.50   13.00    9.00    0.00    0.00    0.00
1132   0.00    7.77  10.11   20.20    8.62    7.62
1141 73.33 40.67  12.00   27.00   17.00   13.00   12.50    8.50    4.50    4.50    9.33
1142   7.33    5.81   9.00   10.57    9.00    5.14
1151 77.33 48.67  22.00   26.00   10.00    0.00    2.66    0.00    0.00    1.00    2.00
1152   2.50    4.00  18.00   22.20   33.00    5.57
1161 70.00 40.00  27.00   24.00   13.50   11.00   28.00   10.00    5.25   14.00   12.50
1162   5.33    7.85   9.00   18.79    7.77   11.71
1171 69.00 41.00  27.00   37.00   15.00    6.33   11.50    9.00    4.50    2.00    7.66
1172   8.33    8.00  11.42   18.66   20.66   17.00
1181 67.33 51.00   6.00    3.75    7.00    1.50    9.00    0.00    4.00    0.00    3.00
1182   3.33    8.75   7.85    8.00    8.60    6.00
1191 76.00 40.00   2.75    5.00    2.00    3.66    3.66    5.66    6.50    2.50   10.00
1192   3.00    9.00  12.57   37.66   23.25    3.63
1201 59.67 37.33   4.00    4.00    8.00    3.00    5.50    5.66    1.50    1.00    2.33
1202  11.00    3.83  47.00    8.33   21.00    6.85
1221 66.67 48.67  24.00   16.00   14.00    4.50    0.00    5.50    6.00    0.00    2.00
1222   0.00  999.99 999.99  999.99  999.99  999.99
1231 59.33 32.00  21.00    8.50    6.00   12.50    6.00    5.00    7.00    7.50    6.00
1232   3.00   10.25  24.25   40.00   27.79   10.75
1261 64.00 40.00   8.33   22.00    8.50    8.00    7.33   10.00    8.00    4.00   12.50
1262  10.50   15.28  84.50   46.00   49.66   15.66
1271 64.00 31.00  21.00   12.00    6.33    4.00   21.00   14.00    6.66    5.33    6.00
1272   0.00    2.18   6.19   10.62   22.00    5.00
1281 52.00 28.00   7.66    7.00    9.50    5.00   23.01    9.50    3.50    7.00    4.50
1282   4.33    8.00  20.79   20.00   19.00    8.89
1291 68.67 36.67   2.75    4.00    2.25    1.79    0.00    3.33    5.25    5.66    6.33
1292  10.50    7.40   4.23    4.23   23.79    9.66
1301 82.00 55.33   2.00    0.00    0.00    0.00    0.00    0.00    1.00    4.00    2.00
1302   0.00    5.25   1.75   75.00   46.33   12.33
1311 63.33 32.33   9.33   13.50    9.33   27.00   14.50   28.00    9.00    9.50    8.66
1312  30.00    6.66  69.50   69.00   68.00    8.50
1321 79.33 38.00  40.00   35.00   32.00   36.00   31.00   29.00   29.00   31.00   26.00
1322  33.00    8.91  30.25   30.79   27.66   13.66
```

Data Set 14 INFANTBP.DAT (*Continued*)

```
---------------------------------------------------------------------------------------
1331 65.33 38.67  12.50   7.66    7.00    4.00    5.25    7.66    5.33    5.33   10.50
1332  3.75  12.16  37.25  30.50  38.00    8.33
1341 61.33 42.00   4.50  11.50  22.00  10.50  23.00  14.50  10.00    0.00  17.50
1342  5.00  65.42  22.79   8.55   6.44    9.75
1351 64.33 40.67   2.00   4.00   3.66    0.00    1.00    4.33    0.00    0.00    4.33
1352  3.79   4.53   6.00  21.60  32.25    6.15
1361 61.33 41.33  10.00   9.00   9.00    7.00    0.00   20.00    6.66    0.00    7.00
1362  4.50   5.00   7.66  39.33  28.75    8.85
1381 73.33 48.67   5.00   3.25   6.50   15.00    6.33    6.00    5.00    9.00   12.66
1382 17.00  12.39  13.42  13.42  58.00    5.50
1391 56.67 38.00  21.00  24.00   9.00   32.00   24.00   27.00   26.00    2.33   37.00
1392 25.00  11.37  32.00  34.50  20.25   15.00
1421 65.00 33.00  17.00  28.00   9.00   17.00   10.00   25.00    8.00    4.00   12.00
1422 10.50   5.85  16.00  11.11  23.00    5.76
1451 81.67 55.33   9.50   6.75  12.00   10.50    7.00    6.33    4.00    4.00    7.00
1452  1.00   4.00   8.00  15.33  18.00    5.00
1461 70.67 47.67   5.00   5.50   5.00    7.00    7.33    4.59   11.00    6.00    1.00
1462  1.00  13.87  32.50  22.20  12.62   11.44
1481 79.33 50.67   9.66  30.00  17.50   25.00   10.66   10.66   24.00   12.66   25.00
1482  9.66  10.33   8.72  11.50  28.20    9.22
1491 63.33 46.00  15.00  15.00   0.00    0.00    0.00   19.00    9.00   34.00   32.00
1492 26.00  17.79 130.00  17.00   7.75    1.33
1501 62.67 43.33  11.00   1.50   3.00    2.00   10.00    4.00    7.00    9.00   12.50
1502  8.00   2.77   6.00   9.10  13.14    4.25
1511 62.67 42.67  17.00  10.50   4.00    4.50   15.00    5.66    2.50    3.25    9.00
1512  3.66   4.83  17.60  12.14  12.87    6.00
1521 65.33 43.33  26.00  13.00  25.00   21.00   21.00   22.00    7.00   11.00   10.50
1522  1.00   3.55  11.85   9.14   7.00    9.00
1541 69.67 47.33   0.00  16.50   5.33    6.00    6.00    5.00    0.00    0.00    1.00
1542  8.50   5.75  11.66  18.33  10.37    6.50
---------------------------------------------------------------------------------------
```

Data Set 15 NEPHRO.DOC

Variable	Column	Code
Study name	1–8	
Study number	10–11	
Endpoint	13	1 = efficacy/2 = nephrotoxicity/3 = ototoxicity
Antibiotic	15	1 = Amikacin/2 = Gentamicin/3 = Netilmicin/ 4 = Sisomycin/5 = Tobramycin
Sample size	17–19	
Number with side effect	21–23	

Data Set 16 NEPHRO.DAT

BODEY	03	2	2	11	0	LOVELESS	23	2	2	16	2
BODEY	03	2	4	11	0	LOVELESS	23	2	3	13	1
BODEY	03	2	5	11	2	MAKI	24	2	2	53	16
KLAST	04	2	2	21	4	MAKI	24	2	4	54	12
KLAST	04	2	4	25	4	SCHNIDER	25	2	2	33	1
WALKER	06	2	2	40	7	SCHNIDER	25	2	3	42	0
WALKER	06	2	5	40	2	BARZA	26	2	1	32	4
FELD	07	2	1	73	11	BARZA	26	2	3	37	3
FELD	07	2	5	77	16	BOCK	27	2	1	29	8
GILBERT	08	2	1	15	2	BOCK	27	2	3	34	13
GILBERT	08	2	2	15	2	LEAL	28	2	2	41	15
LAU	09	2	1	149	12	LEAL	28	2	4	39	9
LAU	09	2	2	130	11	SMITH80	29	2	2	72	19
SMITH77	10	2	1	62	9	SMITH80	29	2	5	74	9
SMITH77	10	2	2	62	11	FONG	31	2	2	102	18
EDEN	11	2	2	20	2	FONG	31	2	5	103	15
EDEN	11	2	3	20	2	HERTIG	32	2	2	362	14
HAHN	12	2	1	33	0	HERTIG	32	2	3	362	8
HAHN	12	2	2	34	0	LORBER	34	2	2	41	4
HAHN	12	2	3	38	0	LORBER	34	2	4	38	3
LERNER78	14	2	1	49	0	WEISS	36	2	2	40	1
LERNER78	14	2	2	52	6	WEISS	36	2	3	40	1
WADE	17	2	2	43	13	BROWN	37	2	2	103	5
WADE	17	2	5	47	11	BROWN	37	2	5	96	2
COX	19	2	2	29	2	FEIG	38	2	2	25	10
COX	19	2	3	29	2	FEIG	38	2	5	29	8
GREENE	20	2	1	16	1	GATELL83	40	2	1	54	7
GREENE	20	2	2	11	2	GATELL83	40	2	5	59	4
GREENE	20	2	3	10	1	HOLM	41	2	1	49	3
GREENE	20	2	5	15	2	HOLM	41	2	2	46	9
KEATING	21	2	1	100	8	LERNER83	42	2	3	116	1
KEATING	21	2	2	105	16	LERNER83	42	2	5	114	5
KEATING	21	2	4	85	19	MATZKE	43	2	2	99	9
LOVE	22	2	1	53	4	MATZKE	43	2	5	97	17
LOVE	22	2	2	58	5	GATELL84	45	2	3	72	12
LOVE	22	2	3	50	2	GATELL84	45	2	5	86	20

Data Set 17 NIFED.DOC

Column	Variable	Code
1–2	ID	
4	Treatment group	N = nifedipine/P = placebo
6–8	Baseline heart rate*	beats/min
10–12	Level 1 heart rate†	beats/min
14–16	Level 2 heart rate	beats/min
18–20	Level 3 heart rate	beats/min
22–24	Baseline systolic bp*	mm Hg
26–28	Level 1 systolic bp	mm Hg
30–32	Level 2 systolic bp	mm Hg
34–36	Level 3 systolic bp	mm Hg

* Immediately prior to randomization.

† Highest heart rate and systolic blood pressure at baseline and each level of therapy, respectively.

Blank values indicate that either

(1) the patient withdrew from the study prior to entering this level of therapy;

(2) the patient achieved pain relief prior to reaching this level of therapy; or

(3) the patient encountered this level of therapy, but this particular piece of data was missing.

Data Set 18 NIFED.DAT

ID	Group	Baseline HR	L1 HR	L2 HR	L3 HR	Baseline SBP	L1 SBP	L2 SBP	L3 SBP
1	P	60	70	64		128	110	120	
2	N	52	64	98		180	156	160	140
3	P	100	94			190	140		
4	N	84	88	96	112	136	126	122	110
5	P	56	70	61	64	230	150	130	150
6	P	105	120			142	150		
7	N	116	116			210	230		
8	N	68	68	72	84	170	150	150	156
9	P	85	88	90	92	150	134	140	154
10	N	64	60			140	120		
11	N	76	90			160	164		
12	N	88	125	140		150	140	140	
13	P	88	78	80	72	130	120	108	118
14	P	96	114		88	152	144		158
15	P	54	60	52	58	100	100	92	110
16	P	60	62	68	60	170	180	206	188
17	N	56	58	56	60	110	112	102	110
18	N	56	60			120	120		
19	N	54	60	78	76	125	120	118	118
20	N	60	60			230	170		
21	P	60	54	60	64	100	120	130	116
22	N	92	100	100	100	124	134	146	180
23	P	72	84	84		168	178	140	
24	N	100	96			110	116		
25	P	100	90	113		150	130	128	
26	N	52	74	88	66	164	144	128	140
27	N	76	76			170	170		
28	P	75	75	75	88	152	152	150	150
29	P	58	58	58	58				
30	N	56	54		59	106	124		120
31	P	70	60			160	180		
32	N	51	66			150	136		
33	P	90	98			180	180		
34	N	90	86			160	140		

Data Set 19 OTO.DOC

Variable	Column	Code
Study name	1–8	
Study number	10–11	
Endpoint	13	1 = efficacy/2 = nephrotoxicity/3 = ototoxicity
Antibiotic	15	1 = Amikacin/2 = Gentamicin/3 = Netilmicin/ 4 = Sisomycin/5 = Tobramycin
Sample size	17–19	
Number with side effect	21–23	

Data Set 20 OTO.DAT

```
KLAST     04 3 2   20    0      LOVE      22 3 3   29    1
KLAST     04 3 4   20    1      MAKI      24 3 2   53    4
WALKER    06 3 2   40    0      MAKI      24 3 4   54    2
WALKER    06 3 5   40    0      SCHNIDER  25 3 2   33    1
LAU       09 3 1  105   21      SCHNIDER  25 3 3   42    0
LAU       09 3 2   96   13      BOCK      27 3 1   23    6
SMITH77   10 3 1   34    2      BOCK      27 3 3   29    2
SMITH77   10 3 2   30    3      LEAL      28 3 2   41    1
EDEN      11 3 2   20    1      LEAL      28 3 4   39    0
EDEN      11 3 3   20    0      SMITH80   29 3 2   47    5
HAHN      12 3 1   10    0      SMITH80   29 3 5   44    5
HAHN      12 3 2   16    1      FONG      31 3 2   11    0
HAHN      12 3 3   16    1      FONG      31 3 5   15    1
HOYME     13 3 3   30    1      HERTIG    32 3 2  166    1
HOYME     13 3 5   31    0      HERTIG    32 3 3  168    1
LERNER78  14 3 1   49    4      LORBER    34 3 2    5    2
LERNER78  14 3 2   52    4      LORBER    34 3 4   11    0
COX       19 3 2   29    0      GATELL83  40 3 1   17    6
COX       19 3 3   29    0      GATELL83  40 3 5   19    8
GREENE    20 3 1    2    0      HOLM      41 3 1   38    3
GREENE    20 3 2    4    0      HOLM      41 3 2   31    5
GREENE    20 3 3    6    1      LERNER83  42 3 3   73    2
GREENE    20 3 5    5    2      LERNER83  42 3 5   84   10
LOVE      22 3 1   29    1      GATELL84  45 3 3   27    2
LOVE      22 3 2   32    2      GATELL84  45 3 5   28    5
```

Data Set 21 SEXRAT.DOC

Variable	Column
Number of children*	1
Sex of children†	3–7
Number of families	9–12

----

* For families with 5+ children, the sex of the first five children are listed. The number of children is given as 5 for such families.
† The sex of successive births is given. Thus, MMMF means that the first three children were males and the fourth child was a female. There were 484 such families.

Data Set 22 SEXRAT.DAT

2	MM	4400	5	MMMFM	478
2	MF	4270	5	MMMFF	495
2	FM	4633	5	MMFMM	524
2	FF	4218	5	MMFMF	469
3	MMM	1651	5	MMFFM	476
3	MMF	1618	5	MMFFF	509
3	MFM	1507	5	MFMMM	472
3	MFF	1403	5	MFMMF	441
3	FMM	1551	5	MFMFM	456
3	FMF	1441	5	MFMFF	453
3	FFM	1610	5	MFFMM	496
3	FFF	1498	5	MFFMF	473
4	MMMM	469	5	MFFFM	495
4	MMMF	484	5	MFFFF	440
4	MMFM	466	5	FMMMM	486
4	MMFF	448	5	FMMMF	447
4	MFMM	398	5	FMMFM	479
4	MFMF	448	5	FMMFF	452
4	MFFM	420	5	FMFMM	492
4	MFFF	425	5	FMFMF	430
4	FMMM	419	5	FMFFM	469
4	FMMF	406	5	FMFFF	429
4	FMFM	406	5	FFMMM	502
4	FMFF	342	5	FFMMF	440
4	FFMM	424	5	FFMFM	445
4	FFMF	428	5	FFMFF	427
4	FFFM	442	5	FFFMM	455
4	FFFF	481	5	FFFMF	461
5	MMMMM	549	5	FFFFM	488
5	MMMMF	516	5	FFFFF	518

Data Set 23 SMOKE.DOC

Variable	Column	Code
ID number	1–3	
Age	4–5	
Gender	6	1 = male/2 = female
Cigarettes/day	7–8	
Carbon monoxide (CO) ($\times 10$)	9–11	
Minutes elapsed since last cigarette	12–15	
Log CO adj* ($\times 1000$)	16–19	
Days abstinent	20–22	Those abstinent less than 1 day were given a value of zero.

* This variable represents adjusted carbon monoxide (CO) values. CO values were adjusted for minutes elapsed since last cigarette smoked using the formula Log_{10} CO (Adjusted) = Log_{10} CO − $(-0.000638) \times (\text{Min} - 80)$, where Min is the number of minutes elapsed since the last cigarette smoked.

Data Set 24 SMOKE.DAT

```
161130270  601419  6        5647240445  751645 64      11537240990  101951  7
354130220 1601393  3        5754135420  551607  3      11822114140  601133  2
467130270  501412365        5939135420  901630365      11959240295  801470  0
556125250  701392 16        6149229245 1501434  0      12047230520  301684  3
854130255  701400 45        6263120300 1201503  2      12152140            30
1452120160 1051220 47       6364120180  551239 74      12237215 7014401713  2
1560111160  951214 36       64331 6 5014401567365      12343220300  971488  2
1751120520 1201742  2       6745250510   51660  1      124282 8280 1101466  2
1963130215  901339365       6866160220  601330 42      12547220290 1031477  2
2236225180  871260365       7158140430  851637365      12630228340  501512365
23721 6 90 7101356365       74252 9110  301009 14      12742150275  601427  7
2468240350 1501589  4       7631230430  551618  3      12830122340  501512 16
2550225380  801580  2       7754210380  301548 12      13228115390  651581 14
2942220145 1431202 26       7835235 4511821356  9      13355120140 6121486  0
3035211150  901182129       7970230310  801491  7      13434113250 1571447  7
3257240270  301399  0       8037223240  851383  3      13525230290  751459157
3329228145  901168 21       8257220115 1121081  2      13644134330  451496  6
3464110190  851282  1       8348150415  451596 11      138432 9 6011151438365
3542220320  601492365       8432118195  901296  7      13928217285 1001468106
3647250315  951508235       8959130420  751620  1      140362 5 70 8401330 35
37442 2 50 285 830  5       9026230450  651644 14      14124120270  601419 14
3850220150  601163 47       9432140280  901454 45      14322244710  751848  4
3927120310         1        95512 7    120      0      14434229170  801230  6
4072240200 1001314  3       9930220270  851435  2      14538230220  721337  2
4132125580  601751  1      10142140415  651608  2      14637130450   51605  2
4433120315  151457 32      10256120260 1651469 47      14746240235  701365  1
4537140540 1801796  5      10476115140  901153  9      14943125310  401466  0
4651230450  131610365      10525215440  971654  1      15057235520  601703  3
5127220130 2251206365      10740250390   81545365      15135120270  621420 90
5254222 65 260 928 15      108402 4 4010801240185      15356220140 1081164 82
5343160450         3       10930220150  151135365      15440240250  601385  0
5450228325  951521 10      11057240195  851293  0      15740230150 1201202 30
5545140300  601464365      11229222375  601561365      15855225285  851458  1
```

Data Set 24 SMOKE.DAT (*Continued*)

```
--------------------------------------------------------------------------------
15927220180 1001268 63      22936140500  101654124      29235230390  661582 17
160391 3 40 205 682365      23128235405  201569 41      29332120275 2401541 20
16126215 4011641294 25      23249218180   11205  8      29462150440  451621 62
16256120140 1251175  3      23327129 80 9451455365      29548140210  401297  3
16430245480  601668  4      23563126290  01411365       29635230350  751541  3
16628120345  651528 21      23624260 4513801483 14      29954229350  101499  5
16824114260  701409  3      23833128480  551665 13      30048123260  901421  1
16953240130 3721300160      23939118190  951288  2      30228240 6010351387311
17043120240  551364  6      24126240485  801686  1      30441130290  451440365
172532 4 50 370 884 90      242282 9150  901182 60      30527120140 1051162 72
174412 7 80 6741282 97      24340220          365        30621215120  601066 71
17559120300  651468 29      244282 5210  261288237      30729135435  101594  4
17644240380  751577 20      24546123 6510201413302      30848120100  70 994  1
17751128500  701693  3      24626233280  801447  1      30933211270  401406365
17847123270  601419  5      24746123255 1201432 11      31139130           45
17929220330  501499  6      24940111265          21      31226210 90 7401375335
18044220300  601464  1      25041218320  101460  5      31337120260  51367365
18537125            30      25146140370   11518 13      31424225100  60 987 16
186491 9170  821232 60      25232133340  101487  3      315322 7 4013201393153
189272 6 70 140 883 14      25335120200  601288  8      31635220 6011101435 54
19037222340  901538 10      25458220185  301235  0      31743230220 1101362 15
19345220 45 9601215 14      25656230 90  35 926 14      31846210 90 6301305244
19735225270 1201457  5      25935225260 1201440 33      32030240320  601492  0
19839215170 1751291  3      26163215100 1101019365      32137123 8514401797 96
19937225330 1051534  0      26231114 95 3301137 35      32226120175  321212 25
20037230     83     33      26342230280   51399  0      32366120150 1951249 12
20156220220  601330  2      26453160240 2101463 14      32539115190 7801725  2
20259215310 1801555  7      26528218475  201638  4      32650210260  451393 21
20325220300 1751538  2      26635160350   11494121      32832125450  701647 20
20528230290  751459365      26753120435  951648 15      32942160160  151163  2
20642230265 1101442365      27248230360 1001569  2      33029116430  601621 15
20736113250  331368 14      27336134390  751588  2      33270113290  151421  1
20833225380   21530 55      27444110 85 2401031  1      33364140185  601254 67
21023218180  661246  3      27550240315 101454252       33437120400  451580  7
21161120200  911308365      27636130410  601600 14      33529214360 1201582  4
21343230260 1051431 21      277271 8180  901262  3      33728225240  751377  3
214531 2 45 9801227  4      27864120125  101052  4      33941117210  701316 11
21546120200 101256365       27927215155  801190 16      34030123270  601419 13
218442 7 50 240 801  4      28344260390   21541  4      34155123265 1301455150
21935215190 1051295 12      28450140 70  90 851365      34349225350  501525 13
22154211           365      28523140240  101336  7      34433110190 1101298  6
22252125360  921564  7      28745225160  901210266      34537120 85 7001325196
22439130320  101460  3      28863120270  301399130      34645225205 1201337  0
22553190190  51231365       29043263100  60 987365      34754120290  601450  7
22630225150 1351211 21      29151130170  451208365      34834222265 1501468 15
--------------------------------------------------------------------------------
```

Data Set 25 VALID.DOC

Column	Variable	Format or code
1–6	ID number	
8–15	Saturated fat-DR	XXXXX.XX
17–24	Saturated fat-FFQ	XXXXX.XX
26–33	Total fat-DR	XXXXX.XX
35–42	Total fat-FFQ	XXXXX.XX
44–51	Alcohol consumption-DR	XXXXX.XX
53–60	Alcohol consumption-FFQ	XXXXX.XX
62–70	Total calories-DR	XXXXXX.XX
72–80	Total calories-FFQ	XXXXXX.XX

Data Set 26 VALID.DAT

100396	33.20	21.20	81.15	53.80	8.26	1.68	1807.00	1242.20
100566	17.73	10.60	53.28	36.60	0.83	0.00	1418.00	907.00
107633	38.73	23.80	83.48	47.20	20.13	15.10	1889.00	786.00
107737	21.57	22.70	49.65	55.30	11.16	7.49	1426.00	1392.50
107744	21.35	30.40	55.18	71.00	7.18	12.84	1253.00	1259.80
107813	28.04	15.10	73.83	41.10	1.76	0.00	1699.00	987.10
107825	23.17	17.80	68.29	49.10	22.66	25.06	1700.00	1189.90
107879	19.73	19.10	58.30	48.90	0.00	0.00	1369.00	1364.10
108618	36.31	23.40	92.58	55.40	0.00	0.00	2163.00	1311.40
109000	20.87	16.00	68.44	44.20	0.00	0.00	1609.00	1200.60
109259	26.40	19.40	69.16	46.60	4.75	1.06	1704.00	1227.40
109565	28.26	27.40	79.42	59.70	4.57	1.81	1884.00	1448.10
109856	19.81	15.80	69.59	50.50	15.12	11.14	1518.00	1367.70
109885	23.48	19.90	65.86	55.60	5.37	2.57	1450.00	1154.40
109908	25.62	17.40	74.22	51.70	0.00	0.00	1702.00	1217.70
110108	23.59	9.50	63.44	23.40	1.89	0.76	1400.00	483.40
110360	14.60	20.90	38.14	46.90	0.78	5.70	910.00	1056.50
110406	20.71	27.70	49.77	70.10	8.09	10.56	1332.00	1265.30
110419	28.63	15.30	80.73	35.30	37.90	35.34	1947.00	890.10
110483	18.12	20.70	53.56	55.70	2.22	0.76	1428.00	1378.50
110579	16.93	17.60	52.75	51.50	20.69	18.22	1557.00	1317.50
110714	25.10	18.90	80.15	57.50	0.00	0.00	1788.00	1257.50
111040	14.04	14.90	39.40	37.00	2.62	1.68	1098.00	1087.80
111220	21.98	32.60	54.91	73.20	9.49	6.49	1362.00	1589.60
111536	24.83	24.00	67.73	63.10	9.69	16.72	1562.00	1422.70
111538	20.72	15.60	55.24	47.60	0.95	1.51	1286.00	1569.60
111799	24.61	24.20	68.11	54.40	3.00	2.74	1529.00	1386.00
111970	32.03	38.20	71.02	98.80	37.28	45.08	1810.00	2220.00
112071	23.49	18.80	63.72	55.80	21.78	33.49	1606.00	1291.40
112087	25.69	23.50	67.99	68.00	15.41	12.91	1363.00	1102.00
112163	23.65	21.30	64.61	46.80	7.85	0.76	1457.00	943.20
112226	23.45	13.90	60.13	31.80	0.39	0.00	1233.00	731.40
112650	46.36	33.30	104.91	88.10	29.81	64.75	2285.00	1878.90
112738	16.04	17.60	44.23	40.60	0.00	0.00	1183.00	1326.60
112812	38.68	17.80	98.14	55.80	6.77	2.57	2176.00	1635.40
112862	24.10	15.90	64.21	41.50	5.97	2.74	1587.00	1035.50
112896	27.92	16.70	91.37	53.50	0.85	0.00	2029.00	1548.30

Data Set 26 VALID.DAT (*Continued*)

112982	26.26	24.10	70.44	61.90	5.23	5.70	1484.00	1376.30
113429	26.46	21.00	76.06	57.70	2.26	2.11	1685.00	1460.50
113436	22.02	19.50	61.13	63.80	7.79	1.06	1440.00	1970.20
113441	24.51	25.90	62.60	60.70	7.75	11.14	1293.00	1350.30
113525	20.06	16.10	55.50	49.90	10.86	10.75	1433.00	1326.30
113613	34.14	38.50	93.04	93.50	22.44	11.38	2333.00	2253.50
113882	24.38	18.20	67.02	48.70	4.49	12.06	1402.00	1139.90
114053	14.38	5.60	41.09	14.80	5.20	4.55	923.00	463.20
114129	17.74	15.20	54.07	48.40	0.38	0.76	1330.00	1498.40
114353	27.83	16.50	78.26	36.70	5.17	3.79	1731.00	946.70
114865	29.17	24.80	84.52	71.10	0.00	0.00	1636.00	1396.70
115013	15.94	12.80	45.89	38.00	22.89	8.64	1311.00	1078.40
115223	28.33	27.20	75.96	88.40	2.93	0.76	1769.00	1996.80
115399	15.18	11.60	49.44	30.70	0.05	0.00	1481.00	1051.90
115512	22.51	31.20	68.28	78.70	4.75	5.70	1649.00	2100.30
115601	46.01	17.40	119.83	49.50	10.12	5.57	2334.00	988.70
115724	27.15	16.90	80.54	37.70	17.90	8.93	1823.00	1025.70
115764	25.47	15.80	75.54	41.60	0.00	1.81	1548.00	1259.60
115806	27.99	17.50	69.13	40.70	3.35	0.76	1456.00	825.10
115822	25.28	17.00	68.36	48.60	10.04	2.87	1535.00	1047.00
115875	26.58	14.80	77.83	40.10	17.27	15.86	1718.00	726.90
115879	28.72	35.80	74.26	76.00	0.00	0.00	1641.00	1961.60
116007	24.62	34.60	66.12	66.00	8.15	13.76	1402.00	1434.40
116164	26.46	18.00	78.20	48.80	27.55	34.42	2143.00	1464.40
116263	29.57	14.40	69.80	33.50	0.60	0.00	1490.00	768.70
116413	20.39	23.70	48.85	56.50	22.31	12.06	1221.00	1698.50
116461	14.68	13.60	52.24	40.30	4.40	2.11	1255.00	758.30
116528	23.43	25.60	71.16	62.40	1.99	0.76	1555.00	1462.00
129518	20.74	23.70	61.82	68.80	0.00	0.00	1374.00	1592.40
129729	21.76	19.30	63.73	56.30	0.38	0.00	1593.00	1661.50
129732	21.58	13.80	64.25	40.90	0.80	1.81	1491.00	1249.30
129743	39.30	32.70	109.83	78.50	0.14	0.76	2505.00	2282.10
129765	21.95	21.00	59.74	52.90	12.03	10.75	1560.00	1152.10
129813	21.90	11.90	67.87	38.30	10.35	3.63	1661.00	979.90
130757	28.65	13.60	78.99	41.00	5.84	0.76	1742.00	1253.00
130786	30.01	33.40	79.13	82.80	17.22	12.99	1680.00	1799.20
130793	29.35	23.90	98.96	75.60	5.97	4.64	2096.00	1873.20
130847	21.56	8.80	72.02	34.60	2.79	0.00	1438.00	811.80
130863	18.83	40.60	54.78	100.00	20.50	37.75	1325.00	1940.40
131023	36.16	18.50	100.56	52.60	8.65	4.55	2051.00	1219.80
131100	31.20	29.30	92.12	70.50	10.41	16.72	2050.00	1644.50
131126	29.06	23.50	80.52	67.90	7.12	6.76	1707.00	1933.50
131129	21.79	14.60	68.42	45.20	12.97	16.06	1502.00	1120.80
132293	29.60	30.30	70.98	78.90	8.75	4.64	1691.00	1470.50
132303	12.71	12.10	35.90	32.30	3.45	1.51	940.00	842.30
132316	22.16	14.80	70.33	45.60	3.11	0.76	1591.00	1156.40
132327	22.61	13.80	65.19	34.60	2.69	0.76	1769.00	1539.80
132328	27.08	18.80	70.95	43.70	49.15	27.75	2054.00	1330.90
132361	17.92	15.20	49.72	41.90	0.57	0.00	1514.00	1509.90
132382	31.41	30.40	90.21	74.10	31.26	11.14	1989.00	1506.50
132390	33.25	25.80	87.69	73.90	11.13	11.38	1968.00	1795.50
132426	32.19	54.90	86.35	133.50	4.67	2.87	1781.00	2263.50
132621	25.17	7.50	70.20	18.40	15.69	11.14	1761.00	513.60

Data Set 26 VALID.DAT (*Continued*)

132622	34.34	19.00	86.68	44.70	23.44	43.75	1639.00	1124.70
132624	25.11	24.20	69.86	63.10	4.60	16.72	1445.00	1340.30
132638	45.31	47.10	106.40	93.80	0.03	0.00	2260.00	2047.80
132662	18.59	12.50	45.37	29.50	19.38	29.92	1107.00	811.10
133002	28.34	35.50	71.45	80.00	0.77	0.00	1878.00	2136.00
133079	18.75	11.80	60.54	36.40	8.49	7.25	1612.00	1251.20
133094	24.37	39.00	67.73	109.30	0.79	0.00	1669.00	3077.30
133317	18.33	19.10	64.53	56.10	11.15	6.63	1533.00	1415.30
133372	17.26	20.00	42.53	41.40	13.11	12.84	1149.00	1055.80
133443	16.73	57.40	54.07	119.00	0.00	0.00	1555.00	2691.00
133497	24.78	11.80	73.18	28.00	0.00	0.00	1470.00	662.40
133507	16.74	5.70	54.84	16.50	4.17	2.57	1620.00	932.50
133533	15.30	17.10	51.96	41.40	6.68	3.63	1122.00	1100.20
133606	27.74	26.60	82.71	70.90	0.40	0.76	1994.00	1530.50
133615	26.40	14.80	80.64	40.80	1.58	0.76	1940.00	902.40
133637	26.76	20.40	70.67	47.00	3.44	0.00	1472.00	954.20
133676	17.73	19.40	52.62	55.10	3.70	3.79	1388.00	1504.70
133692	25.97	24.30	65.65	52.00	4.88	1.98	1683.00	1333.60
133833	18.57	14.00	53.18	37.40	7.31	4.64	1226.00	798.20
133925	26.48	17.30	72.45	37.90	25.74	28.06	1679.00	912.30
133987	26.98	23.20	64.35	60.90	32.64	28.85	1583.00	1210.50
134013	33.26	27.60	80.79	62.30	1.14	2.57	2229.00	1637.10
134021	19.03	11.20	47.56	29.00	0.90	0.00	1321.00	1254.60
134047	16.00	13.60	49.35	37.70	2.25	0.00	1162.00	1046.10
134103	27.59	17.00	72.49	34.80	16.61	10.80	1761.00	1009.60
134119	22.32	13.20	64.53	31.00	1.27	1.85	1376.00	730.60
134129	26.83	19.90	77.98	49.50	2.20	1.06	1689.00	1130.70
134144	36.44	36.80	94.53	89.30	16.51	11.86	2010.00	1870.10
134223	26.25	43.40	73.74	89.80	6.92	10.32	1815.00	2425.40
134350	30.92	20.20	82.98	51.80	17.76	15.86	1779.00	1187.80
134385	27.50	9.80	74.55	25.50	1.84	0.76	2038.00	782.30
134431	34.25	18.80	77.12	41.90	40.11	46.39	1973.00	1187.90
134486	23.79	28.70	62.49	67.80	32.72	56.39	1870.00	2166.60
134495	23.56	36.90	72.11	105.00	1.89	2.57	1936.00	2497.00
134519	20.46	7.50	58.42	21.10	27.86	43.32	1619.00	757.50
134550	24.16	21.80	69.11	60.40	13.62	19.74	1757.00	1297.60
134553	23.42	31.50	72.08	90.40	9.92	7.55	1582.00	1837.60
134566	26.40	14.50	72.99	36.10	5.04	4.64	1670.00	956.90
134598	24.09	14.90	61.17	35.20	12.95	9.30	1454.00	878.40
134611	25.25	20.50	56.16	42.60	9.43	12.08	1616.00	1144.20
134622	23.34	41.10	66.51	113.20	0.00	0.00	1371.00	2921.60
134835	16.61	21.60	49.65	56.80	27.26	27.00	1388.00	1227.60
134838	39.36	22.70	99.10	45.80	18.69	39.26	2425.00	891.80
134839	23.66	18.30	55.86	41.00	28.28	25.90	1280.00	842.60
134845	40.59	48.40	108.72	107.50	6.99	6.63	2376.00	2539.50
135002	34.04	37.40	82.11	91.10	0.00	0.00	1885.00	2389.50
135009	18.94	37.00	56.34	90.70	12.48	12.91	1302.00	1713.90
135026	21.41	23.40	52.20	60.20	4.21	4.42	1016.00	936.60
135126	18.00	24.10	46.44	61.20	12.18	11.86	963.00	1307.60
135130	21.04	24.80	58.92	67.00	0.00	0.00	1501.00	1466.60
135146	11.82	11.50	38.59	28.90	2.18	0.76	1026.00	991.70
135172	27.39	20.80	79.63	58.10	1.76	0.00	1820.00	1532.90
135182	34.61	34.30	98.94	104.60	2.17	2.87	2087.00	2279.80

Data Set 26 VALID.DAT (*Continued*)

135201	19.79	22.60	57.70	62.10	1.43	4.64	1299.00	1228.10
135258	21.34	16.70	57.30	33.90	14.23	6.76	1524.00	854.30
135262	18.79	25.70	57.33	76.60	1.39	2.87	1450.00	1975.80
135313	18.63	6.60	61.11	20.70	2.98	5.70	1340.00	601.10
135324	20.39	16.90	62.54	48.00	3.77	1.06	1634.00	1140.40
135330	18.64	14.90	53.56	44.20	8.10	11.14	1608.00	1520.60
135351	15.39	12.10	50.40	33.70	13.79	5.57	1649.00	922.90
135372	31.35	24.90	91.65	72.80	28.29	39.26	2013.00	1394.80
135391	23.62	23.30	62.62	59.30	15.60	10.75	1626.00	1706.90
135465	27.51	42.80	81.29	112.30	6.97	6.63	1732.00	2447.50
135472	41.01	18.30	114.85	41.70	0.00	0.00	2518.00	1323.70
135488	15.99	20.60	48.25	50.20	8.18	6.63	1561.00	1399.80
135530	17.48	18.80	50.43	46.50	0.51	0.00	1485.00	1739.30
135547	31.35	35.30	76.92	75.10	7.82	12.08	1767.00	1627.50
135559	17.64	17.80	45.39	38.50	6.60	11.14	1006.00	922.50
135588	39.43	26.40	98.34	77.90	28.79	52.07	2085.00	1602.70
135596	30.98	28.80	73.80	70.30	3.82	1.81	1588.00	1570.30
135619	32.86	41.20	84.85	92.50	0.00	0.00	1682.00	2074.90
135769	26.24	26.40	72.73	60.70	8.31	2.91	2121.00	2039.10
135829	20.07	25.90	65.41	87.60	8.22	13.76	1568.00	1958.80
135834	14.42	11.30	40.64	29.60	0.96	0.76	1216.00	1043.60
135848	29.59	29.90	71.41	67.60	0.63	0.00	1612.00	1327.60
135869	28.02	15.50	77.82	40.50	15.72	16.81	1827.00	1026.90
136104	25.85	23.70	70.89	75.00	23.97	12.43	1721.00	1584.60
136377	20.20	20.30	51.99	52.00	7.28	6.63	1467.00	1342.40
136378	36.19	18.10	95.43	47.10	4.60	7.25	1976.00	1204.30
136407	22.42	10.90	62.15	32.00	2.56	1.81	1731.00	981.60
136421	16.98	20.40	42.85	47.50	13.91	8.64	1033.00	1128.30
137461	23.98	18.50	67.43	45.90	5.39	7.19	1585.00	1247.50
184093	21.39	19.00	56.70	40.40	0.60	3.49	1320.00	994.00

ANSWERS
TO SELECTED
PROBLEMS

CHAPTER 2

2.1–2.2 The basic statistics are given in Table A.1.

2.3 Groups in increments of 100 g for heart weight and 10 kg for body weight will be used. This yields the

frequency distributions and statistics as given in Tables A.2–A.4.

2.4 The histograms of the preceding variables are shown in Figures A.1–A.4.

TABLE A.1 Basic statistics for left heart disease and normal males

| | Left heart disease males | | Normal males | |
	THW (g)	BW (kg)	THW (g)	BW (kg)
Sample size	11	11	10	10
Mean	450	55.61	317	56.23
Median	450	54.6	305	56.15
Variance	19,415	133.44	2217.78	133.13
Standard deviation	139.34	11.55	47.09	11.54
Range	475(285,760)	34.2(41.1,75.3)	160(245,405)	34.4(40.5,74.9)
Coefficient of variation	30.96%	20.77%	14.85%	20.52%

TABLE A.2 Frequency distribution for left heart disease males

THW			BW		
Interval	Midpoint	Frequency	Interval	Midpoint	Frequency
200–299	249.5	1	40–49.9	44.95	3
300–399	349.5	3	50–59.9	54.95	5
400–499	449.5	5	60–69.9	64.95	1
500–599	549.5	0	70–79.9	74.95	2
600–699	649.5	1			
700–799	749.5	1			

TABLE A.3 Frequency distribution for normal males

THW			BW		
Interval	Midpoint	Frequency	Interval	Midpoint	Frequency
200–299	249.5	3	40–49.9	44.95	3
300–399	349.5	6	50–59.9	54.95	3
400–499	449.5	1	60–69.9	64.95	3
			70–79.9	74.95	1

TABLE A.4 Grouped statistics for left heart disease and normal males

	Left heart disease males		Normal males	
	THW	BW	THW	BW
Grouped mean	449.5	56.77	329.5	56.95
Grouped variance	20,000	116.4	4000	106.7
Grouped standard deviation	141.4	10.8	63.2	10.3

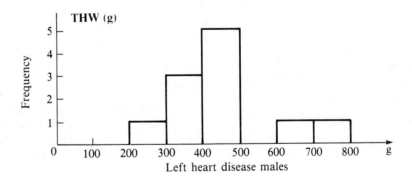

FIGURE A.1

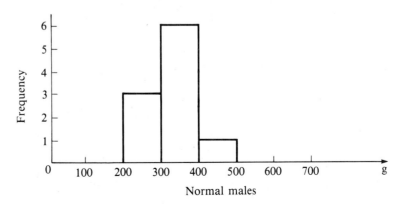

FIGURE A.2

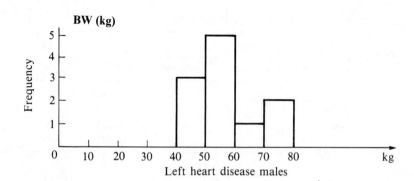

FIGURE A.3   Left heart disease males

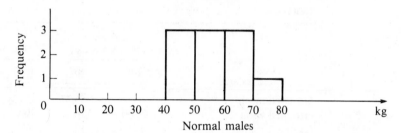

FIGURE A.4   Normal males

2.5 We have the following stem-and-leaf plots.

Total heart weight (g) ($=$STEM.LEAF $\times 10^2$)

Left heart disease					Normal					
2	85				2	45	70	90		
3	25	75	10		3	50	40	00	10	00	60
4	50	95	50	60	25	4	05			
5					5				
6	15				6				
7	60				7				

Body weight (kg)($=$STEM.LEAF $\times 10^1$)

Left heart disease						Normal			
4	46	11	17			4	08	75	05
5	46	03	81	15	97	5	33	12	90
6	13					6	74	22	55
7	35	53				7	49		

2.6 From Problems 2.1–2.5 it appears that body weight is comparable in the two groups, but heart weight is higher and more variable in the diseased group.

2.7 There appears to be a positive relationship between THW and BW in the normal group. No relationship seems to exist in the diseased group. (See Figure A.5.)

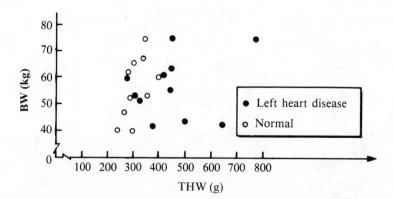

FIGURE A.5   THW (g)

2.8 The principal differences between the groups are that the left heart disease males have larger and more variable heart weights than the normal males. However, the body weights of the two groups are comparable. Finally, there appears to be a relationship between total heart weight and body weight among normal males but not among diseased males.

2.16–2.19 Changing the scale by a factor c will multiply each data value x_i by c, changing it to cx_i (Figure A.6). Again the same individual's value will be at the median and the same individual's value will be at the mode, but these values will be multiplied by c. The geometric mean will be multiplied by c also, as can easily be shown:

$$\text{Geometric mean} = [(cx_1)(cx_2)\cdots(cx_n)]^{1/n}$$
$$= (c^n x_1 \cdot x_2 \cdots x_n)^{1/n}$$
$$= c(x_1 \cdot x_2 \cdots x_n)^{1/n}$$
$$= c \times \text{old geometric mean}$$

The range will also be multiplied by c.

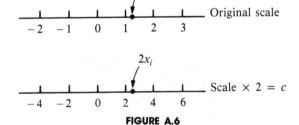

FIGURE A.6

3.1 Yes   **3.2** No   **3.3** .3

3.4 $A \cup C = \{\text{serum cholesterol} \leqslant 299\}$.

3.5 $A \cap C = \{250 \leqslant \text{serum cholesterol} \leqslant 280\}$.

3.6 $B \cup C = \{\text{serum cholesterol} \leqslant 280 \text{ or} \geqslant 300\}$.

3.7 $B \cap C$ is the empty set; that is, it can never occur.

3.8 Yes

3.9 $\bar{B} = \{\text{serum cholesterol} < 300\}$. $Pr(\bar{B}) = .9$.

3.25 .18   **3.26** .30

3.30 Probability $= .30 + (1 - .30) \times .25 = .475$.

3.31 10%

3.32 The mortality incidence per month $= 0.25/5$ years $= 0.25/60$ months $= 0.42\%$ per patient month.

3.33 No. For comparison, mortality data on a control group of patients with the same clinical condition as the patients who underwent open heart surgery but who did not have the operation are needed.

3.50 .0167   **3.51** 180

3.56 The probability that both siblings are affected is

$$\tfrac{1}{2} \times \tfrac{1}{2} = \tfrac{1}{4}$$

3.57 The probability that exactly one sibling is affected is

$$(2)(\tfrac{1}{2}) \times (\tfrac{1}{2}) = \tfrac{1}{2}$$

3.58 The probability that neither sibling will be affected is

$$\tfrac{1}{2} \times \tfrac{1}{2} = \tfrac{1}{4}$$

3.59 The probability that the younger child is affected should not be influenced by whether or not the older child is affected. Thus, the probability of the younger child being affected remains at $\tfrac{1}{2}$.

3.60 The events A, B are independent because whether or not a child is affected does not influence the outcome for other children in the family.

3.70 Bayes' theorem is used here. Dominant is denoted by DOM, autosomal recessive by AR, and sex linked by SL. Let A be the event that two male siblings are affected. We have

$$Pr(\text{DOM}|A) =$$
$$\frac{Pr(A|\text{DOM}) \times Pr(\text{DOM})}{Pr(A|\text{DOM})Pr(\text{DOM}) + Pr(A|\text{AR})Pr(\text{AR}) + Pr(A|\text{SL})Pr(\text{SL})}$$

We also know that

$$Pr(\text{DOM}) = Pr(\text{AR}) = Pr(\text{SL}) = \tfrac{1}{3}$$

from the conditions stated in the problem. Thus,

$$Pr(\text{DOM}|A) = \frac{Pr(A|\text{DOM})}{Pr(A|\text{DOM}) + Pr(A|\text{AR}) + Pr(A|\text{SL})}$$

Finally, we know from Problems 3.56, 3.61, and 3.67 that

$$Pr(A|\text{DOM}) = \tfrac{1}{4} \quad Pr(A|\text{AR}) = \tfrac{1}{16} \quad Pr(A|\text{SL}) = \tfrac{1}{4}$$

Thus,

$$Pr(\text{DOM}|A) = \frac{1/4}{1/4 + 1/16 + 1/4} = \frac{1/4}{9/16} = \frac{4}{9}$$

Similarly,

$$Pr(AR|A) = \frac{Pr(A|AR)}{Pr(A|DOM) + Pr(A|AR) + Pr(A|SL)}$$

$$= \frac{1/16}{9/16} = \frac{1}{9}$$

$$Pr(SL|A) = \frac{Pr(A|SL)}{Pr(A|DOM) + Pr(A|AR) + Pr(A|SL)}$$

$$= \frac{1/4}{9/16} = \frac{4}{9}$$

Thus, the dominant and sex-linked modes of inheritance are the most likely, with the autosomal recessive mode being less likely.

3.71 Let $B = \{$exactly one of two male siblings is affected$\}$. From Problems 3.57, 3.62, and 3.68,

$$Pr(B|DOM) = \tfrac{1}{2} \qquad Pr(B|AR) = \tfrac{3}{8} \qquad Pr(B|SL) = \tfrac{1}{2}$$

Thus, from Bayes' theorem,

$$Pr(DOM|B) = \frac{Pr(B|DOM)}{Pr(B|DOM) + Pr(B|AR) + Pr(B|SL)}$$

$$= \frac{1/2}{1/2 + 3/8 + 1/2} = \frac{1/2}{11/8} = \frac{4}{11}$$

$$Pr(AR|B) = \frac{Pr(B|AR)}{Pr(B|DOM) + Pr(B|AR) + Pr(B|SL)}$$

$$= \frac{3/8}{11/8} = \frac{3}{11}$$

$$Pr(SL|B) = \frac{Pr(B|SL)}{Pr(B|DOM) + Pr(B|AR) + Pr(B|SL)}$$

$$= \frac{1/2}{11/8} = \frac{4}{11}$$

Here the three genetic types are about equally likely.

3.72 Let $C = \{$both one male and one female sibling are affected$\}$. The sex of the siblings is only relevant for sex-linked disease. Thus, from Problems 3.56, 3.61, and 3.67,

$$Pr(C|DOM) = \tfrac{1}{4} \qquad Pr(C|AR) = \tfrac{1}{16} \qquad Pr(C|SL) = 0$$

Thus,

$$Pr(DOM|C) = \frac{Pr(C|DOM)}{Pr(C|DOM) + Pr(C|AR) + Pr(C|SL)}$$

$$= \frac{1/4}{1/4 + 1/16} = \frac{1/4}{5/16} = \frac{4}{5}$$

$$Pr(AR|C) = \frac{Pr(C|AR)}{Pr(C|DOM) + Pr(C|AR) + Pr(C|SL)}$$

$$= \frac{1/16}{5/16} = \frac{1}{5}$$

$$Pr(SL|C) = 0$$

3.73 Let $D = \{$male sibling affected, female sibling not affected$\}$.

$$Pr(D|DOM) = \tfrac{1}{2} \times \tfrac{1}{2} = \tfrac{1}{4} \qquad Pr(D|AR) = \tfrac{1}{4} \times \tfrac{3}{4} = \tfrac{3}{16}$$

$$Pr(D|SL) = \tfrac{1}{2} \times 1 = \tfrac{1}{2}$$

Notice that the event D is not the same as the event that exactly one sibling is affected, since we are specifying which of the two siblings is affected. We have

$$Pr(DOM|D) = \frac{Pr(D|DOM)}{Pr(D|DOM) + Pr(D|AR) + Pr(D|SL)}$$

$$= \frac{1/4}{1/4 + 3/16 + 1/2} = \frac{1/4}{15/16} = \frac{4}{15}$$

$$Pr(AR|D) = \frac{Pr(D|AR)}{Pr(D|DOM) + Pr(D|AR) + Pr(D|SL)}$$

$$= \frac{3/16}{1/4 + 3/16 + 1/2} = \frac{3/16}{15/16} = \frac{1}{5}$$

$$Pr(SL|D) = \frac{Pr(D|SL)}{Pr(D|DOM) + Pr(D|AR) + Pr(D|SL)}$$

$$= \frac{1/2}{15/16} = \frac{8}{15}$$

Thus, in this situation the sex-linked mode of inheritance is the most likely.

3.75

$$Pr(LOW) = Pr(LOW \cap <20 \text{ weeks})$$

$$+ Pr(LOW \cap 20\text{--}27 \text{ weeks})$$

$$+ Pr(LOW \cap 28\text{--}36 \text{ weeks})$$

$$+ Pr(LOW \cap >36 \text{ weeks})$$

$$= Pr(LOW| <20 \text{ weeks})Pr(<20 \text{ weeks})$$

$$+ Pr(LOW|20\text{--}27 \text{ weeks})Pr(20\text{--}27 \text{ weeks})$$

$$+ Pr(LOW|28\text{--}36 \text{ weeks})Pr(28\text{--}36 \text{ weeks})$$

$$+ Pr(LOW| >36 \text{ weeks})Pr(>36 \text{ weeks})$$

$$= (.540)(.0004) + (.813)(.0059)$$

$$+ (.379)(.0855) + (.035)(.9082)$$

$$= .069$$

3.76 The dependence of the events $\{\leqslant 27 \text{ weeks}\}$ and $\{\text{LOW}\}$ can be shown by establishing that

$$Pr\{\leqslant 27 \text{ weeks} \cap \text{LOW}\} \neq Pr\{\leqslant 27 \text{ weeks}\} \times Pr\{\text{LOW}\}$$

We have

$$Pr\{\leqslant 27 \text{ weeks} \cap \text{LOW}\} = Pr\{\text{LOW} \cap < 20 \text{ weeks}\}$$
$$+ Pr\{\text{LOW} \cap 20\text{-}27 \text{ weeks}\}$$
$$= .540(.0004) + .813(.0059)$$
$$= .0050 \text{ (from Problem 3.75)}$$

Similarly, from Problem 3.75

$$Pr(\text{LOW}) = .069$$
$$Pr(\leqslant 27 \text{ weeks}) = .0004 + .0059 = .0063$$

Thus,

$$Pr(\text{LOW}) \times Pr(\leqslant 27 \text{ weeks})$$
$$= .00043 \approx \tfrac{1}{10} Pr\{\leqslant 27 \text{ weeks} \cap \text{LOW}\} = .0050$$

Thus, the two events are dependent.

3.77 $Pr(\leqslant 36 \text{ weeks}|\text{LOW})$ must be computed. Bayes' theorem is used as follows:

$$Pr(\leqslant 36 \text{ weeks}|\text{LOW})$$
$$= \frac{Pr(\leqslant 36 \text{ weeks} \cap \text{LOW})}{Pr(\text{LOW})}$$
$$= \frac{Pr(\text{LOW} \cap < 20 \text{ weeks}) + Pr(\text{LOW} \cap 20\text{-}27 \text{ weeks}) + Pr(\text{LOW} \cap 28\text{-}36 \text{ weeks})}{Pr(\text{LOW})}$$
$$= \frac{(.540)(.0004) + (.813)(.0059) + (.379)(.0855)}{.069}$$
$$= \frac{.0374}{.069} = .542$$

CHAPTER 4

4.1 We have the distribution

x	0	1	2
Pr(X = x)	.80	.18	.02

4.2 0.22   **4.3** 0.212

4.4 We have the distribution

x	<0	≥0, <1	≥1, <2	≥2
F(x)	.00	.80	.98	1.00

4.13 $50 \times 49 \times 48 \times 47 \times 46 = 2.543 \times 10^8$.

4.14 $_{50}C_5 = 2.119 \times 10^6$.

4.19 X = number of hypertensives over a lifetime is binomially distributed with parameters $n = 20$, $p = .2$; that is,

$$Pr(X = x) = {}_{20}C_x(.2)^x(.8)^{20-x} \qquad x = 0, 1, \dots, 20$$

or, from Table 1 in Appendix 1,

TABLE A.5

x	Pr(X = x)	x	Pr(X = x)
0	.0115	7	.0546
1	.0576	8	.0222
2	.1369	9	.0074
3	.2054	10	.0020
4	.2182	11	.0005
5	.1746	12	.0001
6	.1091	13–20	.0000

4.26 .1042   **4.27** .2148

4.28 Expected value = 4.0; variance = 4.0.

4.56 The distribution of the number of gonorrhea cases that occurred over a 3-month period is approximated by a Poisson distribution with parameter $\mu = np = 10,000 \times (50/100,000) = 5.0$. From the Poisson tables, compute

$$Pr(X \geqslant 10|\mu = 5) = .0181 + .0082 + .0034 + .0013$$
$$+ .0005 + .0002$$
$$= .032$$

Thus, the number of gonorrhea cases in this county over this time period is unusual, since the probability of observing at least 10 cases is quite small.

4.57 The probability that exactly 2 out of 50 men will die over a 3-year period is given by

$$\binom{50}{2}(.1)^2(.9)^{48}$$

The probability that not more than two men will die out of 50 $= Pr(X \leqslant 2)$ is given by

$$\sum_{k=0}^{2} \binom{50}{k}(.1)^k(.9)^{50-k}$$

This expression can be evaluated from the recursion rule for binomial probabilities.

$$Pr(0) = (.9)^{50} = .00515$$

$$Pr(1) = \frac{50}{1} \times \frac{.1}{.9} \times Pr(0) = .02861$$

$$Pr(2) = \frac{49}{2} \times \frac{.1}{.9} \times Pr(1) = .07788$$

Thus, $Pr(X \leqslant 2) = .00515 + .02861 + .07788 = .112$.

4.76 The probability that a hypertensive is being treated appropriately and is complying with the treatment is

Pr(hypertensive is told he or she has high blood pressure)

$\times Pr$(adequately treated|told)

$\times Pr$(complying|adequately treated)

$= (\frac{1}{2})^3 = .125$

Thus, we want

$$Pr(X \geqslant 5) = 1 - Pr(X \leqslant 4)$$

$$= 1 - \sum_{k=0}^{4} \binom{10}{k}(.125)^k(.875)^{10-k}$$

$$Pr(0) = (\tfrac{7}{8})^{10} = .26308$$

$$Pr(1) = (\tfrac{10}{1})(\tfrac{1}{7})(.26308) = .37583$$

$$Pr(2) = (\tfrac{9}{2})(\tfrac{1}{7})(.37583) = .24161$$

$$Pr(3) = (\tfrac{8}{3})(\tfrac{1}{7})(.24161) = .09204$$

$$Pr(4) = (\tfrac{7}{4})(\tfrac{1}{7})(.09204) = .02301$$

Thus, $Pr(X \leqslant 4) = .996$, $Pr(X \geqslant 5) = 1 - .996 = .004$.

4.77 Pr(hypertensive knows he or she has high blood pressure) $= \frac{1}{2}$. We want

$$Pr(X \geqslant 7) = \binom{10}{7}\left(\frac{1}{2}\right)^{10} + \binom{10}{8}\left(\frac{1}{2}\right)^{10} + \binom{10}{9}\left(\frac{1}{2}\right)^{10}$$

$$+ \binom{10}{10}\left(\frac{1}{2}\right)^{10}$$

Refer to the binomial tables (Table 1 of Appendix 1) with $n = 10$, $p = .50$ and find that

$$Pr(X = 10) = .0010$$

$$Pr(X = 9) = .0098$$

$$Pr(X = 8) = .0439$$

$$Pr(X = 7) = .1172$$

Thus, $Pr(X \geqslant 7) = .0010 + .0098 + .0439 + .1172 = .172$.

4.78 If the rates are each decreased to 40%, then $(.6)^3 = .216 = 21.6\%$ of hypertensives will be appropriately treated as opposed to $(.5)^3 = .125 = 12.5\%$. Thus, if the current annual mortality rate for *untreated* hypertensives is x and for *treated* hypertensives is $.8x$, then the *current overall* mortality rate for hypertensives is

$$.125(.8x) + .875x = .975x$$

The *new overall mortality* rate would be

$$.216(.8x) + .784x = .957x$$

Thus, the ratio of the new mortality to the current mortality rate is

$$.957x/.975x = 0.982 = 98.2\%$$

Thus, the overall mortality rate among hypertensives would be reduced by 1.8%.

4.82 We have

$$Pr(k \text{ positives}) = \binom{5}{k}(.05)^k(.95)^{5-k}$$

Thus, $Pr(1 \text{ or more } +) = 1 - Pr(0+)$

$$Pr(0+) = (.95)^5 = .77$$

Thus, $Pr(1 \text{ or more} +) = .23$

4.83 $Pr(3 \text{ or more}+) = 1 - Pr(2 \text{ or less}+)$

$$Pr(2 \text{ or less}+) = Pr(0) + Pr(1) + Pr(2)$$

$$Pr(0) = (.95)^{100} = .0059$$

$$Pr(1) = \frac{100}{1}\left(\frac{.05}{.95}\right)Pr(0) = .0311$$

$$Pr(2) = \frac{99}{2}\left(\frac{.05}{.95}\right)Pr(1) = .0810$$

Hence, $Pr(2 \text{ or less}+) = .118$ and $Pr(3 \text{ or more}+) = .882$.

4.84 We know that X can only take on the values 0, 1, or 2.

$Pr(0) = Pr(2 \text{ negatives}) = Pr(\text{negative at time } 0)$
$\qquad \times Pr(\text{negative at time } 1 | \text{negative at time } 0)$
$\qquad = (.95)(1 - .042) = (.95)(.958) = .910$

$Pr(1) = Pr(1 \text{ positive})$
$\qquad = Pr(\text{negative at time } 0 \cap \text{positive at time } 1)$
$\qquad + Pr(\text{positive at time } 0 \cap \text{negative at time } 1)$
$\qquad = Pr(\text{negative at time } 0)$
$\qquad \times Pr(\text{positive at time } 1 | \text{negative at time } 0)$
$\qquad + Pr(\text{positive at time } 0)$
$\qquad \times Pr(\text{negative at time } 1 | \text{positive at time } 0)$
$\qquad = (.95)(.042) + (.05)(.80) = .080$

$Pr(2) = Pr(2 \text{ positives}) = Pr(\text{positive at time } 0)$
$\qquad \times Pr(\text{positive at time } 1 | \text{positive at time } 0)$
$\qquad = (.05)(.20) = .010$

Thus,

X	$Pr(X)$
0	.910
1	.080
2	.010

4.85 Mean of $X = E[X] = 0(.910) + 1(.080) + 2(.010)$
$= 0.100$.

4.86 Variance of $X = E[X^2] - E^2[X]$

$E[X^2] = 0(.910) + 1(.080) + 4(.010) = 0.120$

$Var(X) = 0.120 - (0.100)^2 = 0.110$

4.87 The probability of living to age 65 given that one is age 21 is given by

$$Pr = \frac{\ell_{22}}{\ell_{21}} \times \frac{\ell_{23}}{\ell_{22}} \times \cdots \times \frac{\ell_{64}}{\ell_{63}} \times \frac{\ell_{65}}{\ell_{64}} = \frac{\ell_{65}}{\ell_{21}} = \frac{64{,}177}{95{,}330}$$
$$= .673$$

4.88 The probability of dying exactly between the ages of 56 and 57 given that one is age 21 in 1960 = the probability that one lives to age 56 and then dies before age 57:

$$\frac{\ell_{56}}{\ell_{21}} \times \left(1 - \frac{\ell_{57}}{\ell_{56}}\right) = \frac{79{,}783}{95{,}330} \times \left(1 - \frac{78{,}451}{79{,}783}\right)$$
$$= .8369 \times (1 - .9833) = .8369 \times .0167 = .014$$

4.89 First compute the probability of a single man dying before age 30 given that he is age 21 in 1960. This

probability is given by

$$Pr = 1 - \frac{\ell_{30}}{\ell_{21}} = 1 - \frac{93{,}826}{95{,}330} = 1 - .984 = .016$$

We are now interested in the distribution of the number of deaths among 100 men which follows a binomial distribution with parameters $n = 100$, $p = .016$. Thus, the probability of k deaths is given by

$$Pr(k) = \binom{100}{k}(.016)^k(.984)^{100-k}$$
$$k = 0, 1, 2, \ldots, 100$$

We specifically want $Pr(k \geqslant 5) = 1 - Pr(k \leqslant 4)$. Use the recursion rule for binomial probabilities:

$$Pr(0) = (.984)^{100} = .1993$$

$$Pr(1) = \frac{100}{1}\left(\frac{.016}{.984}\right)(.1993) = .3241$$

$$Pr(2) = \frac{99}{2}\left(\frac{.016}{.984}\right)(.3241) = .2609$$

$$Pr(3) = \frac{98}{3}\left(\frac{.016}{.984}\right)(.2609) = .1386$$

$$Pr(4) = \frac{97}{4}\left(\frac{.016}{.984}\right)(.1386) = .0547$$

Thus,

$$Pr(k \leqslant 4) = .1993 + .3241 + .2609 + .1386 + .0547$$
$$= .978$$
$$Pr(k \geqslant 5) = 1 - .978 = .022$$

4.91 The questions *cannot* be answered if we do not assume that the P_x's remain constant. Current life tables for each succeeding year after 1960 would be needed. For example, in computing the probability that a man will live to age 23 given that he is age 21 in 1960, the probabilities that he will reach age 22 in 1961 must be multiplied by the probability that he will reach age 23 in 1962 given that he was age 22 in 1961. We must assume that the latter probability is the same as the probability that he will reach age 23 in 1961 given that he was age 22 in 1960, which is not necessarily the case. This assumption is especially risky if events happening over a large number of years are being considered, since death rates have tended to go down. The probability that 65-year-old people will live 1 extra year given that they were 65 in the year 2025 is likely to be quite different from the corresponding probability if they were 65 in the year 1960. Life tables that follow the *same* group of people over time are called *cohort* life tables as opposed to the *current* life table data given in this example. These cohort life tables are especially useful in providing baseline mortality data for epidemiologic studies of mortality in high-risk groups.

CHAPTER 5

5.1 .6915   **5.2** .3085   **5.3** .7745

5.4 .0228   **5.5** .0440

5.6 We have Table A.6:

TABLE A.6

x	$\Phi(x)$	x	$\Phi(x)$
-1.28	.10	0.25	.60
-0.84	.20	0.52	.70
-0.52	.30	0.84	.80
-0.25	.40	1.28	.90
0.00	.50		

5.7 We have Table A.7:

TABLE A.7

x	-0.67	0.00	0.67
$\Phi(x)$	.25	.50	.75

5.19 .010

5.20 The number of breast cancer cases (X) is binomially distributed with parameters $n = 10,000$, $p = .010$. This distribution is approximated by a normal distribution (Y) with mean $= np = 100$, variance $= npq = 99$. We have

$$Pr(X \geqslant 120) \approx Pr(Y \geqslant 119.5)$$

$$= 1 - \Phi[(119.5 - 100)/\sqrt{99}]$$

$$= 1 - \Phi(1.96) = .025$$

5.32 If $X = $ birthweight, then $X \sim N(3400, 700^2)$. Thus,

$$Pr(X \leqslant 2500) = \Phi\left(\frac{2500 - 3400}{700}\right) = \Phi(-1.29)$$

$$= 1 - \Phi(1.29) = 1 - .9015 = .0985$$

5.33

$$Pr(X \leqslant 2000) = \Phi\left(\frac{2000 - 3400}{700}\right) = \Phi(-2.00)$$

$$= 1 - \Phi(2.00) = 1 - .9772 = .0228$$

5.34 Let $X = $ number of low-birthweight deliveries. X is binomially distributed with parameters $n = 3$, $p = .0985$. We wish to compute $Pr(X \geqslant 2)$.

$$Pr(X \geqslant 2) = \binom{3}{2}(.0985)^2(.9015) + \binom{3}{3}(.0985)^3$$

$$= .0262 + .0010 = .0272$$

5.37 Let X be the number of people positive for bacteriuria out of 500. X follows a binomial distribution with parameters $n = 500$ and $p = .05$. Compute $Pr(X \geqslant 50)$. X will be approximated by a normal random variable Y with mean $= 500 \times .05 = 50$ and variance $= 500 \times .05 \times .95 = 23.75$. Compute

$$Pr(X \geqslant 50) \approx Pr(Y \geqslant 49.5) = 1 - \Phi\left(\frac{49.5 - 25}{\sqrt{23.75}}\right)$$

$$= 1 - \Phi(5.03)$$

$$= \Phi(5.03) = 1.0$$

5.42 Let $X = $ serum cholesterol and $Y = \ln X$. Compute

$$Pr(X \leqslant 150) = Pr(Y \leqslant \ln 150) = Pr(Y \leqslant 5.01)$$

$$= \Phi\left(\frac{5.01 - 5.39}{0.23}\right)$$

$$= \Phi(-1.65) = 1 - \Phi(1.65)$$

$$= 1 - .9505 = .0495$$

5.43 We want

$$Pr(X \geqslant 250) = Pr(Y \geqslant \ln 250) = Pr(Y \geqslant 5.52)$$

$$= 1 - \Phi\left(\frac{5.52 - 5.39}{0.23}\right)$$

$$= 1 - \Phi(0.57) = 1 - .7157 = .2843$$

5.44 We want

$$Pr(X \geqslant 300) = Pr(Y \geqslant \ln 300) = Pr(Y \geqslant 5.70)$$

$$= 1 - \Phi\left(\frac{5.70 - 5.39}{0.23}\right)$$

$$= 1 - \Phi(1.35) = 1 - .9115 = .0885$$

The proportion of the subpopulation with abnormally high levels, which this represents, is .0885/.2843, or 31%.

5.45 From Problem 5.44, .0885, or 9%.

5.52 Suppose n people are sampled for the study. The number of hypertensives X ascertained from this procedure will be binomially distributed with parameters n and $p = .10$. For large n, from the normal approximation to the binomial, it follows that the distribution of X can be approximated by a normal distribution Y with mean $.1n$ and variance $n \times .1 \times .9 = .09n$. Thus, we want n to be large enough so that $Pr(X \geqslant 100) = .8$. We have

$$Pr(X \geqslant 100) \approx Pr(Y \geqslant 99.5) = 1 - Pr(Y \leqslant 99.5)$$

$$= 1 - \Phi\left(\frac{99.5 - .1n}{\sqrt{.09n}}\right) = .8$$

or

$$\Phi\left(\frac{99.5 - .1n}{\sqrt{.09n}}\right) = .2$$

However, from the normal tables, $\Phi(0.84) = .8$ or $\Phi(-0.84) = .2$. Thus,

$$\frac{99.5 - .1n}{\sqrt{.09n}} = -0.84$$

or

$$99.5 - .1n = -0.84\sqrt{.09n}$$

or

$$.1n - 0.252\sqrt{n} - 99.5 = 0$$

If z^2 is substituted for n, this equation can be rewritten as $.1z^2 - 0.252z - 99.5 = 0$. The solution of this quadratic equation is given by

$$z = \sqrt{n} = \frac{0.252 \pm \sqrt{(0.252)^2 + 4(99.5)(.1)}}{2(.1)}$$

$$= \frac{0.252 \pm \sqrt{39.864}}{.2} = \frac{0.252 \pm 6.314}{.2}$$

$$= -30.310 \quad \text{or} \quad 32.830$$

or $n = z^2 = (-30.310)^2$ or $(32.830)^2 = 918.70$ or 1077.81. Since n must be larger than $100/.1 = 1000$, we have $n = 1077.81$, or 1078. Thus, 1078 people need to be sampled to be 80% sure of recruiting 100 hypertensives.

5.53 The same approach is used as in Problem 5.52. Find n such that

$$Pr(Y \geqslant 99.5) = 1 - \Phi\left(\frac{99.5 - .1n}{\sqrt{.09n}}\right) = .9$$

or

$$\Phi\left(\frac{99.5 - .1n}{\sqrt{.09n}}\right) = .1$$

From the normal tables, $\Phi(1.28) = .9$ or $\Phi(-1.28) = .1$. Thus,

$$\frac{99.5 - .1n}{\sqrt{.09n}} = -1.28$$

or

$$99.5 - .1n = -0.384\sqrt{n}$$

or

$$.1n - 0.384\sqrt{n} - 99.5 = 0$$

Substitute z^2 for n and obtain the quadratic equation

$$.1z^2 - 0.384z - 99.5 = 0$$

The solution is given by

$$z = \frac{0.384 \pm \sqrt{(0.384)^2 + 4(.1)(99.5)}}{2(.1)}$$

$$= \frac{0.384 \pm \sqrt{39.947}}{.2} = \frac{0.384 \pm 6.320}{.2}$$

$$= -29.68 \quad \text{or} \quad 33.52$$

Thus,

$$n = z^2 = (-29.68)^2 = 880.90 \quad \text{or} \quad (33.52)^2 = 1123.59$$

Again use the root such that $n > 1000$, and thus $n = 1123.59$, or 1124 if rounded up to the nearest integer. Thus, 1124 people need to be sampled to be 90% sure of recruiting 100 hypertensives.

CHAPTER 6

6.1 The treatments will be labeled as A and B and patients will be assigned to treatment A if the random digit is from 0 to 4 inclusive and to treatment B if it is from 5 to 9 inclusive. Note that this method is one among many possible randomization schemes that could be used. The treatment assignments in Table A.8 are obtained if one starts at the 28th row of Table A.4.

6.2 There are 9 patients assigned to treatment A and 11 patients assigned to treatment B. Ten patients would be expected in each treatment group.

6.3 The treatments will be labeled as A, B, C, and D. Since 10 is not divisible by 4 but 100 is, two random digits will be used to generate the treatment assignments. If the number formed by two successive random digits is between 00 and 24, the patient will be assigned to treatment A; if between 25 and 49, to treatment B; if between 50 and 74, to treatment C; if between 75 and 99, to treatment D. The treatment assignments in Table A.9 are obtained starting at the 12th row of Table 4.

TABLE A.8

Patient number	Random digit	Treatment assignment
1	6	B
2	9	B
3	6	B
4	4	A
5	4	A
6	3	A
7	7	B
8	1	A
9	9	B
10	8	B
11	0	A
12	0	A
13	0	A
14	2	A
15	8	B
16	9	B
17	8	B
18	1	A
19	9	B
20	5	B

6.4 There are 12 patients assigned to treatment A, 7 patients to treatment B, 8 patients to treatment C, and 13 patients to treatment D. Ten patients would be expected to be assigned to each treatment group.

6.11 2.583 **6.12** -1.313 **6.13** 2.365

6.20 9.24 **6.21** 11.34

6.22 The upper 2.5 percentile $= \chi^2_{2,.975} = 7.38$. The lower 2.5 percentile $= \chi^2_{2,.025} = 0.0506$.

6.23 The chi-square distribution with n df is approximated by a normal distribution with mean n and variance $2n$. Therefore, the upper and lower 2.5 percentiles are given by

$$140 \pm z_{.975}\sqrt{280} = 140 \pm 1.96(16.733)$$

$$= 140 \pm 32.80$$

$$= 107.20 \text{ (lower 2.5 percentile)}, 172.80 \text{ (upper 2.5 percentile)}$$

6.28 Case women, $\frac{89}{283} = .314$; control women, $\frac{640}{3833} = .167$.

TABLE A.9

Patient number	Random digit	Treatment assignment	Patient number	Random digit	Treatment assignment
1	16	A	21	05	A
2	87	D	22	20	A
3	29	B	23	73	C
4	47	B	24	67	C
5	49	B	25	02	A
6	44	B	26	56	C
7	01	A	27	80	D
8	24	A	28	41	B
9	88	D	29	04	A
10	84	D	30	98	D
11	65	C	31	78	D
12	41	B	32	80	D
13	98	D	33	77	D
14	70	C	34	92	D
15	92	D	35	43	B
16	78	D	36	13	A
17	59	C	37	72	C
18	69	C	38	98	D
19	15	A	39	12	A
20	12	A	40	22	A

6.29 A 95% confidence interval for case women is given by

$$.314 \pm 1.96\sqrt{(.314)(.686)/283}$$

$$= .314 \pm 1.96(.0276) = .314 \pm .054$$

$$= (.260, .368)$$

A 95% confidence interval for control women is given by

$$.167 \pm 1.96\sqrt{(.167)(.833)/3833}$$

$$= .167 \pm 1.96(.0060)$$

$$= .167 \pm .012 = (.155, .179)$$

6.54 sem $= 500/\sqrt{20} = 111.8$.

6.55 The standard deviation is a measure of variability for the birthweight of *one* infant. The standard error of the mean is a measure of variability for the *mean* birthweight of a group of n infants (in this case $n = 20$). The standard error will always be smaller than the standard deviation because a mean of more than one birthweight will be less variable in repeated samples than an individual birthweight.

6.66 The best point estimate is $\hat{p} = 6/46 = .130$.

6.67 If a normal approximation is used, then the lower confidence limit is

$$c_1 = \hat{p} - 1.96\sqrt{\frac{\hat{p}\hat{q}}{n}}$$

$$= .130 - 1.96\sqrt{\frac{(.130)(.870)}{46}} = .033$$

The upper confidence limit is

$$c_2 = \hat{p} + 1.96\sqrt{\frac{\hat{p}\hat{q}}{n}}$$

$$= .130 + 1.96\sqrt{\frac{(.130)(.870)}{46}} = .227$$

6.68 Since 10% is within the 95% confidence interval, we would conclude that it is possible that the two drugs are equally effective (i.e., have the same failure rate).

6.73 A 95% confidence interval is given by

$$\hat{p} \pm 1.96\sqrt{\hat{p}\hat{q}/n} = \frac{64}{750} \pm 1.96\sqrt{\left(\frac{64}{750}\right)\left(1 - \frac{64}{750}\right)\bigg/750}$$

$$= .085 \pm 1.96\sqrt{(.085)(.915)/750}$$

$$= .085 \pm .020 = (.065, .105)$$

6.74 The rate of .10 is compatible with these data, since it falls within the 95% confidence interval in Problem 6.73. Thus, we *cannot* conclude from these data that jogging 10 miles per week prevents death from cardiovascular disease.

6.75 We assume that $x_1, \ldots, x_{25} \sim N(\mu, \sigma^2)$, where μ, σ^2 are unknown, and find that $\bar{x} = 7.0$, $s^2 = 4.0$. Thus, a two-sided 95% confidence interval for the mean with unknown variance is given by

$$\left(\bar{x} - t_{n-1, .975}\frac{s}{\sqrt{n}}, \bar{x} + t_{n-1, .975}\frac{s}{\sqrt{n}}\right)$$

$$= [7.0 - 2.064(2)/5, 7.0 + 2.064(2)/5]$$

$$= (6.17, 7.83)$$

6.76 A two-sided 99% confidence interval for the unknown variance σ^2 is given by

$$\left(\frac{(n-1)s^2}{\chi^2_{n-1, .995}}, \frac{(n-1)s^2}{\chi^2_{n-1, .005}}\right) = \left(\frac{24(4)}{45.56}, \frac{24(4)}{9.89}\right)$$

$$= (2.11, 9.71)$$

6.77 The length of the 95% confidence interval in Problem 6.75 is given by

$$2t_{n-1, .975}\frac{s}{\sqrt{n}}$$

and n should be large enough so that this value is 0.5. Thus,

$$2t_{n-1, .975}(2)/\sqrt{n} = 0.5$$

or $\quad \sqrt{n} = 8t_{n-1, .975}, \; n = 64t^2_{n-1, .975}$

For large n we could assume $t_{n-1, .975} = z_{.975} = 1.96$. Thus, $n = 64(1.96)^2 = 245.9$, and an adequate sample size would be 246. This estimate could be improved by using conventional t tables, since we could establish by trial and error that $n > 121$, in which case

$$t_{n-1, .975} < t_{120, .975} = 1.98$$

Thus, $\quad\quad n < 64(1.98)^2 = 250.9$

Thus, we know that $245.9 < n < 250.9$. A conservative estimate of the necessary sample size would be 251. This estimate could be improved with more extensive t tables, but this level of accuracy is probably adequate for an estimate of sample size.

6.84 The concern here is with the variability of two methods of measuring flow, a manual method and a digitizer method. A 95% confidence interval must be constructed for

$$\sigma^2_{\text{manual differences}}$$

We have $\qquad s_{\text{diff}} = 0.0779, n = 10$

The interval is given by

$$\left(\frac{(n-1)s^2}{\chi^2_{n-1,.975}}, \frac{(n-1)s^2}{\chi^2_{n-1,.025}}\right) = \left(\frac{9(0.0779)^2}{\chi^2_{9,.975}}, \frac{9(0.0779)^2}{\chi^2_{9,.025}}\right)$$

$$= \left(\frac{9(0.0779)^2}{19.02}, \frac{9(0.0779)^2}{2.70}\right)$$

$$= (0.0029, 0.0202)$$

So, the 95% confidence interval for

$$\sigma^2_{\text{manual}} = (0.0029, 0.0202)$$

and thus the 95% confidence interval for

$$\sigma_{\text{manual}} = (0.054, 0.142)$$

6.85 The same method can be used for the digitizer differences. We have that $s^2 = 0.00083$. Thus, the 95% confidence interval is given by

$$\left(\frac{(n-1)s^2}{\chi^2_{9,.975}}, \frac{(n-1)s^2}{\chi^2_{9,.025}}\right) = \left(\frac{9(0.00083)}{19.02}, \frac{9(0.00083)}{2.70}\right)$$

$$= (0.00039, 0.00277)$$

Finally, the 95% confidence interval for

$$\sigma_{\text{dig}} = (0.020, 0.053)$$

6.86 $d_i = |\text{manual diff}_i| - |\text{digitizer diff}_i|$. A 95% confidence interval for μ_d comes from the usual t statistic formulation

$$\bar{d} \pm t_{9,.975}\left(\frac{s_d}{\sqrt{10}}\right) = 0.0440 \pm 2.262\left(\frac{0.0353}{\sqrt{10}}\right)$$

$$= (0.0187, 0.0693)$$

6.87 The digitizer method appears to be significantly less variable than the manual method, as shown in Problem 6.86, since the 95% confidence interval for μ_d does not include 0.

6.108 The random samples can be selected in many different ways. The random numbers in Table 4 in Appendix 1 will be used, starting in row 1, where each unique set of three digits identifies a unique delivery in Table 6.2. The random numbers and corresponding random samples of birthweights are identified in Tables A.10 and A.11, respectively. The mean birthweights for the six samples are 103.2, 109.8, 116.2, 115.2, 103.4, and 119.2.

TABLE A.10 Random numbers

Sample	Sample point				
	1	2	3	4	5
1	329	242	232	418	125
2	090	775	463	290	374
3	941	434	929	588	720
4	430	359	708	183	373
5	217	271	190	441	513
6	316	538	098	570	799

TABLE A.11 Random samples of birthweights (oz)

Sample	Sample point				
	1	2	3	4	5
1	88	132	86	97	113
2	118	81	128	114	108
3	116	109	121	108	127
4	125	113	119	121	98
5	118	108	91	95	105
6	118	114	140	91	133

6.109 The standard deviation is given by

$$s = \sqrt{\frac{\sum_{i=1}^{n} x_i^2 - \left(\sum_{i=1}^{n} x_i\right)^2 / n}{(n-1)}} = \sqrt{\frac{74,379.96 - (667)^2/6}{5}}$$

$$= 6.81$$

6.110 The standard deviation from the collection of six third points is

$$\sqrt{\frac{80,463 - (685)^2/6}{5}} = 21.25.$$

6.111 Theoretically, the standard deviation in Problem 6.109 is an expression of the variability of the mean of five sample points $= \sigma/\sqrt{5}$, whereas the standard deviation in Problem 6.110 is an expression of the variability of individual sample points $= \sigma$. Thus, we should approximately have

$$\frac{s \text{ in Problem } 6.110}{s \text{ in Problem } 6.109} \approx \sqrt{5} = 2.24$$

6.112 Indeed,

$$\frac{s \text{ in Problem 6.110}}{s \text{ in Problem 6.109}} = \frac{21.25}{6.81} = 3.12$$

This result shows that the variability of individual sample points is far greater than that of the mean of five sample points.

CHAPTER 7

7.1 Test the hypothesis $H_0: \mu = 15 = \mu_0$ versus $H_1: \mu < 15$. Reject H_0 if $\bar{x} < \mu_0 + z_\alpha \sigma/\sqrt{n} = 15 + z_{.05}(4)/\sqrt{10} = 15 - 1.645(4)/\sqrt{10} = 12.92$ and accept H_0 if $\bar{x} \geqslant 12.92$. Since $\bar{x} = 13 > 12.92$, accept H_0 at the 5% level.

7.2 The p-value is given by $\Phi[(\bar{x} - \mu_0)/(\sigma/\sqrt{n})] = \Phi[(13-15)/(4/\sqrt{10})] = \Phi(-2.0/1.265) = \Phi(-1.58) = 1 - \Phi(1.58) = 1 - .9429 = .057$.

7.3 Test the hypothesis $H_0: \mu = 24 = \mu_0$ versus $H_1: \mu < 24$. Reject H_0 if $\bar{x} < \mu_0 + z_\alpha \sigma/\sqrt{n} = 24 + z_{.01}(11)/\sqrt{8} = 24 - 2.326(11)/\sqrt{8} = 24 - 9.05 = 14.95$ and accept H_0 otherwise. Since $\bar{x} = 11 < 14.95$, reject H_0 at the 1% level.

7.4 The p-value is given by $\Phi[(\bar{x} - \mu_0)/(\sigma/\sqrt{n})] = \Phi[(11-24)/(11/\sqrt{8})] = \Phi(-13/3.889) = \Phi(-3.34) = 1 - \Phi(3.34) = 1 - .9996 = .0004$.

7.5 Since H_0 was rejected at the 1% level in Problem 7.3, p must be $<.01$, which is indeed the case.

7.6 The hypotheses to be tested are $H_0: \mu = 15$ versus $H_1: \mu \neq 15$. This alternative is two-sided in contrast to the one-sided alternative in Problem 7.1.

7.7 The rejection region is given by $\bar{x} < \mu_0 + z_{\alpha/2}\sigma/\sqrt{n} = c_1$ or $\bar{x} > \mu_0 + z_{1-\alpha/2}\sigma/\sqrt{n} = c_2$. We have that $c_1 = 15 + z_{.025}(4)/\sqrt{10} = 15 - 1.96(4)/\sqrt{10} = 15 - 2.48 = 12.52$, $c_2 = 15 + 2.48 = 17.48$. Since $\bar{x} = 13 > 12.52$, accept H_0 using a two-sided test at the 5% level. The exact p-value is given by $2 \times \Phi[(\bar{x} - \mu_0)/(\sigma/\sqrt{n})] = 2 \times \Phi[(13-15)/(4/\sqrt{10})] = 2 \times \Phi(-2.0/1.265) = 2 \times \Phi(-1.58) = 2 \times [1 - \Phi(1.58)] = 2 \times (1 - .9429) = .114$.

7.11 $p = 2 \times Pr(t_7 < -1.52) = 2 \times Pr(t_7 > 1.52)$. Since $t_{7,.9} = 1.415$, $t_{7,.95} = 1.895$, and $1.415 < 1.52 < 1.895$, it follows that $2 \times (1 - .95) < p < 2 \times (1 - .9)$, or $.1 < p < .2$.

7.14 Use a one-sample t test. We have the test statistic $t = (\bar{x} - \mu_0)/(s/\sqrt{n}) = (13 - 15)/(6/\sqrt{10}) = -2/1.897 = -1.05 \sim t_9$ under H_0. Since a two-sided test is being performed, reject H_0 if $t < t_{9,.025} = -2.262$ or $t > t_{9,.975} = 2.262$ and accept H_0 otherwise. Since $-2.262 \leqslant -1.05 \leqslant 2.262$, it follows that H_0 is accepted. The exact p-value is given by $2 \times Pr(t_9 < [(\bar{x} - \mu_0)/(s/\sqrt{n})] = 2 \times Pr(t_9 < -1.05) = 2 \times Pr(t_9 > 1.05)$. Since $t_{9,.8} = 0.883$, $t_{9,.85} = 1.100$, and $0.883 < 1.05 < 1.100$, it follows that $2 \times (1 - .85) < p < 2 \times (1 - .8)$, or $.3 < p < .4$.

7.21 We wish to test the hypothesis $H_0: \mu = \mu_0$ versus $H_1: \mu \neq \mu_0$. Use the following power formula:

$$\text{Power} = \Phi[z_{\alpha/2} + (\mu_1 - \mu_0)\sqrt{n}/\sigma]$$
$$= \Phi(z_{.025} + 0.10\sqrt{100}/0.54)$$
$$= \Phi(-1.96 + 1.852) = \Phi(-0.108)$$
$$= 1 - \Phi(0.108) = 1 - .54 = .46$$

Thus, the study has 46% power.

7.22

$$\text{Power} = \Phi(z_{.025} + 0.20\sqrt{100}/0.54)$$
$$= \Phi(-1.96 + 3.704)$$
$$= \Phi(1.744) = .96$$

7.23 The required sample size is given by

$$n = \frac{\sigma^2(z_{1-\alpha/2} + z_{1-\beta})^2}{(\mu_1 - \mu_0)^2}$$
$$= \frac{(0.54)^2(z_{.975} + z_{.80})^2}{(0.10)^2}$$
$$= 29.16(1.96 + 0.84)^2 = 228.6$$

Thus, 229 people need to be studied to achieve an 80% power.

7.28 Test the hypothesis $H_0: \sigma_1^2 = \sigma_0^2$ versus $H_1: \sigma_1^2 \neq \sigma_0^2$. Use the one-sample χ^2 test for the variance of a normal distribution to test these hypotheses.

7.29 We have the test statistic

$$X^2 = \frac{(n-1)s^2}{\sigma_0^2} = \frac{19(15)^2}{(20)^2}$$
$$= 10.69 \sim \chi_{19}^2 \text{ under } H_0$$

Since $\chi_{19,.05}^2 = 10.12$, $\chi_{19,.10}^2 = 11.65$, and $10.12 < 10.69 < 11.65$, it follows that $2 \times .05 < p < 2 \times .10$, or $.10 < p < .20$. Thus, there is no significant difference between the variances using the two methods.

7.30 Test the hypothesis $H_0: \sigma^2 = \sigma_0^2 = (0.020)^2$ versus $H_1: \sigma^2 \neq \sigma_0^2$. Reject H_0 if $s^2 < \sigma_0^2 \chi_{n-1,\alpha/2}^2/(n-1) = c_1$ or $s^2 > \sigma_0^2 \chi_{n-1,1-\alpha/2}^2/(n-1) = c_2$ and accept H_0 otherwise. We have

$$c_1 = (0.020)^2 \chi_{19,.025}^2/19$$

$$= (0.020)^2(8.91)/19 = 0.000188$$

$$c_2 = (0.020)^2 \chi_{19,.975}^2/19$$

$$= (0.020)^2(32.85)/19 = 0.000692$$

Since $s^2 = (0.016)^2 = 0.000256$ and $c_1 < s^2 < c_2$, it follows that H_0 is accepted at the 5% level and we conclude that the variances of the two methods are not significantly different.

7.31 We have the test statistic

$$X^2 = \frac{(n-1)s^2}{\sigma_0^2} = \frac{(19)(0.000256)}{(0.000400)}$$

$$= 12.16 \sim \chi_{19}^2 \text{ under } H_0$$

Since $\chi_{19,.10}^2 = 11.65$, $\chi_{19,.25}^2 = 14.56$, and $11.65 < 12.16 < 14.56$, it follows that $2 \times .10 < p < 2 \times .25$, or $.20 < p < .50$.

7.49 Test the hypothesis $H_0: \mu = 130 = \mu_0$ versus $H_1: \mu \neq 130$, where σ is assumed to be 20 mm Hg. Use the test statistic

$$z = \frac{\bar{x} - \mu_0}{\sigma/\sqrt{n}} = \frac{135 - 130}{20/\sqrt{85}} = 2.30 \sim N(0, 1) \text{ under } H_0$$

Thus, the two-tailed p-value equals $2 \times [1 - \Phi(2.30)]$ $= 2 \times (1 - .9893) = .021$. Therefore, there is a significant association between glaucoma and high blood pressure.

7.50 Perform the one-sample t test here, since the standard deviation is not assumed known. Use the test statistic

$$t = \frac{\bar{x} - \mu_0}{s/\sqrt{n}} = \frac{135 - 130}{22/\sqrt{85}} = 2.10 \sim t_{84} \text{ under } H_0$$

Since $t_{60,.975} = 2.000$, $t_{60,.99} = 2.390$,

$$t_{120,.975} = 1.980, \quad t_{120,.99} = 2.358$$

it follows that if there were 60 or 120 df, then

$$2 \times (1 - .99) < p < 2 \times (1 - .975)$$

or

$$.02 < p < .05$$

Since there are 84 df, we must also have $.02 < p < .05$. Thus, there is a significant association between glaucoma and high blood pressure in this case as well.

7.55 Use a one-sample test for binomial proportions. We have the hypotheses

$$H_0: p = p_0 = .3$$

$$H_1: p \neq p_0$$

Since $\quad np_0 q_0 = 200(.3)(.7) = 42 \geqslant 5$

the normal approximation can be used. Use the test statistic

$$z = \frac{\hat{p} - p_0}{\sqrt{\dfrac{p_0 q_0}{n}}} = \frac{\frac{110}{200} - .3}{\sqrt{\dfrac{(.3)(.7)}{200}}} = \frac{.55 - .3}{\sqrt{\dfrac{(.3)(.7)}{200}}} = 7.72$$

Since this statistic is distributed as $N(0, 1)$, the p-value $= 2 \times [1 - \Phi(7.72)] \ll .001$. This result is very highly significant.

7.56 Test the hypothesis $H_0: p = p_0$ versus $H_1: p > p_0$. A one-sided alternative is more appropriate here, since we wish to detect only if hair dyes increase the risk of breast cancer rather than decrease the risk.

7.57 Since $np_0 q_0 = 1000(.007)(.993) = 6.95 \geqslant 5$, use the large-sample test. We have the test statistic

$$z = \frac{\hat{p} - p_0}{\sqrt{\dfrac{p_0 q_0}{n}}}$$

$$= \frac{\frac{20}{1000} - .007}{\sqrt{\dfrac{.007 \times .993}{1000}}} = \frac{.013}{.0026} = 5.00 \sim N(0, 1) \text{ under } H_0$$

The p-value is given by $p = 1 - \Phi(z) = 1 - \Phi(5.00) < .001$. Thus, the results are very highly statistically significant, and we conclude that extensive occupational exposure to hair dyes significantly increases the risk of breast cancer.

7.62 Test the hypothesis $H_0: \mu = \mu_0 = 230$ versus $H_1: \mu \neq \mu_0$, where σ^2 is assumed unknown. Use a one-sample t test with test statistic

$$t = \frac{\bar{x} - \mu_0}{s/\sqrt{n}} = \frac{175 - 230}{35/\sqrt{24}} = \frac{-55}{7.144} = -7.70 \sim t_{23}$$

Clearly, since $t_{23,.9995} = 3.767$ and $|t| > 3.767$, it follows that $p < 2 \times .0005 = .001$.

7.63 A two-sided 95% confidence interval for μ_0 is given by

$$\bar{x} \pm t_{n-1,1-\alpha/2}\frac{s}{\sqrt{n}} = 175 \pm t_{23,.975}\left(\frac{35}{\sqrt{24}}\right)$$

$$= 175 \pm 2.069\left(\frac{35}{\sqrt{24}}\right)$$

$$= 175 \pm 14.78 = (160.22, 189.78)$$

Since the 95% confidence interval does not contain 230, we can again conclude that the underlying cholesterol level for macrobiotics is significantly lower than 230.

7.64 The hypothesis test tells precisely how significant the results are ($p < .001$). The confidence interval gives a range of values within which the true mean cholesterol for macrobiotics is likely to fall.

7.68 Test the hypothesis $H_0: p = .10 = p_0$ versus the alternative $H_1: p > .10$. Use the normal approximation test, since $np_0q_0 = 100(.1)(.9) = 9 \geqslant 5$. We have the test statistic

$$z = \frac{(\hat{p} - p_0)}{\sqrt{\dfrac{p_0q_0}{n}}}$$

which should be distributed as $N(0,1)$ under H_0. We have $\hat{p} = .13$, $p_0 = .10$, $n = 100$. Thus,

$$z = \frac{(.13 - .10)}{\sqrt{\dfrac{.1(.9)}{100}}} = \frac{.03}{.03} = 1.0$$

The p-value $= 1 - \Phi(1.0) = 1 - .8413 = .159$. This result is not significant using a one-sided hypothesis test at the 5% level, and the null hypothesis is accepted. Thus, this finding is not indicative of anything about cholesterol, since it is quite possible that the underlying rate for this group of 100 men is .1.

7.69 Perform the same test here with $n = 1000$. We have

$$z = \frac{(\hat{p} - p_0)}{\sqrt{\dfrac{p_0q_0}{n}}} = \frac{.13 - .10}{\sqrt{\dfrac{.1(.9)}{1000}}} = \frac{.03}{.0095} = 3.16$$

We see from the normal tables that the one-sided p-value $= 1 - \Phi(3.16) < .001$. Thus, the null hypothesis that $p = .1$ is rejected and the alternative that the underlying rate for the high cholesterol group is larger than .10 is accepted.

7.70 Refer to the sample-size formula for one-sided alternatives given in (7.40). Use $\alpha = .05$, $\beta = 1 - .80 = .20$, $p_0 = .10$, $q_0 = .90$, $p_1 = .13$, $q_1 = .87$. We have

$$n = \frac{p_0q_0[z_{1-\alpha} + z_{1-\beta}\sqrt{(p_1q_1)/(p_0q_0)}]^2}{(p_1 - p_0)^2}$$

$$= \frac{(.10)(.90)[z_{0.95} + z_{0.80}\sqrt{(.13 \times .87)/(.10 \times .90)}]^2}{(.13 - .10)^2}$$

$$= \frac{(.09)(1.645 + 0.84\sqrt{1.257})^2}{.0009}$$

$$= 100(1.645 + 0.942)^2 = 100(6.693) = 669.3$$

Thus, 670 men are needed in the sample to have an 80% chance of finding a significant difference using a one-sided test with $\alpha = .05$.

7.86 Use a two-sided test since there is no reason to expect self-reported weights to be higher or lower than actual weights.

7.87 Use a one-sample t test to test the hypothesis $H_0: \mu = 0$ versus $H_1: \mu \neq 0$, where μ is the true mean for in self-reported weight − measured weight.

7.88 We have $\bar{x} = -2.5$, $s = 3.27$. The rejection region is given by $\bar{x} > c_2$ or $\bar{x} < c_1$ where

$$c_2 = \mu_0 + t_{n-1,1-\alpha/2}\frac{s}{\sqrt{n}} = 0 + t_{9,.975}\frac{(3.27)}{\sqrt{10}}$$

$$= \frac{2.262(3.27)}{\sqrt{10}} = 2.34$$

$$c_1 = -2.34$$

Since $\bar{x} = -2.50 < c_1$, H_0 is rejected at the 5% level.

7.89 Since $\bar{x} < \mu_0 = 0$, the p-value is obtained from

$$p = 2 \times Pr\left(t_9 < \frac{\bar{x} - \mu_0}{s/\sqrt{n}}\right) = 2 \times Pr\left(t_9 < \frac{-2.5 - 0}{3.27/\sqrt{10}}\right)$$

$$= 2 \times Pr\left(t_9 < \frac{-2.5}{1.034}\right)$$

$$= 2 \times Pr(t_9 < -2.418) = 2 \times Pr(t_9 > 2.418)$$

Since $t_{9,.975} = 2.262$, $t_{9,.99} = 2.821$, and $2.262 < 2.418 < 2.821$, it follows that $2 \times (1 - .99) < p < 2 \times (1 - .975)$, or $.02 < p < .05$.

7.90 There are 7 reported weights out of 10 that end in 0 or 5 ($\hat{p} = .7$) compared with 2 expected ($p_0 = .2$). Use a one-sided hypothesis test of the form $H_0: p = p_0$ versus $H_1: p > p_0$, since p would only be expected to be larger than p_0 due to digit preference. Since $np_0q_0 = 10 \times .2 \times .8 = 1.6 < 5$, use the exact method. The p-value is given by

$$p = \sum_{k=7}^{10} {}_{10}C_k(.2)^k(.8)^{10-k}$$

Refer to Table 1 in Appendix 1 under $n = 10$, $p = .2$ and obtain $p = .0008 + .0001 + .0000 + .0000 = .0009$. Thus, there is a significant excess of 0's and 5's among the self-reported weights.

7.105 Use the one-sample t test to test the hypothesis $H_0: \mu = 0$ versus $H_1: \mu \neq 0$ where μ is the true mean change in blood pressure upon taking potassium supplements.

7.106 We have the test statistic

$$t = \frac{\bar{x} - \mu_0}{s/\sqrt{n}}$$

$$= \frac{-3.2}{8.5/\sqrt{20}}$$

$$= \frac{-3.2}{1.90} = -1.68 \sim t_{19} \text{ under } H_0$$

Note from the t tables that $t_{19,.95} = 1.729$, $t_{19,.90} = 1.328$. Since $1.328 < 1.68 < 1.729$, it follows that $2 \times (1 - .95) < p < 2 \times (1 - .90)$, or $.1 < p < .2$. Thus, there is not a significant change in blood pressure based on the pilot study results.

7.107 A two-sided 95% confidence interval is given by

$$\bar{x} \pm t_{19,.975}s/\sqrt{n} = -3.2 \pm 2.093(8.5/\sqrt{20})$$

$$= -3.2 \pm 2.093(1.90) = -3.2 \pm 4.0$$

$$= (-7.2, 0.8)$$

The 95% confidence interval includes the null value for μ (0), which is consistent with the nonsignificant results in Problem 7.106.

7.108 Use the sample-size formula

$$n = \frac{\sigma^2(z_{1-\alpha/2} + z_{1-\beta})^2}{\Delta^2}$$

In this case, $\sigma = 8.5$, $\alpha = .05$, $\beta = 1 - .80 = .20$, $\Delta = 3.2$. Therefore,

$$n = \frac{(8.5)^2(z_{.975} + z_{.80})^2}{(3.2)^2} = \frac{(8.5)^2(1.96 + 0.84)^2}{(3.2)^2}$$

$$= 55.3$$

Thus, 56 subjects are needed in order to have an 80% chance of detecting a significant difference using a two-sided test with α level $= .05$.

<u>CHAPTER 8</u>

8.1 Use the paired t test because each person is used as his or her own control.

8.2 We have the test statistic

$$t = \frac{\bar{d}}{s_d/\sqrt{n}} = \frac{0.02}{0.04}$$

$$= 0.50 \sim t_{89} \text{ under } H_0$$

Since $0.50 < t_{120,.975} = 1.980 < t_{89,.975}$, it follows that H_0, that there is no significant visual field loss over 1 year, is accepted. Furthermore, since $0.50 < t_{120,.75} = 0.677 < t_{89,.75}$, it follows that $p > 2 \times (1 - .75) = .50$.

8.3 We have the test statistic $t = 0.08/0.05 = 1.60 \sim t_{89}$ under H_0. Since $1.60 < 1.980$, H_0, that there is no significant visual field loss over 2 years, is accepted. Furthermore, if there were 60 df, then since $t_{60,.9} = 1.296 < 1.60 < t_{60,.95} = 1.671$, it follows that $2 \times (1 - .95) < p < 2 \times (1 - .90)$, or $.10 < p < .20$. Similarly, if there were 120 df, then since $t_{120,.9} = 1.289 < 1.60 < t_{120,.95} = 1.658$, it would also follow that $.10 < p < .20$. Finally, since there are 89 df and we reached the same decision with 60 or 120 df, then it follows that $.10 < p < .20$.

8.4 We have the test statistic $t = 0.14/0.07 = 2.000$. The critical value $= t_{89,.975} < t_{60,.975} = 2.000$. Since $t >$ critical value, H_0 is rejected at the 5% level. Since $2.000 < 2.358 = t_{120,.99} < t_{89,.99}$, it follows that $p > 2 \times (1 - 0.99) = .02$. Therefore, $.02 < p < .05$.

8.9 $F_{14,7,.025} = 1/F_{7,14,.975} = 1/3.38 = 0.296$.

8.12 $F_{50,10,.025} = 1/F_{10,50,.975}$. This percentile is not given in the table, so interpolation methods must be used. The following 97.5 percentiles are given in Table A.12.

TABLE A.12

		Numerator df	
		8	12
Denominator df	40	2.53	2.29
	60	2.41	2.17

First compute $F_{10,40,.975}$ and $F_{10,60,.975}$ as follows:

$$F_{10,40,.975} = \frac{(1/10 - 1/12)(2.53) + (1/8 - 1/10)(2.29)}{(1/8 - 1/12)}$$

$$= \frac{(0.0167)(2.53) + (0.0250)(2.29)}{0.0417} = 2.386$$

$$F_{10,60,.975} = \frac{(0.0167)(2.41) + (0.0250)(2.17)}{0.0417}$$

$$= 2.266$$

Now compute $F_{10,50,.975}$ by interpolation as follows:

$$F_{10,50,.975}$$

$$= \frac{(1/50 - 1/60)(2.386) + (1/40 - 1/50)(2.266)}{(1/40 - 1/60)}$$

$$= \frac{(0.0033)(2.386) + (0.0050)(2.266)}{0.0083}$$

$$= 2.314$$

Finally,

$$F_{50,10,.025} = 1/2.314 = 0.432$$

8.15 Test the hypothesis $H_0: \sigma_1^2 = \sigma_2^2$ versus H_1: $\sigma_1^2 \neq \sigma_2^2$. We have the test statistic $F = s_2^2/s_1^2 = (0.76/0.64)^2 = 1.410 \sim F_{39,24}$ under H_0. Since $F < F_{\infty,30,.975} = 1.79 < F_{39,24,.975}$, it follows that H_0 is accepted at the 5% level and we conclude that there is no significant difference between the variances.

8.16 Since H_0 was accepted in Problem 8.15, use the two-sample t test for independent samples with equal variances.

8.17 First compute the pooled variance estimate:

$$s^2 = \frac{(n_1 - 1)s_1^2 + (n_2 - 1)s_2^2}{n_1 + n_2 - 2}$$

$$= \frac{24(0.64)^2 + 39(0.76)^2}{63} = \frac{32.357}{63} = 0.514$$

Compute the test statistic

$$t = \frac{\bar{x}_1 - \bar{x}_2}{\sqrt{s^2(1/n_1 + 1/n_2)}}$$

$$= \frac{6.56 - 6.80}{\sqrt{0.514(1/25 + 1/40)}} = \frac{-0.24}{0.183} = -1.311$$

The critical value is given by $t_{63,.975} > t_{120,.975} = 1.980 > |t|$. Therefore, accept H_0 at the 5% level.

8.18 The p-value is given by $2 \times Pr(t_{63} < -1.311) = 2 \times Pr(t_{63} > 1.311)$. If there were 60 df, then since

$t_{60,.9} = 1.296$, $t_{60,.95} = 1.671$, and $1.296 < 1.311 < 1.671$, it would follow that $2 \times (1 - .95) < p < 2 \times (1 - .9)$, or $.10 < p < .20$. If there were 120 df, then since $t_{120,.9} = 1.289$, $t_{120,.95} = 1.658$, and $1.289 < 1.311 < 1.658$, it would follow that $.10 < p < .20$. Since we reach the same conclusion with either 60 or 120 df and $60 < 63 < 120$, it follows that $.10 < p < .20$.

8.19 The 95% confidence interval is given by $\bar{x}_1 - \bar{x}_2 \pm t_{63,.975}\sqrt{s^2(1/n_1 + 1/n_2)}$. We interpolate to obtain $t_{63,.975}$.

$$t_{63,.975} = \frac{(1/63 - 1/120)(2.000) + (1/60 - 1/63)(1.980)}{(1/60 - 1/120)}$$

$$= \frac{(0.0075)(2.000) + (0.0008)(1.980)}{0.0083} = 1.998$$

Therefore, the 95% confidence interval is given by $-0.24 \pm 1.998(0.183) = -0.24 \pm 0.366 = (-0.61, 0.13)$.

8.33 We have the sample-size formula

$$n = \frac{(\sigma_1^2 + \sigma_2^2)(z_{1-\alpha/2} + z_{1-\beta})^2}{\Delta^2}$$

$$= \frac{[(0.64^2 + 0.76^2)(z_{.975} + z_{.80})^2]}{(0.24)^2}$$

$$= \frac{0.9872(1.96 + 0.84)^2}{0.0576} = 134.4$$

Thus, 135 girls in each group or 270 girls in total need to be recruited.

8.34 Use the sample-size formula

$$n = \frac{(0.64^2 + 0.76^2)(z_{.95} + z_{.80})^2}{(0.24)^2}$$

$$= \frac{(0.9872)(1.645 + 0.84)^2}{0.0576} = 105.8$$

Thus, 106 girls in each group or 212 girls in total need to be recruited using a one-sided test.

8.35 Use the following sample-size formula for the below-poverty-level group (group 1)

$$n_1 = \frac{(\sigma_1^2 + \sigma_2^2/2)(z_{1-\alpha/2} + z_{1-\beta})^2}{\Delta^2}$$

$$= \frac{(0.64^2 + 0.76^2/2)(z_{.975} + z_{.80})^2}{0.24^2}$$

$$= \frac{(0.6984)(1.96 + 0.84)^2}{0.0576} = 95.1$$

Thus, 96 women in the below-poverty-level group and $2 \times 96 = 192$ women in the above-poverty-level group need to be recruited.

8.45 The basic statistics for total heart weight (THW) and body weight (BW) in the two groups are given in Table A.13.

TABLE A.13 Total heart weight and body weight in diseased and normal males

	Left heart disease			Normals		
	Mean	sd	n	Mean	sd	n
THW	450.0	139.34	11	317.0	47.09	10
BW	55.61	11.55	11	56.23	11.54	10

THW Perform the F test for the equality of two variances as follows:

$$F = s_1^2/s_2^2 = (139.34/47.09)^2 = 8.76 \sim F_{10,9} \text{ under } H_0$$

We have

$$F_{10,9,.995} < F_{8,9,.995} = 6.69 < F$$

Thus, the p-value for F is $<2(.005) = .01$, and the variances are significantly different. The t test with unequal variances must be used. We have the following test statistic:

$$t = \frac{\bar{x}_1 - \bar{x}_2}{\sqrt{\dfrac{s_1^2}{n_1} + \dfrac{s_2^2}{n_2}}} = \frac{450.0 - 317.0}{\sqrt{\dfrac{(139.34)^2}{11} + \dfrac{(47.09)^2}{10}}}$$

$$= \frac{133.0}{44.57} = 2.98$$

The appropriate df (d') must be computed.

$$d' = \frac{\left[\dfrac{139.34^2}{11} + \dfrac{47.09^2}{10}\right]^2}{\left(\dfrac{139.34^2}{11}\right)^2 \Big/ 10 + \left(\dfrac{47.09^2}{10}\right)^2 \Big/ 9}$$

$$= \frac{3,947,392}{317,006} = 12.45$$

Thus, there are 12 df. Since $|t| > t_{12,.975} = 2.179$, it follows that H_0 is rejected at the 5% level.

8.46 BW Again perform the F test for the equality of two variances.

$$F = s_1^2/s_2^2 = (11.55/11.54)^2 = 1.002 \sim F_{10,9}$$

which is clearly not statistically significant. The t test with equal variances is therefore used. We have

$$s^2 = \frac{(n_1 - 1)s_1^2 + (n_2 - 1)s_2^2}{n_1 + n_2 - 2} = \frac{10(11.55)^2 + 9(11.54)^2}{19}$$

$$= 133.29$$

$$t = \frac{\bar{x}_1 - \bar{x}_2}{\sqrt{s^2\left(\dfrac{1}{n_1} + \dfrac{1}{n_2}\right)}} = \frac{55.61 - 56.23}{\sqrt{133.29\left(\dfrac{1}{11} + \dfrac{1}{10}\right)}} = \frac{-0.62}{5.044}$$

$$= -0.12 \sim t_{19}$$

which is not statistically significant. Thus, there is no significant difference in body weight between the two groups.

8.55 The distinction between a one-sided and two-sided test in this case is that for a one-sided test we would test the hypothesis $H_0: \mu_1 = \mu_2$ versus $H_1: \mu_1 > \mu_2$ or, alternatively, $H_0: \mu_1 = \mu_2$ versus $H_1: \mu_1 < \mu_2$, where μ_1 represents mean systolic bp sitting upright and μ_2 represents mean systolic bp lying down. For a two-sided test we would test the hypothesis $H_0: \mu_1 = \mu_2$ versus $H_1: \mu_1 \neq \mu_2$.

8.56 A two-sided test is appropriate here, since we have no preconceived notions as to the relative orderings of μ_1 and μ_2 and would be equally interested in the outcomes $\mu_1 < \mu_2$ and $\mu_1 > \mu_2$.

8.57 Since each person is serving as his or her own control, we are dealing with highly dependent samples and must use the paired t test. We test the hypothesis $H_0: \mu_d = 0$ versus $H_1: \mu_d \neq 0$, where d_i = sitting bp − lying bp for the ith person and $d_i \sim N(\mu_d, \sigma_d^2)$. We have the following set of within-pair differences: $-12, -6, +2, -8, -8, 0, -12, +4, 0, -16$. Compute the test statistic

$$t = \bar{d}/(s_d/\sqrt{n}) = -5.60/(6.786/\sqrt{10}) = -5.60/2.146$$

$$= -2.61.$$

Under H_0, $t \sim t_9$ and we have from Table 5 in Appendix 1 that $t_{9,.975} = 2.262 < |t|$.

Therefore, H_0 would be rejected at the 5% level and the hypothesis that the position affects bp, with the upright position having the lower blood pressure, would be accepted.

8.58 We have $t_{9,.975} = 2.262$, $t_{9,.99} = 2.821$. Since $t_{9,.99} = 2.821 > |t|$, $.02 < p < .05$.

8.59 Test the hypotheses $H_0: \mu_d = 0$ versus $H_1: \mu_d \neq 0$, where μ_d represents the mean difference in 1-hour concentration (drug A − drug B) in a specific person. Use a paired t test to test these hypotheses. An independent samples t test cannot be used in this case, since the two samples are from the same people and are not independent.

8.60 The assumption behind this test is that $d_i =$ difference in 1-hour concentration for the ith person is normally distributed with mean μ_d and variance σ_d^2.

8.61 We have the d_i in Table A.14.

TABLE A.14 Difference in 1-hour urine concentration between type A and type B aspirin

Person (i)	1	2	3	4	5	6	7	8	9	10
d_i	2	6	3	7	0	−2	2	6	5	7

It follows that $\bar{d} = 3.60$, $s_d = 3.098$. Thus,

$$t = \bar{d}/(s_d/\sqrt{n}) = 3.60/(3.098/\sqrt{10})$$
$$= 3.60/0.980 = 3.67 \sim t_9$$

Since $t_{9,.995} = 3.250$, $t_{9,.9995} = 4.781$ based on a two-sided test, it follows that $.001 < p < .01$, and H_0 is rejected and we conclude that aspirin A has a significantly higher concentration in urine specimens than aspirin B does.

8.62 The best point estimate is $\bar{d} = 3.60$ mg%.

8.63 A 95% confidence interval is given by

$$\bar{d} \pm t_{9,.975}(s_d/\sqrt{10}) = 3.60 \pm 2.262(3.098)/\sqrt{10}$$
$$= (1.38, 5.82)$$

8.64 If the test result had been significant at the 5% level, then the confidence interval would have excluded 0; otherwise, it would have included 0. The former possibility is what actually occurred.

8.78 Test the hypotheses $H_0: \sigma_1^2 = \sigma_2^2$ versus $H_1: \sigma_1^2 \neq \sigma_2^2$. Use the F test with test statistic

$$F = \frac{s_1^2}{s_2^2} = \left(\frac{7.3}{2.7}\right)^2 = 7.31 \sim F_{39,39} \text{ under } H_0$$

Since $F_{39,39,.975} < F_{24,30,.975} = 2.14 < F$, it follows that $p < .05$, and the variances (and thus the standard deviations) of the two groups are significantly different.

8.79 Test the hypothesis $H_0: \mu_1 = \mu_2$, $\sigma_1^2 \neq \sigma_2^2$ versus $H_1: \mu_1 \neq \mu_2$, $\sigma_1^2 \neq \sigma_2^2$. Use the two-sample t test with unequal variances because H_0 was rejected in Problem 8.78. We have the test statistic

$$t = \frac{\bar{x}_1 - \bar{x}_2}{\sqrt{\dfrac{s_1^2}{n_1} + \dfrac{s_2^2}{n_2}}} = \frac{11.6 - 6.9}{\sqrt{\dfrac{(7.3)^2}{40} + \dfrac{(2.7)^2}{40}}} = \frac{4.7}{1.231} = 3.82$$

Compute the appropriate df (d') as follows:

$$d' = \frac{\left(\dfrac{7.3^2}{40} + \dfrac{2.7^2}{40}\right)^2}{\left(\dfrac{7.3^2}{40}\right)^2 \Big/ 39 + \left(\dfrac{2.7^2}{40}\right)^2 \Big/ 39} = \frac{2.294}{0.046} = 49.9$$

Thus, there are 49 df. Since $t = 3.82 > t_{40,.975} = 2.021 > t_{49,.975}$, it follows that $p < .05$ and H_0 is rejected at the 5% level. Thus, there is a significant difference between the CO concentrations in the two working environments.

8.80 The standard deviations of the two groups were significantly different in Problem 8.78, and thus a two-sample t test with *unequal* variances was used. If the standard deviations were not significantly different, then a two-sample t test with *equal* variances would have been used.

8.106 The hypotheses to be tested are $H_0: \mu_d = 0$ versus $H_1: \mu_d > 0$, where $\mu_d =$ underlying mean change in temperature for a specific patient after taking aspirin. A one-sided alternative is chosen here, since it is only plausible that aspirin could reduce temperature.

8.107 A type I error here is the probability of declaring that aspirin reduces temperature given that it, in fact, has no effect on temperature.

8.108 The power of the test versus this alternative is the probability of declaring that aspirin reduces temperature given that the underlying mean reduction in temperature is 1 degree.

8.109 The power would increase because the alternative mean is further away from 0.

8.110 We have the power formula

$$\text{Power} = \Phi\left(z_\alpha + \frac{\Delta\sqrt{n}}{\sigma_d}\right)$$

In this case, $n = 12$, $\alpha = .05$, and σ is estimated by $s_d = 0.869$. Compute the power for $\Delta = 1$ and 2.

$$\text{Power } (\Delta = 1) = \Phi\left(z_{0.05} + \frac{\sqrt{12}}{0.869}\right)$$

$$= \Phi(-1.645 + 3.986)$$

$$= \Phi(2.34) = 0.990$$

$$\text{Power } (\Delta = 2) = \Phi\left(z_{0.05} + \frac{2\sqrt{12}}{0.869}\right)$$

$$= \Phi(-1.645 + 7.973)$$

$$= \Phi(6.33) = 1.0$$

8.111 Perform a paired t test to test the hypotheses in Problem 8.106. We have the set of within-pair differences given in Table A.15. It follows that $\bar{d} = 1.75$, $s_d = 0.869$.

TABLE A.15 The temperature reduction 1-hour after taking aspirin for 12 5-year-old girls

i	d_i	i	d_i
1	+2.8	7	+1.5
2	+3.1	8	+3.0
3	+1.7	9	+2.1
4	+1.9	10	+1.2
5	+1.4	11	+0.6
6	+0.5	12	+1.2

The paired t statistic is given by

$$t = \frac{\bar{d}}{(s_d/\sqrt{n})} = \frac{1.75}{(0.869/\sqrt{12})} = \frac{1.75}{0.251} = 6.97 \sim t_{11}$$

under H_0. Since $t_{11,.9995} = 4.437$, the one-tailed p-value $<.0005$, indicating that aspirin intake has a very highly significant effect on reducing temperature.

8.141 Since each person is being used as his or her own control, the paired t test is the appropriate test procedure, where the hypothesis $H_0: \mu_d = 0$ versus $H_1: \mu_d \neq 0$ and $\mu_d =$ true mean change in urinary protein over the 8-week period is tested.

8.142 Calculate the difference scores in the raw scale as follows in Table A.16:

TABLE A.16

i	d_i	i	d_i
1	15.5	6	7.2
2	11.3	7	−0.3
3	10.4	8	0.7
4	7.0	9	3.1
5	1.7	10	1.8

Then compute the test statistic:

$$t = \frac{\bar{d}}{s_d/\sqrt{n}}$$

$$= \frac{5.84}{5.294/\sqrt{10}} = \frac{5.84}{1.674} = 3.49 \sim t_9 \text{ under } H_0$$

Since $t_{9,.995} = 3.250$ and $t_{9,.9995} = 4.781$, and $3.250 < 3.49 < 4.781$, it follows that $2 \times (1 - .9995) < p < 2 \times (1 - .995)$, or $.001 < p < .01$. Calculate the difference scores in the log scale as follows in Table A.17:

TABLE A.17

Person (i)	d_i	Person (i)	d_i
1	0.93	6	2.43
2	1.09	7	−0.05
3	1.05	8	0.14
4	1.12	9	0.94
5	0.23	10	0.49

Then compute the test statistic:

$$t = \frac{\bar{d}}{s_d/\sqrt{n}}$$

$$= \frac{0.837}{0.708/\sqrt{10}} = \frac{0.837}{0.224} = 3.74 \sim t_9 \text{ under } H_0$$

Since $t_{9,.995} = 3.250$, $t_{9,.9995} = 4.781$, and $3.250 < 3.74 < 4.781$, it follows that $2 \times (1 - .9995) < p < 2 \times (1 - .995)$, or $.001 < p < .01$. Thus, there is a significant decline in $\log_e$ (urinary protein) over the 8-week period. Since the difference scores appear to be related to the initial urinary protein, the log scale is preferable here.

8.143 The best estimate of mean $[\log_e(X_{1i}) - \log_e(X_{2i})] = \bar{d} = 0.837$ from Problem 8.142 where $X_{1i} =$ urinary protein at baseline for the ith person, $X_{2i} =$ urinary protein at 8 weeks for the ith person. However, $\log_e(X_{1i}) - \log_e(X_{2i}) = \log_e(X_{1i}/X_{2i})$, and, therefore, the best estimate of $X_{1i}/X_{2i} = e^{\bar{d}} = e^{0.837} = 2.309$ or $X_{2i}/X_{1i} = e^{-0.837} = 0.433$. Thus, the best estimate is that urinary protein has declined by 56.7% (i.e., $100\% - 43.3\%$) over the 8-week period.

8.144 A 95% confidence interval for μ_d in the log scale is given by

$$\bar{d} \pm t_{9,.975} s_d/\sqrt{n}$$

$$= 0.837 \pm 2.262\,(0.224)$$

$$= 0.837 \pm 0.507 = (0.330, 1.344)$$

Thus, a 95% confidence interval for the percentage change from time 1 to time $2 = (e^{-1.344}, e^{-0.330}) = (0.261, 0.719)$. This is equivalent to a decline from $100\% - 71.9\% = 28.1\%$ to $100\% - 26.1\% = 73.9\%$. Thus, the 95% confidence interval for percentage decline from time 1 to time 2 is $(28.1\%, 73.9\%)$.

CHAPTER 9

9.1 Use the sign test. If the people who have remained the same are ignored, then 27 people have either improved or declined. Thus, the normal theory version of the test can be used. We have that $C = $ number of patients improved $= 20$, $n = 27$. Reject H_0 if either

$$C > \frac{n}{2} + \frac{1}{2} + z_{1-\alpha/2}\sqrt{n/4} = c_2$$

or

$$C < \frac{n}{2} - \frac{1}{2} - z_{1-\alpha/2}\sqrt{n/4} = c_1$$

and accept H_0 otherwise. We have that $\alpha = .05$. Therefore,

$$c_2 = \frac{27}{2} + \frac{1}{2} + z_{.975}\sqrt{27/4}$$

$$= 14 + 1.96(2.598) = 14 + 5.09$$

$$= 19.09$$

$$c_1 = 14 - 5.09 = 8.91$$

Since $C = 20 > 19.09$, reject H_0 at the 5% level. The exact p-value is given by

$$p = 2 \times \left[1 - \Phi\frac{(C - n/2) - 0.5}{\sqrt{n/4}}\right]$$

$$= 2 \times \left[1 - \Phi\left(\frac{(20 - 27/2) - 0.5}{\sqrt{27/4}}\right)\right]$$

$$= 2 \times \left[1 - \Phi\left(\frac{6.5 - 0.5}{2.598}\right)\right] = 2 \times \left[1 - \Phi\left(\frac{6.0}{2.598}\right)\right]$$

$$= 2 \times (1 - .9896) = .021.$$

9.2 In the subgroup of patients with better visual acuity, 13 patients have either improved or declined. Thus, the exact binomial test must be used. We have $C = 8$, $n = 13$. Refer to the exact binomial tables under $n = 13$, $p = .5$ to compute

$$p = 2 \times \sum_{k=8}^{13} \binom{13}{k}\left(\frac{1}{2}\right)^{13}$$

$$= 2 \times (.1571 + .0873 + .0349 + .0095 + .0016 + .0001)$$

$$= 2 \times .2905 = .581$$

Thus, there is no significant change in VA in the subgroup of patients with VA 20-40 or better.

9.3 Fourteen patients have either improved or declined. Thus, the exact binomial test must again be used. We have $C = 12$, $n = 14$. Refer to the exact binomial tables under $n = 14$, $p = .5$ to compute

$$p = 2 \times \sum_{k=12}^{14} \binom{14}{k}\left(\frac{1}{2}\right)^{14}$$

$$= 2 \times (.0056 + .0009 + .0001)$$

$$= 2 \times .0066 = .013$$

Thus, the visual acuity of persons with VA of worse than 20-40 at baseline has significantly improved.

9.8 The Wilcoxon signed rank test.

9.9 First rank the data by absolute value of the change score (d_i) as follows in Table A.18:

TABLE A.18

$|d_i|$	Negative d_i	Frequency	Positive d_i	Frequency	Total frequency	Range of ranks	Average rank
3	−3	2	3	4	6	18–23	20.5
2	−2	2	2	5	7	11–17	14.0
1	−1	4	1	6	10	1–10	5.5
		8		15	23		
0	0	5					

Since there are 23 pairs with a non-zero d_i, the normal theory test can be used. We have that the rank sum of the positive differences $(R_1) = 4 \times 20.5 + 5 \times 14.0 + 6 \times 5.5 = 185.0$. The test statistic is given by

$$T = \frac{\left| R_1 - \frac{n(n+1)}{4} \right| - .5}{\sqrt{\frac{n(n+1)(2n+1)}{24} - \sum_{i=1}^{g} \frac{(t_i^3 - t_i)}{2}}}$$

$$= \frac{\left| 185 - \frac{23(24)}{4} \right| - .5}{\sqrt{\frac{23(24)(47)}{24} - \frac{(6^3 - 6) + (7^3 - 7) + (10^3 - 10)}{2}}}$$

$$= \frac{|185 - 138| - .5}{\sqrt{1081 - \frac{210 + 336 + 990}{2}}} = \frac{46.5}{\sqrt{1081 - 768}}$$

$$= \frac{46.5}{17.692} = 2.628 \sim N(0, 1) \text{ under } H_0$$

The p-value is obtained from

$$p = 2 \times [1 - \Phi(2.628)]$$
$$= 2 \times (1 - .9957) = .009$$

Thus, the periodontal status of the patients has significantly improved over time.

9.12 The exact tables for the signed rank test must be used, since the number of untied pairs is less than 16. Refer to Table 9 in the Appendix under $n = 15$ and note that the upper critical value for $\alpha = .10 = 90$, whereas the upper critical value for $\alpha = .05$ is 95. Since $R_1 = 90 \geqslant 90$ and $90 < 95$, we have $.05 \leqslant p < .10$ using a two-sided test.

9.15 The normal theory test can be used, since $\min(n_1, n_2) = 12 \geqslant 10$. The test statistic is given by

$$T = \frac{\left| R_1 - \frac{n_1(n_1 + n_2 + 1)}{2} \right| - .5}{\sqrt{\frac{n_1 n_2 (n_1 + n_2 + 1)}{12}}}$$

$$= \frac{\left| 220 - \frac{12(28)}{2} \right| - .5}{\sqrt{\frac{12(15)(28)}{12}}} = \frac{|220 - 168| - .5}{20.494}$$

$$= \frac{52.0 - .5}{20.494}$$

$$= \frac{51.5}{20.494} = 2.513 \sim N(0, 1) \text{ under } H_0$$

The p-value is given by

$$p = 2 \times [1 - \Phi(2.513)] = 2 \times (1 - .9940) = .012$$

Thus, there is a significant difference between the two groups.

9.22 The distribution of length of stay in a hospital is notoriously very skewed and far from being normal. This is due to the relatively short stays of most patients and the very long stays of a relatively small number of patients. To use the t test, we would have to assume underlying normality of the length-of-stay distribution or at the very least that mean length of stay $(\bar{x})$ was normally distributed for moderate sample sizes, which is unlikely to be the case here.

9.23 The Wilcoxon rank sum test can instead be used to test if the median length of stay is significantly different in the two hospitals. First rank the length of stay in the combined sample, as given in Table A.19.

Next compute the rank sum for hospital 1 as follows:

$$R_1 = 1.0 + 2.0 + 3.5 + 5.0 + 6.0 + 7.0 + 9.0 + 10.0$$
$$+ 11.0 + 13.5 + 15.5 = 83.5$$

We will assume that R_1 is normally distributed. Under H_0 we know that

$$E(R_1) = n_1(n_1 + n_2 + 1)/2 = 11(25)/2 = 137.5$$

$Var(R_1)$

$$= \frac{n_1 n_2}{12} \left[n_1 + n_2 + 1 - \frac{\sum_{i=1}^{q} t_i (t_i^2 - 1)}{(n_1 + n_2)(n_1 + n_2 - 1)} \right]$$

$$= \frac{11(13)}{12} \left[25 - \frac{2(3) + 2(3) + 2(3)}{24(23)} \right] = 297.528$$

$$sd(R_1) = 17.249$$

Thus, compute the test statistic

$$T = (|R_1 - E(R_1)| - .5)/sd(R_1)$$
$$= (|83.5 - 137.5| - .5)/17.249$$
$$= 53.5/17.249 = 3.10 \sim N(0, 1)$$

The two-sided p-value is given by

$$2 \times [1 - \Phi(3.10)] = 2 \times (1 - .9990) = .002$$

Thus, there is a significant difference in length of stay between the two hospitals, with hospital 2 patients staying longer. The patient characteristics in the two hospitals would have to be assessed before concluding that this difference was due to procedural variations between the two hospitals.

TABLE A.19 Data layout for length-of-stay data for Wilcoxon rank sum test

Value	Frequency, hospital 1	Frequency, hospital 2	Total frequency	Rank range	Average rank
5	1	0	1	1	1.0
8	1	0	1	2	2.0
10	1	1	2	3–4	3.5
13	1	0	1	5	5.0
21	1	0	1	6	6.0
26	1	0	1	7	7.0
27	0	1	1	8	8.0
29	1	0	1	9	9.0
32	1	0	1	10	10.0
33	1	0	1	11	11.0
35	0	1	1	12	12.0
44	1	1	2	13–14	13.5
60	1	1	2	15–16	15.5
68	0	1	1	17	17.0
73	0	1	1	18	18.0
76	0	1	1	19	19.0
86	0	1	1	20	20.0
87	0	1	1	21	21.0
96	0	1	1	22	22.0
125	0	1	1	23	23.0
238	0	1	1	24	24.0
Total	11	13	24		

9.37 The hypotheses being tested are H_0: median duration of effusion of breast-fed babies = median duration of effusion of bottle-fed babies versus H_1: median duration of effusion of breast-fed babies < median duration of effusion of bottle-fed babies.

9.38 A nonparametric test would be useful because the distribution of duration of effusion is very skewed and the assumptions about normality of the underlying distribution are unlikely to hold.

9.39 The Wilcoxon signed rank test should be used here because the breast- and bottle-fed babies are matched on age, sex, socioeconomic status, and type of medications and thus form two paired samples.

9.40 Apply the signed rank test to these data. First compute the difference (d_i) in duration of effusion between the breast- and bottle-fed babies in the matched pairs, as given in Table A.20.

Now separate the positive and negative differences and order the differences by absolute value (Table A.21).

Then count the number of people with the same absolute value and assign an average rank to each

TABLE A.20 Difference in duration of effusion between breast-fed and bottle-fed babies

i	d_i	i	d_i
1	+2	13	+13
2	−24	14	−1
3	−4	15	−9
4	−158	16	+2
5	+1	17	−1
6	−5	18	−12
7	−165	19	−12
8	0	20	−2
9	−18	21	−1
10	−59	22	+3
11	−169	23	+6
12	−17	24	+5

absolute value, as shown in Table A.21. Since the number of non-zero differences = 23 ⩾ 16, the normal approximation test in **(9.6)** can be used. Compute the

TABLE A.21 Data layout for duration of effusion for Wilcoxon signed rank test

$|d_i|$	Negative d_i	f_i	Positive d_i	f_i	Number of persons with same absolute value	Range of ranks	Average rank
169	−169	1	169	0	1	23	23.0
165	−165	1	165	0	1	22	22.0
158	−158	1	158	0	1	21	21.0
59	−59	1	59	0	1	20	20.0
24	−24	1	24	0	1	19	19.0
18	−18	1	18	0	1	18	18.0
17	−17	1	17	0	1	17	17.0
13	−13	0	13	1	1	16	16.0
12	−12	2	12	0	2	14–15	14.5
9	−9	1	9	0	1	13	13.0
6	−6	0	6	1	1	12	12.0
5	−5	1	5	1	2	10–11	10.5
4	−4	1	4	0	1	9	9.0
3	−3	0	3	1	1	8	8.0
2	−2	1	2	2	3	5–7	6.0
1	−1	3	1	1	4	1–4	2.5
0	0	1					

rank sum of the negative differences as follows:

$$R_1 = 3(2.5) + 1(6.0) + \cdots + 1(23.0)$$

$$= 215$$

The expected value and variance of the rank sum are given as follows:

$$E(R_1) = \frac{n(n+1)}{4} = \frac{23(24)}{4} = 138$$

$$Var(R_1) \doteq \frac{n(n+1)(2n+1)}{24} - \frac{\sum_{i=1}^{g}(t_i^3 - t_i)}{2}$$

$$= \frac{23(24)(47)}{24}$$

$$- \frac{(4^3 - 4) + (3^3 - 3) + (2^3 - 2) + (2^3 - 2)}{2}$$

$$= 1081 - \frac{96}{2} = 1033$$

$$sd(R_1) = 32.14$$

The test statistic is then obtained from

$$T = \frac{215 - 138 - 1/2}{32.14} = 2.38 \sim N(0, 1) \text{ under } H_0$$

It follows that

$$p = 1 - \Phi(2.38) = 1 - .9913 = .009$$

Thus, breast-fed babies have significantly shorter effusions than bottle-fed babies do.

9.49 A nonparametric statistical test would be useful because the change in redness or itching cannot be numerically quantified; only which eye has improved more can be assessed.

9.50 The degree of redness or itching is probably similar in two eyes of the same person at baseline. Thus, this procedure would eliminate the substantial between-person variability that can occur if the drug were administered to one group of people and the placebo to another group.

9.51 The sign test should be used here because we are comparing two paired samples consisting of alternate eyes from the same people and we only know which eye did better but not how much better.

9.52 There are 5 +'s and 11 −'s for redness. Since there are < 20 non-zero differences, the exact test must be used. In particular, the p-value is given by

$$p = 2 \times \sum_{k=0}^{5} \binom{16}{k} \left(\frac{1}{2}\right)^{16}$$

From the binomial tables (Table 1 in the Appendix 1) using $n = 16$, $p = .50$,

$$Pr(0) = .0000 \qquad Pr(1) = .0002$$
$$Pr(2) = .0018 \qquad Pr(3) = .0085$$
$$Pr(4) = .0278 \qquad Pr(5) = .0667$$

Thus,

$$p = 2 \times (.0000 + .0002 + .0018 + .0085$$
$$+ .0278 + .0667)$$
$$= .210$$

Thus, there is no significant difference in redness between the two sets of treated eyes.

9.53 There are 12 +'s and 3 −'s for itching. Again use the exact test, calculated from the binomial tables, with the p-value given by

$$p = 2 \times \sum_{k=0}^{3} \binom{15}{k}\left(\frac{1}{2}\right)^{15}$$

$$= 2 \times (.0000 + .0005 + .0032 + .0139) = .035$$

Thus, there is a significant difference in itching between the drug-treated and placebo-treated eyes. This is what is expected, since drug A is only supposed to be effective against itching.

9.54 For redness there are 22 +'s and 3 0's. We can use the sign test using the normal approximation, since we have 22 untied pairs. The test statistic is given by

$$z = \frac{\left|C - \dfrac{n}{2}\right| - .5}{\sqrt{\dfrac{n}{4}}} = \frac{|22 - 11| - .5}{\sqrt{\dfrac{22}{4}}} = \frac{10.5}{2.345}$$

$$= 4.48 \sim N(0, 1) \text{ under } H_0$$

The p-value is

$$2 \times [1 - \Phi(z)] = 2 \times [1 - \Phi(4.48)] < .001$$

There is clearly overwhelming evidence that drug B is effective against redness.

9.55 For itching there are 7 +'s and 6 −'s. We use the exact binomial version of the sign test. The p-value is obtained from the binomial tables as follows:

$$p = 2 \times \sum_{k=0}^{6} \binom{13}{k}\left(\frac{1}{2}\right)^{13}$$

$$= 2 \times (.0001 + .0016 + .0095 + .0349$$
$$+ .0873 + .1571 + .2095)$$
$$= 1.0$$

There is clearly no significant difference in itching between treated and untreated eyes with drug B. These results are again consistent with our expectations for drug B, which is only supposed to be effective against redness.

9.56 For redness there are 23 +'s and 2 0's. This result is even more extreme than that in Problem 9.54, and if we use the sign test with the normal approximation, we surely will find $p < .001$. Thus, the combination drug is very effective vs. redness.

9.57 For itching there are 14 +'s and 3 −'s. We use the exact binomial version of the sign test as follows:

$$p = 2 \times \sum_{k=0}^{3} \binom{17}{k}\left(\frac{1}{2}\right)^{17}$$

$$= 2 \times (.0000 + .0001 + .0010 + .0052)$$

$$= .013$$

Thus, the combination drug is significantly better than the placebo for itching as well, but not as strongly as for redness.

9.58 The results of the three experiments agree with our prior hypotheses. Indeed, drug A is significantly better than the placebo for itching ($p = .035$) but not for redness; drug B is significantly better than the placebo for redness ($p < .001$) but not for itching; the combination drug is significantly better than the placebo for both redness ($p < .001$) and itching ($p = .013$).

CHAPTER 10

10.1 Test the hypothesis $H_0: p_1 = p_2$ versus $H_1: p_1 \neq p_2$. The test statistic is given by

$$Z = \frac{\hat{p}_1 - \hat{p}_2}{\sqrt{\hat{p}\hat{q}(1/n_1 + 1/n_2)}}$$

where $\hat{p}_1 = 9/199 = .0452$, $\hat{p}_2 = 13/97 = .1340$, $\hat{p} = (9 + 13)/(199 + 97) = 22/296 = .0743$, $\hat{q} = 1 - \hat{p} = 0.9257$.

$$Z = \frac{.0452 - .1340}{\sqrt{(.0743)(.9257)(1/199 + 1/97)}}$$

$$= \frac{-.0888}{.0325} = -2.7323 \sim N(0, 1) \text{ under } H_0$$

Since $Z < -1.96$, reject H_0 at the 5% level.

10.2 The observed table is given by:

TABLE A.22 12-month mortality status

	Dead	Alive	
Streptokinase	9	190	199
Control	13	84	97
	22	274	296

The expected cell counts are obtained from the row and column margins as follows:

$$E_{11} = \frac{199 \times 22}{296} = 14.79$$

$$E_{12} = \frac{199 \times 274}{296} = 184.21$$

$$E_{21} = \frac{97 \times 22}{296} = 7.21$$

$$E_{22} = \frac{97 \times 274}{296} = 89.79$$

These values can be displayed in Table A.23:

TABLE A.23 12-month mortality status

	Dead	Alive	
Streptokinase	14.79	184.21	199
Control	7.21	89.79	97
	22	274	296

10.3 Compute the Yates-corrected chi-square statistic as follows:

$$X^2 = \frac{(|9 - 14.79| - .5)^2}{14.79} + \cdots + \frac{(|84 - 89.79| - .5)^2}{89.79}$$

$$= \frac{5.29^2}{14.79} + \cdots + \frac{5.29^2}{89.79}$$

$$= 1.892 + 0.152 + 3.881 + 0.312$$

$$= 6.24 \sim \chi_1^2 \text{ under } H_0$$

Since $\chi_{1,.95}^2 = 3.84 < X^2$, reject H_0 at the 5% level.

10.4 The decisions reached in Problems 10.1 and 10.3 were the same (reject H_0 at the 5% level). If a chi-square test had been used without continuity correc-

tion, the results would have been identical, since $X^2 = (5.79)^2/14.79 + \cdots + (5.79)^2/89.79 = 7.47 = (2.7323)^2 = Z^2$ and the same p-value is obtained whether Z is compared to an $N(0, 1)$ distribution or $X^2 = Z^2$ to a χ_1^2 distribution.

10.21 Form the following observed 2×2 table:

TABLE A.24 12-month mortality status

	Dead	Alive	
Streptokinase	2	13	15
Control	4	15	19
	6	28	34

The smallest expected value $= E_{11} = (15 \times 6)/34 = 2.65 < 5$. Thus, Fisher's exact test must be used.

10.22

TABLE A.25

0	15	1	14	2	13	3	12
6	13	5	14	4	15	3	16

4	11	5	10	6	9
2	17	1	18	0	19

10.23 Use the recursion rule as follows:

$$Pr(0) = 1 \times Pr(0)$$

$$Pr(1) = \frac{6 \times 15}{1 \times 14} Pr(0) = 6.429 \, Pr(0)$$

$$Pr(2) = \frac{5 \times 14}{2 \times 15} Pr(1) = 15.001 \, Pr(0)$$

$$Pr(3) = \frac{4 \times 13}{3 \times 16} Pr(2) = 16.251 \, Pr(0)$$

$$Pr(4) = \frac{3 \times 12}{4 \times 17} Pr(3) = 8.603 \, Pr(0)$$

$$Pr(5) = \frac{2 \times 11}{5 \times 18} Pr(4) = 2.103 \, Pr(0)$$

$$Pr(6) = \frac{1 \times 10}{6 \times 19} Pr(5) = 0.184 \, Pr(0)$$

Thus,

$$Pr(0) \times (1 + 6.429 + \cdots + 0.184) = 1$$

or

$$Pr(0) \times 49.571 = 1$$

or

$$Pr(0) = \frac{1}{49.571} = .0202$$

Furthermore,

$$Pr(1) = .130, \quad Pr(2) = .303, \quad Pr(3) = .328,$$

$$Pr(4) = .174, \quad Pr(5) = .042, \quad Pr(6) = .004$$

10.24 Our table is the "2" table. Therefore, the two-tailed p-value is given by

$$p = 2 \times \min[Pr(0) + Pr(1) + Pr(2),$$
$$Pr(2) + Pr(3) + \cdots + Pr(7)]$$

$$= 2 \times \min(.453, .851) = 2 \times .453 = .906$$

Clearly, there is no significant difference in 12-month mortality status between the two treatment groups for females.

10.29 87   **10.30** 13

10.31 Since there are less than 20 discordant pairs, the exact binomial version of McNemar's test must be used. Refer to the exact binomial tables (Table 1 in Appendix 1) with $n = 13$, $p = .50$ to evaluate

$$p = 2 \times \sum_{k=9}^{13} {}_{13}C_k (1/2)^{13}$$

$$= 2 \times (.0873 + .0349 + .0095 + .0016 + .0001)$$

$$= 2 \times .1334 = .267$$

Therefore, there is no significant difference between the effectiveness of the two drugs for males.

10.41 Test the hypothesis $H_0: p_A = p_B$ versus $H_1: p_A \neq p_B$, where

$p_A = Pr(\text{first-born child has asthma in a type A family})$

$p_B = Pr(\text{first-born child has asthma in a type B family})$

The observed and expected 2×2 tables are shown in Table A.26 and Table A.27, respectively. The χ^2 test for 2×2 tables will be used, since the expected table has no *expected* value <5. We have the following Yates-corrected chi-square statistic

$$X^2 = \frac{(|15 - 6.3| - .5)^2}{6.3} + \cdots + \frac{(|196 - 187.3| - .5)^2}{187.3}$$

$$= 17.04 \sim \chi_1^2$$

TABLE A.26 Observed asthma status of first-born child

		Observed asthma		
		+	−	
Type of family	A	15	85	100
	B	4	196	200
		19	281	300

TABLE A.27 Expected asthma status of first-born child

		Expected asthma		
		+	−	
Type of family	A	6.3	93.7	100
	B	12.7	187.3	200
		19.0	281.0	300

+ = asthma
− = no asthma

The p-value for this result is $<.001$, since

$$\chi_{1,.999}^2 = 10.83 < 17.04$$

Thus, there is a highly significant association between the type of family and the asthma status of the child.

10.42 The 2×2 table is shown in Table A.28.

TABLE A.28

		Nonasthmatic respiratory disease status		
		+	−	
Type of family	A	3	97	100
	B	2	198	200
		5	295	300

+ = nonasthmatic respiratory disease
− = no nonasthmatic respiratory disease

There are two expected values <5; in particular,

$$E_{11} = \frac{5(100)}{300} = 1.7 \qquad E_{21} = \frac{5(200)}{300} = 3.3$$

Thus, Fisher's exact test must be used to analyze this table. Write all possible tables with the same margins as the observed table, as follows:

TABLE A.29

0	100
5	195

1	99
4	196

2	98
3	197

3	97
2	198

4	96
1	199

5	95
0	200

Use the recursion rule to compute the probability of each table.

$$Pr(0) = 1 \times Pr(0)$$

$$Pr(1) = Pr(0) \times \frac{5 \times 100}{1 \times 196} = 2.551 \, Pr(0)$$

$$Pr(2) = Pr(1) \times \frac{4 \times 99}{2 \times 197} = 2.564 \, Pr(0)$$

$$Pr(3) = Pr(2) \times \frac{3 \times 98}{3 \times 198} = 1.269 \, Pr(0)$$

$$Pr(4) = Pr(3) \times \frac{2 \times 97}{4 \times 199} = 0.309 \, Pr(0)$$

$$Pr(5) = Pr(4) \times \frac{1 \times 96}{5 \times 200} = 0.030 \, Pr(0)$$

Thus,

$$Pr(0)(1 + 2.551 + 2.564 + 1.269 + 0.309 + 0.030) = 1$$

$$Pr(0) = \frac{1}{7.723} = .1295$$

$$Pr(1) = .330$$
$$Pr(2) = .332$$
$$Pr(3) = .164$$
$$Pr(4) = .040$$
$$Pr(5) = .004$$

Thus, since the observed table is the "3" table, the two-tailed p-value is given by

$$2 \times \min(.164 + .040 + .004, .164 + .332 + .330 + .1295) = 2 \times .208 = .416$$

The results are not statistically significant and indicate that there is no significant difference in the prevalence of nonasthmatic respiratory disease among households in which the parents do or do not have asthma.

10.45 The observed table is shown in Table A.30.

TABLE A.30 Association between Oracon use and endometrial cancer

	Use Oracon		
	Yes	No	
Cases	6	111	117
Controls	8	387	395
	14	498	512

The smallest expected value $= 14 \times 117/512 = 3.20 < 5$. Thus, Fisher's exact test must be used to analyze this table. Construct all tables with the same row and column margins as the observed table:

TABLE A.31

0	117
14	381

1	116
13	382

2	115
12	383

3	114
11	384

4	113
10	385

5	112
9	386

6	111
8	387

7	110
7	388

8	109
6	389

9	108
5	390

10	107
4	391

11	106
3	392

12	105
2	393

13	104
1	394

14	103
0	395

Use the recursion rule to compute the exact probability of each table.

$$Pr(0) = 1Pr(0)$$

$$Pr(1) = \frac{14 \times 117}{1 \times 382} Pr(0) = 4.288\ Pr(0)$$

$$Pr(2) = \frac{13 \times 116}{2 \times 383} Pr(1) = 8.442\ Pr(0)$$

$$Pr(3) = \frac{12 \times 115}{3 \times 384} Pr(2) = 10.113\ Pr(0)$$

$$Pr(4) = \frac{11 \times 114}{4 \times 385} Pr(3) = 8.235\ Pr(0)$$

$$Pr(5) = \frac{10 \times 113}{5 \times 386} Pr(4) = 4.822\ Pr(0)$$

$$Pr(6) = \frac{9 \times 112}{6 \times 387} Pr(5) = 2.093\ Pr(0)$$

$$Pr(7) = \frac{8 \times 111}{7 \times 388} Pr(6) = 0.684\ Pr(0)$$

$$Pr(8) = \frac{7 \times 110}{8 \times 389} Pr(7) = 0.169\ Pr(0)$$

$$Pr(9) = \frac{6 \times 109}{9 \times 390} Pr(8) = 0.031\ Pr(0)$$

$$Pr(10) = \frac{5 \times 108}{10 \times 391} Pr(9) = 0.0043\ Pr(0)$$

$$Pr(11) = \frac{4 \times 107}{11 \times 392} Pr(10) = 0.00043\ Pr(0)$$

$$Pr(12) = \frac{3 \times 106}{12 \times 393} Pr(11) = 0.000029\ Pr(0)$$

$$Pr(13) = \frac{2 \times 105}{13 \times 394} Pr(12) = 1.2 \times 10^{-6}\ Pr(0)$$

$$Pr(14) = \frac{1 \times 104}{14 \times 395} Pr(13) = 2.3 \times 10^{-8}\ Pr(0)$$

Thus,

$$Pr(0)(1 + 4.288 + \cdots + 2.3 \times 10^{-8}) = 1$$

or

$$Pr(0) = \frac{1}{39.882} = .0251$$

It follows that

$$Pr(1) = .108$$

$$Pr(2) = .212$$

$$Pr(3) = .254$$

$$Pr(4) = .207$$

$$Pr(5) = .121$$

$$Pr(6) = .053$$

$$Pr(7) = .017$$

$$Pr(8) = .004$$

$$Pr(9) = .001$$

$$Pr(10) = 1.1 \times 10^{-4}$$

$$Pr(11) = 1.1 \times 10^{-5}$$

$$Pr(12) = 7.3 \times 10^{-7}$$

$$Pr(13) = 3.0 \times 10^{-8}$$

$$Pr(14) = 5.8 \times 10^{-10}$$

The observed table is the "6" table. Thus, the one-tailed p-value is given by

$$\sum_{i=6}^{14} Pr(i) = .053 + .017 + \cdots + 5.8 \times 10^{-10} = .075$$

and the two-tailed p-value is given by

$$2 \times \min\left[\sum_{i=0}^{6} Pr(i), \sum_{i=6}^{14} Pr(i)\right] = 2 \times \min(.980, .075)$$

$$= 2(.075) = .150$$

Thus, there is no significant association between the use of Oracon and the development of endometrial cancer. These results differ from those in the article, in which the authors reported a significant association between these two variables. One explanation is that the authors found this significant association after adjusting for age and certain other variables, whereas the results here are unadjusted.

10.52 The 2×2 contingency table is shown in Table A.32.

TABLE A.32

| | Disease status | |
Community	+	−
A	5	99,995	100,000
B	1	199,999	200,000
	6	299,994	300,000

The expected table is formed as follows:

TABLE A.33

2.0	99,998
4.0	199,996

Since two of the four cells have expected values < 5, Fisher's exact test rather than the χ^2 test must be used. Thus, the following seven tables are formed:

TABLE A.34

0	100,000	1	99,999	2	99,998
6	199,994	5	199,995	4	199,996

3	99,997	4	99,996	5	99,995
3	199,997	2	199,998	1	199,999

6	99,994
0	200,000

We have

$$Pr(0) = \qquad\qquad 1.000\ Pr(0)$$

$$Pr(1) = \frac{6 \times 100,000}{1 \times 199,995}\ Pr(0) = 3.000\ Pr(0)$$

$$Pr(2) = \frac{5 \times 99,999}{2 \times 199,996}\ Pr(1) = 3.750\ Pr(0)$$

$$Pr(3) = \frac{4 \times 99,998}{3 \times 199,997}\ Pr(2) = 2.500\ Pr(0)$$

$$Pr(4) = \frac{3 \times 99,997}{4 \times 199,998}\ Pr(3) = 0.937\ Pr(0)$$

$$Pr(5) = \frac{2 \times 99,996}{5 \times 199,999}\ Pr(4) = 0.187\ Pr(0)$$

$$Pr(6) = \frac{1 \times 99,995}{6 \times 200,000}\ Pr(5) = 0.016\ Pr(0)$$

Thus, $Pr(0)(1 + 3.000 + 3.750 + 2.500 + 0.937 + 0.187 + 0.016) = 1$

$$Pr(0) = .0878$$

$$Pr(1) = .263$$

$$Pr(2) = .329$$
$$Pr(3) = .219$$
$$Pr(4) = .082$$
$$Pr(5) = .016$$
$$Pr(6) = .001$$

The one-sided p-value associated with the "5" table is $.016 + .001 = .017$. The two-sided p-value is $.034$. Thus, there is a significant difference in the incidence rates.

10.56 Because the samples are not independent, McNemar's test is used. The distribution of matched pairs is shown in Table A.35.

TABLE A.35 Association between computer and attending physician diagnoses

		Computer +	Computer −	
Attending physician	+	48	2	50
	−	20	9930	9950
		68	9932	10,000

We have

$$n_A = 2 \qquad n_B = 20 \qquad n_D = 22$$

Since $n_D/4 \geqslant 5$, the normal approximation can be used. Under H_0,

$$X^2 = \frac{\left(\left|n_A - \dfrac{n_D}{2}\right| - \dfrac{1}{2}\right)^2}{\dfrac{n_D}{4}} = \frac{\left(|2 - 11| - \dfrac{1}{2}\right)^2}{5.5}$$

$$= 13.14 \sim \chi_1^2$$

Since $\chi_{1,.999}^2 = 10.83 < 13.14 = X^2$, it follows that $p < .001$. The computer seems more likely to diagnose this type of viral infection than the attending physician.

10.62 A two-sample test is needed here, since samples of men who survived and died, respectively, from a first heart attack are being compared.

10.63 The observed table is shown in Table A.36. The smallest expected value is

$$\frac{40 \times 100}{200} = 20 \geqslant 5$$

Thus, the χ^2 test for 2×2 contingency tables can be used here.

TABLE A.36 Observed table relating sudden death from a first heart attack and previous physical activity

		Physical activity		
		Active	Inactive	
Mortality status	Survived	30	70	100
	Died	10	90	100
		40	160	200

10.64 The test statistic is given by

$$X^2 = \frac{n\left[|ad - bc| - \dfrac{n}{2}\right]^2}{(a+b)(c+d)(a+c)(b+d)}$$

$$= \frac{200(|30 \times 90 - 10 \times 70| - 100)^2}{(100)(100)(40)(160)}$$

$$= \frac{200(1900)^2}{(100)(100)(40)(160)} = 11.28 \sim \chi_1^2 \text{ under } H_0$$

Since $\chi_{1,.999}^2 = 10.83 < X^2$, it follows that $p < .001$, and we can conclude that there is a significant association between physical activity and survival after an MI.

10.65 The odds ratio in favor of prior physical activity for MI survivors vs. MI deceased is given by

$$\widehat{OR} = \frac{30 \times 90}{70 \times 10} = \frac{2700}{700} = 3.86.$$

10.66 Use the test-based method to obtain 95% confidence limits. The 95% confidence interval is given by (OR_1, OR_2), where

$$OR_1 = \widehat{OR}^{1-(1.96/\sqrt{X^2})},$$
$$= 3.86^{1-(1.96/\sqrt{11.28})}$$
$$= 3.86^{1-0.58} = 3.86^{0.42} = 1.76$$
$$OR_2 = \widehat{OR}^{1+(1.96/\sqrt{X^2})}$$
$$= 3.86^{1+0.58} = 3.86^{1.58} = 8.45$$

Thus, (1.76, 8.45) is a 95% confidence interval for OR. This interval excludes 1, as it must, since the p-value is less than .05.

10.73 This is a classic example illustrating the use of McNemar's test for correlated proportions. There are two groups of patients, one receiving lithium and one receiving placebo, but the two groups are matched on age, sex, and clinical condition and thus represent dependent samples. Let a type A discordant pair be a pair of people such that the lithium member of the pair has a manic depressive episode and the placebo member does not. Let a type B discordant pair be a pair of people such that the placebo member of the pair has a manic depressive episode and the lithium member does not. Let p = probability that a discordant pair is of type A. Then test the hypothesis

$$H_0: p = \tfrac{1}{2} \text{ vs. } H_1: p \neq \tfrac{1}{2}$$

10.74 There are 2 type A discordant pairs and 10 type B discordant pairs, giving a total of 12 discordant pairs. Since $npq = 12(\tfrac{1}{2})(\tfrac{1}{2}) = 3 < 5$, we *cannot* assume that the normal approximation to the binomial holds and an exact binomial test must be used. Under H_0,

$$Pr(k \text{ type A discordant pairs}) = \binom{12}{k}\left(\frac{1}{2}\right)^{12}$$

In particular, from Table 1 in Appendix 1,

$$Pr(k \leq 2) = \left(\frac{1}{2}\right)^{12}\left[\binom{12}{0} + \binom{12}{1} + \binom{12}{2}\right]$$
$$= .0002 + .0029 + .0161 = .0192$$

Since a two-sided test is being performed,

$$p = 2 \times .0192 = .038$$

Thus, H_0 is rejected and we conclude that the placebo patients are more likely to have manic depressive episodes when the results differ in the two members of a pair.

10.76 Set up a 2 × 2 contingency table to display the data (Table A.37).

TABLE A.37 Observed table relating age to 3-year survival for cancer of the pancreas

		Age group					
		<45	45–54	55–64	65–74	75+	Total
Outcome	Survive	20	14	27	18	13	92
	Die	236	696	1321	1750	1279	5282
		256	710	1348	1768	1292	5374
	Survival rates	.08	.02	.02	.01	.01	.017

Use the chi-square test for trend, as outlined in **(10.24)**. In this regard, use scores $1, 2, \ldots, k$ for the k groups. Compute the test statistic $X^2 = A^2/B$, where

$$A = \left(\sum_{i=1}^{k} x_i S_i\right) - x\left(\sum_{i=1}^{k} n_i S_i\right)\bigg/N$$

$$= [20(1) + 14(2) + \cdots + 13(5)]$$

$$- \frac{92[256(1) + 710(2) + \cdots + 1292(5)]}{5374}$$

$$= 266 - \frac{92 \times 19{,}252}{5374}$$

$$= 266 - 329.58 = -63.58$$

$$B = \bar{p}\bar{q}\left[\left(\sum_{i=1}^{k} n_i S_i^2\right) - \left(\sum_{i=1}^{k} n_i S_i\right)^2\bigg/N\right]$$

$$= \frac{92}{5374} \times \frac{5282}{5374}$$

$$\times \left[256(1^2) + 710(2^2) + \cdots + 1292(5^2)\right.$$

$$\left. - \frac{19{,}252^2}{5374}\right]$$

$$= .0168(75{,}816 - 68{,}969.02) = .0168 \times 6846.98$$

$$= 115.03$$

Thus, $\quad X^2 = \dfrac{(-63.58)^2}{115.03} = 35.14 \sim \chi_1^2$

Since $\chi_{1,.999}^2 = 10.83$, it follows that $p < .001$. Thus, there is a highly significant inverse relationship between 3-year survival and age.

10.92 The data are in the form of a 2×2 table, so the chi-square test may be an appropriate method of analysis if the expected cell counts are large enough. The smallest expected value is given by $(12 \times 22)/50 = 5.28 > 5$. Thus, this is a reasonable method of analysis.

10.93 The observed table is given in Table A.38.

TABLE A.38 Association between working status and health status

	Ill	Well	
Worked	10	12	22
Did not work	2	26	28
	12	38	50

Compute the following chi-square statistic:

$$X^2 = \frac{n\left(|ad - bc| - \dfrac{n}{2}\right)^2}{(a+b)(c+d)(a+c)(b+d)}$$

$$= \frac{50(|10 \times 26 - 2 \times 12| - 25)^2}{(22)(28)(12)(38)}$$

$$= \frac{50(211)^2}{(22)(28)(12)(38)} = 7.92 \sim \chi_1^2$$

Referring to the χ^2 tables, we find that

$$\chi_{1,.995}^2 = 7.88, \quad \chi_{1,.999}^2 = 10.83$$

Thus, $.001 < p < .005$. The authors found a chi-square of 7.8, $p = .01$, and thus our results are somewhat more significant than those claimed in the article.

10.94 The t test is *not* a reasonable test to use in comparing binomial proportions from two independent samples. Instead, either the chi-square test for 2×2 tables with large expected values or Fisher's exact test for tables with small expected values should be used.

10.95 The 2×2 table is given in Table A.39.

TABLE A.39 Association between salad consumption and health status

		Ill	Well	
Ate salad	Yes	25	8	33
	No	3	6	9
		28	14	42

The smallest expected value $= (14 \times 9)/42 = 3.0 < 5$, which implies that Fisher's exact test must be used. First rearrange the table so that the smaller row total is in row 1 and the smaller column total is in column 1, as in Table A.40.

TABLE A.40

		Well	Ill	
Ate salad	No	6	3	9
	Yes	8	25	33
		14	28	

Now enumerate all tables with the same row and column margins as follows:

TABLE A.41

0	9
14	19

1	8
13	20

2	7
12	21

3	6
11	22

4	5
10	23

5	4
9	24

6	3
8	25

7	2
7	26

8	1
6	27

9	0
5	28

Now use the recursion rule to compute the exact probability of each table.

$$Pr(0) = 1\, Pr(0)$$

$$Pr(1) = \frac{14 \times 9}{1 \times 20} Pr(0) = 6.300\, Pr(0)$$

$$Pr(2) = \frac{13 \times 8}{2 \times 21} Pr(1) = 15.600\, Pr(0)$$

$$Pr(3) = \frac{12 \times 7}{3 \times 22} Pr(2) = 19.855\, Pr(0)$$

$$Pr(4) = \frac{11 \times 6}{4 \times 23} Pr(3) = 14.244\, Pr(0)$$

$$Pr(5) = \frac{10 \times 5}{5 \times 24} Pr(4) = 5.935\, Pr(0)$$

$$Pr(6) = \frac{9 \times 4}{6 \times 25} Pr(5) = 1.424\, Pr(0)$$

$$Pr(7) = \frac{8 \times 3}{7 \times 26} Pr(6) = 0.188\, Pr(0)$$

$$Pr(8) = \frac{7 \times 2}{8 \times 27} Pr(7) = 0.012\, Pr(0)$$

$$Pr(9) = \frac{6 \times 1}{9 \times 28} Pr(8) = 0.00029\, Pr(0)$$

Thus,

$$Pr(0)(1 + 6.300 + 15.600 + 19.855 + 14.244 + 5.935$$
$$+ 1.424 + 0.188 + 0.012 + 0.00029) = 1$$

or

$$Pr(0) = \frac{1}{64.558} = .0155$$

$$Pr(1) = .098$$
$$Pr(2) = .242$$
$$Pr(3) = .308$$
$$Pr(4) = .221$$
$$Pr(5) = .092$$
$$Pr(6) = .022$$
$$Pr(7) = .003$$
$$Pr(8) = .0002$$
$$Pr(9) = 4.50 \times 10^{-6}$$

Since our observed table is the "6" table, the two-sided p-value is given by

$$p = 2 \times \min\left[\sum_{i=0}^{6} Pr(i), \sum_{i=6}^{9} Pr(i) \right]$$

$$= 2 \times \min(.999, .0252) = .050$$

Thus, the results are on the margin of being statistically significant ($p = .05$) as opposed to the p-value of .01 given in the paper.

10.136 Form the following 2×2 table to assess age effects among women with a negative history:

TABLE A.42

		Low birthweight		
		Yes	No	
Age	≥30	8	217	225
	<30	30	876	906
		38	1093	1131

The smallest expected cell count $= E_{11} = (38 \times 225)/1131 = 7.56 \geqslant 5$. Therefore, the Yates-corrected chi-square test can be used.

10.137 The test statistic is given by

$$X^2 = \frac{1131(|8 \times 876 - 30 \times 217| - 1131/2)^2}{38 \times 1093 \times 906 \times 225}$$

$$= \frac{1131(498 - 565.5)^2}{8.467 \times 10^9}$$

$$= \frac{5.153 \times 10^6}{8.467 \times 10^9} = 0.00061 \sim \chi_1^2 \text{ under } H_0$$

Clearly, since $\chi_{1,.50}^2 = 0.45$ and $X^2 < 0.45$, it follows that $p > .50$, and there is no significant effect of age on low-birthweight deliveries in this strata.

10.138 Form the following 2×2 contingency table among women with a positive history:

TABLE A.43

		Low birthweight		
		Yes	No	
Age	≥ 30	6	82	88
	> 30	2	151	153
		8	233	241

The smallest expected value $= (8 \times 88)/241 = 2.92 < 5$. Therefore, Fisher's exact test must be used to perform the test.

10.139 First form all possible tables with the same row and column margins, as follows:

TABLE A.44

0	88	1	87	2	86	3	85	4	84
8	145	7	146	6	147	5	148	4	149

5	83	6	82	7	81	8	80
3	150	2	151	1	152	0	153

Now use the recursion rule to compute the exact probabilities of these tables.

$$Pr(0) = 1 \, Pr(0)$$

$$Pr(1) = \frac{8 \times 88}{1 \times 146} Pr(0) = 4.822 \, Pr(0)$$

$$Pr(2) = \frac{7 \times 87}{2 \times 147} Pr(1) = 9.988 \, Pr(0)$$

$$Pr(3) = \frac{6 \times 86}{3 \times 148} Pr(2) = 11.608 \, Pr(0)$$

$$Pr(4) = \frac{5 \times 85}{4 \times 149} Pr(3) = 8.278 \, Pr(0)$$

$$Pr(5) = \frac{4 \times 84}{5 \times 150} Pr(4) = 3.709 \, Pr(0)$$

$$Pr(6) = \frac{3 \times 83}{6 \times 151} Pr(5) = 1.019 \, Pr(0)$$

$$Pr(7) = \frac{2 \times 82}{7 \times 152} Pr(6) = 0.157 \, Pr(0)$$

$$Pr(8) = \frac{1 \times 81}{8 \times 153} Pr(7) = 0.010 \, Pr(0)$$

Thus,

$$Pr(0)[1 + 4.822 + \cdots + 0.010] = 1$$

or

$$Pr(0) = 1/40.591 = .0246$$

Thus, it follows that

$$Pr(1) = .119, \; Pr(2) = .246, \; Pr(3) = .286, \; Pr(4) = .204,$$

$$Pr(5) = .091, \; Pr(6) = .025, \; Pr(7) = .004, \; Pr(8) = .0002.$$

Since our table is the "6" table, a two-sided p-value is computed as follows:

$$p = 2 \times \min\left[\sum_{k=0}^{6} Pr(k), \sum_{k=6}^{8} Pr(k)\right]$$

$$= 2 \times \min(.025 + \cdots + .025, .025 + .004 + .0002)$$

$$= 2 \times \min(.996, .0292) = .058$$

Thus, for women with a negative history, there is a trend toward significance, with older women having a higher incidence of low-birthweight deliveries.

10.140 Let p_{ij} = probability of a low-birthweight delivery in the ith age group and jth history group, $i = 1, 2$, $j = 1, 2$. Test the hypothesis $H_0: \delta = 0$ versus H_1 $\delta \neq 0$, where $\delta = (p_{11} - p_{21}) - (p_{12} - p_{22}) = p_{11} - p_{12} - p_{21} + p_{22}$. The basic approach is to estimate δ by $\hat{\delta} = \hat{p}_{11} - \hat{p}_{12} - \hat{p}_{21} + \hat{p}_{22}$ and perform the hypothesis test by computing the test statistic

$$x = \hat{\delta}/se(\hat{\delta}) \sim N(0, 1) \text{ under } H_0.$$

where

$$Var(\hat{\delta}) = \frac{\hat{p}_{11}\hat{q}_{11}}{n_{11}} + \frac{\hat{p}_{12}\hat{q}_{12}}{n_{12}} + \frac{\hat{p}_{21}\hat{q}_{21}}{n_{21}} + \frac{\hat{p}_{22}\hat{q}_{22}}{n_{22}}$$

If $i = 1$ corresponds to ≥ 30, $i = 2$ to < 30, and $j = 1$ to history = yes, $j = 2$ to history = no, then

$$\hat{\delta} = (.0682 - .0356 - .0131 + .0331 = .0526$$

$$Var(\hat{\delta}) = \frac{.0682 \times .9318}{88} + \frac{.0356 \times .9644}{225}$$

$$+ \frac{.0131 \times .9869}{153} + \frac{.0331 \times .9669}{906}$$

$$= .000722 + .000153 + .000084 + .000035$$

$$= .000994$$

$$se(\hat{\delta}) = .0315$$

The test statistic is given by

$$x = \frac{0.0526}{0.0315} = 1.670 \sim N(0, 1) \text{ under } H_0$$

This yields a two-tailed p-value of $p = 2 \times [1 - \Phi(1.67)] = 2 \times (1 - .9525) = .095$. Thus, there is a trend toward significance, but we cannot definitely state that the effect of age is different in the two subgroups.

CHAPTER 11

11.1

$$L_{xx} = \sum x_i^2 - \frac{(\sum x_i)^2}{n}$$

$$= 54,749 - \frac{(1137)^2}{27}$$

$$= 54,749 - 47,880.33 = 6868.67$$

$$L_{xy} = \sum x_i y_i - \frac{(\sum x_i)(\sum y_i)}{n}$$

$$= 262.93 - \frac{(1137)(6.05)}{27}$$

$$= 262.93 - 254.77 = 8.16$$

$$b = \frac{L_{xy}}{L_{xx}} = \frac{8.16}{6868.67} = 0.0012$$

$$a = \frac{\sum y_i - b \sum x_i}{n}$$

$$= \frac{6.05 - 0.0012(1137)}{27} = \frac{6.05 - 1.364}{27}$$

$$= \frac{4.686}{27} = 0.174$$

Thus, the regression line is given by $y = 0.174 + 0.0012x$.

11.2 The expected LVEF $= 0.174 + 0.0012(45) = 0.174 + 0.054 = 0.228$.

11.3 First compute L_{yy}.

$$L_{yy} = \sum y_i^2 - \frac{(\sum y_i)^2}{n}$$

$$= 1.522 - \frac{(6.05)^2}{27} = 1.522 - 1.356$$

$$= 0.166$$

Now compute the regression and residual sum of squares and mean square:

$$\text{Reg SS} = \frac{L_{xy}^2}{L_{xx}} = \frac{(8.16)^2}{6868.67} = 0.0097 = \text{Reg MS}$$

$$\text{Res SS} = 0.166 - 0.0097 = 0.156$$

$$\text{Res MS} = \frac{0.156}{25} = 0.0062$$

Finally, the F statistic is given by $F = \text{Reg MS}/\text{Res MS} = 0.0097/0.0062 = 1.56 \sim F_{1,25}$ under H_0. Since $F_{1,25,.95} > F_{1,30,.95} = 4.17 > 1.56$, it follows that $p > .05$. Therefore, H_0 is accepted and we conclude that there is no significant relationship between LVEF and age.

11.4 $R^2 = \text{Reg SS}/\text{Total SS} = 0.0097/0.166 = 0.058$.

11.5 It means that only 5.8% of the variance in LVEF is explained by age.

11.6 Compute the standard error of the estimated slope as follows:

$$se(b) = \sqrt{\frac{\text{Res MS}}{L_{xx}}} = \sqrt{\frac{0.0062}{6868.67}}$$

$$= 0.0010$$

Thus, the t statistic is given by

$$t = \frac{b}{se(b)} = \frac{0.0012}{0.0010}$$

$$= 1.20 \sim t_{25} \text{ under } H_0$$

Since $t_{25,.975} = 2.060 > t$, it follows that $p > .05$, and H_0 is accepted at the 5% level.

11.7 $se(b) = 0.0010$ as given in Problem 11.6. The standard error of the intercept is given by

$$se(a) = \sqrt{\text{Res MS}\left(\frac{1}{n} + \frac{\bar{x}^2}{L_{xx}}\right)}$$

$$= \sqrt{0.0062\left[\frac{1}{27} + \frac{(1137/27)^2}{6868.67}\right]}$$

$$= \sqrt{0.0062(0.0370 + 0.2582)}$$

$$= \sqrt{(0.0062)(0.2952)} = 0.043$$

11.8 A 95% confidence interval for the slope is given by

$$b \pm t_{25,.975}se(b) = 0.0012 \pm 2.060(0.0010)$$

$$= 0.0012 \pm 0.0021$$

$$= (-0.0009, 0.0033)$$

A 95% confidence interval for the intercept is given by

$$a \pm t_{25,.975}se(a) = 0.174 \pm 2.060(0.043)$$

$$= 0.174 \pm 0.089 = (0.085, 0.263)$$

11.20 Refer to Table 11 in Appendix 1 under $r = .34$ to obtain $z = 0.354$.

11.26 The test statistic is given by

$$t_s = \frac{r_s\sqrt{n-2}}{\sqrt{1-r_s^2}} = \frac{.4\sqrt{28}}{\sqrt{1-.4^2}}$$

$$= \frac{.4(5.292)}{\sqrt{.84}} = \frac{2.117}{.917}$$

$$= 2.309 \sim t_{28} \text{ under } H_0$$

A one-tailed test is being performed. Thus, the critical value $= t_{28,.95} = 1.701 < t_s$. Therefore, H_0 is rejected at the 5% level and we conclude that the rank correlation is significantly greater than 0.

11.27 Since $t_{28,.975} = 2.048$, $t_{28,.99} = 2.467$, and $2.048 < 2.309 < 2.467$, the one-tailed p-value is given by $1 - .99 < p < 1 - .975$, or $.01 < p < .025$.

11.28 Since $n < 10$, the exact tables for the Spearman rank correlation coefficient must be used. Refer to Table 12 in Appendix 1 under $n = 8$ and note that the two-tailed critical value for $\alpha = .10$ and $.05$ are $.643$ and $.738$, respectively. Since $.643 < .7 < .738$, it follows that the two-tailed p-value is given by $.05 \leqslant p < .10$. Therefore, the one-tailed p-value is given by $.025 \leqslant p < .05$.

11.33 Let $y =$ duration of hospitalization and $x =$ age. Compute

$$L_{xx} = \sum_{i=1}^{n}(x_i - \bar{x})^2 = \sum_{i=1}^{n}x_i^2 - \frac{\left(\sum_{i=1}^{n}x_i\right)^2}{n}$$

$$= 52,217 - \frac{(1031)^2}{25} = 9698.56$$

$$L_{xy} = \sum_{i=1}^{n}(x_i - \bar{x})(y_i - \bar{y}) = \sum_{i=1}^{n}x_iy_i - \frac{\left(\sum_{i=1}^{n}x_i\right)\left(\sum_{i=1}^{n}y_i\right)}{n}$$

$$= 9869 - \frac{(1031)(215)}{25} = 1002.40$$

The least-squares coefficients are then given by

$$b = \frac{L_{xy}}{L_{xx}} = \frac{1002.40}{9698.56} = 0.103$$

$$a = \frac{\sum_{i=1}^{n}y_i - b\sum_{i=1}^{n}x_i}{n} = \frac{215 - 0.103(1031)}{25} = 4.35$$

Thus, the regression line is $y = 4.35 + 0.103x$.

11.34 To test for the significance of this relationship, assume an underlying model of the form

$$y_i = \alpha + \beta x_i + e_i \qquad \text{where } e_i \sim N(0, \sigma^2)$$

Furthermore, under H_0: $\beta = 0$, whereas under H_1: $\beta \neq 0$. Set up the ANOVA table as follows:

$$\text{Regression SS} = \frac{L_{xy}^2}{L_{xx}} = \frac{(1002.40)^2}{9698.56} = 103.60$$

$$\text{Total SS} = L_{yy} = \sum_{i=1}^{n}(y_i - \bar{y})^2 = \sum_{i=1}^{n}y_i^2 - \frac{\left(\sum_{i=1}^{n}y_i\right)^2}{n}$$

$$= 2633 - (215)^2/25 = 784$$

$$\text{Residual SS} = \text{Total SS} - \text{Regression SS}$$

$$= 784 - 103.6 = 680.4$$

TABLE A.45 ANOVA table

	SS	df	MS	F statistic
Regression	103.60	1	103.60	3.50
Residual	680.40	23	29.58	
	784.00	24		

Test for significance by noting that

$$F = \text{Regression MS/Residual MS}$$

$$= 3.50 \sim F_{1,23} \text{ under } H_0.$$

Since

$$F_{1,23,.95} > F_{1,30,.95} = 4.17 > F$$

it follows that $p > .05$. Also, since

$$F_{1,23,.90} < F_{1,20,.90} = 2.97 < F$$

it follows that $p < .10$. Thus, the p-value is given by $.05 < p < .10$, and there is only a trend toward significance in this case.

11.35 R^2 is defined as Regression SS/Total SS $= 103.6/784 = .132$.

11.40 Compute the correlation coefficient between THW and BW in each of the two groups. We first have the following summary statistics ($x = $ THW, $y = $ BW):

Left heart disease

$$\sum_{i=1}^{11} x_i = 4950$$

$$\sum_{i=1}^{11} x_i^2 = 2,421,650$$

$$\sum_{i=1}^{11} y_i = 611.70$$

$$\sum_{i=1}^{11} y_i^2 = 35,350.49$$

$$\sum_{i=1}^{11} x_i y_i = 280,031.50$$

$$L_{xx} = 194,150$$

$$L_{yy} = 1334.41$$

$$L_{xy} = 4766.50$$

$$r = L_{xy}/(L_{xx}L_{yy})^{1/2}$$

$$= .296$$

$$t = r\sqrt{n-2}/\sqrt{1-r^2}$$

$$= .296(3)/\sqrt{1-.296^2}$$

$$= 0.93 \sim t_9, \text{ NS}$$

Normal

$$\sum_{i=1}^{10} x_i = 3170$$

$$\sum_{i=1}^{10} x_i^2 = 1,024,850$$

$$\sum_{i=1}^{10} y_i = 562.3$$

$$\sum_{i=1}^{10} y_i^2 = 32,816.33$$

$$\sum_{i=1}^{10} x_i y_i = 181,462$$

$$L_{xx} = 19,960$$

$$L_{yy} = 1198.20$$

$$L_{xy} = 3212.9$$

$$r = L_{xy}/(L_{xx}L_{yy})^{1/2}$$

$$= .657$$

$$t = r\sqrt{n-2}/\sqrt{1-r^2}$$

$$= .657\sqrt{8}/\sqrt{1-.657^2}$$

$$= 2.465 \sim t_8,$$

$$.02 < p < .05$$

Thus, there is a significant association between total heart weight and body weight in the normal group but

not in the left heart disease group. This confirms what was shown graphically in Problem 2.7.

11.80 Test the hypothesis $H_0: \rho = 0$ versus $H_1: \rho \neq 0$. Use the test statistic

$$t = \frac{r\sqrt{n-2}}{\sqrt{1-r^2}} \sim t_{n-2} \text{ under } H_0$$

In this case

$$r = \frac{L_{xy}}{\sqrt{L_{xx}L_{yy}}}$$

We have

$$L_{xx} = 451,350 - \frac{(2980)^2}{20} = 7330$$

$$L_{yy} = 351,350 - \frac{(2620)^2}{20} = 8130$$

$$L_{xy} = 390,825 - \frac{(2980)(2620)}{20} = 445$$

Thus, $\quad r = \dfrac{445}{\sqrt{7330 \times 8130}} = \dfrac{445}{7719.64} = .058$

The test statistic

$$t = \frac{.058\sqrt{18}}{\sqrt{1-(.058)^2}} = \frac{.246}{.998} = .246 \sim t_{18}$$

Since $t_{18,.975} = 2.101$, this is clearly not statistically significant, and there is no significant correlation between the mother's and father's bp, which is what would be expected on purely genetic grounds.

11.81 Compute

$$r_{yt} = \frac{L_{yt}}{\sqrt{L_{yy} \times L_{tt}}}$$

We have

$$L_{tt} = 210,850 - \frac{(2030)^2}{20} = 4805$$

$$L_{yt} = 269,550 - \frac{(2620)(2030)}{20} = 3620$$

and $\quad r = \dfrac{3620}{\sqrt{8130 \times 4805}} = \dfrac{3620}{6250.17} = .579$

Test the null hypothesis $H_0: \rho = .5$ vs. the alternative $H_1: \rho \neq .5$. Use the Fisher z test, which yields

$$z = \tfrac{1}{2}[\ln(1+r) - \ln(1-r)] = \tfrac{1}{2}[\ln(1.579) - \ln(0.421)]$$

$$= \tfrac{1}{2}(0.4568 + 0.8651) = 0.661$$

Also, from Table 11 in Appendix 1 $z_0 = 0.549$. Use the test statistic

$$\lambda = (z - z_0)\sqrt{n-3} \sim N(0, 1) \text{ under } H_0$$

Thus,

$$\lambda = \sqrt{17}(0.661 - 0.549) = 0.462 \sim N(0, 1)$$

The p-value $= 2 \times [1 - \Phi(0.462)] = .64$. Thus, accept H_0 that $\rho = .5$.

11.82 The mother and first-born child were both living in the same environment, which might explain all or part of the observed correlation of blood pressure.

11.83 Use the model $t = \alpha + \beta y + e$. Test the null hypothesis that $\beta = 0$ versus the alternative hypothesis that $\beta \neq 0$.

$$b = \frac{L_{ty}}{L_{yy}}$$

where $L_{ty} = 3620$

$$L_{yy} = 8130 \qquad \text{from Problems 11.80 and 11.81}$$

Thus, $\qquad b = \dfrac{3620}{8130} = 0.445$

Furthermore,

$$a = \frac{\displaystyle\sum_{i=1}^{20} t_i - b \sum_{i=1}^{20} y_i}{20} = \frac{2030 - 0.445(2620)}{20} = 43.2$$

We thus have the linear relation $t = 43.2 + 0.445y$.

11.84 Test the hypothesis $H_0: \beta = 0$ versus $H_1: \beta \neq 0$. Use the test statistic

$$F = \frac{\text{Reg MS}}{\text{Res MS}} \sim F_{1,n-2} \text{ under } H_0$$

We have

$$\text{Reg MS} = \frac{(L_{ty})^2}{L_{yy}} = \frac{(3620)^2}{8130} = 1611.86 = \text{Reg SS}$$

$$\text{Res MS} = \frac{L_{tt} - \text{Reg SS}}{18} = \frac{L_{tt} - 1611.86}{18}$$

Thus, $\quad \text{Res MS} = \dfrac{4805 - 1611.86}{18} = 177.40 = s_{t \cdot y}^2$

and $\qquad F = \dfrac{1611.86}{177.40} = 9.09 \sim F_{1,18} \text{ under } H_0$

Since $F_{1,18,.99} = 8.29$, $F_{1,18,.995} = 10.22$, $.005 < p < .01$, and there is a significant relationship between the mother's and child's bp.

11.85 The child's bp can be predicted from the mother's bp from the relationship in Problem 11.83. $E(t) = 43.2 + 0.445 \times 130 = 101.1$ mm Hg.

11.86 In this case $E(t) = 43.2 + 0.445 \times 150 = 110.0$ mm Hg.

11.87 In this case $E(t) = 43.2 + 0.445 \times 170 = 118.9$ mm Hg.

11.88 The standard errors are given by the formula

$$se = \sqrt{s_{t \cdot y}^2 \left[\frac{1}{n} + \frac{(y - \bar{y})^2}{L_{yy}} \right]} = \sqrt{177.40 \left[\frac{1}{20} + \frac{(y - \bar{y})^2}{8130} \right]}$$

We have $\qquad \bar{y} = \dfrac{2620}{20} = 131$ mm Hg

Thus, for $y = 130$, 150, and 170 mm Hg, we have, respectively,

$$se(130) = \sqrt{177.40 \left[\frac{1}{20} + \frac{(130 - 131)^2}{8130} \right]} = 2.98$$

$$se(150) = \sqrt{177.40 \left[\frac{1}{20} + \frac{(150 - 131)^2}{8130} \right]} = 4.09$$

$$se(170) = \sqrt{177.40 \left[\frac{1}{20} + \frac{(170 - 131)^2}{8130} \right]} = 6.49$$

11.89 The three standard errors are not the same because the standard error increases the further the value of the mother's bp is from the mean bp for all mothers (131 mm Hg). Thus, the se corresponding to 130 mm Hg is smallest, whereas the se corresponding to 170 mm Hg is largest.

11.90 To test for the independent effect of the father's bp on the child's bp after controlling for the effect of the mother's bp, test the hypothesis $H_0: \beta_1 = 0$, $\beta_2 \neq 0$ versus $H_1: \beta_1 \neq 0$, $\beta_2 \neq 0$. Similarly, to test for the independent effect of the mother's bp on the child's bp after controlling for the effect of the father's bp, test the hypothesis $H_0: \beta_2 = 0$, $\beta_1 \neq 0$ versus $H_1: \beta_2 \neq 0$, $\beta_1 \neq 0$.

11.91 Compute the following test statistics:

Father: $\quad t_1 = b_1/se(b_1) = 0.4150/0.1248$

$\qquad\qquad = 3.33 \sim t_{17}, .001 < p < .01$

Mother: $\quad t_2 = b_2/se(b_2) = 0.4226/0.1185$

$\qquad\qquad = 3.57 \sim t_{17}, .001 < p < .01$

Thus, each parent's bp is significantly associated with the child's bp after controlling for the other spouse's bp.

11.92 From Problems 11.80 and 11.81,

$$s_t^2 = \frac{L_{tt}}{n-1} = \frac{4805}{19} = 252.89$$

$$s_x^2 = \frac{L_{xx}}{n-1} = \frac{7330}{19} = 385.79$$

$$s_y^2 = \frac{L_{yy}}{n-1} = \frac{8130}{19} = 427.89$$

The standardized regression coefficients are given by

Father: $b_s = b \times s_x/s_t$

$$= 0.4150 \times \sqrt{\frac{385.79}{252.89}} = 0.513$$

Mother: $b_s = b \times s_y/s_t$

$$= 0.4226 \times \sqrt{\frac{427.89}{252.89}} = 0.550$$

The standardized coefficients represent the increase in the number of standard deviations in the child's bp that would be expected per standard deviation increase in the parent's bp after controlling for the other spouse's bp.

11.93 The correlation is given by

$$r = \frac{L_{xy}}{\sqrt{L_{xx} \times L_{yy}}}$$

We have

$$L_{xy} = -4.12 - \frac{(2.38)(-15.55)}{8} = 0.506$$

$$L_{xx} = 1.31 - \frac{(2.38)^2}{8} = 0.602$$

$$L_{yy} = 30.71 - \frac{(-15.55)^2}{8} = 0.485$$

Thus, $\quad r = \dfrac{0.506}{\sqrt{(0.602)(0.485)}} = .936$

11.94 Use the test statistic

$$t = \frac{r\sqrt{n-2}}{\sqrt{1-r^2}} \sim t_{n-2} \text{ under } H_0$$

In this case

$$t = \frac{.936\sqrt{6}}{\sqrt{1-.936^2}} = \frac{2.293}{.352} = 6.51 \sim t_6 \text{ under } H_0$$

Refer to the t table and find that $t_{6,.9995} = 5.959 < t$, which implies that $p < .001$. Thus, there is a highly significant association between the lung cancer mor-

tality rate and average cigarette consumption over a 40-year period.

11.95 It is also of interest to fit a regression line to these data of the form $y = a + bx$, where

$$b = \frac{L_{xy}}{L_{xx}} = \frac{0.506}{0.602} = 0.841$$

$$a = \frac{\sum_{i=1}^{8} y_i - b \sum_{i=1}^{8} x_i}{8} = \frac{-15.55 - 0.841(2.38)}{8} = -2.19$$

Thus, the regression line is $y = -2.19 + 0.841x$.

11.96 No. It is unnecessary, since the t test in Problem 11.94 based on the correlation coefficient and the F test based on the regression coefficient are equivalent.

11.97 The expected $\log_{10}$(mortality rate) $= -2.19 + 0.841 \times \log_{10}(1) = -2.19$. Thus, the expected mortality rate $= 10^{-2.19} = 0.00646$, or 646 deaths per 100,000.

11.98 The variables are expressed in the log scale because the relationship between mortality rate and cigarette consumption is most likely to be linear when each is expressed in this scale rather than in the raw scale.

CHAPTER 12

12.1 Bartlett's test for homogeneity of variance.

12.2 The test statistic is given by

$$X^2 = \sum_{i=1}^{3} [(n_i - 1) \ln(s^2/s_i^2)]/C \sim \chi_2^2 \text{ under } H_0$$

where s^2 is the pooled variance estimate.

$$s^2 = \frac{9(16)^2 + 9(16)^2 + 9(9)^2}{27} = \frac{5337}{27} = 198$$

$$C = 1 + \frac{1}{3(2)}\left(\frac{1}{9} + \frac{1}{9} + \frac{1}{9} - \frac{1}{27}\right)$$

$$= 1 + \frac{1}{6}\left(\frac{8}{27}\right) = 1.049$$

$$X^2 = \frac{9\ln\left(\frac{198}{256}\right) + 9\ln\left(\frac{198}{256}\right) + 9\ln\left(\frac{198}{81}\right)}{1.049}$$

$$= \frac{9(-0.257 - 0.257 + 0.894)}{1.049} = \frac{9(0.380)}{1.049}$$

$$= \frac{3.420}{1.049} = 3.26 \sim \chi_2^2 \text{ under } H_0$$

Since $\chi^2_{2,.95} = 5.99 > X^2$, it follows that H_0 is accepted at the 5% level. Since $\chi^2_{2,.75} = 2.77$, $\chi^2_{2,.90} = 4.61$, and $2.77 < 3.26 < 4.61$, it follows that $1 - .90 < p < 1 - .75$, or $.10 < p < .25$.

12.3 Since the variances are not significantly different, a one-way ANOVA can be used.

12.4 The test statistic is given by $F = $ Between MS/ Within MS $\sim F_{k-1,n-k}$ under H_0.

Between SS $= 10(74)^2 + 10(56)^2 + 10(55)^2$

$$- \frac{[10(74) + 10(56) + 10(55)]^2}{30}$$

$$= 54{,}760 + 31{,}360 + 30{,}250 - 114{,}083.33$$

$$= 116{,}370 - 114{,}083.33 = 2286.7$$

Between MS $= \dfrac{2286.7}{2} = 1143$

Within MS $= s^2 = 198$ (from Problem 12.2)

Therefore,

$$F = \frac{1143}{198} = 5.77 \sim F_{2,27} \text{ under } H_0$$

Since $F_{2,27,.95} < F_{2,20,.95} = 3.49 < F$, it follows that $p < .05$, and H_0 is rejected at the 5% level.

12.5 Use the test statistic

$$t = \frac{\bar{y}_1 - \bar{y}_2}{\sqrt{s^2(1/n_1 + 1/n_2)}} \sim t_{n-k}$$

under H_0. The results are given in Table A.46 for each pair of groups.

12.6 This contrast is a comparison of the general vegetarian population with the general nonvegetarian population. Compute the test statistic

$$t = \frac{L}{se(L)} \sim t_{n-k} \text{ under } H_0$$

$$= \frac{0.7(56) + 0.3(55) - 74}{\sqrt{198[(0.7)^2/10 + (0.3)^2/10 + (-1)^2/10]}}$$

$$= \frac{-18.3}{\sqrt{31.284}} = \frac{-18.3}{5.593} = -3.27 \sim t_{27} \text{ under } H_0$$

Since $t_{27,.995} = 2.771$, and $t_{27,.9995} = 3.690$, it follows that $.001 < p < .01$, and we reject H_0 and conclude that among premenopausal women, the general vegetarian population has a significantly lower protein intake than the general nonvegetarian population.

12.13 Refer to Table 13 in Appendix 1 under $\alpha = .01$, $c = 5$, $d = 30$ to obtain $q_{5,30,.99} = 5.05$.

12.14–12.15 First compare groups 1–3, where $1 = $ STD, $2 = $ LAC, $3 = $ VEG. We have the studentized range statistic

$$q = \frac{74 - 55}{\sqrt{\dfrac{198}{2} \times \left(\dfrac{1}{10} + \dfrac{1}{10}\right)}}$$

$$= \frac{19}{4.450} = 4.27 > q_{3,24,.95} = 3.53 > q_{3,27,.95}$$

Therefore, $p < .05$, and we conclude that some of the means are significantly different. We now compare groups 1–2 and 2–3. We have the following computations:

Groups 1–2: $q = \dfrac{74 - 56}{4.450} = \dfrac{18}{4.450} = 4.04 > q_{2,24,.95}$

$$= 2.92 > q_{2,27,.95}$$

TABLE A.46

Groups compared	Test statistic		p-value
STD, LAC	$t = \dfrac{74 - 56}{\sqrt{198(1/10 + 1/10)}}$	$= \dfrac{18}{6.293} = 2.86 \sim t_{27}$	$.001 < p < .01$
STD, VEG	$t = \dfrac{74 - 55}{6.293}$	$= \dfrac{19}{6.293} = 3.02 \sim t_{27}$	$.001 < p < .01$
LAC, VEG	$t = \dfrac{56 - 55}{6.293}$	$= \dfrac{1}{6.293} = 0.16 \sim t_{27}$	NS

Groups 2–3: $q = \dfrac{56-55}{4.450} = \dfrac{1}{4.450} = 0.22 < q_{2,30,.95}$

$\qquad = 2.89 < q_{2,27,.95}$

Therefore, we conclude that groups 1 and 2 are significantly different, whereas groups 2 and 3 are not. In summary, we conclude that for premenopausal women the protein intake of the standard American diet group is significantly higher than the lactovegetarian and strict vegetarian groups, whereas there is no significant difference between the latter two groups.

12.21 The two-way analysis of variance.

12.22 Use the method of unweighted means since max $n_{ij}/\text{min } n_{ij} = 10/6 = 1.7 \leqslant 2$. The two-way table of means by dietary group and menopausal status is presented in Table A.47.

The row and column sums of squares and mean squares are given by

Row SS $= \dfrac{149^2 + 113^2 + 102^2}{2} - \dfrac{(149 + 113 + 102)^2}{3 \times 2}$

$\qquad = 22{,}687 - 22{,}082.67 = 604.33$

Row MS $= \dfrac{604.33}{2} = 302.17$

Col SS $= \dfrac{185^2 + 179^2}{3} - \dfrac{(185 + 179)^2}{3 \times 2}$

$\qquad = 22{,}088.67 - 22{,}082.67 = 6.00 = $ Col MS

To compute the Error MS, compute n_h, where

$$1/n_h = \frac{(1/10 + 1/10 + \cdots + 1/6)}{6} = 0.1111$$

$\qquad n_h = 9.0$

Thus,

$$\text{Error MS} = \frac{9(16)^2 + 9(9)^2 + \cdots + 5(17)^2}{50(9)}$$

$$= \frac{9032}{450} = 20.071$$

Compute following the test statistics to identify significant row and column effects.

$$F_{\text{ROW}} = \frac{302.17}{20.071} = 15.06 \sim F_{2,50} \text{ under } H_0$$

$$F_{\text{COL}} = \frac{6.00}{20.071} = 0.30 \sim F_{1,50} \text{ under } H_0$$

Since $F_{\text{ROW}} = 15.06 > F_{2,40,.95} = 3.23 > F_{2,50,.95}$, it follows that $p < .05$ and there are significant row effects. Since $F_{\text{COL}} = 0.30 < F_{1,60,.95} = 4.00 < F_{1,50,.95}$, it follows that $p > .05$ and there are no significant column effects. In summary, there are significant differences in protein intake among different dietary groups but not between different menopausal groups.

TABLE A.47

			Menopausal status		
			Premenopausal	Postmenopausal	$y_{i.}^*$
	STD	Mean	74	75	149
		sd	16	9	
		n	10	10	
	LAC	Mean	56	57	113
Dietary		sd	16	13	
group		n	10	10	
	VEG	Mean	55	47	102
		sd	9	17	
		n	10	6	
		$y_{.j}^*$	185	179	364

12.23

$$\text{Int SS} = 74^2 + 75^2 + \cdots + 47^2$$

$$- \frac{(74 + 75 + \cdots + 47)^2}{6} - 604.33 - 6.00$$

$$= 22{,}720 - \frac{364^2}{6} - 610.33$$

$$= 22{,}720 - 22{,}082.67 - 610.33 = 27.00$$

$$\text{Int MS} = \frac{27.00}{2} = 13.50$$

Furthermore,

$$F_{\text{INT}} = \frac{13.50}{20.071} = 0.67 \sim F_{2,50} \text{ under } H_0$$

Since $F_{\text{INT}} = 0.67 < F_{2,60,.95} = 3.15 < F_{2,50,.95}$, it follows that $p > .05$ and there are no significant interaction effects.

12.24 In this instance an interaction effect, if it were present, would indicate that the differences in protein intake among diet groups are not the same for premenopausal and postmenopausal women. Since the interaction effect was not significant, it indicates that the diet group differences are similar for premenopausal and postmenopausal women.

12.25 A one-way ANOVA is appropriate here because the three groups are classified according to only one variable—alcohol consumption.

12.26 First perform Bartlett's test for homogeneity of variance to ensure that the overall F test for a one-way ANOVA can validly be used. We have the test statistic

$$X^2 = \frac{\lambda}{c} \sim \chi^2_{k-1} \text{ under } H_0$$

where $\quad \lambda = \sum_{i=1}^{k} (n_i - 1) \ln\left(\frac{s^2}{s_i^2}\right)$

$$c = 1 + \frac{1}{3(k-1)}\left[\left(\sum_{i=1}^{k} \frac{1}{n_i - 1}\right) - \frac{1}{n-k}\right]$$

We have

$$s^2 = \text{Within MS} = \frac{22(25.3)^2 + 14(21.9)^2 + 11(30.3)^2}{47}$$

$$= \frac{30{,}895.51}{47} = 657.35$$

Thus,

$$\lambda = 22 \ln\left[\frac{657.35}{(25.3)^2}\right] + 14 \ln\left[\frac{657.35}{(21.9)^2}\right] + 11 \ln\left[\frac{657.35}{(30.3)^2}\right]$$

$$= 0.585 + 4.413 - 3.675 = 1.323$$

$$c = 1 + \frac{1}{3(2)}\left(\frac{1}{22} + \frac{1}{14} + \frac{1}{11} - \frac{1}{47}\right) = 1.031$$

$$X^2 = \frac{1.323}{1.031} = 1.28 \sim \chi^2_2 \text{ under } H_0$$

Since $\chi^2_{2,.95} = 5.99 > X^2$, it follows that $p > .05$. Thus, the variances are not significantly different, and the F test can validly be used to test for overall differences between means. Use the test statistic

$$F = \frac{\text{Between MS}}{\text{Within MS}} \sim F_{k-1, n-k} \text{ under } H_0$$

We have

Between MS

$$= \frac{(205.6)^2 23 + (182.7)^2 15 + (199.8)^2 12}{2}$$

$$- \frac{[(205.6)(23) + (182.7)15 + (199.8)12]^2}{50(2)}$$

$$= \frac{1{,}951{,}971.11 - 1{,}947{,}114.31}{2}$$

$$= \frac{4856.80}{2} = 2428.40$$

Within MS $= s^2 = 657.35$

Thus,

$$F = \frac{\text{Between MS}}{\text{Within MS}} = \frac{2428.40}{657.35} = 3.69 \sim F_{2,47} \text{ under } H_0$$

Since $F_{2,47,.95} < F_{2,40,.95} = 3.23 < F$, it follows that $p < .05$, and there is an overall significant difference between groups.

12.27 First order the three groups from highest to lowest mean cholesterol level as given in Table A.48.

TABLE A.48 Association between cholesterol level and alcohol consumption

Group number	Group name	Mean	n
1	Nondrinkers	205.6	23
2	Heavy drinkers	199.8	12
3	Light drinkers	182.7	15

Then compare groups 1–3 using the q statistic in (12.15) as follows:

$$q = \frac{205.6 - 182.7}{\sqrt{\frac{657.35}{2}\left(\frac{1}{23} + \frac{1}{15}\right)}} = \frac{22.9}{6.02} = 3.80$$

Compare this value to $q_{3,47,.95}$, since three groups are being compared and s^2 is estimated with 47 df. We have that $q_{3,47,.95} < q_{3,40,.95} = 3.44 < 3.80$. Thus, groups 1 and 3 are significantly different. Now compare each pair of groups, as given in Table A.49.

Thus, neither of these comparisons is significant, and we can only say that nondrinkers (group 1) have significantly higher cholesterol levels than light drinkers (group 3).

12.40 A one-way ANOVA should be used on these data because there is only one effect that distinguishes the groups, namely, the mother's smoking status during pregnancy.

12.41 Table A.51 shows the basic statistics for the four groups.

Test for the homogeneity of the variances using Bartlett's test. We have the test statistic

$$X^2 = \sum_{i=1}^{k} (n_i - 1)\ln(s^2/s_i^2)/c \sim \chi^2_{k-1} \text{ under } H_0$$

$$s^2 = \frac{6(0.925) + 4(0.833) + 6(1.299) + 7(0.518)}{23}$$

$$= 0.883$$

$$X^2 = \left[6\ln\left(\frac{0.883}{0.925}\right) + 4\ln\left(\frac{0.883}{0.833}\right) + 6\ln\left(\frac{0.883}{1.299}\right) \right.$$
$$\left. + 7\ln\left(\frac{0.883}{0.518}\right) \right] \bigg/ c = 1.372/c$$

$$c = 1 + \frac{1}{3(3)}\left[\frac{1}{6} + \frac{1}{4} + \frac{1}{6} + \frac{1}{7} - \frac{1}{23}\right] = 1 + \frac{1}{9}(0.683)$$

$$= 1.076$$

TABLE A.49 Comparison of specific groups using the method of multiple comparisons

Groups compared	Test statistic		Reference value	Statistically significant (yes/no)
1–2	$q = \dfrac{205.6 - 199.8}{\sqrt{\frac{657.35}{2}\left(\frac{1}{23} + \frac{1}{12}\right)}}$	$= \dfrac{5.8}{6.46} = 0.90$	$q_{2,47,.95} = 2.85^*$	No
2–3	$q = \dfrac{199.8 - 182.7}{\sqrt{\frac{657.35}{2}\left(\frac{1}{12} + \frac{1}{15}\right)}}$	$= \dfrac{17.1}{7.02} = 2.44$	2.85	No

$^* q_{2,47,.95}$ is approximated by $\dfrac{(\frac{1}{47} - \frac{1}{60})q_{2,40,.95} + (\frac{1}{40} - \frac{1}{47})q_{2,60,.95}}{\frac{1}{40} - \frac{1}{60}}$

$$= \frac{(0.0046)(2.86) + (0.0037)(2.83)}{(0.0083)} = 2.85$$

TABLE A.50 Relationship between smoking habit during pregnancy and birthweight

Group	Mean	sd	Variance	n
Nonsmoker (NON)	7.59	0.962	0.925	7
Ex-smoker (EX)	7.24	0.913	0.833	5
Light current (CUR < 1)	6.33	1.140	1.299	7
Heavy current (CUR ⩾ 1)	6.01	0.720	0.518	8
	6.73			27

$$X^2 = \frac{1.372}{1.076} = 1.275 \sim \chi_3^2 \text{ under } H_0$$

Since $\chi_{3,.95}^2 = 7.81 > X^2$, it follows that $p > .05$. Thus, the null hypothesis, that the four within-sample population variances are the same, is accepted.

12.42 Perform a one-way ANOVA.

$$\text{Between SS} = \frac{\sum_{i=1}^{4} y_{i.}^2}{n_i} - \frac{y_{..}^2}{n}$$

$$= \frac{53.1^2}{7} + \frac{36.2^2}{5} + \frac{44.3^2}{7} + \frac{48.1^2}{8} - \frac{181.7^2}{27}$$

$$= 1234.446 - 1222.774 = 11.672$$

$$\text{Total SS} = \sum_{i=1}^{4} \sum_{j=1}^{n_i} y_{ij}^2 - \frac{y_{..}^2}{n}$$

$$= 7.5^2 + \cdots + 5.4^2 - \frac{181.7^2}{27}$$

$$= 1254.750 - 1222.774 = 31.976$$

Within SS = Total SS − Between SS

$$= 31.976 - 11.672 = 20.304$$

Thus, the ANOVA table is given in Table A.51.

Since

$$F_{3,23,.975} < F_{3,20,.975} = 3.86 < 4.41$$

it follows that $p < .025$. Also, since

$$F_{3,23,.99} > F_{3,30,.99} = 4.51 > 4.41$$

it follows that $p > .01$. Therefore, $.01 < p < .025$, and the alternative hypothesis, that the four population means are different, is accepted.

12.43 Test for each difference between two groups by computing the appropriate t statistic as follows:

$$t = \frac{\bar{x}_i - \bar{x}_j}{\sqrt{s^2\left(\frac{1}{n_i} + \frac{1}{n_j}\right)}}$$

These are given in Table A.52.

TABLE A.51 ANOVA table

	SS	df	MS	F stat	p-value
Between	11.672	3	3.891	4.41	$.01 < p < .025$
Within	20.304	23	0.883		
	31.976	26			

TABLE A.52 Comparison of specific groups relating smoking habit and birthweight (*t* tests)

Groups compared	Test statistic		p-value
NON, EX	$t = \dfrac{7.59 - 7.24}{\sqrt{0.883(\frac{1}{7} + \frac{1}{5})}}$	$= \dfrac{0.35}{0.550} = 0.64 \sim t_{23}$	NS
NON, CUR < 1	$t = \dfrac{7.59 - 6.33}{\sqrt{0.883(\frac{1}{7} + \frac{1}{7})}}$	$= \dfrac{1.26}{0.502} = 2.51 \sim t_{23}$	$.01 < p < .02$
NON, CUR $\geqslant 1$	$t = \dfrac{7.59 - 6.01}{\sqrt{0.883(\frac{1}{7} + \frac{1}{8})}}$	$= \dfrac{1.58}{0.486} = 3.25 \sim t_{23}$	$.001 < p < .01$
EX, CUR < 1	$t = \dfrac{7.24 - 6.33}{\sqrt{0.883(\frac{1}{5} + \frac{1}{7})}}$	$= \dfrac{0.91}{0.550} = 1.65 \sim t_{23}$	NS
EX, CUR $\geqslant 1$	$t = \dfrac{7.24 - 6.01}{\sqrt{0.883(\frac{1}{5} + \frac{1}{8})}}$	$= \dfrac{1.23}{0.536} = 2.29 \sim t_{23}$	$.02 < p < .05$
CUR < 1, CUR $\geqslant 1$	$t = \dfrac{6.33 - 6.01}{\sqrt{0.883(\frac{1}{7} + \frac{1}{8})}}$	$= \dfrac{0.32}{0.486} = 0.66 \sim t_{23}$	NS

Thus, there are significant differences between the nonsmokers and each group of current smokers but not between nonsmokers and ex-smokers. There are also significant differences between ex-smokers and heavy current smokers but not between ex-smokers and light current smokers. Finally, there are no significant differences between the two groups of current smokers.

12.44 First use the procedure in **(12.15)** to look at the group of four means using the test statistic

$$q = \frac{7.59 - 6.01}{\sqrt{\frac{0.883}{2}\left(\frac{1}{7} + \frac{1}{8}\right)}} = \frac{1.58}{0.344} = 4.593$$

We have $q_{4,23,.95} < q_{4,20,.95} = 3.96 < q$. Thus, some of the means are significantly different. Now look at subgroups of three means. In particular, look at the subgroup (CUR $\geqslant$ 1, CUR $<$ 1, EX) using the test statistic

$$q = \frac{7.24 - 6.01}{\sqrt{\frac{0.883}{2}\left(\frac{1}{5} + \frac{1}{8}\right)}} = \frac{1.23}{0.379} = 3.245$$

Since $q_{3,23,.95} > q_{3,24,.95} = 3.53 > q$, we declare these three means as not significantly different. Next look at the subgroup of means (CUR $<$ 1, EX, NON) using the test statistic

$$q = \frac{7.59 - 6.33}{\sqrt{\frac{0.883}{2}\left(\frac{1}{7} + \frac{1}{7}\right)}} = \frac{1.26}{0.355} = 3.549$$

Note that

$$q_{3,23,.95} \approx \frac{(\frac{1}{23} - \frac{1}{24})q_{3,20,.95} + (\frac{1}{20} - \frac{1}{23})q_{3,24,.95}}{(\frac{1}{20} - \frac{1}{24})}$$

$$= \frac{(0.0018)(3.58) + (0.0065)(3.53)}{0.0083} = 3.541$$

Thus, since $q > q_{3,23,.95}$, it follows that the group of three means is significantly different.

However, we already know that (CUR $<$ 1 and EX) are not significantly different, because they are part of the subgroup (CUR $\geqslant$ 1, CUR $<$ 1, EX), which was tested previously. The only remaining exercise is to test if EX and NON are significantly different, using the following test statistic

$$q = \frac{7.59 - 7.24}{\sqrt{\frac{0.883}{2}\left(\frac{1}{7} + \frac{1}{5}\right)}} = \frac{0.35}{0.389} = 0.900$$

Since $q_{2,23,.95} > q_{2,24,.95} = 2.92 > q$, these two means are not significantly different. The results can be summarized in Figure A.7.

| CUR $\geqslant$ 1 | CUR $<$ 1 | EX | NON |

FIGURE A.7

Thus, the only significant differences are between nonsmokers and each of the groups of current smokers.

12.45 The only difference between the results using the multiple comparison procedures in **(12.16)** and the pairwise t tests in **(12.13)** is that the ex-smokers and heavy current smokers are significantly different ($.02 < p < .05$) with the t tests but not with the multiple comparisons.

12.46 Look at the contrast

$$L = 0\bar{y}_2 + 0.5\bar{y}_3 + 1.3\bar{y}_4$$

However, the sum of the coefficients here adds up to 1.8 rather than 0. If $1.8/3 = 0.6$ is subtracted from each coefficient, then we will have an appropriate contrast of the form

$$L = -0.6\bar{y}_2 + (0.5 - 0.6)\bar{y}_3 + (1.3 - 0.6)\bar{y}_4$$
$$= -0.6\bar{y}_2 - 0.1\bar{y}_3 + 0.7\bar{y}_4$$
$$= -0.6(7.24) - 0.1(6.33) + 0.7(6.01) = -0.77$$

Test the hypothesis H_0: $E(L) = 0$ versus H_1: $E(L) \neq 0$, where under H_0: $E(L) = 0$ and

$$Var(L) = s^2\left[\frac{(0.6)^2}{n_2} + \frac{(0.1)^2}{n_3} + \frac{(0.7)^2}{n_4}\right]$$

$$= 0.883\left(\frac{0.36}{5} + \frac{0.01}{7} + \frac{0.49}{8}\right) = 0.119$$

$$se(L) = 0.345$$

Thus, we have the test statistic

$$t = \frac{L}{se(L)} = \frac{-0.77}{0.345} = -2.232 \sim t_{23} \text{ under } H_0$$

Since $t_{23,.975} = 2.069$, $t_{23,.99} = 2.500$, it follows that $.02 < p < .05$. Thus, there is a significant trend among ever smokers, with a lower mean birthweight as the current number of cigarettes smoked per day increases.

12.47 Use Scheffé's multiple comparisons test. Compare t from Problem 12.46 with $c_1 = -\sqrt{3F_{3,23,.95}}$.

Since $t = -2.232 > -\sqrt{3F_{3,30,.95}}} = -\sqrt{3(2.92)}$

$= -2.96 > c_1 = -\sqrt{3F_{3,23,.95}}$,

and $t < c_2 = \sqrt{3F_{3,23,.95}}$,

it follows that $p > .05$, and the results are not statistically significant using Scheffé's test. Notice the difference in interpretation if the method of multiple comparisons is used, where significant results were not obtained vs. if the t test approach in Problem 12.46 is used, where significant results were found. Which is more appropriate will depend on whether this contrast was, or was not planned before looking at the data.

12.62 A two-way ANOVA is appropriate here because the 80 patients can be classified by two parameters: (1) instruction for weight reduction (yes/no) and (2) instruction for meditation (yes/no).

12.63 Use the method of unweighted means. First, rearrange the data in the format of a 2 × 2 table (Table A.53) giving the sum of the means in each row $(y_{i.}^{*})$ and each column $(y_{.j}^{*})$ in the margins.

Now compute the Row, Column, and Interaction SS and MS as follows:

$$\text{Row SS} = \frac{13.9^2 + 6.0^2}{2} - \frac{(13.9 + 6.0)^2}{2 \times 2}$$

$$= 114.605 - 99.003 = 15.60 = \text{Row MS}$$

$$\text{Column SS} = \frac{13.5^2 + 6.4^2}{2} - \frac{(13.5 + 6.4)^2}{2 \times 2}$$

$$= 111.605 - 99.003 = 12.60$$

$$= \text{Column MS}$$

Interaction SS $= (8.6)^2 + \cdots + (1.1)^2$

$$-\frac{(8.6 + \cdots + 1.1)^2}{4} - 15.60 - 12.60$$

$$= 127.27 - \frac{(19.9)^2}{4} - 28.20$$

$$= 127.27 - 99.00 - 28.20$$

$$= 0.07 = \text{Interaction MS}$$

Finally, compute the Error MS to perform the significance tests. Since the sample size of each group is the same, we have $n_h = 20$. Thus,

$$\text{Error MS} = \frac{19[(6.2)^2 + (5.4)^2 + (7.0)^2 + (6.5)^2]}{76(20)}$$

$$= \frac{(6.2)^2 + (5.4)^2 + (7.0)^2 + (6.5)^2}{4(20)} = 1.986$$

Test for row (weight reduction instruction) effects using the following test statistic:

$$F_{\text{ROW}} = \frac{\text{Row MS}}{\text{Error MS}} = \frac{15.60}{1.986} = 7.85 \sim F_{1,76} \text{ under } H_0$$

Since $F_{1,76,.99} < F_{1,60,.99} = 7.08 < 7.85$

it follows that $p < .01$. Similarly, since

$$F_{1,76,.995} > F_{1,120,.995} = 8.18 > 7.85$$

it follows that $p > .005$. Thus, $.005 < p < .01$, and there is a significant effect of weight reduction instruction on reduction of blood pressure after controlling for the effect of meditation.

TABLE A.53 Effect of counseling for weight reduction and/or meditation on the change in diastolic bp

			Instruction in meditation		
			Yes	No	$y_{i.}^{*}$
		Mean	8.6	5.3	13.9
	Yes	sd	6.2	5.4	
		n	20	20	
Instruction in weight reduction					
		Mean	4.9	1.1	6.0
	No	sd	7.0	6.5	
		n	20	20	
		$y_{.j}^{*}$	13.5	6.4	19.9

12.64 Test for column (meditation instruction) effects using the following test statistic:

$$F_{COLUMN} = \frac{\text{Column MS}}{\text{Error MS}} = \frac{12.60}{1.986}$$

$$= 6.34 \sim F_{1,76} \text{ under } H_0$$

Since $\quad F_{1,76,.975} < F_{1,60,.975} = 5.29 < 6.34$

it follows that $p < .025$. Similarly, since

$$F_{1,76,.99} > F_{1,120,.99} = 6.85 > 6.34$$

it follows that $p > .01$. Thus, $.01 < p < .025$, and there is a significant effect of meditation instruction on reduction of blood pressure after controlling for the effect of weight reduction.

12.65 Test for an interaction effect between weight reduction instruction and meditation instruction. Use the test statistic

$$F_{INT} = \frac{\text{Interaction MS}}{\text{Error MS}} = \frac{0.07}{1.986}$$

$$= 0.04 \sim F_{1,76} \text{ under } H_0$$

Since $F_{1,76,.95} > F_{1,120,.95} = 3.92 > 0.04$, it follows that $p > .05$, and there is no interaction effect between the two types of instruction. Thus, the effect of weight reduction is similar in patients who do and do not receive meditation instruction.

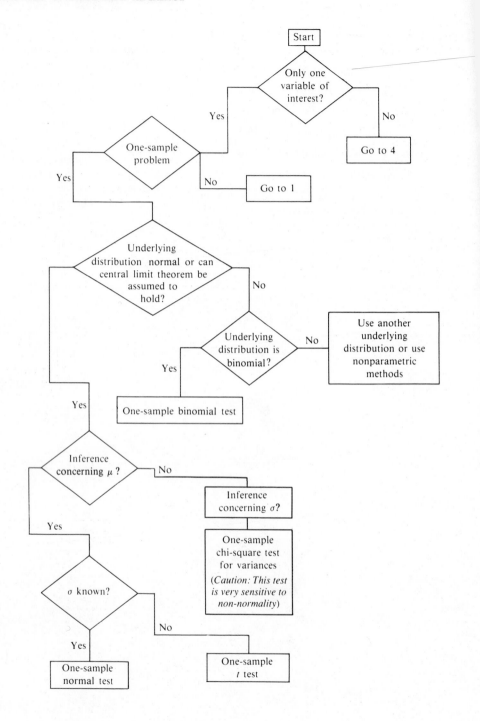

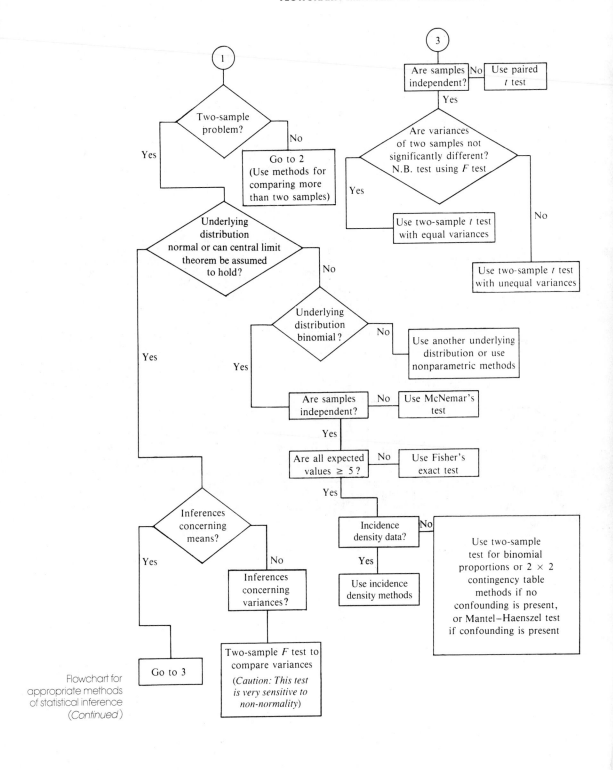

Flowchart for
appropriate methods
of statistical inference
(Continued)

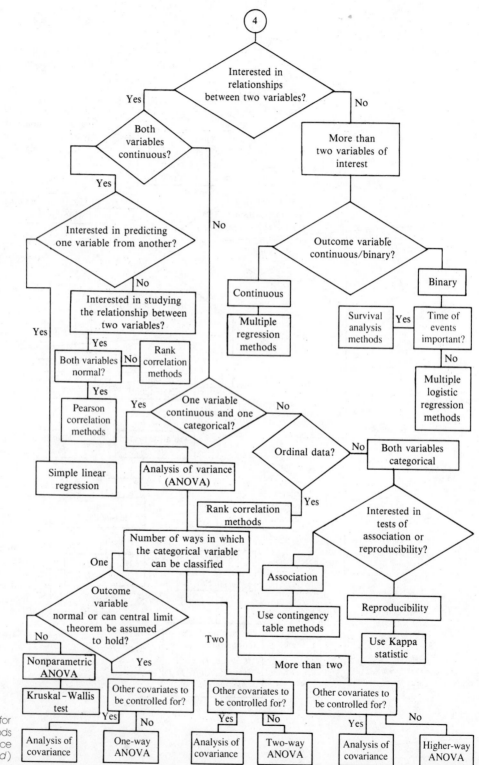

Flowchart for
appropriate methods
of statistical inference
(Continued)

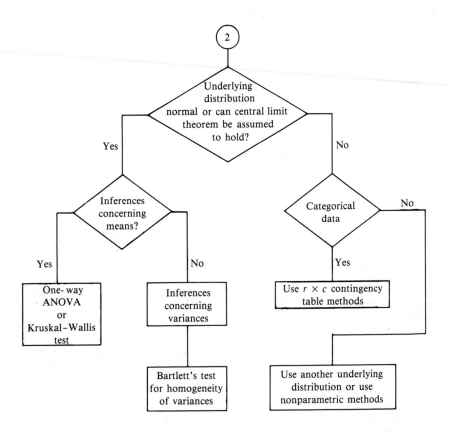

Flowchart for
appropriate methods
of statistical inference
(*Continued*)

INDEX

INDEX OF APPLICATIONS

INDEX OF APPLICATIONS

ISBN 0-534-91973-1